# Beyond the Key

## Locksmithing, Security Systems, and the Future of Access Control



**Richard Skiba**

AFTER MIDNIGHT
PUBLISHING

This publication is designed to provide accurate and authoritative information in regard to the subject matter covered. While the publisher and author have used their best efforts in preparing this book, they make no representations or warranties with respect to the accuracy or completeness of the contents of this book and specifically disclaim any implied warranties of merchantability or fitness for a particular purpose. No warranty may be created or extended by sales representatives or written sales materials. The advice and strategies contained herein may not be suitable for your situation. You should consult with a professional when appropriate. Neither the publisher nor the author shall be liable for any loss of profit or any other commercial damages, including but not limited to special, incidental, consequential, personal, or other damages.

Skiba, Richard (Author)

Beyond the Key: Locksmithing, Security Systems, and the Future of Access Control

ISBN 978-1-7644453-3-7 (Paperback) 978-1-7644453-4-4 (eBook)

Non-fiction

# Contents

# Introduction

Locks have existed for thousands of years, evolving alongside human society's need to protect property, safeguard information, and control access to physical spaces. From simple wooden mechanisms in ancient civilizations to today's electronically connected smart locks, the development of locking systems reflects an ongoing contest between protection and intrusion. Each improvement in lock design has been matched by new techniques developed to bypass, manipulate, or defeat those mechanisms. As a result, locksmithing has never been a static trade. It is a profession defined by adaptation, learning, and the continuous refinement of security technologies.

Traditionally, locksmiths were viewed primarily as skilled tradespeople who worked with mechanical devices. Their work centred on installing locks, cutting keys, repairing worn mechanisms, and restoring access when keys were lost or locks failed. These technical skills remain fundamental to the profession. However, the security landscape has changed significantly. Modern buildings and infrastructure now rely on integrated systems that combine physical locks with electronic access control, surveillance technologies, alarms, and digital credential management. In such environments, the effectiveness of a lock cannot be understood in isolation but must be considered as part of a broader security system that includes people, policies, technology, and physical infrastructure.

This transformation has expanded the role of the locksmith. Today's professionals are not only installers and repairers of mechanical hardware but also advisers who help organisations and individuals design practical and resilient security systems. Locksmiths must understand how doors, frames, locks, access control systems, surveillance technologies, and administrative procedures interact. They must evaluate risk, anticipate how security systems might fail, and recommend solutions that balance protection, usability, safety, and operational needs.

At the same time, the profession is becoming increasingly multidisciplinary. Modern locksmiths encounter electronic locks, biometric authentication systems, wireless communication technologies, cloud-managed access platforms, and cybersecurity considerations. These technologies bring new opportunities for flexible and intelligent access control, but they also introduce new vulnerabilities and responsibilities. As security systems become more connected, locksmiths must work alongside IT professionals, security consultants, and building managers to ensure that physical and digital security measures operate together effectively.

This book explores locksmithing from both its traditional foundations and its modern evolution. It examines how locks function, how they are designed and manufactured, and how they can be defeated through various attack methods. Understanding these mechanisms is essential not only for installation and servicing but also for diagnosing vulnerabilities and selecting appropriate security solutions. At the same time, the book addresses the broader context in which locks operate, including access control systems, human behaviour, regulatory obligations, and emerging technologies.

Throughout the chapters, readers will encounter a range of topics that illustrate how locksmithing extends beyond mechanical craftsmanship. These include the analysis of lock mechanisms, forensic examination of tool marks, ethical responsibilities in providing security services, the role of regulation and licensing, the integration of digital security systems, and the importance of continuous learning in a rapidly evolving field. Each topic highlights how modern locksmithing combines practical skill with analytical thinking and professional judgment.

A central theme of the book is the shift from thinking about locks as isolated devices to understanding security as an interconnected system. Effective protection rarely depends on a single component. Instead, it emerges from the interaction between physical barriers, electronic controls, monitoring technologies, and the behaviour of the people who use them. Recognising these relationships allows locksmiths to move beyond simple hardware replacement and toward designing security solutions that function reliably within real-world environments.

The chapters that follow are intended for readers at various stages of their journey into locksmithing and security. For newcomers, the material introduces the fundamental principles that underpin lock mechanisms and access control. For practitioners, it provides a deeper examination of how locks interact with broader security systems and how the profession continues to evolve. For those interested in security more broadly, the book offers insight into the technical and strategic role locksmiths play in protecting people, property, and critical infrastructure.

Ultimately, locksmithing is a profession built on trust. Locksmiths are granted access to the systems that protect homes, businesses, and institutions. With that access comes responsibility: to act ethically, to maintain professional competence, and to ensure that security solutions genuinely protect those who rely on them. As technology advances and security challenges become more complex, the locksmith’s role will continue to evolve, from a specialist in locks to a professional capable of designing and maintaining integrated security systems.

# Part 1

# Rethinking Locks and Security

# Chapter 1

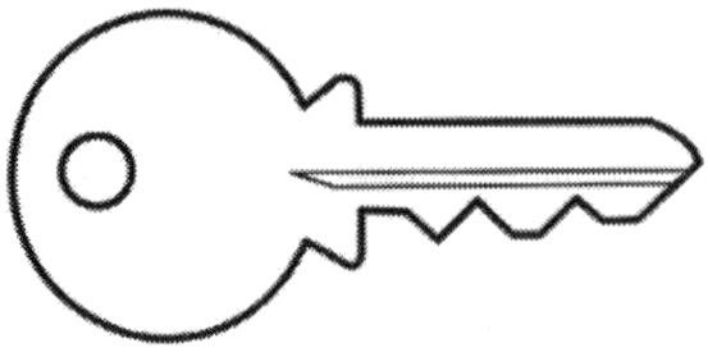

# The Philosophy of Security and Access

This chapter invites you to rethink what security really means in a modern world where access is no longer defined solely by keys and locks, but by systems, data, and human behaviour. Rather than presenting locksmithing as a purely technical trade, it frames it as a discipline shaped by competing priorities, psychological factors, evolving technologies, and ethical responsibilities. From the constant tension between security and convenience, to the influence of trust, perception, and human behaviour on real-world outcomes, this chapter establishes a critical foundation: effective security is not simply engineered, it is experienced, interpreted, and enacted by people. As you move through the sections that follow, the focus is not just on how locks work, but on why they succeed or fail in practice, and how modern locksmithing must adapt to remain relevant in an increasingly complex and interconnected security landscape.

## Security Vs Convenience: The Eternal Trade-Off

Every time a person unlocks a door, they make an unconscious decision about what matters more in that moment—speed or safety. Whether it is a homeowner using a key, an employee tapping a swipe card, or a user relying on a smartphone to gain entry, the same underlying question is being answered: how much security is enough, and how much inconvenience is acceptable? This everyday interaction sits at the heart of modern locksmithing and access control. Understanding this balance is essential, because it shapes not only how locks are designed, but how they are used, trusted, and ultimately whether they succeed or fail in protecting people and property.

The enduring tension between security and convenience is a central theme in locksmithing and access control, spanning traditional mechanical devices through to cloud-based smart access platforms. Security is fundamentally concerned with restricting access through processes of verification and control, whereas convenience prioritises rapid, effortless entry for authorised users. Across the spectrum of locking systems—from basic padlocks to integrated smart-lock ecosystems—designers must position solutions along a continuum that balances these competing priorities. This conceptual framing is widely supported in the literature on access control, smart devices, and IoT security, which consistently emphasises that increased friction in system use often corresponds with stronger verification, while enhanced usability can introduce new vulnerabilities or misuse if not carefully designed [1-4].

High-security configurations typically introduce greater friction into the access process. This may involve multi-factor or multi-step authentication, robust but slower locking mechanisms, restricted credential duplication, tightly controlled access rights, and delays during entry or exit. In contrast, convenience-oriented systems prioritise minimal user input, seamless unlocking methods such as swipe, tap, or proximity access, easily shareable credentials, and simplified interfaces. The fundamental paradox remains that reducing friction tends to weaken security, whereas strengthening security often diminishes usability. This is not a design flaw but a core principle of access control, where verification processes, physical safeguards, and user discipline inherently require time and effort [1, 2, 5, 6].

Evidence from both research and practice indicates that layered security is commonly employed to manage this trade-off. Layered approaches combine simple and usable access mechanisms with additional controls activated when risk levels increase. This includes user-centred and context-aware design strategies that align protection levels with specific environments such as residential, commercial, or high-security settings. Additionally, the balance between fail-safe and fail-secure design illustrates how safety requirements can be integrated without compromising overall control when implemented appropriately [2, 6, 7].

A consistent finding across the security literature is that failures are more often attributable to human behaviour than to mechanical or digital deficiencies. Users frequently bypass security measures for convenience, such as leaving doors ajar, sharing credentials, disabling alarms, or propping open secure access points in workplace environments. These behaviours demonstrate that systems perceived as overly inconvenient are likely to be circumvented, reinforcing the importance of designing security measures that account for human behaviour and real-world usage patterns. Minimising perceived inconvenience while maintaining essential verification is therefore critical to ensuring that systems remain effective in practice [1, 4, 5, 8].

Different environments occupy distinct positions along the security–convenience continuum. Residential systems typically prioritise ease of daily use while maintaining an adequate level of protection. In contrast, environments such as hospitals and airports require controlled access that supports both security and operational efficiency. High-security facilities, however, emphasise strict verification and access control, often accepting reduced convenience as a necessary trade-off for enhanced protection. These distinctions reflect varying risk profiles and inform decisions related to authentication strength, credential management, and system architecture [2, 6, 7].

The integration of digital technologies into locking systems has significantly increased both convenience and complexity. Smart locks enable features such as auto-unlock and networked credential management, enhancing user experience but introducing cybersecurity risks. Biometric systems eliminate the need for physical keys, yet raise concerns related to privacy, reliability, and the difficulty of revoking compromised biometric data. Mobile credentials further simplify access but create dependencies on device security and network connectivity, introducing potential vulnerabilities associated with digital ecosystems. As highlighted in the literature, these advancements expand attack surfaces and require locksmiths and system designers to consider not only mechanical robustness but also digital security, device integrity, and client risk tolerance [5, 6, 9-14].

*Figure 1: Nuki smart lock, controlled with iPhone. Nuki Smart Lock (Flickr User), CC BY 2.0, via Wikimedia Commons.*

Smart locks and connected access systems are particularly vulnerable to cyber threats, especially when deployed in open or heterogeneous network environments. Key risks include weaknesses in credential management, relay and replay attacks, insecure APIs, and vulnerabilities in mobile or cloud-based platforms. Research also highlights risks associated with biometric systems, including template leakage, sensor spoofing, and replay attacks. Effective mitigation strategies include strong device authentication, tokenisation, contextual monitoring, multi-factor authentication, and hardware-based protections that limit exposure of sensitive data [5, 11, 13-17].

For practitioners, several key principles emerge. Layered security design remains essential, combining accessible entry methods with additional controls such as tamper detection, time-based access policies, and revocable credentials to align protection with risk and usability requirements [2, 3, 6, 7]. User-centred design is equally critical, ensuring systems are intuitive and predictable to reduce misuse or circumvention [1, 4, 5]. In high-risk environments, strict verification, auditing, and credential management may justifiably take precedence over convenience [2, 6, 7]. Additionally, the growing reliance on digital technologies necessitates comprehensive cyber-physical risk assessments, incorporating considerations of privacy, device security, and network vulnerabilities [3, 5, 9, 11, 14, 15, 17].

Emerging technologies continue to reshape the balance between security and convenience. Biometric authentication offers significant usability advantages but introduces concerns regarding privacy, reliability, and irreversibility of compromised data [3, 5, 10]. Mobile and cloud-based access systems provide flexibility and efficiency but require robust device security, secure communication channels, and resilient token management to prevent misuse [6, 9, 12-14, 17]. Broader IoT ecosystems further emphasise the need for multi-layered security, coordinated governance, and cross-device trust models to manage interdependencies and evolving threats [2, 6, 7, 18-20].

Several nuanced tensions persist within access control design. Privacy versus convenience remains a key concern, particularly in biometric systems where data breaches may have irreversible consequences [3, 5, 10]. Similarly, security versus performance is evident in fail-safe and fail-secure decisions, where systems must balance safety requirements with controlled access during normal operations [2, 7]. Human factors continue to play a critical role, as users may bypass or misuse systems unless they are designed to align with predictable behaviours and practical needs [1, 4, 5].

The trade-off between security and convenience in locking and access control systems is a well-established and evidence-based principle that remains highly relevant as technologies evolve. Across mechanical, biometric, mobile, and IoT-enabled systems, the most effective solutions are those that achieve a deliberate balance between robust protection and practical usability. Layered, context-aware, and user-centred design—supported by privacy-preserving technologies, secure authentication methods, and comprehensive risk management—provides a practical pathway forward. As systems continue to evolve, integrating behavioural insights, digital security considerations, and system-wide resilience will be essential to maintaining both security and usability in diverse operational contexts [1, 2, 4, 5, 9-12].

## Impact on Locksmithing Practices

The integration of digital technologies into modern locking systems has fundamentally changed the scope and practice of locksmithing. No longer confined to mechanical devices and physical keys, contemporary locksmithing now operates within a broader landscape of interconnected security systems that combine physical hardware with digital platforms. As such, an understanding of both traditional lock mechanisms and emerging technologies is essential for anyone seeking to engage with the field.

Smart locks, biometric systems, and mobile-based access solutions have introduced significant gains in convenience, allowing users to unlock doors automatically, manage access remotely, and eliminate the need for physical keys. However, these advancements also introduce new layers of complexity and risk. Unlike purely mechanical systems, digital locking solutions are vulnerable to cyber threats such as credential interception, relay and replay attacks, and weaknesses in software or network configurations. This means that security is no longer determined solely by the strength of a lock's physical components, but also by the integrity of the devices, networks, and data that support it.

For the locksmith, this shift expands the role from that of a tradesperson to that of a security professional. It is no longer sufficient to understand how to install, repair, or bypass a lock; practitioners must also be capable of evaluating digital vulnerabilities, advising on secure system configurations, and understanding how access credentials are generated, stored, and managed. In many cases, locksmiths are now expected to recommend integrated solutions that combine mechanical and electronic elements, requiring a working knowledge of both domains.

The importance of layered security becomes particularly evident in this context. Effective systems often combine simple and accessible entry methods with additional safeguards such as time-based access controls, audit trails, tamper detection, and revocable digital credentials. This layered approach allows security to be adapted to different risk environments while maintaining usability for the end user. Equally important is the principle of user-centred design, as systems that are overly complex or inconvenient are more likely to be bypassed or misused, thereby undermining their intended purpose.

As locking systems become increasingly reliant on mobile devices, cloud services, and interconnected networks, locksmiths must also consider factors beyond the physical lock itself. Device security, network reliability, data privacy, and system integration all play a role in determining overall security effectiveness. This introduces the need for a cyber-physical perspective, where both digital and physical risks are assessed together rather than in isolation.

This highlights an important reality: locksmithing is no longer solely about locks, but about managing access in a complex and evolving security environment. Developing an understanding of how digital technologies influence security—alongside traditional mechanical knowledge—will be critical to remaining effective, adaptable, and professionally relevant in modern practice.

In the end, the challenge for modern locksmithing is not to choose between security and convenience, but to understand how to balance them intelligently in a world where both expectations continue to rise. The most successful systems will be those that recognise this tension not as a limitation, but as a design opportunity, creating solutions that are secure enough to resist real threats, yet intuitive enough to be used correctly every day. As technology continues to evolve, so too must the thinking of those who design, install, and manage these systems. By embracing this balance, locksmiths are not just protecting doors, they are shaping how people experience security itself.

## The Psychology of Trust in Locking Systems

Before considering the technical performance of any lock, it is worth pausing to reflect on a simple but often overlooked question: what actually makes a lock effective in the real world? While it is easy to assume that strength, complexity, or advanced technology alone determine security, everyday experience suggests otherwise. Doors are left unlocked, systems are bypassed, and even the most sophisticated devices fail, not because they are weak, but because they are misunderstood, mistrusted, or misused. This section explores a critical shift in thinking within modern security practice: that the true effectiveness of locking systems lies not only in their engineering, but in the way people perceive, trust, and interact with them.

The argument that the effectiveness of locking systems rests as much on user trust as on mechanical strength or digital sophistication is well aligned with a cross-disciplinary literature base spanning security engineering, human–computer interaction, and security sociology. Trust operates as a cognitive and affective resource that shapes how people interact with locks, how consistently they engage with access procedures, and how faithfully they adhere to security protocols. Empirical and theoretical work across domains consistently underscores that perceived security, rather than actual cryptographic strength alone, governs real-world security outcomes. For instance, studies of security devices and authentication systems emphasize that users' mental models, perceived control, and confidence in a system determine whether they will adopt, correctly use, and persist with security measures [21-23]. In the lock-and-access domain, this translates into the crucial insight that a lock's protective value is mediated by user trust in its operation and in the institutions that manage it [24, 25].

Perceived security versus actual security is a central tension in locking practice. Heaviest or most formally verified devices can still be trusted less if users do not understand how they function or if the operation remains opaque. Conversely, devices that are technically robust may evoke distrust if users cannot observe or interpret their functioning, leading to suboptimal use or circumvention. In studies of authentication and access-control technology, perceived trust often aligns with tangible signals, feedback, transparency of operation, and visible safeguards, more than with theoretical security guarantees. This is echoed in analyses of user trust in security interfaces, where factors such as ease of verification, clarity of feedback, and the sense of ownership over access rights substantially influence adoption and proper use [21, 23]. The literature on IoT security and physical locking systems also demonstrates how trust signals, cryptographic assurances, auditable logs, and visible security features, shape user cooperation and perceived risk, which in turn affects actual protective behaviour [23, 24] . In this way, perceived security acts as a mediator between the technical properties of a lock and its real-world effectiveness.

Familiarity and the psychology of trust play a substantial role in shaping acceptance of locking technologies. Systems that resemble traditional key-based mechanisms tend to be more readily adopted and trusted, because they align with existing user schemas about access (possession equating to access) and afford straightforward mental models of risk and control. When new modalities such as biometrics or mobile-access solutions are introduced, trust often hinges on users' understanding of how these systems work and their perceived sense of control over the process. If users perceive opaqueness or dependency on external infrastructure (e.g., cloud services, remote permissions), their confidence can wane, irrespective of the underlying cryptographic robustness. This dynamic, familiar systems engender greater trust, whereas unfamiliar, invisible, or remotely managed systems may suffer from reduced perceived control and trust, has been highlighted in discussions of biometric versus physical-key access and in analyses of how users form judgments about "secure enough" configurations [21, 23, 24].

Control and ownership emerge as pivotal drivers of trust. A classic physical key provides a direct, tangible locus of control: possession confers access. Digital systems, by contrast, often introduce a sense of external dependency—on apps, cloud services, or remote authorization processes—that can erode perceived control and, hence, trust. When users worry that access could be altered remotely without their knowledge, they may resist adoption or fail to use advanced features as intended. This tension between direct control and remote control mechanisms is a recurrent theme in security literature addressing user trust, with implications for locksmithing and access-management design [21, 23, 24]. The practical takeaway for locksmiths and security professionals is to emphasize not only cryptographic strength but also feedback-rich interfaces and governance features that reassure users about how access decisions are made and who can alter them.

Reliability and consistency are core determinants of trust in locking systems. The psychology of trust in security emphasizes that repeated positive interactions, where a system unlocks as expected, logs access events, and provides clear confirmations, solidify the belief that the system is dependable. Conversely, intermittent failures or opaque operation undermine trust even when the system is technically robust. In high-use environments, reliability becomes a cornerstone of psychological trust, reinforcing correct usage patterns and reducing risky workarounds [21, 23, 25]. The locksmithing literature thus supports a dual emphasis: design for predictable operation and ensure transparent, user-friendly feedback mechanisms that reinforce trust through observable reliability.

The social environment also moderates trust in locking systems. Social norms—such as colleagues modelling secure behaviour, or organizational enforcement of security protocols—shape how individuals perceive the importance and legitimacy of the locking system. Observed enforcement (badge checks, monitored access points) tends to reinforce trust in the system's purpose and effectiveness, while lax practices (prop open doors, sharing access codes) can erode perceived importance and compliance. This social dimension has empirical support in studies of security cultures and organizational behaviour, where social reinforcement and observed security practices influence user decisions and behaviour around access control [21, 23, 25]. For locksmiths and security professionals, integrating these social cues—clear policies, visible access-control practices, and education about proper usage—can enhance trust-directed behaviours and reduce insecure workarounds.

Finally, the literature acknowledges nuanced disagreements and contextual variation. While there is broad agreement that trust is essential, there is also recognition that trust is not monolithic; it can be domain-specific, modality-specific (e.g., biometrics vs. token-based access), and contingent on environmental factors such as the visibility of monitoring or the perceived threat level. Some sources emphasize the potential for familiar, low-friction interfaces to sustain trust, while others highlight the risks of complacency when trust is built on surface cues rather than deep understanding. For instance, analyses of user perceptions of security tools underscore that ease of use, cognitive burden, and perceived trustworthiness collectively influence adoption decisions, and that messaging about security features must align with actual user experiences to avoid misplaced trust [21-23]. In IoT and smart-lock contexts, reports of vulnerabilities and user perceptions of security risk further illustrate that trust must be navigated alongside technical safeguards, with a focus on actionable transparency and robust, user-centred design [24, 25].

From a design and practice perspective, these insights yield concrete implications for locksmithing and access-control projects. Prioritize feedback-rich designs that provide immediate and understandable confirmation of successful or failed access attempts, and couple these with clear logs, audit trails, and user education to reinforce a shared security narrative. Favor access-control configurations that preserve the user's sense of ownership and control, for example by enabling user-managed keys or revocable credentials, while maintaining strong cryptographic protections and resistant, auditable infrastructures. Recognizing that trust is socially reinforced, implement visible security practices (badge checks, monitored doors) and foster a security-aware culture through training and consistent reinforcement of expected behaviours. These practices are supported by cross-referencing evidence that links trust signals, perceived control, reliability, and social reinforcement to the real-world effectiveness of locking systems, complemented by discussions of authentication frameworks and user perceptions of security tools in related domains [21-24].

In sum, the literature converges on a multidisciplinary understanding: the effectiveness of locking systems rests on a triad of technical robustness, perceptual trust, and social-context factors. A technically excellent lock that elicits low trust or that operates opaquely is unlikely to achieve secure outcomes in daily use;

conversely, a well-designed system that communicates its functioning clearly, aligns with users' mental models, and leverages social norms and education to reinforce secure behaviour can realize substantial protective benefits beyond what its hardware specifications alone would predict. For locksmiths and security professionals, this means designing and recommending systems that integrate perceptible feedback, user education, controlled access, transparent governance, and alignment with local security cultures as much as selecting the strongest mechanical or electronic components [21-24].

## Impact on Locksmithing Practices

The relevance of this discussion lies in its redefinition of what it means for a locking system to be truly effective. While locksmithing has traditionally emphasised mechanical strength, resistance to attack, and technical performance, contemporary understanding shows that real-world effectiveness depends equally on how users perceive, trust, and interact with the system. A lock that is technically robust but poorly understood or mistrusted may be misused or bypassed, whereas a system that is trusted, intuitive, and consistently used can deliver far greater practical security.

This has direct implications for locksmithing practice. The selection and installation of locks are no longer purely technical tasks; they require a human-centred approach that considers how users will engage with the system. Locksmiths must evaluate whether a system is understandable, whether users feel a sense of control, and whether it is likely to be used correctly over time. Systems that are overly complex or unclear can lead to behaviours such as credential sharing, feature avoidance, or complete circumvention, ultimately undermining security. As a result, the locksmith's role increasingly involves designing and recommending solutions that align with user behaviour and expectations.

The distinction between perceived security and actual security is particularly significant in this context. Users often assess security based on visible and tangible cues—such as the physical presence of a lock, audible or visual feedback, and clarity of operation—rather than on technical specifications alone. Effective locksmithing therefore requires not only the provision of robust systems but also ensuring that those systems clearly communicate their function and reliability. Features such as strong physical design, intuitive feedback, and visible access records can enhance user confidence and encourage correct use.

User familiarity and psychological factors further influence how locking systems are accepted and used. Traditional key-based systems are often trusted because they are familiar and provide a straightforward model of control. In contrast, newer technologies such as biometric or mobile-based access may be perceived as less trustworthy if their operation is not well understood or if users feel a loss of control. This places an additional responsibility on locksmiths to educate clients, explaining how systems function, what risks exist, and how control is maintained. Without this understanding, even highly secure systems may be rejected or used improperly.

Control and ownership are central to building trust in locking systems. Physical keys offer a clear and direct form of control, whereas digital systems may introduce uncertainty due to reliance on applications, cloud services, or remote permissions. Where users perceive control as external or unclear, trust can diminish. Locksmiths must therefore prioritise systems that provide transparent and user-understandable control mechanisms, such as revocable credentials, clearly defined permissions, and accessible access logs.

Reliability also plays a critical role in shaping user trust and behaviour. Systems that operate consistently and provide clear confirmation of access reinforce correct usage patterns and build confidence over time. In contrast, systems that behave unpredictably or lack clear feedback can lead to frustration, mistrust, and the development of insecure workarounds. For this reason, locksmiths must consider not only the security of a system but also its dependability and clarity in everyday use, particularly in high-frequency or high-risk environments.

In commercial and organisational settings, the broader social environment further influences the effectiveness of locking systems. Security practices are shaped by workplace culture, visible enforcement, and shared behaviours. Locksmiths operating in these contexts must consider how systems will function within this environment, including the presence of policies, training, and visible security measures such as monitored access points or credential checks. The success of a system depends not only on its design but also on how it is supported and reinforced within the organisation.

Overall, this perspective highlights that locksmithing has evolved into a multidisciplinary field that integrates engineering, human behaviour, and organisational considerations. Trust is not uniform; it varies according to the system, the context, and the user's experience. Consequently, locksmiths must adopt flexible, context-sensitive approaches when selecting and designing security solutions.

In practice, this means prioritising user-centred design, ensuring clear and transparent system operation, acting as both technician and advisor, and integrating mechanical, digital, and behavioural elements into a cohesive solution. Ultimately, the effectiveness of a locking system is determined not solely by its technical strength, but by the extent to which it is trusted, understood, and used correctly in real-world conditions.

Ultimately, this reframing of locksmithing invites you to look beyond the lock itself and consider security as a living system shaped by people, perception, and context. The most effective solutions are not necessarily the most complex or technologically advanced, but those that are understood, trusted, and consistently used as intended. In this sense, locksmithing becomes as much about insight and design as it is about tools and mechanisms. By recognising that true security emerges from the alignment of technical strength, human behaviour, and social context, practitioners are better equipped to create systems that not only resist attack, but also function reliably in the realities of everyday use.

## "Perceived Security" Vs Actual Security

Imagine standing in front of two doors: one fitted with a large, heavy lock that *looks* secure, and another with a sleek, almost invisible smart system that quietly manages access in the background. Which one would you trust more? For most people, the answer is immediate—and often wrong. This simple scenario highlights a powerful truth at the heart of modern locksmithing: what people *believe* about security can differ greatly from what actually protects them. Understanding this gap between perception and reality is essential, because it shapes not only how systems are chosen, but how they are used, trusted, and ultimately how effective they are in practice.

The distinction between perceived security and actual security is a critical lens through which to examine why some locking and access-control systems perform well in practice while others falter, despite possessing robust technical specifications. Perceived security refers to the subjective sense of protection that users experience, which is often shaped by visible, tangible cues and psychological factors. Visual features such as lock size, the presence of a keypad, alarm stickers, surveillance cameras, and warning signage can generate a reassuring impression of safety even when the system's underlying protections are comparatively weak or easily bypassed [26, 27]. This alignment or misalignment between appearance and capability has important consequences: if users rely on perceptual signals rather than technical realities, they may develop false confidence that promotes lax security behaviours or premature relaxation of more substantive protections [26, 27]. Conversely, the mere presence of familiar mechanical cues—such as tactile feedback from traditional locks—may strengthen perceived security even when contemporary systems offer superior, but less tangible, protections such as encryption or fortified ingress controls [26, 28]. In this sense, perceived security acts as a heuristic that can both deter casual intruders and lull determined adversaries into misjudging the level of risk, highlighting the divergence between what is believed and what is enforceable in practice [26, 27].

Actual security, by contrast, rests on engineering robustness, rigorous design, and demonstrable resistance to attack. It encompasses material strength, the quality of internal lock mechanisms, resistance to picking, drilling, bypass techniques, and, for digital systems, encryption standards and secure integration within the overall system. Importantly, actual security is typically validated through controlled testing, certifications, and evidence of real-world performance under attempted breaches, which are often invisible to non-specialists who would otherwise rely on intuitive visual cues [26, 28, 29]. The divergence between perceived and actual security can thus yield two distinct pathologies. A system with high perceived but low actual security creates false confidence and a brittle security posture susceptible to subversion by determined actors; a dummy deterrent such as a fake camera or a low-grade lock may deter casual observation yet fail under targeted inspection or sophisticated attack, illustrating the risk of overemphasizing perceptual signals at the expense of substantive protection [26, 27]. Conversely, a system with impressive actual security but modest perceptual salience may be underutilized or mistrusted if users cannot experience its strength, reducing the likelihood of proper engagement or adherence to security procedures [27, 28].

The tension between perception and reality is particularly evident in modern access-control ecosystems where traditional mechanical devices tend to convey robust security through tactile engagement and visible engagement, even when their protections lag behind more advanced digital systems. Smart locks, encrypted credentials, and networked access controls may deliver superior actual security by leveraging modern cryptographic protections and rigorous access management but can appear opaque or impalpable to users who base trust on straightforward, visible cues. In such cases, security professionals face a twofold obligation: to maintain or elevate actual security through robust design and testing, and to foster user trust and accurate mental models by making the security properties more visible or understandable without compromising system integrity [26, 28, 29]. The psychological dimension of security is not merely supplementary; it directly shapes how systems are adopted, used, and maintained in practice, reinforcing the need for a design philosophy that intentionally harmonizes perception with measurable protection [26, 27, 30].

Human behaviour amplifies the perception-actual security gap, as individuals frequently base decisions on what they perceive rather than on technical indicators. Visual deterrents, prior experiences, and generic assurances can dominate risk assessment, guiding choices in ways that may either reinforce or undermine protective measures. This tendency highlights a fundamental challenge for locksmiths and security professionals: how to align user beliefs with actual capabilities while preserving or enhancing the deterrent effect of perceptual cues. Strategies to achieve this alignment include incorporating visible indicators of strength that are truthful proxies for robustness, providing clear feedback mechanisms (such as audible confirmations or visual indicators) that reflect successful operation, and, crucially, educating users about how the system works and why its security level is appropriate for the context. When users understand the relationship between the physical or digital hardening of a system and its demonstrated protective performance, the likelihood of genuine adherence to security practices increases, reducing the risk of both false confidence and misplaced scepticism [26, 27, 30].

In the domain of physical security, the interplay between perception and protection has practical ramifications for installation choices, policy, and practice. CCTV and surveillance technologies exemplify how visible security apparatus can influence perceived safety even when their effectiveness is contingent on broader operational factors such as placement, functionality, and integration with response protocols. Research on CCTV presence in business contexts indicates that stakeholders' perceptions of safety and deterrence are significantly shaped by the visibility and perceived reliability of surveillance, with the location and usability of cameras shaping confidence and behaviour alongside actual deterrent effects [27, 30-32]. This body of work suggests that the perceptual aura surrounding surveillance can function as a psychological shield, reducing fear and improving perceived security, while simultaneously pushing organizational emphasis toward ensuring that the installed measures are aligned with concrete security capabilities and processes. The practical takeaway for practitioners is that deploying perceived deterrents should be complemented by strengthening the concrete value of the security program through proper maintenance, testing, and user education to avoid the mismatch that undermines effective protection [30-32].

A specificity that underscores the need for harmonization across perception and performance emerges in the context of modern access-control and deterrence systems. While traditional mechanical locks sustain a sense of reassurance through familiar tactile feedback, modern encrypted access systems promise higher actual security through cryptography and integrated control architectures, yet may have their credibility undermined by intangible or technocratic interfaces. This tension can be addressed by combining visible strength cues with transparent, audibly or visually verifiable feedback that communicates successful operation, and by building user trust through clear explanations of how the system resists common attack vectors such as bypass, picking, or tampering. In practice, such alignment requires thoughtful design that preserves the deterrent value of perceptual signals while simultaneously ensuring robust, certifiable protection underneath. The broader literature on perceived versus actual security, including experimental work on security perceptions and the deterrent effect of surveillance, supports the conclusion that effective security design must integrate human factors with engineering rigor to realize the true protective potential of any locking or access-control solution [26, 27, 30, 32].

The strategic imperative for locksmiths and security professionals is thus to craft systems in which what users believe about a system mirrors its real protective capability. This involves not only selecting robust hardware and sophisticated software protections but also engineering user experiences that accurately convey strength and reliability. Practical approaches include deploying visible indicators of system integrity, ensuring feedback mechanisms faithfully reflect system status, and, crucially, educating users about the nature of the protections in place and how they work together to deter and withstand attack. When perception and reality converge, security outcomes improve: users feel secure, and they are genuinely protected against unauthorized access or manipulation. The literature on perception versus reality in security contexts, including empirical studies on randomized security measures and the deterrent impact of surveillance, provides a rich evidence base for developing such integrated strategies and underlines the necessity of aligning human cognition with engineering performance to achieve truly effective protection [26, 27, 30-32].

## Impact on Locksmithing Practices

This distinction between perceived security and actual security is highly relevant to anyone interested in locksmithing because it directly explains why some locking systems succeed in the real world while others fail—regardless of their technical quality. Locksmithing is not just about installing the strongest possible lock; it is about ensuring that the lock is both technically secure and correctly used. If users misunderstand a system, overestimate its strength, or fail to trust it, the effectiveness of even the most advanced locking solution can be significantly reduced. This means that understanding perception is not optional—it is central to achieving real security outcomes.

In practical locksmithing, this insight reshapes how systems are selected, installed, and presented to clients. Locksmiths must recognise that users often judge security based on visible cues and intuitive signals, such as the size of a lock, the presence of keypads or cameras, or the feedback a system provides when operated. These cues influence behaviour. A system that *looks* secure may encourage compliance and deter casual intrusion, while a system that appears weak or confusing may be ignored, bypassed, or misused, even if it is technically superior. As a result, locksmiths must balance engineering strength with perceptual clarity, ensuring that the system both performs well and communicates its strength effectively.

This also influences how locksmiths approach modern technologies. Digital and smart locking systems often provide higher levels of actual security through encryption and controlled access management, but they can appear opaque or intangible to users. Without clear feedback or understanding, users may distrust these systems or fail to use them properly. Locksmiths therefore take on an expanded role as educators and advisors, explaining how systems work, what protections are in place, and how users retain control. This helps align user perception with actual capability, reducing both scepticism and misuse.

The concept also affects how deterrence is designed in real environments. Visible security measures, such as robust-looking locks, cameras, or access control points, can influence behaviour by increasing perceived risk for potential intruders and reinforcing correct behaviour among authorised users. However, locksmiths must ensure that these visible measures are backed by genuine security capability, as relying on appearance alone can create dangerous gaps. A system that only looks secure may deter casual threats but will fail against determined ones.

Human behaviour is a critical factor in this relationship. People tend to act based on what they believe is secure, not what is technically proven. This means locksmiths must design and recommend systems that guide behaviour in the right direction—through clear feedback, intuitive operation, and appropriate visibility of security features. When perception and reality are aligned, users are more likely to follow procedures, maintain systems properly, and avoid risky shortcuts.

This concept reinforces a key principle for modern locksmithing practice: security is both a technical and psychological outcome. Effective locksmiths do not simply install locks, they design security solutions where what users see, feel, and believe accurately reflects the system's true level of protection. By aligning perceived security with actual security, locksmiths can create systems that not only resist attack, but are also trusted, used correctly, and effective in everyday conditions.

In the end, the true measure of a locking system is not just how well it resists attack, but how accurately it communicates its strength to those who rely on it. When perception and reality are aligned, security becomes both visible and genuine, users act with confidence, systems are used as intended, and protection is reinforced rather than undermined. For the modern locksmith, this represents a shift from simply installing hardware to shaping understanding and behaviour. By bridging the gap between what people *see* and what actually *protects* them, locksmiths play a critical role in ensuring that security is not only engineered, but truly realised in everyday practice.

## Why Locks Fail: Human Behaviour Over Mechanical Failure

It is often assumed that when a lock fails, the problem lies in the metal, the mechanism, or the technology behind it. Yet, in many real-world situations, the failure begins long before any tool touches the lock, it begins with human behaviour. Doors left unsecured, systems ignored, shortcuts taken in the name of convenience, these everyday actions quietly undermine even the most carefully engineered protections. To understand why locks truly succeed or fail, it is necessary to look beyond the hardware and examine the human element at the centre of every security system.

In real-world locking systems, failures are rarely caused by fundamental flaws in mechanical design or raw materials. Instead, a substantial portion of breaches and ineffective security arises from how people interact with locks and access control technologies. The literature across security usability, human-centered design, and usable authentication consistently shows that human behaviour, whether inconsistent use, shortcutting, miscomprehension, or social dynamics, often undermines engineered protections. This synthesis integrates evidence from diverse sources to argue that locksmiths and security professionals must design systems that anticipate and accommodate human behaviour, rather than assume perfect user compliance. The synthesis also identifies practical design, training, and policy implications supported by multiple, converging sources.

A foundational point across the literature is that user behaviour can nullify even robust technical protections. Usability-focused security research demonstrates that the effectiveness of security controls depends on user interaction patterns as much as on cryptographic strength or mechanical resilience. When users fail to engage a lock, mismanage credentials, or disable features perceived as inconvenient, the system's protective value deteriorates substantially. This observation is echoed by studies of secure system design and human-centric security approaches, which emphasize that usability and security are deeply interdependent and that complex authentication or access controls often degrade compliance and security outcomes if not designed with users in mind [33-35]. The broader security community consistently notes that human factors are a critical determinant of whether security measures succeed or fail, necessitating design choices that reduce friction while preserving security properties [34, 35] .

Inconsistent engagement with locks is identified as a leading cause of security failures. When doors are left unlocked for convenience or treated as "safe enough" due to a perception of low risk, the protective effect of even the strongest mechanical or electronic locking system is compromised. This pattern is frequently observed in residential and familiar workplace settings where repeated safe experiences create complacency and diminish vigilance over time [33, 36, 37]. The literature on usable security further supports this point by showing that users often bypass or override security controls to save time or effort, such as propping doors open, sharing access credentials, or disabling features considered burdensome or inconvenient. Such shortcut behaviours can invalidate multiple layers of engineered security with a single act of convenience, highlighting the importance of designing for natural work patterns and behavioural economics [33, 35, 36].

A lack of user understanding about how a locking system works is a major contributor to misuse and non-use, particularly for more advanced technologies such as electronic or smart locks. Misunderstandings can lead to failure modes such as neglecting secondary locking mechanisms, mismanaging digital credentials, or ignoring system alerts. Trust and comprehension are critical; when users do not trust or understand a system, misuse and avoidance follow. This emphasizes the need for clear feedback, intuitive interfaces, and educational components integrated into the lock design and deployment process [34, 35, 37]. The research on human-centered development reinforces that usable design must consider users' cognitive load and mental models to support correct operation and interpretation of feedback signals from security devices [34, 37].

Security practices do not occur in a vacuum. Group norms and observed behaviours strongly influence individual actions, such that if peers routinely ignore protocols, others are more likely to mimic that non-compliant behaviour. Over time, this normalization of non-compliance erodes even well-designed systems and increases breach likelihood. This social dimension of security behaviour is well-documented in human factors and security literature, underscoring that education and policy must address not only individual users but also the social environment in which security decisions are made [34, 37, 38].

Paradoxically, the absence of incidents can breed complacency. When security systems function without issue for extended periods, users may underestimate risks and reduce attention to procedures or checks. This tendency to rely on routine and infer safety from a lack of past problems is a well-known driver of non-adherence to security protocols and an important consideration in the design of training and reinforcement strategies for locksmithing and access control systems [35, 36, 38].

Mechanisms for physical lock failure, wear, environmental exposure, or maintenance lapses, are often predictable and addressable through inspection and replacement. By contrast, human behaviour is less predictable and cannot be engineered out entirely; it requires proactive management through design, training, and reinforcement strategies that acknowledge human limitations and variability. This distinction emphasizes a preventative security philosophy centred on human-centered design and ongoing user education, rather than relying solely on improvements in hardware [33-35].

Implications for lock design and practice: strategies for human-centered security Given the central role of human behaviour in lock system effectiveness, security professionals should pursue design strategies that reduce the opportunity for human error and support correct usage. Practical strategies found in the literature include:

- Simplifying operation and reducing cognitive load: locks and access control mechanisms should be intuitive, with minimal steps required to engage or disengage security features. This reduces the likelihood of user error and inconsistent engagement, aligning with human-centered design principles that prioritize usability alongside security goals [34, 35].
- Incorporating automatic locking features and clear feedback: automatic engagement after door closure or when a user leaves a space, along with immediate, obvious feedback when a lock is engaged or an alarm is triggered, helps close gaps created by inconsistent user action and misinterpretation of system state [34, 35].
- Layered security and non-reliance on single user actions: implementing multiple protective layers that do not rely on a single user action (e.g., separate hardware, software, and procedural controls) can mitigate the impact of shortcut behaviour or forgetfulness. This approach resonates with broader design patterns in secure systems that balance usability and risk across multiple controls [33-35].
- Education and continuous awareness: training that recognizes diverse learning styles and roles, combined with ongoing reinforcement and evaluation of security culture, strengthens user engagement with security practices and reduces non-compliance due to complacency or misunderstanding [34, 38, 39].
- Designing for accessibility and inclusive use: ensuring that authentication and access controls accommodate users with diverse cognitive or physical abilities reduces the likelihood of non-use or misuse due to inaccessibility and cognitive barriers. This is echoed in research exploring accessibility considerations in authentication and security usability [35, 37].

The convergence of findings across disparate domains strengthens the case for prioritizing human-centered approaches in locksmithing and security engineering. Studies on human-centered cybersecurity training argue for integrating usability, education, and engagement to foster a culture of security within organizations. These frameworks emphasize aligning security practices with human capabilities and organizational realities, which is directly applicable to the design and deployment of locks and access control systems in real-world contexts [34, 39] . Related analyses on secure system design highlight the need to balance usability with core security principles, recognizing that user experience strongly influences the effectiveness of security controls and that trade-offs between ease-of-use and security must be navigated with careful, evidence-based design considerations [33, 35]. In the realm of authentication and access control technologies, work on password usability and feedback mechanisms demonstrates that human factors can both strengthen and weaken security, depending on how feedback and usability are implemented. These insights support the assertion that lock systems should guide users toward correct behaviour without imposing excessive cognitive or operational burden [35, 40].

While there is broad consensus on the centrality of human factors, some literature notes nuance in balancing usability and security. For example, discussions about the inevitable trade-offs between strong authentication and user convenience highlight that perfect usability and perfect security cannot be achieved simultaneously; instead, the goal is to achieve an acceptable equilibrium that minimizes risk while preserving workflow efficiency [33, 35] . Other sources emphasize the importance of continuous evaluation and adaptation, noting that security culture and user behaviours evolve with technology, policies, and organizational changes. This implies that locksmiths and security teams must engage in ongoing assessment, training, and redesign to maintain alignment with user needs and threat landscapes [34, 38, 39].

Practical recommendations for locksmiths and security professionals

- Prioritize user-centred design in lock selection and installation, choosing mechanisms that align with typical user workflows and reduce opportunities for error.
- Integrate automatic locking and explicit feedback mechanisms, ensuring users clearly perceive when a lock is engaged and when access has been granted or denied.
- Implement multi-layered security strategies that do not place sole responsibility on a single user action, thereby reducing risk from inconsistent use or shortcut behaviour.
- Develop and deliver targeted education programs that accommodate diverse learner profiles and emphasize the importance of consistent engagement with locking mechanisms.
- Assess accessibility and inclusivity in authentication and locking interfaces, addressing cognitive and physical barriers that could lead to misuse or non-use.
- Monitor social norms within organizations and communities, actively promoting compliant behaviours through leadership, peer modelling, and reinforcement of protocols.
- Establish regular maintenance and inspection routines for mechanical components while recognizing that human factors may require changes in training, signage, and feedback to sustain effectiveness over time.

Across locksmithing, security engineering, and human-centered design, the weight of evidence indicates that human behaviour is the dominant determinant of lock system performance in real-world settings. While mechanical wear and defects can contribute to failures, these factors are comparatively predictable and manageable. The unpredictable nature of human interaction with locks—encompassing inconsistent engagement, shortcutting, misunderstandings, social dynamics, and complacency—demands a deliberate, user-centred approach to design, deployment, and education. By embracing designs that minimize user error, provide clear feedback, and reinforce secure behaviour through layered controls and ongoing training, locksmiths and security professionals can enhance the resilience of locking systems in the face of diverse human factors. The literature consistently supports this integrated approach, demonstrating that usability, accessibility, and human-centered strategies are not merely add-ons but essential components of effective physical and digital access security [33-35, 37, 38, 40].

## Impact on Locksmithing Practices

Traditionally, locksmithing has been grounded in mechanical strength, precision engineering, and resistance to physical attack. However, the evidence presented here demonstrates that even the most technically robust lock can fail if it is not used correctly, consistently, or as intended. In practice, this means that security is not determined solely by the quality of the hardware, but by the interaction between the system and the people using it. For a locksmith, this shifts the focus from purely technical performance to real-world effectiveness, where human behaviour becomes a central consideration.

The influence on locksmithing practice is significant. It expands the role of the locksmith beyond installation and repair into that of a security designer and advisor. When selecting and installing locking systems, locksmiths must now consider how users will interact with them: whether the system is intuitive, whether it aligns with daily routines, and whether it reduces the likelihood of error or misuse. For example, a highly secure system that is complex or inconvenient may encourage users to bypass it entirely, whereas a slightly less complex but well-designed system may achieve better security outcomes through consistent use.

This perspective also places strong emphasis on usability and human-centred design in locksmithing decisions. Locks and access control systems must be easy to understand, require minimal effort to operate, and provide clear feedback to the user. Features such as automatic locking, audible or visual confirmation, and simple interfaces are no longer optional enhancements—they are critical components of effective security. Locksmiths must therefore evaluate not only how resistant a system is to attack, but also how likely it is to be used correctly under normal conditions.

Another key influence is the growing importance of education and communication. Locksmiths are increasingly required to explain systems to clients, ensuring they understand how the lock works, what its limitations are, and how to use it properly. Without this understanding, even well-designed systems can be misused or ignored. This positions the locksmith as an educator who helps align user behaviour with the intended security function of the system, reducing the gap between design and real-world use.

The concept also affects how locksmiths approach risk and system design through layered security. Recognising that human error is inevitable, effective locksmithing incorporates multiple layers of protection that do not rely on a single user action. This might include combining mechanical locks with electronic controls, alarms, or monitoring systems to ensure that if one layer is compromised through human behaviour, others remain in place. This layered approach reflects a shift toward resilience rather than reliance on perfect compliance.

Social and environmental factors further influence locksmithing practice. In workplaces or shared environments, behaviour is shaped by group norms and organisational culture. Locksmiths working in these contexts must consider how systems will be used collectively, not just individually. This may involve recommending solutions that reinforce correct behaviour through visibility, accountability, and ease of use, as well as supporting policies and training that promote consistent engagement with security measures.

Finally, this understanding reinforces that locksmithing is increasingly a multidisciplinary profession, integrating engineering, psychology, and behavioural science. The effectiveness of a lock is no longer judged solely by its resistance to physical attack, but by how well it fits into human systems of use, trust, and routine. As a result, modern locksmithing practices are shaped by the need to design systems that anticipate human limitations, guide behaviour, and support consistent, correct use.

In practical terms, this means that successful locksmithing is not about creating the strongest possible lock in isolation, but about delivering a complete security solution where hardware, user behaviour, and environment work together. By recognising the dominant role of human behaviour in security outcomes, locksmiths can design and implement systems that are not only technically sound, but also reliable and effective in everyday use.

Overall, the strength of a lock is only part of the story—the real test lies in how it performs in the hands of the people who use it every day. A perfectly engineered system can be undone by a moment of distraction, habit, or convenience, while a well-designed, user-focused system can reinforce secure behaviour without resistance. This shifts the focus of locksmithing from simply building stronger locks to designing smarter systems that work with human nature rather than against it. By recognising that people are not the weakest link but a central component of security, locksmiths can create solutions that are not only resilient in theory, but reliable in the realities of everyday use.

## Ethical Locksmithing in Modern Society

Before a locksmith ever touches a cylinder or programs a smart credential, they are being trusted with something far more valuable than hardware: access. In a single job, a locksmith may be handed the keys to a family's safety, a business's assets, or an organisation's sensitive systems—often in moments of urgency, vulnerability, or high stress. That reality makes locksmithing one of the most trust-dependent trades in modern society, where a small ethical lapse can have consequences as serious as a technical failure. This section therefore begins from a simple premise: the true measure of professional locksmithing is not only what a practitioner *can* do, but what they *choose* to do—lawfully, transparently, and with the client's security and privacy at the centre of every decision.

Ethical locksmithing in modern society extends far beyond technical skill. Locksmiths operate at a unique intersection of trust, security, privacy, and legal responsibility, often being granted access to homes, businesses, vehicles, and sensitive systems. This position places them in a role where their actions can either protect or compromise the safety of individuals and organisations. As a result, ethical practice is not optional—it is fundamental to the profession. The ethical framework guiding locksmith practice must therefore encompass confidentiality, authorised access, privacy protections, transparency in service delivery, safety advocacy, and strict legal compliance across evolving technologies and regulatory contexts [41-46].

At the core of ethical locksmithing is the principle of authorised access only. Locksmiths must ensure that any service they provide, such as gaining entry, rekeying locks, or installing systems, is performed only for individuals who have a legitimate right to access the property. This requires careful verification of identity and authority, even in distressed or urgent situations. Failing to confirm authorisation can result in unlawful access, legal consequences, and harm to others. The emphasis on verification aligns with broader data protection and privacy norms that treat consent, legitimacy, and purpose limitation as central to lawful data processing and access control in security-relevant environments [41, 42]. While the literature on locksmithing-specific authorisation is sparse, the general consensus across privacy- and security-focused scholarship is that access should be mediated by documented authority and verifiable identity checks to reduce misappropriation and harm [41-43]. In analogous high-trust occupations, recourse to transparent procedures, dual custody where feasible, and verifiable credentials are routinely recommended to safeguard both clients and practitioners [44, 45].

Closely related to this is the responsibility to protect privacy and confidentiality. Locksmiths often become aware of security vulnerabilities, access codes, key systems, or personal circumstances during their work. Ethical practice requires that this information is never disclosed or exploited. For example, knowledge of a client's locking system or security weaknesses must not be shared, as doing so could directly endanger the client. Maintaining confidentiality builds trust and upholds the integrity of the profession. The primacy of confidentiality in professional ethics is reinforced across disciplines, including data protection law and professional codes, which describe confidentiality as a foundational pillar of trust and professional integrity [41, 44]. Additionally, data protection frameworks emphasise that handling personal data—such as access credentials or system configurations—requires safeguards against disclosure, access by unauthorized parties, and inappropriate use, thereby supporting the locksmith's obligation to keep such information secure [42, 46]. In practice, confidentiality obligations are strengthened by combining technical measures (limited data collection, encryption, access controls) with organisational policies and staff training that address the specific risks of locksmith work [41-43].

Modern locksmithing also involves increasing engagement with digital and smart security systems, which introduces additional ethical considerations. Access control systems, smart locks, and mobile credentials often involve user data, access logs, and network connectivity. Locksmiths must handle this information responsibly, ensuring compliance with privacy laws and cybersecurity principles. This includes safeguarding login credentials, avoiding unauthorized data access, and ensuring systems are configured securely. The expansion of digital security modalities necessitates a privacy-by-design approach, where privacy considerations are integrated into the system design, configuration, and aftercare. Hardware-based security mechanisms and trusted execution environments provide a model for how locksmiths can support privacy and security in cloud- or device-centric contexts by establishing secure execution domains and auditable privacy controls within devices and networks [43]. Moreover, systematic reviews of cloud- and data-protection practices under LGPD and related regulations emphasize that transparent, privacy-preserving configurations, coupled with robust encryption and access management, are essential for sustaining client trust in technologically advanced security deployments [42, 46]. In this sense, the locksmith's advisory role increasingly resembles that of a security consultant, recommending configurations and controls that balance usability with risk reduction and privacy preservation [44].

**LGPD and Related Regulations**

The LGPD, or *Lei Geral de Proteção de Dados* (General Data Protection Law), is Brazil's primary data protection legislation. It regulates how personal data is collected, processed, stored, and shared, and is broadly aligned with international frameworks such as the General Data Protection Regulation (GDPR) in the European Union. Introduced under Law No. 13,709/2018 and effective from 2020, the LGPD was designed to strengthen individual privacy rights while establishing clear responsibilities for organisations handling personal information.

At its core, the LGPD aims to protect the privacy and personal data of individuals in Brazil, regulate how organisations manage both personal and sensitive data, and establish enforceable rights for individuals—referred to as data subjects. It also imposes obligations and penalties on organisations that misuse or inadequately protect data. Importantly, the law has extraterritorial reach, applying not only to organisations operating within Brazil but also to any entity processing the personal data of individuals located in Brazil, regardless of where the organisation itself is based.

A key feature of the LGPD is its distinction between general personal data and sensitive personal data. Personal data includes any information capable of identifying an individual, such as names, email addresses, phone numbers, IP addresses, or location data. Sensitive personal data, however, is subject to stricter protections and includes categories such as health information, biometric data, and religious, political, or genetic details. This distinction is particularly significant in modern security contexts, where technologies such as biometric access systems and smart locks routinely process highly sensitive data.

Under the LGPD, organisations must have a lawful basis for processing personal data. Common legal bases include obtaining consent from the individual, fulfilling a legal obligation, performing a contract, or relying on legitimate interests—provided appropriate safeguards are in place. These principles closely mirror those found in the GDPR, reinforcing a global convergence toward structured, accountable data governance.

The law also grants individuals a range of enforceable rights over their data. These include the right to access their personal information, correct inaccuracies, request deletion, withdraw consent, and understand how their data is being used. Collectively, these rights shift control toward individuals and require organisations to be transparent and responsive in their data handling practices.

To comply with LGPD, organisations are required to implement robust security measures to protect data, establish clear and accessible privacy policies, report data breaches, and in many cases appoint a Data Protection Officer (DPO). They must also adhere to principles such as data minimisation, ensuring that only the necessary data is collected and retained. Non-compliance can result in significant penalties, including fines of up to 2% of a company's revenue in Brazil, subject to a statutory cap.

Enforcement of the LGPD is overseen by the Autoridade Nacional de Proteção de Dados (ANPD), Brazil's national data protection authority. The ANPD is responsible for issuing regulatory guidance, investigating data breaches, and applying administrative sanctions, playing a central role in shaping how the law is interpreted and enforced in practice.

LGPD exists within a broader global movement toward stronger data protection regulation. Comparable frameworks include the Privacy Act 1988 in Australia, which incorporates the Australian Privacy Principles; the California Consumer Privacy Act (CCPA) in the United States, which emphasises transparency and consumer rights; and China's Personal Information Protection Law (PIPL), which imposes strict controls on data processing and cross-border transfer. Together, these regulations reflect a shared international recognition of the importance of safeguarding personal information in an increasingly digital world.

The relevance of LGPD is particularly evident in modern locksmithing and security system design. Contemporary security solutions—such as smart locks, biometric authentication systems, cloud-based access control, and CCTV—routinely collect and process personal data, including entry logs, user identities, and biometric credentials. As a result, locksmiths and security providers may be classified as data controllers or processors under the law, bringing legal obligations alongside technical responsibilities.

In practical terms, this means that systems must be designed not only for physical security but also for data protection. This includes implementing encryption and secure access controls, ensuring transparency so users understand what data is being collected, and maintaining user control, such as the ability to revoke access or request data deletion. For example, installing a fingerprint-based access system in Brazil would require explicit user consent, secure storage of biometric data, and mechanisms to delete that data upon request.

Ultimately, the LGPD and similar global regulations signal a fundamental shift in how security is understood. Security is no longer limited to controlling physical access; it now extends to protecting the data generated by access systems themselves. For modern security professionals, including locksmiths, this requires a dual focus on physical and digital protection, as well as an ability to balance security, usability, and legal compliance in an increasingly interconnected environment.

Mapping the LGPD to smart locks, access control systems, and CCTV installations requires translating legal principles into practical decisions about system design, installation, and day-to-day operation. For locksmiths and security professionals, this means developing a clear understanding of how each technology collects, stores, and processes personal data, and ensuring that compliance is maintained across the entire lifecycle of the system—from installation through to ongoing use and eventual decommissioning.

Smart locks, particularly those that are biometric or app-based, are among the most directly affected technologies under LGPD because they routinely process both personal and sensitive data. Personal data may include user names, email addresses, phone numbers, device identifiers, IP addresses, and detailed access logs that record when and how a person enters a space. Where biometric authentication is used, such as fingerprints, facial recognition, or voice identification, the data becomes classified as sensitive and subject to higher levels of protection. Under LGPD, this requires a clear lawful basis for processing, most commonly explicit and informed consent. Users must understand what data is being collected, how it is used, and where it is stored. Systems must also adhere to data minimisation principles, ensuring that only necessary data is collected, and that biometric information is stored securely—typically as encrypted templates rather than raw images. In practice, locksmiths must prioritise systems that support encryption, allow users to delete or revoke credentials, and provide transparency about data use. They also play a key role in educating clients about managing users, revoking access, and maintaining compliance over time.

Access control systems, whether in residential complexes or commercial environments, introduce additional complexity due to the involvement of multiple users, centralised databases, and continuous monitoring. These systems typically process personal data such as employee or tenant names, identification numbers, card or fob credentials, and detailed access logs showing entry and exit times. Where biometrics are used, the same heightened protections apply. Under LGPD, organisations must ensure that data processing has a lawful basis, often legitimate interest or legal obligation in workplace settings, while still maintaining transparency and proportionality. Data must only be used for its intended purpose, such as security and access management, and not for unrelated monitoring without disclosure. Strong governance is required, including role-based access for administrators, audit trails for system activity, and clearly defined data retention periods. Locksmiths and system installers must therefore configure systems to enforce these controls, ensure that clients understand data ownership and responsibilities, and select technologies that support auditability, secure storage, and breach response processes.

CCTV and surveillance systems are also heavily regulated under LGPD due to their ability to capture identifiable visual data, often continuously and in shared or public environments. Personal data in this context includes video footage, facial images, and associated metadata such as time and location. In some cases, footage may indirectly reveal sensitive information, such as behavioural patterns or personal characteristics. The lawful basis for CCTV is typically legitimate interest in security; however, this must be balanced against individual privacy rights. Transparency is critical, requiring clear signage that informs individuals they are being recorded and explains the purpose of surveillance. Data minimisation principles require careful camera placement to avoid unnecessary or intrusive monitoring, and retention limits must be established so that footage is not stored indefinitely. Access to recordings must be restricted to authorised personnel, with logs maintained to track usage. For locksmiths, this translates into practical responsibilities such as installing cameras in appropriate locations, ensuring compliant signage is in place, configuring automatic deletion periods, and selecting systems with secure storage and controlled access features.

Across all these technologies, several core LGPD principles apply consistently. Security by design requires that systems incorporate encryption, strong authentication, and secure update mechanisms from the outset. Privacy by design emphasises limiting data collection, providing user control, and ensuring that systems default to privacy-protective settings. Accountability is also a key consideration, as locksmiths and installers influence system design and may share responsibility if insecure or non-compliant systems are implemented. Effective data lifecycle management is essential, covering every stage from data collection and storage through to use, retention, and eventual deletion, all of which must be controlled and documented.

The broader implication is that LGPD fundamentally changes the nature of locksmithing practice. No longer confined to physical hardware, the profession now requires an understanding of how security systems generate and manage data. Installing a lock or access system involves not only ensuring physical protection but also ensuring legal and ethical compliance with data protection requirements. This includes selecting systems that support encryption, consent management, and data deletion, as well as designing solutions that balance security with user privacy.

Another key ethical obligation is honesty and transparency in service delivery. Clients often rely on locksmiths in urgent or vulnerable situations, such as lockouts or security breaches. Ethical locksmiths provide clear explanations of the work required, fair pricing, and appropriate solutions without exploiting the situation. Recommending unnecessary upgrades, overcharging, or using fear-based tactics to sell products undermines both the client's trust and the reputation of the industry. This emphasis on fair dealing and transparency is consistent with broader ethical and legal expectations in service industries that handle sensitive information and critical safety functions. Privacy- and data-protection-focused literature stresses that informed consent, transparent data practices, and avoidance of deceptive practices are central to sustaining trust and minimizing legal risk in high-stakes environments [41, 42, 46]. When communicating with clients about security needs, locksmiths should offer evidence-based assessments, present trade-offs clearly, and document agreed-upon actions to support accountability and prevent misunderstandings or disputes [44, 45].

Ethical locksmithing also includes a duty to promote safety rather than merely provide access. This means advising clients on appropriate security measures, identifying weaknesses in existing systems, and recommending improvements that align with the client's needs and risk profile. In this sense, the locksmith acts not just as a tradesperson but as a security advisor, helping clients make informed decisions about protecting their property and assets. The broader literature on privacy, compliance, and security highlights the professional obligation to translate technical findings into practical risk mitigation recommendations, balancing access needs with resilience against potential threats. This advisory role is reinforced by data-protection regimes and cybersecurity standards, which emphasise the need for protective controls, threat awareness, and risk-based decision making when dealing with sensitive security information [42-44, 46].

Legal compliance is another critical component. Locksmiths must operate within relevant laws, licensing requirements, and industry standards. In Australia, this includes adhering to state and territory regulations governing security work, as well as broader obligations related to privacy and consumer protection. While the Australian specifics are not exhaustively covered in the sources provided, the principle that professionals must stay informed about and comply with applicable privacy, consumer protection, and security regulations is consistently echoed across data-protection literature and regulatory analyses from multiple jurisdictions [41, 42, 45, 46]. Ethical practice involves ongoing professional development to ensure that licensing, credentialing, and security-service procedures meet current legal expectations, and that incidents or breaches are handled in accordance with regulatory requirements and professional codes [44]. This commitment to legal and regulatory alignment supports public trust, professional legitimacy, and the orderly functioning of the locksmithing field in a rapidly evolving security landscape [42, 44, 46].

Ethical locksmithing also requires an ongoing commitment to professional integrity and continuous improvement. As security technologies evolve, locksmiths must update their skills and knowledge to ensure they can provide safe and effective solutions. This includes understanding emerging threats, such as cyber vulnerabilities in smart locks, and adapting practices to address them responsibly. The literature on privacy, security, and professional ethics emphasises that continuous education, adoption of privacy-preserving design principles, and engagement with evolving regulatory frameworks are essential to maintain high standards and public confidence in professional practice [41-44, 46]. The integration of hardware-based security concepts into practical locksmith workflows demonstrates a concrete pathway for maintaining data privacy and system integrity in increasingly connected environments, reinforcing the need for ongoing training and adherence to best practices in security engineering [43].

Ethical locksmithing requires a holistic approach that integrates the protection of confidentiality, the verification of authorisation, responsible handling of digital security data, transparent client communication, safety-oriented advisory roles, and rigorous legal compliance. The convergence of traditional physical security with digital and cloud-enabled security systems makes these ethical commitments more important than ever, ensuring that the locksmith's trusted role contributes to the safety and privacy of individuals and organisations while upholding the integrity and reputation of the profession [41-44, 46].

### Impact on Locksmithing Practices

Locksmiths are granted access to some of the most sensitive environments—homes, businesses, vehicles, and increasingly digital systems—and this access comes with significant ethical, legal, and social responsibility. Understanding ethical locksmithing ensures that practitioners recognise the weight of this trust and the potential consequences of misuse. It reinforces that locksmithing is fundamentally a profession built on integrity, where technical competence alone is not sufficient; ethical judgement and accountability are equally critical to protecting clients and maintaining public confidence.

In practical terms, this directly influences how locksmiths approach everyday decision-making. For example, the requirement for authorised access means locksmiths must develop and follow strict verification procedures before undertaking work. This may include checking identification, confirming ownership or tenancy, and documenting consent. Rather than acting quickly based solely on a client's request, ethical practice requires a deliberate process that prioritises legitimacy over convenience. This reduces the risk of unlawful entry, protects vulnerable individuals, and safeguards the locksmith from legal liability.

The emphasis on confidentiality and privacy also reshapes locksmithing practice. Locksmiths frequently encounter sensitive information, such as key systems, access codes, security weaknesses, and client routines. Ethical practice requires that this information is protected at all times. In modern contexts, this extends beyond physical knowledge to digital data, including access logs, user credentials, and system configurations. As a result, locksmiths must adopt secure handling practices, limit data exposure, and ensure that any information obtained during their work is not disclosed or misused. This elevates the role from mechanical service provider to trusted custodian of security information.

The integration of digital technologies and data protection laws such as LGPD further expands the scope of ethical responsibility. Smart locks, access control systems, and CCTV installations generate and process personal data, meaning locksmiths may be involved in systems that fall under privacy legislation. This influences practice by requiring locksmiths to understand consent, data minimisation, secure storage, and user rights such as data access and deletion. It also means recommending systems that comply with these principles and advising clients on their obligations. In effect, locksmiths are no longer just installers—they are increasingly acting as security and data advisors, bridging physical and digital protection.

Ethical locksmithing also affects how professionals engage with clients through honesty and transparency. Clients often seek locksmith services in urgent or stressful situations, which creates a power imbalance. Ethical practice requires clear communication about costs, options, and limitations, avoiding unnecessary upselling or exploitation. This builds long-term trust and supports the reputation of the industry. It also aligns with broader consumer protection expectations and reinforces the idea that security services must be delivered with fairness and professionalism.

Another key influence is the shift toward a safety-focused advisory role. Ethical locksmiths do not simply provide access or install hardware; they actively assess risks, identify vulnerabilities, and recommend appropriate solutions. This may involve advising on lock upgrades, access control strategies, or integration with surveillance systems. By doing so, locksmiths contribute to proactive risk management rather than reactive problem-solving, enhancing overall security outcomes for clients.

Legal compliance is also a central influence on practice. Locksmiths must remain informed about licensing requirements, privacy laws, and industry standards across jurisdictions. Ethical practice requires ongoing professional development to ensure that services remain compliant as technologies and regulations evolve. This includes understanding obligations related to data breaches, record-keeping, and system security. Failure to comply not only risks legal penalties but can also damage professional credibility and trust.

This perspective reinforces that locksmithing is an evolving, multidisciplinary profession. It now sits at the intersection of mechanical engineering, cybersecurity, privacy law, and human behaviour. Ethical locksmithing requires continuous learning, adaptation to new technologies, and a commitment to best practice. It encourages practitioners to think beyond individual jobs and consider the broader impact of their work on safety, privacy, and trust.

Ultimately, the influence on locksmithing practice is profound: it transforms the role from one of technical execution to one of responsible security stewardship. A locksmith is not just someone who opens doors or installs locks, but a professional who protects access, safeguards information, and upholds trust. In this context, ethical practice is not an added consideration—it is the foundation that determines whether security systems truly serve the people they are meant to protect.

Ethical locksmithing is what turns a set of tools and technical knowledge into a profession worthy of trust. In a world of smart locks, biometric credentials, cloud platforms, and increasingly strict privacy regulation, the locksmith's reputation will be shaped as much by how they verify authority, protect client information, configure systems, and explain risks as by how cleanly they cut a key or fit a cylinder. The most effective locksmiths will be those who treat every job as both a security task and a stewardship task—leaving behind not just a working lock, but a safer system, a better-informed client, and a clear record of lawful, responsible practice. In that sense, ethics is not a constraint on locksmithing; it is the craft's modern edge, the thing that ensures security remains real, privacy remains respected, and trust remains earned every time a door is opened.

## Chapter 1 Review Questions

**Question 1**

## Beyond the Key

A company installs a high-security mechanical lock on a staff entrance door. Within weeks, employees begin leaving the door propped open because unlocking it repeatedly during busy periods slows down their work.

What is the most likely cause of the security failure?

A. The lock was incorrectly installed
B. The lock mechanism is defective
C. The security design did not consider user convenience and workflow
D. The door frame is too weak

**Question 2**

A school installs electronic keypad locks on several staff doors. After a few months, it is discovered that most teachers share the keypad code with colleagues to avoid repeatedly entering the code.

What principle of security design does this situation demonstrate?

A. Mechanical locks are always superior to electronic locks
B. Access control systems must align with real user behaviour
C. Electronic systems should never be used in schools
D. Codes should never be used in security systems

**Question 3**

A hospital installs strong locks on a restricted medication room but fails to monitor who enters the area. Staff frequently borrow keys from one another.

What is the primary weakness in this system?

A. The locks are not strong enough
B. The room should not be locked
C. Poor key control and access management
D. Lack of CCTV

**Question 4**

A retail store installs an advanced biometric access system for its stockroom. However, when the system temporarily fails during a network outage, employees are unable to access inventory needed for customers.

What security concept is illustrated in this situation?

A. Security systems must be as complex as possible
B. All locks should be biometric

C. Security systems must consider reliability and backup access methods
D. Mechanical locks should never be replaced

**Question 5**

A residential building replaces traditional keys with smartphone-based access credentials. Some residents struggle to use the mobile system and frequently request assistance.

What key issue has been overlooked?

A. Smartphone access is always insecure
B. User capability and accessibility considerations
C. Mobile access is illegal in residential buildings
D. Mechanical keys are obsolete

**Question 6**

A factory installs high-security locks on external doors but does not upgrade the door frames or hinges. During a break-in, the intruder forces the door frame rather than attacking the lock.

What does this scenario demonstrate?

A. Locks are ineffective
B. Criminals always pick locks
C. Security must consider the entire physical structure, not just the lock
D. Strong locks are unnecessary

**Question 7**

An office installs a complex access system requiring multiple authentication steps. Employees complain that it slows down entry during busy mornings.

What risk may arise if the system is not redesigned?

A. Employees may create unsafe workarounds
B. The building will become inaccessible
C. The locks will fail mechanically
D. The access system will automatically reset

**Question 8**

A company installs security cameras and strong locks but fails to establish procedures for managing staff access when employees leave the organisation.

What type of security weakness does this represent?

A. Mechanical failure
B. Procedural or administrative weakness
C. Electronic malfunction
D. Structural weakness

**Question 9**

A property manager chooses a simpler lock system instead of a highly complex one because it is easier for staff to operate and maintain.

What principle of security design is being applied?

A. Convenience should always replace security
B. Simplicity can improve compliance and security effectiveness
C. Complex locks are unnecessary
D. Mechanical locks are obsolete

**Question 10**

A building installs new electronic locks but does not provide training for staff. Many users repeatedly enter incorrect codes or forget how to operate the system.

What lesson does this scenario illustrate?

A. Electronic locks should not be used
B. Training and user education are essential parts of security implementation
C. Mechanical locks are always better
D. Security systems should be replaced annually

# Chapter 2

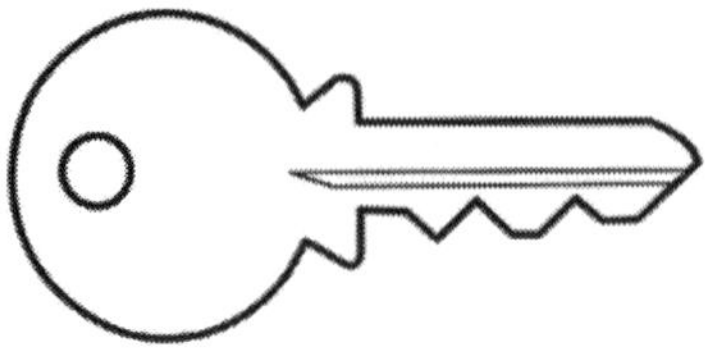

# The Evolution of Locking as a Social Technology

Chapter 1 established that security is never purely technical. It demonstrated that every locking system exists within a constant tension between security and convenience, and that its real-world effectiveness depends not only on mechanical strength or digital sophistication, but also on human behaviour, trust, and perception. Locks succeed or fail not simply because of how they are engineered, but because of how people understand, use, and sometimes bypass them. This foundation reframes locksmithing as a discipline that sits at the intersection of technology, psychology, and everyday practice, where design decisions must account for both physical protection and human interaction

Building on this understanding, Chapter 2 shifts the focus outward—from the individual lock and user interaction to the broader social systems in which locks operate. If Chapter 1 asked how locks work and why people trust or misuse them, this chapter asks a deeper question: what role do locks play in shaping society itself? It explores how locking systems function as social technologies that define boundaries, organise access, and reflect structures of power, control, and privacy. In doing so, it moves beyond mechanisms and materials to examine the social meaning embedded in every act of locking and unlocking.

This transition is critical. Once we recognise that security is influenced by human behaviour, it becomes clear that locks cannot be understood in isolation. They are part of larger systems of governance, culture, and technology that determine who has access, under what conditions, and with what consequences. Whether in a household, workplace, or digital network, locks act as instruments that both enforce and communicate rules about inclusion, exclusion, and authority.

Chapter 2 therefore expands the lens of locksmithing. It examines how access control evolves from simple physical barriers into complex, policy-driven systems involving identity, permissions, and data. It introduces concepts such as permissioning, delegation, and identity management, showing how modern access is no longer static but dynamic, contextual, and continuously managed. At the same time, it highlights how these systems raise important questions about privacy, accountability, and trust in increasingly connected environments.

In doing so, this chapter invites the reader to reconsider what a lock truly represents. It is not merely a device that resists entry, but a mechanism that encodes decisions about power, responsibility, and personal boundaries. As locking systems become more integrated with digital platforms and societal structures, the role of the locksmith evolves accordingly—from a technician working with hardware to a professional engaged in designing and managing access within complex social and technological systems.

Overall, this chapter provides the conceptual bridge between understanding locks as tools and understanding them as systems of control embedded within society. It sets the stage for examining how technology, culture, governance, and ethics converge in modern access control, and why the future of locksmithing depends as much on understanding people and systems as it does on understanding mechanisms.

## Locks as Tools of Power, Control, and Privacy

Before we think of a lock as a cylinder, a shackle, or a line of code, it helps to see it for what it really is: a decision made visible. Every lock draws a boundary, between public and private, insider and outsider, trusted and untrusted, and that boundary is never neutral. It reflects who is allowed in, who must stay out, who gets to decide, and what happens when that decision is challenged. From palace gates and prison cells to apartment foyers, office turnstiles, and encrypted mobile credentials, locking systems have always done more than resist force: they organise daily life, distribute authority, and shape privacy. With that in mind, the story of locks becomes less about metal and mechanisms and more about how societies manage power, control, and personal space.

The conventional view of locks as merely mechanical devices underestimates their broader social significance. Locks function as social technologies that encode and enforce how power, control, and privacy are organized within a society. Historical evidence shows that access to spaces and resources was contingent on the possession of keys and the right to entry, a dynamic that reveals who commands authority and who is excluded. In contemporary contexts, this logic extends to digital and organizational environments where access control systems regulate entry to workplaces, data repositories, and critical infrastructure. Contemporary "keys" include keycards, biometric credentials, and digital tokens, and the issuers or revokers of these credentials exercise institutional power. These observations are supported by studies of access governance in both traditional and modern systems, which emphasize how access control structures reflect governance arrangements and social inequality [47-49].

At the workplace, residential settings, and public spaces, locking and surveillance systems shape not only safety but also routines, behaviour, and social norms, demonstrating that control is as much about shaping conduct as it is about restricting access [47, 48, 50]. The encryption and privacy dimensions of modern locks—ranging from encrypted digital keys to privacy-preserving access logs—further illustrate how locks mediate personal boundaries and information flow in a networked world [51-53].

### Power, Control, and Access Governance

Locks historically reinforced hierarchy by conferring access to those in authority: palace attendants, treasury guardians, wardens of prisons, and administrators controlled entry points and valuable assets, thereby encoding social order in physical form. In modern organizations, access control systems extend this logic into digital domains, where authorization decisions determine who may access sensitive data, facilities, or systems. The governance logic embedded in access control is captured in studies of permissioning and delegation in Cyber-Physical Systems (CPS) and the Internet of Things (IoT), where models of authorization govern who can operate, monitor, or audit systems [48].

In Cyber-Physical Systems (CPS) and the Internet of Things (IoT), permissioning and delegation are fundamental mechanisms that determine who can do what, when, and under what conditions across interconnected physical and digital environments. These concepts extend traditional access control beyond simple lock-and-key models into dynamic, distributed systems where users, devices, and services continuously interact. As a result, access is no longer a fixed state but an actively managed process shaped by policies, context, and system design.

Permissioning refers to the structured assignment of access rights within a system. It defines the rules that govern how users, devices, and applications interact with both physical assets—such as doors, vehicles, and machinery—and digital resources, including data, logs, and system controls. In CPS and IoT environments, permissioning becomes significantly more complex than in traditional systems. Access is often multi-layered, spanning device, network, application, and data levels. The entities involved are not limited to people but also include automated systems and interconnected devices. Furthermore, permissions are frequently dynamic, changing based on contextual factors such as time, location, or system state.

In practical terms, this means permissioning extends far beyond simply granting entry. For example, in a smart building, a staff member may be allowed access to certain areas only during working hours, while a maintenance system may be authorised to override locks during emergencies. At the same time, a cloud-based platform may have permission to collect and analyse access data. Permissioning, therefore, governs not only physical access but also the broader interaction between physical infrastructure and digital systems. Common models used to manage this complexity include Role-Based Access Control (RBAC), where permissions are assigned according to defined roles; Attribute-Based Access Control (ABAC), where access depends on contextual conditions such as time or location; and policy-based control systems, which apply centralised rules across the entire environment.

Delegation complements permissioning by enabling the transfer or sharing of access rights between entities. It allows access to be granted temporarily or conditionally, reflecting the realities of everyday use where access needs frequently change. In CPS and IoT systems, delegation is essential for accommodating scenarios such as guests, contractors, or service providers, without compromising overall security. It also reflects the evolving role of users, who increasingly act as access managers rather than passive holders of credentials.

Examples of delegation are common in modern systems. A homeowner may grant temporary access to a cleaner through a smart lock application, a facility manager may allow a contractor entry to specific areas for a limited period, or a system administrator may delegate control of part of a system to another operator. Delegation can take several forms, including time-bound access that expires automatically, scope-limited access restricted to certain areas or devices, and conditional access that depends on specific triggers such as presence or approval. For delegation to be secure, it must always be revocable, auditable, and traceable, ensuring that access can be withdrawn, actions can be logged, and responsibility can be clearly identified.

The importance of permissioning and delegation in CPS and IoT lies in their direct impact on security, usability, and trust. Poorly designed permission structures or uncontrolled delegation can lead to serious vulnerabilities, including unauthorized access, privilege escalation, and system misuse. For instance, a delegated credential that is not properly revoked may remain active indefinitely, creating a hidden security risk. At the same time, overly complex or rigid permissioning can reduce usability, encouraging users to share credentials or bypass controls altogether. Effective delegation provides a structured and secure way to share access, reducing the likelihood of these behaviours.

Accountability is another critical factor. Modern systems must maintain clear records of who accessed what, when access occurred, and who authorised that access. This level of auditability is essential for incident investigation, regulatory compliance, and maintaining organisational trust. Without it, even technically secure systems may fail to meet operational or legal expectations.

For locksmiths and security professionals, permissioning and delegation represent a significant shift from traditional, static access control to dynamic access management. The role is no longer limited to installing locks and issuing keys; it now involves configuring user roles, defining access hierarchies, and ensuring that systems support secure delegation through features such as time-limited access, instant revocation, and comprehensive audit trails. Locksmiths increasingly act as advisors, helping clients design access strategies that balance security with usability, and understanding how digital credentials—such as mobile apps, biometrics, and tokens—integrate with or replace physical keys.

In practice, installing a modern access control system involves more than hardware deployment. It requires defining who has authority to grant access, establishing time-based permissions, ensuring that access can be revoked immediately when required, and configuring systems to record and monitor all access events. This reflects a broader transformation in the profession, where technical skill must be combined with an understanding of system design, user behaviour, and data governance.

Ultimately, in CPS and IoT environments, access is no longer a fixed possession but a managed and evolving relationship between users, devices, and systems. Permissioning defines the boundaries of that relationship, while delegation enables it to function flexibly in real-world conditions. Together, they transform security from a static barrier into a dynamic, policy-driven system in which control is continuously defined, monitored, and enforced across both physical and digital domains.

The broader governance implications of access control are echoed in discussions of identity and access management (IAM) as a unifying framework for physical and digital security, underscoring how centralized or distributed control structures shape risk, accountability, and trust [47, 53]. Research on attribute-based access control (ABAC) and related cryptographic mechanisms demonstrates how control policies can be expressed in formal terms and enforced through trusted computation, illustrating how governance considerations are inseparable from technical design choices [50, 52, 54].

Identity and Access Management (IAM), Attribute-Based Access Control (ABAC), and their supporting cryptographic and trusted computation mechanisms form the backbone of modern security in digital, physical, and cyber-physical systems. Together, they define how identities are established, how access decisions are made, and how those decisions are securely enforced in environments where users, devices, and services interact continuously.

Identity and Access Management (IAM) is the overarching framework used to identify entities and control their access to systems and resources. An "identity" in IAM is not limited to a person; it can also represent devices, applications, or services within a network. IAM systems manage the full lifecycle of these identities, including their creation, authentication, authorisation, and eventual deactivation. Authentication verifies that an entity is who or what it claims to be, using mechanisms such as passwords, biometrics, tokens, or multi-factor authentication. Authorisation then determines what that authenticated entity is allowed to do, based on defined policies.

In modern environments, IAM extends beyond traditional IT systems into Cyber-Physical Systems (CPS) and IoT. For example, a smart building may authenticate a user via a mobile credential, then use IAM policies to determine which doors they can access, at what times, and under what conditions. IAM therefore acts as the central coordination layer that links identity, policy, and enforcement across both digital and physical domains.

Attribute-Based Access Control (ABAC) is an advanced authorisation model within IAM that makes access decisions based on attributes rather than fixed roles. Attributes can relate to the user (e.g., job role, clearance level), the resource (e.g., door type, sensitivity), the environment (e.g., time of day, location), or the action being requested. Instead of assigning static permissions, ABAC evaluates policies dynamically at the moment of access.

For example, a technician may be allowed to access a server room only if they are on-site, during working hours, and assigned to a relevant task. If any of these attributes change—such as the time or location—the access decision changes automatically. This makes ABAC highly flexible and scalable, particularly in complex environments where static role-based models (RBAC) become too rigid. However, it also introduces greater complexity in policy design, requiring careful definition of attributes and conditions to avoid unintended access outcomes.

Underlying IAM and ABAC are cryptographic mechanisms that ensure identities, credentials, and communications are secure. Cryptography provides the trust foundation for verifying identities and protecting data. Common mechanisms include encryption, which protects data at rest and in transit; hashing, which securely stores credentials such as passwords; and digital signatures, which verify the integrity and origin of messages or transactions. Public Key Infrastructure (PKI) plays a central role by enabling secure key management and certificate-based authentication, allowing devices and users to establish trust without sharing secret keys directly.

In IoT and CPS environments, cryptography is particularly important because many interactions occur over networks and involve distributed devices. Secure communication protocols (such as TLS) protect data exchanges, while device identity certificates ensure that only trusted devices can participate in the system. In access control systems, cryptographic tokens or credentials—such as encrypted mobile keys or smart cards—replace traditional physical keys, enabling secure and revocable access.

Trusted computation mechanisms ensure that access control decisions and cryptographic operations are executed securely, even in potentially untrusted environments. These mechanisms rely on hardware-based security features such as Trusted Execution Environments (TEEs), secure enclaves, and hardware security modules (HSMs). A TEE, for example, creates an isolated area within a processor where sensitive computations can occur without exposure to the rest of the system. This protects critical processes such as credential validation, biometric matching, and key management from tampering or malware.

In practice, trusted computation enforces IAM and ABAC policies by ensuring that decisions are made and executed in a secure, verifiable manner. For instance, when a user presents a credential to a smart lock, the device may use a secure enclave to verify the credential, evaluate access policies, and unlock the door only if all conditions are met. Similarly, in cloud-based systems, HSMs protect encryption keys and ensure that sensitive operations cannot be accessed or altered by unauthorised processes.

The integration of IAM, ABAC, cryptography, and trusted computation creates a layered security model. IAM defines identities and high-level policies, ABAC enables fine-grained and context-aware decision-making, cryptography secures identities and communications, and trusted computation ensures that all these processes are enforced reliably and cannot be compromised. This layered approach is essential in modern systems where threats can arise from both external attackers and internal misuse.

For locksmithing and security practice, these concepts represent a significant evolution. Traditional locksmithing focused on physical keys and mechanical access, where identity was implicit in possession. In contrast, modern systems require explicit identity verification, dynamic policy enforcement, and secure digital credentials. Locksmiths must now understand how identities are managed, how access decisions are made based on attributes, and how cryptographic systems protect those processes. They are increasingly responsible for configuring IAM policies, managing digital credentials, and ensuring that systems are both secure and usable.

IAM, ABAC, and their supporting technologies transform access control from a static, object-based model into a dynamic, identity-driven system. Access is no longer defined solely by possession of a key, but by a continuously evaluated set of conditions, securely enforced through cryptographic trust and trusted computation. This shift reflects the broader transition from mechanical security to integrated cyber-physical security, where identity, context, and trust are central to protecting both physical spaces and digital assets.

In contemporary networks, the shift from static to dynamic access control intensifies governance challenges. For example, blockchain-enabled data governance and multi-authority CP-ABE schemes are proposed to strengthen fine-grained access control while preserving policy privacy and resisting illegal authorization, reflecting a governance concern with transparency, auditability, and resilience against single points of failure [55]. Likewise, anonymous and secure access control frameworks for dynamic environments (e.g., vehicular systems and distributed digital twins) illustrate how smart contracts and cryptographic constructs can operationalize governance without centralized bottlenecks, highlighting the interplay between technology, policy, and institutional authority [49, 56]. In security-critical domains such as civil registration or healthcare, secure, auditable, and privacy-preserving access control mechanisms become essential to maintain legitimacy and public trust in governance processes [51, 57].

## Privacy as a Dimension of Lock Design

Privacy emerges from the capacity to exclude others from personal space, information, and digital footprints. A locked physical space or a secured device creates boundaries around a person's life, and digital locks extend this boundary to data and metadata—who can see access logs, when, and under what conditions. Modern locksmithing expands beyond doors to include data privacy in smart locks, access analytics, and surveillance systems, where the design choices about data collection and retention influence autonomy and exposure. Privacy-enhancing access controls and privacy-preserving authentication are central concerns in contemporary security design, with scholarly attention given to how cryptographic techniques (e.g., CP-ABE, zero-knowledge approaches, and confidential computing) can decouple access authorization from policy disclosure and reduce leakage of sensitive attributes [52, 54, 55].

The literature on privacy in blockchain-enabled contexts further emphasizes the tension between transparency and privacy; while blockchains provide immutable audit trails and accountability, they can also expose policy details or activity histories if not carefully architected, necessitating design choices that protect identity and policy privacy [49, 55, 56]. In this sense, a "lock" in modern systems is a dual barrier: a physical or logical gate that prevents entry and a privacy-preserving boundary that protects personhood and sensitive information.

### Balancing Control and Privacy in Practice

A persistent design tension in locking and access systems concerns balancing legitimate control with the protection of individual autonomy and privacy. Systems designed to maximize monitoring and enforcement (e.g., extensive access logs, CCTV in workplaces, centralized databases) can erode privacy if data collection outpaces governance safeguards or if data is misused. Conversely, privacy-preserving approaches can complicate oversight and accountability if access policies become opaque or unverifiable. The scholarship on delegated and self-contained authorization in CPS/IoT highlights this tension by classifying models of access governance according to policy delegation, trust assumptions, and auditability, thereby illuminating the trade-offs between central authority and local control [48].

Related discussions in privacy-centric data-sharing frameworks using blockchain and CP-ABE show how to achieve fine-grained access control while mitigating illegal authorization and privacy leakage, albeit at the cost of increased system complexity and performance considerations [55, 56]. The emergence of privacy-preserving digital identities and decentralized credentials—e.g., anonymous credentials and zero-knowledge proof-based mechanisms—addresses the need to verify access rights without exposing underlying attributes or identities, thereby helping reconcile control with privacy [52, 56]. In healthcare and civil registry contexts, privacy-preserving access control is essential for maintaining public trust and legal compliance, reinforcing that technical design choices must align with ethical and legal norms as well as social expectations [51, 57].

### Implications for Locksmiths and Security Professionals

For practitioners, the perspective of locks as social technologies reframes the role of locksmiths and security professionals from mere installation to the broader consideration of access governance, privacy, and ethical design. Decisions about who gets access, how access is granted, and what data is produced or stored are social choices with implications for trust, power dynamics, and social equity. This reframing invites practitioners to engage with governance questions, such as how to implement ABAC or CP-ABE in distributed environments, how to balance auditability with privacy, and how to design access systems that support fair and transparent processes while preserving individual autonomy [50, 54-56]

The literature on IAM and integrated physical-digital access control demonstrates practical pathways for achieving such alignment through unified identity management, biometric-enabled authentication, and layered privacy controls that withstand regulatory scrutiny while maintaining usability [47, 53]. In domains like healthcare IoT and civil registration, security professionals must anticipate the socio-ethically charged consequences of access policies and data retention practices, ensuring that systems are robust against misuse while respectful of privacy rights and public trust [51, 52, 57].

Evolution of Locks and the Shape of Society As locking systems evolve from purely mechanical devices to complex, networked access platforms, their role in shaping social boundaries, governance, and information flows becomes more pronounced. The literature suggests that the evolution toward blockchain-enabled access management, smart contracts, and privacy-preserving cryptographic techniques will intensify the intersection of security with governance and civil liberties. For locksmiths and security professionals, staying informed about advances in ABAC, CP-ABE, zero-knowledge approaches, and distributed identity mechanisms is essential to designing systems that are not only secure but also fair, transparent, and aligned with the values and needs of the communities they serve [49, 50, 54, 55]. The integration of digital credentials with physical access points—such as smart locks integrated into IAM platforms or CPS/I IoT environments—offers a compelling vision of future practice where access control is both technically rigorous and socially responsible [47, 48, 53].

Locks function as more than hardware; they are social technologies that embody and propagate patterns of power, control, and privacy. The shift from traditional mechanical locks to digital and networked access systems amplifies the social and ethical dimensions of access governance, requiring practitioners to consider not only the strength of cryptographic protections and physical security but also the fairness, transparency, and autonomy of the people who interact with these systems. By acknowledging that decisions about who gains access and what data is produced are social and ethical decisions, locksmiths and security professionals can design systems that are secure, privacy-preserving, and aligned with societal values. The literature consistently shows that the most robust and ethically sound designs balance legitimate control with respect for privacy, harnessing both technical innovations (e.g., ABAC, CP-ABE, digital identities, and smart contracts) and governance considerations to build systems that serve people and communities as well as assets and spaces [47, 48, 53-56].

### Impact on Locksmithing Practices

Understanding locks as social technologies, rather than purely mechanical devices, is highly relevant to anyone interested in locksmithing because it fundamentally reshapes how the profession is perceived and practiced. It moves locksmithing beyond the narrow task of opening doors or installing hardware and positions it within a broader system of power, governance, privacy, and human behaviour. For a modern locksmith, this perspective is not abstract theory; it directly influences day-to-day decisions, system design, and professional responsibility.

At a foundational level, this concept highlights that locksmiths are not just technicians—they are gatekeepers of access and, by extension, participants in the distribution of power. Every time a locksmith installs a lock, programs a credential, or configures an access system, they are helping to determine who is included and who is excluded. In traditional contexts, this might involve issuing keys to property owners or restricting access to secure areas. In modern environments, however, it extends to configuring access permissions in workplaces, managing digital credentials, and enabling or revoking access to sensitive systems. This elevates the role of the locksmith into one that directly intersects with organisational authority, governance structures, and even social equity.

This perspective also has a strong influence on how locksmiths approach system design and installation. Rather than viewing locks as isolated devices, locksmiths must now understand them as part of broader access ecosystems that include Identity and Access Management (IAM), permissioning models, and data flows. Decisions about who gets access are no longer static; they are dynamic and policy-driven. For example, configuring an access control system requires defining roles, setting conditions (such as time or location), and enabling delegation. This requires locksmiths to think in terms of systems, policies, and user behaviour—not just hardware specifications.

The shift toward Cyber-Physical Systems (CPS) and IoT further reinforces this relevance. In these environments, access is not simply about possession of a key but about managing relationships between users, devices, and systems. Permissioning and delegation become central tasks, requiring locksmiths to configure who can access what, under which conditions, and for how long. This introduces new responsibilities, such as ensuring access can be revoked, actions are auditable, and permissions are not overly broad. Poorly configured systems can create serious vulnerabilities, meaning that locksmiths must now consider cybersecurity principles alongside physical security.

Another critical influence is the growing importance of privacy. Modern locks—particularly smart locks, access control systems, and CCTV—generate and process personal data. This means locksmiths are increasingly involved in systems that affect not only physical security but also informational boundaries. Decisions about data collection, storage, and access logs directly impact user privacy. As a result, locksmiths must understand concepts such as data minimisation, secure storage, and user consent, especially in jurisdictions governed by privacy laws. The lock is no longer just a barrier to entry; it is also a mechanism that produces data about people's movements, behaviours, and routines.

This understanding also changes how locksmiths engage with clients. Instead of simply responding to requests (e.g., "install a lock" or "open a door"), locksmiths must take on an advisory role. They need to explain trade-offs between security and convenience, help clients design appropriate access policies, and ensure that systems align with both operational needs and ethical considerations. For example, recommending a system with strong audit capabilities may improve accountability, but it may also raise privacy concerns if not properly managed. The locksmith must navigate these tensions and guide clients toward balanced solutions.

Ethical responsibility becomes more pronounced under this framework. Because locks encode control and influence behaviour, locksmiths must consider the broader implications of their work. Installing surveillance systems, configuring access hierarchies, or enabling monitoring capabilities can shape how people behave within a space. This means locksmiths must think about fairness, proportionality, and transparency, ensuring that systems do not unnecessarily infringe on privacy or create unjust power imbalances. Ethical locksmithing, therefore, involves not only doing the job correctly but also considering the social impact of the system being implemented.

The influence on professional skills is also significant. Traditional locksmithing skills—such as key cutting, lock picking, and mechanical installation—remain important, but they are no longer sufficient on their own. Modern locksmiths must understand digital systems, networked devices, encryption, and access control models such as RBAC and ABAC. They must be able to configure software, manage user identities, and interpret system logs. This represents a shift toward a hybrid profession that combines mechanical expertise with elements of cybersecurity, data governance, and systems engineering.

Finally, this perspective prepares locksmiths for the future of the industry. As access control continues to evolve—through technologies such as biometrics, mobile credentials, blockchain-based systems, and privacy-preserving authentication—the social and governance dimensions of locking systems will become even more significant. Locksmiths who understand locks as social technologies will be better equipped to adapt to these changes, design systems that are both secure and ethical, and remain relevant in an increasingly complex security landscape.

Seen this way, the evolution of locks is also the evolution of responsibility. As cylinders become credentials, doors become networks, and "keys" become policies, the locksmith's work increasingly shapes not only who gets in, but how safety, privacy, and trust are experienced in everyday life. The future of the trade will belong to practitioners who can move comfortably between metal and software, between convenience and control, and between strong security and respectful governance—designing access systems that are defensible, auditable, and fair. In that sense, modern locksmithing is not simply keeping pace with technology; it is helping to decide what kind of access society will normalise, what kinds of monitoring it will tolerate, and how personal boundaries will be protected. When locks are understood as social technologies, the craft becomes more than problem-solving—it becomes stewardship of the spaces, data, and people those systems are meant to protect.

## Cultural Differences in Locking Practices Globally

Before a lock is ever installed, it is already shaped by something far less visible than metal or code: culture. The simple act of securing a door carries different meanings across societies—signalling trust or suspicion, independence or community, privacy or shared living. What feels like "normal" security in one place may appear excessive, intrusive, or even unnecessary in another. By looking at how different cultures approach locking, we begin to see that security is not a universal standard but a reflection of deeply held social values, lived experiences, and collective expectations about risk, responsibility, and belonging.

Cultural variation in locking practices is a consequential lens through which to understand security as a social practice rather than a purely technical artifact. The function of locks, to control access, remains constant across contexts, but the meanings, endorsements, and configurations of locking vary with trust norms, community structures, and perceived risk. In settings with strong social cohesion and high trust, locking may be minimal or symbolic, reflecting an expectation that interpersonal accountability substitutes for physical barriers. Rural and high-trust contexts in some East Asian and Nordic environments illustrate this tendency, where homes, bicycles, or small businesses are left unlocked because security is believed to emerge from communal norms rather than hardware alone. This suggests that security is as much a social contract as a mechanical device, with the "lock" existing in shared norms and mutual respect as much as in the tangible components of a locking mechanism [58-61]. Conversely, environments characterized by lower social trust or higher inequality tend to deploy more layered physical and electronic barriers, illustrating how social and economic structures shape the perceived necessity and design of security systems [58, 62, 63].

Economic conditions exert a powerful influence on the modality of locking. In lower-income regions, affordability and simplicity often steer security choices toward basic padlocks, improvised mechanisms, or reinforced barriers because they are accessible, maintainable, and understandable for a broad segment of the population. In contrast, wealthier, urbanized societies frequently deploy more sophisticated access controls, including electronic entry systems, biometrics, and integrated surveillance. Yet higher technological adoption does not universally translate into greater perceived security: it can reflect higher perceived risk or lower tolerance for uncertainty within a given culture. This pattern, economic constraints shaping the granularity of security technology, coupled with cultural interpretations of risk, emerges in cross-national analyses of biometric and digital security adoption, where policy, privacy norms, and institutional trust interact with technical choices to determine what technologies are acceptable or desirable in a given locale [58, 62, 64-67].

Cultural attitudes toward privacy are central to how locking practices are configured and perceived. In Western contexts, there is a pronounced emphasis on individual privacy and personal space, which is typically reflected in widespread use of locks on residential entry points, personal spaces, and even interior doors. In collectivist contexts that emphasize shared living arrangements and communal access, the raison d'être for internal locks may be more limited, with locking practices following from communal norms that regulate access rather than from individual ownership alone. As digital systems increasingly mediate access, these cultural distinctions extend into perceptions of data privacy, surveillance, and the acceptability of monitoring technologies such as CCTV. Comparative work on biometric privacy, privacy laws, and data protection demonstrates that cultural expectations significantly colour the acceptance and regulation of security technologies, including the use of biometric authentication and data collection across borders [58, 61, 62, 65-67]. Disparities in privacy values are not merely theoretical; they shape regulatory responses, the design of consent mechanisms, and the social acceptability of data-sharing arrangements that underpin digital access systems [41, 58, 60, 62, 64].

Historical and architectural factors intersect with cultural norms to shape locking regimes. Traditional building designs frequently incorporate passive security features, such as courtyard layouts, elevated living spaces, and community-oriented spatial configurations, that facilitate natural surveillance and controlled access without heavy reliance on locking hardware. By contrast, modern urban environments characterized by high population density and anonymity tend to rely more heavily on formal access control systems, surveillance, and electronic credentials. The evolution of locks in different regions thus tracks the way spaces are designed, used, and imagined to be secure, revealing how architecture and culture jointly determine the prevalence and form of locking technologies [58, 59, 63, 67]. When new digital paradigms, including smart locks and app-based access, arrive, they often interact with local architectural tendencies and cultural expectations—some communities embracing mobile credentials and biometrics, others resisting digital data collection on privacy or trust grounds [58, 62, 64, 67].

Legal and regulatory frameworks further shape the adoption and configuration of locking systems. Jurisdictions with robust privacy protections and clear data governance regimes tend to impose specific requirements on the installation and operation of locks, access control systems, and surveillance technologies. Signage, data retention limits, and biometric data protections become salient design constraints, influencing not only what technologies are used but how they are configured in practice. In other settings with looser privacy protections or more permissive surveillance regimes, broader use of monitoring technologies and flexible access arrangements may be permitted. Across jurisdictions, academic and policy analyses consistently show that legal norms interact with cultural expectations and economic conditions to determine the built security environment, including the alignment or tension between physical security measures and privacy rights [41, 58, 59, 62, 67]. This underscores the need for security professionals to anticipate regulatory design beyond technical performance, incorporating privacy-by-design principles and culturally informed consent and notification practices when deploying access control and surveillance systems [41, 58, 61, 62].

Religion and cultural beliefs can also influence how locks are perceived and used. In some contexts, there may be spiritual or moral beliefs that reduce reliance on physical security measures, or alternatively elevate the sense of safeguarding property as a communal obligation. Ritual practices or traditional norms can coexist with or partially substitute for modern locking systems, illustrating that security is not solely material. The literature on privacy regulation and biometrics points to a complex interplay of cultural values, religious norms, and socio-political structures that shape how people experience and accept security technology, including biometrics and data collection protocols [41, 58, 61, 66]. In some settings, religious or moral frameworks may reinforce or challenge the acceptability of monitoring and data collection practices that underpin contemporary access-control ecosystems [41, 62, 65].

The following are practical, scenario-based case studies that clearly demonstrate how cultural context shapes locking practices. Each scenario is written to reflect real-world conditions and highlight implications for locksmiths and security professionals.

**Case Study 1: Rural Japan – Security Through Social Trust**

**Scenario:** A locksmith is called to a small rural town in Japan to install locks on a newly built home owned by a returning expatriate. The homeowner requests high-security deadbolts, window locks, and a CCTV system.

**Context:** Most homes in the area traditionally rely on minimal locking. Neighbours know each other, crime rates are extremely low, and community surveillance (informal observation) is strong.

**Outcome:** Neighbours question the installation, interpreting it as a signal of distrust. The homeowner later scales back visible security measures to better align with local norms, retaining only discreet internal locks.

**Insight:** Here, security is socially enforced rather than mechanically enforced. Over-securing can create social friction.

**Implication for locksmiths:** Recommend low-visibility, culturally sensitive solutions rather than overt security.

**Case Study 2: Scandinavian City – Minimal Locking, High Digital Trust**

**Scenario:** A property manager in Copenhagen installs smart locks in an apartment complex to enable keyless entry and remote access for tenants.

**Context:** Residents are comfortable with shared spaces and often leave bicycles unlocked. However, they are highly aware of digital privacy and data protection.

**Outcome:** Tenants accept the smart locks but demand strict controls over data collection, access logs, and third-party access.

**Insight:** Physical security expectations are low, but data privacy expectations are extremely high. Implication: Locksmiths must prioritise privacy-by-design, not just physical security.

**Case Study 3: Urban United States – Layered Security Culture**

**Scenario:** A locksmith is contracted to upgrade security in a suburban home after a nearby burglary.

**Context:** The homeowner requests multiple layers: deadbolts, alarm system, CCTV, smart locks, and motion sensors.

**Outcome:** The system becomes highly complex, but family members begin sharing PIN codes and disabling alarms due to inconvenience.

**Insight:** High perceived risk leads to over-securitisation, but poor usability leads to security bypass.

**Implication:** Locksmiths must balance security with usability, or systems will fail due to human behaviour.

**Case Study 4: Informal Settlement (Developing Economy) – Visible Security Over Advanced Systems**

**Scenario:** A locksmith works in a low-income urban area installing security for small shops.

**Context:** Clients cannot afford electronic systems. Instead, they request heavy padlocks, steel bars, and reinforced doors.

**Outcome:** Security relies on deterrence through visibility rather than sophisticated access control.

**Insight:** Security is shaped by economic constraints—simple, robust, and repairable solutions dominate.

**Implication:** Locksmiths must prioritise cost-effective, durable, and maintainable solutions over high-tech systems.

**Case Study 5: Middle Eastern Compound Housing – Controlled Access and Social Hierarchy**

**Scenario:** A locksmith installs access control systems in a gated residential compound.

**Context:** Different access levels exist: residents, domestic workers, security staff, and visitors. Movement is tightly controlled.

**Outcome:** The system uses hierarchical access permissions, restricting movement based on role and status.

**Insight:** Locks encode social hierarchy and authority structures.

**Implication:** Locksmiths must understand power dynamics and configure systems that reflect organisational or cultural hierarchy.

**Case Study 6: European Office (GDPR Environment) – Privacy-Constrained Security**

**Scenario:** A company in Germany installs CCTV and access control systems.

**Context:** Strict privacy laws (GDPR) limit surveillance and data retention.

**Outcome:** Cameras cannot monitor certain areas, access logs must be limited, and employees must be informed of all data use.

**Insight:** Legal frameworks strongly shape how locks and surveillance systems are implemented. **Implication:** Locksmiths must ensure legal compliance, not just technical installation.

**Case Study 7: Multi-Generational Home in India – Shared Access Culture**

**Scenario:** A locksmith installs locks in a large family home with multiple generations living together.

**Context:** Internal doors are rarely locked. Access is shared, and privacy is secondary to family cohesion.

**Outcome:** The family rejects internal locking systems but installs a strong perimeter lock.

**Insight:** Security focuses on external threats, not internal boundaries.

**Implication:** Locksmiths must design systems that respect collectivist living arrangements.

**Case Study 8: Smart Apartment in South Korea – High-Tech, High Adoption**

**Scenario:** A developer installs biometric smart locks in a high-rise apartment complex.

**Context:** Residents are comfortable with advanced technology and biometric systems.

**Outcome:** High adoption and acceptance, but residents expect seamless integration and convenience.

**Insight:** Technology acceptance is culturally influenced—convenience and efficiency drive adoption.

**Implication:** Locksmiths must prioritise integration, speed, and user experience.

**Case Study 9: Remote Australian Community – Hybrid Security Approach**

**Scenario:** A locksmith installs security in a remote community with strong social ties but increasing external risks.

**Context:** Traditionally low locking, but recent incidents have increased concern.

**Outcome:** A hybrid approach is adopted: basic locks combined with community awareness and selective surveillance.

**Insight:** Security evolves as risk perception changes, even in high-trust environments.
**Implication:** Locksmiths must design adaptive systems that respect culture while addressing emerging risks.

**Case Study 10: Global Corporate Office – Standardisation vs Local Culture**

**Scenario:** A multinational company rolls out a global access control system across offices in Europe, Asia, and the Middle East.

**Context:** Corporate policy requires uniform systems, but local expectations differ.

**Outcome:** Some regions resist biometric systems (privacy concerns), while others demand stricter controls.

**Insight:** Global systems must be locally adapted to succeed.

**Implication:** Locksmiths and security designers must balance standardisation with cultural flexibility.

Globalisation and technology are gradually harmonising practices while leaving distinctive local footprints. Innovations such as smart locks, mobile credentials, and IoT-based access systems are increasingly adopted worldwide, but rollout, acceptance, and governance remain contingent on local norms around trust, control, and privacy. Biometric authentication, in particular, often experiences differential uptake across countries and cultures, with privacy concerns and trust in data handling playing decisive roles in adoption and regulatory response. Studies comparing privacy expectations and the social acceptability of biometric technologies across borders emphasize that techno-scientific progress must be paired with culturally informed governance to achieve durable, legitimate security outcomes [41, 58, 59, 61, 62, 64-67].

For locksmiths and security professionals, these findings imply that effective security design must be deeply context-sensitive. A system that is well received in one cultural and regulatory environment may be inappropriate or rejected in another if it conflicts with local expectations about privacy, trust, and social responsibilities. Practitioners should integrate assessments of user behaviour, cultural norms, and social acceptability into the technical evaluation of lock systems, access controls, and monitoring technologies. This means prioritizing user-centred design, transparent communication about data collection and purpose, and governance structures that reflect local ethical and legal norms while maintaining core security objectives. The literature repeatedly demonstrates that security is a social practice as much as a technical one, and that cultural and regulatory contexts shape both the design and the acceptance of locking systems in ways that extend beyond hardware performance alone [41, 58, 59, 61, 62, 64, 65, 67].

Locking practices are not universal; they crystallize the values, histories, risks, and social structures of their respective cultures. The cross-cultural analysis of privacy norms, biometrics, and access-control governance reveals that security constitutes a social contract as much as a technical solution. As digital technologies merge with traditional locking practices, the necessity for context-sensitive, culturally aware approaches becomes ever more important for practitioners who seek to protect both property and personal information in a globally connected world. The literature indicates that when security design integrates local norms around privacy, trust, and social responsibility with rigorous technical performance, security outcomes are more robust and more socially legitimate across diverse cultural settings [41, 58, 59, 61, 62, 64-67].

### Impact on Locksmithing Practices

Understanding locks as social technologies, rather than purely mechanical devices, is highly relevant to anyone interested in locksmithing because it fundamentally reshapes how the profession is perceived and practiced. It moves locksmithing beyond the narrow task of opening doors or installing hardware and positions it within a broader system of power, governance, privacy, and human behaviour. For a modern locksmith, this perspective is not abstract theory; it directly influences day-to-day decisions, system design, and professional responsibility.

At a foundational level, this concept highlights that locksmiths are not just technicians—they are gatekeepers of access and, by extension, participants in the distribution of power. Every time a locksmith installs a lock, programs a credential, or configures an access system, they are helping to determine who is included and who is excluded. In traditional contexts, this might involve issuing keys to property owners or restricting access to secure areas. In modern environments, however, it extends to configuring access permissions in workplaces, managing digital credentials, and enabling or revoking access to sensitive systems. This elevates the role of the locksmith into one that directly intersects with organisational authority, governance structures, and even social equity.

This perspective also has a strong influence on how locksmiths approach system design and installation. Rather than viewing locks as isolated devices, locksmiths must now understand them as part of broader access ecosystems that include Identity and Access Management (IAM), permissioning models, and data flows. Decisions about who gets access are no longer static; they are dynamic and policy-driven. For example, configuring an access control system requires defining roles, setting conditions (such as time or location), and enabling delegation. This requires locksmiths to think in terms of systems, policies, and user behaviour—not just hardware specifications.

The shift toward Cyber-Physical Systems (CPS) and IoT further reinforces this relevance. In these environments, access is not simply about possession of a key but about managing relationships between users, devices, and systems. Permissioning and delegation become central tasks, requiring locksmiths to configure who can access what, under which conditions, and for how long. This introduces new responsibilities, such as ensuring access can be revoked, actions are auditable, and permissions are not overly broad. Poorly configured systems can create serious vulnerabilities, meaning that locksmiths must now consider cybersecurity principles alongside physical security.

Another critical influence is the growing importance of privacy. Modern locks—particularly smart locks, access control systems, and CCTV—generate and process personal data. This means locksmiths are increasingly involved in systems that affect not only physical security but also informational boundaries. Decisions about data collection, storage, and access logs directly impact user privacy. As a result, locksmiths must understand concepts such as data minimisation, secure storage, and user consent, especially in jurisdictions governed by privacy laws. The lock is no longer just a barrier to entry; it is also a mechanism that produces data about people's movements, behaviours, and routines.

This understanding also changes how locksmiths engage with clients. Instead of simply responding to requests (e.g., "install a lock" or "open a door"), locksmiths must take on an advisory role. They need to explain trade-offs between security and convenience, help clients design appropriate access policies, and ensure that systems align with both operational needs and ethical considerations. For example, recommending a system with strong audit capabilities may improve accountability, but it may also raise privacy concerns if not properly managed. The locksmith must navigate these tensions and guide clients toward balanced solutions.

Ethical responsibility becomes more pronounced under this framework. Because locks encode control and influence behaviour, locksmiths must consider the broader implications of their work. Installing surveillance systems, configuring access hierarchies, or enabling monitoring capabilities can shape how people behave within a space. This means locksmiths must think about fairness, proportionality, and transparency, ensuring that systems do not unnecessarily infringe on privacy or create unjust power imbalances. Ethical locksmithing, therefore, involves not only doing the job correctly but also considering the social impact of the system being implemented.

The influence on professional skills is also significant. Traditional locksmithing skills—such as key cutting, lock picking, and mechanical installation—remain important, but they are no longer sufficient on their own. Modern locksmiths must understand digital systems, networked devices, encryption, and access control models such as RBAC and ABAC. They must be able to configure software, manage user identities, and interpret system logs. This represents a shift toward a hybrid profession that combines mechanical expertise with elements of cybersecurity, data governance, and systems engineering.

Finally, this perspective prepares locksmiths for the future of the industry. As access control continues to evolve—through technologies such as biometrics, mobile credentials, blockchain-based systems, and privacy-preserving authentication—the social and governance dimensions of locking systems will become even more significant. Locksmiths who understand locks as social technologies will be better equipped to adapt to these changes, design systems that are both secure and ethical, and remain relevant in an increasingly complex security landscape.

Ultimately, these cultural patterns remind us that a "good lock" is not just one that resists attack, but one that fits the people and the place it serves. The most sophisticated system can fail if it clashes with local expectations about privacy, trust, status, or everyday convenience, just as a simple padlock can be highly effective when it aligns with community norms and realistic risks. For locksmiths, this is the deeper craft: reading the environment as carefully as the door, listening for what clients value as much as what they fear, and designing security that feels legitimate, usable, and respectful. In a globalised world where the same smart lock can be installed in Tokyo, Copenhagen, Melbourne, or Dubai, cultural awareness becomes a form of technical competence—because the future of access control will be decided not only by engineering, but by what societies are willing to accept, share, and protect.

## Historical Failures that Reshaped Lock Design

Historical failures in locking systems have played a crucial role in shaping how locks are designed, tested, and trusted today. Rather than evolving purely through innovation, lock technology has often advanced in response to being defeated, sometimes publicly, sometimes catastrophically. Each major failure has exposed weaknesses not just in mechanisms, but in assumptions about users, attackers, and the nature of security itself.

### 1. The Bramah Challenge Lock (18th–19th Century)

The history of lock design provides a compelling case study in the importance of adversarial testing and empirical validation of security claims. The narrative of Joseph Bramah's "unpickable" lock and its eventual defeat by Alfred Charles Hobbs represents a pivotal moment in the development of security technology and the broader understanding that theoretical design complexity does not guarantee practical security. This account demonstrates fundamental principles about security validation that remain relevant to contemporary security practices.

In 1784, Joseph Bramah introduced a lock mechanism that represented a significant advancement in lock design technology [68]. The lock was prominently advertised as "unpickable," reflecting the confidence of its designer in the sophistication of its internal mechanisms [68]. To substantiate these claims and generate public interest, Bramah displayed a challenge lock in London with an accompanying notice offering a substantial reward, two hundred guineas, to anyone who could successfully pick it [68]. This public challenge was intended to serve as definitive proof of the lock's security properties, transforming the lock into a symbol of technological achievement and reliability.

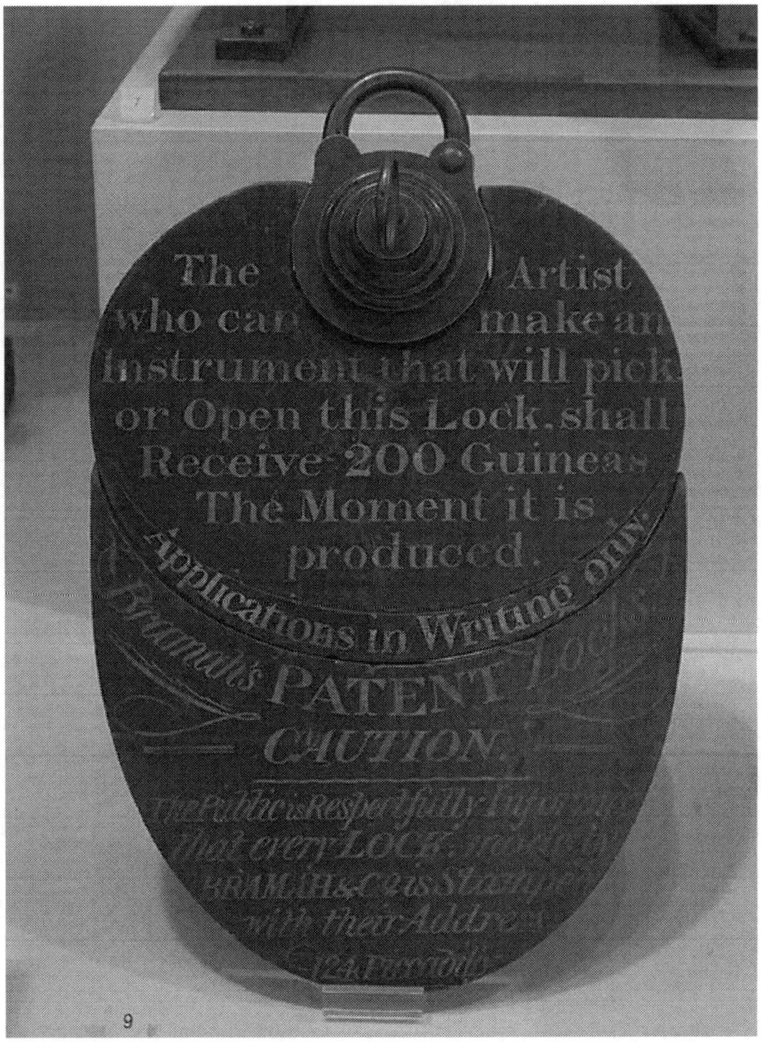


*Figure 2: Photo of the Bramah challenge lock. Geni, CC BY-SA 4.0, via Wikimedia Commons.*

The Bramah challenge lock remained unpicked for an extended period, which served to reinforce the manufacturer's claims of invulnerability [68]. For decades, the lock stood as a testament to the effectiveness of Bramah's design, accumulating prestige and market value through its apparent impermeability to attack [68]. The lock's continued resistance to picking attempts created a powerful marketing narrative: the longer it remained unopened, the more credible the claims of its security became. This extended period of success created what might be termed a "security myth"—a widely accepted belief in the lock's unpickability based on the absence of successful attacks rather than on rigorous testing under controlled conditions.

The narrative of the Bramah lock's invulnerability was fundamentally disrupted in 1851 when Alfred Charles Hobbs, an American locksmith representing the firm Day & Newell, successfully picked the challenge lock [68]. Hobbs's achievement was not instantaneous; rather, it required 51 hours of sustained, focused work [68]. This extended effort underscores an important principle: the time required to defeat a security mechanism under controlled conditions may be substantially longer than the time available to an opportunistic attacker, yet the mechanism remains vulnerable to determined, skilled adversaries with adequate resources.

Hobbs's success at the Great Exhibition of 1851 captured significant public attention and generated considerable press coverage [68]. The event became known as the "Great Lock Controversy," and Hobbs's conquest of the Bramah lock—along with his simultaneous defeat of the Chubb & Son detector lock, originally patented in 1818—demonstrated that locks previously considered inviolable could indeed be defeated through systematic, skilled attack [68]. The public spectacle of these lock-picking demonstrations was particularly significant; it provided a visible, tangible demonstration of security failure before crowds of onlookers, transforming abstract claims about security into concrete evidence of vulnerability [68].

Impact on Lock Design and Security Philosophy

**Destruction of the "Unpickable" Myth:** Hobbs's successful picking of the Bramah lock fundamentally undermined the security mythology that had accumulated around the device [68]. The lock's defeat demonstrated that marketing claims of invulnerability, no matter how long sustained by the absence of successful attacks, could not substitute for rigorous adversarial testing. The myth of the "unpickable" lock was not merely challenged; it was definitively destroyed through practical demonstration [68].

**Validation Through Adversarial Testing:** The lock-picking competitions of the 1850s and 1860s established a new paradigm for security validation [68]. Rather than relying on theoretical analysis or the manufacturer's confidence in design complexity, security claims would henceforth be tested through actual attack by skilled adversaries [68]. These competitions provided "a spectacular forum for marketing new security commodities, which promised to serve as a uniquely credible means of demonstrating the utility of these technically advanced products" [68]. The competitions transformed security validation from a matter of design assertion into a matter of empirical demonstration.

**Influence on Lock Design Evolution:** The defeat of previously "unpickable" locks prompted manufacturers to develop more sophisticated internal mechanisms and to implement tighter manufacturing tolerances [68]. The security industry responded to the challenge posed by skilled lock-pickers by continuously advancing the complexity of lock designs. However, this escalation in design complexity was now informed by a more realistic understanding of the threat posed by determined, skilled attackers rather than by theoretical assumptions about security.

**Emergence of the Security Industry:** The lock-picking competitions and challenges of the 1850s and 1860s played a significant role in the commodification of security and the emergence of the security industry as a distinct economic sector [68]. The competitions demonstrated that security could be marketed as a commodity, with credibility established through public demonstration of effectiveness against skilled attack. This transformation had lasting effects on how security products and services would be developed, marketed, and validated in subsequent centuries [68].

Key Lessons for Security Validation

### Beyond the Key

The Bramah lock narrative illustrates several fundamental principles about security that extend far beyond the specific domain of lock design:

**Security claims must be validated under real attack conditions:** The Bramah lock's decades of unopened status created a false sense of security. The lock had never been subjected to systematic attack by a skilled, determined adversary with adequate time and resources. Only when such conditions were created, through Hobbs's deliberate, sustained effort. did the lock's vulnerabilities become apparent. This principle applies across all domains of security: theoretical design complexity and the absence of known attacks do not constitute proof of security.

**Design complexity alone does not ensure security:** The Bramah lock represented a sophisticated design for its era, yet this sophistication proved insufficient against systematic attack. The lock's internal mechanisms, while advanced, contained exploitable weaknesses that a skilled attacker could identify and exploit given adequate time and resources. This demonstrates that security cannot be achieved through complexity alone; rather, security must be validated through rigorous testing against realistic threat models.

**Adversarial testing reveals vulnerabilities that theoretical analysis may miss:** The lock-picking competitions provided a mechanism for identifying security weaknesses through practical attack rather than through theoretical analysis. This approach proved more effective at revealing vulnerabilities than any amount of design review or theoretical analysis could have achieved. The principle of adversarial testing, deliberately attempting to defeat security mechanisms, remains a cornerstone of modern security validation practices.

**Security is a process, not a state:** The Bramah lock's defeat did not represent the end of lock security; rather, it initiated a continuous process of security improvement driven by the knowledge that locks could be picked. Manufacturers responded by developing more sophisticated designs, which in turn prompted lock-pickers to develop more sophisticated techniques. This dynamic process of attack and defence continues to drive security innovation.

The history of the Bramah lock and its defeat by Alfred Charles Hobbs provides a foundational case study in security validation. The lock's decades of unopened status created a powerful but ultimately false sense of security, demonstrating that the absence of successful attacks does not constitute proof of security. Only through deliberate, adversarial testing under realistic conditions did the lock's vulnerabilities become apparent. This lesson—that security claims must be validated through rigorous adversarial testing rather than assumed from design complexity—remains as relevant to contemporary security practices as it was to nineteenth-century lock design. The emergence of the security industry and the development of more sophisticated lock designs in response to Hobbs's challenge demonstrate how adversarial testing can drive security innovation and improvement. The Bramah lock narrative thus stands as a historical precedent for the principle that effective security requires continuous validation through realistic attack scenarios, not merely confidence in design sophistication.

## 2. The Rise and Defeat of Early Pin Tumbler Locks

The pin tumbler lock represents a foundational security mechanism with deep historical roots. While the pin tumbler principle itself dates back to ancient times, with evidence suggesting its use in ancient civilizations

[69], the modern pin tumbler lock was refined and popularized in the 19th century [69, 70]. The introduction of Western-style pin tumbler locks fundamentally transformed global lock design, eventually displacing traditional lock mechanisms in many regions [70]. This widespread adoption reflects the design's practical advantages and relative security compared to earlier locking mechanisms [69, 70].

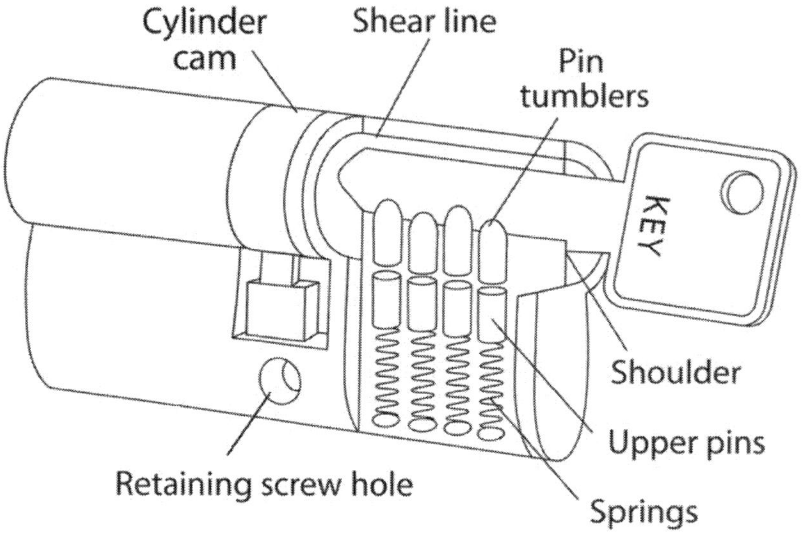


*Figure 3: Pin Tumbler Lock.*

The pin tumbler lock's dominance in the global market stems from its effectiveness as a security device [71]. However, despite its widespread use and refinement over more than a century, the design contains inherent vulnerabilities that have become increasingly apparent through both academic study and criminal exploitation [71, 72].

The image illustrates the internal operation of a pin tumbler lock, one of the most widely used locking mechanisms in the world. It shows four key stages that explain how the lock transitions from a locked to an unlocked state.

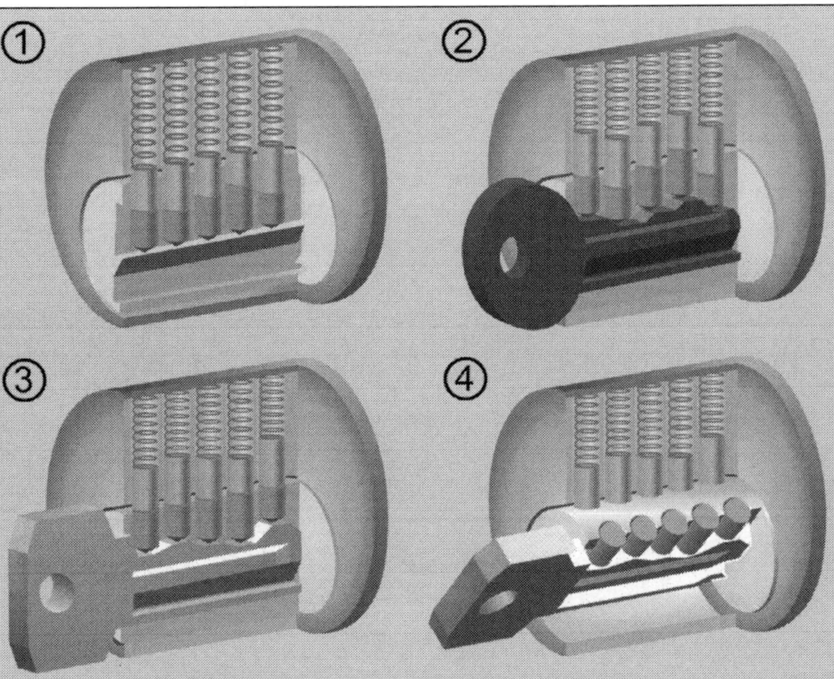


*Figure 4: Explanation of how a tumbler key lock operates. PNG: User:Wapcaplet; SVG: User:Pbroks13; this combined image: User:Crisco 1492., CC BY-SA 3.0, via Wikimedia Commons.*

**1. Locked State (No Key Inserted)**

In the top-left image, the lock is in its default, secure position.

- The lock contains two sets of pins in each chamber:
  - Key pins (bottom, red)
  - Driver pins (top, blue/purple), pushed down by springs
- These pins sit across the boundary between the plug (the part that turns) and the housing (outer shell).
- Because the pins block this boundary—called the shear line—the plug cannot rotate.

**Key idea:** The lock is secure because the pins physically prevent movement.

**2. Incorrect Key Inserted**

In the top-right image, a wrong key is inserted.

- The key lifts the pins, but not to the correct heights.

- Some pins still cross the shear line, meaning the boundary is still blocked.
- The plug remains unable to turn.

**Key idea:** Even small differences in key shape prevent alignment, keeping the lock secure.

**3. Correct Key Inserted (Alignment Achieved)**

In the bottom-left image, the correct key is inserted.

- Each cut on the key lifts the key pins to precise heights.
- This causes the gap between key pins and driver pins to align exactly at the shear line.
- Now, no pin crosses the boundary between plug and housing.

**Key idea:** The correct key creates a clean separation at the shear line.

**4. Plug Rotates (Unlocking)**

In the bottom-right image, the lock is turned.

- With all pins aligned at the shear line, the plug is free to rotate.
- Turning the key rotates the plug, which:
    - Retracts the locking mechanism (e.g., latch or bolt)
    - Opens the lock

**Key idea:** Alignment enables movement, allowing the lock to function.

The effectiveness of the pin tumbler lock lies in its reliance on precision, variation, and mechanical constraint. Each pin within the lock must be lifted to an exact height for the mechanism to function correctly. Even a slight deviation in the position of a single pin is enough to prevent the lock from operating. This requirement for precise alignment ensures that only a correctly cut key can successfully manipulate all pins simultaneously.

In addition to precision, the use of multiple pin stacks significantly increases the number of possible key combinations. Each pin chamber represents a variable, and together they create a complex system that is difficult to replicate without the correct key. This multiplicity enhances security by making unauthorised access more challenging.

The design also relies on mechanical blocking. When the pins are not correctly aligned, they physically obstruct the rotation of the plug, preventing the lock from opening. This means the lock does not depend on electronics or external systems—its security is built into the physical arrangement of its components.

A useful way to understand this mechanism is to imagine a series of barriers blocking a rotating wheel. If even one barrier is out of place, the wheel cannot turn. Only when all the barriers are perfectly aligned does the wheel rotate freely. In the same way, a pin tumbler lock will only operate when every pin is positioned exactly at the shear line, allowing the plug to turn and the lock to open.

Vulnerabilities and the Shear Line Weakness

The fundamental vulnerability of pin tumbler locks lies in their mechanical operation at the shear line, the critical interface where pins must align to permit cylinder rotation [71, 72]. This design feature, while necessary for the lock's basic function, creates an exploitable weakness that has been extensively documented in both forensic and security literature [73, 74].

Lock picking represents one of the most well-documented methods of exploiting pin tumbler vulnerabilities. The "Foil Impressioning" kit exemplifies modern picking techniques that target the shear line mechanism [74]. This technique involves using shaped aluminium foil to create impressions of the pin positions, allowing attackers to manipulate the lock without the original key [74]. The effectiveness of such techniques demonstrates that the shear line vulnerability remains a persistent security concern despite the lock's widespread adoption [73, 74].

Beyond traditional picking methods, pin tumbler locks have proven vulnerable to unconventional attack vectors. Recent research has documented the use of nitric acid as an attack method against pin tumbler locks in European burglaries [75]. In experimental testing, 65% nitric acid successfully compromised three out of four standard pin tumbler locks, with the fastest compromise occurring in approximately 22 minutes [75]. This vulnerability reveals that the mechanical integrity of the lock's internal components—the pins and springs—can be compromised through chemical degradation, representing a significant security gap in conventional designs [75].

Impact on Lock Design: Iterative Security Improvements

The recognition of pin tumbler vulnerabilities has driven substantial innovation in lock design and manufacturing. These improvements operate across multiple dimensions: material selection, manufacturing precision, and mechanical design modifications.

**Security Pin Development:** The introduction of security pins represents a direct response to picking vulnerabilities [71]. Security pins, including spool and serrated variants, were specifically engineered to complicate the picking process by creating additional binding points and resistance to manipulation [71]. These pins exploit the principle that traditional pins can be manipulated individually at the shear line, whereas security pins create multiple contact points that resist conventional picking techniques [71].

The Mul-T-Lock cylinder exemplifies advanced security pin design through its "pin-in-pin" telescoping tumbler design [76]. This innovation significantly complicates picking attempts, as traditional picking tools can become stuck between the inner and outer pins [76]. The specialized H&M Mul-T-Lock picking tool was subsequently developed to overcome this specific security feature, demonstrating the ongoing arms race between lock designers and those seeking to defeat locks [76].

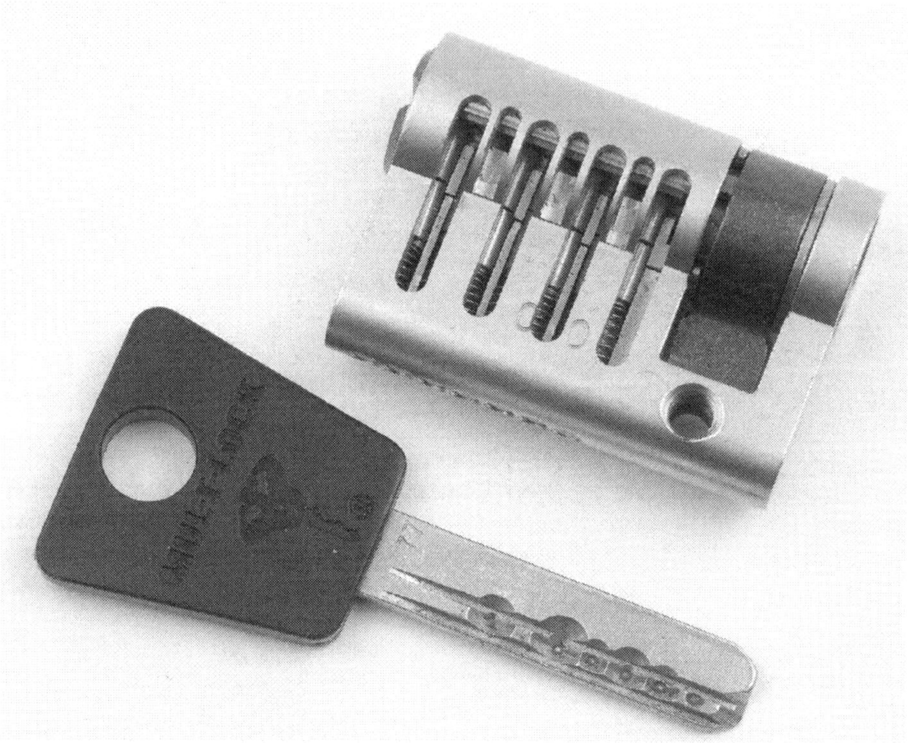

*Figure 5: Mul-T-Lock Cylinder and Key. Dennis van Zuijlekom, CC BY-SA 2.0, via Flickr.*

Mul-T-Lock is a high-security locking system developed by the Israeli company Mul-T-Lock, engineered to provide advanced protection against picking, drilling, bumping, and unauthorized key duplication. It is widely deployed in residential, commercial, and critical infrastructure settings where conventional locking systems are considered insufficient to manage modern security risks.

At the core of the system is telescopic pin technology, which distinguishes Mul-T-Lock from standard pin tumbler locks. In a conventional lock, each chamber contains a single pin stack that must align at the shear line to allow rotation. In contrast, Mul-T-Lock uses a dual-pin arrangement in each chamber, consisting of an inner and outer pin. Both pins must align precisely at the shear line simultaneously for the lock to operate. This effectively creates two levels of precision within each chamber, dramatically increasing the number of possible combinations and significantly enhancing resistance to manipulation.

The key design reflects this complexity. Mul-T-Lock keys are typically dimple keys rather than traditional edge-cut keys. Instead of cuts along the edge, the key contains precisely drilled dimples at varying depths. Each dimple is engineered to lift both the inner and outer pins to their exact required positions. In more advanced systems, additional features such as side pins or interactive elements are incorporated, further increasing security. As a result, the system demands extremely high precision—any deviation, even slight, will prevent the lock from opening.

This design enables several important security features. Pick resistance is greatly enhanced because an attacker must align multiple shear lines across both inner and outer pins simultaneously, making manipulation highly complex. Drill resistance is achieved through the use of hardened steel inserts that protect critical components of the cylinder, preventing damage from drilling attacks. The telescopic pin structure also disrupts traditional bumping techniques, providing strong bump resistance. In addition, Mul-T-Lock employs strict key control measures: keys are patented and restricted, and duplication is only possible through authorized locksmiths with the correct security credentials, typically verified באמצעות a security card.

More advanced variants of Mul-T-Lock systems incorporate additional layers of security. For example, Interactive+ systems include a spring-loaded pin embedded within the key itself, adding a dynamic element to the locking process. The MT5+ platform extends this further by combining telescopic pins with a sidebar locking mechanism and additional features such as magnetic or moving components. These innovations introduce secondary locking layers that go beyond traditional pin-based systems, creating a multi-dimensional security architecture.

Mul-T-Lock systems are commonly used in environments where high security is essential. These include high-end residential properties, commercial facilities, government buildings, and critical infrastructure such as utilities and data centres. In such contexts, the combination of mechanical complexity, controlled key duplication, and resistance to physical attack provides a robust and reliable security solution.

The effectiveness of Mul-T-Lock lies in its layered approach. It integrates mechanical sophistication through telescopic pins, enforces strict key control through restricted duplication systems, incorporates physical protection against forced entry, and introduces additional security layers such as sidebars and interactive elements. Together, these features create a system that is significantly more resilient than standard locking mechanisms.

A useful way to understand this system is through analogy. A conventional lock can be thought of as requiring a single correct value for each pin—similar to entering a single password. In contrast, Mul-T-Lock requires multiple correct values per pin, along with additional verification steps, making it comparable to needing several passwords at once, all entered correctly and simultaneously.

For locksmiths and security professionals, Mul-T-Lock represents a shift from basic mechanical security toward engineered, high-precision security systems. Working with these locks requires specialized tools, proper authorization for key cutting, and a solid understanding of multi-layered pin configurations and advanced access control principles.

Below is a technical comparison table that contrasts Mul-T-Lock, standard pin tumbler locks, and other high-security systems such as Medeco and ASSA ABLOY (Abloy disc detainer systems). This focuses on engineering design, security mechanisms, and practical locksmith considerations.

*Table 1: Technical Comparison of Locking Systems*

| Feature | Standard Pin Tumbler | Mul-T-Lock | Medeco | ASSA ABLOY (Abloy Disc Detainer) |
|---|---|---|---|---|
| **Core Mechanism** | Single pin tumbler (top + bottom pins) | Telescopic pin tumbler (inner + outer pins per chamber) | Angled pin tumbler + sidebar | Rotating discs (no pins) |
| **Shear Line Complexity** | Single shear line | Dual shear lines per chamber | Dual requirement: pin height + rotation | No shear line (disc alignment gates) |
| **Key Type** | Edge-cut key | Dimple key (multi-directional cuts) | Chisel-tip / angled cut key | Flat key with cuts controlling disc rotation |
| **Additional Locking Elements** | None (basic designs) | Optional sidebar, interactive pins, magnetic elements | Sidebar that engages when pins rotate correctly | Sidebar engaged by aligned discs |
| **Picking Resistance** | Low–moderate | Very high (multi-layer pin alignment) | Very high (must lift + rotate pins) | Extremely high (no traditional pins to pick) |
| **Bump Resistance** | Low (vulnerable) | High (telescopic pins disrupt bumping) | High (angled pins prevent bump transfer) | Immune (no pins) |
| **Drill Resistance** | Low–moderate | High (hardened inserts) | High (hardened components) | Very high (hardened steel discs) |
| **Key Control** | Low (easy duplication) | High (patented, restricted keys) | High (restricted keyways) | High (restricted systems) |
| **Tolerance Precision** | Moderate | Very high (dual pin alignment) | Very high (lift + rotation precision) | Extremely high (disc alignment precision) |
| **Master Key Capability** | Basic | Advanced (complex pin stacks) | Advanced (sidebar + pin angles) | Advanced (disc combinations) |
| **Wear & Durability** | Moderate (springs wear over time) | High (precision components) | High | Very high (no springs, minimal wear) |
| **Maintenance Complexity** | Low | Moderate–high | High | Low–moderate |
| **Typical Use Cases** | Residential, low-security | High-security residential, | Government, military, commercial | High-security facilities, harsh environments |

| Feature | Standard Pin Tumbler | Mul-T-Lock | Medeco | ASSA ABLOY (Abloy Disc Detainer) |
|---|---|---|---|---|
| | | commercial, critical infrastructure | | |

The relationship between mechanical complexity and security is central to understanding how modern locking systems have evolved. Standard pin tumbler locks rely on a single variable—pin height—meaning that each pin must be lifted to the correct position to align at the shear line. Mul-T-Lock systems significantly increase this complexity by introducing telescopic pins, where each chamber contains both an inner and outer pin. This effectively creates two variables per chamber, requiring dual alignment for the lock to operate. Medeco systems extend this further by adding rotational alignment, meaning that pins must not only be lifted to the correct height but also rotated to the correct angle. In contrast, Abloy disc detainer systems eliminate pins altogether, relying instead on the precise rotational alignment of multiple discs. As mechanical complexity increases across these systems, the nature of potential attacks shifts from relatively simple manipulation techniques to highly precise, engineering-intensive challenges.

Another important feature that enhances security in advanced locking systems is the use of a sidebar mechanism. Standard locks typically do not include a sidebar, relying solely on pin alignment. However, in systems such as Mul-T-Lock (in advanced variants), Medeco, and especially disc detainer designs, the sidebar introduces a secondary locking condition. Even if an attacker successfully manipulates the primary locking elements, the lock will not open unless the sidebar also aligns correctly. This means multiple independent conditions must be satisfied simultaneously, significantly increasing resistance to picking and other forms of attack.

Each locking system also reflects a distinct philosophy of attack resistance. Standard locks are generally designed as a basic deterrent, often described as "keeping honest people honest." Mul-T-Lock systems focus on increasing the number of variables and the precision required, creating layered mechanical complexity. Medeco adopts a different approach by adding directional control, requiring both correct positioning and rotation of components. Abloy disc detainer systems take a more radical approach by eliminating traditional weaknesses entirely, removing pins and springs and replacing them with rotating discs that are inherently more resistant to conventional attack methods.

Modern high-security locks also extend beyond mechanical design into what can be described as a broader security ecosystem. Systems such as Mul-T-Lock and Medeco incorporate patented key control, meaning that key duplication is tightly restricted. Keys can typically only be copied by authorised locksmiths and often require a security card or formal verification. This shifts the concept of security from being purely physical to a combination of physical and procedural control, where administrative processes play a critical role in maintaining system integrity.

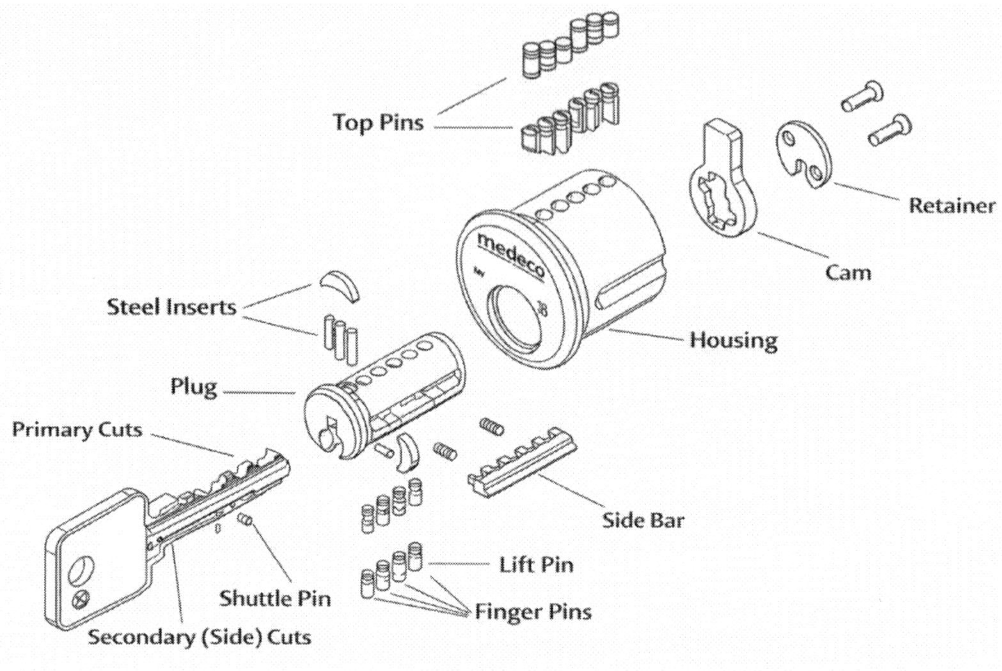


*Figure 6: Medeco Lock Cylinder exploded view.*

Durability and environmental performance are also key considerations in lock design. Traditional pin-based systems rely on springs and small moving components that can wear over time or be affected by environmental conditions such as dust, moisture, or temperature extremes. Disc detainer systems, by contrast, do not rely on springs and are therefore more robust in harsh environments, making them particularly suitable for industrial, outdoor, or high-exposure applications.

A simple way to understand these differences is through analogy. A standard lock can be thought of as requiring one correct number per position. Mul-T-Lock systems require two correct values per position, increasing complexity. Medeco systems add another layer by requiring both the correct number and the correct direction, similar to aligning a dial precisely. Abloy systems, on the other hand, resemble a set of rotating combination wheels that must all align perfectly to allow movement.

For locksmiths, this comparison highlights a clear evolution in the profession—from working with simple mechanical alignment systems to managing multi-dimensional, controlled, and policy-driven security solutions. Modern practice requires not only an understanding of mechanical principles but also the ability to manage key control systems, integrate physical and digital access technologies, and advise clients on the balance between security, usability, and cost.

**Manufacturing Tolerance Improvements:** Tighter manufacturing tolerances represent another critical response to picking vulnerabilities [71]. Enhanced precision in manufacturing reduces the mechanical play within the lock mechanism, making it more difficult for attackers to manipulate pins through the shear line

[71, 77]. Research on the mechanical characteristics of cylinder pin tumbler locks has established methodologies for assessing resistance to non-destructive breaking methods, providing quantitative frameworks for evaluating lock security [77]. These testing methodologies enable manufacturers to verify that their products meet specified resistance standards and maintain consistent security levels across production batches [77].

**Anti-Pick Features and Restricted Keyways:** The development of anti-pick features and restricted keyways constitutes a third major design response [71]. Restricted keyways limit the types of keys and tools that can be inserted into the lock, thereby preventing the use of standard picking implements [71]. These design modifications work in conjunction with security pins to create multiple layers of resistance against unauthorized access [71].

### Systemic Weaknesses and Iterative Improvement

The history of pin tumbler lock development illustrates a fundamental principle in security design: even widely adopted systems can harbor systemic weaknesses that require continuous iterative improvement [71, 72]. The pin tumbler lock's dominance in the global market did not prevent the emergence of picking techniques, chemical attack methods, and other vulnerabilities [73-75]. Rather than rendering the design obsolete, these vulnerabilities prompted systematic improvements in manufacturing, materials, and mechanical design [71, 77].

The recognition of these vulnerabilities has also spurred exploration of alternative security approaches. Modern research has investigated opto-electronic ignition locks as potential replacements for mechanical pin tumbler designs in automotive applications, suggesting that even in contexts where pin tumbler locks have been standard for over a century, security concerns continue to drive innovation toward fundamentally different technological approaches [78]. Additionally, contemporary research has explored IoT-based smart door lock systems as alternatives to conventional pin tumbler locks, acknowledging that traditional designs remain susceptible to lock-picking techniques and offer limited access control capabilities [72].

The pin tumbler lock exemplifies how security systems evolve in response to identified vulnerabilities. From its refinement in the 19th century to its current status as a globally dominant design, the pin tumbler lock has undergone continuous modification to address weaknesses at the shear line and other critical points [69-71]. The introduction of security pins, manufacturing tolerance improvements, and anti-pick features demonstrates that security design is inherently iterative, with each innovation prompting new attack methods and subsequent defensive responses [71, 76, 77]. This dynamic process continues to shape lock design, with emerging technologies offering potential alternatives to mechanical pin tumbler systems [72, 78]. The pin tumbler lock's history thus serves as a case study in how systemic vulnerabilities in widely adopted technologies drive sustained innovation and improvement.

## 3. Master Keying Failures and Cross-Key Vulnerabilities

Master keying systems represent a fundamental compromise in mechanical access control design, wherein the convenience of hierarchical key systems introduces significant security vulnerabilities. These systems, which allow multiple keys to open the same lock through the incorporation of additional shear lines, have been subject to extensive scrutiny within both security research and practical building operations. The tension between operational convenience and security integrity remains a critical challenge in physical access control design.

### Master Key System Design and Vulnerabilities

Master key systems function by introducing additional shear lines within pin tumbler locks, enabling multiple keys with different cuts to operate the same mechanism (Lad, 2026). While this architectural approach provides substantial operational benefits for large facilities requiring hierarchical access control, the design introduces inherent vulnerabilities that have been documented across multiple domains of security research.

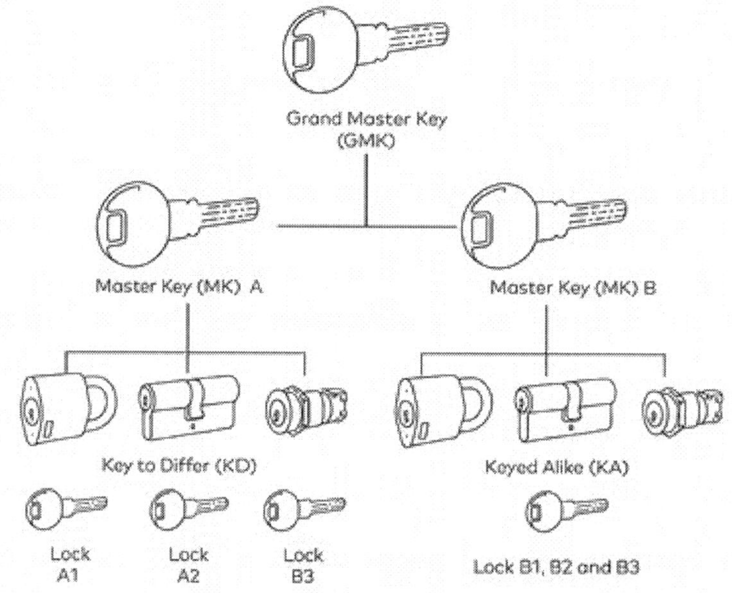


*Figure 7: Master Key System.*

The fundamental vulnerability in master key systems stems from the mathematical relationship between the number of shear lines and the potential for unintended key combinations. When additional shear lines are introduced to accommodate master keys, the system creates what is known as cross-keying—a condition where keys not intended to open a particular lock may nonetheless do so due to the geometric relationships between the shear lines [79]. This phenomenon represents a direct consequence of the physical constraints inherent to pin tumbler lock design.

### Cross-Keying and Unintended Access Paths

Cross-keying vulnerabilities emerge from the geometric properties of master key systems and represent a critical security concern in mechanical access control. The problem manifests when the combination of shear lines required to accommodate multiple master keys creates unintended alignment possibilities [79]. These unintended access paths can be exploited by individuals possessing keys that were never intended to open specific locks, yet whose geometric properties align with the multiple shear lines present in the system.

The research on self-impressioning master keys demonstrates the practical exploitation of master key system vulnerabilities [80]. Self-impressioning master keys operate by creating an impression of the lock's internal pin configuration, allowing unauthorized individuals to fabricate functional keys that can open locks for which they lack authorization [80]. This technique has become increasingly prevalent in criminal activity, with manufactured self-impressioning master keys readily available through commercial channels and extensively documented in instructional materials [80].

### Impact on Lock Design and Access Control Standards

The recognition of master key vulnerabilities has fundamentally altered the trajectory of lock design and access control system development. Stricter design standards have been implemented to limit the vulnerabilities inherent in mechanical master key systems [79]. These standards impose constraints on the number of shear lines that can be incorporated into a single lock mechanism, thereby reducing the potential for cross-keying while maintaining some degree of hierarchical access capability [79].

The operational management of keying systems in large facilities demonstrates the practical implementation of these design constraints. In a comprehensive case study of a high-rise residential building project, the keying system was structured with deliberate limitations on the master key hierarchy to prevent uncontrolled access proliferation [79]. The system employed a two-track architecture consisting of a building-controlled hierarchy with a Grand Master key and seven operational subcategories, combined with Single-Keyed-Different (SKD) unit keys maintained at the individual unit level [79]. This approach represents a deliberate design choice to limit the scope of master key authority and reduce the potential for cross-keying vulnerabilities [79].

The implementation of such constrained hierarchies requires rigorous operational controls to maintain security integrity. The case study documented the necessity for disciplined governance mechanisms, including tagged tracking systems linking opening identity, room logic, keying intent, and subcontractor coding [79]. Installation-phase breakdowns were systematically categorized and addressed, including cylinder-lock interface incompatibility, finish-driven cylinder length misfit, and unit-level traceability slips that could result in uncontrolled key swaps [79]. These operational controls demonstrate that security in master key systems depends not solely on mechanical design but equally on administrative discipline and verifiable identity control [79].

Electronic Access Control as a Replacement for Mechanical Hierarchies

The vulnerabilities inherent in mechanical master key systems have driven substantial adoption of electronic access control systems as alternatives to traditional hierarchical key arrangements. Electronic locking systems offer several advantages over mechanical systems, including more flexible access rights management, the ability to revoke physical keys, and the theoretical resistance of electronic keys to cloning [81]. These features address fundamental limitations of mechanical master key systems by eliminating the geometric constraints that create cross-keying vulnerabilities.

However, the transition to electronic access control introduces new security considerations. Analysis of widely deployed electronic locking systems has revealed that while the system concepts may be well-designed, implementation flaws can compromise security [81] Specifically, vulnerabilities in electronic locking systems have been identified that allow extraction of system-wide master secrets through brute force attacks or Differential Power Analysis attacks on electronic keys [81]. Additionally, weaknesses in random number generation have been documented that permit door opening without breaking cryptography under certain circumstances [81].

The security challenges in electronic access control extend beyond individual lock vulnerabilities to encompass broader system architecture concerns. Smart lock systems utilizing RSA encryption and one-time passwords represent advances in electronic security, yet they introduce new attack vectors including man-in-the-middle attacks where digital keys can be duplicated [82]. The integration of electronic access control with cloud-based logging and breach detection systems adds complexity to the security architecture while potentially introducing additional vulnerabilities through network-based attack surfaces [82].

Key Lesson: The Convenience-Security Tradeoff

The history of master key system development illustrates a fundamental principle in security design: convenience and security exist in tension, and poorly managed convenience can introduce hidden vulnerabilities that undermine the entire system. Master key systems were developed to provide operational convenience, the ability to grant hierarchical access without requiring multiple distinct keys for each access level [79]. This convenience, however, comes at the cost of introducing geometric relationships within the lock mechanism that create potential for unintended access.

The principle extends beyond mechanical systems to electronic access control. Electronic systems provide convenience through flexible access management and remote revocation capabilities, yet these conveniences can introduce implementation flaws or architectural vulnerabilities if not carefully controlled [81]. The security of access control systems, whether mechanical or electronic, depends fundamentally on the discipline with which convenience features are implemented and managed.

The operational case study of the high-rise building project demonstrates that security in master key systems is achieved not through increasingly complex mechanical hierarchies, but rather through disciplined governance and verifiable identity control [79]. The project's success in managing approximately 512 hollow metal openings with a constrained master key hierarchy and rigorous operational controls illustrates that security effectiveness depends on the integration of design constraints, administrative procedures, and verification mechanisms [79].

Master keying failures and cross-key vulnerabilities represent a persistent challenge in physical access control design. The fundamental issue—that convenience in the form of hierarchical key systems introduces geometric vulnerabilities—has driven evolution in both mechanical lock design standards and the adoption of electronic access control systems. However, the transition to electronic systems introduces new vulnerabilities that require equally rigorous attention to implementation details and system architecture. The key lesson from this evolution is that security cannot be achieved through design alone; rather, it requires the integration of constrained design, strict standards, rigorous operational controls, and verifiable administrative procedures. Whether mechanical or electronic, access control systems must balance the legitimate operational need for hierarchical access with the security imperative to prevent unintended access paths.

### 4. Bumping Attacks on Modern Locks (Early 2000s)

Lock bumping represents a significant vulnerability in mechanical security systems that emerged prominently in the early 2000s. This attack method exploits fundamental design weaknesses in pin tumbler locks, enabling rapid unauthorized access with minimal skill and expertise. The discovery and widespread dissemination of lock bumping techniques fundamentally altered the landscape of physical security, prompting substantial changes in lock design philosophy and consumer awareness of lock vulnerabilities.

Lock bumping is a technique that exploits the mechanical properties of pin tumbler locks through the use of a specially modified key and a striking force. The attack operates by inserting a "bump key", a key cut to the maximum depth at all positions, into the lock cylinder and then striking it with a light tap while applying rotational pressure [83]. This mechanical action causes the pins within the lock to jump momentarily, potentially allowing the cylinder to rotate and the lock to open [83].

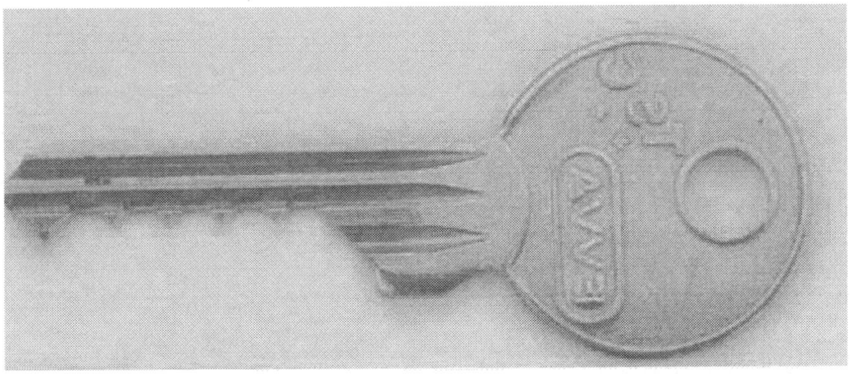

*Figure 8: Bumping Key. Madox at Polish Wikipedia, Public domain, via Wikimedia Commons.*

The attack is particularly concerning because it requires minimal technical skill to execute. Unlike traditional lock picking, which demands considerable practice and understanding of lock mechanics, bumping can be performed by individuals with little to no specialized training [83]. Furthermore, the attack leaves minimal forensic evidence, making it difficult to detect whether a lock has been compromised through bumping or opened with a legitimate key [83].

Impact on Lock Design and Security Architecture

**Development of Anti-Bump Mechanisms:** The widespread recognition of lock bumping vulnerabilities prompted manufacturers to develop countermeasures specifically designed to prevent or mitigate bumping attacks. These anti-bump mechanisms represent a fundamental shift in lock design philosophy, moving beyond traditional pin tumbler architecture to incorporate additional security features [83]. The development of these mechanisms reflects the security industry's response to a low-skill, high-impact attack vector that could be deployed at scale.

**Increased Consumer Awareness:** The discovery of lock bumping in the early 2000s significantly elevated public consciousness regarding the limitations of standard pin tumbler locks [83]. This awareness extended beyond security professionals to include homeowners, business operators, and policy makers. The realization that widely-used locks could be compromised through a simple technique with minimal evidence created a market demand for higher-security alternatives and prompted consumers to reassess their physical security assumptions.

**Shift Toward Higher-Security and Restricted Key Systems:** In response to bumping vulnerabilities, the lock industry experienced a notable shift toward higher-security lock systems and restricted key architectures [83]. These systems employ more complex internal mechanisms, tighter manufacturing tolerances, and proprietary key designs that are more difficult to duplicate or modify into bump keys. Restricted key systems limit key duplication to authorized locksmiths and manufacturers, reducing the availability of keys that could be modified for bumping attacks [83].

High-quality locks can, somewhat counterintuitively, be more susceptible to bumping unless they incorporate specific anti-bump features. This is largely due to their precise manufacturing tolerances. In well-engineered cylinders, the internal components fit together with minimal variation, allowing energy from a bump key strike to transfer efficiently through the pin stacks. Because there is less friction and less dissipation of force, the pins move more freely and align more easily at the shear line. In contrast, lower-quality locks often have rougher tolerances, which can absorb or misdirect this energy and sometimes cause the mechanism to bind or jam. Similarly, locks constructed from hardened steel are less likely to deform under repeated impacts, meaning they continue to function smoothly during a bumping attempt, whereas cheaper materials may fail or seize.

To address these vulnerabilities, standards and countermeasures have been developed. In the United Kingdom, the TS007 Kitemark standard provides a benchmark for resistance to snapping and bumping, with 1-star cylinders offering a minimum resistance of around one minute and 3-star cylinders providing significantly higher protection, typically resisting attack for at least ten minutes. While security pins, such as spool or mushroom pins, can introduce irregular movement and make bumping more difficult, they do not fully prevent it.

The TS007 Kitemark is a British security standard developed to address a specific and widespread vulnerability in Euro-profile cylinders—particularly lock snapping, but also resistance to bumping, picking, drilling, and other forms of forced entry. It is part of a broader UK initiative led by the British Standards Institution (BSI) and the door and hardware industry to improve real-world burglary resistance, especially in residential properties.

TS007 is not just a generic quality mark—it is a performance-based testing standard. Locks and door security products are physically attacked under controlled conditions to determine how long they can resist common burglary techniques.

These include:

- Lock snapping (breaking the cylinder to access the cam)
- Bumping (using a bump key to momentarily align pins)
- Picking and manipulation
- Drilling and forced entry

The goal is to simulate real attack scenarios, not just laboratory conditions.

TS007 uses a star-based rating system (0 to 3 stars) to indicate the level of protection. The two most commonly referenced levels are 1-star and 3-star, which relate directly to resistance time and overall security performance.

**★ 1-Star Cylinders**

- Provide basic enhanced security over standard cylinders
- Tested to resist common attacks for a minimum duration (approximately 1 minute)
- Must include some anti-snap features (e.g., sacrificial sections or strengthened core)
- Often do not provide full protection on their own

Important: A 1-star cylinder is typically not considered sufficient by itself for high-security applications.

**★★★ 3-Star Protection (Two Pathways)**

A door can achieve 3-star protection in one of two ways:

**Option 1: 3-Star Cylinder Alone**

- A high-security cylinder independently meets the full TS007 3-star requirements
- Provides strong resistance (typically 10 minutes or more) against attack methods
- Includes:
    - Advanced anti-snap design

- Anti-bump and anti-pick features
- Drill-resistant materials
- Reinforced structure

**Option 2: 1-Star Cylinder + 2-Star Door Furniture**

- A 1-star cylinder combined with 2-star rated door hardware (e.g., security escutcheons or handles)
- The external hardware protects the cylinder from being accessed or snapped

This reflects a key design philosophy: Security is a system, not just a lock.

The concept of "time to defeat" is central to understanding modern lock security. Most burglaries are opportunistic and highly time-sensitive, meaning that attackers are unlikely to persist if entry takes too long or becomes too difficult. As resistance time increases, so too does the likelihood that the intruder will abandon the attempt, be detected, or move on to a more vulnerable target. The difference between one minute and ten minutes of resistance is not merely incremental; it represents a significant escalation in deterrence. A lock that resists for one minute may slow an attacker, but it may not be enough to stop a determined attempt. In contrast, a lock that withstands attack for ten minutes often exceeds the practical threshold for most break-ins, making the risk far less acceptable to the intruder.

To achieve these higher levels of resistance, TS007-rated cylinders incorporate a range of engineered security features. Anti-snap lines are designed to control how a lock fails under force, protecting the internal mechanism even if part of the cylinder is broken. Hardened steel inserts are used to resist drilling attacks, while more complex pin systems reduce the effectiveness of bumping and picking techniques. Reinforced cams and plugs help prevent forced rotation after damage, and tighter manufacturing tolerances combined with stronger materials ensure that the lock maintains structural integrity under attack. Together, these features transform the lock from a simple mechanical device into a layered defensive system.

For locksmiths and security professionals, the TS007 standard represents a shift in how security is evaluated and recommended. Rather than selecting locks based solely on brand reputation or cost, there is an increasing emphasis on independently tested performance against real-world attack methods. In practice, this means recommending 3-star rated solutions for residential doors where appropriate, recognising that door hardware—such as handles and escutcheons—plays a critical role alongside the cylinder itself, and educating clients that a "high-quality" lock is not necessarily a "high-security" lock unless it has been formally tested and certified.

It is also important to note that key control systems do not inherently protect against bumping. A bump key only needs to match the keyway profile of the lock, meaning that even restricted or registered key systems can be vulnerable if a compatible blank or modified key is used. Although legitimate blanks may be controlled, improvised bump keys can often be created by filing down standard keys.

More advanced mechanical countermeasures include trap pins, which are designed to engage and jam the cylinder if improper manipulation occurs, and shallow-drilled pin chambers, where certain pins are positioned too high for a bump key to effectively engage. These features disrupt the energy transfer required for a successful bump attack.

Some lock designs eliminate the vulnerability entirely by removing the spring-loaded pin mechanism that bumping relies on. Systems that use sidebars without top pins, such as BiLock, are inherently resistant, as are time locks, combination locks, electronic and electromagnetic locks, and disc tumbler locks, all of which operate on fundamentally different principles. However, it is worth noting that electronic locks with mechanical key overrides may still be vulnerable through their backup cylinders. Warded locks, while immune to bumping, can be defeated using a different method involving skeleton keys.

Finally, adjustments to internal components, such as increasing spring tension, can also reduce susceptibility to bumping. Stronger springs alter the timing and movement of the pins, disrupting the "ricochet" effect required for the attack to succeed. This illustrates that resistance to bumping is not determined by a single feature, but by a combination of design choices that influence how energy is transferred and controlled within the lock.

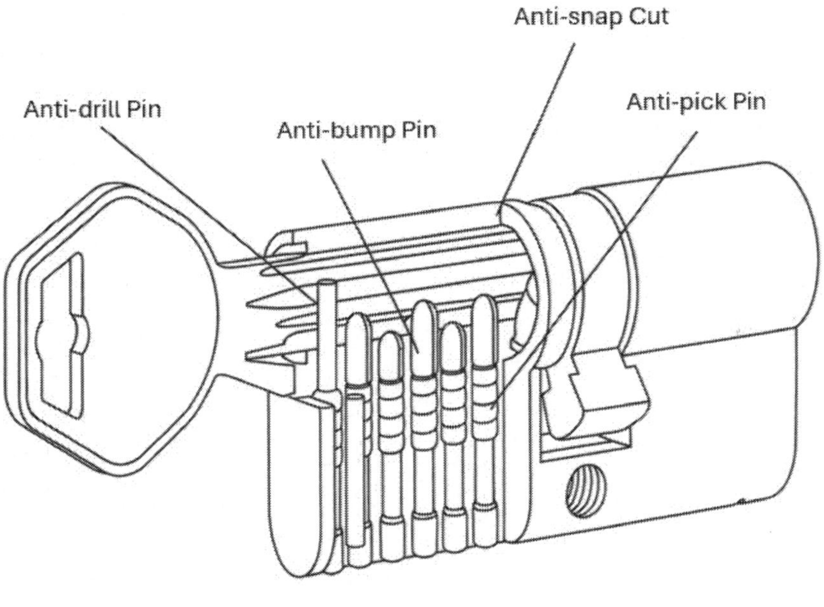


*Figure 9: Anti-bump cylinder.*

The diagram above illustrates a modern anti-bump euro cylinder, an advanced evolution of the standard pin tumbler lock designed to resist common forms of attack such as bumping, picking, drilling, and snapping. While it is built on the same fundamental operating principle as a traditional lock, it incorporates additional engineered features that disrupt and resist these attack methods. The result is a locking system that retains familiar functionality while significantly enhancing security.

At its core, the anti-bump cylinder still operates using the conventional pin tumbler mechanism. Inside the cylinder are bottom pins (key pins) and top pins (driver pins), both held under spring pressure. In the absence of a key, these pins sit across the shear line, preventing the plug from rotating. When the correct key is inserted, it lifts each pin to a precise height, aligning all pin stacks at the shear line. This allows the plug to rotate freely and the lock to open. Although this principle remains unchanged, the introduction of advanced pin designs and structural features makes manipulation far more difficult.

One of the primary enhancements is the use of anti-bump pins. These pins are specially shaped—often spool, mushroom, or otherwise modified—and are designed to disrupt the transfer of kinetic energy during a bumping attack. In a typical bump attack, a specially cut key is struck to momentarily force all pins upward to the shear line. Anti-bump pins interfere with this process by absorbing or dispersing the impact energy and moving in less predictable ways. In some cases, they create false shear lines, meaning that even if the pins appear aligned, the plug will not rotate. This makes the precise "jump" required for bumping unreliable or ineffective.

Anti-pick pins further enhance security by resisting manipulation during lock picking. Unlike standard smooth pins, these have irregular shapes that create false feedback when tension is applied. As a result, they can bind within the plug and give the impression that they are correctly set when they are not. This forces the person attempting to pick the lock to overcome multiple false sets, significantly increasing the time, skill, and effort required to open the lock.

To counter destructive attacks, anti-drill pins and hardened inserts are incorporated into critical areas of the cylinder. These components are typically made from hardened steel and are positioned to protect the pin chambers, shear line, and plug. When a drill bit contacts these inserts, the material resists cutting and may even deflect or spin against the bit, making drilling slow and difficult. This ensures that attempts to bypass the lock through force are significantly impeded.

Another important feature is the anti-snap cut, also known as a sacrificial section. This is a deliberately engineered weak point within the cylinder body designed to control how the lock behaves under forced attack. In a snapping attempt, where force is applied to break the cylinder, this section is intended to break in a controlled manner. Crucially, the internal locking mechanism—including the cam and core—remains intact and protected. Even after the outer portion is broken, the attacker cannot easily access or operate the lock.

The effectiveness of the anti-bump cylinder lies in this layered approach to security. The pin system still controls rotation, but anti-bump and anti-pick pins prevent manipulation of that system. Hardened inserts resist drilling, while the anti-snap design protects against forced entry. Instead of relying on a single line of defence, the lock requires multiple independent conditions to be overcome simultaneously, greatly increasing resistance to attack.

Compared to traditional locks, which rely on relatively simple and predictable mechanical alignment, anti-bump cylinders introduce multiple layers of engineered resistance. They disrupt energy transfer during bumping, create false feedback during picking, resist physical destruction through hardened materials, and maintain security even when subjected to force. This transforms the lock from a basic mechanical device into a sophisticated, multi-layered security system.

A useful way to understand this is through analogy. A standard lock can be thought of as a single row of pins that must align once for the mechanism to work. An anti-bump cylinder, by contrast, behaves more like a system where the pins do not respond predictably, some elements intentionally mislead the user, and the structure itself resists damage. Even if part of the system is compromised, it does not necessarily result in access.

For locksmiths and security professionals, this represents a clear shift in practice. The focus moves away from simple mechanical alignment toward understanding and mitigating specific attack methods. It requires knowledge of techniques such as bumping, picking, and snapping, as well as the ability to select cylinders based on tested performance standards such as TS007 or EN ratings. It also involves educating clients that true security is not defined by brand or material alone, but by the design features and certified resistance of the system.

Ultimately, an anti-bump cylinder does not change the fundamental principle of how a lock works. Instead, it enhances that principle by introducing precision, complexity, and engineered resistance at every stage. This makes both manipulation and forced entry significantly more difficult, time-consuming, and often impractical, reflecting the evolution of locksmithing into a discipline that combines mechanical design with strategic security engineering.

## Critical Security Lesson: Low-Skill, High-Impact Attacks

The lock bumping phenomenon illustrates a fundamental principle in security design: attacks that require minimal skill but produce significant impact pose particular dangers because they scale easily across large populations [83]. Unlike sophisticated attacks that require specialized knowledge and equipment, bumping can be learned from publicly available information and executed with common tools. This scalability means that a single vulnerability can be exploited by numerous actors simultaneously, creating widespread security risks across entire classes of locks.

The ease of learning and executing bumping attacks, combined with the difficulty of detecting such attacks, creates a particularly dangerous security scenario. Traditional security models often assume that attackers possess significant technical expertise; however, bumping demonstrates that this assumption may be invalid. The attack's accessibility means that individuals with minimal motivation or capability can compromise physical security, fundamentally challenging assumptions about the effectiveness of standard locks as security barriers.

Lock bumping attacks represent a watershed moment in the understanding of mechanical lock security. The early 2000s discovery of this vulnerability revealed fundamental weaknesses in widely-deployed pin tumbler lock designs and prompted significant changes in lock manufacturing, consumer behaviour, and security awareness. The incident underscores the critical importance of designing security systems that resist not only sophisticated, skilled attacks but also simple, low-skill attacks that can be executed and scaled by large populations. The lessons learned from lock bumping continue to inform security design principles across both physical and digital domains, emphasizing the necessity of anticipating and defending against attacks that combine ease of execution with significant impact.

### 5. Hotel Keycard System Hacks

Electronic locks in hotels have undergone significant evolution as security technologies, yet their implementation has revealed critical vulnerabilities that challenge conventional assumptions about digital security. What was once perceived as a highly secure advancement in access control has been demonstrated to be susceptible to inexpensive hacking devices capable of reading or spoofing credentials. This vulnerability has prompted substantial changes in lock design philosophy and implementation practices across the hospitality industry.

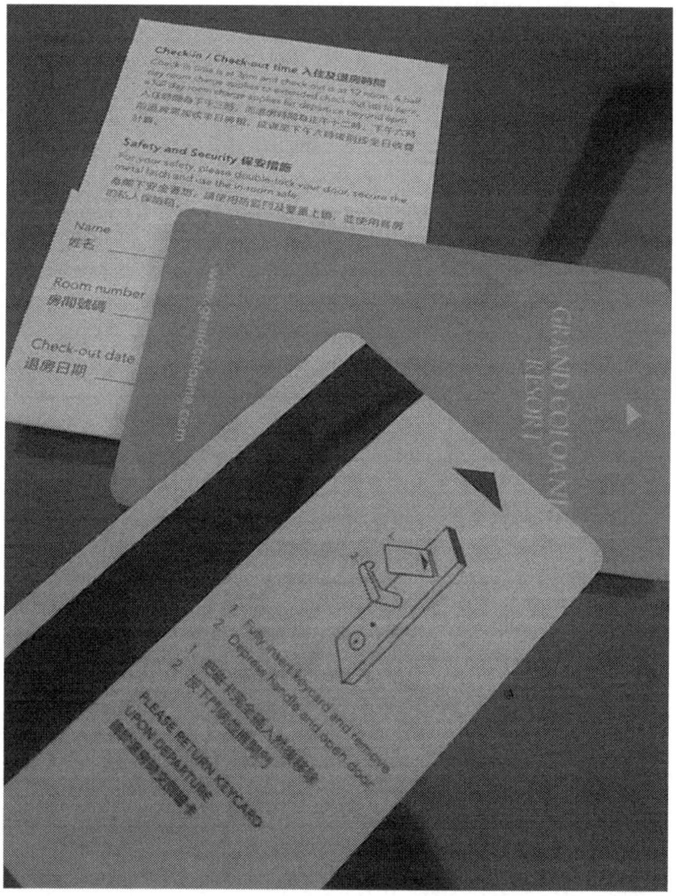


*Figure 10: MC Macau hotel room plastic doorcard - Grand Coloane Resort 1918 Estrada de Hac Sa. Sonia Hung awingt, CC BY-SA 4.0 , via Wikimedia Commons.*

<u>Vulnerabilities in Hotel Electronic Lock Systems</u>

Electronic locks deployed in hotel environments have demonstrated multiple security weaknesses that extend beyond traditional mechanical vulnerabilities. Research has identified that smart locks, including those used in hospitality settings, are inherently hackable, allowing untrusted users to gain remote access to secured spaces [84]. The vulnerability landscape encompasses both physical and digital attack vectors, with smart lock systems combining mechanical and electronic technologies that create multiple points of potential compromise [85].

The specific vulnerabilities in hotel lock systems stem from several technical factors. Connected locks can be exploited through various attack methodologies, including the use of inexpensive hacking devices that can read or spoof digital credentials [84]. These vulnerabilities are particularly concerning in hotel environments where temporary access credentials must be issued to numerous guests and service personnel. The Internet of Things (IoT) protocols commonly used in smart lock systems, such as MQTT, have been identified as presenting significant security risks when inadequately secured, potentially allowing attackers to leak sensitive lock data and remotely control door locks through compromised cloud infrastructure [86].

MQTT (Message Queuing Telemetry Transport) is a lightweight messaging protocol specifically designed to enable efficient and reliable communication between devices, particularly in environments where bandwidth, power, or network stability are limited. It has become a foundational technology in the Internet of Things (IoT), including smart homes, industrial automation, and modern access control systems. Unlike traditional communication methods, MQTT is optimised for small data packets and low overhead, making it ideal for devices such as smart locks, sensors, and embedded systems that must operate continuously with minimal resource consumption.

At its core, MQTT uses a publish–subscribe model, which differs significantly from the traditional client–server (request–response) approach. Instead of devices communicating directly with one another, they interact through a central intermediary known as a *broker*. Devices (called clients) publish messages to the broker under specific "topics," while other devices subscribe to those topics to receive relevant information. This architecture decouples devices from each other—they do not need to know who is receiving the data, only the topic they are publishing to or subscribing from. This makes systems more flexible, scalable, and easier to manage.

In practical operation, the process is straightforward. A device such as a smart lock or mobile app first connects to an MQTT broker (for example, Mosquitto or a cloud-based service like AWS IoT). Once connected, the device can either publish messages, subscribe to topics, or both. Topics are structured like labels—for instance, *home/door/status* or *building/floor1/access*. When a device publishes a message (e.g., "door locked"), the broker immediately distributes that message to all devices subscribed to the corresponding topic. This allows real-time communication across the system without direct device-to-device connections.

In access control and locksmithing contexts, MQTT plays a critical role in enabling modern, connected systems. It is commonly used for transmitting lock status updates (locked/unlocked), sending remote access commands, maintaining audit logs, and integrating with broader building management systems. For example, when a user presses "unlock" on a smartphone app, the app publishes a command to a topic such as *home/door/command*. The smart lock, which is subscribed to that topic, receives the command and executes the action almost instantly. This creates a seamless interaction between physical security devices and digital control systems.

One of MQTT's key strengths is its efficiency. The protocol is extremely lightweight, meaning it uses minimal bandwidth and processing power. This is particularly important for battery-powered devices like smart locks, which must operate for long periods without frequent recharging. MQTT is also designed to function reliably over unstable or intermittent networks, making it suitable for real-world environments where connectivity may not always be consistent. Additionally, it supports near real-time communication and can scale to support thousands of connected devices through a single broker.

MQTT includes configurable Quality of Service (QoS) levels, which allow system designers to balance speed and reliability. At QoS 0, messages are delivered "at most once," prioritising speed with no delivery guarantee. QoS 1 ensures messages are delivered "at least once," meaning duplicates may occur but delivery is guaranteed. QoS 2 provides the highest level of reliability, ensuring messages are delivered "exactly once," though with increased overhead. In security-critical applications such as lock and unlock commands, QoS 1 or 2 is typically used to ensure reliability.

Despite its simplicity, MQTT requires careful attention to security. By default, the protocol does not enforce encryption or authentication, so additional measures must be implemented. These include Transport Layer Security (TLS) for encrypted communication, authentication mechanisms such as usernames/passwords or digital certificates, and strict access control policies that define which devices can publish or subscribe to specific topics. Without these protections, an attacker could potentially intercept messages or send unauthorized commands, such as unlocking a door.

Compared to traditional protocols like HTTP, MQTT is far more efficient for IoT applications. HTTP relies on a request–response model, which introduces higher overhead and is less suited to real-time, device-to-device communication. MQTT's publish–subscribe architecture enables continuous, low-latency data exchange with minimal resource usage, making it a preferred choice for connected security systems.

A useful way to understand MQTT is to think of it as a radio broadcasting system. The broker acts like a radio tower, devices publish messages like radio broadcasts, and other devices "tune in" by subscribing to specific channels (topics). Devices do not communicate directly with each other—they simply listen to the channels relevant to them.

For modern locksmiths and security professionals, MQTT represents a significant shift in the nature of access control. Locks are no longer purely mechanical devices; they are networked endpoints within a larger digital ecosystem. This means that security is no longer فقط about physical resistance—it also involves network communication, authentication, encryption, and system configuration. As a result, professionals in this field must develop an understanding of both physical security and cybersecurity principles to design, install, and manage effective modern access control systems.

Furthermore, research has demonstrated that smart locks present vulnerabilities at the intersection of multiple convenience-focused devices. For example, adversaries can leverage vulnerabilities in network webcams and voice interfaces to activate conversational devices trusted to control physical infrastructure, thereby subverting the intended security benefits of electronic locks [84]. The August Smart Lock case study revealed specific vulnerabilities that hackers could exploit to open the lock [87], illustrating that even commercially available smart lock systems contain exploitable security flaws.

<u>Impact on Lock Design and Implementation Practices</u>

The discovery of vulnerabilities in electronic hotel locks has catalysed significant changes in how these systems are designed, deployed, and maintained. The recognition that digital locks require ongoing security management has fundamentally altered the approach to lock system lifecycle management.

**Improved Encryption and Authentication Protocols:** In response to identified vulnerabilities, lock manufacturers and hospitality technology providers have implemented enhanced encryption and authentication mechanisms. Research on authentication encryption mechanisms for intelligent door locks has demonstrated the effectiveness of advanced cryptographic approaches, such as those based on the SM9 algorithm, in protecting against unauthorized access attempts [86]. These improved protocols aim to prevent the leakage of sensitive information that could be exploited by attackers using inexpensive hacking devices.

Modern smart lock systems now incorporate multiple layers of authentication. Two-factor authentication has emerged as a critical security measure, with recommendations to protect voice channels and other communication pathways using contextual monitoring and voice source embedding [84]. Additionally, blockchain-based access control systems have been proposed to provide enhanced security for smart lock credentials, offering a distributed and tamper-resistant approach to managing access permissions [88].

The implementation of improved authentication protocols extends to the credentials themselves. Rather than relying solely on static digital keys, modern hotel lock systems employ temporary and revocable digital credentials that can be issued to guests for specific time periods [84]. This approach reduces the risk associated with credential compromise, as stolen or intercepted credentials have limited temporal validity.

**Regular Firmware Updates and Patching Practices:** The vulnerability of electronic locks to hacking has established firmware updates and security patching as essential maintenance practices. Research emphasizes that manufacturers may patch vulnerabilities in smart lock systems, though the only way to prevent Internet-enabled exploits is to remove the lock from a wide-area network [84]. This recognition has led to the establishment of regular update protocols as standard practice in hotel lock system management.

Smart home security research has identified that device owners must update their firmware to the most recent version as a primary defence against cybersecurity threats [8]. For hotel operators managing large numbers of electronic locks across multiple properties, this requirement has necessitated the development of centralized firmware management systems that can deploy security patches across distributed lock installations. The importance of maintaining current firmware versions is underscored by the fact that security vulnerabilities are continuously discovered in IoT devices, requiring ongoing vigilance and rapid response capabilities [8].

**Recognition of Ongoing Lifecycle Security Management:** Perhaps the most significant impact of electronic lock vulnerabilities has been the recognition that digital locks are not "set and forget" security devices but rather require continuous lifecycle security management. This paradigm shift represents a fundamental change from the traditional approach to mechanical locks, which, once installed, required minimal ongoing security maintenance beyond occasional physical servicing.

The lifecycle security management approach encompasses multiple dimensions. First, it requires continuous monitoring of lock system performance and security status. IoT-based smart lock systems now incorporate real-time monitoring capabilities that enable hotel operators to track access events, identify suspicious patterns, and respond to potential security breaches [89]. Second, it necessitates regular security assessments and vulnerability testing. Forensic examination methodologies for smart lock systems have been developed to identify potential attack vectors and evaluate the effectiveness of security measures [85].

Third, lifecycle management includes user education and awareness. Research on smart home security emphasizes that user awareness and education are critical components of maintaining the security of smart-home environments, including electronic locks [90]. Hotel staff and guests must understand the proper use of electronic lock systems and the security implications of their actions. Fourth, it requires collaboration among multiple stakeholders. Ensuring the security of smart lock systems demands collaboration between lock manufacturers, connected device manufacturers, internet service providers, and property owners [8].

### Technical Implementations Reflecting Security Lessons

The lessons learned from electronic lock vulnerabilities have been reflected in contemporary smart lock implementations for hotels. Modern systems integrate multiple security technologies to address identified weaknesses. For example, IoT-based smart door lock systems now commonly incorporate two-factor authentication using touchscreen keypads and RFID scanners, enabling users to generate temporary PIN codes with predetermined validity periods [91]. These systems maintain detailed access logs that record exactly who accessed the door and when, providing accountability and forensic capabilities [91].

Additionally, contemporary hotel lock systems have adopted cloud-based architectures with enhanced security features. Mobile key technology, as implemented by major hotel chains, has demonstrated significant operational benefits while maintaining security through encrypted credential transmission [92]. The implementation of mobile key systems has decreased check-in times by 70% and reduced physical key card costs by 85%, while simultaneously improving security through the use of encrypted-steganography codes sent via mobile applications [92, 93].

### Implications for Hotel Operations and Guest Experience

The recognition that electronic locks require ongoing lifecycle security management has implications extending beyond technical security measures. Hotels must now allocate resources to continuous monitoring, regular updates, and security training as part of their operational budgets. However, when properly implemented, these security measures enhance both guest safety and operational efficiency. Research demonstrates that IoT-enabled smart lock systems have reduced security incidents by 60% while improving guest access convenience [92].

The integration of electronic locks with broader IoT hotel systems has enabled additional security and operational benefits. Smart lock systems can be integrated with access control systems for multiple hotel areas, including elevators, parking lots, and restaurants, providing comprehensive security management

[94]. Furthermore, the ability to issue temporary credentials and revoke access remotely provides hotels with flexibility in managing guest access and responding to security incidents [84].

The vulnerabilities discovered in electronic hotel locks have fundamentally transformed the approach to lock system design, implementation, and management. Rather than viewing digital locks as static security devices, the hospitality industry has adopted a lifecycle security management perspective that recognizes the need for continuous monitoring, regular updates, and ongoing security assessment. The implementation of improved encryption and authentication protocols, combined with regular firmware updates and comprehensive security management practices, has enabled hotels to address the vulnerabilities that inexpensive hacking devices could previously exploit. This evolution demonstrates that digital security is not a destination but rather an ongoing process requiring sustained attention, resources, and collaboration among all stakeholders in the hotel technology ecosystem.

## 6. Smart Lock and IoT Vulnerabilities

The proliferation of Internet of Things (IoT) devices in smart homes has fundamentally transformed residential security infrastructure, introducing unprecedented convenience alongside significant cybersecurity challenges. Smart locks, as critical components of home automation systems, exemplify the complex security landscape where physical security mechanisms must now contend with digital vulnerabilities. This synthesis examines the vulnerabilities inherent in modern smart lock systems, the attack vectors exploited by adversaries, and the design paradigms that have emerged in response to these threats.

### Vulnerabilities in Smart Lock Systems

**Encryption and Protocol Weaknesses:** Modern smart locks have demonstrated susceptibility to compromise through multiple attack vectors. Weak encryption implementations represent a fundamental vulnerability in many deployed systems [95]. The taxonomy of cyber-physical threats in smart homes identifies specific attack scenarios affecting smart locks, including Bluetooth Low Energy (BLE) automated door opening through compromised wireless protocols and backdoor PIN code injection attacks [95]. These vulnerabilities stem from both protocol-level weaknesses and implementation flaws that fail to adequately protect credential transmission and device authentication.

Bluetooth Low Energy, widely adopted in smart lock implementations due to its low power consumption, presents particular security challenges [96]. The protocol's simplified design, while enabling efficient operation on resource-constrained devices, introduces vulnerabilities in encryption key exchange mechanisms [96]. Attackers can exploit poor implementation of encryption and custom key exchange methods to decrypt communications, as the BLE protocol's security relies heavily on proper implementation rather than inherent protocol strength [96]. Furthermore, BLE bonding credentials stored on companion mobile devices can be accessed by malicious applications installed on the same platform, enabling unauthorized access to paired smart locks [96].

**Insecure APIs and Remote Access Vulnerabilities:** Beyond wireless protocol vulnerabilities, smart locks face threats through insecure application programming interfaces (APIs) and cloud service integrations. The integration of smart locks with cloud-based management systems and mobile applications introduces additional attack surfaces [97]. While OAuth2 authentication and HTTPS encryption are recommended security measures for API communication, inadequate implementation of these standards leaves systems vulnerable to unauthorized access and data breaches [97]. The potential for attackers to unlock doors remotely or intercept credentials represents a critical threat to physical security [95].

Voice-activated smart lock systems present an emerging vulnerability class. Spoofed voice commands can activate smart locks through compromised voice assistant integrations, exploiting the trust relationship between voice interfaces and physical access control systems [95]. This attack vector demonstrates how the convergence of multiple IoT devices creates secondary vulnerabilities not present in isolated systems [84].

**Authentication and Credential Management Weaknesses:** Poor authentication mechanisms constitute a fundamental vulnerability in smart lock deployments. Weak or default credentials, inadequate multi-factor authentication, and insufficient credential lifecycle management expose systems to brute force attacks and credential compromise [98]. The lack of standardized authentication protocols across heterogeneous smart lock implementations exacerbates this challenge, as manufacturers implement proprietary solutions with varying security rigor [90].

Node capture attacks represent a severe physical threat to smart lock authentication schemes [99]. When adversaries compromise a single device node, they can potentially extract cryptographic keys and authentication credentials, compromising the security of the entire authentication system [99]. This vulnerability highlights the need for authentication schemes specifically designed to resist physical device compromise.

### Impact on Smart Lock Design and Security Architecture

**Adoption of End-to-End Encryption and Secure Protocols:** In response to identified vulnerabilities, the smart lock industry has increasingly adopted end-to-end encryption as a foundational security principle [97]. End-to-end encryption ensures that credential data and access commands remain encrypted throughout transmission, preventing interception by network-level attackers [97]. This approach extends beyond simple transport-layer encryption (HTTPS) to implement application-level encryption where the lock device itself maintains exclusive access to decryption keys.

Secure protocol implementation has become essential for smart lock security. Rather than relying on proprietary protocols, industry leaders have begun adopting standardized, well-vetted cryptographic protocols [90]. The integration of post-quantum cryptographic algorithms represents an emerging trend, as organizations prepare for potential threats from quantum computing that could compromise current encryption standards [100]. These advanced cryptographic approaches, while computationally intensive, can be optimized for resource-constrained smart lock devices through careful implementation [100].

**Integration with Identity and Access Management (IAM) Systems:** Modern smart lock architectures increasingly incorporate formal Identity and Access Management systems, moving beyond simple credential verification to comprehensive access control frameworks [97]. IAM integration enables granular permission management, allowing property owners to grant temporary or revocable access credentials to specific individuals with defined time windows and usage restrictions [84]. This approach provides significant advantages over traditional physical keys, which cannot be easily revoked or time-limited.

IAM systems integrated with smart locks support mutual authentication, ensuring that both the user and the lock device verify each other's identity before granting access [99]. This bidirectional authentication prevents impersonation attacks where adversaries masquerade as legitimate users or devices. Additionally, IAM frameworks enable comprehensive audit logging, creating forensic records of all access attempts and credential usage [95].

**Secure-by-Design and Privacy-by-Design Principles:** The security failures in early smart lock deployments have catalysed a fundamental shift toward secure-by-design and privacy-by-design methodologies [90]. Rather than treating security as an afterthought or optional feature, manufacturers now integrate security considerations throughout the entire product development lifecycle [90]. This approach encompasses threat modelling during initial design phases, security-focused code review processes, and comprehensive vulnerability testing before product release.

Privacy-by-design principles ensure that smart lock systems minimize data collection and retention, implementing data minimization strategies where only essential information necessary for lock operation is processed (Vardakis et al., 2024). Personal data such as access patterns, location information, and user behaviour patterns are protected through privacy-preserving techniques including differential privacy and federated learning approaches [101]. These methodologies prevent unauthorized profiling and behavioural inference based on smart lock usage data.

## Emerging Security Solutions and Best Practices

**Multi-Layered Security Approaches:** Contemporary smart lock security implementations employ multi-layered protection strategies that address vulnerabilities at multiple system levels [95]. These approaches combine strong encryption, robust authentication mechanisms, secure communication protocols, and intrusion detection capabilities [102]. The integration of IoT-based intrusion detection systems enables real-time monitoring of smart lock activity, identifying suspicious access attempts and unauthorized connection attempts [102].

Machine learning algorithms applied to smart lock usage patterns can detect anomalies indicating potential security breaches [102]. By establishing baseline patterns of normal access behaviour, these systems can identify deviations suggesting compromised credentials or unauthorized access attempts, triggering automated responses including access denial and alert notifications [102].

**Secure Communication and Encryption Standards:** The implementation of advanced encryption standards specifically tailored for IoT environments has become standard practice in smart lock design [100]. While AES-256 encryption provides strong symmetric encryption for data protection, the challenge of secure key management in resource-constrained devices requires specialized approaches [100]. Hardware-based key storage, secure enclaves, and trusted execution environments provide protected spaces for cryptographic key storage, preventing extraction even if the device is physically compromised [99].

Secure communication protocols must address the unique constraints of IoT devices, which often operate with limited computational resources and battery power [100]. Lightweight cryptographic implementations, such as optimized versions of established algorithms, enable strong security without excessive power consumption [100]. The selection of appropriate cryptographic algorithms requires careful analysis of the threat model, computational constraints, and expected operational lifetime of the device.

**Authentication Framework Innovations:** Advanced authentication schemes for smart locks incorporate multiple verification factors beyond simple password or PIN entry. Biometric authentication, including fingerprint recognition and facial recognition, provides user verification resistant to credential theft [103]. Physically unclonable functions (PUFs) leverage inherent manufacturing variations in semiconductor devices to create unique, unclonable device identifiers that cannot be forged or extracted [103].

Dynamic key lifecycle management ensures that cryptographic keys used for authentication are regularly rotated and updated, limiting the window of vulnerability if a key is compromised (Obaid et al., 2025). Blockchain-based revocation mechanisms enable secure credential revocation without requiring centralized management infrastructure, supporting decentralized smart home architectures [103]. Proxy re-encryption and threshold secret sharing techniques enable secure credential recovery if a user loses access to their primary authentication device, without requiring complete re-registration [103].

## Security Challenges and Ongoing Vulnerabilities

**Firmware and Software Update Challenges:** Despite advances in smart lock security design, the challenge of maintaining security through the device's operational lifetime remains significant [98]. Outdated firmware and software represent critical vulnerabilities, as manufacturers may cease security updates for older device models [98]. The heterogeneous landscape of smart lock implementations, with devices from numerous manufacturers using different platforms and architectures, complicates coordinated security patching [90].

The tension between security updates and device usability creates practical challenges. Users may resist security updates that require device downtime or reconfiguration, leaving vulnerable devices in operation [98]. Manufacturers must balance the need for frequent security updates against user convenience, potentially through automatic background updates that minimize disruption [98].

**Privacy and Data Protection Concerns:** Smart lock systems collect sensitive data regarding access patterns, user location, and behavioural routines [104]. Unauthorized access to this data enables identity impersonation, routine analysis for burglary planning, and behavioural profiling [104]. The integration of smart locks with broader smart home ecosystems and cloud services multiplies the potential for data exploitation, as information flows through multiple systems and organizations [104].

Third-party integrations with cloud services, voice assistants, and mobile applications introduce additional privacy risks [97]. Data sharing agreements and privacy policies often lack transparency regarding how access data is used, retained, and potentially monetized [104]. Users frequently lack awareness of security implications when accepting terms and conditions for smart lock systems, creating a gap between actual data practices and user expectations [104].

**Standardization and Interoperability Gaps:** The absence of standardized security protocols across smart lock implementations creates fragmentation in the security landscape [90]. Different manufacturers implement proprietary security solutions with varying levels of rigor, making it difficult for users to assess relative security posture [90]. This fragmentation also complicates security research and vulnerability disclosure, as researchers must understand numerous proprietary implementations rather than analysing standardized protocols [105].

Interoperability challenges arise when users attempt to integrate smart locks from different manufacturers into unified smart home systems [90]. The need to support multiple communication protocols and authentication mechanisms increases system complexity and introduces potential security gaps at integration points [90].

### Key Lesson: Software and Network Security as Critical as Physical Mechanisms

The evolution of smart lock security demonstrates a fundamental shift in how physical security must be conceptualized in the digital age. Traditional lock security relied primarily on physical mechanisms, the difficulty of picking locks, the strength of materials, and the complexity of mechanical designs [84]. Modern smart locks have revealed that physical security mechanisms alone provide insufficient protection when devices connect to networks and rely on software for operation.

Security now depends as much on software and networks as on physical mechanisms [84]. A smart lock with an impeccable physical design can be compromised through weak encryption, insecure APIs, or poor firmware implementation [95]. Conversely, even a mechanically simple lock can provide strong security if protected by robust cryptographic protocols and secure software architecture [97]. This paradigm shift requires security professionals, manufacturers, and users to understand that smart lock security encompasses the entire system, from cryptographic algorithms to network protocols to firmware update mechanisms [90].

The integration of smart locks into broader IoT ecosystems amplifies this lesson. A single compromised smart lock can serve as an entry point for attacks on other home devices sharing the same network [95]. Conversely, security measures implemented at the network level, such as network segmentation and intrusion detection, can protect smart locks even if individual device security is imperfect [102]. This systemic perspective recognizes that smart lock security cannot be achieved through device-level measures alone but requires coordinated security across the entire smart home infrastructure [90].

Smart lock vulnerabilities exemplify the broader challenges facing IoT security in connected environments. Weak encryption, insecure APIs, and poor implementation have enabled attackers to compromise physical access control systems through digital means. The industry response, adoption of end-to-end encryption, integration with IAM systems, and implementation of secure-by-design principles, reflects recognition that modern physical security depends fundamentally on software and network security. As smart homes continue to proliferate, the security of smart locks will increasingly depend on the robustness of cryptographic implementations, the security of cloud integrations, and the vigilance of manufacturers in maintaining security throughout device lifecycles. The convergence of physical and digital security in smart locks demonstrates that future security architectures must treat these domains as inseparable, with equal emphasis on both physical resilience and digital protection.

## 7. The TSA Master Key Leak (2014)

The TSA Master Key Leak of 2014 represents a significant security breach in physical access control systems. Photographs of Transportation Security Administration master keys were published online, enabling individuals to create functional copies using three-dimensional printing technology [106]. This incident compromised the security of millions of luggage locks worldwide and fundamentally challenged assumptions about the security of universal master key systems.

The vulnerability exploited in the TSA Master Key Leak exemplifies a critical distinction in security analysis: the difference between mechanical security and information security. While the physical locks themselves were mechanically sound, the security of the entire system depended upon the secrecy of the master key design. Once photographic evidence of the key geometry became publicly available, the mechanical strength of the locks became irrelevant. This incident demonstrates that security systems relying on universal master keys face inherent risks when key designs are not adequately protected from information disclosure [106].

The ability to reproduce functional keys from published photographs using three-dimensional printing technology represented a novel threat vector that traditional lock manufacturers had not adequately anticipated. The accessibility of both the information and the manufacturing technology (3D printing) created a perfect storm for security compromise [106].

### Impact on Lock Design and Security Philosophy

**Reduced Trust in Universal Master Key Systems:** The TSA Master Key Leak fundamentally altered industry perspectives on universal master key systems. The incident revealed that systems designed for convenience and centralized access control inherently create single points of failure when key information is compromised. Organizations worldwide began reassessing their reliance on master key systems, recognizing that the administrative convenience they provided came at the cost of distributed vulnerability [106].

The widespread deployment of electronic locking systems has partially addressed these concerns by offering alternatives to mechanical master keys. Electronic locking systems provide more flexible access rights management and claim that electronic keys cannot be cloned as easily as their mechanical counterparts [107]. However, as subsequent research has demonstrated, electronic systems introduce their own security challenges that require careful cryptographic implementation [107].

**Greater Emphasis on Key Control and Secrecy:** Following the TSA Master Key Leak, the security industry placed substantially greater emphasis on key control protocols and the protection of key design information. Organizations recognized that master key designs must be treated as highly sensitive intellectual property, subject to strict access controls and compartmentalization [106].

This shift in perspective extended beyond physical keys to encompass cryptographic key management more broadly. Comprehensive key management infrastructure designs have become essential for protecting sensitive systems, with emphasis on key generation, secure storage, key revocation, and controlled distribution [108]. The principle that keys must be protected against unauthorized disclosure, misuse, alteration, or loss has become foundational to modern security architecture [108].

**Awareness of Risks from Information Leakage:** The TSA Master Key Leak elevated awareness throughout the security industry regarding the risks posed by information leakage. The incident demonstrated that even information that might seem innocuous, photographs of physical objects, could pose catastrophic security risks when combined with modern manufacturing capabilities [106].

This awareness has influenced security practices across multiple domains. In the context of integrated circuit design and intellectual property protection, for example, researchers have developed logic locking techniques to prevent reverse-engineering and unauthorized reproduction, recognizing that design information itself represents a critical security asset [81, 109]. The principle that security can be compromised through information exposure, not merely through mechanical or cryptographic failure, has become widely accepted [106].

### Key Lesson: Security Beyond Mechanics

The fundamental lesson from the TSA Master Key Leak is that security cannot be evaluated solely through mechanical or cryptographic analysis. Information security represents an equally critical dimension. A system may be mechanically sound and cryptographically robust, yet remain vulnerable if the information necessary to bypass it becomes publicly available [106].

This principle has broad implications for security system design. Electronic locking systems, while offering advantages over purely mechanical locks, must still protect their cryptographic primitives and design information. Research analysing electronic locking systems has revealed that implementation flaws can allow extraction of system-wide master secrets through brute force attacks or side-channel analysis, even when the underlying system concepts are well-designed [107]. This underscores that information security, protecting the details of how systems work, is as important as the security of the systems themselves.

The incident has also influenced approaches to hardware security. Techniques such as logic locking for integrated circuits incorporate security measures specifically designed to prevent reverse-engineering and information extraction, recognizing that protecting design information is essential to overall system security [81, 109]. Similarly, scan-based security techniques have been developed to prevent unauthorized access to chip internals through test interfaces, acknowledging that information about system internals can be weaponized by attackers [110, 111].

### Broader Implications for Security Architecture

The TSA Master Key Leak has contributed to a more holistic understanding of security that encompasses not only the technical properties of locks and keys but also the information security practices surrounding them. Modern security systems increasingly incorporate multiple layers of protection: mechanical or cryptographic security for the primary function, combined with information security practices to protect design details and operational secrets [106].

This multi-layered approach is evident in contemporary smart lock systems, which combine electronic authentication mechanisms with secure key management protocols [112, 113]. The recognition that information leakage represents a critical vulnerability has led to the development of more sophisticated access control systems that limit the distribution of sensitive information and implement compartmentalization of security-critical knowledge [106].

**Overarching Insights**

Across these examples, several consistent themes become clear. One of the most significant is that failure is a primary driver of innovation. Locks rarely evolve in isolation; instead, they improve most rapidly after vulnerabilities are exposed or systems are successfully defeated. Each failure reveals weaknesses in design, materials, or assumptions, prompting the development of stronger, more resilient solutions. In this way, the history of lock design is closely tied to the history of how those locks have been broken.

Another recurring theme is the importance of human factors. Many security failures are not purely technical but arise from how systems are used, managed, or misunderstood. Convenience often leads to compromised practices, such as sharing keys, reusing codes, or bypassing security features altogether. Even the most advanced locking systems can fail if they are poorly implemented or if users prioritise ease of access over proper security protocols.

The tension between transparency and secrecy also plays a critical role. When vulnerabilities are publicly exposed—whether through research, demonstration, or real-world incidents—it can initially appear to weaken security. However, this transparency often leads to stronger designs, as flaws are acknowledged, tested, and addressed. In contrast, relying on secrecy alone can create a false sense of security, allowing weaknesses to persist unnoticed until they are exploited.

Finally, these examples reinforce the reality that security is never absolute. No lock is permanently secure; every system exists within a constantly evolving threat landscape where new tools, techniques, and technologies emerge over time. What is considered high security today may become vulnerable tomorrow. As a result, effective security is not a fixed state but an ongoing process of adaptation, improvement, and reassessment.

Each historical failure reminds us that a lock is not a static object but part of a living system shaped by attackers, users, and context. What appears secure today may be vulnerable tomorrow—not because the materials are weak, but because assumptions have changed. The evolution of locking systems is therefore not just a story of better mechanisms, but of learning from failure, adapting to new threats, and continually renegotiating the balance between trust, control, and resilience.

**Impact on Locksmithing Practices**

Understanding these historical failures is highly relevant to anyone interested in locksmithing because they fundamentally shape how the profession is practiced today. Locksmithing is no longer just about fitting keys to locks or repairing mechanisms—it is about understanding how and why locks fail, and using that knowledge to make informed decisions about security. These examples demonstrate that effective locksmithing is grounded in both mechanical knowledge and an awareness of real-world attack methods.

One of the most important influences is the shift toward evidence-based security selection. The story of the Bramah lock shows that claims of security cannot be trusted without real-world testing. For modern locksmiths, this translates into selecting products based on certified performance standards (such as TS007 or EN ratings), rather than relying on brand reputation or perceived quality. It reinforces the need to think like an attacker—considering how a lock might actually be defeated, not just how it is designed to work.

These failures also highlight the importance of understanding attack techniques as part of everyday practice. The evolution of pin tumbler locks, the rise of bumping, and the development of anti-pick and anti-bump features demonstrate that locksmiths must be familiar with both constructive and destructive entry methods. This knowledge directly influences how locks are installed, serviced, and recommended. A locksmith is not just a technician—they are a risk assessor who must match the right level of security to the threat environment.

Another key influence is the recognition that security is a system, not a single component. Master keying failures and TS007 standards clearly show that a lock cannot be considered in isolation. Door hardware, installation quality, key control policies, and user behaviour all affect overall security. This has shifted locksmithing practice toward a more holistic approach, where professionals assess entire access control systems rather than just individual cylinders or keys.

The examples also demonstrate the growing importance of key control and administrative security. Incidents like cross-keying vulnerabilities and the TSA master key leak show that even well-designed locks can fail if key management is poor. As a result, locksmiths must now advise on restricted key systems, duplication controls, and organisational procedures. This extends the role of the locksmith beyond physical hardware into governance and policy.

With the emergence of smart locks and IoT systems, these historical lessons now extend into cybersecurity awareness. Modern locksmiths must understand that vulnerabilities are no longer purely mechanical. Protocols such as MQTT, wireless communication, and cloud-based access systems introduce new risks that require knowledge of encryption, authentication, and network security. This represents a major shift in the profession—from purely mechanical expertise to a hybrid of physical and digital security.

These failures also reinforce the importance of client education. Many vulnerabilities arise from human factors—poor key management, convenience-driven decisions, or misunderstanding of security levels. Locksmiths play a critical role in explaining that "high quality" does not necessarily mean "high security," and that certified, tested systems provide measurable protection. Educating clients helps bridge the gap between technical capability and real-world use.

Finally, these examples shape a core professional mindset: security is dynamic and never absolute. Locksmithing practices must continuously adapt to new tools, techniques, and threats. What was considered secure a decade ago may now be vulnerable. This requires ongoing learning, staying current with standards and technologies, and regularly reassessing security solutions.

In practice, this means that modern locksmithing is no longer reactive but strategic. It involves anticipating threats, selecting systems based on proven resistance, integrating physical and digital security, and guiding clients toward informed decisions. The influence of historical failures is therefore profound—they have transformed locksmithing from a trade focused on mechanisms into a profession centred on risk, systems thinking, and continuous improvement.

## Case Study: Prisons, Banks, and Domestic Locks

Security requirements vary significantly depending on the environment, and this is clearly illustrated when comparing prisons, banks, and domestic (residential) locking systems. Each setting reflects a different balance between risk, usability, cost, and control, and each has influenced locksmithing practices in distinct ways.

### Prisons: Maximum Control, Minimal Tolerance for Failure

Prison locking systems represent the highest level of physical security infrastructure, where failure is fundamentally unacceptable. The design philosophy underlying these systems reflects the unique operational environment of correctional facilities, where security personnel must contend with individuals who possess significant motivation, extended time, and potentially sophisticated tools to circumvent access controls. This demanding context necessitates locking mechanisms that are substantially larger, more mechanically complex, and considerably more durable than those found in commercial or residential applications [81].

The fundamental challenge in prison security design stems from the adversarial nature of the environment. Unlike conventional access control systems designed for cooperative users, prison locking systems must function reliably against determined, sustained, and deliberate attacks from individuals with both the motivation and opportunity to exploit vulnerabilities [81]. This reality shapes every aspect of system design, from the selection of mechanical components to the integration of electronic control infrastructure.

Prison locks typically employ heavy-duty lever mechanisms engineered to withstand extraordinary mechanical stress. These mechanisms are designed with substantial material thickness, reinforced pivot points, and hardened steel components that resist both direct force attacks and sophisticated manipulation techniques [81]. The mechanical complexity of prison locks serves a dual purpose: it increases the difficulty of unauthorized manipulation while simultaneously providing redundancy through multiple independent locking points.

The durability requirements for prison locking systems exceed those of virtually all other security applications. Locks must maintain their integrity and functionality despite exposure to harsh environmental conditions, including moisture, temperature fluctuations, and corrosive substances that inmates may deliberately introduce [81]. This requirement for extreme durability is reflected in the material selection, manufacturing tolerances, and quality control procedures employed in their production.

## Beyond the Key

Modern prison facilities increasingly supplement or entirely replace mechanical locks with advanced electronic locking systems that enable centralized control from dedicated control rooms [114]. These electronic systems provide significant operational advantages over purely mechanical approaches, including the ability to rapidly secure or release multiple access points simultaneously, the capacity to implement time-based access restrictions, and the capability to create detailed audit trails of all access events [114].

The integration of electronic locking systems into prison infrastructure represents a fundamental shift in security management philosophy. Rather than relying on the physical properties of individual locks to prevent unauthorized access, electronic systems distribute security responsibility across multiple layers: the cryptographic security of the electronic keys, the integrity of the communication protocols, the robustness of the central control system, and the procedural controls governing access authorization [81].

A defining characteristic of prison security is the implementation of extraordinarily strict key control and accountability procedures. Keys used in prison facilities are typically uniquely coded and individually tracked through comprehensive documentation systems (Bennett, 2023). The issuance of keys is governed by tightly controlled procedures that document the identity of the individual receiving the key, the specific lock or locks the key opens, the date and time of issuance, and the authorized duration of possession [114].

This rigorous approach to key management reflects the recognition that physical keys represent a critical vulnerability in any mechanical locking system. A single lost or compromised key could potentially provide unauthorized access to secure areas, necessitating the replacement of locks and the reissuance of keys to all authorized personnel. The administrative burden of managing this scenario has driven the adoption of electronic systems in many modern facilities, where the revocation of access rights can be accomplished instantaneously through software commands rather than through the time-consuming and expensive process of physical lock replacement [81].

Electronic locking systems substantially reduce reliance on physical keys by implementing access control through electronic credentials, typically radio frequency identification (RFID) cards, electronic fobs, or biometric identifiers, that can be remotely disabled or revoked [81]. This capability provides a significant security advantage, as it eliminates the possibility of unauthorized key duplication and enables rapid response to security breaches through the immediate revocation of compromised credentials [81].

For locksmiths and security professionals operating in prison environments, the paramount concern extends beyond the ability of individual locks to resist picking or forced entry. Prison locking systems must demonstrate exceptional reliability under extreme operational conditions, including sustained mechanical stress, environmental exposure, and deliberate sabotage attempts [81]. The system must function flawlessly throughout its operational lifetime, as any failure could compromise facility security and potentially endanger both staff and inmates.

This requirement for absolute reliability has driven the adoption of redundant systems and layered security approaches. Electronic locking systems typically incorporate redundant power supplies, backup communication pathways, and mechanical override mechanisms that function independently of electronic components [114]. These redundancies ensure that even in the event of partial system failure, the facility can maintain secure control of critical access points.

The integration of locking systems with broader security protocols represents another critical dimension of prison security architecture. Locking systems do not operate in isolation but rather function as components of comprehensive security frameworks that include surveillance systems, alarm systems, access control systems, and personnel procedures [114]. The effectiveness of the overall security posture depends on the seamless integration of these various components and the consistent application of security procedures by trained personnel.

Beyond the technical specifications of locks and electronic systems, prison security depends fundamentally on the consistent application of rigorous operational procedures. These procedures govern the authorization of access, the issuance and revocation of credentials, the maintenance and testing of locking systems, and the investigation of any security incidents or anomalies [114]. The effectiveness of these procedures depends critically on the training, competence, and professional discipline of prison personnel.

The lesson derived from prison security practice is that high-security environments demand a comprehensive, multi-layered approach that combines robust physical security infrastructure, advanced electronic control systems, strict procedural controls, and rigorous personnel training [114]. No single element, whether mechanical durability, electronic sophistication, or procedural discipline, is sufficient in isolation. Rather, security effectiveness emerges from the integration of all these elements into a coherent system where each component reinforces the others and where failure in any single component is compensated for by redundancy and backup systems [81].

The principle of zero tolerance for uncontrolled access represents the fundamental organizing principle of prison security. This principle acknowledges that in a correctional environment, any unauthorized access, whether to secure areas, to other inmates, or to the outside world, represents a potential threat to facility security, staff safety, and public security. Consequently, prison security systems are designed with the assumption that any potential vulnerability will eventually be discovered and exploited by determined adversaries, and therefore all systems must be designed with sufficient redundancy and robustness to maintain security even in the face of sophisticated attacks [81].

**Banks: Layered Security and Controlled Access**

Banks operate within a distinctive security paradigm that fundamentally differs from other high-security environments such as correctional facilities. While both require robust physical protection mechanisms, banks must simultaneously facilitate controlled, authorized access for diverse stakeholders including staff members and customers [115]. This operational requirement has necessitated the development of sophisticated, multi-layered security systems that balance security imperatives with accessibility needs [115, 116].

The modern bank security infrastructure represents a coordinated integration of multiple security components rather than isolated protective measures. Electronic security systems (ESS) in banking institutions comprise several interconnected subsystems, including closed-circuit television (CCTV) monitoring systems, Access Control Systems (ACS), Security Alarm Systems (SAS), and Fire Alarm Systems (FAS) [115]. These components function synergistically to provide prevention, detection, response to incidents, and recovery capabilities following security disturbances [115].

Contemporary bank security extends beyond traditional mechanical approaches to incorporate advanced technological solutions. Real-time monitoring capabilities, intelligent control systems, and immediate emergency action protocols have become essential components of modern banking security infrastructure [116]. The integration of Internet of Things (IoT) technologies with traditional security measures has enabled banks to implement comprehensive surveillance and access control systems that provide enhanced situational awareness and rapid response capabilities [116].

Bank vaults and safes employ highly specialized locking mechanisms designed to prevent unauthorized access while maintaining operational functionality. These mechanisms include time locks, combination locks, and dual-control systems requiring authorization from two individuals to access secure areas. The implementation of dual-control systems represents a critical procedural safeguard that prevents any single person from exercising complete control over sensitive areas or systems.

Access segmentation constitutes another fundamental principle in bank security design. By distributing access privileges across multiple individuals and systems, banks ensure that no single person possesses comprehensive control over all areas or security systems. This architectural approach significantly reduces the risk of internal threats and unauthorized access.

Modern banking institutions have increasingly adopted biometric authentication technologies to enhance access control mechanisms. Fingerprint-based door locking systems utilize individuals' unique physical characteristics for authentication, capturing and storing biometric data to permit access exclusively to recognized fingerprints [117]. Such systems prove particularly suitable for high-security banking environments, demonstrating practical applications of Internet of Things technology in security solutions [117].

Multi-factor authentication approaches have emerged as best practices in contemporary bank security. Advanced systems integrate face recognition technology, one-time password (OTP) verification, and traditional physical key access to create robust three-level authentication systems. These layered authentication mechanisms minimize unauthorized access risks and enhance overall security reliability.

Surveillance systems constitute essential components of comprehensive bank security architectures. Real-time video monitoring through IP cameras and CCTV systems enables continuous observation of critical areas, with recorded data providing valuable evidence in security incidents [115]. The integration of motion detection capabilities with surveillance systems allows for automated recording initiation upon detecting suspicious activity [118].

Remote monitoring capabilities have become increasingly important in modern banking security. Systems utilizing ESP32-CAM modules and Telegram-based communication enable security personnel to receive real-time image analysis and remote verification of access attempts [116]. This technological integration facilitates rapid response to potential security threats regardless of personnel location [116].

**ESP32-CAM Modules and Telegram-Based Communication**

The ESP32-CAM is a compact and cost-effective microcontroller module that integrates Wi-Fi connectivity, processing capability, and a camera interface into a single device. Built around the ESP32 chip and commonly paired with the OV2640 camera sensor, it enables real-time image capture, basic video streaming, and wireless communication. This combination makes it particularly well suited to embedded and IoT applications where space, cost, and power efficiency are important considerations.

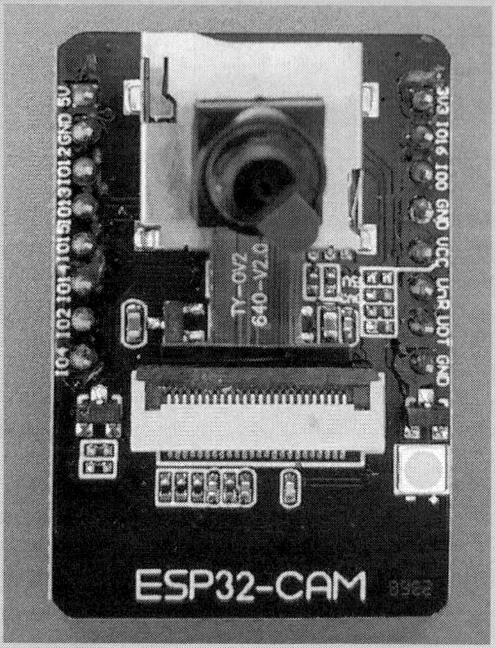


*Figure 11: ESP32-CAM. Nowforever, CC BY-SA 4.0, via Wikimedia Commons.*

Unlike traditional CCTV systems, which typically rely on dedicated recording hardware or servers, the ESP32-CAM functions as a self-contained edge device. This means it can capture, process, and transmit data independently, without requiring a full computer or centralised processing system. As a result, it supports decentralised architectures where intelligence is distributed across devices rather than concentrated in a single location.

Technically, the ESP32-CAM includes a dual-core processor capable of handling lightweight image processing tasks, along with integrated Wi-Fi and Bluetooth for network communication. It features a camera interface that supports JPEG image capture, as well as GPIO pins that allow integration with external sensors such as motion detectors or door switches. In addition, the module supports microSD card storage, enabling local data logging or image storage when required.

In practice, the ESP32-CAM is widely used across a range of applications, including smart surveillance systems, remote monitoring for homes, warehouses, and vehicles, access control and security systems, and IoT-based alert and automation platforms. Its versatility and low cost make it an attractive option for both hobbyist projects and professional deployments.

Communication with the ESP32-CAM is achieved over Wi-Fi using standard internet protocols. These include HTTP or HTTPS for sending images or data to web servers or APIs, MQTT for lightweight and efficient messaging in IoT environments, and WebSockets for real-time, bidirectional communication. Through these protocols, the device can capture an image or detect an event, process or compress the data, and transmit it to a remote system such as a cloud platform, application, or monitoring service. In security contexts, this enables instant alerts, remote monitoring, and automated responses to events.

Telegram-based communication further extends the functionality of the ESP32-CAM by transforming it into an interactive security device. Instead of requiring a custom-built mobile application or dashboard, the system uses a Telegram Bot as the interface between the device and the user. The ESP32 communicates with the bot via the Telegram Bot API, while users interact with the system through the familiar Telegram messaging application.

This approach creates a simple, secure, and widely accessible control interface. Users can receive alerts, images, or status updates directly within a chat, and can also send commands back to the device. In effect, Telegram acts as both a notification system and a control panel, significantly reducing development complexity while maintaining real-time interaction and usability.

**How It Works (Step-by-Step)**

**1. Create a Telegram Bot**

Using Telegram's BotFather, a bot is created and assigned a unique API token.

**2. ESP32 Connects to Wi-Fi**

The module connects to the internet and is programmed with the bot token.

**3. Event Occurs**

Example: motion detected, door opened, or manual trigger.

**4. ESP32 Sends Data to Telegram API**

- Captures an image
- Sends it via HTTPS request to Telegram servers

**5. User Receives Message**

The user instantly receives:

- Photo or video
- Text alert (e.g., "Motion detected at front door")

**6. User Sends Command (Optional)**

The user can reply with commands such as:

- /capture → take a photo
- /status → check system state
- /unlock → trigger a smart lock

**7. ESP32 Executes Command**

The device reads the message and performs the action.

In modern locksmithing and security system applications, the integration of ESP32-CAM modules with Telegram-based communication enables a significant shift toward intelligent, connected access control. This setup allows for remote visual verification before granting access, meaning users can see who is at a door in real time before unlocking it. It also supports instant alerts for unauthorised entry attempts, ensuring that potential security breaches are identified and responded to immediately. When integrated with smart locks, users can send commands such as lock or unlock directly from their mobile device, while Telegram chat history provides a simple but effective audit trail of events, actions, and system interactions.

A typical workflow illustrates how seamlessly this system operates. For example, a motion sensor connected to the ESP32-CAM detects activity and triggers the camera to capture an image. This image is then sent via Wi-Fi to a Telegram bot, which delivers it instantly to the user's phone. The user can visually verify the visitor and, if appropriate, send a command such as "/unlock." The system receives this command and remotely unlocks the door. This process effectively merges physical security mechanisms, such as locks and sensors, with digital communication technologies, including IoT networks and messaging platforms.

From a technical perspective, this approach offers several advantages. It is significantly more cost-effective than traditional CCTV and access control systems, making it accessible for a wide range of applications. There is no need to develop or maintain a dedicated mobile application, as Telegram serves as the user interface. Communication benefits from Telegram's built-in encryption, enhancing data security during transmission. The system also supports real-time interaction with minimal latency, allowing immediate responses to events. Additionally, it is highly scalable, with multiple devices able to connect to and be managed through a single Telegram bot.

However, these benefits come with important security considerations. Because the system relies on internet connectivity and API-based communication, it must be carefully configured to prevent vulnerabilities. API tokens used to connect the ESP32-CAM to the Telegram bot must be securely stored and never exposed. User authentication must be enforced to ensure that only authorised individuals can send commands to the system. Communication should be secured using HTTPS or TLS protocols, and the underlying Wi-Fi network must be properly protected. Without these safeguards, an attacker could potentially intercept camera feeds or issue unauthorised commands, such as unlocking a door.

A useful way to understand this system is through analogy. The ESP32-CAM functions as both the camera and the on-site guard, capturing and reporting events. The Wi-Fi and internet connection act as the communication radio, transmitting information between locations. Telegram serves as the control room interface, allowing the user to monitor and control the system remotely. Instead of requiring a dedicated control centre, the user's smartphone effectively becomes the command hub.

Ultimately, the combination of ESP32-CAM modules and Telegram-based communication represents a broader transformation in security practice. It moves locksmithing beyond standalone mechanical devices toward integrated, networked systems where monitoring, access control, and decision-making occur in real time. For modern locksmiths and security professionals, this highlights the growing importance of understanding not only physical security hardware but also digital communication, system integration, and cybersecurity principles.

While technological systems provide essential security infrastructure, procedural safeguards and human oversight remain critical components of effective bank security. Access control systems require careful configuration to prevent both unauthorized access and denial of legitimate access to authorized personnel [119]. The complexity of managing multiple zones, numerous users, and thousands of access-control rules necessitates rigorous configuration management to prevent security misconfigurations [119].

Data logging and audit trail capabilities enable banks to maintain comprehensive records of system access and personnel movement. These records document which individuals accessed specific areas and at what times, providing accountability and facilitating investigation of security incidents [120]. The integration of data logging with access control systems creates a complete audit trail of all security-relevant activities [120].

The banking security model demonstrates that effective security cannot be achieved through isolated protective measures. A lock, regardless of its mechanical sophistication, represents only one component within a comprehensive security system [116]. The effectiveness of any locking mechanism depends fundamentally on its integration with alarms, surveillance systems, access protocols, and human procedures [116].

This systems-level approach to security design emphasizes the importance of coordinated implementation across all security components. Electronic access control systems must function in concert with physical barriers, surveillance capabilities, alarm mechanisms, and procedural protocols to achieve comprehensive protection [115]. The failure or misconfiguration of any single component can compromise the entire security architecture [119].

Bank security systems exemplify the principles of layered, coordinated security design. Through the integration of specialized locking mechanisms, electronic access control, real-time surveillance, biometric authentication, and procedural safeguards, banks create comprehensive security environments that protect assets while maintaining operational functionality. This multi-layered approach demonstrates that security effectiveness depends not on individual components in isolation, but rather on their coordinated integration into cohesive systems designed to address multiple threat vectors simultaneously.

## Domestic Locks: Balancing Security, Cost, and Convenience

Domestic locks operate within a fundamentally different context than high-security applications such as prisons or banks. The design and implementation of residential locking systems must navigate a complex interplay of competing demands: security effectiveness, affordability, ease of use, and aesthetic integration with the home environment. This balance reflects the distinct threat landscape facing residential properties, where security measures must deter opportunistic intruders rather than withstand prolonged, sophisticated attacks [121-123].

### Pin Tumbler Cylinders and Anti-Defeat Features

Most residential properties continue to rely on pin tumbler cylinder locks as their primary security mechanism [121]. However, contemporary residential security has evolved significantly beyond basic pin tumbler designs. Modern domestic locks increasingly incorporate anti-bump, anti-pick, and anti-snap features to address the most common burglary methods [121]. These technological enhancements represent a practical response to documented vulnerabilities in conventional lock designs, where bump keys, picking tools, and forced entry techniques pose realistic threats to residential security [124].

The development of these enhanced features reflects a broader trend in residential security toward layered protection strategies. Research on Crime Prevention Through Environmental Design (CPTED) principles demonstrates that access control and target hardening measures, including improved door lock systems, significantly reduce burglary intentions [125]. Furthermore, evidence from burglary prevention programs indicates that households upgrading from basic to enhanced security configurations, including improved locks with deadbolts, experience substantially reduced victimization risk [122].

### Time to Defeat and Opportunistic Attack Deterrence

Unlike institutional or commercial security contexts, domestic environments face a distinctly different threat profile. Residential properties are typically exposed to opportunistic attacks rather than highly skilled or prolonged attempts by professional criminals [121]. This fundamental difference in threat assessment makes the concept of "time to defeat" particularly important in residential lock design [121].

The principle underlying this approach is straightforward: locks are designed to delay entry long enough to deter the intruder, who is likely to move on to an easier target [121]. Research on active security measures supports this logic, demonstrating that visible security features and functional barriers significantly reduce criminal intent [125]. The effectiveness of this deterrent approach is reinforced by evidence showing that even modest security improvements can substantially reduce burglary risk, as intruders typically seek properties with minimal resistance [122].

### Residential Lock Technologies and Features

Contemporary residential locking systems have expanded considerably beyond traditional mechanical designs. Smart door locks with biometric authentication, wireless communication, and remote access capabilities have emerged as increasingly popular alternatives [123, 126, 127]. These systems offer enhanced convenience through features such as fingerprint recognition, facial recognition, and smartphone-based access control [123, 126, 128, 129]

However, the adoption of advanced technologies in residential settings introduces new security considerations. Research on smart home security adoption reveals that consumers balance perceived security benefits against concerns about ease of use and system reliability [126]. Digital door locks utilizing fingerprint recognition technology, for instance, must address vulnerabilities such as fingerprint residue exposure while maintaining practical usability [123]. Similarly, Bluetooth-enabled door lock systems offer wireless convenience but require careful implementation to prevent unauthorized access through compromised wireless signals [130, 131].

The diversity of residential lock technologies reflects the heterogeneous needs of domestic users. Some systems employ multi-factor authentication combining multiple access methods—such as RFID cards, keypads, fingerprint sensors, and facial recognition—to enhance security while providing fallback options if any single authentication method fails [129, 132]. Others utilize IoT integration to enable remote monitoring and control through smartphone applications, allowing homeowners to verify door lock status and grant access to authorized visitors from any location [133-135].

### Risk-Based Recommendations and Client Education

For locksmiths and security professionals working in residential contexts, the emphasis differs markedly from institutional security work. Residential work emphasizes risk-based recommendations and client education, requiring professionals to help clients understand the practical and economic trade-offs inherent in security decisions [121].

This educational role encompasses several key areas. First, locksmiths must help clients understand the difference between standard and high-security cylinders, explaining how enhanced features such as anti-pick and anti-bump mechanisms provide improved protection against common attack methods [121]. Second, they must educate clients about the importance of complementary security hardware, including escutcheons and strike plates, which significantly influence overall door security [121]. The quality of door hardware extends beyond the lock cylinder itself; properly installed strike plates and reinforced door frames substantially increase resistance to forced entry [122, 125].

Third, locksmiths must familiarize clients with certified product standards and their implications. TS007-rated locks, for example, represent a recognized standard for residential lock security in certain markets, providing clients with objective criteria for evaluating lock quality [121]. Understanding these standards enables clients to make informed decisions about security investments.

### Balancing Security, Budget, and Practical Needs

A critical aspect of residential locksmith work involves balancing security improvements with the client's budget and practical needs [121]. Unlike institutional settings where security budgets may be substantial and security is the paramount concern, residential clients typically operate under financial constraints and must prioritize security alongside other home improvement needs [121].

This balancing act requires locksmiths to assess each property's specific risk profile and recommend proportionate security measures. Research on burglary prevention programs demonstrates that even modest security improvements—such as installing proper locks and deadbolts—can substantially reduce victimization risk [122]. This evidence supports a tiered approach to residential security recommendations, where basic improvements provide significant protection at reasonable cost, with optional enhancements available for clients with higher security concerns or greater financial resources.

The practical needs dimension of residential security extends beyond the lock mechanism itself. Homeowners must consider factors such as ease of use for all household members, compatibility with existing door hardware, aesthetic integration with home décor, and maintenance requirements [121, 126]. A lock that provides excellent security but proves difficult for elderly residents or children to operate may ultimately compromise household security if users circumvent the system or fail to engage it consistently.

Domestic locks represent a distinct category of security hardware, fundamentally different from institutional or commercial applications. The residential context demands a careful balance among security effectiveness, affordability, ease of use, and aesthetic considerations [121]. Modern residential locks increasingly incorporate advanced features such as anti-bump and anti-pick mechanisms, biometric authentication, and IoT integration [121, 123, 126, 127]. However, the core principle remains constant: residential locks are designed to deter opportunistic intruders by creating sufficient delay to encourage them to seek easier targets [121].

Effective residential security requires not only appropriate lock hardware but also complementary measures such as proper door hardware installation and client education [121, 122]. Locksmiths and security professionals working in residential contexts must balance technical security considerations against practical constraints of cost, usability, and client needs [121]. This nuanced approach, grounded in understanding the distinct threat landscape facing residential properties, enables homeowners to achieve meaningful security improvements proportionate to their circumstances and resources.

### Comparative Insight and Industry Impact

Across these three environments, a number of consistent insights emerge that have direct implications for modern locksmithing practice. First and foremost, security is highly context-dependent. What is appropriate in a high-risk environment such as a prison, where absolute control and containment are essential, is neither practical nor desirable in a residential setting. Locksmiths must therefore move beyond simply installing hardware and instead assess the level of risk, the nature of the environment, and the expectations of the user before recommending a solution.

Another key insight is that layered security systems consistently outperform single-point solutions. Prisons and banks demonstrate that combining mechanical locks with electronic controls, surveillance systems, and strict procedural measures creates a far more resilient security framework. No single device provides complete protection; rather, security emerges from the interaction of multiple systems working together. This reinforces the need for locksmiths to think in terms of integrated systems rather than isolated components.

Human factors also play a critical role in determining the effectiveness of any security system. Even the most advanced locks can be undermined by poor key management, weak access control policies, or user behaviour that prioritises convenience over security. In both prison and banking contexts, strict key control and accountability processes are essential to maintaining system integrity. For locksmiths, this highlights the importance of advising clients not only on hardware, but also on procedures, user practices, and governance.

In residential environments, the emphasis shifts toward time and deterrence rather than absolute resistance. Domestic locks are not designed to withstand prolonged, determined attacks; instead, they aim to delay entry long enough to discourage opportunistic intruders or increase the likelihood of detection. This reinforces the importance of selecting locks and systems that balance security with usability, ensuring that they are both effective and practical for everyday use.

Finally, technology is reshaping all sectors of security, from institutional to domestic settings. Electronic locking systems, smart home devices, and integrated access control platforms are becoming increasingly common, requiring locksmiths to develop skills that extend beyond traditional mechanical expertise. Understanding digital systems, communication protocols, and cybersecurity principles is now essential. As a result, the modern locksmith must operate at the intersection of physical and digital security, adapting to a rapidly evolving landscape where access is managed as much by code and networks as it is by keys and cylinders.

These case studies illustrate that locksmithing is not about applying a single solution to every situation. Instead, it requires adapting security strategies to the specific environment, threat level, and user needs. Whether working on a prison, a bank, or a residential property, the locksmith's role is to design and implement systems that balance security, usability, and risk, drawing on lessons learned from both historical failures and modern innovations.

### Impact on Locksmithing Practices

The differences between prisons, banks, and domestic environments show that there is no single "best" lock or solution. Instead, effective locksmithing requires the ability to assess the environment, identify threats, and design appropriate, proportionate security solutions.

One of the most important influences on locksmithing practice is the shift from a product-based mindset to a risk-based approach. In a prison, the expectation is absolute control with zero tolerance for failure, which drives the use of heavy-duty mechanisms, redundancy, and strict procedures. In contrast, residential settings prioritise affordability and convenience, where the goal is deterrence rather than absolute resistance. For locksmiths, this means developing the ability to evaluate risk levels and recommend solutions that align with the client's environment rather than simply offering the most complex or expensive lock.

The case studies also reinforce the importance of layered security design, which has become a core principle in modern locksmithing. Banks, in particular, demonstrate that locks alone are insufficient; they must be integrated with surveillance, alarms, access control systems, and procedural safeguards. This has influenced locksmithing practice by expanding the role of the locksmith into that of a security integrator, responsible not only for mechanical hardware but also for how different systems interact to provide comprehensive protection.

Another major influence is the recognition that human factors are often the weakest link in security systems. Prison and banking environments highlight how critical key control, access management, and staff procedures are to maintaining security. This has shifted locksmithing practice toward client education and policy advice, where locksmiths guide users on key management, access control protocols, and the risks associated with poor security habits. It is no longer enough to install a secure lock if the system around it is poorly managed.

The comparison also highlights the growing importance of time, deterrence, and behavioural psychology in residential locksmithing. Domestic locks are designed not to be invulnerable, but to delay entry and encourage intruders to abandon the attempt. This has influenced locksmiths to focus on practical improvements—such as upgrading cylinders, reinforcing doors, and improving visibility—rather than pursuing unrealistic notions of absolute security. Understanding how intruders think and behave is now just as important as understanding lock mechanisms.

Perhaps the most significant impact on locksmithing practice is the increasing integration of electronic and digital technologies. Prisons and banks have long used electronic access control, and these technologies are now entering residential environments through smart locks, IoT devices, and systems like ESP32-CAM with remote communication. This evolution requires locksmiths to develop skills in networking, access control systems, and cybersecurity, fundamentally changing the profession from purely mechanical work to hybrid physical–digital security expertise.

Finally, these environments demonstrate that effective locksmithing requires a systems-thinking approach. Locks do not operate in isolation—they are part of broader ecosystems that include people, procedures, technology, and physical infrastructure. The locksmith's role is therefore to design and maintain these systems so that each component supports the others, ensuring that security is resilient even if one element fails.

In practical terms, this means that a modern locksmith must be able to:

- Assess different risk environments and tailor solutions accordingly
- Design layered security systems rather than relying on single devices
- Advise on key control, access policies, and user behaviour
- Balance security with usability, cost, and client needs
- Integrate mechanical locks with electronic and digital systems

These case studies show that locksmithing has evolved into a professional discipline centred on security strategy, not just hardware. Understanding how prisons, banks, and domestic systems differ—and why—equips locksmiths to make informed decisions, provide better advice, and deliver solutions that are both effective and appropriate for the real-world context in which they are used.

# The Future of "Ownership" in a Keyless World

The fundamental concept of ownership is undergoing a profound transformation as physical keys are progressively replaced by digital, networked access systems. This shift represents more than a technological upgrade; it constitutes a reconceptualization of what ownership means in an increasingly connected world. Understanding this transition requires examining how traditional notions of control and possession are being redefined through cloud-based and Internet of Things (IoT) technologies.

## The Transition from Physical to Digital Ownership

In traditional locksmithing, ownership was straightforward and tangible: possession of a physical key directly conferred control over access [136]. This model was characterized by clear, binary relationships, one either possessed the key or did not. Keys could be handed over, duplicated, lost, or access could be revoked by changing the lock entirely. Ownership and control were inextricably linked to a physical object, making the relationship between possession and authority unambiguous [136].

The transition to keyless environments fundamentally alters this relationship. In modern keyless systems, access is no longer defined by possession of a physical object but rather by permissions, identities, and system-level controls [136]. Access is granted through digital credentials such as PIN codes, RFID cards, smartphone applications, or biometric identifiers [136]. Critically, these credentials are not "owned" in the traditional sense; instead, they are issued, managed, and revoked by a system [136]. This represents a fundamental shift in the locus of control: the person or entity controlling the system, whether a homeowner, business, or cloud service provider, effectively controls access [136].

## Programmable and Granular Ownership

One of the most significant innovations in keyless systems is the introduction of programmable ownership (Khurshid et al., 2018). In digital systems, access can be tailored with remarkable precision and flexibility. Rather than the binary nature of traditional keys, access in modern systems becomes granular, dynamic, and conditional [136]. For example, a cleaning service might be granted access only on specific days and times, such as every Tuesday between 9 am and 12 pm, while a contractor might receive temporary access that automatically expires [136]This level of control was impossible with physical keys but is routine in digital systems.

A critical distinction emerges between ownership and possession in keyless environments. Historically, giving someone a physical key meant transferring ongoing access until the key was physically returned or the lock was changed [136]. In contrast, digital systems enable access to be granted remotely without transferring any physical object, and access can be revoked instantaneously [136]. This separation is particularly valuable in environments with frequent access changes, such as rental properties, hotels, and workplaces, where flexibility and rapid updates are essential [136].

**Platform Dependency and Cloud-Based Control**

The rise of cloud-based and Internet of Things (IoT)-enabled systems introduces a new and complex layer to ownership: platform dependency [136]. In many smart lock systems, access is managed through mobile applications and cloud services rather than through local, hardware-based mechanisms [136]. This architecture raises fundamental questions about the true nature of ownership: Does the user own access, the device manufacturer, or the service provider? [136]

The integration of IoT devices with cloud computing platforms has become increasingly prevalent, with cloud providers offering scalable infrastructure for managing access control systems [137, 138]. However, this integration introduces significant security and control challenges. If a cloud platform is discontinued, experiences a security breach, or becomes locked behind subscription requirements, users may lose control over their own locks [136]. In this sense, ownership becomes partially dependent on digital infrastructure and service agreements rather than being solely determined by hardware ownership [136].

Research on cloud-based IoT systems emphasizes the importance of access control mechanisms. In cloud-enabled IoT environments, access control is typically achieved through Discretionary Access Control (DAC) and Mandatory Access Control (MAC) models, including Access Control Lists (ACLs), Role-Based Access Control (RBAC), and Capability Systems [136]. In DAC models, the data owner can modify access permissions, which aligns with the concept of programmable ownership [136]. However, the delegation of access control to cloud providers introduces questions about who ultimately controls these permissions and under what conditions they can be modified or revoked.

**Security Implications and Expanded Responsibilities**

Security risks significantly complicate the concept of ownership in keyless systems. In physical systems, duplicating a key requires physical access and considerable effort (Khurshid et al., 2018). In digital systems, credentials can potentially be copied, intercepted, or shared instantaneously through various attack vectors [136]. This fundamental difference means that ownership in digital systems must now encompass identity verification, encryption, and comprehensive cybersecurity practices [136]. The responsibility of ownership expands far beyond safeguarding a physical key to include managing passwords, devices, and network security [136].

Cloud-based IoT systems face multiple security challenges that directly impact ownership and control. Research identifies four major categories of security concerns in IoT-cloud integration: data security, network and service security, application security, and people-related security issues [137, 139]. Data security concerns include unauthorized access, data breaches, and data loss [137, 139]. Network security challenges involve securing communications between IoT devices and cloud platforms, protecting against man-in-the-middle attacks, and ensuring secure data transmission [137, 139]. These security challenges mean that true ownership in a cloud-connected system requires ongoing vigilance and technical expertise [136].

To address security concerns, modern keyless systems employ advanced encryption and access control mechanisms. End-to-end encryption, such as Advanced Encryption Standard (AES) 256-bit encryption, is used to protect data during transmission and storage [140]. Additionally, attribute-based encryption (ABE) and other cryptographic approaches enable fine-grained access control, allowing system administrators to specify precisely who can access what resources under which conditions [141]. These technical measures are essential for maintaining ownership and control in cloud-connected environments, but they also require technical knowledge and ongoing management [136].

### Professional and Practical Implications

For locksmiths and security professionals, this transformation has direct and profound implications for professional practice [136]. The traditional role of cutting keys and installing mechanical locks is evolving into managing complex access control ecosystems [136]. Modern security professionals must now set permissions, integrate diverse devices, secure networks, and advise clients on digital risks [136]. This requires understanding not only how to secure a physical door but how to secure the entire system that controls that door [136].

The integration of access control systems with broader IoT platforms adds another layer of complexity. IoT platforms serve as middleware solutions that connect different hardware devices, access points, and networks [142]. These platforms must provide user-friendly interfaces while maintaining security and reliability [142]. The role of security professionals now includes understanding how virtual objects (digital representations of physical entities) are managed within these platforms and how access to these virtual objects is controlled [142].

### Broader Societal and Economic Implications

Keyless systems enable entirely new models of ownership and access that were previously impossible with physical keys [136]. Shared access and temporary ownership have become practical realities. For example, Airbnb hosts can grant time-limited access to guests without issuing physical keys, and businesses can manage large numbers of users without the logistical challenges of key distribution [136]. These systems create more flexible and efficient access management, but they also challenge traditional ideas of control, privacy, and responsibility [136].

The flexibility provided by digital access systems offers significant advantages. Organizations can rapidly grant or revoke access, modify permissions in real-time, and maintain detailed audit trails of who accessed what and when [136]. This is particularly valuable in dynamic environments where access requirements change frequently. However, this flexibility comes with the responsibility of managing complex permission systems and ensuring that access controls are properly configured and maintained [136].

### The Future of Ownership in Keyless Systems

Ultimately, the future of ownership in a keyless world is defined by a fundamental shift from objects to access rights [136]. Ownership is no longer about holding a physical key, it is about controlling permissions, managing identities, and maintaining secure systems [136]. This transformation does not eliminate the need for locksmithing or security expertise; rather, it expands these fields into more complex domains where physical security and digital control are inseparably linked [136].

The convergence of physical and digital security requires integrated approaches. Cloud-based IoT platforms must provide comprehensive security frameworks that address data protection, network security, application security, and user authentication [137-139]. These frameworks must balance security requirements with usability and cost-effectiveness [138]. The responsibility for security is increasingly shared between users, device manufacturers, and cloud service providers, requiring clear agreements about who is responsible for what aspects of security [139].

Emerging technologies such as blockchain are being explored to enhance security and transparency in cloud-based IoT systems. Blockchain-based approaches can provide distributed access control and auditable records of access permissions, reducing dependence on centralized authorities [143, 144]. These technologies offer potential solutions to the platform dependency problem by enabling decentralized management of access control [143, 144].

The concept of ownership is undergoing a fundamental transformation as keyless, cloud-connected access systems replace traditional physical keys. This shift moves ownership from the possession of physical objects to the control of digital permissions and access rights. While this transformation enables unprecedented flexibility and efficiency in access management, it also introduces new complexities related to platform dependency, cybersecurity, and the distribution of responsibility among users, manufacturers, and service providers. For security professionals and society more broadly, this evolution requires new skills, new frameworks for understanding control and responsibility, and new approaches to ensuring that digital access systems are secure, reliable, and aligned with users' interests and values. The future of ownership in a keyless world will be defined not by what we hold, but by what we control, and by our ability to maintain that control in an increasingly complex digital ecosystem.

**Impact on Locksmithing Practices**

This shift from physical keys to digital, permission-based access is highly relevant to anyone interested in locksmithing because it fundamentally changes what the profession is, what skills are required, and how security is delivered. It is not simply a new type of lock, it is a transformation of the entire concept of access control.

At a practical level, locksmithing is no longer centred on hardware alone. Traditionally, the locksmith’s role was to manage physical objects, keys, cylinders, and mechanical systems. In a keyless world, the focus moves to managing access rights. This means configuring who can enter, when they can enter, and under what conditions. As a result, locksmiths increasingly act as access control managers, not just installers of locks. The work involves setting permissions, programming devices, and ensuring systems operate securely rather than simply ensuring a key turns in a cylinder.

This transformation also expands the required skill set. A modern locksmith must understand digital systems, networking, and cybersecurity alongside mechanical principles. Knowledge of concepts such as encryption, authentication, cloud platforms, and IoT communication is becoming essential. For example, installing a smart lock is no longer just a physical task, it requires configuring apps, securing API access, and ensuring the device is protected from unauthorised digital control. This represents a shift from mechanical expertise to hybrid technical expertise.

Another major influence on locksmithing practice is the rise of programmable and flexible access. In traditional systems, access was static, once a key was issued, control was difficult to change. In digital systems, access is dynamic and can be modified instantly. This changes how locksmiths design solutions. Instead of asking "who needs a key," the question becomes "what level of access should each user have?" Locksmiths must now think in terms of roles, schedules, and permissions, similar to IT system administrators.

The concept of platform dependency also introduces new responsibilities. Locksmiths must consider not just the lock itself, but the ecosystem it operates in, apps, cloud services, firmware, and manufacturers. This affects product selection and advice to clients. A locksmith must evaluate questions such as: What happens if the service is discontinued? How secure is the platform? Who controls the data? This shifts practice toward system evaluation and risk assessment, rather than simple product installation.

Security practices are also significantly affected. In mechanical systems, security risks were primarily physical (picking, bumping, forced entry). In digital systems, risks include hacking, credential theft, and network vulnerabilities. This means locksmiths must adopt a broader security mindset that includes:

- Securing user authentication (who can send commands)
- Protecting credentials (PINs, tokens, digital keys)
- Ensuring encrypted communication (HTTPS, TLS)
- Advising on secure network configurations

In other words, locksmithing now overlaps with cybersecurity, and failure to address digital risks can completely undermine physical security.

The shift also changes how locksmiths interact with clients. There is a greater emphasis on education and ongoing support. Clients must understand how to manage digital access, update systems, and maintain security over time. Unlike a traditional lock, which is largely "fit and forget," digital systems require continuous management. Locksmiths therefore take on a consultative role, guiding clients on best practices, updates, and risk mitigation.

From a business perspective, this evolution creates new opportunities and challenges. Locksmiths can expand into areas such as smart home integration, commercial access control systems, and IoT security solutions. However, it also means increased competition from IT professionals and technology providers. To remain relevant, locksmiths must position themselves as integrated security specialists, combining physical and digital expertise.

Finally, this shift reinforces a deeper professional insight: security is no longer about a single device, it is about a connected system of control. A lock is now just one component within a broader architecture that includes software, networks, users, and policies. Effective locksmithing practice therefore requires a systems-thinking approach, where the goal is not just to secure a door, but to secure the entire access environment.

The move to keyless systems is directly shaping locksmithing into a more advanced, interdisciplinary profession. It changes the focus from keys to permissions, from hardware to systems, and from physical protection to integrated security. For anyone entering the field, understanding this shift is essential, as it defines both the future direction of the industry and the competencies required to operate within it.

## Chapter 2 Review Questions

### Question 1

A property manager installs smart locks in an apartment complex in a city where residents are comfortable with shared spaces but are highly concerned about digital privacy. Tenants accept the locks but demand strict control over access logs and personal data.

What is the locksmith's most important design consideration?

A. Increasing the number of mechanical locks
B. Minimising data collection and protecting digital privacy
C. Installing stronger deadbolts on all doors
D. Removing remote access features entirely

### Question 2

A homeowner requests multiple layers of security after a nearby burglary: alarms, cameras, smart locks, and PIN entry systems. After installation, the family begins sharing access codes and disabling alarms because the system feels inconvenient.

What key security principle does this situation demonstrate?

A. More technology always improves security
B. Complex systems eliminate user error
C. Security systems must balance protection with usability
D. Mechanical locks should always replace electronic locks

### Question 3

A locksmith installs heavy padlocks and steel bars for a small shop in a low-income neighbourhood where advanced electronic systems are unaffordable.

Why is this solution appropriate?

A. Visible deterrence and durability suit the economic context
B. Electronic locks are illegal in small shops
C. Steel bars eliminate all security risks
D. High-tech systems only work in offices

**Question 4**

A residential compound uses an access control system that gives different permissions to residents, domestic workers, visitors, and security staff.

What does this system primarily represent?

A. A failure of mechanical security
B. A hierarchical access structure reflecting social roles
C. A purely technical security upgrade
D. A temporary emergency access solution

**Question 5**

A European company installs CCTV and electronic access control but must limit monitoring and restrict data storage because of strict privacy laws.

What factor most influences the system design?

A. Hardware quality
B. Legal and regulatory requirements
C. Cost of installation
D. User convenience

**Question 6**

A locksmith recommends replacing a widely used pin-tumbler lock system with an anti-bump design after learning that lock bumping can be performed quickly using simple tools.

Why is this change necessary?

A. The lock is difficult to manufacture
B. Low-skill attacks with high impact create widespread vulnerabilities
C. Only professional locksmiths should open locks
D. Pin tumbler locks are obsolete

**Question 7**

A hotel installs digital keycards for guests. Researchers later demonstrate that inexpensive devices can copy or spoof these credentials.

What lesson does this scenario illustrate?

A. Electronic locks are always safer than mechanical locks
B. Digital systems can introduce new attack vectors
C. Hotels should never use electronic locks
D. Mechanical keys cannot be copied

**Question 8**

A landlord replaces traditional keys with mobile-app access for rental properties. Access permissions can be granted remotely and revoked instantly when tenants move out.

What advantage does this system provide compared with physical keys?

A. Keys cannot be duplicated
B. Access permissions can be dynamically managed
C. Mechanical security becomes unnecessary
D. Doors no longer require hardware

**Question 9**

A locksmith observes that many security failures occur not because locks are weak, but because users misunderstand how systems work or share access credentials.

What role should the locksmith take in response?

A. Focus only on installing stronger locks
B. Educate clients about correct security practices
C. Remove all electronic access systems
D. Avoid advising clients

**Question 10**

A modern smart lock system communicates with other devices through a lightweight IoT messaging protocol to send lock and unlock commands.

What new responsibility does this create for locksmiths?

A. Designing new mechanical keys
B. Understanding cybersecurity and secure network communication
C. Eliminating electronic systems
D. Replacing digital credentials with physical keys

# Part 2

# Advanced Mechanical Systems Beyond the Basics

# Chapter 3

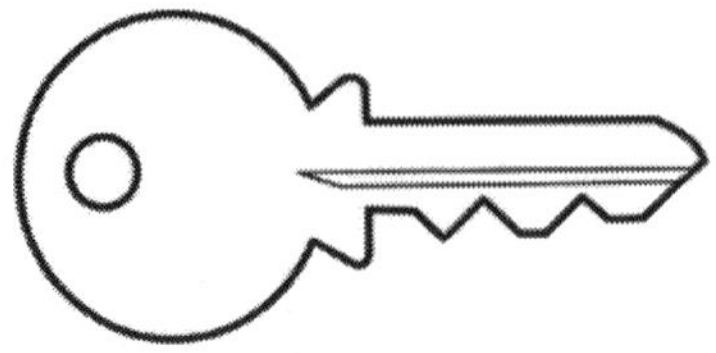

# Micro-Mechanics of Modern Lock Systems

Every component of a lock, from the shape of the keyway to the composition of internal parts, has been deliberately designed to control access, resist attack, and ensure long-term reliability.

This chapter explores the critical foundations that underpin contemporary lock and key systems. It begins by examining how keyways function not only as physical guides for keys but as engineered security features that influence access control and resistance to manipulation. It then considers the role of materials and manufacturing processes, highlighting how choices such as brass, hardened steel, and advanced alloys directly impact durability, wear resistance, and performance in different environments.

Building on this foundation, the chapter examines key cutting as a precision process rather than a purely mechanical task. Techniques such as decoding, code cutting, and the use of depth charts and gauges demonstrate how modern locksmithing relies on measurement, accuracy, and system knowledge. These principles become even more critical when working with high-security systems, where complex mechanisms—such as sidebars, rotating pins, sliders, and disc detainers—require exact alignment and interaction with specially designed keys.

Finally, the chapter introduces high-security lock designs from leading manufacturers, illustrating how innovation in key control, mechanical complexity, and material engineering has transformed the industry. These systems reflect a shift away from simple pin tumbler mechanisms toward multi-layered security solutions that integrate precision machining, restricted keyways, and advanced locking technologies.

Together, these topics highlight a key theme: locksmithing is no longer just a trade of replication, but a profession centred on analysis, design, and controlled precision. Understanding these principles is essential for anyone seeking to work effectively with modern lock systems, diagnose faults accurately, and implement secure, reliable solutions in a wide range of environments.

## Tolerances, Wear Patterns, and Failure Points

Understanding the micro-mechanics of modern lock systems, particularly tolerances, wear patterns, and failure points, is critical for explaining why locks behave the way they do in real-world conditions. While locks are often described in terms of their visible components (pins, plugs, keys), their true performance is governed by microscopic interactions between materials, clearances, and forces. For locksmiths and security professionals, this level of detail separates basic installation knowledge from deep diagnostic and forensic expertise.

### Tolerances: The Invisible Geometry of Security

Tolerances represent the allowable deviations in size, alignment, and positioning of components within a lock mechanism [145]. In pin tumbler systems specifically, tolerances encompass multiple critical dimensions including the diameter of pins, the spacing of chambers, the fit between plug and housing, and the exact height each pin must reach to align at the shear line [145]. These dimensional specifications are fundamental to both the functional and security characteristics of mechanical locks.

At a microscopic level, no component can be manufactured to perfect specifications, every part exists within a tolerance range, often measured in microns [145]. These tiny gaps and deviations from nominal dimensions are not merely manufacturing imperfections but are essential design features [145]. Without such tolerances, the internal components of a lock would seize and become immobile, rendering the mechanism non-functional [145]. However, these necessary operational spaces create what can be characterized as the "space" that attackers exploit when attempting to compromise lock security [145].

The relationship between tolerance tightness and lock security presents a nuanced and sometimes counterintuitive picture. In high-quality locks manufactured to tighter tolerances, several mechanical advantages emerge [145]:

- Pins move more smoothly and predictably through their chambers
- Friction and mechanical "slop" are minimized
- Alignment at the shear line becomes more precise

However, this precision creates a paradoxical security situation. The very smoothness and predictability that characterize high-quality locks can sometimes make certain attacks, particularly lock bumping, more effective [145]. This occurs because energy is transferred efficiently through well-aligned components with minimal friction losses [145].

Conversely, cheaper locks manufactured with looser tolerances present a different security profile [145]. These locks may resist certain attacks unintentionally because friction and irregularities in component alignment absorb energy that would otherwise be transmitted through the mechanism [145]. However, this same lack of precision makes these locks more vulnerable to picking attacks, as the inconsistent feedback from poorly-aligned components provides less reliable information to an attacker attempting to manipulate the pins [145].

In advanced lock systems, particularly high-security cylinders, tolerances are deliberately engineered not merely for functional performance but specifically for security behaviour [145]. This represents a sophisticated approach to lock design where tolerance specifications are used as active security features [145]. These engineered tolerances may introduce controlled binding of pins, provide false feedback to potential attackers, or establish multi-axis alignment requirements that complicate unauthorized manipulation [145].

The role of tolerances in lock security extends far beyond simple manufacturing specifications. Tolerances fundamentally shape both the operational characteristics and the security properties of mechanical locks, creating a complex relationship where precision can enhance certain vulnerabilities while loose tolerances create different security challenges. Understanding this relationship is essential for both lock designers seeking to optimize security and for security researchers analysing lock vulnerabilities.

**Wear Patterns: The Mechanical Memory of Use**

Every time a key is inserted and turned within a lock, microscopic friction occurs between the key and the pin surfaces, the pins and the chamber walls, and the plug and the housing interface. Although these interactions are not visible to the naked eye, they gradually alter the internal components of the lock. Over time, this repeated contact produces wear patterns that effectively act as the lock's "memory," recording how it has been used throughout its service life.

As wear develops, several changes begin to occur within the mechanism. Pins may become rounded or flattened, reducing the precision required for correct shear line alignment. Keys themselves can wear down, subtly altering the heights at which pins are lifted and, in some cases, allowing partially correct or poorly cut keys to operate the lock. The internal chambers may also enlarge slightly, increasing tolerances and making manipulation easier. In addition, surfaces can become polished through repeated movement, reducing friction and allowing components to move more freely than originally intended.

From a security perspective, these changes can have significant consequences. Increased smoothness and reduced resistance can make locks easier to pick, as the tactile feedback becomes clearer and more predictable. Similarly, smoother pin movement can increase susceptibility to bumping, as kinetic energy transfers more efficiently through the system. In some cases, excessive wear may even result in accidental operation by similar or worn keys, undermining the reliability of the locking mechanism.

For locksmiths, wear patterns provide valuable diagnostic insight. They can reveal how frequently a lock has been used, whether poor key duplication practices have been employed, or whether the lock has been installed with misalignment that accelerates internal degradation. In more advanced applications, such as forensic analysis, these patterns may also indicate whether a lock has been subjected to manipulation or attack, offering clues about the methods used and the history of the device.

**Failure Points: Where Systems Break Down**

Failure points are the locations within a lock where stress, fatigue, or attack forces concentrate, eventually leading to malfunction or compromise.

These can be grouped into three categories:

**1. Mechanical Failure Points**

These arise from normal use and material fatigue:

- **Springs** → lose tension over time, causing pins to fail to reset
- **Pins** → deform or jam due to wear or contamination
- **Plug-housing interface** → develops excessive play, reducing security

A common example is a lock that begins to "feel loose" or inconsistent—often a sign that tolerances have increased beyond safe limits.

### 2. Structural Weak Points

These are inherent to the lock's design and are often targeted in attacks:

- **Shear line** → primary target for picking and bumping
- **Cylinder body (euro profile)** → vulnerable to snapping if not reinforced
- **Cam or tailpiece** → can be forced or twisted under torque

Modern high-security locks reinforce these areas with:

- Hardened steel inserts
- Anti-snap sacrificial sections
- Reinforced cam mechanisms

### 3. Attack-Induced Failure Points

These occur when external forces are applied deliberately:

- **Drilling** → targets pin chambers and shear line
- **Bumping** → exploits kinetic transfer through pins
- **Picking** → exploits binding tolerances
- **Snapping** → exploits structural weakness in cylinder design

Each attack method targets a different micro-mechanical vulnerability. For example:

- Bumping relies on low friction and precise tolerances
- Picking relies on binding caused by slight misalignments
- Drilling bypasses tolerances entirely by destroying components

### System Interaction: Why These Factors Matter Together

Tolerances, wear, and failure points do not operate independently within a lock mechanism; rather, they interact dynamically and continuously influence one another. The internal condition of a lock is therefore not static, but evolves over time as these factors combine to affect performance, reliability, and security.

In a newly manufactured lock with tight tolerances and minimal wear, the mechanism operates with high precision. However, this same precision can allow forces from certain attacks, such as bumping, to transfer more efficiently through the system. Conversely, as tolerances increase and wear accumulates, the behaviour of the lock becomes less predictable. Looser internal clearances and worn components can make the lock easier to manipulate, as pins may bind less consistently and provide clearer feedback during picking.

The introduction of reinforced failure points further alters this balance. Features such as anti-snap sections or hardened inserts are designed to resist specific attack methods, but in doing so, they can redirect force or attention toward other, less protected vulnerabilities within the lock. This demonstrates that strengthening one area of a system often shifts, rather than eliminates, the point of weakness.

As a result, a lock must be understood as an evolving system rather than a fixed product. Its security characteristics change over time, meaning real-world performance can differ significantly from original factory specifications. Maintenance practices, frequency of use, environmental conditions, and key management all play a direct role in shaping how secure the lock remains throughout its lifespan.

### Practical Insight for Locksmiths

For modern locksmithing practice, an understanding of micro-mechanics has direct and practical implications across installation, maintenance, diagnosis, and security planning. During installation, even minor misalignment can introduce uneven stresses within the lock, accelerating wear and creating unintended failure points. What may appear to be a small fitting error can, over time, significantly reduce both performance and security.

Maintenance plays an equally critical role in preserving the integrity of a lock's internal components. Proper lubrication and regular inspection help maintain designed tolerances, reduce friction, and prevent premature degradation. Without this care, microscopic wear accumulates, gradually altering how the lock behaves and increasing its susceptibility to manipulation or failure.

From a diagnostic perspective, recognising wear patterns allows locksmiths to make informed decisions about whether a lock can be serviced or should be replaced entirely. These patterns provide insight into usage frequency, potential misuse, and underlying mechanical issues that may not be visible externally. This level of analysis moves the locksmith's role beyond reactive repair toward proactive risk assessment.

Security advice is also shaped by micro-mechanical understanding. Rather than selecting locks solely based on advertised features or initial performance, locksmiths must consider how a lock is likely to degrade and fail over time. The way a lock fails—whether predictably, securely, or catastrophically—is just as important as how it functions when new.

At the micro-mechanical level, a lock is not a static device but a dynamic system shaped by precision, friction, and time. Tolerances determine how it operates, wear patterns reveal how it has been used, and failure points dictate how it will ultimately fail. Understanding these hidden mechanisms enables locksmiths and security professionals to move beyond surface-level assessments and engage with the true engineering reality of locks, where outcomes are often determined at the scale of microns.

The following are case study examples that demonstrate how tolerances, wear patterns, and failure points influence lock behaviour in practice.

**Case Study 1: High-Tolerance Cylinder and Unexpected Bumping Vulnerability**

A commercial office installed a premium high-security pin tumbler system marketed for its tight manufacturing tolerances and precision engineering. Shortly after installation, a penetration test revealed that the lock was unexpectedly vulnerable to bumping.

Investigation showed that the extremely tight tolerances allowed pins to move with minimal friction. When a bump key was used, kinetic energy transferred efficiently through the pin stacks, causing consistent alignment at the shear line. Rather than resisting the attack, the precision of the system actually enhanced energy transmission.

This case highlights the paradox of tolerances: while tight tolerances improve reliability and smooth operation, they can increase susceptibility to certain attack methods. For locksmiths, the lesson is that lock selection must consider *attack behaviour*, not just build quality. High precision does not automatically equate to higher security.

**Case Study 2: Worn Residential Lock Allowing Incorrect Key Operation**

A residential client reported that their front door lock could occasionally be opened using a similar, but incorrect, key. The lock had been in service for over 12 years with frequent daily use.

Upon disassembly, the locksmith identified significant wear patterns:

- Pins were rounded and slightly shortened
- The key showed visible wear along the cuts
- Chambers had minor enlargement from long-term friction

These changes increased internal tolerances and reduced the precision required to reach the shear line. As a result, a near-matching key could lift the pins sufficiently to allow rotation.

This case demonstrates how wear patterns act as a "mechanical memory" of use. For locksmiths, it reinforces the importance of recognising when a lock has degraded beyond safe tolerances and must be replaced rather than serviced.

**Case Study 3: Misaligned Installation Leading to Premature Failure**

A newly installed mortice lock in a commercial property began failing within six months. Users reported stiffness, inconsistent key operation, and eventual jamming.

Inspection revealed that the lock had been installed slightly out of alignment with the strike plate. This caused:

- Uneven pressure on the plug during operation
- Increased friction between pins and chambers
- Accelerated wear on internal components

The misalignment introduced abnormal stress, effectively altering internal tolerances and creating premature wear patterns. Eventually, pins began to bind, and springs lost effectiveness.

This case highlights how installation errors directly affect micro-mechanics. For locksmiths, precision during installation is critical, not only for function but for long-term security and durability.

**Case Study 4: Euro Cylinder Snap Attack Exploiting Structural Failure Point**

A burglary occurred at a residential property fitted with a standard euro-profile cylinder. Entry was gained without picking or drilling.

Forensic analysis showed the cylinder had been snapped at its central fixing point—an inherent structural weakness. Once broken, the cam was exposed and easily manipulated to unlock the door.

This case illustrates a design-based failure point, unrelated to tolerances or wear. The lock functioned correctly but failed structurally under attack.

Modern anti-snap cylinders address this by:

- Incorporating sacrificial break sections
- Reinforcing critical areas with hardened inserts

For locksmiths, this reinforces the importance of understanding *how a lock fails*, not just how it operates.

**Case Study 5: Progressive Wear Increasing Pickability in a Commercial Lock**

A locksmith conducting a security audit on a retail premises found that older locks were significantly easier to pick than newly installed ones of the same model.

Closer inspection revealed:

- Polished pin surfaces from repeated use
- Reduced friction inside chambers
- Slightly increased tolerances

These changes improved tactile feedback during picking, making it easier to identify binding pins and manipulate them accurately.

This case demonstrates how wear can improve attack conditions, even in locks that were originally secure. For locksmiths, it highlights the importance of periodic security reassessment—not just functional checks.

**Case Study 6: Reinforced Lock Redirecting Attack Method**

A high-security lock with anti-drill pins and hardened inserts was installed in a small business. During an attempted break-in, the attacker abandoned drilling and instead applied torque to the cam mechanism.

The lock resisted drilling as designed, but the attack shifted to a less protected component—the cam/tailpiece, which eventually failed under force.

This demonstrates a critical principle: reinforcing one failure point often redirects attack forces elsewhere. Security is not eliminated, it is redistributed.

For locksmiths, this reinforces the need for holistic security design, ensuring all components—not just primary ones, are appropriately protected.

**Key Takeaways Across Case Studies**

Across these examples, several important insights emerge:

- Tolerances influence attack behaviour, sometimes in unexpected ways
- Wear patterns degrade security over time, not just functionality
- Failure points determine how a lock is defeated, not just if it is secure
- Installation and maintenance directly shape micro-mechanical performance
- Security is dynamic, evolving with use, environment, and attack methods

These case studies clearly demonstrate that locksmithing is not just about hardware—it is about understanding the microscopic engineering reality inside every lock. This level of insight enables better installation, more accurate diagnosis, and more effective security recommendations in real-world practice.

## Impact on Locksmithing Practices

Locks are often assumed to function correctly as long as they are fitted into a door, but micro-mechanical understanding reveals that even slight misalignment can distort internal tolerances. Over time, this leads to uneven wear, binding, or premature failure. A locksmith who understands tolerances will ensure correct alignment, minimise stress on components, and preserve the designed mechanical relationships within the lock. This transforms installation from a basic fitting task into a precision-based activity that directly impacts long-term security.

It also has a strong influence on maintenance practices. Locks are not static devices; they degrade through use. Without an understanding of wear patterns, maintenance becomes reactive—only addressing problems after failure occurs. With micro-mechanical insight, locksmiths can recognise early signs of degradation, such as subtle changes in key feel, resistance, or inconsistency. This allows for preventative maintenance, such as lubrication, adjustment, or timely replacement, which preserves both functionality and security.

From a diagnostic and troubleshooting perspective, micro-mechanics is essential. Many lock issues—such as intermittent operation, key sticking, or unexpected opening—cannot be understood by looking at external components alone. They are the result of microscopic changes inside the lock. By understanding how wear alters pin geometry or how tolerances influence binding and movement, a locksmith can accurately diagnose the root cause rather than applying temporary fixes. This elevates the role from technician to problem-solver and mechanical analyst.

Micro-mechanical understanding also plays a critical role in security assessment and advice. Locks are often marketed based on features, materials, or brand reputation, but these factors do not fully determine real-world security. A locksmith who understands tolerances and failure points can evaluate how a lock will respond to attack methods such as picking, bumping, drilling, or snapping—not just when new, but after years of use. This leads to more informed recommendations, where security is assessed based on behaviour over time rather than initial specifications.

Importantly, this knowledge highlights that security is dynamic, not fixed. A lock that is secure when installed may become vulnerable as wear increases tolerances or as certain components degrade. Similarly, reinforcing one part of a lock may shift attack forces to another area. For locksmiths, this means security must be viewed as an evolving condition that requires ongoing assessment, not a one-time installation decision.

In modern locksmithing practice, these insights expand the profession beyond mechanical work into areas such as forensic analysis and risk management. Wear patterns can indicate misuse, poor key control, or attempted manipulation. Failure points can reveal how a lock was attacked or why it failed. This is particularly relevant in insurance, legal, and investigative contexts, where understanding the *cause* of failure is just as important as fixing it.

The influence of micro-mechanics on locksmithing practice is profound. It changes the role of the locksmith from someone who works *on* locks to someone who understands the *engineering system within* the lock. This enables more precise installation, proactive maintenance, accurate diagnosis, and higher-quality security advice. For anyone entering or advancing in the field, this level of understanding is what distinguishes basic competency from true expertise.

## Locking System Operation Fundamentals

Locks used in residential and commercial environments span a wide spectrum of designs, ranging from purely mechanical systems to fully integrated electronic and smart access solutions. At their core, all locks perform the same fundamental function—controlling access—but they achieve this in different ways depending on the level of security required, the environment in which they are used, and the balance between convenience and risk. Traditional mechanical locks rely on physical components such as pins, bolts, and levers to restrict movement, while modern systems increasingly incorporate electronics, sensors, and network connectivity to manage access dynamically and remotely.

### Deadbolts

Deadbolts are widely regarded as the standard for securing external doors in residential settings due to their simplicity and strength. Unlike spring-loaded latch mechanisms, a deadbolt uses a solid metal bolt that extends deep into the door frame, making it highly resistant to forced entry. They are commonly available in single-cylinder configurations, where a key is used externally and a thumbturn internally, or double-cylinder configurations, where a key is required on both sides. While double-cylinder deadbolts provide increased protection against attacks such as glass break-and-reach, they introduce safety concerns in emergency situations where rapid egress is required.

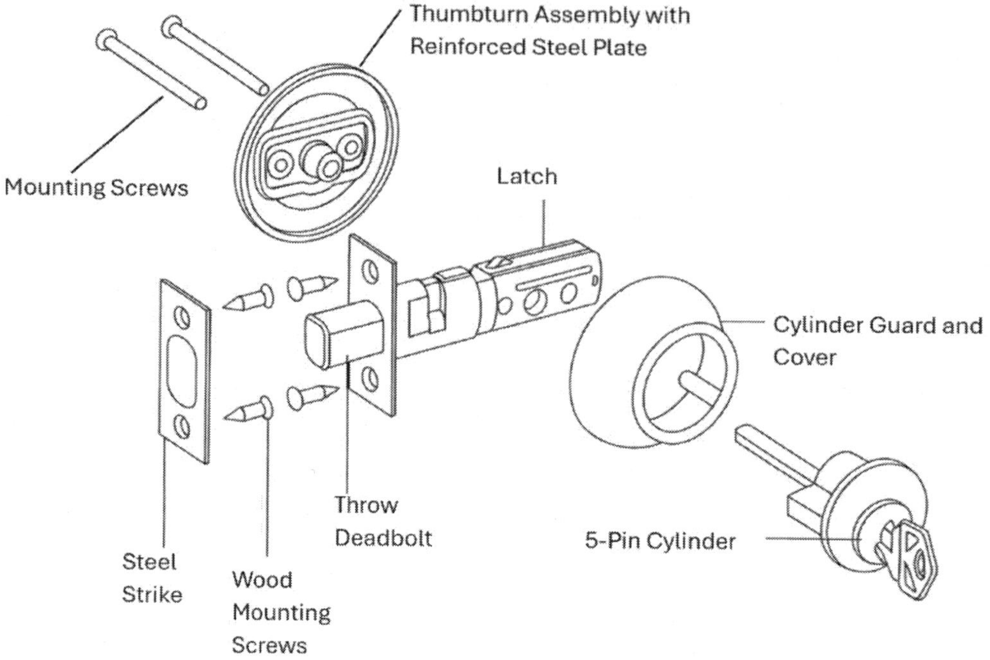


*Figure 12: Example of Typical Deadlock Component Assembly.*

Deadlocks (commonly referred to as deadbolts) are one of the most widely used locking mechanisms in both residential and light commercial environments. Their primary purpose is to provide secure, tamper-resistant locking that cannot be easily bypassed by spring manipulation (unlike latch-based locks).

In residential settings, deadlocks are typically installed on external entry doors, such as front doors, back doors, and garage access doors. They are often used in combination with a knob or lever lock, creating a two-point locking system, one for convenience (latch) and one for security (deadbolt). Double-cylinder deadlocks may also be used on doors with nearby glass panels to prevent reaching inside and unlocking the door.

*Figure 13: Door knob (lockable) with deadbolt fitted above. Brian Katt, CC BY-SA 3.0, via Wikimedia Commons.*

In commercial environments, deadlocks are commonly found on offices, storerooms, and low-to-medium security external doors, particularly where a simple but reliable mechanical solution is required. They may also be integrated into access control systems, working alongside electric strikes or electronic credentials.

Despite their simplicity, deadlocks are subject to a range of mechanical, environmental, and user-related failures.

One of the most common issues is misalignment between the bolt and the strike plate. This can occur due to door sagging, building movement, or poor installation. When misaligned, the bolt may not fully extend or retract, leading to sticking or incomplete locking. Over time, this also accelerates wear on internal components.

Another frequent problem is key or cylinder wear. As keys and pins wear down, the precision required for proper operation is reduced. This can result in difficulty turning the key, intermittent operation, or even the lock accepting incorrect or worn keys.

Environmental factors also play a significant role. Temperature changes, moisture, and dirt ingress can cause internal corrosion or swelling of door materials. Timber doors, for example, may expand in humid conditions, placing pressure on the bolt and making operation difficult.

Mechanical failures can occur internally, including:

- Spring fatigue (in the cylinder)
- Cam or tailpiece wear or breakage
- Bolt mechanism jamming

In more severe cases, deadlocks may be compromised through forced entry attempts, such as snapping (in euro cylinders), drilling, or excessive torque applied to the cylinder or cam.

**Repair Methods**

Repairing a deadlock depends on accurately diagnosing the root cause, which often requires both mechanical understanding and practical experience.

For alignment issues, the most common solution is adjusting the strike plate or repositioning the door hardware. This may involve enlarging the strike plate opening, tightening hinges, or correcting door sag. In some cases, the door frame itself may need reinforcement.

When wear is the issue, locksmiths may rekey or replace the cylinder. Rekeying involves changing the internal pin configuration to suit a new key, restoring proper tolerances without replacing the entire lock. If wear is extensive, full cylinder replacement is often more effective.

For internal mechanical faults, such as a broken cam or jammed bolt, the lock is typically disassembled and either repaired or replaced. In many modern locks, replacement is more cost-effective than attempting intricate repairs.

Environmental problems are addressed through cleaning and lubrication, typically using graphite or specialist lock lubricants (not oil-based products, which attract dirt). Weather seals and door alignment may also be improved to prevent recurrence.

In cases of forced damage or security compromise, the recommended approach is usually complete lock replacement, often upgrading to a higher-security model (e.g., anti-snap or TS007-rated cylinders).

### Mortice Locks

Mortice locks represent a more robust and complex locking solution, typically found in commercial buildings or high-end residential applications. These locks are installed within a cavity, or mortise, cut into the door itself, allowing for a larger and more durable internal mechanism. Mortice locks often combine multiple functions, such as a latch and a deadbolt, within a single unit, and they are designed to withstand high usage and greater mechanical stress. Their strength and reliability make them suitable for environments where both security and durability are critical.

Mortice locks are typically used in timber doors where a higher level of strength, durability, and integration is required. Unlike surface-mounted locks, they are installed בתוך a pocket (mortice) cut into the door, which allows the mechanism to sit securely within the door structure.

In residential settings, mortice locks are commonly found on front doors, back doors, and high-quality internal doors, particularly in older or architecturally traditional buildings. They are often used as sash locks (combining latch and deadbolt) or deadlocks for enhanced security.

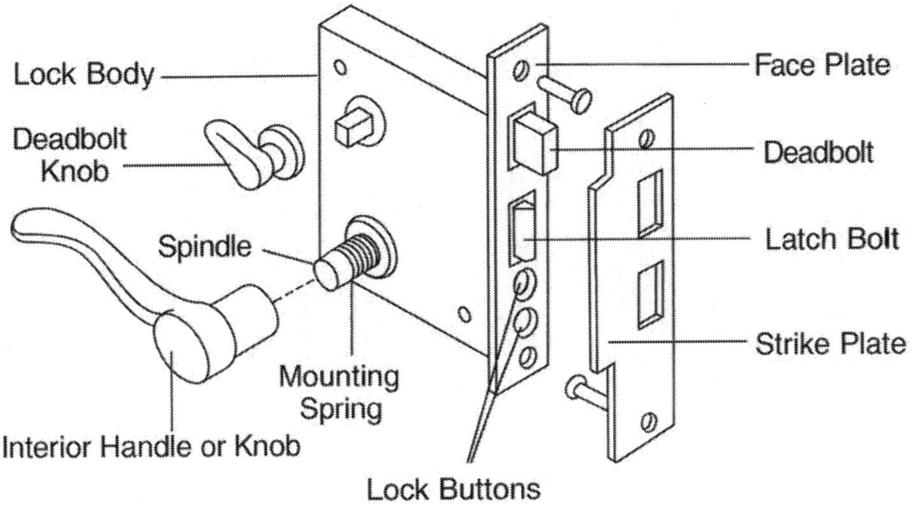


*Figure 14: Example of a Typical Mortice Lock Component Assembly.*

In commercial environments, mortice locks are widely used on office doors, schools, hotels, and institutional buildings, where they provide a robust and long-lasting locking solution. Their internal placement makes them more resistant to tampering and better suited for high-traffic doors.

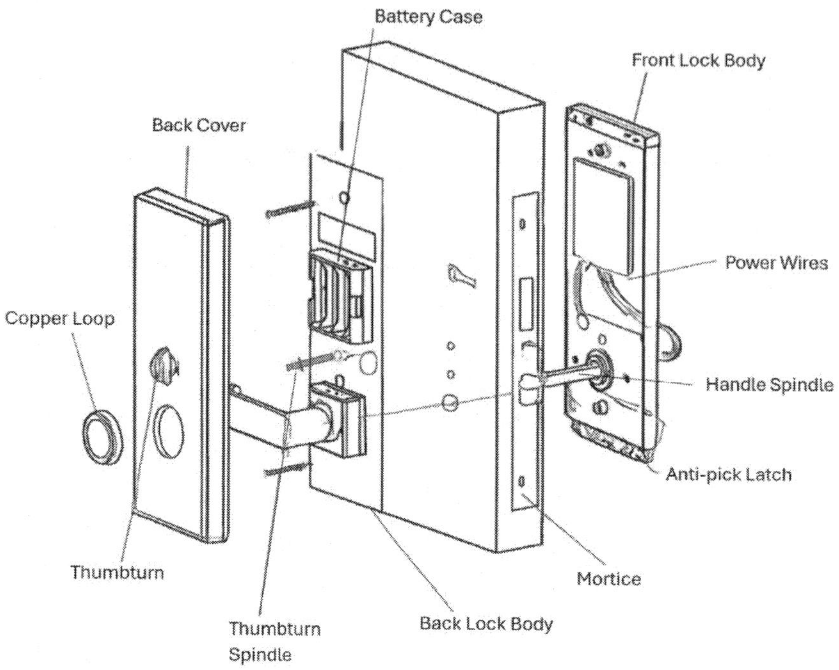


*Figure 15: Diagram of a Mortise Style Latch with Deadbolt Hotel Lock.*

They are also frequently used where door furniture (handles, escutcheons, cylinders) needs to be integrated cleanly, supporting both functionality and aesthetics.

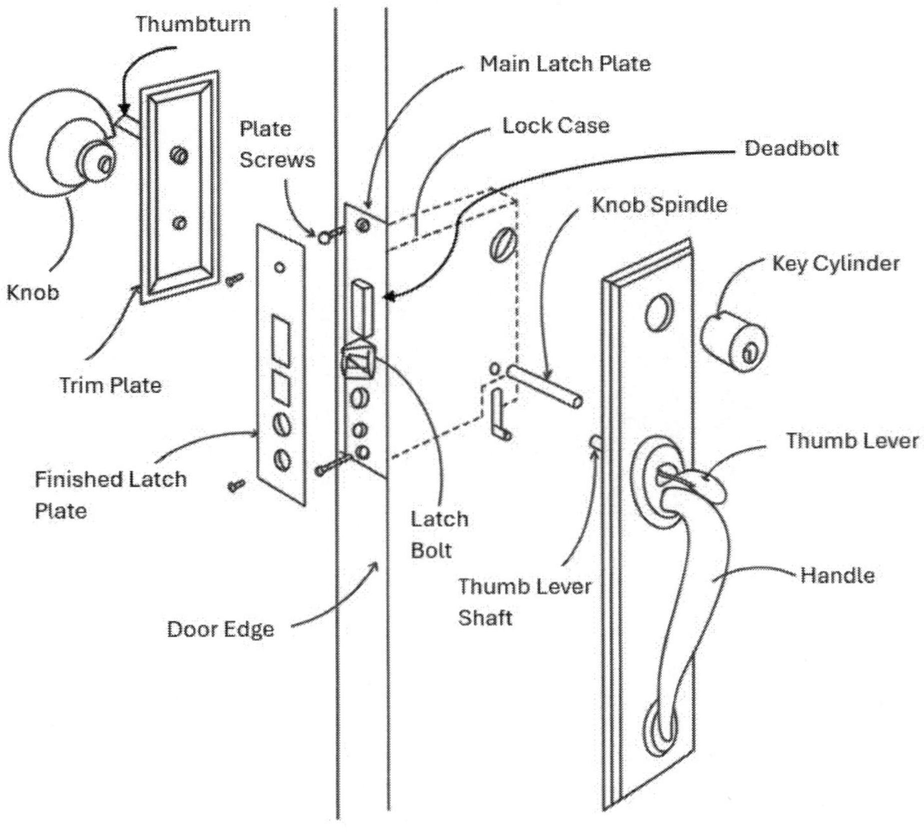


*Figure 16: Mortice Lock Installation.*

Although mortice locks are durable, their internal complexity and embedded design mean that faults can develop over time, often hidden within the door.

One of the most common issues is misalignment between the lock and the strike plate. This may result from door sagging, hinge wear, or building movement. Because mortice locks rely on precise alignment of both latch and bolt, even slight movement can cause sticking, jamming, or incomplete locking.

Internal wear is another major factor. Mortice locks often contain levers (in lever locks), springs, and moving linkages, all of which are subject to fatigue. Over time:

- Springs may weaken or break
- Levers may wear or deform
- Internal linkages may loosen

This can lead to symptoms such as keys not turning smoothly, the latch failing to retract, or the bolt not engaging fully.

Environmental factors also contribute. Timber doors can expand or contract with humidity and temperature, placing pressure on the lock case and internal components. Dirt, dust, and lack of lubrication can further increase friction داخل the mechanism.

In older mortice locks, key wear or poor duplication can reduce accuracy, particularly in lever systems, allowing incorrect keys to partially operate the lock or causing inconsistent behaviour.

Finally, because mortice locks are installed within the door, structural damage to the door itself (e.g., splitting or weakening around the mortice cut-out) can compromise both function and security.

**Repair Methods**

Repairing a mortice lock typically requires removal of the lock case from the door, as the mechanism is enclosed within the mortice cavity. This makes diagnosis and repair more involved than surface-mounted locks.

For alignment issues, locksmiths will usually adjust the strike plate, hinges, or door position. This may involve repositioning the strike, packing hinges, or correcting door sag to ensure smooth operation of both latch and bolt.

If the issue is internal wear or failure, the lock must be disassembled. Minor faults such as broken springs or debris can sometimes be repaired, but in many cases—especially with older or heavily worn locks—full lock case replacement is more practical and reliable.

For cylinder-based mortice locks (e.g., euro cylinder mortice locks), the cylinder itself can be replaced or rekeyed without removing the entire lock case. This is a common solution when keys are lost or compromised.

If the door has been affected (e.g., swelling or structural weakening), repairs may involve:

- Trimming or adjusting the door
- Reinforcing the mortice cavity
- Reinstalling the lock to restore proper fit and alignment

Lubrication and cleaning are also essential maintenance steps, typically using dry lubricants to reduce friction without attracting debris.

### Smart and Keyless Locks

Smart and keyless locks reflect the ongoing shift toward digital access control. These systems eliminate or supplement traditional keys by allowing entry through PIN codes, smartphones, RFID credentials, or biometric identifiers such as fingerprints. In addition to convenience, they offer advanced features such as remote access, audit trails, and integration with broader security systems. However, they also introduce new considerations related to cybersecurity, power supply, and system reliability, requiring users and locksmiths to think beyond purely mechanical security.

## Beyond the Key

Smart and keyless locks introduce a broader range of potential failure points because they combine mechanical, electronic, and networked components. Failures can occur in any of these layers.

*Figure 17: Example of a Smart Lock. Liadmrm, CC BY-SA 4.0, via Wikimedia Commons.*

One of the most common issues is power failure, particularly in battery-operated locks. If batteries are depleted, the lock may become unresponsive, preventing normal entry. While most systems include low-battery warnings or backup power options, failure to maintain power is a frequent real-world issue.

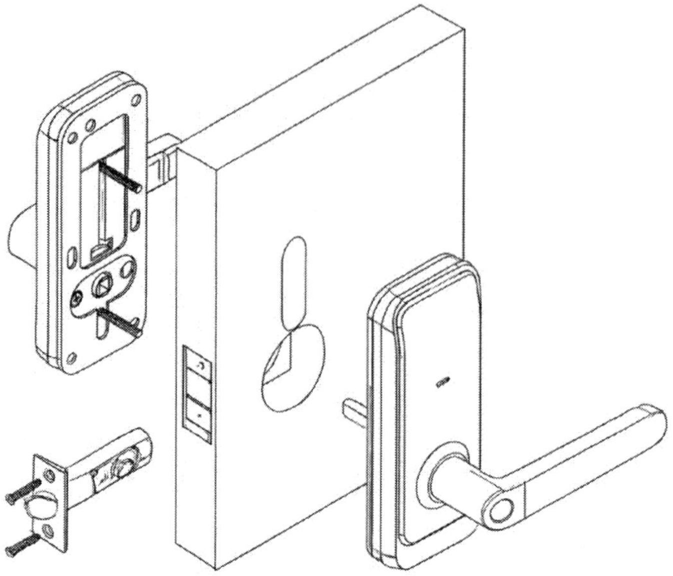

*Figure 18: Typical Smart Lock installation.*

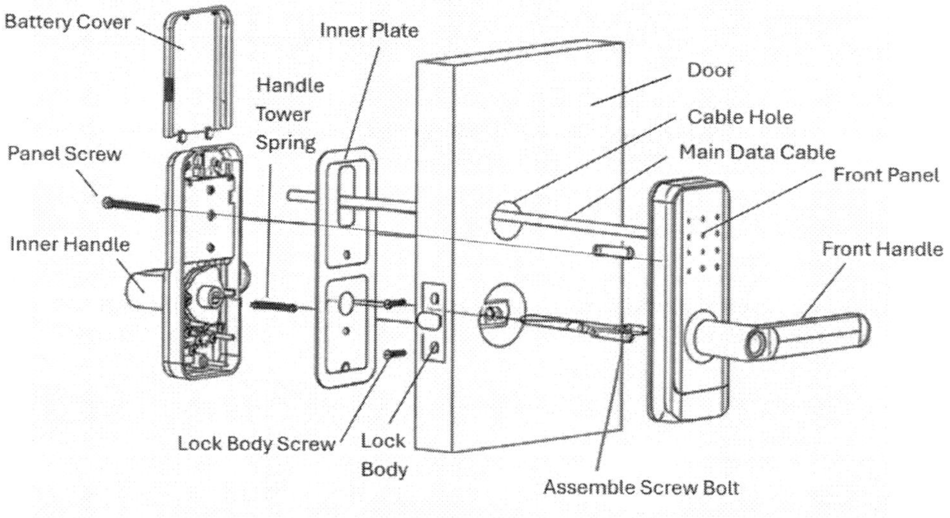


*Figure 19: Example Keyless Lock Installation.*

Connectivity problems are another major concern. Smart locks often rely on Bluetooth, Wi-Fi, or cloud services, and disruptions can prevent communication between the lock and the user's device. This may result in delayed commands, inability to unlock remotely, or system desynchronisation.

Authentication failures can also occur. Keypads may become worn or unresponsive, fingerprint sensors may fail due to dirt, moisture, or poor calibration, and software errors may reject valid credentials. In some cases, users are locked out due to app errors, expired permissions, or incorrect configuration.

Hardware faults can develop within the lock itself. Internal motors, actuators, or circuit boards may fail due to wear, manufacturing defects, or environmental exposure such as moisture or temperature extremes. Unlike purely mechanical locks, these failures can be less predictable and harder to diagnose without specialised tools.

Cybersecurity risks represent a unique category of failure. If systems are poorly configured, vulnerabilities may allow:

- Unauthorized remote access
- Interception of communication signals
- Exploitation of weak passwords or exposed API tokens

Finally, there are human-factor issues, such as users forgetting codes, sharing credentials insecurely, or mismanaging access permissions. In many cases, system failure is not purely technical but results from poor configuration or user behaviour.

**Repair Methods**

Repairing smart and keyless locks requires a hybrid approach, combining traditional locksmithing skills with electronics, software, and networking knowledge.

For power-related issues, the solution is often straightforward: battery replacement or restoring power supply. Many locks also include emergency power terminals or backup keys, which allow temporary access while power is restored.

Connectivity problems are typically addressed through system diagnostics, including checking Wi-Fi or Bluetooth connections, resetting the device, or updating firmware. In some cases, locks must be repaired through software reconfiguration, such as reconnecting to the network or re-pairing with user devices.

Authentication issues may require reprogramming user credentials, recalibrating biometric sensors, or replacing worn keypads. Cleaning sensors and ensuring proper environmental conditions can also restore functionality.

If internal electronic components fail, such as motors or circuit boards, the repair often involves module replacement rather than component-level repair, as many smart locks are designed as sealed units. This is a key difference from traditional locks, where individual parts can often be repaired.

Cybersecurity-related issues require re-securing the system, including:

- Changing passwords or credentials
- Updating firmware
- Restricting user access permissions
- Securing API keys and communication channels

In many cases, locksmiths must also ensure that fail-safe mechanisms are operational. Most smart locks include a mechanical override cylinder, allowing access if the electronic system fails. Ensuring this backup system is functional is critical.

### Multipoint Locking Systems

Multipoint locking systems provide enhanced security by engaging multiple locking points along the length of a door frame, typically controlled by a single handle or key action. Commonly used in uPVC, aluminium, and patio doors, these systems distribute force across several points rather than relying on a single bolt, making forced entry significantly more difficult. They also improve door sealing and alignment, contributing to both security and energy efficiency.

Multipoint locking systems are widely used in modern residential and light commercial applications where enhanced security, weather sealing, and door stability are required. They are most commonly found on uPVC doors, aluminium-framed doors, composite entry doors, and patio or sliding doors. Unlike traditional single-point locks, a multipoint system secures the door at several locations, typically at the top, centre, and bottom, using hooks, rollers, bolts, or a combination of these. All locking points are engaged simultaneously through a central gearbox, usually operated by lifting the handle and turning a key.

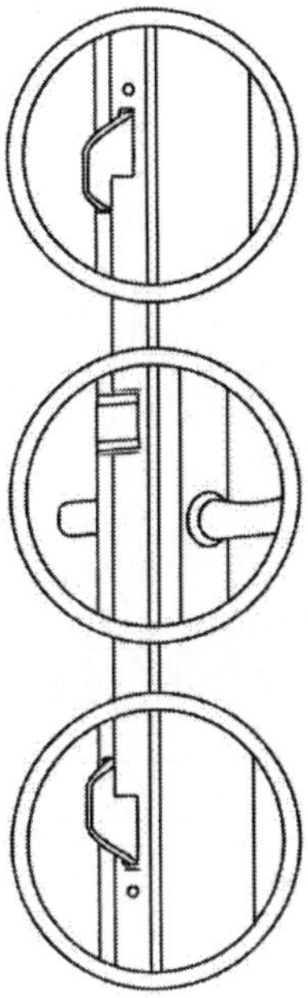

*Figure 20: Example Security Door Multi-point Locking.*

This design provides several advantages. It distributes locking forces along the entire length of the door, improving resistance to forced entry such as prying or kicking. It also pulls the door tightly into the frame, enhancing insulation, reducing draughts, and improving acoustic sealing. For these reasons, multipoint systems are standard in many modern building designs and are often required to meet energy efficiency and security standards.

However, the complexity of multipoint locking systems also introduces a range of potential failure points. One of the most common issues is misalignment between the door and frame. Over time, doors may drop due to hinge wear, building movement, or improper installation. When this occurs, the locking points may no longer align correctly with their keeps (strike plates), making the handle difficult to lift or preventing the lock from engaging fully. This is often experienced by users as stiffness, grinding, or an inability to lock the door.

Another frequent problem lies within the central gearbox, which acts as the control mechanism for all locking points. Repeated use, lack of lubrication, or internal component wear can cause the gearbox to fail. When this happens, the handle may become loose, fail to return to its original position, or no longer engage the locking points. In some cases, the key may turn without actually locking the door, indicating internal failure of the mechanism.

Individual locking components, such as hooks, rollers, or bolts, can also become worn, bent, or jammed. Dirt, corrosion, or debris entering the mechanism may restrict movement, particularly in external doors exposed to weather. Additionally, the euro cylinder (commonly used with multipoint systems) can develop its own issues, including key wear, pin failure, or vulnerability to snapping if not reinforced.

**Repair Methods**

Repairs to multipoint locking systems typically begin with diagnosis of alignment and mechanical condition. In many cases, issues can be resolved by adjusting the door hinges to restore proper alignment between the locking points and keeps. This is often the first and simplest corrective action. If the problem lies in the locking strip or individual components, these may require cleaning, lubrication, or replacement.

When the gearbox fails, it usually cannot be repaired internally and must be replaced as a complete unit. This involves removing the handle, cylinder, and faceplate to access and replace the central mechanism. If the euro cylinder is faulty or insecure, it can be replaced independently, often with an upgraded anti-snap or high-security version.

In more severe cases—such as extensive wear, corrosion, or structural damage—the entire multipoint locking strip may need to be replaced. This requires careful matching of dimensions, backset, and locking point configuration to ensure compatibility with the door.

### Knob and Lever Locks

Knob and lever locks are among the most common locking mechanisms for internal doors and low-security applications. While knob locks are widely used in residential settings, lever handles are increasingly preferred due to their ergonomic design and compliance with accessibility standards. Lever locks require less force and dexterity to operate, making them suitable for environments where ease of use is important, such as public buildings or facilities requiring universal access.

*Figure 21: Typical Knob and Lever Lock. rrafson, CC BY-SA 3.0, via Wikimedia Commons.*

Knob and lever locks are among the most commonly used locking mechanisms, particularly in residential and light commercial environments. They are typically installed on internal doors such as bedrooms, bathrooms, and offices, where the primary function is privacy or basic access control rather than high security. In some cases, they are also used on external doors, although this is generally not recommended unless combined with a secondary locking device such as a deadbolt. Lever locks are especially common in commercial buildings, healthcare facilities, and public spaces because they are easier to operate and meet accessibility requirements, including those outlined in standards for people with disabilities.

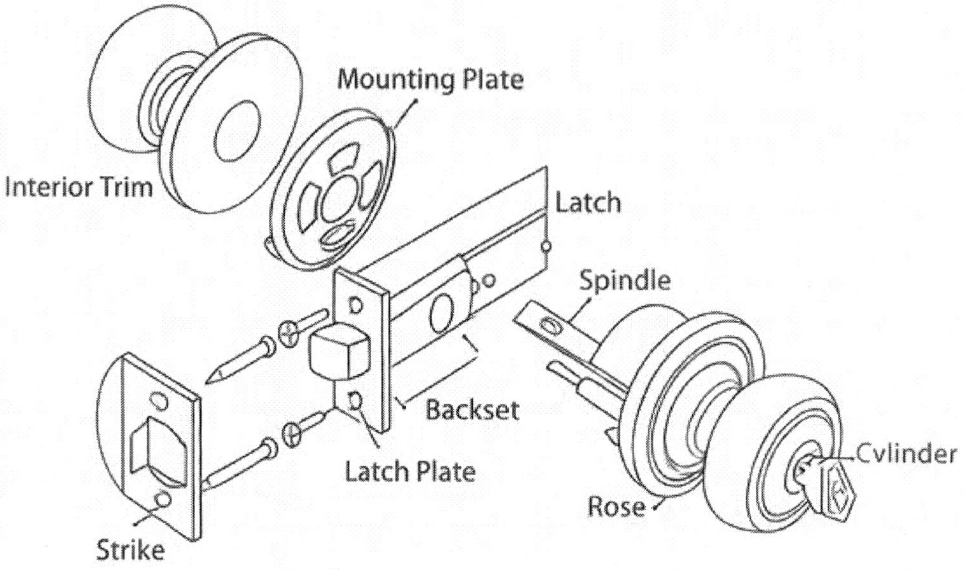


*Figure 22: Typical Knob and Lever Lock arrangement.*

Despite their widespread use, knob and lever locks have several inherent limitations. One of the most common issues is mechanical wear over time. Internal components such as springs, latches, and return mechanisms can weaken or fail, leading to handles that feel loose, do not return properly, or fail to retract the latch when operated. In knob locks, the spindle mechanism can wear or slip, causing the knob to turn without engaging the latch. Lever handles may sag or drop due to worn springs or internal breakage, which can prevent the door from opening or closing correctly.

Another common problem is misalignment between the latch and the strike plate. This can occur due to door sagging, building movement, or poor initial installation. When misalignment occurs, the latch may not fully engage or may bind against the strike plate, making the door difficult to close or lock. Over time, this can accelerate wear on both the lock and the door frame. In external applications, knob and lever locks are also vulnerable to forced entry, as the relatively lightweight construction can be overcome with torque or impact.

Key-related issues can also arise in keyed versions of these locks. Worn or poorly cut keys may fail to operate the cylinder smoothly, while dirt, debris, or lack of lubrication inside the cylinder can cause sticking or jamming. In lower-quality locks, internal tolerances may degrade quickly, increasing the likelihood of malfunction or accidental operation by similar keys.

**Repair Methods**

Repairing knob and lever locks depends on the nature and severity of the fault. Minor issues, such as stiffness or sticking, can often be resolved through cleaning and lubrication of the internal components or cylinder. If the problem is related to misalignment, adjusting the strike plate, tightening hinges, or repositioning the door may restore proper function. Loose handles or knobs can typically be fixed by tightening mounting screws or replacing worn spindles.

In cases where internal components have failed, such as broken springs or worn latch mechanisms, the lock may need to be disassembled and repaired if parts are available. However, due to the relatively low cost of these locks, complete replacement is often more practical and cost-effective than repair. For security upgrades, especially on external doors, locksmiths will typically recommend replacing knob or lever locks with more robust systems, such as deadbolts or multipoint locks, to provide a higher level of protection.

**Rim Cylinders**

Rim cylinders are surface-mounted locking components typically used in conjunction with night latches. Unlike mortise cylinders, which are embedded within the door, rim cylinders are attached to the surface and connect to a locking mechanism on the interior side. They are commonly found on entrance doors, particularly in older or retrofitted systems, and provide a practical solution where internal door modification is limited.

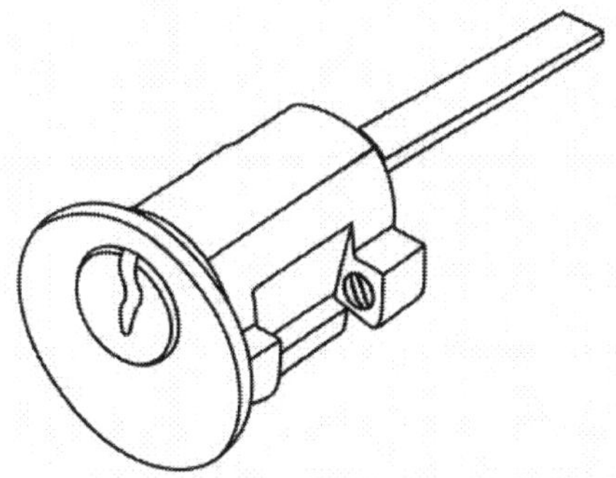

*Figure 23: Night Latch Rim Cylinder.*

In terms of function, a rim cylinder operates by turning a tailpiece (cam) at the rear of the cylinder, which engages the internal locking mechanism. Because the cylinder is externally mounted and directly accessible, it becomes a critical security component—and also a potential vulnerability if not properly specified or installed.

Several issues can arise with rim cylinders over time. One of the most common problems is wear within the cylinder mechanism, particularly in the pins, springs, and plug. This can lead to stiff operation, keys becoming difficult to insert or turn, or the cylinder failing to operate smoothly. Worn keys or poorly cut duplicates can worsen this issue by accelerating internal wear or causing misalignment of the pins.

Another common issue is tailpiece misalignment or incorrect sizing. If the tailpiece is too long, too short, or not correctly aligned with the internal lock body, the lock may fail to engage properly or may feel loose and inconsistent during operation. Over time, repeated stress on a poorly aligned tailpiece can lead to failure of the internal mechanism or damage to the connecting components.

From a security perspective, rim cylinders can be vulnerable to forced entry attacks, particularly if they are not reinforced. Because they are mounted on the surface of the door, they may be exposed to techniques such as snapping, pulling, or drilling. Lower-quality cylinders without hardened inserts or anti-drill protection are especially susceptible. In addition, if the cylinder protrudes too far from the door surface, it becomes easier for an attacker to grip and apply force.

Environmental factors can also affect performance. Exposure to moisture, dust, and temperature changes can lead to corrosion or contamination inside the cylinder, resulting in sticking or jamming. In external applications, this is a common cause of failure if regular maintenance is not carried out.

**Repair Methods**

Repairing rim cylinders depends on the specific fault. Minor operational issues, such as stiffness or sticking, can often be resolved by cleaning the cylinder and applying appropriate lubricant (typically a dry graphite or specialist lock lubricant rather than oil-based products). If the issue is related to key wear, cutting a new key to the correct specification may restore proper function.

If the problem lies with the tailpiece, the cylinder may need to be removed and adjusted or replaced with a correctly sized unit. Ensuring proper alignment between the cylinder and the internal lock body is critical for reliable operation. In cases where the cylinder has been damaged, worn excessively, or compromised by an attempted attack, replacement is usually the most effective solution.

From a security upgrade perspective, locksmiths often recommend replacing standard rim cylinders with high-security versions that include features such as anti-drill pins, reinforced housings, and restricted key systems. Additionally, ensuring that the cylinder does not protrude excessively and is paired with a robust internal locking mechanism significantly improves overall security.

### Padlocks

Padlocks offer a portable and flexible security solution, commonly used for gates, sheds, lockers, and chains. Their design separates the locking mechanism from the object being secured, allowing them to be applied across a wide range of applications. Padlocks vary significantly in security level, from basic designs suitable for low-risk environments to high-security versions constructed from hardened steel with advanced anti-cut, anti-drill, and anti-pick features.

*Figure 24: Solex 99 30 padlock with keys. Trougnouf, CC BY 4.0, via Wikimedia Commons.*

Padlocks are portable locking devices used wherever a fixed lock cannot be permanently installed. They are commonly found on gates, sheds, storage units, toolboxes, lockers, chains, containers, and industrial enclosures. In residential settings, they are often used for securing backyard gates or garden sheds. In commercial and industrial environments, padlocks play a critical role in asset protection and safety systems, including lockout/tagout (LOTO) procedures, where equipment must be isolated to prevent accidental operation during maintenance.

Their flexibility is their greatest strength: a single padlock can be moved between locations, used with chains or hasps, and adapted to a wide variety of security situations. Padlocks range from basic brass models for low-risk applications to hardened steel, high-security variants designed to resist cutting, drilling, and picking.

A padlock is a self-contained locking mechanism that secures an object by trapping a movable metal loop called the shackle. Although designs vary, most padlocks operate using a similar internal principle: a locking mechanism holds the shackle in place until the correct key or code releases it.

In a typical key-operated padlock, the process begins when the key is inserted into the cylinder. Inside the lock body, there are components such as pins, wafers, or discs (depending on the design) that block movement. When the correct key is inserted, it aligns these internal components so that a shear line is created—this allows the internal plug to rotate.

When the plug turns, it activates a locking mechanism—usually a set of locking bars, ball bearings, or spring-loaded latches—that are holding the shackle in place. Once released, the shackle pops open under spring pressure. When the padlock is closed again, the shackle is pushed back into the body, and the locking mechanism automatically re-engages, securing it without needing to turn the key (in most designs).

In simple terms, a padlock works by:

- Blocking movement of the shackle with internal locking components
- Using a key or code to align internal parts
- Releasing the shackle when alignment is correct

Most padlocks, regardless of type, include:

- **Shackle** – the U-shaped bar that secures the object
- **Body** – the outer casing that houses the mechanism
- **Cylinder/Mechanism** – where the key or code operates
- **Locking mechanism** – holds or releases the shackle (e.g., ball bearings, latches)
- **Spring** – pushes the shackle open when unlocked

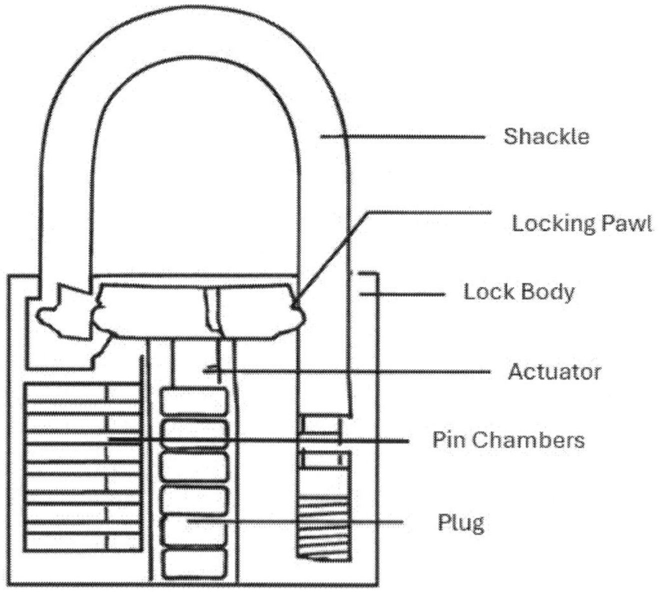


*Figure 25: Padlock general internal design.*

<u>Variations in Padlocks</u>

## Beyond the Key

### 1. Keyed Padlocks (Pin Tumbler, Wafer, Disc)

These are the most common types and use a physical key to align internal components.

- **Pin tumbler padlocks**: Use pins and springs (similar to door locks)
- **Wafer locks**: Simpler and cheaper, often used in low-security applications
- **Disc detainer locks**: Use rotating discs instead of pins (high security, resistant to picking and dirt)
- **Dimple locks**: Use keys with drilled dimples for higher precision

Security level depends on internal complexity and build quality.

### 2. Combination Padlocks

These do not use a key but instead rely on a code.

*Figure 26: A Master Lock brand combination padlock. Thegreenj, CC BY-SA 3.0, via Wikimedia Commons.*

Inside, rotating wheels (called tumblers) must align at the correct positions. When the correct combination is entered:

- The wheels align a gap or notch
- A locking bar drops into place
- The shackle is released

Common in lockers, luggage, and low-to-medium security applications.

### 3. Shrouded and High-Security Padlocks

These are designed to resist physical attacks.

- **Shrouded shackles**: Most of the shackle is covered to prevent cutting
- **Ball-bearing locking**: Uses hardened steel balls instead of spring latches (resists pulling and shimming)
- **Hardened materials**: Resist drilling, sawing, and impact

*Figure 27: Shrouded Padlock.*

Used in industrial, commercial, and high-risk environments.

### 4. Smart / Electronic Padlocks

Modern padlocks replace keys with digital credentials.

They may use:

- Bluetooth or Wi-Fi (smartphone control)
- Keypads (PIN entry)
- Biometrics (fingerprint recognition)

Internally, instead of mechanical alignment:

- An electronic signal activates a motor or solenoid
- The locking mechanism releases the shackle

Often used in smart homes, shared access systems, and logistics.

Smart or electronic padlocks represent a significant evolution from traditional mechanical locking systems. Instead of relying on a physical key to align internal pins, these devices use digital control, electronic authentication, and electromechanical actuation to determine access. In essence, the act of unlocking is no longer based on mechanical alignment, but on verifying a user's identity and then electronically releasing the locking mechanism.

*Figure 28: Smart NFC padlock. Jing jsb, CC BY-SA 4.0, via Wikimedia Commons.*

At a fundamental level, a smart padlock still performs the same core function as a conventional padlock—securing or releasing the shackle. However, the way it decides when to release that shackle is entirely different. Inside the lock, three primary subsystems operate together: authentication, control, and actuation. The authentication system is responsible for identifying the user and may involve technologies such as Bluetooth or Wi-Fi via a smartphone app, a numeric PIN entered through a keypad, biometric verification such as a fingerprint, or RFID/NFC cards and tags. When a user attempts access, the system checks whether the provided credentials are valid.

Once credentials are received, the control system, typically a small embedded microcontroller, processes the request. It verifies the input either locally or through a connected application or cloud service, evaluates permissions such as time-based access, and determines whether access should be granted or denied. This component effectively acts as the "brain" of the lock, making real-time decisions based on programmed rules and user data.

If access is approved, the actuation system is triggered. The microcontroller sends a signal to an actuator, usually a small motor or solenoid, which physically releases the locking mechanism. This may involve retracting a locking bar or disengaging ball bearings that secure the shackle. Once released, the shackle opens, replacing the traditional action of turning a key with an electronically controlled movement.

In practical use, the operation of a smart padlock follows a clear sequence. A user initiates access by opening an app, entering a PIN, or using biometric input. The credentials are transmitted via Bluetooth, Wi-Fi, or direct input. The system then verifies the identity, either locally or through a cloud service, and makes an access decision. If approved, the microcontroller activates the actuator, the shackle is released, and the lock opens. At the same time, the event is recorded, creating a digital log of access within the system's memory or associated application.

Internally, these locks are composed of several integrated components, including a battery as the power source, a microcontroller for processing logic, communication modules such as Bluetooth, Wi-Fi, or NFC, and input interfaces like fingerprint readers or keypads. They also include an actuator for physical movement and a locking mechanism, typically based on ball bearings or latch systems, to secure the shackle.

*Figure 29: Smart padlock (S35) for access control . Chole Wu, CC BY-SA 4.0, via Wikimedia Commons.*

## Beyond the Key

One of the defining advantages of smart padlocks is their functionality. They enable remote access control, allowing users to unlock devices or grant access from anywhere via the internet. They also support time-based permissions, meaning access can be scheduled or limited to specific time windows. Additionally, they generate audit trails, recording who accessed the lock, when it was opened, and how access was granted. This provides a level of accountability that is not possible with traditional keys. The absence of physical keys also eliminates risks associated with duplication or loss, although it introduces new digital vulnerabilities.

Communication in smart padlocks is achieved through various technologies. Bluetooth is commonly used for short-range, direct control via smartphones, while Wi-Fi enables remote access through cloud-based systems. NFC and RFID allow for quick, contact-based unlocking, and some systems integrate with IoT protocols such as MQTT for broader system connectivity. Many devices combine multiple communication methods to ensure reliability and redundancy.

However, this increased functionality also introduces new security considerations. Unlike mechanical locks, which are primarily vulnerable to physical attacks, smart padlocks operate within a cyber-physical environment. Risks include weak authentication methods, unencrypted communication channels, compromised user accounts, and exposed API tokens or firmware vulnerabilities. To mitigate these risks, systems must implement strong encryption protocols such as TLS or AES, enforce secure authentication methods including two-factor authentication or biometrics, and maintain regular firmware updates and access control policies. Ultimately, the security of a smart lock depends on both its physical integrity and its digital protection.

When compared to traditional mechanical padlocks, the differences are substantial. Mechanical locks rely solely on physical keys and local control, with failure typically resulting from wear or damage. In contrast, smart padlocks use digital credentials, allow both local and remote control, and introduce new failure modes such as battery depletion, software errors, or network issues. Key duplication is replaced by digital sharing, shifting the concept of access from possession to permission.

*Figure 30: Passive Electronic Lock and Smart KEY. Chole Wu, CC BY-SA 4.0, via Wikimedia Commons.*

For locksmiths and security professionals, this represents a fundamental shift in practice. The role is no longer limited to cutting keys or installing hardware; it now involves managing access permissions and understanding integrated systems. Professionals must develop knowledge in basic electronics, wireless communication, and cybersecurity. Diagnostic work has also evolved, requiring the ability to identify issues related to battery performance, connectivity, and software configuration, in addition to traditional mechanical faults.

Padlocks are exposed devices, often used outdoors, which makes them particularly vulnerable to both environmental and physical stress.

One of the most common issues is corrosion and contamination. Exposure to rain, dust, salt air, or industrial pollutants can cause internal components—such as springs and pins—to corrode or seize. Over time, this leads to stiff operation, keys not turning, or the shackle failing to release. In colder climates, moisture inside the lock can freeze, temporarily rendering the padlock inoperable.

Another frequent problem is mechanical wear. Repeated use gradually wears down keyways, pins, and shackles. This can result in sloppy key movement, difficulty locking or unlocking, or the lock opening with worn or incorrect keys. In cheaper padlocks, softer materials accelerate this process, reducing both lifespan and security.

Padlocks are also common targets for physical attacks. These include:

- Cutting attacks using bolt cutters or grinders, especially on exposed shackles

- Shimming attacks, where thin tools are inserted to bypass the locking mechanism in lower-quality designs
- Picking or decoding, particularly in simple pin tumbler or wafer-based padlocks
- Pulling or twisting, where force is applied to break internal components

In addition, key-related issues can arise. Lost keys, poorly cut duplicates, or key breakage inside the cylinder can render the padlock unusable. For combination padlocks, users may forget the code or the internal mechanism may drift out of calibration.

**Repair Methods**

The approach to repairing a padlock depends on its value, construction, and the nature of the fault. In many cases, especially with low-cost padlocks, replacement is more economical than repair. However, higher-quality or specialized padlocks can often be serviced.

For minor issues such as stiffness or sticking, cleaning and lubrication is usually the first step. Dirt and debris can be flushed out using appropriate cleaning agents, followed by application of a dry lubricant (such as graphite or PTFE). This restores smoother movement of internal components.

If corrosion is present, the lock may be partially restored through cleaning, but severe corrosion—especially affecting springs or locking mechanisms—typically requires replacement, as internal damage is often irreversible.

In cases of key problems, locksmiths may:

- Extract broken keys using specialised tools
- Cut new keys if the lock code is known or can be decoded
- In some cases, re-pin the lock (for serviceable, higher-end padlocks)

For mechanical failures, such as a jammed shackle or failed locking mechanism, repair is often limited. Some high-security padlocks are designed to be serviceable, allowing internal components to be replaced, but many are sealed units and must be replaced entirely.

When a padlock has been attacked or compromised, it should always be replaced rather than repaired, as its structural integrity and security can no longer be guaranteed.

### Dimple Locks

Dimple locks represent a more advanced variation of the traditional pin tumbler system. Instead of vertical cuts along the edge of the key, dimple keys feature small indentations or “dimples” on a flat surface, which interact with pins arranged in multiple directions. This design allows for more complex keying and often provides higher resistance to picking and unauthorized duplication. Dimple locks are commonly used in high-security applications where key control and advanced protection are required.

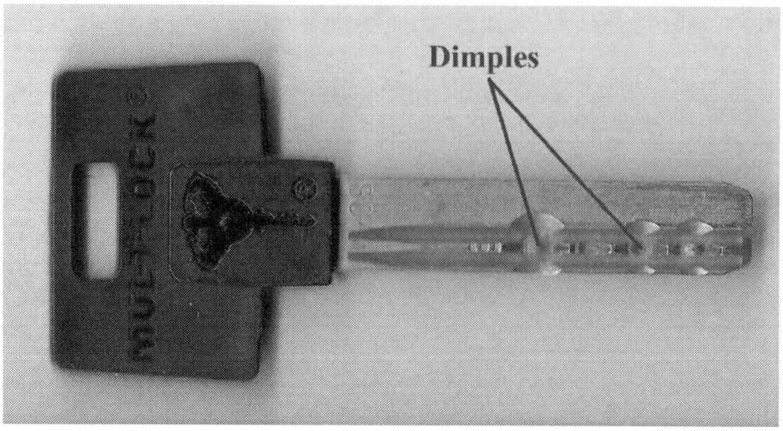


*Figure 31: MUL-T-LOCK Standard Interactive Keyway key has only dimple indentations and no floating pins. Kim Meyrick, CC-BY-SA-3.0, via Wikimedia Commons.*

In residential settings, dimple locks are often installed in high-end homes or apartments where owners want improved resistance to picking and unauthorised key duplication. In commercial environments, they are frequently used in offices, retail premises, and restricted-access areas because they support master key systems and controlled key distribution. They are also common in institutional and industrial settings, such as hospitals, schools, and warehouses, where access needs to be managed across multiple users and zones. In many cases, they are integrated into euro cylinders, mortise locks, or access control systems.

Their popularity stems from a combination of compact design, high key variation (many possible combinations), and the ability to incorporate advanced features such as side pins, sliders, or interactive elements that increase security.

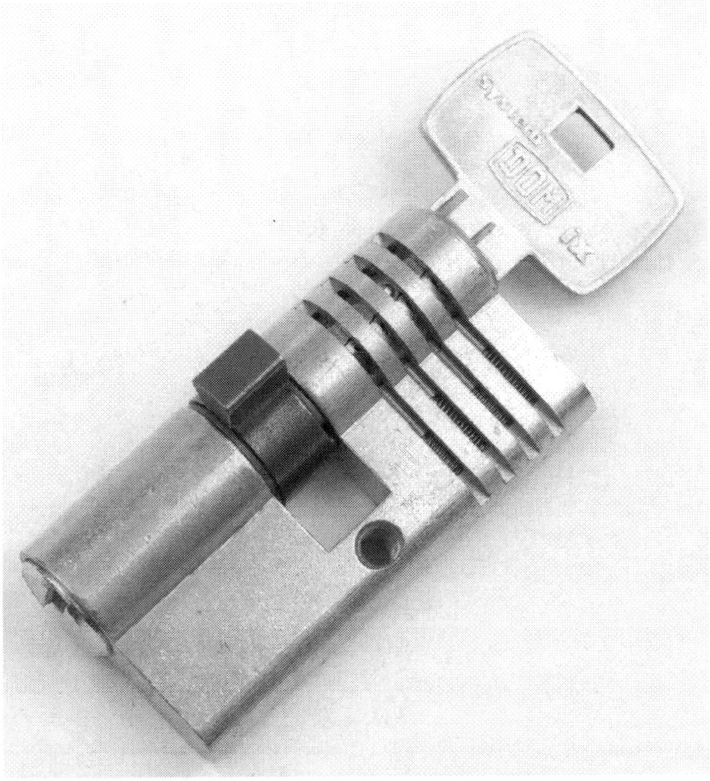

*Figure 32: Dimple lock and key - DOM ix cylinder refers to a high-security lock cylinder manufactured by DOM Sicherheitstechnik (Germany). Dennis van Zuijlekom, CC BY-SA 2.0, via Flickr.*

**Dimple Lock Example**

A DOM ix cylinder refers to a high-security lock cylinder manufactured by DOM Sicherheitstechnik in Germany, using the "ix" key system—a sophisticated dimple key design combined with a multi-level pin mechanism. The term "key turned" simply describes the operational state of the lock in which the correct key has been inserted and rotated, allowing the cylinder to actuate and either lock or unlock the door.

What distinguishes a DOM ix cylinder from standard pin tumbler locks is its advanced internal design and security features. Instead of a traditional vertical keyway, it uses a horizontal dimple key with precisely positioned indentations. Internally, the lock contains multiple rows of pins, including top, bottom, and sometimes side pins, as well as secondary locking elements such as sidebars or check pins. These components are manufactured to very tight tolerances and are often protected by patent-restricted key profiles, limiting unauthorised duplication. As a result, DOM ix cylinders offer strong resistance to common attack methods such as picking, bumping, drilling, and unauthorised key copying.

When the correct key is inserted and turned, a series of complex micro-mechanical actions occur simultaneously. The dimples on the key lift multiple pins—often in different directions—to precise heights so that they align exactly at the shear line. In addition to the primary pins, secondary elements such as side pins or control pins must also align correctly. This multi-axis alignment requirement makes the mechanism significantly more complex than a standard lock.

At the same time, secondary security elements come into play. In many DOM ix systems, sidebars or similar control components must retract into the body of the lock before rotation can occur. Special check pins may also verify the correct key profile. Even if the main pins are correctly aligned, the lock will not operate unless these additional mechanisms are also satisfied, adding another layer of security.

Once all internal components are correctly aligned, the plug is free to rotate within the housing. As the user turns the key, the plug begins to move, transferring motion to the cam or tailpiece at the rear of the cylinder. This action ultimately retracts a bolt in a mortise lock or releases a latch mechanism, completing the unlocking process.

In locksmithing practice, the phrase "key turned" carries important diagnostic meaning. A smoothly turning key generally indicates that the internal components are correctly aligned, tolerances are within acceptable limits, and there are no significant obstructions or wear issues. Conversely, if a key will not turn, only turns partially, or feels loose, it can indicate problems such as misalignment, worn pins, debris within the mechanism, or excessive internal tolerances.

For locksmiths, observing a DOM ix cylinder successfully turning confirms that a complex, multi-layered system is functioning as intended. It demonstrates correct key coding, proper alignment of pins and sidebars, and the overall integrity of the internal mechanism. However, due to the high precision of these systems, even minor wear, contamination, or poor-quality key duplication can prevent correct operation. As a result, diagnosing faults in these cylinders often requires specialised knowledge and tools, and repairs may involve re-pinning, replacing internal components, or in some cases, replacing the entire cylinder.

Despite their advanced design, dimple locks are still mechanical systems and are subject to wear, misuse, and attack. Problems generally fall into mechanical, operational, and security-related categories.

From a mechanical perspective, internal wear is one of the most common issues. Over time, the pins that align with the dimples on the key can become rounded or worn, reducing precision at the shear line. The key itself may also wear, particularly if it is frequently used or poorly duplicated, which can lead to inconsistent operation or the lock accepting incorrect keys. Dirt, dust, or debris can accumulate in the horizontal keyway, especially in exposed environments, causing pins to stick or jam.

Operational issues often arise from poor key management or incorrect duplication. Because dimple keys require high precision, low-quality copies can result in partial alignment of pins, making the lock feel stiff or unreliable. In some cases, users may force the key, leading to internal damage or misalignment of components. Misinstallation of the cylinder can also create binding between the plug and housing, accelerating wear and reducing performance.

From a security standpoint, while dimple locks are more resistant to picking than standard pin tumbler locks, they are not immune. Skilled attackers using specialised tools can manipulate dimple pins, particularly if the lock has worn over time and provides clearer feedback. Lower-quality dimple locks without additional security features (such as sidebars or security pins) may still be vulnerable to picking, bumping (in some designs), or drilling attacks.

**Repair Methods**

Repairing a dimple lock requires both mechanical skill and an understanding of precision components. The approach depends on the nature of the fault.

For minor issues such as stiffness or inconsistent operation, cleaning and lubrication are often sufficient. The lock cylinder is flushed with an appropriate cleaner to remove debris, followed by application of a suitable lubricant (typically graphite or a dry PTFE-based product) to restore smooth pin movement. Care must be taken to avoid oil-based lubricants that can attract dirt.

If the issue is related to worn or damaged keys, the first step is to cut a new key from the original code or a verified template rather than duplicating a worn key. This restores correct pin alignment and often resolves operational problems without further intervention.

In cases of internal wear or damage, the cylinder may need to be disassembled. A locksmith can inspect the pins, springs, and chambers for signs of wear, deformation, or contamination. Worn pins and springs can be replaced, and the lock can be re-pinned to restore proper tolerances. In higher-security systems, this process must be done carefully to maintain key control and system integrity.

If the lock has been compromised by an attempted attack, such as drilling or forced entry, repair may not be viable. In these situations, the cylinder is typically replaced entirely. For security upgrades, locksmiths may recommend higher-grade dimple systems with additional features such as sidebars, anti-drill inserts, and patented key profiles.

### Gun Safe Locks

Gun safe locks are specialised locking systems designed to secure firearms against theft, unauthorised access, and accidental use. Unlike standard door locks, they are engineered with a strong emphasis on tamper resistance, reliability, and compliance with legal and safety requirements. In many jurisdictions, including Australia, gun safes must meet specific standards, meaning the lock is not just a convenience feature, it is a critical safety component.

*Figure 33: Example Gun Safe Locking Mechanisms.*

Mechanical Dial Combination Locks

Mechanical dial locks are one of the most traditional and trusted systems used on gun safes. They operate using a rotating dial connected to a spindle and a set of internal wheels (known as a wheel pack). Each wheel corresponds to a number in the combination. When the correct sequence is dialled, all wheels align at a common point called the "gate," allowing a fence to drop and the lock bolt to retract.

These locks are highly valued for their durability and independence from power sources. Because they are purely mechanical, they are not affected by battery failure or electronic faults. However, they require precise operation and can be slower to open, especially under stress. Over time, internal wear or incorrect dialing technique can cause alignment issues, requiring adjustment or servicing.

Electronic Keypad Locks

Electronic keypad locks are now very common on modern gun safes. Instead of a dial, users enter a PIN code on a keypad. Internally, a circuit board verifies the code and activates a solenoid or motor to release the locking bolt.

These locks offer fast access, user-friendly operation, and the ability to change codes easily. Many models include features such as multiple user codes, time delays, and audit logs. However, they depend on batteries and electronic components, which introduces new failure modes such as flat batteries, keypad wear, or circuit faults. If not properly maintained, these locks may fail unpredictably, making regular testing and battery replacement essential.

Biometric (Fingerprint) Locks

Biometric locks use fingerprint recognition to grant access. A sensor scans the user's fingerprint and compares it to stored templates. If a match is found, the system releases the locking mechanism.

These locks are designed for rapid access, particularly in situations where time is critical. They eliminate the need to remember combinations or codes. However, their reliability can be affected by factors such as dirty sensors, worn fingerprints, moisture, or low battery power. From a security perspective, lower-quality biometric systems may be more vulnerable to spoofing or failure to recognise authorised users.

Key Override Systems

Many electronic and biometric gun safe locks include a key override system as a backup. This is typically a concealed mechanical cylinder that allows access if the electronic system fails.

While this adds redundancy, it can also introduce a potential vulnerability. If the override cylinder is of low quality or poorly protected, it may become the weakest point in the system. High-quality safes mitigate this risk by using restricted keyways or shielding the cylinder from attack.

Gun safe locks, like all systems, are subject to failure. Mechanical locks may suffer from wear in the wheel pack, misalignment, or jammed components. Electronic locks can fail due to battery depletion, damaged keypads, wiring faults, or software errors. Biometric systems may experience sensor degradation or recognition issues.

Environmental factors also play a role. Dust, humidity, corrosion, and temperature fluctuations can all affect performance. In some cases, user-related issues—such as incorrect code entry, forgotten combinations, or poor maintenance—are the primary cause of lock failure.

**Repair Methods**

Repair methods depend on the type of lock. Mechanical dial locks may require disassembly, cleaning, lubrication, and re-calibration of the wheel pack. Electronic locks often involve battery replacement, keypad or circuit board replacement, or reprogramming. Biometric locks may need sensor cleaning, recalibration, or full module replacement.

In high-security or locked-out situations, locksmiths may need to use specialised techniques such as safe manipulation (for mechanical locks) or controlled drilling to access and replace the lock without compromising the integrity of the safe more than necessary. This work requires advanced training due to the risk of damaging internal components or voiding compliance.

Gun safe locks represent a balance between security and accessibility. They must prevent unauthorised access while still allowing the authorised user to gain entry quickly when required. For locksmiths and security professionals, this means understanding not only how each lock type works, but also how it fails.

Choosing the right lock involves considering factors such as reliability, speed of access, environmental conditions, and regulatory requirements. Increasingly, professionals must also understand both mechanical and electronic systems, as modern gun safes often combine elements of both.

**Other Features**

In addition to lock types, several common features influence how locks function and how they are used. Double-cylinder mechanisms, for example, enhance security by requiring a key on both sides but must be carefully considered in relation to fire safety regulations. Thumbturns provide quick and convenient internal operation, improving usability in residential settings. Electric strikes, commonly used in commercial access control systems, replace traditional strike plates and allow doors to be unlocked electronically, often in conjunction with keypads, card readers, or intercom systems.

A double-cylinder mechanism is a type of lock, most commonly used in deadbolts, where both sides of the door require a key to operate. Unlike single-cylinder locks, which use a key on the outside and a thumbturn on the inside, double-cylinder locks eliminate the internal thumbturn and replace it with a second key-operated cylinder. While the mechanism still relies on the traditional pin tumbler principle, the addition of a key interface on both sides connected to a shared internal cam or tailpiece fundamentally changes how access and control are managed.

Inside a double-cylinder lock, there are two keyways, one on each side of the door, both connected to a common internal mechanism. When a key is inserted into either side, it interacts with a set of pins that must align precisely at the shear line. When the correct key is turned, it lifts the key pins and driver pins to their correct positions, clearing the shear line and allowing the plug to rotate. This rotation drives a cam or tailpiece, which in turn retracts or extends the bolt, typically a deadbolt. Because both sides are mechanically linked, turning the key from either side will operate the same locking bolt. In some designs, one side may override or disengage the other depending on configuration, but critically, the lock cannot be operated from either side without a key.

The internal configuration of a double-cylinder system includes two independent plugs, each with its own set of pin stacks aligned by their respective keys. These plugs are connected by a shared cam or tailpiece that controls the central bolt mechanism. In higher-security versions, each cylinder may incorporate additional protective features such as anti-pick pins, hardened anti-drill inserts, and reinforced cam systems. This arrangement also allows for flexible keying options, meaning both sides can be operated by the same key or by different keys, depending on security requirements.

Double-cylinder locks are typically used in situations where internal access must be tightly controlled, even if an intruder gains entry through another means. A common example is doors with nearby glass panels, where a single-cylinder lock could be defeated by breaking the glass and reaching inside to turn the thumbturn. They are also used in higher-risk residential environments, commercial premises requiring controlled internal access, and storage or restricted areas. The primary advantage of this design is that it prevents unauthorised unlocking from the inside without a key.

However, these locks also introduce several risks and potential failure points. The most significant concern is safety during emergencies. Because a key is required on both sides, occupants may be delayed in exiting during a fire or other emergency if a key is not immediately available. For this reason, many building codes restrict or regulate their use. Key dependency is another issue, as lost or misplaced keys can result in occupants being locked in or out, and leaving keys in the lock can create additional security vulnerabilities.

Mechanical wear is also a factor, as both cylinders are subject to regular use. Over time, this can lead to pin wear, misalignment, stiffness, and difficulty turning the key. Installation issues can further exacerbate these problems. If the door or cylinder is not properly aligned, the force applied during key rotation may not transfer efficiently to the cam, resulting in binding, partial operation, or accelerated wear.

Although double-cylinder locks improve resistance to certain attack methods, such as reach-through attacks, they are still vulnerable to conventional techniques including picking, drilling, and, in some designs, snapping. Ultimately, their effectiveness depends on the quality of the cylinder and the inclusion of appropriate security features.

Overall, the diversity of lock types and features reflects the need to tailor security solutions to specific environments and risks. For locksmiths and security professionals, understanding these variations is essential for selecting appropriate hardware, ensuring compliance with safety standards, and delivering solutions that balance security, functionality, and user needs.

## Impact on Locksmithing Practices

Understanding the wide range of lock types, from traditional mechanical systems to modern smart and integrated access solutions, is fundamentally important for anyone entering locksmithing because it defines the scope of the profession itself. Locksmithing is no longer limited to working with keys and cylinders; it now involves selecting, installing, maintaining, and advising on a diverse ecosystem of security devices. Each lock type represents a different engineering approach to solving the same problem, controlling access, and each introduces unique behaviours, risks, and failure modes that a competent locksmith must understand.

From a practical perspective, this knowledge directly influences how locksmiths assess security in real-world environments. A residential front door, a commercial office, and an industrial facility all require different solutions, even though they may appear similar at a glance. For example, a deadbolt may be perfectly adequate for a residential setting, while a commercial door may require a mortice lock integrated with an access control system. This reinforces a key professional principle: locksmithing is not about installing hardware in isolation, but about assessing risk, usage patterns, and environmental conditions to determine the most appropriate solution.

The diversity of lock types also shapes diagnostic and repair practices. Mechanical locks such as deadbolts, mortice locks, and padlocks require an understanding of internal components like pins, levers, springs, and cams, as well as how wear, alignment, and environmental factors affect their performance. In contrast, smart and keyless locks introduce entirely new diagnostic layers, including power supply issues, connectivity problems, software faults, and cybersecurity vulnerabilities. As a result, modern locksmiths must be capable of troubleshooting both mechanical failures and digital system issues, often within the same device.

Another critical influence on locksmithing practice is the need to balance security with safety and usability. The example of double-cylinder deadbolts highlights this clearly. While they provide increased protection against certain attacks, they can create serious risks in emergency situations. This means locksmiths must understand not only how locks function, but also how they interact with building codes, fire safety regulations, and human behaviour. Advising a client therefore becomes a complex task that involves evaluating trade-offs between security, convenience, and legal compliance.

The evolution toward smart and networked locking systems has significantly expanded the skill set required in the profession. Locksmiths must now understand basic electronics, wireless communication, access control systems, and cybersecurity principles. Tasks such as programming credentials, managing user permissions, updating firmware, and securing communication channels are becoming as important as traditional skills like key cutting and cylinder replacement. This shift transforms the locksmith's role from a purely mechanical tradesperson to a hybrid security technician.

In addition, the wide variation in lock design influences how locksmiths approach security planning and system integration. High-security environments often rely on layered systems that combine mechanical locks, electronic access control, and procedural controls. Understanding how different lock types interact, such as deadbolts working alongside electric strikes or smart locks integrated into broader IoT systems, enables locksmiths to design more effective and resilient security solutions.

Finally, this knowledge reinforces the importance of lifecycle thinking in locksmithing practice. Locks are not static devices; they wear, degrade, and become vulnerable over time. Different lock types fail in different ways, and understanding these failure patterns allows locksmiths to provide better maintenance advice, anticipate issues before they occur, and recommend timely upgrades. This moves the profession beyond reactive repair toward proactive risk management.

## Advanced Pin Dynamics and Security Pins

Advanced pin dynamics refers to the microscopic behaviour of pins and springs inside a pin-tumbler lock during normal key operation or when manipulation is attempted. Rather than being simple mechanical components, pins respond to very small forces, friction changes, and alignment shifts within the lock.

Security pins are specifically engineered to exploit these behaviours by introducing misleading feedback, false stopping points, and unstable pin positions, making controlled manipulation significantly more difficult.

In a standard pin-tumbler lock, each pin stack consists of a key pin at the bottom, a driver pin above it, and a spring applying downward pressure. When the correct key is inserted, the interface between the key pin and driver pin aligns precisely at the shear line, the boundary between the plug and the housing, allowing the plug to rotate. However, when torque is applied, either by a key or another tool, a series of subtle physical effects begin to influence how the pins behave.

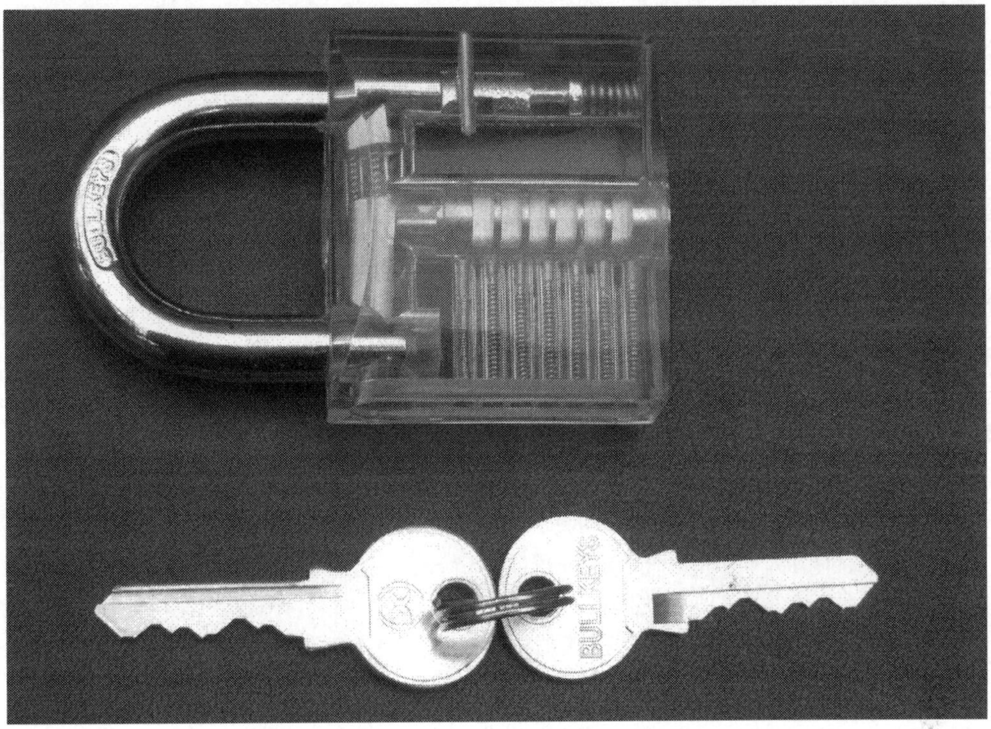


*Figure 34: Clear padlock and keys showing pins and springs. zaphad1, CC BY 2.0, via Flickr.*

One of the most important of these effects is binding order. Because no lock is perfectly manufactured, small tolerances cause the plug to sit slightly off-centre within the housing. When rotational force is applied, this misalignment causes one pin chamber to bind before the others. As a result, pins tend to set one at a time in a predictable sequence. In lower-quality locks with larger tolerances, this binding may feel inconsistent and vague. In contrast, higher-quality locks with tighter tolerances often produce sharper and more repeatable feedback, which can sometimes make interpretation easier.

Friction also plays a critical role in pin behaviour. Pins do not move smoothly; instead, they experience static friction, known as stiction, which resists initial movement, followed by lower kinetic friction once movement begins. This leads to a stick–slip motion, where pins move in small, incremental jumps. These micro-movements are often perceived as clicks or subtle shifts, forming the basis of tactile feedback. Security pins are designed to manipulate these friction effects, making feedback less reliable and more difficult to interpret.

Spring behaviour further influences how pins respond. Springs not only push pins into position but also affect resistance and responsiveness. Stronger springs increase the force required to move pins and can reduce accidental oversetting, but they may also create sharper feedback. Additionally, pins can exhibit slight bounce when released, particularly in low-friction environments or under higher torque. This behaviour is relevant not only to dynamic attacks but also to the way security pins create false feedback conditions.

As torque is applied and the plug rotates slightly, another important effect emerges: false alignment. Pins may momentarily appear to align at the shear line even when they are not correctly positioned. Security pins are specifically designed to take advantage of this effect by catching on the edge of the plug, creating convincing but misleading feedback that suggests progress has been made.

Security pins function by deliberately introducing complexity into these interactions. Their purpose is not to make a lock impossible to manipulate, but to disrupt interpretation by creating false sets, counter-rotation, spongy feedback, and overset traps. A false set occurs when the plug rotates slightly, giving the impression that the lock is close to opening, while in reality, multiple pins may still be incorrectly positioned.

Spool pins are one of the most common types of security pins. Shaped like a dumbbell, with a narrow centre and wider ends, they create a distinctive false set when the narrow section aligns at the shear line. As further force is applied, the wider section catches, forcing the plug to rotate slightly in the opposite direction—a phenomenon known as counter-rotation. This requires careful control of torque, as excessive force can prevent the pin from moving, while insufficient force can allow previously set pins to drop.

Serrated pins take a different approach by incorporating multiple grooves along their length. As the pin is lifted, each groove can catch at the shear line, producing a series of small clicks that simulate progress. This creates multiple false feedback points and increases the likelihood of oversetting, as it becomes difficult to distinguish between a true set and a temporary catch.

Mushroom pins operate in a similar manner to spool pins but have a more rounded profile. This results in smoother, less obvious feedback, making them more difficult to identify by feel. Tapered pins, which gradually narrow along their length, alter the way pins bind and reduce the clarity of contact at the shear line, contributing to ambiguous feedback. More advanced designs, such as T-pins or hybrid "Christmas tree" pins, combine multiple features to create highly complex interactions, producing deep false sets, continuous feedback, and a strong requirement for precise control.

These pin designs rely on the interaction between false sets and counter-rotation as a deliberate security strategy. The lock is made to feel as though it is nearly open, encouraging incorrect assumptions while maintaining internal misalignment. Most security features are incorporated into driver pins, although some systems may also use modified key pins, specialised springs, or restrictive keyways to increase complexity.

The behaviour of security pins is also heavily influenced by tolerances and wear. In tightly manufactured locks, security pins tend to produce clean and predictable feedback, including pronounced counter-rotation. In contrast, in worn or loosely manufactured locks, feedback can become inconsistent or "mushy," sometimes making manipulation easier due to increased internal play. Wear and lubrication further modify these effects. Smooth, polished surfaces reduce friction and can make pin movement more fluid, while contamination or corrosion can create erratic binding and even prevent normal operation.

For locksmiths, understanding advanced pin dynamics has direct practical importance. It influences the selection and specification of locks, as choosing a cylinder involves more than simply selecting the number of pins. Consideration must be given to pin types, tolerances, key control, and the intended operating environment. In diagnostic work, recognising how pins behave can help identify issues such as worn keys, degraded pin stacks, debris within the mechanism, or weakened springs.

When rekeying or rebuilding cylinders, pin selection becomes a balance between usability and security. Aggressive security pins may increase resistance to manipulation but can also make locks more difficult for legitimate users to operate. Long-term wear and maintenance requirements must also be considered. At a higher level, observing pin and plug behaviour can provide insight into whether a fault is due to normal wear, poor key duplication, misalignment, or other mechanical issues.

In practical field situations, certain patterns can provide useful indicators. A plug that rotates deeply but does not open may indicate a false set caused by spool or mushroom pins. A series of small, repeated clicks with little progress may suggest serrated pins. A lock that feels very stiff followed by a sudden release of movement often reflects strong binding combined with security pin interaction.

In pin-tumbler locks, security pins such as spool (mushroom) pins and serrated pins are specially designed driver pins that make picking more difficult by manipulating the way the lock "feels" under tension. They don't stop picking completely, but they introduce misleading feedback, false positions, and instability that require greater skill to overcome.

**Spool (Mushroom) Pins:** Spool pins (sometimes called mushroom pins depending on their exact shape) have a narrowed middle section with wider ends, resembling a dumbbell or spool.

When tension is applied during picking, the narrow centre section can align at the shear line before the rest of the pin is correctly set. This allows the plug to rotate slightly, creating what is known as a false set—it feels like the lock is opening, but it isn't fully unlocked.

As the picker continues lifting the pin, the wider section of the spool catches on the edge of the plug. To set the pin correctly, the plug must rotate slightly backward, a movement called counter-rotation. This behaviour forces the operator to carefully control tension.

From a practical locksmithing perspective, spool pins:

- Create deep, convincing false sets
- Require precise tension control to overcome
- Provide strong tactile feedback (counter-rotation)
- Are commonly used in mid- to high-security cylinders

Mushroom pins work in a similar way but tend to have a smoother, rounded profile. This makes the feedback more subtle and sometimes harder to interpret than a standard spool pin.

**Serrated Pins:** Serrated pins have a series of small grooves or notches cut along their length. Instead of a single clean edge, they present multiple "steps" that can catch at the shear line.

When tension is applied and the pin is lifted, each groove can momentarily align with the shear line, producing a series of small clicks. This creates the illusion that the pin is being set correctly multiple times, even when it is not.

In practice, serrated pins:

- Produce multiple false sets or "micro-sets"
- Generate frequent clicking feedback
- Increase the likelihood of oversetting (lifting the pin too far)
- Make it difficult to distinguish real progress from false feedback

Unlike spool pins, serrated pins don't usually create strong counter-rotation. Instead, they rely on confusion and repetition, forcing the operator to carefully evaluate each movement.

Understanding security pins such as spool, mushroom, and serrated pins is essential in locksmithing because they directly influence both the security performance of a lock and how it behaves during servicing. These components are not just design features; they play a practical role in how locks are selected, maintained, and evaluated in real-world applications.

When selecting and specifying locks, locksmiths must consider the type and configuration of security pins used within a cylinder. Higher-security locks often incorporate combinations of spool, serrated, and other advanced pin designs to increase resistance to manipulation. This means that choosing a lock is not simply about the number of pins, but about the quality, geometry, and interaction of those pins within the system.

From a diagnostic perspective, security pins can also influence how a lock feels during normal operation. Irregular key movement, stiffness, or intermittent operation may not always indicate a major fault but can instead be related to worn, damaged, or contaminated security pins. Recognising these subtle behaviours allows locksmiths to identify underlying issues more accurately and avoid unnecessary replacement of otherwise serviceable components.

During rekeying and rebuilding, the selection of pin types becomes a balance between security and usability. While aggressive security pins can significantly increase resistance to picking, they may also make the lock feel less smooth for everyday users. A skilled locksmith must therefore choose pin configurations that provide adequate security without compromising reliable and user-friendly operation.

Finally, the presence of security pins is an important factor in security assessment. Locks that incorporate these features generally offer greater resistance to casual or opportunistic picking attempts. For locksmiths, this knowledge supports informed recommendations to clients, helping them select locking systems that align with their specific risk level and operational needs.

Overall, advanced pin dynamics reveal that lock behaviour is governed not just by visible components, but by complex interactions at a microscopic level. Understanding these interactions allows locksmiths to move beyond basic operation and develop deeper expertise in diagnosis, maintenance, and security assessment.

## Rekeying and Rebuilding Cylinders

Rekeying focuses on changing access by altering key compatibility, while rebuilding focuses on restoring or improving the internal condition of the lock. Rekeying is generally quicker and less invasive, whereas rebuilding requires deeper inspection and component replacement.

In practice, the two processes are often combined. For example, a locksmith may rekey a lock while also replacing worn pins and springs, effectively performing a partial rebuild.

### Rekeying a Cylinder (Changing the Key)

Rekeying and rebuilding cylinders are core locksmithing tasks that involve altering or restoring the internal pin configuration of a lock so that it operates with a different key or returns to proper working condition. While both processes deal with the internal components of a cylinder, they serve different purposes: rekeying changes key access, while rebuilding restores mechanical performance and reliability.

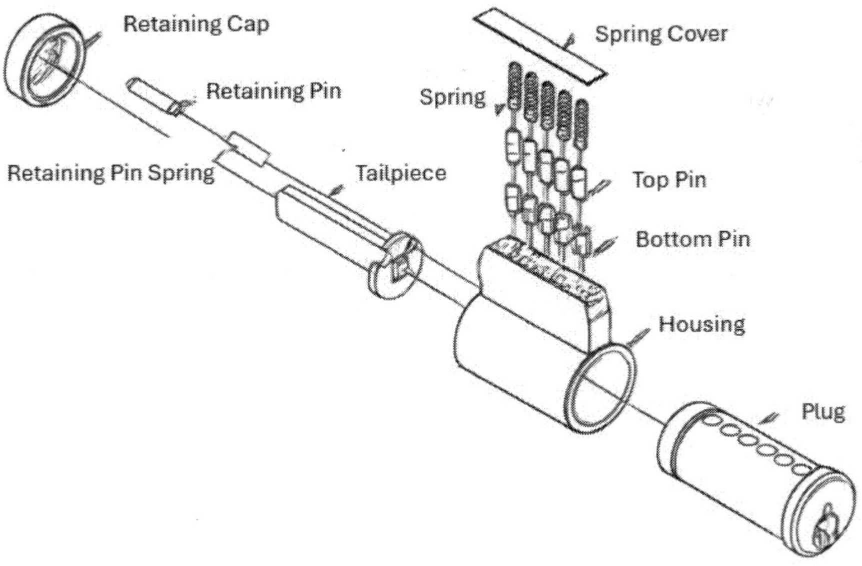


*Figure 35: Key in Knob Cylinder.*

Rekeying is the process of changing the internal pin configuration so that an existing lock works with a new key, without replacing the entire lock. This is commonly done when keys are lost, stolen, or when access control needs to change.

Inside a pin-tumbler cylinder, each pin stack consists of a key pin and a driver pin. The key pin height corresponds directly to the cuts on the key. When rekeying, the locksmith replaces the key pins so that they match a new key's profile while keeping the existing driver pins and springs (unless worn or damaged).

The process typically begins with removing the cylinder from the door and extracting the plug using a follower tool. The follower is critical because it holds the driver pins and springs in place while the plug is removed, preventing them from ejecting under spring pressure.

Once the plug is exposed, the old key pins are removed from each chamber. The locksmith then inserts the new key into the plug and selects replacement key pins that align perfectly with the new key's cuts. Each pin must be chosen so that, when the key is inserted, the top of the key pin aligns exactly at the shear line.

After all pins are replaced, the plug is reinserted into the cylinder housing, again using the follower to maintain control of springs and driver pins. The cylinder is then tested to ensure smooth operation with the new key and that no pins bind or protrude incorrectly.

Rekeying is efficient, cost-effective, and widely used in residential and commercial settings where hardware replacement is unnecessary.

### Rebuilding a Cylinder (Restoring Function)

Rebuilding goes further than rekeying. It involves restoring or replacing worn, damaged, or contaminated internal components so that the lock functions reliably and securely.

Over time, locks experience wear due to friction, dirt ingress, poor lubrication, and environmental exposure. Common issues include worn or flattened pins, weakened springs, debris buildup, and enlargement of internal chambers. These issues can lead to symptoms such as stiff key operation, inconsistent turning, key jamming, or even security vulnerabilities.

The rebuilding process starts similarly to rekeying, with disassembly of the cylinder and removal of the plug. However, instead of simply replacing key pins, the locksmith inspects all internal components in detail. Worn pins are replaced, springs are checked for tension and replaced if fatigued, and the plug and housing are examined for scoring, corrosion, or excessive tolerance.

Cleaning is a critical step in rebuilding. Debris, old lubricant residues, and contaminants are removed using appropriate cleaning agents. Once cleaned, the cylinder is lightly lubricated using suitable lock lubricants (such as graphite or PTFE-based products) to restore smooth operation without attracting dirt.

If the cylinder design allows, the locksmith may also upgrade components—for example, installing new springs, higher-quality pins, or even security pins during the rebuild process. In cases where structural damage is present (such as cracked housings or excessive wear), the cylinder may be deemed beyond repair and require replacement.

After reassembly, the lock is tested thoroughly to ensure correct pin alignment, smooth operation, and reliable locking/unlocking.

### Impact on Locksmithing Practices

Advanced pin dynamics and the use of security pins are highly relevant to anyone interested in locksmithing because they reveal that lock behaviour is not simply mechanical—it is governed by subtle, microscopic interactions that directly affect security, performance, and serviceability. For a locksmith, this understanding marks the transition from basic lock handling to true technical expertise.

At a practical level, this knowledge fundamentally shapes how locksmiths select and specify locks. A cylinder is not defined solely by its size or number of pins, but by the behaviour of those pins under real-world conditions. Understanding how spool, serrated, or mushroom pins interact with torque, friction, and tolerances allows a locksmith to recommend appropriate security levels. For example, environments requiring higher resistance to manipulation may benefit from aggressive security pin configurations, while high-use environments may require a balance between security and smooth operation.

This concept also has a strong influence on diagnostic practices. Many lock faults are not immediately visible and cannot be understood without recognising how pins behave internally. Issues such as intermittent key operation, stiffness, or inconsistent feedback may be caused by worn pins, debris, weakened springs, or tolerance changes rather than obvious mechanical failure. A locksmith who understands pin dynamics can interpret these subtle signs and identify root causes more accurately, reducing unnecessary replacement and improving repair outcomes.

In rekeying and rebuilding, advanced pin knowledge becomes even more critical. Selecting pin types is not just about matching a key—it involves deciding how the lock will behave over time. Aggressive security pins can improve resistance to picking but may introduce increased friction, more complex feedback, and reduced usability for the client. Locksmiths must therefore balance security requirements with user experience, durability, and maintenance considerations, especially in high-traffic or commercial environments.

From a security assessment perspective, advanced pin dynamics provide insight into how resistant a lock is to manipulation and how that resistance may change over time. A lock with well-designed security pins and tight tolerances may perform very differently from one that is worn or poorly manufactured. This allows locksmiths to move beyond generic advice and provide tailored recommendations based on actual mechanical behaviour, risk level, and environmental conditions.

These principles also influence fieldwork and practical decision-making. Recognising patterns such as deep plug rotation (false sets), repeated clicking (serrated pins), or counter-rotation (spool pins) helps locksmiths quickly interpret what is happening inside a lock without invasive inspection. This is particularly valuable when diagnosing faults on-site, dealing with lockouts, or assessing whether a lock is functioning correctly or has been compromised.

More broadly, understanding advanced pin dynamics expands the role of the locksmith from a tradesperson to a security specialist. It supports informed decisions about maintenance, upgrades, and system design, and it underpins the ability to work with both traditional and high-security cylinders. It also reinforces the importance of precision in key cutting, proper lubrication, and correct installation, as small deviations can significantly affect internal behaviour.

## Sidebar Mechanisms and Secondary Locking Systems

Sidebar mechanisms and secondary locking systems are advanced security features used in high-security locks to provide an additional layer of protection beyond the standard pin-tumbler design. Their purpose is to ensure that even if the primary pin system is manipulated or compromised, the lock will still not operate unless a second, independent set of conditions is satisfied.

### Sidebar Mechanisms (Primary Secondary Locking Feature)

A sidebar is a solid metal bar positioned along the side of the plug within a lock cylinder. Unlike conventional pin-tumbler systems, where driver pins align vertically at the shear line to allow rotation, the sidebar functions in a different orientation. It operates horizontally (or radially), engaging with specially designed grooves or slots within the plug. This additional component acts as a secondary locking barrier, preventing rotation unless very specific conditions are met.

In a lock equipped with a sidebar, the mechanism is normally in a locked state with the sidebar extended outward into the cylinder housing. In this position, it physically blocks the plug from turning. For the lock to operate, more than just the main pin stacks must be aligned. Additional components, such as side pins, sliders, or specially shaped key features, must also be correctly positioned.

When the correct key is inserted, it performs multiple functions simultaneously. It lifts the primary pin stacks to the correct heights, aligning them at the shear line as in a standard lock. At the same time, it interacts with secondary elements within the lock, positioning them so that grooves or gates within the plug align precisely with the sidebar. Once this alignment is achieved, the sidebar is able to retract into the body of the plug.

With the sidebar retracted, the obstruction is removed and the plug can rotate freely, allowing the locking mechanism to actuate. The critical point is that both systems, the primary pin stacks and the sidebar mechanism—must align at the same time. If either system is incorrect, even slightly, the lock will not operate.

Several well-known high-security locking systems use sidebar mechanisms in different ways. For example, Medeco systems utilise angled key cuts that not only lift pins but also rotate them to align sidebar gates. Mul-T-Lock systems employ telescopic pin designs combined with interactive elements to achieve multi-layered alignment. DOM systems, particularly the ix series, incorporate side pins and additional checking mechanisms to control sidebar engagement. Each of these approaches uses the sidebar concept to introduce an additional level of precision and security beyond standard pin-tumbler designs.

**Medeco Sidebar System:** Medeco cylinders are well known for introducing rotational pin technology in addition to standard pin lifting. The key contains angled cuts, which do not simply raise the pins to a height but also rotate each pin to a precise orientation.

Inside the lock, each pin has a small slot or gate. When the correct key is inserted, two things must occur simultaneously:

- The pins must be lifted to the correct height (as in a normal pin tumbler)
- The pins must also be rotated so that their gates align with the sidebar

Only when both conditions are satisfied can the sidebar retract, allowing the plug to rotate.

This dual requirement makes Medeco locks highly resistant to picking and bumping because an attacker must control both vertical movement and rotational alignment, which is significantly more complex than standard systems.

**Mul-T-Lock Sidebar System:** Mul-T-Lock systems use a different approach based on telescopic pins, where each pin stack consists of an outer pin and an inner pin. This effectively doubles the number of pin elements that must be aligned.

When the key is inserted:

- The outer pins are lifted to one height
- The inner pins are lifted independently to another height
- Additional features, such as interactive elements (e.g., a moving pin within the key), may also need to engage correctly

These aligned components allow the sidebar (or similar locking element) to disengage.

This design creates multi-dimensional complexity: instead of just one shear line, there are effectively multiple internal alignment points, making manipulation extremely difficult. Mul-T-Lock systems are also known for strong key control, meaning keys are difficult to duplicate without authorisation.

**DOM (ix Series) Sidebar / Secondary System:** DOM ix cylinders represent a modern European high-security approach, combining dimple key technology with side pins and check systems.

In these locks:

- The key uses dimple cuts to control multiple rows of pins (often top, bottom, and side pins)
- Additional check pins or control elements verify the correct key profile
- Side pins interact with a sidebar or similar locking component

For the lock to operate:

- All pin rows must align precisely

- Secondary elements (such as check pins) must also match
- Only then can the sidebar retract and allow rotation

This creates a layered verification system, where multiple independent conditions must be satisfied. DOM systems are known for tight tolerances and precision engineering, making them highly resistant to manipulation and unauthorised key duplication.

*Table 2: Key Comparison of Sidebar Systems.*

| Feature | Medeco | Mul-T-Lock | DOM (ix series) |
|---|---|---|---|
| Key Type | Angled-cut key | Dimple key with telescopic pins | Dimple key (multi-row) |
| Unique Mechanism | Pin rotation | Telescopic (dual) pins | Side pins + check systems |
| Sidebar Function | Engages with rotated pin gates | Engages after multi-pin alignment | Engages after multi-row + check alignment |
| Complexity Type | Rotational + vertical | Multi-layer vertical (inner/outer pins) | Multi-directional + verification layers |
| Key Control | High | Very high | Very high (often patented) |
| Picking Resistance | High (requires rotation control) | Very high (multiple shear lines) | Very high (multi-system alignment) |

**Secondary Locking Systems (Beyond the Main Pins)**

Secondary locking systems are additional mechanisms within a lock that must be correctly aligned before the lock will operate. They function alongside, or independently from, the primary pin-tumbler system. While a traditional cylinder relies on vertical pin alignment at the shear line, secondary systems introduce extra conditions that must be satisfied simultaneously. This layered approach significantly increases resistance to manipulation and unauthorised access because more than one mechanical interaction must be correct before rotation can occur.

One common type of secondary system involves side pins, sometimes referred to as finger pins. These components are positioned perpendicular to the main vertical pin stacks and interact with grooves or tracks cut into the side of the key. When the correct key is inserted, it not only lifts the main pins but also aligns these side pins so that they no longer block rotation. Often, side pins engage with a sidebar mechanism, meaning that both the vertical pins and the side pins must align before the plug can turn.

Another widely used secondary mechanism is the slider or gate system. Sliders are flat components that move horizontally within the cylinder. Each slider contains a gate, or slot, that must align with a sidebar or locking bar. When the correct key is inserted, it pushes each slider into a precise position. Only when all gates are aligned can the sidebar retract. This system is common in high-security dimple locks and creates a very controlled and precise interaction between key and cylinder.

Telescopic pins, also known as pin-in-pin systems, represent another form of secondary locking. In this design, one pin is nested inside another, effectively creating two separate alignment requirements within a single pin chamber. Both the inner and outer pins must reach their correct positions for the lock to function. This doubles the alignment conditions and increases key variation. Systems such as those produced by Mul-T-Lock use this technology to create high resistance to picking and unauthorised duplication.

Rotating pin systems add a different dimension of complexity. In these locks, pins must not only be lifted to the correct height but also rotated to a specific angle. The key's angled cuts cause the pins to turn as they are raised. If the rotation is incorrect, a sidebar will remain engaged and prevent movement. This method, used in systems such as Medeco, introduces both vertical and rotational alignment requirements, greatly increasing manipulation difficulty.

More advanced secondary systems incorporate magnetic or electronic verification elements. Some high-security locks use small magnetic components embedded within the key that interact with concealed magnets inside the cylinder. Others may integrate electronic authentication layers, where mechanical alignment alone is insufficient without additional verification. These hidden mechanisms function as secondary authentication checks, adding another level of security beyond purely mechanical design.

Together, these secondary locking systems illustrate how modern high-security cylinders move beyond simple pin alignment. By requiring multiple independent conditions to be satisfied at once, they increase complexity, improve resistance to attack, and significantly influence both lock performance and locksmith servicing practices.

### Impact on Locksmithing Practices

Understanding sidebar mechanisms and secondary locking systems is highly relevant to anyone entering locksmithing because it marks the transition from basic mechanical knowledge to advanced, security-focused practice. At a foundational level, many people learn locksmithing through standard pin-tumbler systems. However, modern security environments increasingly rely on high-security cylinders that incorporate multiple layers of protection. Without an understanding of these systems, a locksmith's ability to diagnose, service, and recommend appropriate locking solutions is significantly limited.

From a practical standpoint, these mechanisms directly influence how locks are selected and specified. A locksmith is not simply choosing a "lock," but rather selecting a security system with defined resistance characteristics. Sidebar-equipped systems such as those used in Medeco, Mul-T-Lock, or DOM cylinders provide enhanced protection against picking, bumping, and unauthorised key duplication. This means that locksmiths must understand not only how these systems function, but also how their design aligns with specific risk environments. For example, a high-risk commercial site or restricted-access facility may justify the complexity and cost of a multi-layered locking system, whereas a standard residential application may not.

These systems also have a significant impact on locksmithing techniques and service delivery. Traditional methods used for standard cylinders—such as basic rekeying, picking, or impressioning—may not be effective, or may require modification, when dealing with sidebar or secondary locking systems. The presence of sidebars, sliders, or rotating pins means that multiple alignment conditions must be satisfied simultaneously, which increases the complexity of both legitimate servicing and any form of manipulation. As a result, locksmiths must develop higher levels of technical skill, precision, and system-specific knowledge.

In diagnostic work, understanding these advanced mechanisms allows locksmiths to interpret faults more accurately. Problems such as a key not turning, partial rotation, or inconsistent operation may not be caused by simple wear or misalignment, but by failure in a secondary system such as side pins, sliders, or check mechanisms. Without this knowledge, a locksmith may misdiagnose the issue, leading to unnecessary replacement or ineffective repairs. Recognising how multiple systems interact within a cylinder enables more efficient troubleshooting and more appropriate solutions.

Rekeying and rebuilding practices are also influenced by these systems. In high-security cylinders, rekeying is often more complex due to the presence of additional components such as telescopic pins, side pins, or interactive elements. In some cases, specialised tools, restricted key blanks, or manufacturer authorisation may be required. This introduces an additional layer of professional responsibility, as locksmiths must adhere to key control protocols and ensure that security is not compromised during servicing. The choice of whether to rekey, rebuild, or replace a cylinder becomes more strategic when advanced security features are involved.

From a security assessment perspective, the presence of sidebar mechanisms and secondary locking systems is a strong indicator of higher resistance to common attack methods. Locksmiths who understand these features are better equipped to evaluate the true security level of a lock, rather than relying on superficial characteristics such as brand or appearance. This knowledge enables them to provide informed advice to clients, balancing factors such as security, usability, cost, and long-term reliability.

Finally, these systems reflect the broader evolution of locksmithing as a profession. The field is no longer limited to basic mechanical repair and key cutting; it now requires an understanding of complex mechanical engineering, precision manufacturing, and, increasingly, integration with electronic and access control systems. Sidebar and secondary locking mechanisms are a clear example of this progression, demonstrating how modern locks rely on layered security principles rather than single-point solutions.

In summary, sidebar mechanisms and secondary locking systems are relevant because they fundamentally change how locks operate, how they fail, and how they must be serviced. For anyone interested in locksmithing, mastering these concepts is essential for developing advanced skills, delivering effective security solutions, and keeping pace with the increasing complexity of modern locking systems.

## Key Cutting

Key cutting is a core locksmithing skill that involves accurately shaping a blank key so that it matches the internal configuration of a lock. While it may appear simple from the outside, effective key cutting requires precision, correct equipment, and an understanding of lock mechanics, tolerances, and key profiles.

At its most fundamental level, a key works by lifting pins (or other internal components) to precise heights so they align at the shear line. Cutting a key is therefore the process of creating those exact depths and spacing along the key blade. Each cut on a key corresponds to a specific pin height inside the lock. If the cuts are too deep, too shallow, or incorrectly spaced, the lock will not operate correctly.

### Key Duplication (Copying an Existing Key)

Key duplication is the most common method used in locksmithing and is widely employed for producing additional keys quickly and efficiently. In this process, the original key is placed into a key cutting machine alongside a blank key that matches the same profile. The machine uses a guide, known as a tracer, to follow the shape and cuts of the original key, while a rotating cutting wheel removes material from the blank key to replicate that exact pattern.

The process begins by inserting the original key into one side of the machine and securing the matching blank key on the opposite side. Both keys must be carefully aligned to ensure that the spacing and depth of the cuts are transferred accurately. Once aligned, the machine is activated, and the tracer follows the contours of the original key while the cutting wheel simultaneously shapes the blank key. After cutting, the new key is deburred and smoothed to remove any sharp edges or metal fragments that could interfere with operation.

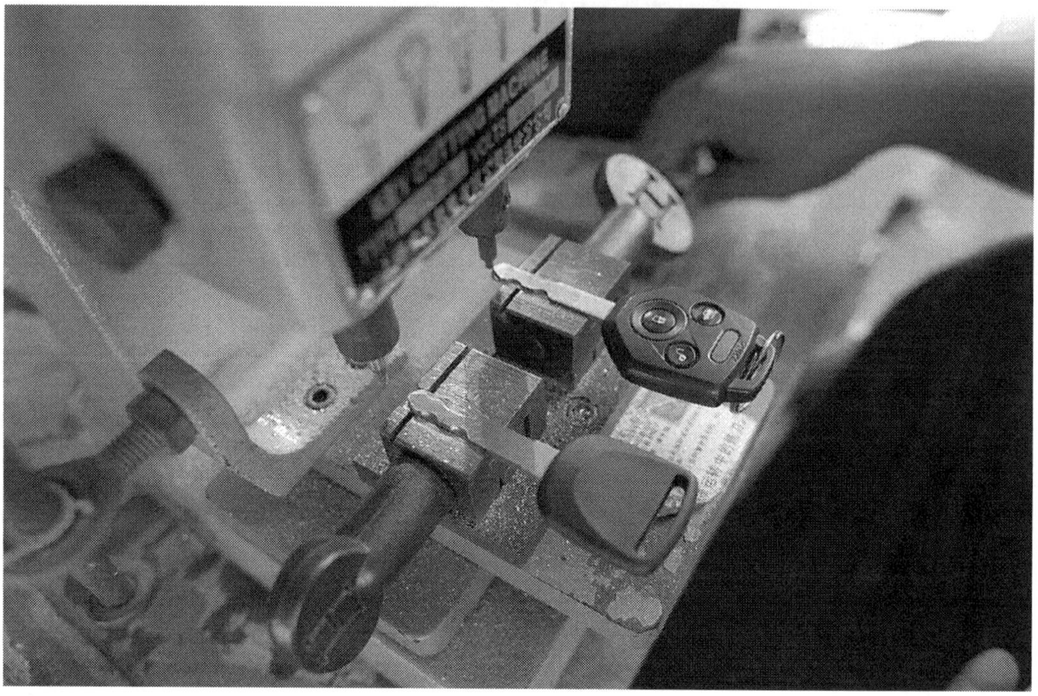

*Figure 36: Car key being duplicated with a key cutting machine. Christian Heimig, CC BY 4.0, via Wikimedia Commons.*

This method is fast, cost-effective, and suitable for most everyday key duplication tasks. However, its accuracy is directly dependent on the condition of the original key. If the original key is worn or damaged, the duplicated key will replicate those imperfections, which can lead to poor performance or unreliable operation in the lock.

**Code Cutting (Cutting from Key Code)**

Key cutting by code is a more precise method than standard duplication because it does not rely on copying an existing key. Instead, the locksmith uses a key code, which is a series of numbers that represent the exact depths and positions of each cut on the key. These codes can be obtained from the lock itself, a manufacturer's key card, or a database.

Once the code is available, it is entered into a key cutting machine or used in conjunction with a depth and spacing chart. The machine then cuts the key directly to the manufacturer's original specifications, ensuring that each cut is accurate and consistent with the intended design.

*Figure 37: The use of Code Cutter duplication machine. Visualeagle, CC BY-SA 4.0, via Wikimedia Commons.*

This method produces a cleaner and more precise key compared to duplicating a worn original. For this reason, it is commonly preferred in high-security systems, restricted keyways, and situations where accuracy and reliability are critical.

**Impressioning (Advanced Technique)**

Impressioning is a manual method used when no key is available.

A blank key is inserted into the lock and turned slightly. Internal pins leave tiny marks on the blank. The locksmith files those marked areas and repeats the process until the key works.

Most beginners are unfamiliar with the concept of lock impressioning and may not fully understand what the process involves. This section introduces the fundamentals of impressioning, outlining how it works, its limitations, and the basic tools required. The focus is on providing a clear and practical overview suitable for those new to locksmithing. As with any lock-related skill, it is essential to emphasise a key ethical principle: impressioning should only ever be carried out on locks that you own or have explicit permission to work on. This aligns with the same professional standards applied to lock picking and other access techniques [146].

The discussion primarily focuses on standard pin tumbler locks, as these provide the most accessible starting point for learning impressioning. While the technique can be applied to other lock types, beginning with simpler mechanisms allows learners to develop the necessary understanding and control before progressing to more complex systems [146].

Lock impressioning is a method used to create a working key for a lock without disassembling or damaging it. Instead of opening the lock through manipulation or force, the locksmith uses a blank key to "read" the internal pin positions [146]. By inserting and applying controlled pressure to the blank, the pins inside the lock leave small marks on the key surface. These marks indicate where material needs to be removed. Over a series of repeated insertions and careful filing, the blank is gradually shaped into a functioning key [146].

This method is often favoured because it is direct and does not require full lock disassembly. While alternative techniques such as picking or bumping can be used to gain access, impressioning focuses on producing a usable key, which can be more practical in situations where long-term access is required. Although it takes time and practice to master, experienced locksmiths can often complete the process relatively quickly once the technique is understood [146].

Impressioning is not limited to professional locksmiths. Anyone with the appropriate permission and a willingness to practise can develop the skill. However, it requires patience, attention to detail, and a good understanding of how pin tumbler mechanisms operate. Compared to lock picking, impressioning can be more methodical, but it still demands consistent practice to achieve reliable results [146].

While many lock types can be impressioned, the practicality of doing so varies. Standard pin tumbler locks are the most suitable for beginners due to their relatively straightforward internal design. More complex systems, such as disc detainer locks or high-security cylinders with sidebars, are significantly more difficult and often impractical to impression. In these cases, alternative methods may be more efficient [146].

The tools required for impressioning are relatively simple and widely available. A suitable lock and a correctly matched key blank form the foundation of the process. The key blank must match the lock's keyway and is typically made of soft brass, which allows clear marking from the pins. A fine-cut file is used to remove small amounts of material from the blank, shaping it to match the internal pin configuration [146]. A key grip is often used to hold the blank securely while applying controlled turning force, improving both accuracy and ease of handling. A magnifying glass is also highly useful, as it allows the locksmith to clearly see the small marks left by the pins and maintain precision during filing [146].

Together, these tools and techniques form the basis of lock impressioning. While the process is simple in principle, its effectiveness relies on careful observation, controlled force, and incremental adjustment. For beginners, developing these skills provides valuable insight into how locks function internally and builds a strong foundation for more advanced locksmithing practices [146].

Impressioning Technique

Impressioning is a technique used to create a working key directly from a lock by using a blank key to "read" the internal pin positions. The process relies on close interaction between the key blank and the pin-tumbler mechanism. A blank key is inserted into the lock and gentle turning pressure is applied. As the key is moved slightly, jiggled, rocked, or withdrawn, the binding pins inside the cylinder leave small marks on the surface of the key blade. These marks indicate where material must be removed to form the correct cuts [146].

After marking, the key is removed and examined. The visible marks are carefully filed to create shallow cuts that correspond to the pin heights inside the lock. Over multiple cycles of inserting, marking, and filing, the key gradually takes shape. When all cuts are correct, the pins align at the shear line and the lock will turn.

While the concept is simple, the process requires patience, precision, and repetition to achieve an accurate result [146].

**Preparation and Setup:** The process begins by preparing a suitable workspace and gathering the required tools. These typically include the lock, correctly matched key blanks (preferably soft brass), a fine file, a key grip, and a magnifying aid. Having multiple blanks available is recommended, as beginners often make mistakes such as filing too deeply, which can render a blank unusable. A stable work surface with good lighting is essential for maintaining accuracy [146].

Before use, the key blank should be prepared by lightly filing the blade surface. Many blanks have a plated finish, and removing this outer layer exposes the softer brass underneath, making it easier for the pins to leave visible marks. Proper preparation significantly improves the clarity of markings and the overall success of the process.

**Marking the Key Blank:** Once prepared, the key blank is inserted fully into the lock. A small amount of turning pressure (torque) is applied, typically using a key grip or by hand. This pressure causes certain pins to bind against the plug. With the key under tension, slight movements are introduced to encourage the pins to mark the blank [146].

There are several common ways to generate these marks. The pulling method involves applying tension and withdrawing the key, allowing pins to scratch the surface. The rocking method involves gently tilting or wiggling the key while inserted, helping pins mark their positions more clearly. The tapping method introduces light impacts to the key bow to create subtle movement. All methods aim to produce visible marks, but they require careful control, too little force produces no marks, while too much can damage the key or lock [146].

**Filing the Marks:** After marking, the key is removed and inspected, often with the aid of a magnifying glass. The marks made by the pins are typically very small and can be difficult to see clearly. Each mark represents a point where material must be removed. Using a fine file, the locksmith carefully removes a small amount of material at each marked location.

It is important to file gradually and only where marks are present. Removing too much material too quickly can overshoot the correct depth and compromise the entire process. Precision and restraint are critical at this stage, as each cut directly affects how the pins align within the lock.

**Repeating the Process:** Impressioning is an iterative process. After the first round of filing, the key is reinserted into the lock and the marking process is repeated. New marks will appear on areas that still require adjustment. With each cycle, the number and intensity of marks will decrease as the key approaches the correct configuration.

This sequence—insert, apply tension, mark, remove, inspect, and file—is repeated until no further marks appear and the key operates the lock smoothly. While beginners may find the process time-consuming and occasionally inconsistent, experience improves both speed and accuracy. Over time, the ability to interpret marks and control filing becomes more refined, making impressioning a reliable and valuable locksmithing skill [146].

**Cutting High-Security Keys**

Cutting high-security keys represents a significant step beyond traditional key duplication, both in terms of complexity and required expertise. Unlike standard edge-cut keys, modern high-security systems use advanced key designs that incorporate additional dimensions, tighter tolerances, and controlled key profiles. These features are specifically engineered to increase resistance to unauthorised duplication and manipulation, meaning that the cutting process must be far more precise and controlled.

One of the most common high-security designs is the dimple key, which uses a series of drilled holes (dimples) rather than cuts along the edge of the blade. These dimples are positioned at varying depths and locations across the flat surface of the key, often on one or both sides. Because the cuts are vertical rather than edge-based, they require specialised vertical milling machines that can accurately control depth and spacing in multiple axes. Even slight deviations can prevent proper pin alignment, particularly in systems with tight manufacturing tolerances.

Another advanced design is the laser-cut or sidewinder key, which features internal grooves running along the centre of the key blade rather than traditional external cuts. These keys are commonly used in automotive and high-security applications. Cutting them requires CNC (computer numerical control) or electronic key cutting machines, which follow digital key codes or precise tracing paths. The internal groove profile must be reproduced with extreme accuracy, as these keys often interact with complex mechanisms such as sidebars or wafer systems.

In addition to the physical design of the key, many high-security systems use restricted keyways. These are patented or controlled profiles that limit who can legally obtain key blanks. Locksmiths must be authorised by the manufacturer or distributor to access these blanks, and duplication often requires proof of ownership or security credentials. This adds an administrative layer to the cutting process, reinforcing key control as part of the overall security system.

The machinery used for cutting high-security keys reflects the need for precision. Unlike basic duplicators, these machines often operate digitally, allowing locksmiths to input key codes directly or scan existing keys. This ensures that the new key is cut to original factory specifications rather than replicating wear or inaccuracies from an existing key. In many cases, this results in a more accurate and reliable key than traditional duplication methods.

From a practical locksmithing perspective, cutting high-security keys demands not only technical skill but also an understanding of the underlying lock system. The locksmith must consider factors such as tolerances, pin configuration, and secondary locking mechanisms. Because these systems are less forgiving than standard locks, errors in cutting can lead to non-functional keys or even damage to the lock.

Ultimately, high-security key cutting represents a shift from purely mechanical duplication toward precision engineering and controlled access. It requires specialised equipment, authorised materials, and a higher level of expertise. For modern locksmiths, mastering these techniques is essential, as high-security systems are increasingly common in residential, commercial, and automotive applications.

## Key Cutting Equipment

Common tools include:

- Manual key cutting machines
- Automatic or electronic duplicators
- CNC key cutting machines
- Key gauges and depth charts
- Deburring brushes or files

Each tool must be calibrated regularly to maintain precision.

### Manual Key Cutting Machine

A manual key cutting machine (often called a key duplicator) copies an existing key by mechanically linking two movements:

- A tracer follows the profile of the original key
- A cutting wheel removes material from a blank key at the same time

*Figure 38: Börkey 954-2 Key Cutting Machine. Dennis van Zuijlekom, CC BY-SA 2.0, via Flickr.*

Because both sides are rigidly linked on a carriage, whatever shape the tracer follows is reproduced on the blank.

Key Components and Their Roles:

- **Original key clamp (vice):** Holds the reference key
- **Blank key clamp (vice):** Holds the uncut key
- **Tracer (guide stylus):** Physically rides along the original key's cuts
- **Cutting wheel:** A hardened rotating wheel that cuts the blank
- **Carriage:** Moves both keys together across the tracer and cutter
- **Depth/spacing alignment system:** Ensures both keys sit at the same height and position
- **Motor:** Drives the cutting wheel

**The Duplication Principle (What's Happening Mechanically):** At a fundamental level, the key duplication process is based on a simple mechanical principle. The original key acts as the reference and defines the geometry that will be copied. As the machine operates, a tracer follows the contours of the original key, physically "reading" the depths and spacing of its cuts. This movement is transferred through the carriage, which ensures that the blank key mirrors the same motion. At the same time, a rotating cutting wheel removes material from the blank key, reproducing the pattern of the original.

Because this process relies on direct mechanical copying rather than measured or coded input, the accuracy of the final key depends on several critical factors. The condition of the original key is particularly important, as any wear, rounding, or damage will be replicated in the duplicate. Machine calibration also plays a significant role, as even minor misalignment between the tracer and cutting wheel can result in incorrect cuts. In addition, the operator's control and consistency during the cutting process influence the quality of the outcome, as uneven pressure or poor handling can affect the precision of the duplicated key.

**Preparation and Setup (High-Level):** Before any cutting takes place, competent practice focuses on preparation:

- **Correct blank selection:** The blank must match the keyway profile and length
- **Visual inspection:** The original key should be checked for wear, bending, or damage
- **Machine condition:** Cutter sharpness, tracer condition, and alignment all affect accuracy
- **Clean clamping surfaces:** Dirt or swarf can introduce misalignment

If the original key is heavily worn, professionals often avoid duplication and instead cut from a code or factory specification to restore accuracy.

## Beyond the Key

**Alignment Fundamentals (Critical Concept):** Accurate duplication depends on indexing—ensuring both keys are positioned identically.

Two key references are used:

- **Shoulder alignment:** The flat stop near the bow is aligned against a guide
- **Tip alignment:** Used for keys without a defined shoulder (e.g., some automotive keys)

If alignment is off—even slightly—the entire key will be incorrect, regardless of how well it is traced.

**Cutting Motion (Conceptual Operation):** During the cutting process, the operator plays a critical role in ensuring accuracy and quality. The carriage must be moved in a controlled and steady manner along the length of the key, allowing the tracer to follow the original key's profile precisely. Consistent contact between the tracer and the original key is essential, as any loss of contact can result in incorrect depths or missed cuts. At the same time, excessive pressure must be avoided, as forcing the machine can distort the cut, wear the cutter prematurely, or even damage the key blank.

As the carriage moves, the cutting wheel progressively removes material from the blank key, shaping it to match the original. This process should be smooth and continuous, rather than rushed or uneven. Good practice in key cutting emphasises using smooth, controlled motion rather than force. The operator should allow the cutter to do the work, maintaining a light but consistent feed. It is also important to avoid "over-tracing" or repeatedly grinding in the same area, as this can remove too much material and result in an inaccurate or weakened key.

**Finishing and Quality Checks:** After cutting, the new key is not immediately ready for use. It must be finished and checked:

*Deburring*

- Cutting leaves sharp edges and burrs
- These are removed with a brush or deburring tool
- This improves smooth insertion and reduces wear on the lock

*Surface Smoothing*

- Light finishing ensures clean edges and consistent profile

*Functional Testing (Authorised Context Only)*

- The key should operate smoothly without forcing
- Any stiffness may indicate:

- Misalignment
- Overcutting
- Worn original key copied inaccurately

**Common Issues and Their Causes:** Understanding common key-related problems is essential for effective locksmithing, as many operational issues can be traced back to the quality of the key itself. When a key works poorly or only intermittently, the cause is often duplication from a worn or damaged original. In these cases, the inaccuracies present in the original key are transferred to the duplicate, resulting in inconsistent pin alignment and unreliable operation.

If a key does not insert fully into the lock, this usually indicates that the wrong key blank has been used or that there is misalignment in the cutting process. Even small differences in keyway profile or shoulder positioning can prevent proper insertion and engagement with the lock mechanism.

Rough or noisy operation when turning the key is commonly caused by burrs, sharp edges, or poor finishing after cutting. These imperfections create additional friction داخل the keyway, leading to a gritty or resistant feel. Proper deburring and smoothing are critical steps to ensure smooth operation.

A loose or “sloppy” feel when the key is inserted or turned typically points to excessive material removal during cutting or the use of a heavily worn original key. This can reduce the precision of the fit between the key and the lock components, affecting both performance and security.

**Safety Considerations:** Manual key cutting machines involve rotating cutters and metal removal, so:

- Eye protection is essential
- Hands must be kept clear of the cutting area
- Loose clothing or jewellery should be avoided
- Machines should only be used with proper training

<u>Automatic or Electronic Duplicators</u>

Automatic or electronic key duplicators are designed to improve accuracy, consistency, and efficiency compared to manual machines. Instead of relying entirely on operator skill to trace and cut, these machines use motors, sensors, and sometimes digital calibration to control the cutting process. While the basic principle remains the same—copying an original key onto a blank—the level of automation reduces human error and produces more precise results.

Before starting, ensure the machine is powered on and has completed any self-check or calibration cycle. Many electronic machines will automatically reference their cutting position or require a quick calibration step using a supplied calibration block. The correct cutter and tracer (or probe) must be installed and in good condition, as worn components will affect accuracy. Select the correct jaw or clamp type depending on the key style (e.g., standard edge-cut, laser/sidewinder, or dimple key).

The original key is placed into the designated clamp (usually the left side), and the blank key is secured in the corresponding clamp (right side). It is critical that both keys are aligned correctly—typically against a shoulder stop or tip stop, depending on the key design. Proper alignment ensures that spacing and depth are replicated accurately. Tighten the clamps firmly to prevent any movement during cutting.

On automatic duplicators, the operator usually selects a duplication mode via a control panel. Some machines are simple “trace and cut” systems, while more advanced units scan the original key electronically using a probe or optical sensor. In scanning systems, the machine first reads the profile of the original key and then calculates the cutting path before engaging the cutter.

Once the process begins, the machine will move the carriage automatically. In tracer-based systems, the probe follows the contours of the original key while the cutter simultaneously removes material from the blank. In fully electronic systems, the cutter follows a programmed path based on the scanned data. The operator’s role during this stage is mainly supervisory—ensuring smooth operation, monitoring for unusual noise or vibration, and being ready to stop the machine if necessary.

After cutting is complete, remove the duplicated key and inspect it carefully. Check that the cuts are clean, consistent, and free from defects. As with manual cutting, deburring is essential. Use a wire brush, deburring tool, or light file to remove sharp edges and metal shavings, particularly along the cuts and edges of the key blade.

The duplicated key should then be tested in the lock. It should insert smoothly, turn without excessive force, and operate the lock consistently. If issues are detected—such as stiffness, incomplete turning, or poor engagement—the cause may be incorrect alignment, a worn original key, or machine calibration issues. In such cases, adjustments or recutting may be required.

Regular maintenance of automatic and electronic duplicators is critical to maintaining accuracy. This includes cleaning metal shavings from the machine, checking cutter sharpness, verifying calibration, and ensuring clamps and guides remain secure and undamaged. Because these machines rely on precision, even minor wear or misalignment can lead to noticeable errors in the finished key.

In practice, automatic and electronic duplicators allow locksmiths to produce high-quality duplicates quickly and consistently, particularly in high-volume or commercial environments. However, they still require a solid understanding of key geometry, correct setup procedures, and careful inspection to ensure reliable results.

<u>CNC Key Cutting Machines</u>

CNC (Computer Numerical Control) key cutting machines represent the highest level of precision and automation in locksmithing. Unlike manual or basic duplicators, CNC machines do not rely on physical tracing. Instead, they use digital measurements, coded input, or electronic scanning to calculate exact cutting paths. This allows locksmiths to produce highly accurate keys, including complex high-security designs such as dimple, laser (sidewinder), and restricted key systems.

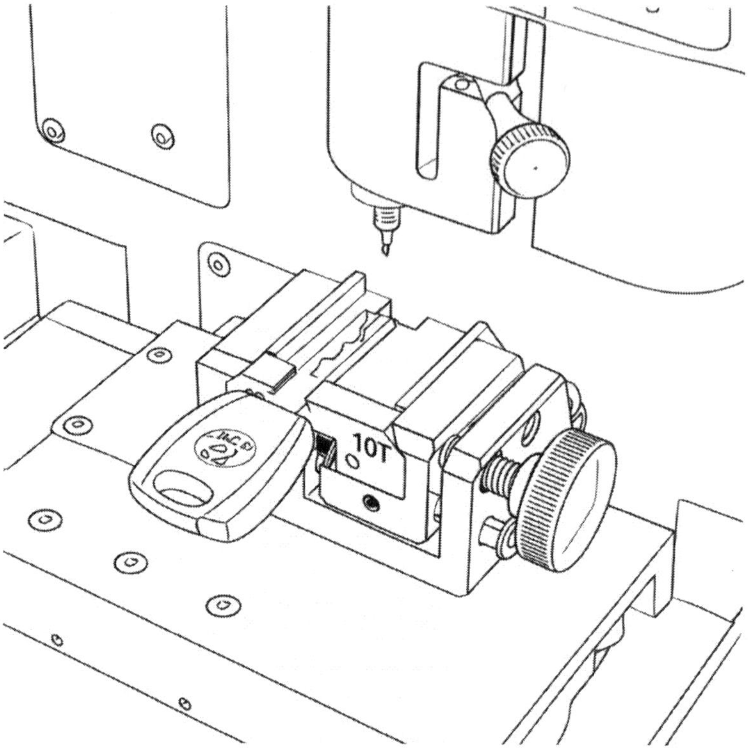


*Figure 39: CNC Key Cutting.*

Before using a CNC key cutting machine, the operator must ensure the machine is properly set up and calibrated. Power on the unit and allow it to complete its startup sequence. Most machines perform an automatic homing process, where the cutting head moves to a reference position. If required, run a calibration routine using the manufacturer's calibration block to confirm that the cutter and probe are aligned correctly. Check that the correct cutting tool (e.g., milling cutter for dimples or sidewinder blade for laser keys) is installed and in good condition, as tool wear directly affects accuracy.

The next step is to select the correct key type and input method. CNC machines typically offer several options: duplicating an existing key by scanning it, cutting from a key code, or selecting a key profile from an internal database. For duplication, the original key is placed into the machine's clamp and scanned using a probe or optical sensor. For code cutting, the operator enters the key code manually or retrieves it from a database. This method is preferred for accuracy because it produces a key to factory specifications rather than copying wear from an existing key.

Once the cutting method is selected, the key blank must be installed. The blank is secured in a precision clamp or vice, and it is critical that it is aligned correctly using the appropriate reference point (shoulder stop or tip stop). Many CNC machines provide visual guides or automatic positioning systems to assist with alignment. The clamp must be tightened securely to prevent any movement during cutting, as even slight shifts can result in an unusable key.

After setup, the operator initiates the cutting cycle through the machine's interface. The CNC system calculates the tool path based on the scanned data or input code and then controls the movement of the cutting head along multiple axes. The cutter removes material in a controlled, step-by-step process, often making multiple passes to achieve precise depths and smooth finishes. During this process, the operator should monitor the machine for abnormal vibration, unusual noise, or tool issues, although direct manual control is minimal.

When the cutting cycle is complete, the machine will return to a safe position, and the finished key can be removed. The key should be inspected carefully for accuracy, paying attention to cut depth, spacing, and surface finish. Any burrs or sharp edges must be removed using a deburring brush or fine file. Even with CNC precision, finishing is still an essential step to ensure smooth operation in the lock.

Testing the key in the corresponding lock is the final verification step. A properly cut CNC key should insert smoothly and operate the lock without resistance. If problems occur, they are typically due to incorrect key blank selection, improper alignment, incorrect code input, or calibration issues within the machine. Troubleshooting may involve rechecking the input data, recalibrating the machine, or verifying that the correct key profile was used.

Routine maintenance is critical for CNC key cutting machines. This includes cleaning metal shavings from the cutting area, regularly replacing or sharpening cutting tools, checking calibration accuracy, and updating the machine's software or key database as required. Because CNC machines operate to very fine tolerances, even small amounts of debris or tool wear can affect performance.

In professional locksmithing practice, CNC machines are essential for working with modern high-security systems. They enable precise duplication, accurate code cutting, and the ability to handle complex key designs that cannot be produced reliably by manual methods. However, they also require a higher level of technical understanding, including familiarity with digital interfaces, key coding systems, and machine calibration, making operator competence just as important as the machine itself.

## Key Gauges and Depth Charts

Understanding how to use key gauges and depth charts is fundamental in locksmithing, particularly when cutting keys to code, diagnosing locks, or verifying accuracy in high-security systems. These tools allow a locksmith to move beyond simple duplication and work with precise, measurable data.

A key gauge, also known as a decoder, is a precision measuring tool used in locksmithing to determine the depth of each cut on a key. Rather than relying on visual estimation, the gauge converts the physical shape of the key into a series of numerical values, commonly referred to as bitting numbers. These numbers represent the exact depth of each cut and form the key's code, which can then be used for accurate duplication or code cutting.

Each position along the key blade corresponds to a specific cut depth defined by the lock manufacturer. The key gauge is designed with calibrated steps or markings that match these standard depth increments. This allows the locksmith to measure each cut consistently and in alignment with the manufacturer's specifications.

When the key is inserted into the gauge, the measuring surface of the tool makes contact with the lowest point of each cut. The depth at that position is then read directly from the scale or indicated by the gauge's markings. This process is repeated across all cut positions along the key.

As each cut is measured, the corresponding values are recorded in sequence, typically forming a series such as 2-4-3-1-5. This sequence represents the complete bitting code of the key, which can then be used to accurately reproduce the key to factory specifications or to analyse the lock's configuration.

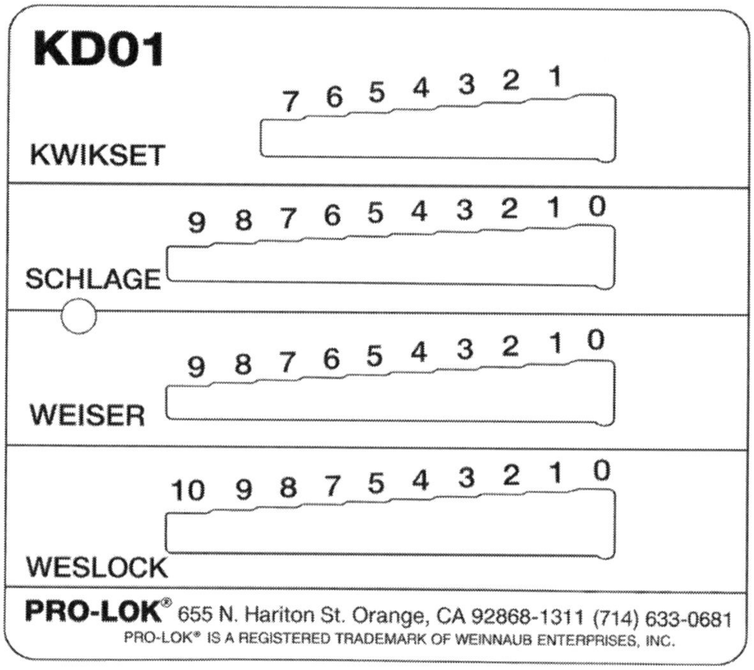


*Figure 40: Key Decoder Guage Card.*

Kwikset, Schlage, Weiser, and Weslock—are all well-known lock manufacturers (brands) that produce residential and light commercial locking hardware. They each have their own keyway designs, pin systems, and keying standards, which is why they appear separately on key gauges and depth charts.

**Kwikset**

Kwikset is a widely used residential lock brand, especially in North America.

In locksmithing terms, Kwikset is known for:

- A simple pin-tumbler system (traditionally 5-pin)
- A distinct keyway profile (KW1 is very common)
- Relatively wide manufacturing tolerances, making keys easier to duplicate

Kwikset also introduced SmartKey technology, which is a rekeyable system that does not use traditional pins in the same way. Because of its popularity, many locksmith tools (like gauges and cutters) include Kwikset standards as a default.

**Schlage**

Schlage is another major manufacturer, often considered a step up in quality compared to basic residential locks.

Key characteristics include:

- Typically 5 or 6 pin systems
- Tighter tolerances than Kwikset
- Common keyways like SC1 (5-pin) and SC4 (6-pin)

Schlage locks are widely used in both residential and commercial settings, and their tighter tolerances often provide better security and more consistent key performance.

**Weiser**

Weiser is very similar to Kwikset and is commonly used in Canada and some international markets.

Important points:

- Often shares design similarities with Kwikset (they are historically related brands)
- Uses similar keyway styles and pinning concepts
- Found mainly in residential applications

For locksmiths, Weiser keys may look similar to Kwikset, but they are not always interchangeable, so correct identification is important.

**Weslock**

Weslock is a smaller brand compared to the others but still commonly encountered in residential hardware.

Features include:

- Standard pin-tumbler mechanisms
- Unique keyway profiles
- Often used in decorative or mid-range residential locks

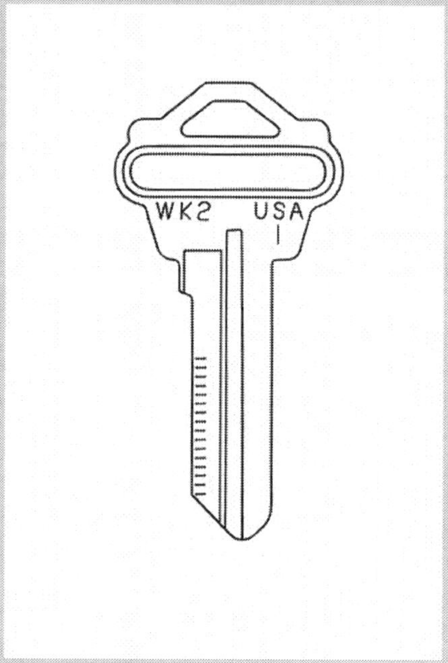


*Figure 41: Weslock key blank example.*

For locksmiths, Weslock requires specific blanks and depth charts, as its key specifications differ from Kwikset or Schlage.

Step-by-Step Use:

**1. Identify the key system**

Ensure you are using the correct gauge for the key type (e.g., Kwikset, Schlage, automotive, dimple).

**2. Insert the key into the gauge**

Position the key so each cut aligns with the measuring probe.

**3. Read each cut depth**

Move along the key from shoulder to tip (or tip to shoulder, depending on system) and record each value.

**4. Record the bitting sequence**

Write down the numbers in order—this becomes your working code.

**5. Verify consistency**

Re-check readings to avoid errors, especially on worn keys.

Key gauges and decoded bitting values have several important practical uses in locksmithing. One of the most common applications is decoding an existing key so that a new key can be cut by code rather than copied. This approach produces a more accurate result because it restores the key to original factory specifications instead of duplicating any wear or damage present on the original.

They are also valuable for identifying worn or incorrect cuts on a key. Over time, keys can become rounded or distorted through use, which can lead to poor operation or intermittent faults. By measuring each cut precisely, a locksmith can compare the actual depths against standard specifications and determine whether the key is the source of the problem.

Another key use is reverse-engineering a key when no original code is available. In situations where documentation has been lost or a code card is missing, a key gauge allows the locksmith to reconstruct the bitting sequence directly from the physical key. This makes it possible to reproduce or replace keys without needing access to manufacturer records.

Key gauges also play an important role in rekeying and master keying systems. Accurate measurement of cut depths ensures that new pin combinations are correctly matched to the intended key system, helping maintain both functionality and security across multiple locks.

A depth chart is a reference table that shows the exact physical depth (usually in millimetres or thousandths of an inch) for each cut number in a specific key system.

Depth charts are an essential reference tool in locksmithing because they translate abstract key codes into precise physical measurements. Each lock manufacturer establishes its own system that defines the number of available depth levels, the exact measurement associated with each depth, and the spacing between each cut position along the key blade. For example, a manufacturer may use a scale such as 0–9 or 1–6, where each number corresponds to a specific depth of cut, with lower numbers typically representing shallower cuts and higher numbers representing deeper cuts.

In addition to depth levels, manufacturers also define the exact physical measurements for each cut. These measurements are usually specified in thousandths of an inch or millimetres and are critical for ensuring that the key aligns correctly with the internal pin stacks of the lock. Even small deviations can result in poor operation or failure of the key to function, which highlights the importance of accuracy when using depth charts.

Spacing is another key element defined by depth charts. Each cut on a key must be positioned at an exact distance from a reference point, usually the shoulder or tip of the key. These spacing measurements ensure that each cut aligns precisely with the corresponding pin chamber inside the lock cylinder. Incorrect spacing, even if the depths are correct, will prevent the lock from operating properly.

A depth chart brings all of this information together by converting a bitting number sequence into exact cutting instructions. For example, a code such as 2-4-3-1-5 is not just a series of numbers—it represents specific cut depths at specific positions along the key. By following the chart, a locksmith or machine can accurately reproduce the key to factory specifications, ensuring reliable operation and maintaining the intended security level of the lock.

*Table 3: Example Concept.*

| Cut Number | Depth (mm) |
|---|---|
| 1 | 0.30 mm |
| 2 | 0.60 mm |
| 3 | 0.90 mm |
| 4 | 1.20 mm |

*(Values vary by manufacturer and system)*

Using depth charts in practice begins with selecting the correct chart for the specific key system being worked on. Each manufacturer defines its own depth and spacing standards, so using the wrong chart will result in an inaccurate key that will not function properly. Ensuring the correct chart is used is a critical first step in maintaining accuracy and compatibility with the lock.

Once the correct chart is obtained, the locksmith converts the bitting numbers into physical cut depths. Each number in the sequence corresponds to a specific depth measurement defined by the manufacturer. These measurements are then used to guide the cutting process, ensuring that each cut matches the intended geometry of the key.

The next stage involves setting up the key cutting machine. Whether using a manual, electronic, or CNC machine, the cutter must be adjusted to the required depth for each position. This setup phase is crucial, as incorrect calibration will result in inaccurate cuts even if the correct data is used.

Cutting the key requires careful attention to both depth and spacing. Each position must be cut precisely according to the chart, maintaining consistent alignment with the reference point (shoulder or tip). Accuracy at this stage directly determines whether the key will operate smoothly or fail to function.

After cutting, tolerances must be checked, particularly for high-security systems. Modern locks, especially those using dimple, laser, or restricted key designs, require extremely tight tolerances. Even minor deviations in depth or spacing can prevent proper operation or reduce security performance.

Key gauges and depth charts are most effective when used together. The key gauge is used to decode an existing key, translating its physical cuts into a numerical bitting code. The depth chart then converts that code into exact measurements for cutting a new key. This combination allows locksmiths to move beyond simple duplication and instead produce keys to precise specifications.

In professional practice, this approach enables several important capabilities. Locksmiths can cut keys without needing the original, restore worn keys back to factory dimensions, and avoid errors that arise from duplicating damaged or inaccurate keys. This is particularly valuable in rekeying, master keying, and high-security applications.

There are several key considerations to keep in mind. Accuracy is critical, as even small errors can prevent a key from working correctly. Tools must always match the specific manufacturer's system, as standards vary significantly. Worn keys can produce misleading readings when decoded, so multiple checks or decoding directly from the lock may be necessary. High-security systems, in particular, demand a high level of precision and attention to detail.

Overall, key gauges and depth charts form the foundation of precision locksmithing. Rather than relying on mechanical copying alone, they allow locksmiths to interpret, measure, and recreate the exact geometry of a key system. This transforms key cutting into a controlled and repeatable process, which is essential for modern locks and advanced security applications.

A practical example helps demonstrate how key gauges and depth charts are used together in real locksmithing work. Consider a scenario where a customer brings in a worn house key for a common residential lock (for example, a Kwikset or similar system), and requests a new key because the existing one works poorly.

The first step is to identify the correct key system and obtain the matching depth chart. The locksmith confirms the keyway profile (e.g., Kwikset) and selects the appropriate depth chart for that manufacturer. This chart will define both the depth values (e.g., 1–7) and the spacing between cuts. Using the wrong chart at this stage would result in a key that may physically fit but will not operate the lock correctly.

Next, instead of duplicating the worn key directly, the locksmith uses a key gauge to decode it. The key is inserted into the gauge, and each cut is measured. Suppose the readings produce a bitting code such as **2–4–3–1–5**. This sequence represents the relative depths of each cut along the key blade. At this point, the locksmith may double-check readings, especially if the key is visibly worn, to ensure accuracy.

Using the depth chart, the locksmith now converts each number into an exact physical depth. For example (illustrative only), the chart may specify:

- Depth 1 = 0.230"
- Depth 2 = 0.220"
- Depth 3 = 0.210"
- Depth 4 = 0.200"

- Depth 5 = 0.190"

The bitting **2–4–3–1–5** is then translated into precise measurements for each position along the key. This step transforms the decoded information into actionable cutting data.

The cutting machine is then set up. If using a manual or electronic machine, the locksmith adjusts the depth settings according to the chart and ensures the spacing guide (shoulder stop or tip stop) is correctly aligned. A suitable key blank is selected that matches the original keyway. Proper alignment here is critical—any offset will shift all cuts and render the key unusable.

The locksmith proceeds to cut each position one at a time, following the specified depths and spacing. Care is taken to maintain consistent positioning and avoid overcutting. Unlike duplication, this process is controlled and deliberate, with each cut made to a known specification rather than copied from a potentially inaccurate original.

After cutting, the key is deburred and cleaned to remove sharp edges. The locksmith then tests the key in the lock. Because the key was cut to factory specifications rather than copied from a worn original, it will often operate more smoothly than the customer's old key. If necessary, minor adjustments can be made, but ideally, the key should function correctly on the first attempt.

This example highlights the advantage of combining key gauges and depth charts. Instead of reproducing errors from a worn key, the locksmith restores the key to its intended design. This approach is particularly valuable in professional settings, such as rekeying systems, creating master keys, or working with high-security locks where precision is essential.

In practice, this method represents a shift from simple duplication to precision engineering. It allows locksmiths to produce reliable, accurate keys, diagnose issues more effectively, and maintain higher standards of workmanship—especially important as modern lock systems become more complex and tolerance-sensitive.

### Common Problems in Key Cutting

Several common problems can arise during or after the key cutting process, and understanding these issues is essential for producing accurate and reliable keys. One of the most frequent problems is duplicating a worn key. When a worn or damaged key is copied, any inaccuracies in the original are transferred to the new key, often compounding existing errors. This can result in poor operation, intermittent function, or a key that only works under certain conditions.

Another issue involves using an incorrect key blank. If the blank does not match the required key profile exactly, it may not fit into the lock at all. Even very small differences in shape or dimensions can prevent proper insertion or alignment within the keyway. Depth errors are also a critical concern. If cuts are made too deep, the pins inside the lock may be overset, preventing proper alignment at the shear line. Conversely, if cuts are too shallow, the pins may not be lifted high enough, also preventing the lock from operating.

Burrs and rough edges left on the key after cutting can create additional problems. These imperfections can cause the key to stick, feel rough during operation, or even damage internal lock components over time. For this reason, deburring and smoothing the key is an essential finishing step. Machine misalignment is another potential source of error. If the cutting machine is not properly calibrated or aligned, it can produce uneven or inaccurate cuts, leading to a key that does not function correctly despite following the correct process.

Quality control and finishing are therefore critical stages in key cutting. After a key is cut, it should be carefully inspected for accuracy and consistency. Any burrs or sharp edges should be removed, and the key should be tested in the lock to confirm proper operation. A correctly cut key should insert smoothly and turn without excessive resistance or force, indicating that the internal components are aligning as intended.

From a practical locksmithing perspective, key cutting is far more than a simple mechanical duplication task. It is a precision process influenced by tolerances, wear, and the specific design of the lock system. Locksmiths must recognise when it is inappropriate to duplicate a worn key and instead use code cutting to restore the key to its original specifications. They must also understand the differences between standard and high-security key systems, maintain their equipment to ensure accuracy, and be able to distinguish whether a problem lies with the key itself or within the lock mechanism.

Ultimately, key cutting sits at the intersection of craftsmanship and engineering. Each cut represents a precise instruction to the internal components of the lock. A skilled locksmith understands that even the smallest deviation—often measured in fractions of a millimetre—can determine whether a lock operates smoothly or fails altogether.

## Impact on Locksmithing Practices

Key cutting is relevant to locksmithing because it sits at the centre of what locksmiths actually *deliver*: controlled access, reliable operation, and security that performs as designed. Every other service—rekeying, master keying, lock repair, lock upgrades, and access control recommendations—depends on keys being made accurately and ethically. If the key is wrong, the lock can appear "faulty," the customer loses confidence, and the security performance of the system can be unintentionally reduced.

A locksmith who understands how keys interact with internal components (pins, wafers, sidebars, sliders) can predict how small cutting errors will show up in real-world symptoms. A key is not just a "shape that fits"—it is a set of precise instructions that positions internal parts at the correct heights (and sometimes rotations or lateral positions) so the plug can turn. That makes key cutting inseparable from lock mechanics, tolerances, and wear. In practice, this knowledge changes how a locksmith approaches everyday jobs, because it helps them choose the *right* cutting method for the situation instead of defaulting to quick duplication.

This is why the distinction between duplication, code cutting, and impressioning matters in professional practice. Duplication is fast and convenient, but it copies whatever is present on the original key—including wear, rounding, and bending. Over time, repeated "copy of a copy" leads to cumulative error, which is a common cause of intermittent operation, stiff turning, and keys that only work if you jiggle them. A locksmith who understands this will often recommend cutting from code (or decoding and cutting to factory depths) when accuracy matters, when the original is worn, or when the lock has tighter tolerances. This improves reliability and reduces call-backs, which directly affects a locksmith's reputation and profitability.

Code cutting is also relevant because it shifts the job from "copying" to "manufacturing to specification." Using depth charts, spacing data, and correct blanks allows the locksmith to restore keys to intended geometry—especially important in commercial systems, restricted keyways, and higher-quality cylinders. That capability influences how locksmiths structure their workflow: maintaining code databases, selecting correct charts and gauges, calibrating machines, and documenting bittings for authorised jobs. It also supports more advanced services like master keying and system audits, where predictable, standardised key geometry is critical.

Impressioning is relevant because it represents problem-solving when standard information isn't available, such as when no key exists and disassembly is undesirable or impractical. Even if a locksmith doesn't use impressioning daily, understanding it deepens practical knowledge of binding, marking, and how locks "communicate" through metal-to-metal contact. It also reinforces professional ethics and legal boundaries, because impressioning, like other access techniques, must be restricted to authorised work.

High-security key cutting has an even bigger influence on locksmithing practice because it adds two additional realities: tighter tolerances and controlled access to blanks. Dimple and sidewinder keys require specialised machines and more careful quality control, and restricted keyways may require authorisation and record-keeping. This changes how locksmiths invest in equipment, how they train staff, how they manage supplier relationships, and how they handle customer expectations (for example, explaining why a restricted key cannot be duplicated without proper authority). It also increases the importance of inspection, deburring, and testing because high-security systems are less forgiving of small errors.

Finally, key cutting influences diagnostic practice. Many "lock problems" are actually key problems—worn keys, wrong blanks, inaccurate duplication, burrs, or spacing errors. A locksmith who treats key cutting as precision work will routinely verify the blank, check alignment (shoulder vs tip indexing), confirm machine calibration, and assess the condition of the original key before cutting. That approach reduces damage to locks, improves customer outcomes, and supports better security advice—because the locksmith can distinguish between a key-related fault and a mechanical fault inside the cylinder.

## Keyway Design as a Security Feature

Keyway design is a security feature because it controls what key (and what tools) can physically enter the lock, how well the key is guided inside the cylinder, and how difficult it is to manipulate the internal components without the proper key. In other words, the keyway is not just a "slot"—it's a deliberately engineered interface that affects access control, pick resistance, durability, and key control.

A keyway is the specifically shaped opening inside a lock cylinder that determines which keys are able to be inserted and operate the lock. Cylinders are designed so that only a key with a matching cross-sectional shape can enter and engage the locking mechanism. This shape is applied along the length of the key blade and can be seen clearly if you look at a key from the front. The profile you see is the keyway (also referred to as the key section), and it acts as the first level of access control by physically preventing incorrect keys from entering the lock.

It is important to understand that the keyway is separate from the cuts or "teeth" on the key. The keyway controls whether a key can be inserted, while the cuts on the key determine whether the lock will turn once the key is inside. A simple way to visualise this is to think of the keyway as a basic shape, such as a circle or triangle. A key with a circular profile will only fit into a circular keyway, and not into a triangular one. However, even if the shape matches, the key will still not operate the lock unless the cuts align correctly with the internal pin configuration.

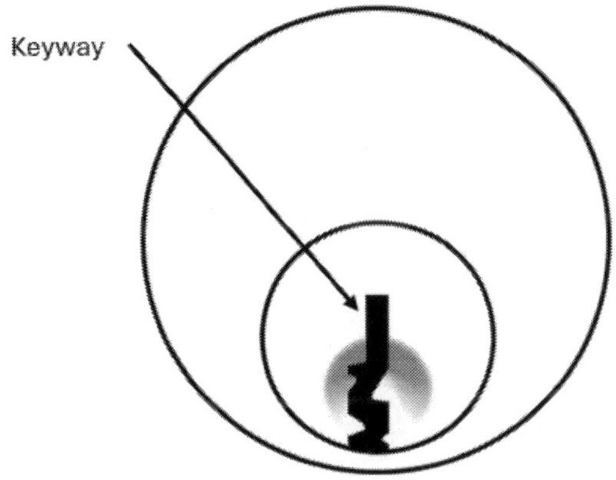


*Figure 42: Keyway on lock cylinder.*

Keyways are created during manufacturing by milling grooves into the key blade, forming the final profile. Some keys have a single groove, while others have multiple grooves that create more complex shapes. The complexity of the keyway can vary depending on the level of security required. Simpler keyways are widely available and not restricted, while more complex designs may require specialised tooling to produce and may be protected or controlled by the manufacturer.

Many lock manufacturers group their designs into what are known as keyway families. Within a single brand, multiple keyways may look similar but are not interchangeable. This allows users to choose different keyways within the same brand to prevent keys from operating neighbouring locks. For example, two properties using the same brand of lock can be supplied with different keyways so that their keys cannot be used interchangeably. This is particularly useful in residential developments, commercial buildings, and large key systems.

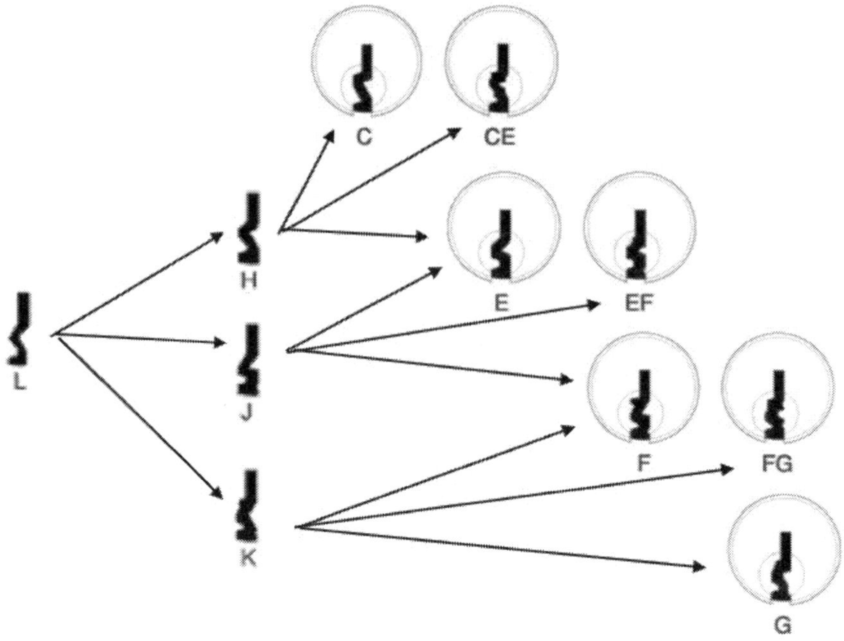


*Figure 43: Schlage 40-100 Multiple Tailpiece Key-in-Knob/Lever Cylinder Classic Key Family example.*

Some keyways are designed to be more versatile through the use of multi-section or multi-milled profiles. These keys are shaped in a way that allows them to fit into more than one keyway within a family. This can be useful in large master key systems that span multiple keyway types, providing flexibility in system expansion while maintaining control over access.

Keyways are also classified as either open or restricted, which relates to how easily keys and blanks can be obtained. An open keyway is widely available and can typically be purchased without special authorisation. Locksmiths and distributors can supply keys and cylinders for these systems directly to customers. Some open keyways may still be limited to certain distributors, but they generally do not require formal approval or documentation to obtain duplicates.

Restricted keyways, on the other hand, provide a higher level of key control. These systems require authorisation from both the manufacturer and the end user before keys or cylinders can be supplied. When a restricted key system is established, formal documentation is created, and any future orders for additional keys must be verified against authorised signatures or credentials. This prevents unauthorised duplication, even if someone has physical access to a key. For example, an individual cannot simply take a restricted key to a distributor and request a copy without proper approval.

Different lock manufacturers use distinct keyway designs to achieve compatibility control, product differentiation, and varying levels of security. For a locksmith, recognising these differences is essential because the keyway determines which blanks to use, how keys are cut, how systems are managed, and what level of security is provided.

## Common Types of Keyway Designs by Manufacturers

### 1. Standard (Open) Keyways

Many manufacturers produce open keyways, which are widely available and easy to duplicate.

**Examples:**

- Kwikset (KW1)
- Schlage (SC1, SC4)
- Weiser (WR series)

**Characteristics:**

- Simple, relatively wide profiles
- Minimal warding (internal grooves)
- Blanks readily available from most suppliers

**Influence on locksmithing:**

- Fast and easy duplication
- Common in residential work
- Lower security due to unrestricted key copying

### 2. Paracentric Keyways

Paracentric keyways are designed with curves and overhangs that obstruct straight-line tool access.

**Used by:**

- Schlage (some commercial profiles)
- Yale (various commercial systems)
- ASSA ABLOY systems

**Characteristics:**

- Twisted or offset internal shape
- Narrow access path
- Forces angled key and tool insertion

**Influence on locksmithing:**

- Increases pick resistance
- Requires more precise key cutting
- Makes servicing and decoding more complex

**3. Restricted Keyways**

Restricted keyways are controlled systems where key blanks are not freely available.

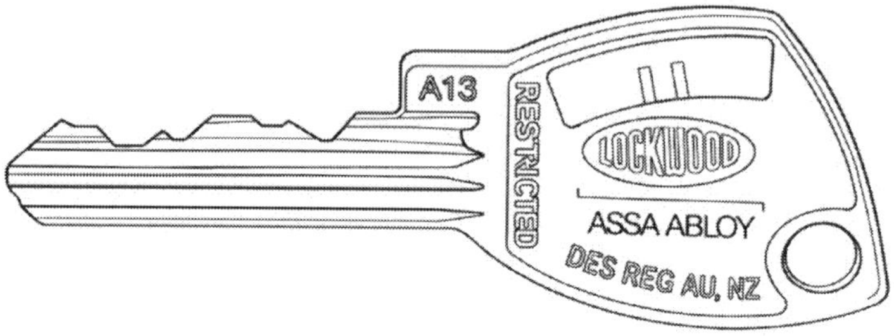


*Figure 44: Restricted key.*

**Used by:**

- Medeco
- ASSA / ASSA ABLOY
- Mul-T-Lock
- Abloy (disc systems)

**Characteristics:**

- Patented or licensed profiles
- Distribution controlled by authorised locksmiths
- Often require key cards or authorisation

**Influence on locksmithing:**

- Requires authorisation to cut keys
- Involves administrative control (not just technical work)
- Common in commercial and government systems

**4. Dimple Keyways**

Dimple systems use flat keys with drilled holes instead of edge cuts.

**Used by:**

- Mul-T-Lock
- KABA
- DOM
- Yale (some models)

**Characteristics:**

- Flat blade inserted horizontally
- Pins engage from multiple directions
- Often combined with sidebars

**Influence on locksmithing:**

- Requires specialised cutting machines (vertical milling)
- Higher precision required
- Common in high-security applications

**5. Laser / Sidewinder Keyways**

These keyways use internal milled grooves rather than traditional cuts.

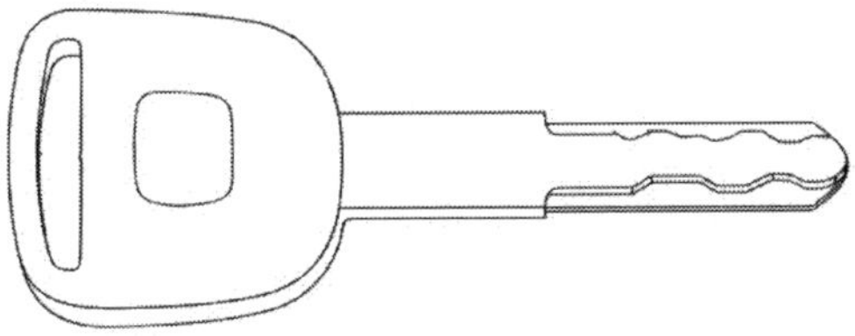

*Figure 45: Sidewinder key.*

**Used by:**

- Automotive manufacturers (e.g., Toyota, BMW, Ford)

- Some high-security residential locks

**Characteristics:**

- Symmetrical or central groove
- Thicker, stronger key blade
- Often reversible (works both ways)

**Influence on locksmithing:**

- Requires CNC or electronic cutters
- More durable and wear-resistant
- Common in automotive locksmithing

**6. Sidebar / Multi-Dimensional Keyways**

These systems use additional locking elements beyond standard pins.

**Used by:**

- Medeco (rotating pins)
- Mul-T-Lock (telescopic pins)
- DOM (side pins and check systems)

**Characteristics:**

- Multiple axes of interaction (lift + rotate or align)
- Keyway precisely guides the key
- Often combined with restricted profiles

**Influence on locksmithing:**

- Requires advanced knowledge and tools
- Extremely tight tolerances
- Errors in cutting are less forgiving

Understanding different keyway designs is essential in locksmithing practice because it directly influences how a locksmith approaches everyday tasks. One of the most fundamental applications is key identification. Accurately recognising the correct keyway ensures that the appropriate key blank is selected. Using an incorrect blank—even if it appears similar—will result in a key that cannot be inserted or operated, leading to wasted time and potential damage to equipment or customer trust.

Keyway design also affects tool selection. Different manufacturers and keyway profiles often require specific cutting machines, gauges, and accessories. For example, tighter or more complex keyways may require precision cutting equipment or specialised cutters to achieve the correct geometry. A locksmith must therefore understand the physical characteristics of the keyway before beginning any cutting or duplication process to ensure accuracy and efficiency.

In terms of security assessment, knowledge of keyway design allows locksmiths to recommend appropriate locking systems based on the level of risk. Open keyways may be suitable for low-security residential applications, while restricted or high-security keyways are more appropriate for commercial or sensitive environments. By understanding how keyways contribute to key control and pick resistance, locksmiths can provide informed advice that aligns with the client's security requirements.

Keyway design is also critical when developing and managing key systems, particularly master key systems. The choice of keyway determines compatibility between cylinders, availability of blanks, and the ability to enforce controlled duplication. Selecting the correct keyway at the design stage ensures the system can be expanded, maintained, and managed effectively over time without compromising security or functionality.

From a troubleshooting perspective, keyway-related issues can often be mistaken for faults within the lock itself. Problems such as incorrect key blanks, burrs on a newly cut key, or debris داخل the keyway can all cause poor operation. A locksmith who understands keyway design can quickly diagnose whether the issue lies with the key, the keyway, or the internal mechanism, avoiding unnecessary repairs or replacements.

Ultimately, keyway design varies significantly between manufacturers because it serves as a core feature for both security and compatibility. For locksmiths, this reinforces the need for a system-specific approach rather than a one-size-fits-all mindset. Each keyway must be understood in terms of its geometry, intended security level, and operational tolerances.

As locksmithing continues to evolve, keyways are becoming increasingly complex. This progression is shifting the profession away from simple mechanical duplication and toward precision engineering, controlled key systems, and high-security design expertise. A modern locksmith must therefore combine practical skills with technical understanding to work effectively across a wide range of locking systems.

In practical terms, this means that restricted keyways significantly enhance security by controlling not only who can use a key, but also who can reproduce it. This is particularly important in commercial environments, government facilities, and any situation where key control is critical. By combining physical keyway design with administrative controls, restricted systems provide a layered approach to security.

Overall, the keyway is a fundamental feature of any lock system. It acts as the first barrier to unauthorised access, determines compatibility between keys and cylinders, and plays a major role in both physical and administrative security. Understanding keyways is essential for locksmiths, as it influences lock selection, key cutting, system design, and long-term key management.

Keyway design plays a critical role in increasing the security of a lock by controlling both physical access and the ability to duplicate keys. One of the primary ways this is achieved is through restricted keyway profiles. In these systems, the key blank is not widely available and may be protected by licensing agreements or patents. Access to these blanks is typically limited to authorised locksmiths or approved facilities. As a result, even if someone temporarily gains possession of a key, producing an unauthorised duplicate becomes significantly more difficult without proper authorisation and access to the correct materials. This form of key control is especially important in commercial master key systems and higher-security residential applications.

Another important security feature is the use of narrow, complex, or paracentric keyway designs. These keyways incorporate curves, offsets, and warding that restrict direct access to the internal components of the lock. Paracentric keyways, in particular, include overhanging sections that prevent tools from moving in a straight line. Heavy warding forces picking tools and tension devices to approach pins at awkward angles, reducing control and feedback. Additionally, tight keyways limit the available space for tool movement, making it more difficult to manipulate pins accurately. In practice, this increases the likelihood of errors such as oversetting pins or losing tactile feedback, thereby raising the level of difficulty for unauthorised entry.

Keyways also serve as precision guiding systems in high-security locks that incorporate secondary locking mechanisms. In these systems, the key does more than lift top pins—it may also interact with side pins, sliders, rotating elements, or sidebar mechanisms. For these components to function correctly, the key must be positioned with a high degree of accuracy. The keyway ensures this by acting like a guide rail, maintaining consistent alignment in both vertical and lateral directions. This precise positioning allows all internal elements to align correctly, improving both the security and reliability of the lock. However, this also means that key cutting must be extremely accurate, as even minor deviations can prevent proper operation.

In addition to controlling duplication and increasing pick resistance, keyway design can also limit the effectiveness of common bypass techniques. Some bypass methods rely on accessing internal components directly or manipulating parts of the lock beyond the standard pin interface. Carefully designed warding within the keyway can block access to the rear of the plug, restrict interaction with sidebars or sliders from the front, and prevent the use of certain tools such as rakes or jigglers. Even if the internal pin configuration is relatively simple, a well-designed keyway can significantly increase resistance to non-destructive entry methods.

Despite these advantages, highly secure keyway designs can introduce practical trade-offs. Tight and complex keyways are often more sensitive to dirt, debris, and poor lubrication, which can lead to binding or reduced performance. They also require a higher level of accuracy in key cutting, as small errors become more noticeable and can affect operation. From a user perspective, keys may feel less smooth when inserting or turning, particularly if they are worn or slightly damaged. Additionally, servicing these locks can be more complex, as locksmiths must have access to the correct blanks, specialised tools, and system-specific knowledge.

For locksmiths, understanding keyway design is essential in everyday practice. When selecting a lock, the keyway must be matched to the security requirements of the environment. For example, residential applications may prioritise convenience and ease of duplication, while rental properties may require controlled key systems, and commercial settings often demand restricted keyways with formal key management procedures. The choice of keyway directly affects how the system is used, maintained, and secured over time.

Keyway design also influences rekeying and system management. It determines compatibility between cylinders, the availability of replacement blanks, and the ability to enforce key control policies. In master key systems, the keyway must support both current needs and future expansion, making long-term planning an important consideration. If a keyway becomes obsolete or restricted, it can complicate maintenance and replacement.

In key cutting and quality control, tighter keyways demand greater precision. Accurate depth and spacing are critical, as even small deviations can result in rough insertion, stiff operation, or intermittent function. These issues can often be mistaken for mechanical faults within the lock itself. For this reason, professional locksmiths frequently prefer code cutting or decoding combined with depth charts rather than duplicating worn keys, particularly when working with high-security systems.

Finally, keyway knowledge is essential for diagnostics and service work. When a lock does not operate smoothly, the problem may not be the lock mechanism but the key or keyway interaction. Issues such as burrs on a newly cut key, incorrect key blanks, debris داخل the keyway, or wear between the key and plug can all affect performance. A clear understanding of keyway design allows locksmiths to identify the root cause quickly and apply the correct solution without unnecessary replacement of components.

Common terminology used in the industry reflects these design features. A restricted keyway refers to a system with controlled blank distribution and authorised duplication. A paracentric keyway includes curves and overhangs that restrict tool access. A high-ward keyway contains additional internal ridges that increase complexity, while a tight tolerance keyway has minimal clearance between the key and the cylinder to improve precision and security. Together, these features demonstrate how keyway design is a fundamental element in modern locksmithing and an essential consideration in both security and practical application.

**Impact on Locksmithing Practices**

Keyway design is highly relevant to anyone interested in locksmithing because it sits at the foundation of how locks control access, how keys are produced, and how security is achieved in real-world applications. While many beginners focus on pins, springs, and key cutting, the keyway is actually the first point of interaction between the user, the key, and the lock. It determines not only whether a key can enter the lock, but also how accurately it is guided and how resistant the lock is to manipulation. This makes it a critical concept that influences nearly every aspect of locksmithing practice.

At a practical level, understanding keyways directly affects lock selection and system design. A locksmith is not simply choosing a lock body—they are selecting a keyway system that matches the client's security needs. For example, a residential client may prioritise convenience and easy key duplication, making an open keyway suitable. In contrast, a commercial facility or government site may require a restricted keyway to control duplication and maintain accountability over who can obtain keys. This means that keyway knowledge influences decisions about security policy, not just hardware.

Keyway design also has a strong influence on key control, which is one of the most important aspects of modern locksmithing. Restricted keyways limit the availability of key blanks and require authorisation for duplication. This introduces an administrative layer of security that goes beyond the physical lock itself.

For a locksmith, this means managing documentation, verifying authority, and maintaining system integrity over time. In practice, this shifts the role from simple key cutting to managing controlled access systems.

From a technical perspective, keyway design affects how locks resist picking and manipulation. Narrow, paracentric, or heavily warded keyways restrict tool access and reduce the effectiveness of picking techniques. For a locksmith, this has two major implications. First, it increases the difficulty of non-destructive entry when authorised access is required. Second, it requires a deeper understanding of tool selection, positioning, and technique. A locksmith working with complex keyways must adapt their approach and cannot rely on basic methods used for simpler locks.

Keyways also play a critical role in guiding the key and ensuring precise alignment within the lock. In high-security systems, the key may interact with multiple components such as side pins, sliders, or sidebar mechanisms. The keyway ensures that the key is positioned exactly where it needs to be for all these elements to align correctly. For locksmiths, this means that accuracy in key cutting becomes far more important. Even small deviations in depth, spacing, or key profile can result in failure, especially in tight-tolerance systems.

This directly influences key cutting practices. Locksmiths must be able to identify the correct keyway, select the correct blank, and ensure precise alignment during cutting. In many cases, especially with high-security or tight keyways, duplicating a worn key is not acceptable because it introduces errors. Instead, locksmiths rely on decoding, depth charts, and code cutting to restore the key to factory specifications. This represents a shift from basic duplication to precision-based work.

Keyway design also affects diagnostics and troubleshooting. When a customer reports that a lock is “sticky” or difficult to use, the issue may not be a mechanical fault inside the lock. It could be caused by an incorrect key blank, poor cutting, burrs, or debris in a tight keyway. A locksmith with a strong understanding of keyways can quickly identify whether the issue lies with the key, the keyway, or the internal mechanism. This improves efficiency and prevents unnecessary replacement of locks.

There are also important trade-offs that influence locksmithing decisions. High-security keyways offer better control and resistance to attack, but they can be more sensitive to dirt, wear, and cutting errors. They may also require specialised tools, authorised blanks, and additional training. Locksmiths must balance these factors against the environment in which the lock will be used, such as high-traffic areas, dusty conditions, or situations requiring frequent key duplication.

Finally, keyway design influences long-term system management, particularly in master key systems. The choice of keyway affects compatibility between cylinders, availability of blanks, and the ability to expand or maintain the system over time. A poorly chosen keyway can create problems years later if it becomes obsolete or difficult to source. For this reason, experienced locksmiths consider not only immediate needs but also future scalability and maintainability.

## Material Science in Lock Durability

Material science constitutes a fundamental discipline in understanding lock performance, durability, and security characteristics. The selection and application of specific materials in lock components directly determines how effectively a locking mechanism can resist wear, environmental degradation, and deliberate attack. This analysis synthesizes current knowledge regarding the role of material properties in lock design and function, with particular emphasis on practical applications for locksmithing professionals.

## Mechanical Strength and Component Selection

The foundational principle of lock durability rests upon the mechanical properties of materials used in critical load-bearing components. Hardened steel represents the primary material choice for components subjected to high stress and potential attack, including shackles, bolts, and locking bars. The selection of hardened steel is justified by its exceptional tensile strength and resistance to mechanical failure modes such as cutting, sawing, and forced entry attempts [147]. This material choice reflects a deliberate engineering decision to prioritize structural integrity under extreme loading conditions.

In contrast to external load-bearing components, internal mechanisms such as pins require fundamentally different material properties. Brass and nickel silver are intentionally selected for pin manufacture despite their lower absolute strength compared to hardened steel [147]. This apparent paradox in material selection reflects a sophisticated understanding of tribological principles: softer metals reduce wear on the key during normal operation and provide smoother mechanical action while maintaining adequate strength for routine use cycles. The deliberate selection of materials with different hardness characteristics demonstrates how material science enables the optimization of competing performance objectives within a single mechanical system.

Understanding which materials are used by specific manufacturers—and why—helps locksmiths make informed decisions about durability, application suitability, and security performance. Below are practical, real-world examples linking materials → lock components → manufacturers → intended use.

<u>ASSA ABLOY / Lockwood (Australia & Global)</u>

### Common Materials Used

- Brass cylinders and pins
- Hardened steel shackles and anti-drill inserts
- Nickel-silver keys (restricted systems)

### Example Products

- Lockwood 001/002 Deadlatch
- Lockwood Twin / Generation Six restricted key systems

### Why These Materials Are Used

Lockwood uses brass extensively for cylinders because it offers excellent machinability, corrosion resistance, and smooth operation—ideal for high-use doors. In restricted systems, nickel-silver keys are preferred because they are harder than brass, reducing wear and maintaining tight tolerances.

**Security Enhancement**

Higher-security ASSA ABLOY systems incorporate hardened steel rods and pins inside the cylinder to resist drilling attacks.

<u>Abloy (Finland – High-Security Disc Detainer Systems)</u>

**Common Materials Used**

- Hardened stainless steel discs and bodies
- Corrosion-resistant alloys (marine-grade)

**Example Products**

- Abloy Protec2 / Protec2 CLIQ
- Abloy PL series padlocks

**Why These Materials Are Used**

Abloy locks are designed for extreme environments (e.g., marine, mining, infrastructure). Stainless steel is used almost exclusively because it resists corrosion, freezing, and contamination far better than traditional brass systems.

**Security Enhancement**

The use of **hardened steel disc stacks** provides exceptional resistance to picking and drilling, while maintaining durability in harsh conditions.

<u>Schlage (Residential & Commercial – USA)</u>

**Common Materials Used**

- Brass pins and plug
- Steel springs and bolts
- Zinc die-cast housings (lower-cost models)

**Example Products**

- Schlage B60/B560 Deadbolts
- Schlage Everest / Primus high-security systems

**Why These Materials Are Used**

Schlage balances cost and durability:

- **Brass** → smooth operation and corrosion resistance
- **Steel bolts** → strength against forced entry
- **Zinc housings** → cost-effective manufacturing for residential use

**Security Enhancement**

Higher-end systems (e.g., Primus) include hardened steel inserts and sidebars, increasing resistance to picking and drilling.

<u>Mul-T-Lock (High-Security – Israel)</u>

**Common Materials Used**

- Hardened steel pins (telescopic pin-in-pin system)
- Brass or nickel-silver cylinders
- Steel anti-drill plates and inserts

**Example Products**

- Mul-T-Lock Interactive+
- Mul-T-Lock MT5+

**Why These Materials Are Used**

Mul-T-Lock combines precision brass machining with hardened steel security elements. The telescopic pin system requires extremely tight tolerances, supported by durable materials.

**Security Enhancement**

Extensive use of hardened steel inserts and complex pin systems provides strong resistance to picking, bumping, and drilling.

<u>Master Lock (Padlocks – Consumer to Industrial)</u>

**Common Materials Used**

- Laminated steel bodies (stacked steel plates)
- Brass cylinders (in many models)
- Hardened boron alloy steel shackles

**Example Products**

- Master Lock No. 1 (brass padlock)
- Master Lock laminated series (e.g., No. 3)

**Why These Materials Are Used**

- **Laminated steel** → cost-effective strength
- **Boron alloy shackles** → high cut resistance
- **Brass internals** → smooth operation

**Security Enhancement**

Higher-end models include anti-shim features and hardened shackles to resist cutting tools.

<u>Yale (Residential & Smart Locks – Global)</u>

**Common Materials Used**

- Brass cylinders and pins
- Zinc alloy housings
- Steel locking bolts

**Example Products**

- Yale Deadbolts
- Yale Smart Locks (Assure series)

**Why These Materials Are Used**

Yale focuses on mass-market durability and affordability:

- Brass ensures reliability
- Zinc alloys reduce cost
- Steel provides structural strength

**Security Enhancement**

In smart locks, electronic components are paired with traditional metal locking mechanisms, requiring both mechanical and electronic durability.

Material selection in lock design varies significantly depending on the intended application and operating conditions. Environmental factors play a major role; for example, locks used in marine or coastal environments typically require stainless steel components to resist corrosion, whereas indoor applications often use brass or zinc alloys due to their cost-effectiveness and adequate durability. The required level of security also influences material choice, with high-security locks incorporating hardened steel inserts or components to resist drilling, cutting, and other forms of attack. In addition, manufacturers must balance cost and durability, meaning consumer-grade hardware may prioritise affordability, while commercial or industrial locks are built with more robust materials to withstand frequent use and harsher conditions.

For locksmiths, these material differences have direct practical implications. Material choice affects product recommendations, such as selecting corrosion-resistant locks for coastal installations or heavy-duty hardware for high-traffic environments. It also influences expectations around service life, as components made from zinc alloys may wear more quickly than those machined from solid brass. Understanding materials is essential for accurate failure diagnosis, helping to identify issues related to corrosion, mechanical wear, or material fatigue. Furthermore, harder materials used in high-security systems require greater precision in key cutting and may demand specialised tooling, reinforcing the importance of matching equipment and techniques to the materials being worked with.

Ultimately, there is no single "best" material in locksmithing—only the most appropriate material for a given application. High-quality lock systems often combine multiple materials, such as brass for smooth operation and hardened steel for enhanced security, achieving a balance between durability, usability, and resistance to attack. For locksmiths, recognising and understanding these material choices is critical to making informed decisions, designing effective security solutions, and ensuring long-term reliability in the systems they install and maintain.

## Wear Resistance and Friction Management

Locks function as mechanical systems characterized by repeated contact cycles between moving components. Over the operational lifetime of a lock, thousands or even millions of engagement cycles occur, creating significant wear potential if materials are not carefully selected. Brass emerges as a particularly suitable material for cylinder construction due to its inherently low friction properties and resistance to seizing compared to harder metallic alternatives [147]. The low friction characteristics of brass reduce the mechanical resistance encountered during key insertion and rotation, thereby extending the operational life of both the key and the cylinder.

Advanced lock systems employ additional strategies to manage friction and wear beyond simple material selection. Specialized coatings and alloy blends represent technological enhancements designed to further reduce friction coefficients and extend service life [147]. These surface treatments modify the tribological interface without requiring wholesale changes to base material composition, offering a cost-effective approach to performance enhancement. The consequences of inadequate wear management are substantial: poor material selection or insufficient friction control leads to rapid wear progression, resulting in sloppy tolerances, key sticking, and eventual mechanical failure [147].

Tribology, the study of friction, wear, and lubrication, provides the theoretical foundation for understanding why surface treatments are critical in lock design [147, 148]. In locking mechanisms, tribological interfaces occur at multiple points: between pins and plugs, along keyways, and within rotating or sliding components such as cams and bolt mechanisms [147]. These interfaces experience continuous mechanical stress, and friction at these surfaces directly determines component longevity and operational reliability.

The fundamental principle underlying coating applications is that most mechanical failures in locks occur at the surface level rather than within the bulk material [147, 148]. When friction coefficients are elevated, components experience accelerated wear, rough operation, and reduced service life [147]. Specialised coatings function by reducing friction coefficients at critical interfaces, thereby enabling smoother movement, minimizing wear rates, and extending the operational lifespan of locking components [147, 148].

Research on tribological systems demonstrates that surface treatment processes can significantly alter wear behaviour and failure mechanisms [147]. For instance, studies on locking pins subjected to fretting wear, a particular concern in mechanical locking systems, show that surface treatments such as plasma nitriding and low-friction coatings substantially improve wear resistance compared to untreated substrates [147]. The wear mechanisms themselves change with surface treatment; while untreated surfaces may experience uniform degradation, treated surfaces often exhibit localized wear patterns that reflect specific friction points or contamination zones [147].

**Low-Friction Surface Coatings: Nickel Plating and Related Technologies:** Nickel plating represents one of the most widely adopted low-friction surface coating strategies in lock manufacturing [148]. Nickel coatings provide a harder, smoother surface than raw brass, the traditional base material for lock components [148]. This enhanced surface hardness reduces wear rates, while the smoother finish improves corrosion resistance—a critical consideration in locks exposed to humid or corrosive environments [148].

The effectiveness of nickel plating in lock applications stems from its dual functionality: it simultaneously reduces friction coefficients and increases surface hardness [147, 148]. Research on metallic coatings applied to steel substrates demonstrates that nickel coatings improve wear resistance across multiple wear mechanisms, including oxidative and adhesive wear [148]. In lock cylinders, nickel-plated pins maintain consistent key feel over extended operational periods, as the coating prevents the surface degradation that would otherwise cause binding or rough operation [148].

Beyond simple nickel plating, electroplating and anodising techniques extend the application of low-friction coatings to external lock components [148]. These processes are particularly valuable for padlocks and external lock bodies, where environmental exposure creates additional wear and corrosion challenges [148]. The cost-effectiveness of these coating approaches, compared to manufacturing entire locks from premium materials, makes them economically attractive for manufacturers seeking to balance performance with affordability [148].

**Polytetrafluoroethylene and Dry Lubricant Coatings:** PTFE (polytetrafluoroethylene), commonly known as "Teflon," represents an advanced category of low-friction coatings designed to provide exceptionally low friction coefficients [148]. These coatings are sometimes applied to internal lock components or used in combination with dry lubricants, creating hybrid systems that minimize binding between pins and chambers [148].

The tribological advantages of PTFE-based coatings are well-established in mechanical systems [147]. PTFE coatings exhibit self-lubricating properties that reduce friction coefficients and energy dissipation at sliding interfaces [147]. However, research indicates that while PTFE coatings excel at friction reduction, they may exhibit higher wear rates than alternative coating systems under certain conditions [147]. This trade-off between friction reduction and wear resistance necessitates careful selection based on specific application requirements [147].

In automotive and high-use commercial lock systems, PTFE-based treatments ensure smooth operation over thousands of operational cycles [148]. The coating's effectiveness in reducing binding—a common failure mode in locks subjected to frequent use, makes it particularly valuable in applications where reliability and consistent performance are paramount [148].

**Hard Coatings for Security and Wear Resistance:** High-security lock manufacturers employ hard coatings specifically designed to resist drilling, picking, and abrasive wear [148]. These coatings typically involve carbide or hardened steel treatments applied to critical security elements such as anti-drill pins or inserts [148]. The strategy differs from low-friction coatings in that it prioritizes hardness and resistance to attack over smoothness of operation [148].

Research on hard coating systems demonstrates that carbide and hardened steel coatings significantly increase resistance to abrasive wear and mechanical attack [147, 148]. In security applications, these coatings serve a dual purpose: they protect against deliberate attack while simultaneously reducing wear from normal operational use [148]. The combination of a brass base material—which provides smooth operation and machinability—with hardened, coated security elements represents an optimized approach to balancing performance with security [148].

The wear mechanisms in hard-coated systems differ from those in low-friction systems [147]. Hard coatings typically experience abrasive and oxidative wear rather than adhesive wear, and the localized nature of wear in coated systems allows manufacturers to identify specific failure points and contamination sources [147].

**Advanced European Approaches: Precision Alloys and Integrated Coating Systems:** European lock manufacturers have developed sophisticated approaches combining precision alloy blends with integrated surface treatments [148]. These systems represent the current state-of-the-art in lock design, where material science and surface engineering converge to optimize multiple performance characteristics simultaneously [148].

DOM cylinders exemplify this integrated approach, featuring specially treated brass alloys combined with hardened and coated steel elements [148]. This combination balances the smooth operation required for consistent key feel with the wear and manipulation resistance demanded by security applications [148]. Similarly, ABLOY's disc detainer systems utilize hardened stainless steel components with protective coatings, ensuring long-term durability in harsh environments such as marine or industrial settings where corrosion and abrasive wear present significant challenges [148].

The effectiveness of these integrated systems reflects a fundamental principle in materials engineering: targeted enhancement of critical surfaces, rather than wholesale material replacement, achieves superior performance at lower cost [148]. By applying hardened and coated elements only where security or wear resistance is critical, manufacturers maintain the operational advantages of traditional materials while addressing specific failure modes [148].

**Alloy Engineering and Material Composition Optimization:** Beyond surface coatings, alloy blending represents a complementary strategy for enhancing lock performance [148]. Rather than using pure metals, manufacturers engineer alloys with optimized compositions to improve specific properties including machinability, wear resistance, and corrosion performance [148].

Brass used in lock cylinders is frequently modified with small quantities of additional elements, such as zinc or lead substitutes, to enhance these properties [148]. These compositional adjustments occur at the material level, affecting the entire component rather than just the surface [148]. Research on alloy systems demonstrates that carefully controlled composition can simultaneously improve wear resistance, corrosion resistance, and machinability, reducing manufacturing costs while enhancing performance [148].

High-security locks often employ nickel-silver alloys for keys, which are harder and more wear-resistant than standard brass [148]. These alloys reduce key deformation over time, maintaining the precision fit required for tight-tolerance keyways [148]. The improved wear resistance of nickel-silver alloys reflects their enhanced hardness and resistance to plastic deformation, properties that become increasingly important as keys are used repeatedly in high-security applications.

**Practical Implications for Locksmithing and Maintenance:** The widespread adoption of coatings and engineered alloys has significant practical implications for locksmiths and lock maintenance professionals [148]. First, coated and alloy-optimised components operate more smoothly and consistently over extended periods, reducing the frequency of maintenance interventions [148]. Second, these enhancements influence maintenance practices; locks with advanced coatings may require specific lubricants, often dry or non-reactive types, to avoid degrading the coating surface [148].

Third, wear pattern analysis becomes more complex with coated components [148]. Untreated materials typically exhibit uniform wear patterns, while coated surfaces often show localized wear reflecting specific friction points or contamination zones [148]. This difference in wear morphology requires locksmiths to develop new diagnostic approaches, as traditional wear pattern interpretation may not apply to coated systems [148].

Research on tribological systems confirms that wear mechanisms vary significantly between coated and uncoated surfaces [147]. Coated surfaces may experience abrasive, oxidative, or adhesive wear depending on coating type and operating conditions [147]. Understanding these mechanisms enables more accurate troubleshooting and more effective maintenance strategies [147, 148].

**Cost-Effectiveness and Performance Optimization:** A central advantage of coating and alloy-based approaches is their cost-effectiveness relative to performance gains [148]. Rather than manufacturing entire locks from premium materials such as solid stainless steel or exotic alloys, manufacturers selectively enhance critical surfaces where friction and wear occur [148]. This targeted approach allows manufacturers to achieve high performance without excessive cost, making advanced lock designs accessible across residential, commercial, and high-security markets [148].

Research on coating systems demonstrates that surface treatments can provide wear resistance comparable to or exceeding that of bulk premium materials, while maintaining the cost advantages of traditional base materials [147, 148]. This economic efficiency explains the widespread adoption of coating technologies across the lock manufacturing industry [148].

**Comparative Performance and Material Selection:** Two locks manufactured from similar base materials can exhibit dramatically different performance characteristics due to variations in coating type and alloy composition [148]. This variability underscores the importance of understanding coating and alloy technologies for locksmiths engaged in product selection and troubleshooting [148].

Research on comparative coating performance indicates that selection of appropriate coating systems depends on specific application requirements [147]. Low-friction coatings optimize smoothness and consistency of operation, while hard coatings prioritize security and resistance to deliberate attack [147, 148]. Integrated systems combining multiple coating types and alloy elements represent the most sophisticated approach, balancing multiple performance objectives simultaneously [148].

Specialised coatings and engineered alloy blends have become fundamental to modern lock design, enabling manufacturers to enhance performance, durability, and wear resistance without fundamentally altering base materials [148]. These technologies are grounded in tribological principles, where surface-level interactions drive component degradation [147]. Through low-friction coatings such as nickel plating and PTFE treatments, hard coatings for security applications, and carefully engineered alloy compositions, manufacturers achieve optimized performance across multiple dimensions [148]. For locksmiths, understanding these technologies supports better product selection, more accurate troubleshooting, and improved long-term outcomes for clients [148]. The continued evolution of coating and alloy technologies promises further enhancements in lock performance, reliability, and cost-effectiveness [148].

### Corrosion Resistance and Environmental Durability

Environmental exposure represents a critical challenge for lock durability, particularly in applications involving moisture, salt air, or chemical exposure. Locks deployed in external environments or harsh conditions must incorporate materials specifically selected for corrosion resistance. Stainless steel, brass alloys, and specially coated metals constitute the primary material options for corrosion-resistant lock construction [147]. The selection between these options depends upon the specific environmental conditions anticipated during the lock's service life.

Marine-grade padlocks exemplify the application of corrosion-resistant materials in demanding environments, typically incorporating stainless steel components throughout their construction to resist rust formation and material degradation [147]. In contrast, locks intended for indoor use may employ plated steel or zinc alloys, which provide adequate corrosion protection for controlled indoor environments at lower cost than stainless steel alternatives. The consequences of inadequate corrosion resistance are severe: corrosion processes lead to internal binding of moving components, weakened structural integrity of load-bearing elements, and complete lock failure if not properly managed through appropriate material selection [147].

**Material Science and Security Performance**

The security characteristics of locks are fundamentally influenced by material science principles and material selection strategies. Hardened steel inserts and anti-drill plates represent specialized material applications designed to resist common attack methodologies including drilling and prying [147]. These reinforced components increase the mechanical resistance encountered by attackers attempting to compromise lock security through direct mechanical attack.

High-security lock designs frequently employ strategic combinations of multiple materials to balance usability with attack resistance [147]. A representative example combines a brass cylinder, selected for its superior operational characteristics and low friction properties, with hardened steel pins or inserts specifically designed to resist drilling attacks. This multi-material approach demonstrates sophisticated engineering that recognizes the fundamental incompatibility between optimal operational performance and maximum attack resistance, requiring compromise solutions that integrate materials with complementary properties.

<u>Core Features of High-Security Locks</u>

**1. Complex Key and Keyway Design**

High-security locks use advanced key profiles that go beyond traditional edge-cut keys. These include dimple keys, laser/sidewinder keys, and restricted keyways with complex warding. The keyway itself is often paracentric (curved and obstructed), limiting tool access and guiding the key with high precision.

Restricted keyways are a major feature, ensuring that key blanks are not freely available. This provides key control, meaning duplication can only occur through authorised channels. For locksmiths, this introduces administrative security alongside physical security.

**2. Secondary Locking Mechanisms (Sidebar Systems)**

Many high-security locks include a secondary locking system such as a sidebar, which must align in addition to the main pins. This creates a dual requirement for operation.

Examples include:

- Sidebars that engage with grooves in pins or sliders
- Rotating pins that must be aligned to specific angles

- Sliders or finger pins that align gates for sidebar retraction

The key insight: the lock will only open when *multiple independent systems* align simultaneously.

**3. Advanced Pin Technologies**

High-security cylinders use specialised pins to resist manipulation:

- Spool pins create false feedback during picking
- Serrated pins catch at multiple points, increasing difficulty
- Telescopic (pin-in-pin) systems require dual alignment
- Anti-bump pins reduce vulnerability to bump keys

These features increase resistance to picking and bumping by disrupting tactile feedback and increasing precision requirements.

**4. Anti-Drill and Physical Reinforcement**

High-security locks incorporate hardened materials to resist destructive entry:

- Hardened steel pins and inserts to blunt drill bits
- Carbide rods or plates protecting critical areas
- Anti-snap and anti-pull features in euro cylinders

These features ensure that even if an attacker bypasses picking resistance, physical attacks remain difficult and time-consuming.

**5. Tight Tolerances and Precision Engineering**

High-security locks are manufactured to very tight tolerances. This means:

- Minimal clearance between key and keyway
- Precise pin alignment requirements
- High sensitivity to incorrect cuts or wear

Result: keys must be cut accurately, and duplication errors are more likely to cause failure.

High-security lock manufacturers each take a distinct engineering approach to improving resistance against picking, drilling, bumping, and unauthorised key duplication. These differences are reflected in the internal mechanisms, key designs, and overall system architecture used by each brand.

Medeco (USA) is well known for its use of angled key cuts that require pins to both lift and rotate into precise positions. This introduces a second dimension of alignment beyond standard pin tumbler systems. In addition, Medeco incorporates a sidebar mechanism that must align with grooves in the pins before the lock will operate. This combination of rotation and sidebar engagement provides strong resistance to picking and bumping, while also supporting controlled key systems.

Mul-T-Lock (Israel) is widely recognised for its telescopic pin design, often referred to as a pin-in-pin system. Each pin stack consists of an outer and inner pin, requiring two levels of alignment for correct operation. Many systems also include interactive elements within the key, adding another layer of complexity. Combined with hardened inserts for drill resistance, these locks are commonly used in commercial and high-risk environments where durability and security are critical.

ASSA ABLOY (global group) encompasses a range of brands, including ASSA and Yale, and offers a broad portfolio of high-security locking solutions. Their systems typically incorporate restricted keyways, sidebar mechanisms, and advanced master keying capabilities. A key strength of ASSA ABLOY products is their integration with access control systems, allowing both mechanical and electronic security solutions to be combined within the same platform.

ABLOY (Finland) takes a fundamentally different approach by using a disc detainer mechanism rather than traditional pins. In this design, a series of rotating discs must be aligned to allow a sidebar to move and the lock to open. Because there are no springs or standard pins, these locks are highly resistant to picking and perform exceptionally well in harsh environments. As a result, ABLOY systems are often used in critical infrastructure and outdoor applications.

DOM (Germany) combines conventional pin tumbler systems with additional side pins and check mechanisms, as seen in products such as the ix series. These secondary systems interact with the key in multiple ways, increasing both security and precision requirements. DOM places a strong emphasis on modular design, allowing locksmiths to adapt and service systems efficiently while maintaining high levels of security.

Kaba (dormakaba) (Switzerland) specialises in dimple key systems that use multiple rows of pins arranged in different orientations. These systems often incorporate advanced sidebar mechanisms and precise pin alignment requirements, along with strong patent protection to enforce key control. The result is a highly controlled and secure system suitable for complex master key applications.

Evva (Austria) uses slider-based locking mechanisms in systems such as EPS and MCS. Instead of traditional pins, these locks rely on movable sliders that must align gates to allow operation. Some models also incorporate magnetic elements, adding a non-mechanical layer of verification. This multi-dimensional interaction between key and lock provides an additional level of security beyond conventional designs.

Overall, these manufacturers demonstrate how high-security lock design can vary significantly. Each system reflects a different balance of mechanical complexity, material strength, and key control, reinforcing the idea that there is no single solution to security—only systems engineered for specific risks and applications.

### Manufacturing Methods and Material Quality

The relationship between material properties and manufacturing methods significantly influences lock performance and durability. Lower-cost lock designs frequently employ zinc die-cast construction, commonly referred to as pot metal, for housing components [147]. While zinc die-casting offers economic advantages through rapid production and low material costs, this manufacturing approach produces housings that are less durable and more prone to cracking under mechanical stress compared to alternatives. The brittleness of zinc die-cast materials limits their suitability for applications requiring high reliability or extended service life.

Higher-quality lock designs utilize machined brass or steel components manufactured through precision machining processes [147]. These manufacturing methods produce components with superior structural integrity and substantially longer service life compared to die-cast alternatives. The precision of manufacturing processes, combined with the inherent properties of the selected materials, directly affects the tolerances achievable in lock assembly and the overall performance characteristics of the completed mechanism. Manufacturing precision and material quality represent interdependent factors that jointly determine lock reliability and durability.

### Temperature and Environmental Conditions

Thermal effects on lock materials represent an important but frequently overlooked aspect of material science in lock design. Metals exhibit thermal expansion and contraction in response to temperature changes, and these dimensional changes directly affect the clearances and tolerances within lock mechanisms [147]. Temperature-induced dimensional changes can alter the fit between components, potentially affecting lock operation and security characteristics.

Extreme environmental conditions impose additional material science challenges. In extreme cold environments, certain materials may become brittle and prone to fracture under mechanical stress [147]. Conversely, in high-temperature environments, lubricants may break down more rapidly, reducing their effectiveness in managing friction and wear. Well-designed locks account for these environmental factors through careful material selection that maintains dimensional and mechanical stability across the range of environmental conditions anticipated during the lock's operational life.

### Impact on Locksmithing Practices

Understanding material science principles directly informs practical locksmithing decisions and professional recommendations. When recommending locks for specific applications, locksmiths must consider multiple environmental and operational factors including indoor versus outdoor deployment, expected usage intensity (high-traffic versus occasional use), and specific security requirements [147]. Material science knowledge enables locksmiths to match lock specifications to application requirements, ensuring appropriate performance and durability.

During lock servicing and maintenance, recognition of material wear patterns provides diagnostic information regarding lock condition and failure mechanisms. Identifying binding caused by corrosion, wear patterns indicating inadequate material selection, or component fatigue resulting from excessive stress enables locksmiths to diagnose problems accurately and recommend appropriate remediation [147]. Material science knowledge guides maintenance practices, including selection of appropriate lubricants and cleaning methods compatible with specific lock materials. Different materials require different maintenance approaches; for example, brass components may require different cleaning and lubrication protocols compared to hardened steel components.

Material science underpins every aspect of lock durability, reliability, and security performance. The choice of materials for lock components determines how effectively a lock resists wear, withstands deliberate attack, performs under environmental stress, and maintains reliability over extended operational periods [147]. For modern locksmithing practice, material science knowledge extends professional capability beyond simple installation and repair toward informed selection of appropriate hardware, system design optimization, and long-term asset management. The integration of material science principles into locksmithing practice represents a critical advancement in professional competency, enabling locksmiths to provide superior service through evidence-based material selection and maintenance recommendations.

## Chapter 3 Review Questions

### Question 1

A locksmith installs a high-precision pin-tumbler cylinder with extremely tight manufacturing tolerances. During a later security test, the lock is found to be unexpectedly vulnerable to bumping.

What explains this vulnerability?

A. Tight tolerances allow energy to transfer efficiently through the pin stacks
B. Loose tolerances prevent pins from moving freely
C. High-security locks cannot resist bumping
D. Bumping only affects worn locks

**Question 2**

A residential lock that has been in service for over a decade begins opening occasionally with a similar but incorrect key.

What is the most likely cause?

A. A manufacturing defect in the lock
B. Excessive lubrication inside the cylinder
C. Wear increasing tolerances inside the lock mechanism
D. A faulty strike plate

**Question 3**

A newly installed mortice lock begins to jam only a few months after installation. Inspection shows that the strike plate and bolt are slightly misaligned.

What is the most likely reason for the failure?

A. The lock contains defective springs
B. Misalignment introduced abnormal stresses and accelerated wear
C. The lock mechanism is incompatible with the door
D. Mortice locks always fail within a year

**Question 4**

A locksmith analysing a lock after a break-in notices that the euro cylinder has snapped cleanly at the centre fixing point.

What type of failure does this represent?

A. Mechanical fatigue
B. Structural weakness targeted by an attack
C. Wear from long-term use
D. A key duplication error

**Question 5**

A security audit reveals that older locks in a retail store are easier to pick than newly installed ones of the same model.

What is the most likely explanation?

A. Wear has reduced friction and improved tactile feedback during picking
B. Older locks contain fewer pins
C. New locks cannot be picked
D. Older locks always have weaker springs

**Question 6**

A locksmith installs reinforced anti-drill pins in a lock. During a burglary attempt, the attacker abandons drilling and instead attacks the cam mechanism.

What principle does this illustrate?

A. Reinforcing one component can redirect attacks to other weaknesses
B. Anti-drill pins prevent all forms of attack
C. Cam mechanisms cannot fail
D. Drilling is the only attack method used against locks

**Question 7**

A homeowner reports that their deadbolt is becoming increasingly difficult to operate during humid weather.

What is the most likely cause?

A. Electronic failure inside the lock
B. Environmental effects causing door expansion and bolt misalignment
C. A broken key inside the cylinder
D. Excessive spring tension inside the lock

**Question 8**

A smart lock stops responding to commands even though the mechanical components appear intact. Investigation reveals that the batteries are depleted.

What category of failure does this represent?

A. Structural failure
B. Electronic or power-supply failure
C. Mechanical wear
D. Keyway obstruction

**Question 9**

A commercial door fitted with a multipoint locking system becomes difficult to lock because the handle cannot lift fully.

What is the most likely cause?

A. Door misalignment preventing locking points from engaging properly
B. The euro cylinder has been drilled
C. The door hinges have no effect on lock operation
D. The locking strip cannot wear over time

**Question 10**

A locksmith servicing a lock notices that the key operates smoothly and the plug rotates freely after insertion of the correct key.

What does this most likely indicate about the lock?

A. Internal pins are worn beyond tolerance
B. Pins and shear line alignment are functioning correctly
C. The lock is vulnerable to snapping
D. The cylinder has been incorrectly installed

# Chapter 4

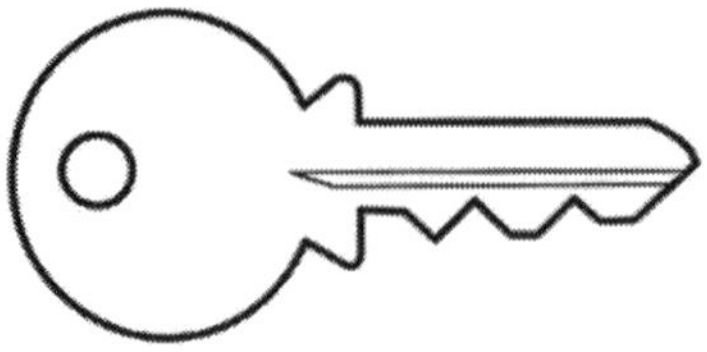

# High-Security Lock Engineering

This chapter explores the engineering principles, technologies, and professional practices that underpin modern high-security locking systems, moving well beyond the fundamentals of traditional lock-and-key mechanisms. It brings together critical concepts such as restricted key systems, patent protections, advanced anti-attack innovations, hybrid magnetic and electronic technologies, and real-world applications in high-risk environments. In doing so, it reflects the evolving nature of locksmithing, from a primarily mechanical trade to a multidisciplinary field grounded in security engineering, system design, and risk management.

The chapter begins by examining how control over key duplication, rather than simple key possession, fundamentally changes security outcomes. It then builds on this foundation by exploring the layered design of modern locks, including anti-pick, anti-bump, and anti-drill innovations developed by leading manufacturers. These technical features demonstrate how contemporary lock design relies on complexity, redundancy, and material science to resist both covert and destructive attack methods.

As the discussion progresses, the chapter introduces hybrid systems that combine mechanical, magnetic, and electronic elements, highlighting a significant shift toward multi-factor authentication and digitally integrated access control. These systems illustrate how security is no longer achieved through a single mechanism, but through the interaction of multiple layers working together to prevent, detect, and respond to threats.

The inclusion of real-world case studies, such as critical infrastructure security, reinforces the practical application of these technologies and emphasises the importance of system-level thinking. In these environments, locks are not standalone devices but part of a broader security ecosystem that includes access control, monitoring, and regulatory compliance.

Finally, the chapter examines reverse engineering as a professional skill, demonstrating how analysing lock design enables locksmiths to understand performance, identify vulnerabilities, and make informed decisions about system selection and maintenance. This analytical approach is central to modern locksmithing practice, where success depends not only on technical skill, but also on the ability to interpret complex systems and apply engineering-level thinking.

Overall, this chapter provides a comprehensive foundation for understanding high-security lock engineering and its direct impact on contemporary locksmithing practice, preparing the reader to engage with increasingly sophisticated security technologies in both theory and application.

## Restricted Key Systems and Patent Protections

Restricted key systems are designed to control not only who can use a key, but, more importantly, who can duplicate it. In standard (open) key systems, key blanks are widely available, meaning anyone with access to a key can usually have it copied. Restricted systems remove this vulnerability by limiting access to key blanks, enforcing authorisation procedures, and often incorporating legal protections such as patents and registered designs.

The following examples illustrate how control over key duplication, rather than just key use, fundamentally changes security outcomes.

**1. Residential Open Key System (Common Hardware Store Locks)**

In a typical residential setting using common keyways such as Kwikset (KW1) or Schlage (SC1), the system is considered *open*. Key blanks are widely available from hardware stores, locksmiths, and even online suppliers.

**Scenario:** A tenant, cleaner, or contractor briefly borrows a house key. They visit a hardware store and have it duplicated in minutes without needing identification or authorisation.

**Outcome:**

- No control over how many copies exist
- Homeowner cannot track duplication
- Security relies entirely on physical possession of keys

This demonstrates the weakness of open systems: anyone with access can duplicate the key freely.

**2. Commercial Office Using a Restricted System (Authorised Duplication Only)**

A commercial office installs a restricted system such as Medeco, Mul-T-Lock, or ASSA ABLOY restricted profiles. Each key is issued with a control card and recorded in a key register.

**Scenario:** An employee attempts to duplicate their office key at a locksmith.

**What happens:**

- The locksmith requests the control card or authorised signature
- Without approval, the locksmith refuses to cut the key

- Even if the employee finds another provider, blanks are not available outside authorised channels

**Outcome:**

- Key duplication is strictly controlled
- Organisation maintains accountability
- Risk of unauthorised copies is significantly reduced

This shows how restricted systems enforce administrative + physical control over duplication.

**3. High-Security Government or Infrastructure Facility**

Critical infrastructure sites (e.g., utilities, defence, transport) often use advanced systems such as **ABLOY**, **Evva**, or **Mul-T-Lock**.

**Scenario:** A contractor temporarily accesses a secure area and attempts to duplicate the key.

**Barriers in place:**

- Key uses patented design (e.g., rotating discs, magnetic elements, telescopic pins)
- Key blanks are legally protected and not commercially available
- Duplication requires manufacturer approval and registered credentials

**Outcome:**

- Duplication is both physically difficult and legally restricted
- Even skilled locksmiths cannot copy the key without authorisation
- Security is maintained even if the key is briefly exposed

This demonstrates layered protection: mechanical complexity + patent law + controlled distribution.

**4. Patent Expiry Scenario (Loss of Restriction Over Time)**

A previously restricted system loses patent protection after its legal term expires.

**Scenario:** A high-security key system that was once protected becomes available to third-party manufacturers.

**What changes:**

- Aftermarket key blanks enter the market

- More locksmiths can cut keys without authorisation
- Duplication control weakens significantly

**Outcome:**

- System transitions toward an "open" state
- Original security advantage is reduced
- Organisations may need to upgrade systems

This highlights the importance of patent protection in maintaining long-term key control.

**5. Master Key System in a Large Facility (Controlled Hierarchy)**

A hospital, university, or large commercial building uses a restricted master key system.

**Scenario:**

- Staff have access only to specific areas
- Managers have broader access
- A master key opens multiple doors

**Duplication control:**

- All keys are logged and issued formally
- Replacement or duplication requires authorisation
- Lost keys trigger system review or rekeying decisions

**Outcome:**

- Controlled access hierarchy
- High accountability
- Reduced risk of system compromise

This demonstrates how restricted systems support complex access control structures.

At the core of a restricted key system is key control. This is achieved by using a unique keyway profile and specialised key blanks that are not available through general retail channels. Instead, these blanks are supplied only to authorised locksmiths or directly through the manufacturer. When a restricted system is established, it is typically accompanied by formal documentation—often including a key control register and authorised signatories. Any request for additional keys must be verified against this record, ensuring that duplication can only occur with proper approval.

Patent protection plays a critical role in strengthening these systems. A patent can protect the design of the key, the keyway profile, or specific mechanical features such as sidebars, sliders, or interactive elements. While a patent is active (commonly up to 20 years, depending on jurisdiction), it legally prevents unauthorised parties from manufacturing compatible key blanks or duplicating the protected design. This means that even if someone has physical access to a key, they cannot legally produce a copy without permission from the patent holder.

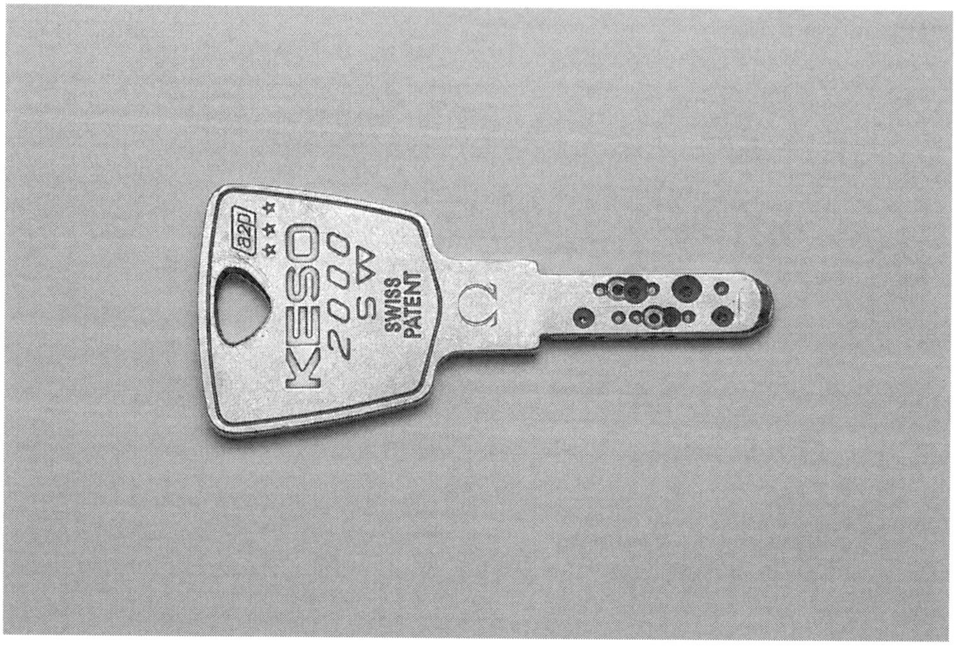


*Figure 46: Assa Abloy Keso 2000 key. D4m1en, CC BY-SA 3.0, via Wikimedia Commons.*

For example, high-security systems often combine restricted keyways with patented features such as angled cuts, side milling, magnetic elements, or multi-dimensional key profiles. These features are not only difficult to replicate physically but are also legally protected, creating both a technical barrier and a legal barrier to duplication. This dual-layer protection significantly enhances overall security.

However, it is important to understand what happens when patents expire. Once a patent lapses, third-party manufacturers may legally produce compatible key blanks. At this point, the system may transition from fully restricted to partially restricted or even effectively open, depending on how widely those blanks become available. To address this, manufacturers frequently develop new patented systems before older ones expire, ensuring continued control and market differentiation.

Restricted key systems are widely used in environments where security and accountability are critical, such as commercial buildings, government facilities, hospitals, universities, and large residential complexes. In these settings, controlling who can duplicate keys is just as important as controlling who can use them. A well-managed restricted system allows organisations to track key issuance, prevent unauthorised copies, and maintain long-term security integrity.

## Beyond the Key

From a locksmithing perspective, working with restricted systems introduces additional responsibilities. Locksmiths must be authorised by the manufacturer, follow strict ordering procedures, and verify customer credentials before supplying keys or cylinders. This requires not only technical skill but also administrative discipline and adherence to legal and contractual obligations.

In practice, restricted key systems also influence how locksmiths design and maintain master key systems. Because key blanks are controlled, system expansion must be carefully planned to ensure future compatibility and availability. Additionally, accurate record-keeping becomes essential, as any errors in documentation can compromise the integrity of the entire system.

### Impact on Locksmithing Practices

Restricted key systems are highly relevant to anyone entering or working in locksmithing because they fundamentally change the role of the locksmith—from a key cutter to a security professional responsible for access control and risk management. In open systems, locksmithing is largely a mechanical task: duplicating keys, replacing locks, and responding to failures. In restricted systems, however, the locksmith becomes part of a controlled security framework that includes legal, administrative, and technical responsibilities.

At the most practical level, restricted systems directly influence how locksmiths interact with customers and keys. A locksmith can no longer simply duplicate a key on request. Instead, they must verify authorisation, check control cards or signatures, and ensure compliance with the system's rules. This introduces a layer of accountability that does not exist in open systems. It also means that locksmiths must understand not just the physical lock, but the entire key management process, including documentation, record keeping, and client-specific restrictions.

These systems also shape product selection and recommendation, which is a core part of locksmithing practice. A locksmith must assess the client's risk profile and determine whether an open or restricted system is appropriate. For example, a residential client may prioritise convenience and low cost, making an open system acceptable. In contrast, a commercial or government client requires strict key control, making restricted systems essential. This decision directly affects long-term security outcomes, meaning locksmiths must think beyond immediate installation and consider future risks such as unauthorised duplication and staff turnover.

Restricted systems significantly influence master key system design, which is a critical area of professional locksmithing. In large facilities, the ability to control duplication ensures that the hierarchy of access remains secure over time. Without restriction, a single unauthorised copy could compromise an entire system. As a result, locksmiths must carefully plan key hierarchies, maintain compatibility of keyways, and ensure that expansion is possible without weakening control. This elevates locksmithing into a form of system engineering, where long-term planning and structure are essential.

Another key influence is on legal and ethical responsibilities. Because restricted systems often involve patents, licensing agreements, and controlled distribution, locksmiths must operate within strict legal frameworks. Supplying or duplicating keys without proper authorisation can breach contractual obligations or intellectual property laws. This requires locksmiths to maintain professional integrity, follow manufacturer agreements, and implement secure processes. In this sense, locksmithing becomes not only a technical trade but also a regulated professional practice.

From a technical perspective, restricted systems also demand a higher level of precision and specialised knowledge. These systems often use complex keyways, sidebars, sliders, or interactive elements that require accurate cutting and system-specific tools. Locksmiths must be familiar with manufacturer standards, authorised blanks, and specialised equipment. This reduces reliance on simple duplication and increases the importance of code cutting, decoding, and precision measurement, reinforcing the shift toward engineering-level accuracy in the trade.

Restricted systems also play a major role in diagnostics and troubleshooting. When issues arise, locksmiths must consider not only mechanical faults but also system integrity. For example, a problem may not be a worn lock, but an unauthorised or incorrect key blank, a poorly cut duplicate, or a breach in key control. Understanding restricted systems allows locksmiths to diagnose issues more effectively and avoid unnecessary replacement of hardware, improving both efficiency and client outcomes.

Importantly, restricted systems influence long-term client relationships and service models. Locksmiths often become ongoing partners in managing a client's security system, handling authorised key orders, system updates, and audits. This creates opportunities for higher-value services but also requires consistent record-keeping, communication, and system knowledge. In practice, this transforms locksmithing from a reactive service into a proactive security management role.

Finally, the concept of patent protection highlights how security evolves over time, which is critical for locksmiths to understand. As patents expire, previously restricted systems may become vulnerable due to the availability of aftermarket blanks. Locksmiths must recognise when a system is no longer secure and recommend upgrades accordingly. This requires staying informed about product lifecycles, manufacturer developments, and emerging technologies—further reinforcing the need for continuous professional development.

## Anti-Pick, Anti-Bump, and Anti-Drill Innovations

Modern lock design has evolved significantly to resist common attack methods such as picking, bumping, and drilling. For a locksmith, understanding these innovations is essential because they directly influence lock selection, installation, servicing, and security assessment. These features are not standalone—they are often layered together to create a comprehensive defence system within a single cylinder.

### Anti-Pick Innovations

## Beyond the Key

Anti-pick features are designed to prevent or disrupt the manipulation of pins using tools such as picks and tension wrenches. Traditional pin tumbler locks rely on aligning pins at the shear line, but anti-pick mechanisms make this process far more difficult by introducing false feedback and additional alignment requirements.

One of the most common innovations is the use of security pins, such as spool, mushroom, and serrated pins. These pins are shaped in a way that causes them to "catch" at the shear line, creating false sets that mislead the person attempting to pick the lock. Instead of smooth, predictable feedback, the operator experiences resistance, counter-rotation, and uncertainty.

Another important feature is the use of paracentric or restricted keyways, which physically limit access to the internal components. These keyways force tools into awkward angles, reducing control and making it difficult to manipulate pins individually.

More advanced systems incorporate secondary locking mechanisms, such as sidebars, sliders, or rotating pins. In these designs, even if the main pins are picked correctly, additional elements must also align before the lock will open. This significantly increases complexity and time required for unauthorised entry.

In practice, anti-pick innovations increase the skill, time, and precision required to manipulate a lock without a key, often deterring opportunistic attacks. Understanding these manufacturer-specific innovations is essential for locksmiths because it directly affects how locks are serviced, diagnosed, and specified.

<u>Medeco (USA) – Rotating Pin & Sidebar System</u>

Medeco is one of the most well-known innovators in anti-pick technology. Its core innovation is the use of angled key cuts that both lift and rotate pins. Unlike standard pin-tumbler locks, where pins only move vertically, Medeco pins must also be rotated to a precise orientation before the lock will open.

These pins have chisel-shaped tips and side grooves. When correctly rotated, these grooves align with a sidebar, allowing it to retract and permit plug rotation. If the pins are not rotated correctly, the sidebar remains engaged and blocks movement.

Anti-pick effect:

- Picking tools must control two variables simultaneously (height + rotation)
- False feedback is common due to partial rotation
- Sidebar prevents opening even if pins are lifted correctly

This makes Medeco locks highly resistant to traditional picking techniques.

<u>Mul-T-Lock (Israel) – Telescopic Pin (Pin-in-Pin) System</u>

Mul-T-Lock uses telescopic pins, where each pin consists of an outer pin and an inner pin. Both components must align correctly at the shear line.

In addition, many Mul-T-Lock systems include:

- Sidebars for secondary locking
- Interactive elements (spring-loaded components in the key)

Anti-pick effect:

- Doubles the number of alignment points
- Requires extremely precise manipulation
- False sets occur frequently

Even if a picker sets one layer correctly, the second layer prevents opening.

ASSA / ASSA ABLOY (Global) – Security Pins & Sidebar Systems

ASSA (part of ASSA ABLOY) uses a combination of advanced pin designs and sidebar mechanisms, particularly in systems like ASSA Twin.

Common anti-pick features include:

- Spool pins (create false set feedback)
- Serrated pins (multiple false shear lines)
- Sidebar systems requiring secondary alignment

Anti-pick effect:

- False feedback misleads the picker
- Multiple shear lines increase difficulty
- Sidebar adds a second locking condition

This layered approach combines deception (false feedback) with mechanical redundancy.

ABLOY (Finland) – Disc Detainer System

ABLOY takes a completely different approach by eliminating traditional pins altogether. Instead, it uses rotating discs that must be aligned to allow a sidebar to retract.

Anti-pick effect:

- No vertical pins → traditional picking tools ineffective
- Requires specialised tools and techniques
- Extremely tight tolerances

This design is inherently resistant because it removes the standard picking interface.

Kaba (dormakaba, Switzerland) – Multi-Row Dimple Pin Systems

Kaba systems use dimple keys with multiple rows of pins, often combined with sidebars.

Features include:

- Pins arranged in multiple directions (vertical + horizontal)
- Tight tolerances and precision alignment

Anti-pick effect:

- Requires multi-directional manipulation
- Limited space for tools
- Increased complexity and reduced feedback

The geometry itself becomes a barrier to picking.

EVVA (Austria) – Slider & Magnetic Systems

EVVA uses slider-based systems instead of traditional pins. These sliders must align gates to allow sidebar movement. Some systems (like MCS) also incorporate magnetic elements.

Magnetic systems rely on embedded magnets in the key to align internal components, meaning traditional picking tools cannot manipulate them.

Anti-pick effect:

- No conventional pin stacks to manipulate
- Magnetic elements cannot be "felt" or lifted
- Multi-dimensional interaction required

This creates both mechanical and non-contact barriers to picking.

DOM (Germany) – Side Pins & Check Systems

DOM systems (e.g., ix series) combine traditional pins with side pins and check systems.

Features include:

- Secondary side pin rows
- Sidebar engagement based on multiple alignments

Anti-pick effect:

- Multiple independent pin systems
- Increased manipulation complexity
- Reduced feedback clarity

Picking requires aligning both primary and secondary systems simultaneously.

Across different manufacturers, anti-pick innovations tend to follow several core design strategies aimed at increasing resistance to manipulation. One common approach is adding additional dimensions to the locking process. For example, systems such as Medeco require both lifting and rotation of pins, while Kaba incorporates multiple rows of pins, and EVVA uses slider mechanisms. These designs move beyond simple vertical pin movement, making manipulation significantly more complex.

Another strategy involves adding redundancy within the locking mechanism. This can be seen in systems like Mul-T-Lock, which uses telescopic (pin-in-pin) configurations, and DOM, which incorporates side pins alongside traditional pin stacks. By requiring multiple components to align simultaneously, these systems increase the number of variables a locksmith or attacker must manage.

Manufacturers also introduce deception as a defensive measure. ASSA, for instance, uses spool and serrated pins that create false feedback during picking attempts. These features are designed to mislead the operator into thinking the lock is correctly set when it is not, increasing the likelihood of errors and prolonging the picking process.

In some cases, manufacturers move away from traditional pin-tumbler mechanisms entirely. Systems such as ABLOY use rotating disc detainers, while EVVA and certain Chubb designs incorporate magnetic or non-traditional elements. By eliminating or fundamentally altering the standard picking interface, these locks render conventional picking techniques largely ineffective.

For locksmiths, these variations have important practical implications. There is no universal picking method that applies to all lock types, meaning that tools and techniques must be carefully matched to the specific system being serviced. Effective diagnosis also depends on understanding the internal mechanism of the lock, as different designs present different failure modes and behaviours. Additionally, as locks become more precise and incorporate secondary systems such as sidebars and sliders, the accuracy of key cutting becomes increasingly critical to ensure proper operation.

## Anti-Bump Innovations

Lock bumping is a technique that uses a specially cut "bump key" and impact force to momentarily separate the pins, allowing the plug to rotate. Anti-bump innovations are specifically designed to absorb, disrupt, or prevent this rapid energy transfer.

One approach is the use of modified pin stacks, including specially shaped driver pins and stronger or varied springs. These components reduce the effectiveness of the kinetic energy transferred during bumping, preventing the pins from aligning simultaneously.

Another method involves trap pins or check pins, which can engage if abnormal movement is detected. Instead of allowing the lock to open, these pins may lock the mechanism further when a bump attempt is made.

Locks that incorporate sidebars or secondary systems are naturally more resistant to bumping. Even if the top and bottom pins are momentarily aligned, the sidebar or secondary mechanism remains engaged unless additional conditions are met.

In real-world terms, anti-bump features ensure that quick, low-skill attacks using bump keys are far less effective, particularly on modern high-security cylinders. Across the industry, several distinct anti-bump strategies have emerged, each implemented differently by specific manufacturers.

### Medeco (USA) – Rotating Pins and Sidebar Systems

Medeco is widely recognised for combining anti-pick and anti-bump protection through its angled key cuts and rotating pin system. Unlike standard vertical pins, Medeco pins must be both lifted and rotated to a precise angle before the sidebar can retract.

From an anti-bump perspective, this is highly effective because bumping relies on vertical energy transfer. The added rotational requirement disrupts this process—pins may momentarily jump, but they will not align rotationally, preventing the lock from opening.

Example systems:

- Medeco Original
- Medeco Biaxial
- Medeco M3

### Mul-T-Lock (Israel) – Telescopic Pins (Pin-in-Pin System)

Mul-T-Lock uses a telescopic pin system, where each pin stack consists of an inner and outer pin. This effectively doubles the number of shear lines that must align.

In bumping, energy is transferred through the pin stack. With telescopic pins:

- Energy is dissipated across multiple components
- Inner and outer pins move independently
- Precise simultaneous alignment becomes extremely difficult

Example systems:

- Mul-T-Lock Classic
- Mul-T-Lock Interactive / Interactive+

ASSA ABLOY / ASSA (Sweden) – Dual Shear Lines & Security Pins

ASSA high-security cylinders often incorporate dual shear lines and advanced driver pins (spool, serrated, barrel).

For bump resistance:

- Multiple shear lines mean pins must align in more than one position
- Security pins absorb and disrupt impact energy
- Sidebar systems require additional alignment beyond pin lifting

Example systems:

- ASSA Twin / Twin Combi
- ASSA Maximum+

Schlage (USA) – Security Pins & Controlled Tolerances

Schlage improves bump resistance primarily through security pins and tighter tolerances.

- Spool and serrated pins disrupt smooth pin movement
- Tighter manufacturing tolerances reduce uncontrolled pin travel
- In higher-end systems (e.g., Primus), side pins and a sidebar add another layer of control

Example systems:

- Schlage Everest

- Schlage Primus / Primus XP

ABLOY (Finland) – Disc Detainer Mechanism (No Pins)

ABLOY takes a completely different approach by eliminating pins altogether. Instead, it uses a disc detainer system.

- Rotating discs must align to allow a sidebar to move
- No spring-loaded pins = no energy transfer path for bumping
- Bump keys are ineffective because the mechanism does not rely on pin stacks

Example systems:

- ABLOY Protec / Protec2
- ABLOY Sentry

EVVA (Austria) – Sliders and Magnetic Elements

EVVA incorporates slider mechanisms and magnetic coding in some systems.

- Sliders must be positioned laterally, not just vertically
- Magnetic elements require correct polarity and positioning
- These systems are unaffected by impact-based attacks like bumping

Example systems:

- EVVA 3KS (springless slider system)
- EVVA MCS (magnetic coding system)

DOM (Germany) – Multi-Component Pin Systems

DOM uses side pins, check pins, and multi-row configurations to increase complexity.

- Additional pin layers absorb and dissipate bump energy
- Alignment requirements extend beyond a single shear line
- Often combined with restricted keyways and tight tolerances

Example systems:

- DOM ix series
- DOM Diamant

Across different manufacturers, anti-bump innovations are based on a number of core engineering strategies designed to interrupt the fundamental mechanics of bumping. One of the most common approaches is disrupting energy transfer within the lock. Systems such as those used by Mul-T-Lock incorporate telescopic (pin-in-pin) designs, while manufacturers like Schlage and ASSA use security pins (e.g., spool and serrated). These features absorb, scatter, or delay the kinetic energy introduced during a bump attack, preventing the pins from aligning simultaneously at the shear line.

Another key strategy involves adding additional alignment requirements beyond simple vertical pin movement. For example, Medeco systems require pins to be rotated as well as lifted, while ASSA and Schlage Primus incorporate sidebar mechanisms that must align independently of the main pin stacks. These added requirements mean that even if bumping briefly lifts the pins, the lock will not open unless all secondary conditions are also met.

Manufacturers also increase resistance by introducing multi-dimensional interaction within the locking mechanism. Systems such as EVVA use sliders that must move laterally, while DOM incorporates side pins and check systems. These designs require movement in more than one plane, making it extremely difficult for impact-based methods like bumping to achieve the precise alignment needed to operate the lock.

In some cases, manufacturers eliminate the traditional pin tumbler system altogether. ABLOY, for example, uses a disc detainer mechanism with rotating discs instead of spring-loaded pins. Because there are no pins to transfer kinetic energy through, bumping is effectively rendered impossible, demonstrating a complete departure from conventional lock design.

For locksmiths, these anti-bump innovations have significant practical implications. There is no universal approach to attack, diagnosis, or servicing, as each lock system behaves according to its own internal design. Techniques that work on one system may be completely ineffective on another.

Precision in key cutting becomes increasingly important, particularly for systems that incorporate sidebars, sliders, or other multi-element mechanisms. Even minor inaccuracies can prevent proper alignment and lead to operational issues that may be mistaken for mechanical faults.

Lock selection must also be informed by the intended application. Factors such as environmental conditions, required security level, and the need for key control all influence the suitability of a particular system. High-security environments demand more advanced anti-bump features, while lower-risk applications may prioritise cost and ease of use.

Ultimately, a thorough understanding of internal lock mechanisms is essential for effective locksmithing practice. Whether diagnosing faults, recommending products, or maintaining systems, locksmiths must be able to interpret how these anti-bump features function in order to provide accurate, reliable, and professional service.

## Anti-Drill Innovations

Drilling is a destructive attack aimed at physically removing or bypassing internal components. Anti-drill innovations focus on using materials and design features that resist or slow down drill penetration.

A common solution is the use of hardened steel inserts or pins placed in critical areas of the lock, such as the front of the cylinder or within the pin chambers. These inserts are much harder than standard drill bits, causing them to dull or break during an attack.

Some high-security locks include rotating hardened discs or plates. When a drill bit makes contact, these components spin freely, preventing the bit from gaining traction and cutting effectively.

Another strategy is strategic material layering, where softer materials are combined with hardened components. This creates unpredictable drilling conditions, reducing efficiency and increasing the time required to breach the lock.

In addition, some cylinders are designed so that even if drilling partially succeeds, secondary mechanisms (like sidebars) still prevent the lock from opening.

From a security perspective, anti-drill features aim to delay forced entry long enough to deter attackers or trigger detection. The following examples illustrate how different manufacturers implement anti-drill innovations in practice.

One of the most widely recognised approaches is the use of hardened steel inserts positioned at critical attack points. Medeco integrates hardened steel rods and inserts within the cylinder plug and housing. These inserts are specifically located in front of and around the pin chambers and shear line. When a drill bit contacts these hardened components, it is deflected or rapidly dulled, significantly slowing penetration. In addition, Medeco often combines this with its sidebar system, meaning that even if partial drilling occurs, the lock will not function without proper pin rotation and sidebar alignment.

Mul-T-Lock employs a layered anti-drill strategy using hardened steel ball bearings and inserts embedded throughout the cylinder. These ball bearings are free-spinning, which means that when a drill bit makes contact, the bearing rotates instead of allowing the bit to bite into the material. This greatly reduces drilling efficiency. Mul-T-Lock also reinforces both the plug and the cylinder body, ensuring that multiple layers must be penetrated before reaching critical components such as the telescopic pins and shear line.

Within the ASSA ABLOY group, high-security systems such as ASSA and Yale commercial cylinders incorporate hardened steel pins, drill-resistant plates, and protective discs. These components are strategically placed at the face of the cylinder and along the pin chambers. In some designs, the front of the cylinder includes a hardened steel shield or cap, which acts as the first line of defence against drilling. Even if the outer layer is breached, internal hardened pins continue to resist further progress.

ABLOY takes a fundamentally different approach by combining material strength with an alternative locking mechanism. Its disc detainer systems are constructed using hardened steel discs and components that are inherently resistant to drilling. Because there are no traditional pin stacks, drilling attacks aimed at pin chambers are ineffective. Additionally, the use of hardened materials throughout the cylinder body increases overall resistance to destructive entry methods.

EVVA integrates anti-drill features into its high-security systems such as MCS and 4KS. These systems include hardened steel inserts and protective elements positioned around critical areas, including sidebars and slider tracks. EVVA also combines these features with complex internal mechanisms (such as sliders and magnetic elements), meaning that even if drilling is partially successful, the lock remains extremely difficult to bypass due to the multi-dimensional alignment required.

Kaba (dormakaba) incorporates hardened steel pins, inserts, and plates within its dimple key systems. In products such as the Kaba expert and pExtra series, drill resistance is enhanced by reinforcing both vertical and horizontal pin chambers. The presence of multiple rows of pins further complicates drilling, as an attacker must accurately target several independent locking elements rather than a single shear line.

DOM (Germany) uses hardened steel rods and inserts in both the plug and housing, often combined with side pin systems. These hardened components are placed to protect not only the main pin stacks but also the secondary locking elements. As a result, drilling must overcome multiple reinforced barriers before any meaningful compromise can occur.

Some manufacturers also employ sacrificial or trap-based designs. In certain high-security cylinders, drilling into the plug triggers internal damage that permanently locks the cylinder, preventing rotation even with the correct key. While not always visible, these features are designed to ensure that a failed drilling attempt results in a denial of entry rather than a successful bypass.

From a locksmithing perspective, these anti-drill innovations significantly influence both product selection and service procedures. Locks intended for high-risk environments—such as commercial premises, critical infrastructure, or government facilities—require robust anti-drill protection, often combining multiple strategies (e.g., hardened inserts, ball bearings, and complex mechanisms).

Servicing and replacement also become more complex. Drilling a high-security cylinder for legitimate access (e.g., during lockout or failure) requires specialised drill bits, precise knowledge of internal construction, and often significantly more time. In many cases, non-destructive entry methods are preferred due to the difficulty and cost associated with drilling these systems.

**Combined Security Approach**

Modern high-security locks rarely rely on a single defence. Instead, they integrate anti-pick, anti-bump, and anti-drill features into a unified system. For example:

- A lock may use security pins (anti-pick)
- Combined with a sidebar mechanism (anti-pick + anti-bump)
- Reinforced with hardened steel inserts (anti-drill)

This layered approach ensures that defeating one mechanism does not compromise the entire lock.

## Impact on Locksmithing Practices

Understanding modern lock security innovations, particularly anti-pick, anti-bump, and anti-drill features, is highly relevant to anyone entering locksmithing because these technologies fundamentally shape how locks are selected, installed, serviced, and assessed. Locksmithing is no longer limited to basic mechanical duplication or simple pin-tumbler systems; it has evolved into a profession that requires a detailed understanding of layered security engineering. These innovations directly influence how a locksmith approaches both everyday tasks and complex security challenges.

From a practical perspective, these features determine how a locksmith gains entry during legitimate work. Traditional techniques that may work on older or basic locks are often ineffective on modern high-security systems. For example, anti-pick mechanisms such as security pins, sidebars, or rotating pins change the feedback and behaviour of a lock, meaning that entry techniques must be adapted or replaced with alternative methods. Similarly, anti-bump features disrupt impact-based entry techniques, and anti-drill protections make destructive entry slower, more difficult, and more resource-intensive. As a result, locksmiths must develop a broader skill set that includes both non-destructive methods and controlled destructive techniques.

These innovations also have a direct impact on lock selection and recommendation. A locksmith must assess the environment, risk level, and client requirements before recommending a suitable system. For example, a residential property may not require advanced anti-drill protection, while a commercial or high-risk site may demand a lock that integrates multiple layers of defence. Understanding how anti-pick, anti-bump, and anti-drill features function allows the locksmith to match the correct product to the application, balancing security, cost, and usability.

In terms of installation and system design, these features influence how locks are specified and integrated into broader security systems. High-security cylinders often require precise alignment, correct hardware pairing, and careful handling during installation. In master key systems, the presence of sidebars, sliders, or multi-dimensional key interactions adds complexity that must be accounted for during system planning. Poor installation or incorrect specification can compromise the effectiveness of these advanced features, even if the lock itself is highly secure.

Servicing and maintenance are also significantly affected. High-security locks are typically manufactured with tighter tolerances and more complex internal components. This means that wear, debris, or incorrect lubrication can have a greater impact on performance. A locksmith must be able to diagnose whether a problem is caused by key wear, internal component misalignment, or environmental factors. In systems with advanced features such as sidebars or sliders, even minor inaccuracies in key cutting can lead to operational issues, making precision and quality control essential.

These innovations further influence troubleshooting and fault diagnosis. A lock that fails to operate correctly may not be mechanically damaged but instead affected by subtle issues such as incorrect key geometry, worn components, or interference within a complex mechanism. Without an understanding of how anti-pick, anti-bump, and anti-drill features interact, a locksmith may misdiagnose the problem and recommend unnecessary replacement. Knowledge of internal design allows for accurate, efficient, and cost-effective solutions.

Another important influence is on tooling and equipment. Modern lock systems often require specialised tools for key cutting, decoding, and servicing. For example, systems with multi-row pins, rotating elements, or restricted keyways demand precise machinery and measurement techniques. Similarly, dealing with anti-drill features during legitimate access may require high-quality drill bits, specific drilling points, and detailed knowledge of internal construction. This shifts locksmithing toward a more technical and equipment-driven profession.

These innovations reinforce the importance of professional judgement and ongoing learning. As manufacturers continue to develop new security features, locksmiths must stay current with evolving technologies and system designs. The integration of multiple defensive layers within a single lock means that no single technique or approach is universally applicable. Instead, locksmithing becomes a process of analysis—understanding the mechanism, selecting the correct method, and applying it with precision.

## Magnetic and Electronic Hybrid Locks

Magnetic and electronic hybrid locks represent one of the most advanced developments in modern locksmithing, combining traditional mechanical security with non-contact or digitally controlled elements. These systems are designed to overcome many of the limitations of purely mechanical locks by adding additional layers of authentication that cannot be manipulated using conventional tools. For a locksmith, understanding these systems is essential because they are increasingly used in high-security environments and require a different approach to installation, servicing, and diagnostics.

### Magnetic Locking Systems (Pure Mechanical with Magnetic Coding)

Magnetic locks use embedded magnets within the key and/or lock to create an additional layer of security that operates without direct physical contact. Unlike traditional pin-tumbler systems, where components must be physically lifted to the correct height, magnetic systems rely on magnetic fields to position internal elements.

In these systems, the key contains small magnets placed at specific positions and polarities. Inside the lock cylinder, corresponding magnetic sensors or movable elements respond to these magnetic fields. When the correct magnetic pattern is presented, internal components such as sliders or sidebars align, allowing the lock to operate. If the magnetic pattern is incorrect, or absent, the lock remains secured, even if other mechanical elements are correctly aligned.

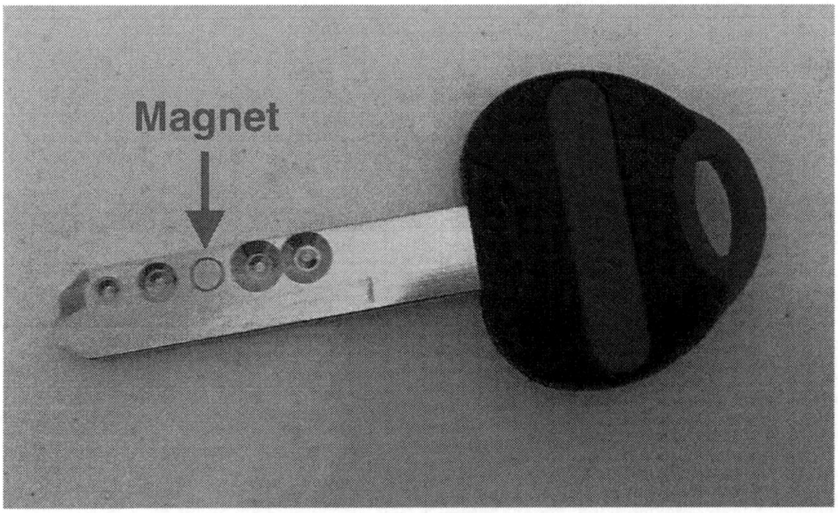


*Figure 47: Avocet ABS Magnetic Key. Willh26, CC BY-SA 4.0, via Wikimedia Commons.*

A well-known example is EVVA's Magnetic Code System (MCS), where magnets in the key interact with internal magnetic rotors. This creates a "hidden" authentication layer that cannot be easily detected or manipulated using traditional locksmith tools. Because there is no direct tactile feedback, magnetic systems are extremely resistant to picking and bumping.

EVVA's Magnetic Code System (MCS) is one of the most advanced high-security locking technologies available, combining magnetic coding with traditional mechanical elements to create a multi-layered security system. It represents a significant shift away from purely mechanical locks toward hybrid systems in which physical interaction and magnetic alignment operate together to control access.

EVVA MCS is a high-security cylinder system that uses permanent magnets embedded within the key to control internal components inside the lock. Unlike conventional systems that rely solely on pins or discs, MCS introduces a contactless magnetic layer of authentication alongside mechanical coding. This results in what EVVA describes as "triple security," combining a magnetic code, a precision key profile (keyway), and additional mechanical locking elements that must all align for the lock to function.

At the core of the system is its magnetic coding mechanism, which represents the primary innovation. The key contains small permanent magnets, typically manufactured from high-grade materials such as samarium cobalt. These magnets are positioned at precise locations along the key and oriented with specific polarity (north and south alignment). Inside the cylinder, corresponding magnetic rotors—free-spinning elements located within the keyway—react to these magnetic fields. When the correct key is inserted, the magnets influence the rotors, causing them to align into the required positions. This alignment is essential before the lock can operate. Importantly, this interaction is entirely contactless, meaning there is no physical contact between the magnets and the rotors, which significantly reduces wear and extends the lifespan of the system.

In addition to the magnetic system, MCS incorporates mechanical coding as a secondary layer of security. This includes a precisely milled key profile that must match the keyway, as well as additional internal mechanical locking elements. These components must align simultaneously with the magnetic system for the lock to open. Even if the magnetic code is correct, the cylinder will not rotate unless the mechanical elements are also correctly positioned, reinforcing the layered nature of the system.

The operation of the lock therefore depends on a multi-layer alignment requirement. The key must physically fit the keyway, the magnetic rotors must align correctly under the influence of the embedded magnets, and the mechanical locking elements must also be positioned accurately. Only when all of these conditions are satisfied at the same time can the cylinder rotate, making unauthorised entry extremely difficult.

From a technical perspective, the system incorporates several key features. Magnetic rotors—typically around eight in number—are positioned around the key channel and remain free-spinning until influenced by the correct magnetic field. The permanent magnets embedded in the key maintain fixed polarity and strength and cannot be easily altered or re-magnetised. The contactless nature of the system results in minimal friction and wear, contributing to a long service life compared to traditional pin-based systems. Additionally, the modular cylinder design allows the system to be adapted to a wide range of door types and supports complex master key configurations.

The security advantages of MCS are substantial. The absence of traditional pins means there is no conventional interface for picking, and the magnetic elements cannot be detected or manipulated using standard tools, providing extreme resistance to picking. Similarly, because there are no spring-loaded pin stacks, bumping techniques are ineffective. The use of hardened materials and a complex internal layout also enhances resistance to drilling. Key duplication is tightly controlled, as keys are restricted and can only be produced through authorised EVVA partners, typically requiring a security card or formal authorisation. The contactless operation further enhances durability, ensuring consistent performance over time.

In practical terms, EVVA MCS is used in environments where both high security and long-term reliability are essential. These include government and defence facilities, critical infrastructure such as utilities and transport networks, high-security commercial buildings, and data centres. In such settings, the combination of strong resistance to attack, strict key control, and extended service life makes MCS an effective and reliable solution.

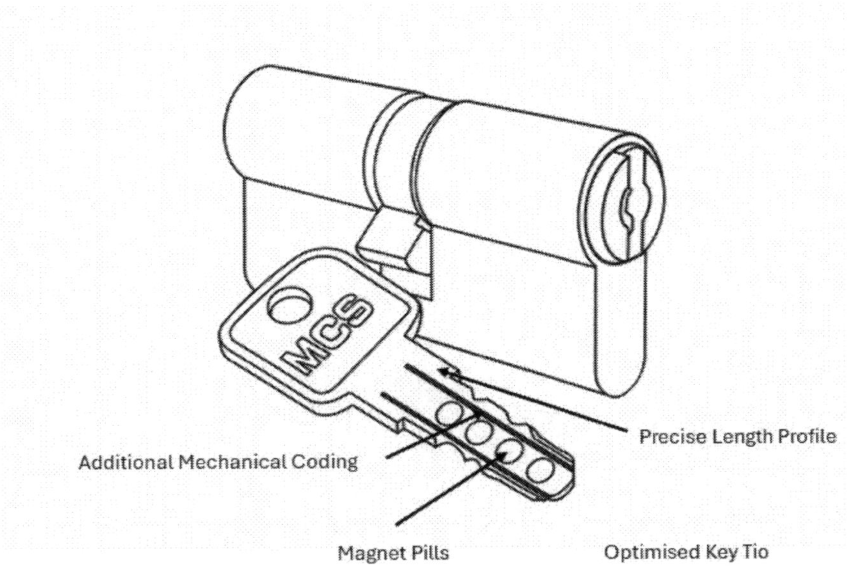


*Figure 48: EVVA MCS Double Cylinder.*

From a locksmithing perspective, magnetic systems introduce several important considerations. Key duplication requires specialised equipment and authorised blanks, as both the physical cuts and magnetic configuration must be reproduced accurately. Diagnostics also become more complex, as faults may involve magnetic degradation, misalignment, or internal component failure rather than visible mechanical wear.

**How Magnetic Key Duplication Works (Conceptual Overview)**

In legitimate locksmithing or manufacturer-controlled environments, duplication is a controlled and highly specialised process.

At a high level, it involves:

**1. Authorisation and Key Control**

Magnetic systems are almost always part of restricted key systems. Before duplication:

- The locksmith must be authorised by the manufacturer
- The customer must provide control cards or registered credentials
- Key blanks are not available on the open market

This is the first and most important barrier—most locksmiths cannot legally obtain the blanks.

**2. Decoding the Magnetic Configuration**

Unlike standard keys, the "code" includes invisible data:

- Magnet positions along the key
- Magnet polarity (north/south orientation)
- Sometimes magnetic strength or interaction patterns

Manufacturers use proprietary systems to determine this, such as:

- Secure databases linked to key codes
- Specialised decoding tools or readers
- Original factory specifications

This information is not typically measurable with standard locksmith tools.

**3. Preparing the Key Blank**

Magnetic key blanks are:

- Precision-machined to match the keyway exactly
- Manufactured with predefined cavities for magnets
- Often patented or restricted

In many systems (e.g., EVVA MCS), blanks are only supplied to:

- Approved locksmiths
- Direct manufacturer channels

**4. Magnetic Encoding Process**

The defining step is inserting magnets into the blank:

- Magnets must be placed in exact positions
- Each magnet must have the correct polarity and orientation
- In some systems, magnets are sealed or bonded into the key

This requires:

- Specialised jigs or fixtures
- Manufacturer-specific tooling
- Strict quality control

Incorrect polarity is one of the most common failure points.

**5. Mechanical Cutting (if applicable)**

Some magnetic keys also include:

- Edge cuts
- Dimple patterns
- Side milling

These are cut using:

- CNC key machines
- Manufacturer-specific cutters

Even if the magnetic coding is correct, poor mechanical cutting will prevent operation.

**6. Testing and Verification**

Final verification involves:

- Testing the key in the lock
- Confirming smooth operation
- Ensuring both magnetic and mechanical systems align

In high-security environments, duplication may also be:

- Logged in a key control register
- Linked to an audit system

### Electronic Locking Systems (Digitally Controlled Access)

Electronic locks replace or supplement mechanical interaction with digital authentication methods. These systems use electronic credentials such as PIN codes, RFID cards, key fobs, or mobile devices to grant access. Instead of relying solely on physical key geometry, access is controlled by a microprocessor that verifies the credential before actuating a locking mechanism.

*Figure 49: Electronic lock with number pad. User:Mattes —Restored by ZooFari, CC BY-SA 3.0, via Wikimedia Commons.*

Internally, electronic locks typically include components such as:

- A microcontroller (decision-making unit)
- Input interface (keypad, RFID reader, Bluetooth module)
- Actuator (motor or solenoid to move the locking bolt or clutch)
- Power source (battery or wired supply)

When a valid credential is presented, the system sends a signal to the actuator, which disengages the lock and allows the door to open. If the credential is invalid, no mechanical movement occurs.

*Figure 50: Electronic lock core. Chole Wu, CC BY-SA 4.0, via Wikimedia Commons.*

For locksmiths, electronic systems shift the skillset beyond purely mechanical work. Installation involves wiring (in some systems), configuration, and programming. Servicing may include battery replacement, firmware updates, and troubleshooting electronic faults. Security assessment also expands to include risks such as hacking, signal interception, or credential management.

Electronic locking systems are widely used to either replace traditional keys entirely or to supplement mechanical locking with digital authentication. These systems rely on credentials such as PIN codes, RFID cards, key fobs, or mobile devices, and are available across residential, commercial, and high-security applications. Below is a structured overview of commercially available systems, with examples from major manufacturers relevant to locksmithing practice.

<u>PIN Code / Keypad Electronic Locks</u>

These locks use a numeric keypad where users enter a PIN code to unlock the door. Many also include a traditional key override or smart connectivity.

Examples:

- Yale Assure Lock series – Keypad + optional Wi-Fi/Bluetooth integration
- Schlage Encode / Sense – Smart deadbolts with keypad and remote access
- Lockwood Digital DX series (Australia) – Common in residential and light commercial installs
- Samsung SHP / Smart Door Locks – Popular in Asia-Pacific markets

**Use case:** Residential homes, apartments, short-term rentals

**Locksmith relevance:** Code management, battery maintenance, fallback entry methods

RFID / Card / Key Fob Access Systems

These systems use contactless credentials such as RFID cards or fobs. They are widely used in commercial and hospitality environments.

*Figure 51: Electronic access control (BSL3 Lab) using magnetic swipe card. Ca.garcia.s, CC BY-SA 4.0, via Wikimedia Commons.*

Examples:

- Salto XS4 / KS systems – Advanced wireless access control
- dormakaba Saflok / Ilco – Hotel locking systems
- HID Global readers – Industry-standard RFID credential systems
- Paxton Net2 / Paxton10 – Integrated building access systems

**Use case:** Hotels, offices, universities, hospitals

**Locksmith relevance:** Credential programming, system integration, audit trails

Smart Locks (Bluetooth / Wi-Fi / Mobile App)

**Beyond the Key**

Smart locks allow access via smartphones, apps, or remote systems. They often integrate with home automation platforms.

Examples:

- August Smart Lock Pro – Retrofit design for existing deadbolts
- Nuki Smart Lock – Popular in European markets
- Ultraloq U-Bolt Pro – Combines PIN, fingerprint, and app access
- Level Lock – Concealed smart lock inside the door

**Use case:** Smart homes, Airbnb properties, remote access control

**Locksmith relevance:** Installation, connectivity troubleshooting, user management

*Figure 52: The second generation silver August smart lock, in an unlocked state. username, CC BY 4.0, via Wikimedia Commons.*

The **August Smart Lock (2nd Generation)** is a retrofit smart lock that mounts on the *inside* of an existing door and works with your current deadbolt. Instead of replacing the entire lock, it automates the turning of the thumb-turn using a motorised mechanism, while adding digital access control through a smartphone app.

**Core Concept: Motorised Control of a Standard Deadbolt**

At its core, the August lock does not change the mechanical lock itself. The original key cylinder, pins, and keyway remain unchanged on the outside of the door.

Internally, the smart lock replaces the thumb-turn and connects to the deadbolt tailpiece using an adapter.

When the lock is activated:

- A motor inside the unit rotates
- This turns the deadbolt tailpiece
- The deadbolt extends (lock) or retracts (unlock)

In simple terms: It is a motor-driven thumb-turn controlled electronically instead of manually.

The internal operation of the smart lock is based on a combination of mechanical and electronic components working together to control the deadbolt. At the centre of the system is a compact electric motor connected to a gear reduction mechanism. This motor converts electrical energy into rotational force, which is then used to turn the deadbolt. The gear train plays a critical role by reducing the speed of the motor while increasing torque, allowing the lock to operate smoothly even when the mechanism is stiff or under load. This ensures consistent and controlled movement rather than abrupt or jerky operation.

To connect the motor to the existing lock, the system uses a tailpiece adapter. Because different deadbolts use different tailpiece shapes—such as flat or D-shaped designs—interchangeable adapters are provided. These adapters act as the physical link between the motor and the lock, transferring rotational movement directly to the deadbolt mechanism. This design allows the smart lock to remain compatible with a wide range of standard residential locks without requiring full replacement of the existing hardware.

Communication between the lock and the user is handled through a Bluetooth Low Energy (BLE) module. This enables a smartphone to send lock and unlock commands directly to the device. The lock receives the signal, verifies it, and executes the command locally, meaning basic operation does not depend on an internet connection. For extended functionality, such as remote access from outside Bluetooth range, the system can be paired with a Wi-Fi bridge, which allows commands to be sent over the internet.

The lock also incorporates sensors that monitor its operational state. These sensors detect whether the lock is engaged or disengaged, measure the rotation angle of the deadbolt, and, when paired with an additional door sensor, determine whether the door itself is open or closed. This information is relayed to the user through the mobile application, enabling accurate status reporting and supporting automated features such as auto-lock and auto-unlock.

Power is supplied by internal batteries, typically CR123 cells, which are selected for their reliability and energy density. These batteries power all electronic components, including the motor, communication module, and sensors. The system is designed for low power consumption, and battery status is continuously monitored and reported through the app to ensure the user is aware of when replacement is required.

Control of the lock is primarily managed through a smartphone application, which serves as the main user interface. Through the app, users can lock or unlock the door, check its current status, manage user permissions, and view activity logs that record when the lock has been operated. When combined with a Wi-Fi bridge, these functions can be performed remotely, adding an additional layer of convenience and control.

One of the most notable features is the auto-unlock function, which uses geofencing technology. When the user's phone leaves a predefined area, the system prepares for automatic operation. As the user returns and approaches the door, Bluetooth communication is re-established, and the lock automatically disengages. This process relies on a combination of GPS data and short-range Bluetooth detection to determine proximity accurately.

Instead of relying solely on physical keys, the system introduces digital access control through virtual keys. These digital credentials can be assigned to different users and configured with varying levels of access, such as permanent, temporary, or scheduled permissions. This shifts access control from a physical model—where keys can be freely duplicated—to a managed digital system where permissions can be granted, modified, or revoked instantly.

From a security perspective, the system maintains the integrity of the original mechanical lock. The existing deadbolt and cylinder remain unchanged, meaning the physical security provided by the lock is not reduced. At the same time, communication between the smartphone and the lock is encrypted, helping to prevent unauthorised interception or replay of signals. Because the device is mounted on the interior side of the door, it is also protected from external tampering or physical attack.

A key feature of the design is its fail-safe capability. If the electronic system loses power or encounters a fault, the lock can still be operated using the original key from the outside, and manually from the inside. This redundancy ensures that access is never entirely dependent on the electronic system, which is an important consideration in both locksmithing practice and general safety design.

Biometric Locks (Fingerprint / Multi-Factor)

These locks use biometric data, most commonly fingerprints, as the primary credential. Many combine biometrics with PIN or app access.

*Figure 53: ZKT-ECO fingerprint scanner in Guangzhou Haizhu. 中少, CC BY-SA 4.0, via Wikimedia Commons.*

Examples:

- Samsung biometric locks – Fingerprint + PIN hybrid systems
- Ultraloq fingerprint models – Multi-method authentication
- ZKTeco biometric access systems – Commercial-grade solutions
- Eufy Smart Lock Touch – Consumer smart biometric lock

**Use case:** High-convenience residential, offices requiring user tracking

**Locksmith relevance:** Sensor calibration, enrolment management, failure diagnosis

Integrated Electronic Access Control Systems

These are large-scale systems that integrate multiple doors, credentials, and monitoring software into a centralised platform.

Examples:

- HID Global + controllers – Enterprise credential systems
- Gallagher security systems – Widely used in infrastructure and government
- Honeywell Pro-Watch – Integrated building security
- Bosch Access Control – Commercial and industrial solutions

**Use case:** Airports, hospitals, corporate buildings, critical infrastructure

**Locksmith relevance:** System design, compliance, integration with mechanical hardware

### Hybrid Magnetic-Electronic Lock Systems

Hybrid locks combine mechanical, magnetic, and electronic elements into a single system, creating multiple layers of authentication. These systems are designed so that defeating one layer does not compromise the entire lock.

A hybrid system may include:

- A traditional mechanical keyway (for physical access)
- Magnetic coding (for hidden authentication)
- Electronic verification (for user credentials or access control)

For example, a high-security electronic cylinder may require:

1. A correctly cut key (mechanical alignment)
2. A valid magnetic pattern (magnetic verification)
3. An authorised electronic credential (digital approval)

Only when all required conditions are met will the lock operate. This multi-factor approach significantly increases security by combining physical, magnetic, and digital barriers.

Some systems also include audit trails, allowing access events to be recorded and monitored. This is particularly valuable in commercial, government, and critical infrastructure environments where accountability is essential.

Below are commercially available examples from leading manufacturers that demonstrate how hybridisation is implemented in practice.

Mechanical + Magnetic Hybrid Systems

**EVVA MCS (Magnetic Code System) – Austria**

EVVA's MCS is one of the most advanced hybrid systems combining magnetic coding with mechanical key profiling.

- **Magnetic layer:** Permanent magnets embedded in the key align internal magnetic rotors (contactless)
- **Mechanical layer:** Precision key profile and additional locking elements
- **Hybrid effect:** Both magnetic and mechanical conditions must be met simultaneously

Widely used in:

- Government facilities
- Critical infrastructure
- Data centres

**Chubb / Union Magnetic Systems (UK – legacy & specialised systems)**

Some high-security Chubb-style systems incorporate magnetic elements alongside mechanical locking.

- Magnetic coding supplements traditional mechanisms
- Often used in custodial or high-control environments

These systems demonstrate early hybridisation before modern digital integration.

Mechanical + Electronic Hybrid Locks

**SALTO Systems (Spain) – XS4 / Neo Cylinders**

SALTO combines mechanical locking hardware with RFID-based electronic access control.

- Mechanical cylinder provides physical locking
- RFID card/mobile credential provides authentication
- Audit trails and access permissions managed digitally

Common in:

- Hotels
- Universities
- Commercial buildings

**SimonsVoss (Germany) – Digital Locking System 3060**

A fully integrated hybrid system combining mechanical locking with digital transponder authentication.

- Electronic transponder replaces traditional key
- Internal motor actuates mechanical locking mechanism
- No wiring required (battery-powered cylinders)

Known for:

- High-end commercial access control
- Retrofit capability

**dormakaba (Switzerland) – Digital Cylinder Series**

dormakaba offers systems combining precision mechanical cylinders with RFID or mobile credentials.

- Electronic authorisation controls access
- Mechanical redundancy remains in some models
- Can integrate with building management systems

Used in:

- Offices
- Healthcare
- Government buildings

**ASSA ABLOY Aperio (Global)**

Aperio integrates wireless electronic access control with mechanical lock bodies.

- Uses RFID, PIN, or mobile credentials
- Connects wirelessly to access control systems
- Mechanical locking hardware remains core to operation

Key advantage:

- Scalable integration into existing security systems

<u>Mechanical + Electronic + Magnetic (Advanced Multi-Layer Hybrid)</u>

**EVVA MCS + Electronic Access Integration**

EVVA MCS cylinders can be integrated into electronic access control environments, creating a triple-layer hybrid system:

- Magnetic coding (primary authentication)
- Mechanical key profile (secondary)
- Electronic system (access control layer)

Result:

- Extremely high security with both physical and digital control

**Kaba (dormakaba) – Mechatronic Systems (e.g., Kaba elolegic / evolo)**

These systems combine:

- Mechanical cylinder precision
- Electronic credential verification (RFID/mobile)
- Optional audit and access control

Known for:

- “Mechatronic” design philosophy (mechanical + electronic integration)

**Gunnebo / Chubb Safes – Electronic + Mechanical Safe Locks**

High-security safes often use dual systems:

- Electronic keypad (PIN authentication)
- Mechanical key or time-lock backup
- Sometimes magnetic or encoded components internally

Used in:

- Banking
- Cash handling
- High-value storage

**Biometric Smart Locks (e.g., Samsung, Yale Assure, Ultraloq)**

Modern residential/commercial locks often combine:

- Mechanical deadbolt
- Electronic keypad or app control
- Biometric fingerprint authentication
- Backup physical key

Example:

- Yale Assure Lock 2 (fingerprint + app + key backup)

**Why This Matters in Locksmithing Practice**

Traditional skills such as key cutting and mechanical diagnosis are still important, but they must be complemented by an understanding of electronics, digital systems, and advanced materials. In practical terms, this means:

- **No single entry method applies**: Mechanical picking, bumping, or decoding may be ineffective against magnetic or electronic layers
- **Key duplication becomes controlled and complex**: Requires authorised blanks and specialised equipment
- **Diagnostics require multi-system analysis**: Faults may be mechanical, magnetic, or electronic
- **Installation involves configuration and programming**: Not just physical fitting
- **Security assessment expands**: Includes both physical and digital vulnerabilities

Magnetic and electronic hybrid locks represent a shift in locksmithing from purely mechanical craftsmanship to integrated security engineering. By combining physical precision with non-contact magnetic systems and digital control, these locks create layered security that is far more resistant to traditional attack methods.

For modern locksmiths, understanding these systems is not optional—it is essential. As security technology continues to evolve, the profession increasingly requires the ability to work across mechanical, electronic, and hybrid domains, ensuring that locks are not only installed correctly but also understood at a systems level.

**Impact on Locksmithing Practices**

Magnetic and electronic hybrid locks are highly relevant to anyone entering locksmithing because they represent the direction in which the industry is rapidly evolving. Traditional locksmithing was centred on mechanical systems—understanding pins, keyways, and physical tolerances. However, hybrid systems introduce non-contact magnetic coding and digital authentication, fundamentally changing how locks are designed, attacked, serviced, and managed. For a modern locksmith, this shifts the role from purely mechanical work to a broader discipline that combines mechanical engineering, electronics, and access control systems.

One of the most important influences on locksmithing practice is the shift from single-layer security to multi-layer authentication. In a traditional lock, if the correct key geometry is replicated or the pins are manipulated, the lock opens. In hybrid systems, this is no longer sufficient. A key may need to satisfy mechanical alignment, magnetic coding, and electronic authorisation simultaneously. This means that common entry techniques such as picking, bumping, or simple key duplication are often ineffective. For locksmiths, this removes the idea of a "universal approach" and replaces it with system-specific knowledge and strategies.

Key control is also fundamentally transformed. In open mechanical systems, duplication is relatively easy because blanks are widely available. In magnetic systems such as EVVA MCS, duplication involves hidden data (magnet position and polarity) and requires authorised blanks and specialised equipment. When electronic layers are added, duplication becomes credential management rather than physical copying. Locksmiths must therefore understand not only how to cut keys, but also how to manage authorisation, access rights, and system records, particularly in commercial or high-security environments.

Diagnostics and servicing become significantly more complex. In purely mechanical locks, faults are typically related to wear, debris, or misalignment. In hybrid systems, failures may occur in multiple domains simultaneously—mechanical (e.g., worn components), magnetic (e.g., misaligned rotors or degraded magnet interaction), or electronic (e.g., battery failure, communication errors, or firmware issues). This requires a structured diagnostic approach where the locksmith must isolate which layer is causing the problem. As a result, troubleshooting becomes more analytical and less purely tactile.

Installation practices are also affected. Instead of simply fitting a lock, locksmiths must often configure and commission systems. Electronic locks may require pairing with mobile apps, programming user credentials, integrating with access control platforms, or installing wireless bridges. Even hybrid mechanical systems require strict adherence to manufacturer specifications to ensure that magnetic and mechanical elements align correctly. This introduces a level of system setup and configuration that goes beyond traditional hardware installation.

Security assessment expands considerably in hybrid environments. A locksmith is no longer evaluating only physical resistance to attack (e.g., picking or drilling), but also digital vulnerabilities, such as unauthorised credential sharing, poor access management, or insecure communication protocols. In high-security settings, the locksmith must consider how mechanical, magnetic, and electronic layers interact to create a defence-in-depth strategy, rather than relying on a single protective mechanism.

Another major influence is the increasing importance of precision and tolerances. Hybrid systems, particularly those involving sidebars, sliders, or magnetic alignment, require extremely accurate key cutting and component positioning. Small deviations that might be tolerated in basic locks can cause complete failure in high-security systems. This reinforces the need for code cutting, CNC machining, and manufacturer-specific tooling, rather than relying on duplication of worn keys.

From a professional perspective, these systems elevate locksmithing into a more specialised field. Locksmiths working with hybrid locks often need authorisation from manufacturers, must follow controlled procedures for key duplication, and may be required to maintain records or comply with security protocols. This introduces legal and administrative responsibilities alongside technical skills, particularly when dealing with restricted systems.

## Case Study: High-Security Locks Used in Critical Infrastructure

High-security locking systems play a vital role in protecting critical infrastructure such as power grids, water treatment plants, transport networks, and data centres. These environments demand continuous operation, strict control over access, and strong resistance to both physical attacks and covert entry methods. Unlike residential or standard commercial settings, a failure in security within these facilities can lead to major service disruptions, safety hazards, and, in some cases, national security implications. As a result, the design and implementation of locking systems in these environments must be both robust and carefully managed.

A typical example can be seen in an electrical utility company operating multiple substations across a region. Each site contains high-voltage equipment, control cabinets, and secured perimeter access points. These substations are often located in remote or unattended areas and are exposed to harsh environmental conditions such as rain, dust, and temperature extremes. This combination of isolation and exposure makes them particularly vulnerable to both opportunistic interference and deliberate attack.

To address these risks, the organisation establishes clear security requirements. These include strict key control to prevent unauthorised duplication, resistance to common attack methods such as picking, bumping, and drilling, and the ability to withstand environmental conditions over long periods. In addition, the system must provide auditability, allowing the organisation to track who accessed specific areas and when, while also being scalable across multiple sites to support operational growth.

Electrical utility companies require sophisticated security systems to protect critical infrastructure, sensitive equipment, and control systems from unauthorized access and potential sabotage. The types of locks employed in these facilities have evolved significantly from traditional mechanical systems to advanced electronic and smart lock technologies. Understanding the specific lock types used in electrical utility environments is essential for maintaining operational security and regulatory compliance.

**Traditional Mechanical Locks in Utility Infrastructure**

Electrical utility companies have historically relied on traditional padlocks and key-based locking mechanisms to secure critical areas. These conventional systems include standard padlocks and key locks that are used to protect access points to vital equipment and control systems [149]. However, traditional padlocks present significant vulnerabilities. They can be compromised through physical tampering, such as gluing lock cores or inserting foreign objects into lock mechanisms [149]. Additionally, key-based systems are susceptible to loss, duplication, and unauthorized access, making them increasingly inadequate for modern security requirements [82].

**Electronic and Smart Lock Systems**

Solenoid-Based Electronic Locks

Modern electrical utility facilities have increasingly adopted electronic locking mechanisms, particularly solenoid locks controlled by microcontroller systems. These systems utilize Arduino microcontrollers or similar processing units to manage access control [150]. The solenoid lock mechanism converts electrical signals into mechanical movement, enabling automated locking and unlocking based on authenticated commands [151]. This technology provides a significant improvement over traditional mechanical locks by enabling remote control and automated access management [150].

RFID-Based Access Control Systems

Radio Frequency Identification (RFID) technology has become a prominent choice for access control in utility environments. RFID-based smart lock systems utilize RFID reader modules integrated with Arduino microcontrollers to authenticate authorized personnel [152, 153]. Users can gain access by presenting registered RFID cards or key fobs to the reader, which then triggers the locking mechanism [152]. These systems offer several advantages over traditional locks, including the ability to record access data in real-time, monitor incoming and outgoing activities, and provide automated access control without physical keys [152]. RFID technology is particularly suitable for critical environments such as engine rooms and platforms in industrial installations [152].

Biometric Authentication Systems

Fingerprint-based smart door locking systems represent an advanced security approach increasingly adopted in high-security utility environments [117]. These systems capture and store biometric data, allowing access only to recognized fingerprints [117]. Fingerprint recognition technology provides a high level of security that eliminates the need to remember PINs, passwords, or carry physical keys [154]. The implementation typically involves an Arduino UNO microcontroller that controls the locking mechanism and manages the fingerprint authentication process [154]. Such systems are particularly valuable in utility facilities where access control must be strictly limited to authorized personnel [117].

**Remote Monitoring and Control Systems**

Network-Based Access Control

Electrical utility companies, particularly substations, have implemented trusted WLAN-based access control systems that enable remote monitoring of locks and access points [155]. These systems connect access control and screen cabinet lock controllers to auxiliary control systems through trusted wireless networks, allowing centralized supervisory control systems to monitor and manage access remotely [155]). This approach addresses the limitations of traditional work permits and on-site safety measures by preventing unauthorized personnel from entering restricted work areas [155].

IoT-Enabled Smart Lock Systems

Internet of Things (IoT) technology has enabled the development of comprehensive smart lock systems for utility facilities. These systems integrate multiple security layers, including RF detection, NFC technology, and networked communication capabilities [151]. The systems are managed by central processing units such as Arduino Uno microcontrollers, which facilitate communication and synchronization between various security components [151]. Network modules enable seamless information exchange across the facility, supporting real-time monitoring and remote operations that allow administrators to manage access permissions from any location [151].

Multi-Factor Authentication Approaches

Modern electrical utility companies increasingly employ multi-factor authentication systems to enhance security. These systems may combine multiple authentication methods, such as one-time passwords (OTP) sent to registered users combined with master codes [82]. Some advanced systems incorporate end-to-end encryption using algorithms such as RSA to protect against man-in-the-middle attacks and unauthorized key duplication [82]. Additionally, breach detection systems with image capture capabilities are integrated to detect forced intrusions and unauthorized access attempts [82].

**Specialized Security Features for Utility Environments**

Vibration and Motion Detection

Advanced lock systems used in utility facilities may incorporate vibration sensors and motion detection capabilities to identify tampering attempts [154]. These sensors can trigger alarms and alert security personnel when unauthorized access is attempted [154]). The integration of vibration detection with biometric authentication provides an additional layer of security against physical attacks on locking mechanisms [154].

Real-Time Monitoring and Notification Systems

Electrical utility companies employ lock systems with real-time monitoring capabilities that send notifications to administrators when access events occur [156]. These systems can integrate with mobile applications and messaging platforms to provide immediate alerts regarding door lock status, unauthorized access attempts, and system anomalies [156]. Short Message Service (SMS) notifications enable rapid response to security incidents [156].

Physical Security Infrastructure

Beyond individual lock mechanisms, electrical utility companies implement comprehensive physical security systems that include locks as one component of a multi-layered approach. Physical security features include protection utilities such as fencing and fence topping, exterior intrusion detectors and alarms, locks and padlocks, and surveillance utilities such as closed-circuit television (CCTV) [157]. This integrated approach ensures that locks function as part of a comprehensive security system rather than as isolated devices [157].

Regulatory and Compliance Considerations

The selection and implementation of lock systems in electrical utility companies must comply with regulatory requirements and industry standards. Security guidelines for utility facilities recommend the selection and implementation of physical security approaches and measures designed to detect, delay, and respond to the presence of intruders [157]. The specific types of locks employed must be chosen based on their ability to meet these regulatory requirements while providing practical security solutions for the facility's operational needs [157].

Electrical utility companies employ a diverse range of lock types, ranging from traditional mechanical padlocks to advanced smart lock systems incorporating biometric authentication, RFID technology, and IoT-enabled remote monitoring capabilities. The evolution from simple key-based systems to sophisticated electronic and networked solutions reflects the increasing security demands of critical infrastructure protection. Modern utility facilities typically implement multi-layered security approaches that integrate various lock technologies with surveillance systems, access control networks, and real-time monitoring capabilities. The selection of specific lock types depends on the facility's security requirements, regulatory compliance needs, operational efficiency considerations, and the level of access control sophistication required to protect critical equipment and personnel.

In response, a layered high-security locking system is implemented. At the mechanical level, high-security cylinders—such as disc detainer or sidebar-based designs—are installed on gates and equipment housings. These systems eliminate or enhance traditional pin mechanisms, making them highly resistant to picking and bumping. Their construction typically includes hardened steel components, which provide strong resistance to drilling and ensure reliable performance in demanding outdoor environments.

Alongside this, a restricted key system is introduced to control duplication. Key blanks are not available on the open market and can only be obtained through authorised channels. Any request for duplication requires formal approval and verification, and all issued keys are recorded in a key register. This ensures that even if a key is temporarily accessed or lost, it cannot be easily copied without proper authorisation, significantly improving overall security control.

A master key hierarchy is also developed to manage access efficiently. Technicians are issued keys that allow entry only to specific substations, while supervisors are granted broader access across multiple sites. An emergency master key is held under strict control for high-level access when required. This structured approach reduces risk by limiting access to only what is necessary, while still maintaining operational flexibility.

At critical points within the infrastructure, such as control rooms, electronic access systems are integrated to form a hybrid security layer. These systems use technologies such as RFID cards or smart credentials and record access events in real time. Permissions can be updated remotely, allowing rapid response to changes in personnel or security conditions. This combination of mechanical and electronic systems provides both physical protection and digital accountability.

When tested against real-world threat scenarios, the effectiveness of this layered approach becomes clear. Attempts at unauthorised key duplication are prevented due to restricted keyways and controlled authorisation processes. Non-destructive entry methods such as picking or bumping are unsuccessful because of the advanced internal mechanisms of the cylinders. Even destructive attacks, such as drilling, are significantly delayed or defeated by hardened materials and protective design features. In each case, the system increases the time, effort, and likelihood of detection, which is the fundamental objective of physical security.

From a locksmithing perspective, this case study highlights several important considerations. Lock selection must be aligned with both the environmental conditions and the security risk level, requiring careful evaluation of materials, mechanisms, and key control systems. The role of the locksmith extends beyond mechanical installation to include management of restricted systems, maintenance of documentation, and integration of electronic access technologies.

Precision and compliance are also critical. High-security systems require accurate installation, correct key cutting—often using code-based methods—and strict adherence to manufacturer and organisational requirements. Maintenance and diagnostics become more complex, as faults may arise from environmental wear, electronic issues, or breaches in key control rather than simple mechanical failure. This requires locksmiths to adopt a broader, systems-based approach to troubleshooting.

Ultimately, this example demonstrates that in critical infrastructure, locks are not standalone devices but components of a comprehensive, layered security strategy. Mechanical strength, controlled key management, and electronic monitoring work together to create a resilient system. For modern locksmiths, this represents a shift from traditional trade skills toward integrated security engineering, where understanding the interaction between systems, assessing risk, and ensuring long-term reliability are just as important as the technical aspects of installation and repair.

## Impact on Locksmithing Practices

For someone interested in locksmithing, this example is highly relevant because it demonstrates how the profession has evolved beyond traditional lock-and-key work into a broader discipline of integrated security engineering. In critical infrastructure environments such as electrical substations, locks are not simply devices that open and close doors—they are part of a coordinated system designed to prevent, detect, and respond to security threats. Understanding this shift is essential for any modern locksmith who aims to work in high-security or industrial sectors.

One of the most important influences on locksmithing practice is the move from single-layer mechanical security to layered security systems. In the past, a locksmith's role may have focused primarily on selecting and installing a suitable mechanical lock. In critical infrastructure, however, security is achieved through a combination of mechanical resistance, restricted key control, and electronic access systems. This means locksmiths must understand how different technologies interact, rather than viewing each lock as an isolated component.

The example also highlights the growing importance of key control and system management. In standard environments, key duplication is often a routine task. In high-security settings, it becomes a tightly controlled process involving authorisation, documentation, and restricted keyways. Locksmiths are therefore required to manage key registers, enforce duplication policies, and ensure compliance with organisational procedures. This shifts part of the role from technical work to administrative and security governance responsibilities.

Another major influence is the increasing need for multi-disciplinary technical skills. Traditional locksmithing focused on mechanical systems such as pin tumblers and wafer locks. In modern infrastructure, locksmiths must also work with electronic systems including RFID access control, biometric devices, networked locks, and remote monitoring platforms. This requires knowledge of electronics, communication protocols, software configuration, and sometimes even basic networking concepts. As a result, the profession now sits at the intersection of mechanical engineering, electronics, and information technology.

The case study also demonstrates how environmental and operational factors shape locksmithing decisions. Locks used in substations must withstand extreme weather, dust, corrosion, and long periods without maintenance. This influences material selection, lock design, and installation techniques. Locksmiths must therefore consider not only security features, but also durability, reliability, and lifecycle performance—particularly in remote or high-risk environments.

From a practical perspective, the example reinforces the importance of precision and compliance in locksmithing work. High-security systems often rely on tight tolerances, complex internal mechanisms, and manufacturer-specific requirements. Errors in installation, key cutting, or configuration can compromise the entire system. Additionally, locksmiths must adhere to regulatory standards and organisational policies, particularly in industries where security breaches can have serious consequences.

The integration of electronic systems introduces a new dimension: diagnostics and troubleshooting across multiple domains. Problems are no longer purely mechanical. A failure may be caused by battery issues, communication faults, software errors, or access control misconfigurations. Locksmiths must be able to systematically diagnose whether a fault is mechanical, electronic, or procedural, and respond accordingly. This requires a more analytical and systems-based approach to problem-solving.

Importantly, this scenario illustrates a shift in the locksmith's role from reactive to proactive security planning. Instead of simply repairing or replacing locks after a failure, locksmiths are increasingly involved in designing access systems, assessing risks, and recommending appropriate security solutions. This includes advising on master key systems, selecting suitable lock technologies, and integrating physical security with digital monitoring systems.

## Reverse Engineering Lock Security Features

Reverse engineering lock security features refers to the process of systematically analysing how a lock is designed and how its internal mechanisms function, with the goal of understanding how it achieves security. In legitimate locksmithing practice, this is not about bypassing locks unlawfully, but about developing a deep technical understanding of lock construction, performance, and vulnerabilities so that systems can be properly selected, installed, maintained, and improved.

At its core, reverse engineering involves breaking a lock system down into its fundamental components and behaviours. A locksmith may examine the keyway geometry, pin configuration, sidebar interaction, materials, tolerances, and any secondary mechanisms such as sliders, discs, or magnetic elements. This analysis can be done through non-destructive methods, such as studying manufacturer documentation and observing lock behaviour, or through controlled disassembly in a workshop environment. The objective is to understand how each component contributes to the overall security of the lock.

One of the key areas of focus is identifying the security features embedded within the design. For example, a locksmith might analyse how spool or serrated pins create false feedback during manipulation, how a sidebar requires precise multi-point alignment, or how a restricted keyway prevents unauthorised duplication. In high-security systems, this may extend to understanding multi-dimensional key interaction, such as rotating pins in Medeco systems, telescopic pins in Mul-T-Lock, or slider mechanisms in EVVA cylinders. By reverse engineering these features, the locksmith gains insight into why certain locks are more resistant to attack than others.

Material science and manufacturing quality also form an important part of this process. Reverse engineering allows the locksmith to evaluate the hardness of components, the presence of anti-drill inserts, and the wear characteristics of moving parts. For instance, identifying hardened steel pins or carbide inserts explains a lock's resistance to drilling, while observing machining precision helps explain its resistance to manipulation. These insights directly influence decisions about product selection and expected service life.

Another critical aspect is understanding failure modes and vulnerabilities. Reverse engineering helps identify where and how a lock might fail, whether due to wear, poor tolerances, environmental degradation, or design limitations. For example, a locksmith might discover that a particular lock is prone to key breakage due to narrow keyways, or that certain components wear prematurely under high usage. This knowledge is essential for troubleshooting, maintenance planning, and recommending more suitable alternatives.

In modern locksmithing, reverse engineering increasingly extends beyond purely mechanical systems into electronic and hybrid locks. This includes analysing how electronic credentials are verified, how actuators engage the locking mechanism, and how communication protocols (e.g., Bluetooth or RFID) interact with physical hardware. Understanding these systems allows locksmiths to diagnose faults, assess cybersecurity risks, and ensure proper integration with broader access control systems.

Importantly, reverse engineering also supports system design and security assessment. By understanding how different locks function internally, a locksmith can compare products, evaluate their resistance to various attack methods, and recommend appropriate solutions based on risk level and environment. This is particularly important in high-security applications, where layered systems combine mechanical, magnetic, and electronic elements.

From a professional standpoint, reverse engineering reinforces the transition of locksmithing from a trade based on manual skill to one grounded in engineering knowledge and analytical thinking. It enables locksmiths to move beyond routine tasks and engage in higher-level activities such as forensic analysis, product evaluation, and security consultancy.

A clear way to demonstrate reverse engineering in locksmithing is to walk through a real, widely used lock and analyse how its security features function, how they are identified, and what that means in practice. A good example is the Schlage Primus cylinder, which is commonly used in commercial and institutional environments and incorporates both traditional and high-security elements.

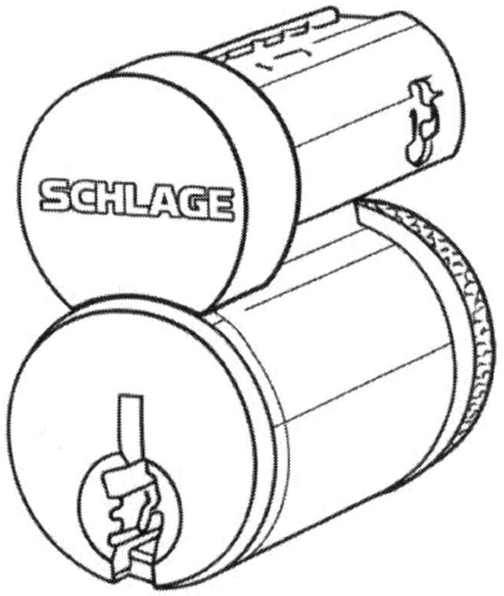


*Figure 54: Schlage Primus High Security Cylinder.*

## Beyond the Key

### 1. Initial Observation (External Features)

The reverse engineering process begins with external inspection—what can be identified without disassembly.

- **Key type:** Appears similar to a standard Schlage key (SC1/SC4 profile)
- **Visible difference:** Additional side milling (track cuts) along the key blade
- **Keyway:** Restricted or controlled profile (not a standard open keyway)

**Initial hypothesis:** This is not a standard pin-tumbler lock—it likely includes a secondary locking mechanism beyond top pins.

### 2. Functional Testing (Behaviour Analysis)

Before opening the lock, behaviour is analysed using controlled testing.

- Key inserts smoothly → confirms correct keyway profile
- Slight resistance before rotation → suggests secondary alignment requirement
- Incorrect key (same profile but no side cuts) → partially inserts but will not turn

**Conclusion:** The lock likely uses dual authentication:

1. Traditional pin alignment
2. Secondary mechanism (triggered by side cuts)

### 3. Controlled Disassembly

The cylinder is disassembled in a workshop environment to expose internal components.

**Primary components identified:**

- Standard pin stacks (top pins, bottom pins, springs)
- A sidebar mechanism running longitudinally
- A set of finger pins aligned with the key's side milling

This confirms the lock is a hybrid mechanical system:

- Conventional pin tumbler
- Sidebar with side-coded interaction

**4. Mechanism Analysis (How It Actually Works)**

**A. Primary System – Pin Tumbler**

- Key lifts bottom pins to the shear line
- This is identical to standard Schlage cylinders

✓If this were the only system → lock would open

**B. Secondary System – Sidebar + Finger Pins**

- Each finger pin is positioned to interact with side milling on the key
- When the correct key is inserted:
  - Finger pins rotate or shift
  - This aligns gates (notches) along the sidebar
- The sidebar retracts into the plug

× If side cuts are incorrect:

- Sidebar remains engaged
- Plug cannot rotate—even if top pins are correct

**Critical insight:** The sidebar acts as a second shear line, requiring simultaneous alignment.

**5. Security Feature Identification**

Through reverse engineering, the following features are confirmed:

**Anti-Pick Features**

- Standard pins may include spool or serrated pins
- Sidebar provides no direct tactile feedback
- Requires dual-axis manipulation (vertical + lateral)

**Anti-Bump Features**

- Even if bumping aligns top pins:
  - Sidebar remains engaged
  - Lock will not open

**Key Control**

- Side milling requires specialised cutting equipment
- Blanks are typically restricted

## 6. Failure Mode Analysis

Reverse engineering also identifies how the lock can fail in legitimate use:

- **Worn side milling** → sidebar misalignment → key stops working
- **Debris in sidebar channel** → sidebar fails to retract
- **Incorrect duplication** → top pins align but sidebar blocks rotation

These are critical for diagnostics and servicing.

## 7. Locksmithing Implications

From a practical standpoint, this analysis changes how a locksmith approaches the lock:

### A. Key Cutting

- Requires:
  - Standard depth cutting (pins)
  - Precision side milling (secondary system)
- Cannot be duplicated on basic machines

### B. Picking / Entry Techniques

- Traditional picking alone is insufficient
- Requires understanding of sidebar behaviour

### C. System Design

- Suitable for:
  - Commercial master key systems
  - Restricted access environments

### D. Troubleshooting

- Must diagnose:
  - Mechanical pin issues
  - Sidebar alignment issues

**8. Security Assessment Outcome**

After reverse engineering, the locksmith can confidently conclude:

- The lock provides layered mechanical security
- It resists:
    - Basic picking
    - Bumping
    - Unauthorised duplication
- It introduces complexity rather than just stronger materials

This makes it appropriate for medium to high-security applications

This worked example demonstrates that reverse engineering in locksmithing is not about defeating or bypassing a lock, but rather about understanding the underlying design logic that governs how the system operates. By analysing the internal structure and interaction of components, a locksmith gains insight into why the lock behaves the way it does, how its security features function, and how it should be correctly serviced, maintained, or specified for particular applications.

In the case of the Schlage Primus system, security is achieved through a multi-dimensional interaction between components. The lock does not rely solely on traditional vertical pin alignment, but also incorporates a lateral sidebar mechanism that must align simultaneously. This dual-layer requirement means that both vertical and horizontal elements must be correctly positioned before the lock can operate, significantly increasing resistance to manipulation and unauthorised entry.

As a result, effective servicing of such systems requires more than basic mechanical skill. Locksmiths must develop a deeper understanding of how multiple subsystems interact within a single lock, including pin stacks, side pins or finger pins, and sidebar engagement. This knowledge is essential for accurate key cutting, fault diagnosis, and system maintenance, particularly when dealing with restricted or high-security applications.

Ultimately, this example highlights a broader shift in locksmithing practice. The profession is moving away from a “single mechanism” mindset—focused only on basic pin-tumbler operation—towards a multi-layer analytical approach. Modern locksmiths must be capable of evaluating complex, integrated systems, where security is achieved through the combination of multiple interacting elements rather than a single line of defence.

Reverse engineering lock security features is a critical skill that allows locksmiths to understand not just how a lock works, but why it works the way it does. This knowledge underpins effective installation, accurate diagnostics, informed product selection, and robust security design, making it an essential component of modern locksmithing practice.

## Impact on Locksmithing Practices

Reverse engineering lock security features is highly relevant to anyone entering or advancing in locksmithing because it builds the depth of understanding required to work with modern, complex locking systems. Locksmithing is no longer limited to simply cutting keys or replacing cylinders; it increasingly requires the ability to analyse, interpret, and make informed decisions about security systems. Reverse engineering develops this capability by teaching the locksmith how and why a lock functions, rather than just how to operate it.

At a foundational level, this process strengthens core technical competence. By examining internal components such as pin stacks, sidebars, sliders, or magnetic elements, a locksmith gains a clear understanding of how different mechanisms achieve security. This directly improves practical skills such as installation, rekeying, and servicing. For example, recognising how a sidebar interacts with side milling—as seen in high-security systems like Schlage Primus—ensures that keys are cut accurately and faults are diagnosed correctly, rather than relying on trial-and-error approaches.

This knowledge also has a significant influence on diagnostic and troubleshooting practices. In traditional locksmithing, a fault might be approached as a simple mechanical issue. However, reverse engineering teaches the locksmith to think in terms of systems and interactions. A problem may not be caused by worn pins alone, but by misalignment in a secondary mechanism, debris affecting a sidebar, or incorrect key geometry. As a result, locksmiths become more efficient, accurate, and reliable in identifying root causes rather than symptoms.

Reverse engineering further influences how locksmiths select and specify locking systems. By understanding the internal design and security features of different locks, a locksmith can match the correct product to the risk level and environment. For instance, recognising the advantages of multi-dimensional systems (e.g., pin + sidebar, disc detainer, or magnetic systems) allows for better recommendations in commercial or high-security settings. This shifts the role of the locksmith from installer to security advisor, capable of assessing threats and designing appropriate solutions.

Another important impact is on key control and duplication practices. High-security locks often rely on restricted keyways, side milling, or specialised elements that require precise cutting and controlled authorisation. Reverse engineering helps the locksmith understand why these controls exist and how they function, reinforcing the importance of compliance with manufacturer requirements and legal obligations. It also highlights the limitations of standard equipment, driving the need for specialised machinery and procedures.

In modern contexts, the relevance extends beyond purely mechanical systems into electronic and hybrid technologies. Reverse engineering principles are applied to understanding how electronic credentials, sensors, actuators, and communication systems integrate with physical locks. This broadens the locksmith's skill set into areas such as access control systems, smart locks, and cybersecurity considerations. As a result, the profession evolves toward integrated security engineering rather than isolated mechanical work.

From a professional development perspective, reverse engineering fosters analytical thinking and continuous learning. Each lock becomes a case study in design, materials, and security strategy. This mindset encourages locksmiths to stay current with emerging technologies, evaluate new products critically, and adapt to changing industry demands. It also supports higher-level roles such as security auditing, forensic analysis, and consultancy.

## Chapter 4 Review Questions

### Question 1

A homeowner using a common hardware-store lock lends their house key to a contractor for a few hours. The contractor secretly duplicates the key at a local store.

Why was this duplication possible?

A. The lock used a restricted key system
B. The system used patented key blanks
C. The lock used an open key system with widely available blanks
D. The locksmith incorrectly installed the lock

### Question 2

A commercial office installs a restricted key system. An employee attempts to duplicate their key but is refused by the locksmith because they cannot provide the system's control card.

What security principle does this demonstrate?

A. Mechanical complexity prevents duplication
B. Key control through authorised duplication procedures
C. Locks cannot be copied once installed
D. Restricted systems eliminate the need for records

### Question 3

A government facility uses a patented key system with specialised key blanks that are unavailable outside authorised suppliers.

What is the primary security advantage of this system?

A. Keys cannot be physically lost
B. Lock cylinders cannot wear out

C. Key duplication is controlled both legally and technically
D. Mechanical attacks become impossible

**Question 4**

A previously restricted key system becomes widely copied after its patent expires and third-party manufacturers begin producing compatible key blanks.

What is the likely consequence?

A. The system becomes more secure
B. Duplication control weakens
C. Locks stop functioning correctly
D. Master key systems stop working

**Question 5**

A university installs a master key system where maintenance staff can open certain rooms while administrators have access to all areas.

What is the main purpose of this system?

A. Eliminating key duplication entirely
B. Creating a structured hierarchy of access permissions
C. Reducing the number of locks in the building
D. Preventing locksmith access to the system

**Question 6**

A burglar attempts to pick a lock containing spool and serrated pins but repeatedly encounters false feedback and cannot align the pins correctly.

What feature is preventing successful picking?

A. Magnetic key coding
B. Anti-pick security pins creating false shear lines
C. Electronic credential verification
D. Hardened steel drill inserts

**Question 7**

An attacker attempts to open a cylinder using a bump key but the lock includes a rotating pin and sidebar system that prevents alignment.

Why does the attack fail?

A. The lock has too many pins
B. The sidebar requires additional alignment beyond vertical pin movement
C. Bump keys only work on worn locks
D. The keyway is too narrow

**Question 8**

During a break-in attempt, a burglar tries drilling a lock but encounters spinning hardened steel ball bearings that prevent the drill bit from gaining traction.

What security feature is responsible?

A. Anti-pick pins
B. Magnetic authentication
C. Anti-drill protection using hardened components
D. RFID authentication

**Question 9**

A building installs a magnetic locking system where magnets embedded in the key align internal rotors inside the cylinder.

Why are these systems highly resistant to picking?

A. Magnetic elements cannot be manipulated using traditional lock-picking tools
B. Magnetic locks contain no internal components
C. Keys cannot be lost or copied
D. The lock has no keyway

**Question 10**

A security system requires a correctly cut key, a magnetic pattern, and an electronic credential before the lock will open.

What type of system is this?

A. A traditional mechanical lock
B. A hybrid multi-factor locking system
C. A basic keypad lock
D. A master key system

# Part 3

# Lock Bypassing, Forensics, and Ethical Boundaries

# Chapter 5

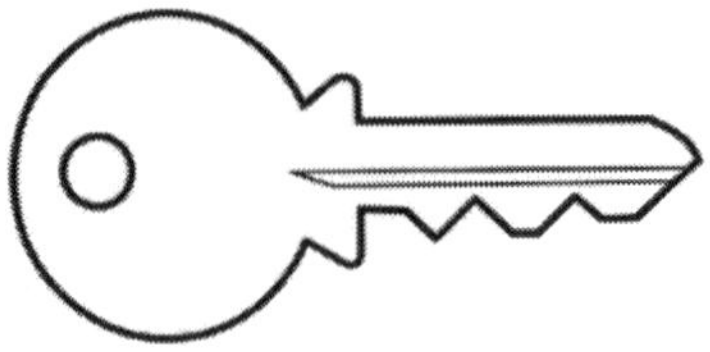

# Forensic Locksmithing and Crime Analysis

This chapter builds on the foundational principles of forensic locksmithing by moving beyond individual techniques and into the structured application of investigative practice. Having explored methods of entry, tool mark identification, and key analysis, the focus then shifts toward integrating these elements into a coherent, defensible investigative process. In real-world scenarios, locksmiths are not simply technicians, they are observers, analysts, and, at times, contributors to legal and insurance outcomes. This requires a disciplined approach that combines technical skill with forensic awareness.

At this stage, the emphasis is on how evidence is interpreted within context. Locks, keys, and access systems do not exist in isolation; they are part of a broader environment that includes human behaviour, security practices, and situational factors. Understanding how to connect physical evidence with contextual information is critical in forming accurate conclusions about events such as break-ins, unauthorised access, or system compromise. The ability to distinguish between genuine incidents and staged or misleading scenarios becomes particularly important.

The chapter also reinforces the importance of methodology. Consistency in examination, documentation, and reasoning ensures that findings are not only accurate but also repeatable and defensible. Whether the work is conducted for investigative purposes, insurance assessments, or legal proceedings, the credibility of the locksmith depends on adherence to structured processes and evidence-based conclusions.

Finally, this chapter highlights the evolving role of the locksmith within forensic and security domains. As locking systems become more advanced and integrated with electronic and digital technologies, the scope of analysis expands. Modern locksmithing increasingly requires an understanding of both mechanical and electronic evidence, as well as the ability to interpret data from multiple sources.

In essence, this chapter serves as a bridge between technical knowledge and professional application, demonstrating how forensic locksmithing operates as a disciplined, analytical practice within real-world investigative environments.

## Identifying Forced Vs Non-Forced Entry

In forensic locksmithing and crime analysis, the ability to distinguish between forced entry and non-forced (covert) entry is fundamental to understanding how an incident occurred. This determination influences not only the technical investigation, but also legal outcomes, insurance claims, and the overall direction of a criminal enquiry. For a locksmith operating in a forensic or investigative capacity, this process requires a structured, evidence-based assessment of both the lock and its surrounding environment.

Forced entry involves the use of physical force to overcome a locking mechanism or barrier. This typically results in visible damage to the lock, door, frame, or surrounding hardware. Non-forced entry, on the other hand, occurs without obvious physical damage. It includes methods such as the use of a legitimate key, unauthorised key duplication, lock picking, decoding, or bypass techniques. These methods are often referred to as *covert entry* because they leave minimal or no visible signs.

### Indicators of Forced Entry

In cases of forced entry, physical evidence is typically present and can often be identified through careful visual inspection. Damage to the lock, door, or surrounding hardware provides clear indicators that force has been applied to gain access.

One of the most common signs is deformation of lock components. This may include bent plugs, damaged keyways, or distorted cylinders, which are often the result of prying or torque-based attacks. Such damage indicates that significant mechanical force has been applied directly to the locking mechanism in an attempt to overcome it.

Drill marks or tool impressions are also strong indicators of forced entry. Circular abrasions, visible penetration points, or the presence of metal shavings suggest that drilling has been used to destroy internal components of the lock. These marks can often be analysed to determine the type of tool used and the method of attack.

In some cases, cylinders may be broken or snapped, particularly in euro-profile locks. Snapping attacks typically leave clear fracture points where the cylinder has been forcibly separated, allowing access to the internal mechanism. This type of damage is usually obvious and provides strong evidence of deliberate forced entry.

*Figure 55: Forced entry. Tim Samoff, CC BY-ND 2.0, via Flickr.*

Damage to the door or frame is another key indicator. Splintered timber, bent or displaced strike plates, and visible pry marks are commonly associated with the use of crowbars or levering tools. This type of damage often extends beyond the lock itself and reflects attempts to bypass the locking mechanism entirely.

Additionally, hardware such as handles, escutcheons, or entire lock bodies may be detached or displaced. This suggests that force has been applied not only to the internal components but also to the external fixtures of the locking system.

From a forensic perspective, these physical signs are critical in reconstructing the event. They can help determine the type of tool used, such as a drill, crowbar, or bolt cutters, as well as the direction and magnitude of force applied. Furthermore, analysis of the damage can provide insight into the sequence of events, indicating whether the attack was attempted but unsuccessful, or whether it ultimately resulted in entry.

**Indicators of Non-Forced (Covert) Entry**

Non-forced entry is significantly more difficult to identify because it typically leaves little or no visible damage to the lock or surrounding structure. In many cases, the lock appears completely intact, which can make it challenging to determine whether unauthorised access has occurred without a more detailed examination.

One of the primary indicators of non-forced entry is the absence of visible damage to the lock, door, or frame. The locking mechanism often appears normal and undisturbed, with no signs of deformation, tool marks, or impact. This can initially suggest that no intrusion has taken place, even when access has been gained.

Another key characteristic is that the lock continues to function normally. The correct key will still operate smoothly, and there are no obvious signs of mechanical failure or resistance. This indicates that no destructive method, such as drilling or prying, has been used to gain entry.

However, more subtle evidence may be present upon closer inspection. Under magnification, minor scratches or fine markings inside the keyway can sometimes be detected. These may indicate the insertion of picking tools or other instruments used to manipulate the internal components without causing visible external damage.

Evidence of key access is also an important consideration. This may include missing keys, unauthorised duplication of keys (particularly in systems with open or unrestricted keyways), or access by individuals who legitimately hold keys, such as staff or contractors. In such cases, the entry may appear legitimate from a physical standpoint but may still be unauthorised.

At a broader level, system-based inconsistencies can provide valuable clues. In electronic or hybrid systems, access logs may reveal unusual entry times, repeated access attempts, or use of credentials outside expected patterns. In purely mechanical systems, discrepancies in key registers or access records may indicate misuse or loss of control over keys.

Common covert entry methods include the use of an authorised or stolen key, unauthorised key duplication, lock picking, and, in lower-security systems, bumping. More advanced techniques such as decoding or impressioning may also be used to create a working key without leaving obvious traces. These methods rely on skill and precision rather than force, making detection more reliant on careful analysis than on visible damage.

Detecting non-forced entry methods requires a combination of physical inspection, behavioural analysis, and, where available, system data. Each method leaves different types of evidence—often subtle—so the locksmith must adopt a structured, forensic approach rather than relying on obvious damage.

**Use of an Authorised or Stolen Key:** This is one of the hardest forms of entry to detect because the lock is operated exactly as designed. There is typically no physical damage or abnormal wear within the lock.

Detection relies primarily on contextual and administrative evidence rather than mechanical signs. Key indicators include discrepancies in key control records, such as missing keys, unreturned keys from former employees or contractors, or access occurring outside authorised times. In environments with electronic systems, access logs may show entry by a valid credential at unusual hours or by a user who should not have been present.

A locksmith may also consider patterns such as repeated unexplained access or reports of doors being found unlocked. In these cases, the conclusion is often reached not by what is visible on the lock, but by eliminating all other entry methods and identifying a breakdown in key control procedures.

**Unauthorised Key Duplication:** Unauthorised duplication can also leave no direct evidence on the lock itself, as the duplicated key may function normally. However, detection focuses on inconsistencies in key management systems and duplication controls.

One of the main indicators is the presence of additional keys in circulation beyond what has been issued or recorded. This may be identified through audits, staff reports, or security breaches. In restricted systems, any duplication without proper authorisation is a clear red flag.

From a mechanical perspective, poorly duplicated keys may leave subtle signs. These can include:

- Slightly rough or inconsistent operation of the lock
- Increased wear on pins due to inaccurate cutting
- Minor scoring or abrasion within the keyway from incorrectly shaped keys

In open keyway systems, the risk is significantly higher because blanks are freely available. A locksmith may recommend transitioning to a restricted system if duplication is suspected but cannot be directly proven.

**Lock Picking:** Lock picking is a non-destructive method that manipulates internal components, and while it leaves minimal visible evidence, careful inspection can reveal indicators.

The most common signs are found inside the keyway and on internal components:

- Fine scratches or scoring on the plug, particularly around pin chambers
- Polishing or abnormal wear patterns inconsistent with normal key use
- Slight deformation or wear on pins (especially security pins)

Externally, the lock will appear intact and functional, but under magnification, these internal marks can suggest tool insertion and manipulation.

Another indicator is inconsistent or intermittent lock behaviour following the incident. For example, pins may feel slightly “rough” or irregular due to minor internal disturbance.

It is important to note that highly skilled picking may leave extremely minimal evidence, so conclusions often depend on experience and exclusion of other methods.

**Bumping (Lower-Security Systems):** Bumping is a covert but more forceful non-destructive method, and it can leave subtle but detectable signs, particularly in lower-security pin tumbler locks.

Key indicators include:

- Impact marks on the face of the plug or keyway, caused by repeated striking of the bump key
- Unusual wear or peening on pin tops and driver pins, due to repeated kinetic impact
- Slight deformation of springs or reduced spring tension over time
- A slightly “looser” or more worn feel when operating the lock

Unlike picking, bumping involves repeated impacts, so evidence may accumulate with multiple attempts. In some cases, very fine metal particles or dust may also be present inside the cylinder.

Locks that have been bumped frequently may show premature wear compared to their age and usage, which can be a strong indicator when assessed alongside the context of the incident.

In practice, no single sign definitively proves a specific method. Instead, locksmiths use a process of elimination and evidence correlation, combining:

- Physical inspection (external and internal)
- Magnified examination of keyways and components
- Functional testing of the lock
- Review of key control records or access logs
- Consideration of the environment and user behaviour

This approach allows the locksmith to form a reasoned professional opinion about how entry was gained.

**Key Forensic Insight**

The absence of damage does not mean no entry occurred—it often indicates a higher level of sophistication. In many cases, non-forced entry suggests:

- Insider involvement
- Poor key control
- Use of professional or knowledgeable methods

Conversely, forced entry is often associated with:

- Opportunistic or time-pressured attacks
- Lower skill levels
- Higher likelihood of noise and detection

### Forensic Analysis Process

Forensic locksmith investigation represents a specialized discipline within forensic science that applies systematic methodologies to determine how unauthorized entry was gained at crime scenes. The process requires meticulous attention to physical evidence, contextual information, and established investigative protocols. While the provided task describes a comprehensive forensic locksmith methodology, the available reference materials do not contain direct sources specifically addressing forensic locksmith practices. However, the references do provide relevant information about broader forensic investigation methodologies, evidence collection standards, and the integration of multiple evidence types in forensic analysis that parallel the structured approach described in the task.

## Scene Assessment and Physical Evidence Examination

The initial phase of forensic locksmith investigation involves thorough scene assessment, examining doors, frames, and surrounding areas for signs of disturbance. This foundational approach aligns with established forensic investigation principles. Research on forensic evidence collection emphasizes the importance of comprehensive scene examination and the integration of multiple evidence types. Forensic investigations should "adhere to stringent standards that consider the possibility of secondary DNA transfer" and that "investigators ought to emphasize thorough evidence collection methods, including gathering witness statements and reviewing surveillance footage to support" physical findings [158]. This principle of combining physical examination with contextual evidence reflects the methodological rigor required in forensic locksmith analysis.

The examination of secondary damage, misaligned hardware, and signs of tampering represents critical components of scene assessment. The identification of tool marks, deformation, and unusual wear patterns on locks requires the same level of systematic documentation and analysis that characterizes other forensic disciplines. The contextual approach to evidence evaluation, as emphasized in forensic science literature, supports the integration of physical lock examination with broader scene analysis.

## Lock Examination and Technical Analysis

Detailed lock examination, including inspection of the keyway, plug, and cylinder for visible and microscopic signs of interference, represents the technical core of forensic locksmith investigation. This examination phase requires distinguishing between normal operational wear and damage caused by manipulation or attack. The methodology parallels approaches used in other forensic disciplines that examine physical evidence for signs of interference or manipulation.

The assessment of functional lock condition and the identification of internal disruption indicators require specialized knowledge and careful documentation. This technical examination phase must be conducted with the same rigor and objectivity applied in other forensic analyses, ensuring that conclusions are based on observable, measurable evidence rather than assumptions.

## Key Control Investigation

The key control investigation phase addresses scenarios where no obvious forced entry is present, requiring examination of legitimate access pathways. This investigative step considers unauthorized key duplication, lost or unreturned keys, and insider involvement. The systematic evaluation of access authorization records and key registers represents a critical component of comprehensive forensic analysis. This approach reflects the principle that forensic investigation must consider multiple potential entry methods and access scenarios, not merely obvious signs of forced entry.

## Mechanism Analysis and Internal Component Examination

The mechanism analysis phase, typically conducted in controlled workshop environments, involves careful disassembly and detailed examination of internal lock components. This stage requires distinguishing between normal wear patterns and damage caused by manipulation or attack, and evaluating how built-in security features may have been bypassed. The controlled environment examination allows for detailed microscopic analysis and systematic documentation of findings.

The evaluation of security features such as security pins, sidebars, and anti-drill components requires specialized knowledge of lock design and security mechanisms. This technical analysis must be conducted with appropriate documentation and photographic evidence to support conclusions about the method of entry.

### Contextual Evidence Integration

The final investigative phase incorporates contextual evidence to validate and support physical findings. Forensic investigations should include "gathering witness statements and reviewing surveillance footage to support" physical evidence findings [158]. The integration of CCTV footage, alarm system data, and electronic access logs represents a critical component of comprehensive forensic analysis.

The use of CCTV footage in forensic investigations has become increasingly important. "CCTV footage is utilized in crime prevention and law enforcement to safeguard and improve public security" and that "images of offenders captured in surveillance are considered evidence in the court of the law" [159]. The systematic review of temporal data, including the timing of the incident and sequence of events, allows forensic locksmiths to construct a comprehensive narrative of how entry was gained.

The combination of physical evidence with contextual information enables the formation of well-reasoned conclusions about entry methods. This integrated approach reflects established forensic science principles that emphasize the importance of multiple lines of evidence in supporting investigative conclusions.

### Impact on Locksmithing Practices

The distinction between forced and non-forced entry is highly significant in professional locksmithing practice, particularly for those working in forensic investigation, insurance assessment, and security consulting roles. One of the most important implications is the need for accurate reporting. Locksmiths are often required to determine whether forced entry has occurred, as many insurance claims depend on clear evidence of physical break-in. An incorrect assessment can result in claims being wrongly approved or denied, making precision and objectivity essential.

This distinction also directly informs security recommendations. Where non-forced entry is identified, it often indicates underlying weaknesses in key control or access management rather than physical security failure. In such cases, locksmiths may recommend the implementation of restricted key systems, tighter key control procedures, or the upgrade to high-security cylinders with advanced anti-duplication and anti-manipulation features. These recommendations are based on an understanding that the breach occurred through access rather than force.

In addition, locksmith findings may carry weight as legal evidence. Reports and expert opinions can be used in court proceedings, insurance disputes, or internal investigations. This requires locksmiths to provide objective, evidence-based conclusions supported by clear documentation, logical reasoning, and, where necessary, photographic or physical evidence. The ability to justify findings in a professional and defensible manner is a critical aspect of forensic locksmithing.

Finally, the issue has a direct impact on professional credibility. Misidentifying the method of entry—such as incorrectly concluding forced entry where none exists, or overlooking signs of covert access—can lead to serious consequences. These may include liability issues, disputes with insurers or clients, and damage to professional reputation. As a result, locksmiths must apply a systematic, analytical approach to every investigation, ensuring that conclusions are accurate, well-supported, and consistent with observed evidence.

## Tool Mark Analysis and Evidence Preservation

### Tool Mark Analysis in Forensic Locksmithing

Tool mark analysis is a core component of forensic locksmithing and involves examining the physical impressions, scratches, cuts, and deformations left on locks, doors, and associated hardware during an attempted or successful entry. Every tool, whether a screwdriver, drill bit, bolt cutter, or pry bar, has unique characteristics such as wear patterns, manufacturing imperfections, and edge geometry. When these tools come into contact with a lock or door, they leave behind identifiable marks that can be analysed to determine how the entry occurred.

At a basic level, tool marks are categorised into three types: impression marks (caused by direct force, such as a crowbar pressing into a frame), abrasion marks (caused by sliding contact, such as a screwdriver slipping in a keyway), and cutting marks (produced by tools like drills or saws removing material). By studying these marks, a forensic locksmith can often determine the type of tool used, the direction of force applied, and whether the attack was controlled or opportunistic.

<u>Impression Marks (Direct Force)</u>

Impression marks are created when a tool applies direct pressure without significant movement, leaving a clear indentation in the material. These marks typically appear as compressed, deformed, or crushed areas where the surface has been forced inward. On softer materials such as timber door frames, the fibres may appear flattened, splintered, or compacted, often retaining the general shape of the tool edge. On metals, the mark may appear as a dented or stamped outline, sometimes with smooth, polished areas where the tool made firm contact.

The defining characteristic of an impression mark is that it often reflects the shape or profile of the tool. For example, a crowbar may leave a broad, curved indentation, while a screwdriver might leave a narrower, more defined imprint. There is usually little to no directional scratching, as the tool did not slide significantly during contact.

Abrasion Marks (Sliding Contact)

Abrasion marks are formed when a tool moves across a surface, creating friction and removing small amounts of material. These marks appear as scratches, scoring lines, or fine striations, often running in a consistent direction that reflects the motion of the tool. In a lock keyway, for example, they may appear as thin, linear scratches inside the plug, sometimes only visible under magnification.

Unlike impression marks, abrasion marks are characterised by their directionality and texture. They may range from light surface scuffing to deeper grooves, depending on the force applied and the hardness of the material. The surface often looks roughened or polished in streaks, and multiple overlapping lines may indicate repeated attempts or tool repositioning.

These marks are particularly important in identifying covert entry methods, such as picking or probing, where tools are inserted and manipulated repeatedly.

Cutting Marks (Material Removal)

Cutting marks are produced when a tool actively removes material, such as with a drill, saw, or cutting wheel. These marks are typically the most obvious and destructive, appearing as holes, grooves, or severed surfaces. A drilled lock, for example, will show a circular hole with spiral or concentric patterns, often accompanied by metal shavings or burrs around the entry point.

The edges of cutting marks are usually sharp, rough, or jagged, depending on the tool used. Drill bits leave distinct rotational patterns, while saws create linear cuts with visible tooth marks. In many cases, the surrounding area may also show signs of heat or friction, such as slight discoloration or melted residues.

The key feature of cutting marks is that they clearly indicate material has been removed, rather than just displaced or scratched. This makes them strong evidence of deliberate, forceful entry using tools designed to penetrate or destroy components.

In more advanced investigations, tool marks may be compared against suspected tools using magnification or microscopy. Fine striations and microscopic patterns can link a specific tool to a specific mark, much like a fingerprint. While this level of analysis is often conducted in forensic laboratories, locksmiths play a crucial role in recognising and preserving these marks at the scene.

**Common Attacks Against Padlocks**

The most common method of attacking a padlock is by cutting the shackle, either partially or completely through. This is typically achieved using bolt cutters and is highly effective against most low- to medium-security padlocks that do not incorporate shrouded shackles. The attack is a form of shearing, where the opposing blades of the cutter apply force until the metal fails. This process leaves clear physical evidence, including compressed and displaced metal at the cut site, often reflecting the shape and pressure points of the cutter jaws [160].

Closer examination of these tool marks may reveal trace materials transferred from the cutting tool. Substances such as paint, grease, dirt, or corrosion products can adhere to the damaged area of the shackle. These residues can be forensically significant, as they may be compared with tools recovered from a suspect to establish a link between the tool and the damage observed [160].

Drilling is another common form of destructive entry and is widely used against a variety of lock types, including by locksmiths when non-destructive methods are unsuccessful. In this process, the cylinder, such as a key-in-knob (KIK) cylinder, is drilled at the shear line to destroy the internal pin stacks, allowing the plug to rotate freely. This method also produces a significant amount of metal debris, including shavings and fragments that may contain evidentiary value, such as remnants of a damaged drill bit. For this reason, openings in drilled locks are often sealed or taped during evidence collection to preserve these materials [160].

The internal surfaces of a drilled lock frequently display distinctive spiral striations left by the drill bit. These markings are analogous to ballistic striations in firearms analysis, as they can reflect the unique characteristics of a specific tool. Such marks can be compared to a suspect drill bit to determine whether it was used in the attack [160].

Impact-based attacks represent another destructive method, particularly effective against padlocks, doors, and windows. Repeated hammering can cause the shackle to fracture, and the direction of the break can provide insight into the angle and force of the attack. The body of the lock often exhibits multiple impact marks, such as crescent-shaped indentations, which can be measured to estimate the type and size of the striking tool. Multiple points of impact may also indicate repeated blows during the attack [160].

Severe damage to the lock face or keyway can indicate the use of prying or levering tools such as screwdrivers or chisels. In such cases, the keyway may be widened or gouged to allow tool insertion. Internal examination may reveal that upper pin chambers have been sheared off entirely, rather than individual pins being manipulated. This type of damage is typically associated with the application of significant torsional force, often through levering tools inserted into the keyway [160].

The internal pins themselves may show heavy deformation, with clear tool marks indicating that a tool was forcibly driven into the plug, sometimes with the assistance of hammering. These marks are generally inconsistent with covert entry techniques such as picking or bumping and instead reflect destructive methods. The distinct nature of these marks may allow for later comparison with tools recovered during an investigation [160].

Euro-profile cylinders present a well-known structural weakness at the central fixing point, where the retaining screw passes through the cylinder. This area is typically the thinnest section of the lock, making it susceptible to snapping attacks. If part of the cylinder protrudes beyond the door face, even slightly, it can be gripped with tools such as pliers and broken through applied force. Tool marks on the exposed face of the cylinder often indicate where gripping tools were applied, and examination of the cam can help determine the direction in which the cylinder was snapped. In some cases, the direction of damage may suggest that the attack could not have occurred while the door was closed, raising the possibility of staged damage [160].

Padlock shackles may also fail at multiple points under extreme force. In some cases, a broken section may remain lodged beneath the locking mechanism. Examination of both the shackle and the lock body can reveal signs of torsion, such as twisting or distortion, indicating that rotational force was applied rather than direct cutting. Tool marks along the edges of the lock body may appear fresh and can provide evidence of the tools used. For example, parallel striations or tooth marks may indicate the use of a wrench or tongue-and-groove pliers, which can later be compared to suspect tools [160].

In some instances, individuals attempt to simulate forced entry, particularly in cases of insurance fraud. These attempts often involve crudely damaging the keyway with tools such as screwdrivers to imitate lockpicking. However, such damage typically results in excessive material removal and large, irregular tool marks concentrated at the front of the cylinder. Unlike genuine covert entry, there is little to no evidence of manipulation deeper within the lock, and the internal pins often show damage inconsistent with picking or bumping techniques [160].

Chemical attacks represent a less common but highly effective method of compromising lock components. Certain chemicals can alter the molecular structure of metals, weakening them and making them more susceptible to further attack. For example, exposure of brass components to nitric acid produces nitrogen dioxide fumes and results in a reddish-brown discolouration as the copper content is affected. Further examination may reveal blue crystalline residues, such as copper nitrate and zinc nitrate, which are byproducts of the chemical reaction. These residues provide strong forensic indicators of chemical use in the attack [160].

### Investigation Tools

Identifying different types of forensic evidence in locks does not require an extensive range of tools; however, it does require careful selection of appropriate equipment and a controlled workspace. The priority in forensic locksmithing is always to preserve existing evidence while avoiding the introduction of false or misleading marks. Proper handling and disassembly of the lock are just as important as the tools themselves, as poor technique can compromise the integrity of the investigation [160].

One of the most important tools used in the disassembly of pin-tumbler cylinders is the plug follower. This device allows the plug to be pushed out of the cylinder while retaining the top pins and springs within their chambers. For forensic purposes, plug followers must be made from non-metallic materials such as wood or plastic. This prevents the creation of artificial tool marks that could interfere with later analysis [160].

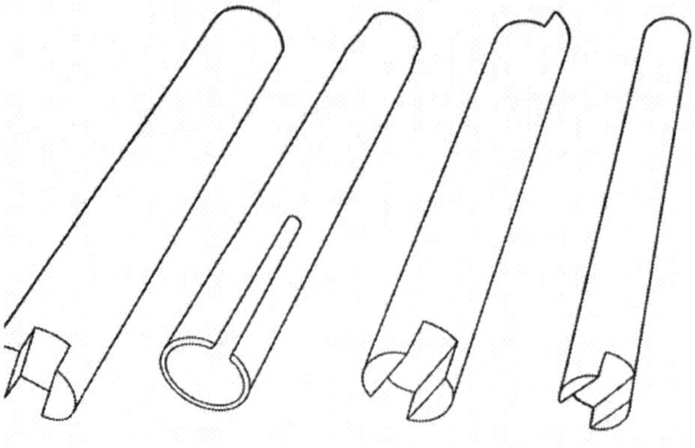

*Figure 56: Plug Followers.*

Shims are another commonly used tool, particularly for opening pin-tumbler locks in a controlled manner. These thin strips of metal are inserted into the rear of the cylinder and used in conjunction with a blank key to separate pin stacks at the shear line. Although shimming is slightly invasive, it is often preferred over destructive methods because it minimises damage to internal components and preserves potential evidence [160].

Handling small lock components requires precision, making tweezers an essential tool. In forensic work, plastic tweezers are strongly recommended over metal ones, as even polished metal can leave marks on softer materials such as brass or nickel-silver pins. Using a variety of tweezer shapes can improve handling efficiency and reduce the risk of damaging evidence [160].

Microscopy plays a critical role in tool mark identification and detailed examination of lock components. For this purpose, a top-lit microscope, such as a metallurgical or jeweller's microscope, is required, rather than a bottom-lit biological microscope. Magnification of at least 20x is necessary, although higher magnification (100x or more) is preferred for detailed analysis. Digital microscopes offer convenience and direct computer integration, while manual microscopes often provide superior image quality. In more advanced investigations, comparison microscopes allow side-by-side analysis of tool marks, and scanning electron microscopes (SEM) can produce extremely high-resolution images of surfaces and residues, although their cost limits accessibility [160].

Photography is another essential component of forensic documentation. Medium to high-quality cameras are suitable, provided they are used correctly. Accessories such as tripods, ring strobes, and filter kits can significantly improve image clarity and consistency. In some jurisdictions, non-digital cameras may still be preferred due to concerns about image manipulation, although this view is increasingly outdated. When photographing through microscopes, appropriate mounting systems are required to ensure stability and alignment [160].

Lighting is critical in revealing fine details such as scratches, residues, and tool marks. Investigators should experiment with different light intensities, angles, and colour temperatures to optimise visibility. Ultraviolet A (UVA) light sources are particularly useful for detecting residues such as inks, fibres, or certain chemical traces that are not visible under normal lighting conditions [160].

Hand tools are generally limited to those required for controlled disassembly or sectioning of components. Fine-tooth hacksaws or jeweller's saws are commonly used for cutting lock plugs during detailed examination. Power tools, such as drills or rotary tools, are used only when destructive disassembly is unavoidable. In such cases, the investigator must understand the lock's internal design and security features to avoid unnecessary damage or safety risks [160].

Additional supporting tools include pinning trays for organising components, measuring instruments such as micrometres and callipers (preferably non-metallic where possible), and holding devices such as vices with protective inserts. Documentation materials, such as notebooks, markers, and cameras, are equally important for recording observations and maintaining a clear chain of evidence. Items such as modelling clay can also assist in stabilising components during microscopic examination [160].

Equally important is the forensic workspace itself. Investigations should be conducted in a clean, controlled environment that is separate from general workshop areas. Dust, debris, and stray components from everyday work can contaminate evidence and compromise results. Ideally, a dedicated clean workspace or "clean room" should be used for forensic examinations [160].

Finally, strict procedural discipline is essential. Investigators should wash their hands and clean all tools before commencing work, avoid using any equipment that could contaminate evidence, and document every step of the process. Components must always be examined before any cleaning is undertaken, and if washing is required, any residue should be retained for further analysis. Through careful tool selection, controlled environments, and disciplined methodology, forensic locksmiths can ensure accurate and defensible identification of evidence [160].

### Lockpicking Evidence

Lockpicking is typically carried out using two primary tools: a tension tool and a pick. The tension tool is used to apply a controlled rotational force to the plug, while the pick is used to manipulate the internal locking components into the correct positions. As tension is applied, individual components, such as pins, begin to bind due to slight manufacturing tolerances. The picker uses the pick to identify which component is binding and then carefully lifts or positions it. Feedback is critical in this process and is obtained through subtle tactile sensations, sounds, or visual cues. Once a component reaches its correct position, the tension tool holds it in place while the process is repeated for the remaining components. When all elements align correctly, the lock can be operated [160].

Because of the forces involved, lockpicking tools are typically made from strong materials such as steel, iron, or aluminium. These tools are thin, often around 0.025 inches for pin-tumbler picks, but must withstand moderate force during manipulation. When these harder tools come into contact with softer lock components, such as brass or nickel-silver pins and plugs, they leave behind characteristic marks. These marks often appear as scratches, gouges, or abrasions and form the basis of forensic evidence. While the most significant evidence is usually found on internal components such as pins and chambers, additional marks may also be present on the plug, housing, bolt, or cam depending on the lock design [160].

The act of picking is inherently invasive, and forensic examination typically reveals evidence of tool interaction. For example, single-pin picking often leaves consistent, localised scratches where individual pins have been lifted. These marks tend to be relatively uniform in direction and depth. In contrast, raking, a technique involving rapid, repeated movement of the pick across multiple pins, produces a wider range of elongated scratches at varying angles and depths. In many real-world cases, a combination of both techniques is evident, with raking used initially to set some pins, followed by single-pin manipulation to complete the process [160].

The characteristics of tool marks can also provide insight into the skill level of the person manipulating the lock. Amateur attempts often result in deep, irregular gouges caused by excessive force applied through both the tension tool and the pick. High tension increases binding forces, requiring greater effort to move components and leading to more pronounced damage. In contrast, skilled operators typically use minimal force, resulting in lighter, more subtle markings. However, even in these cases, careful forensic examination can still reveal evidence of manipulation, particularly under magnification and controlled lighting [160].

More complex lock designs, such as telescopic pin systems or specialised security pins, also exhibit distinctive signs of tool interaction. For instance, telescopic pins, such as those found in some high-security cylinders, may show marks on both inner and outer pin elements, as well as within internal chambers. Similarly, uniquely shaped pins, such as spool or "nippled" designs, may display wear patterns along specific contours where tools have engaged them. In high-security systems that use specialised tools, such as those designed to interact with sidebars or rotating pins, marks may also be found in secondary channels or non-traditional contact surfaces [160].

Evidence of lockpicking is not limited to pins alone. The movement of tools through the keyway often leaves marks along the internal walls of the plug. These scratches are typically irregular in direction and differ from the smooth, consistent wear caused by normal key use. Certain areas of the lock, such as the top of the keyway, are particularly valuable for forensic examination because they are not normally contacted by a key. Marks in these "clean" areas strongly indicate tool insertion. Additional damage may be observed on warding surfaces, inside pin chambers, or along the plug walls where picks have been inserted, repositioned, or levered [160].

Pin chambers themselves may also show signs of manipulation. Marks in these areas may appear as localized deformations or patterned scratches caused by repeated lifting and releasing of pin stacks. In some cases, the shape and direction of these marks can indicate the technique used, such as the characteristic up-and-down motion associated with searching for the shear line. Similarly, the sides of pins may display scratches caused by incidental contact during manipulation, particularly in locks with challenging pin configurations where adjacent pins must be lifted to different heights [160].

The tension tool also leaves identifiable forensic evidence. As torque is applied, the tool can create shear marks or gouges at the point of contact with the plug. These marks are often found at the bottom or top of the keyway, depending on tool placement. Inexperienced users may leave deeper, more obvious marks, while skilled users may leave only faint traces that require careful lighting to detect. In some cases, additional scratches may be present from repositioning the tension tool or from incidental contact with the pick during operation [160].

Finally, evidence may also extend to components such as the cam or rear of the plug. During manipulation, particularly with less controlled techniques, tools may contact areas that are not normally engaged by a key, leaving scratches or impact marks. Amateur attempts may result in extensive damage in these areas, including deep scoring or repeated impacts, while more controlled attempts may leave only minor but still detectable marks [160].

Overall, forensic examination of lockpicking relies on identifying and interpreting these subtle but characteristic signs of tool interaction. By analysing the location, depth, and pattern of marks across multiple components, a forensic locksmith can determine not only whether a lock has been manipulated, but also gain insight into the methods used and, in some cases, the level of skill of the individual involved [160].

### Pick Gun Evidence

Pick guns are covert entry tools designed to manipulate pin-tumbler locks by rapidly separating pin pairs at the shear line, allowing the plug to rotate. They are available in both manual and electric (vibrational) forms, and although they operate differently in terms of mechanics, both rely on the transfer of kinetic energy to momentarily disengage the locking components. In this respect, pick guns function on principles similar to bump keys, using impact or vibration rather than precise individual pin manipulation [160].

Manual pick guns are spring-loaded devices that resemble a small handheld tool with a trigger mechanism and an interchangeable picking blade, commonly referred to as a "needle." To operate the tool, the needle is inserted into the keyway beneath the pin stacks. A separate tension tool is used to apply light rotational force to the plug, similar to traditional lockpicking. When the trigger is pulled, the needle strikes upward against the bottom pins. This impact transfers kinetic energy through the pin stack, causing the top pins to momentarily jump above the shear line. If this occurs simultaneously across all pin stacks, the plug can rotate, allowing the lock to open [160].

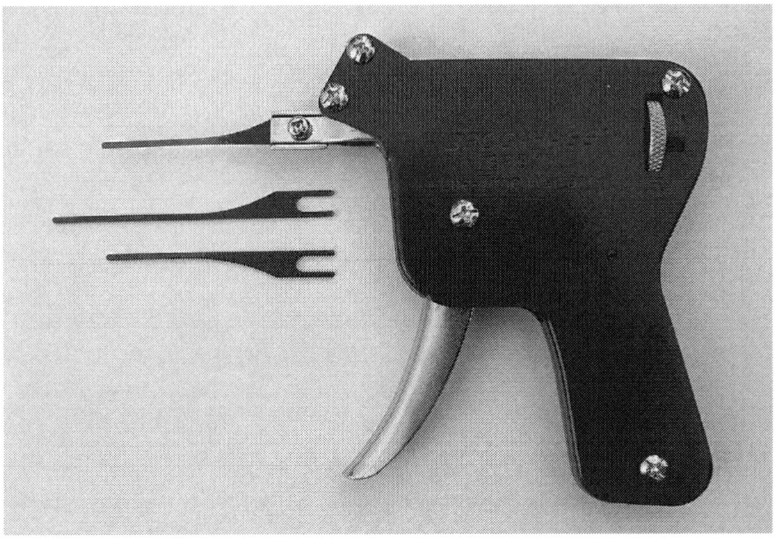

*Figure 57: Lock pick gun. Photograph: wdwd, CC BY-SA 4.0, via Wikimedia Commons.*

Electric or vibrational pick guns operate on the same fundamental principle but use rapid oscillation rather than discrete impacts. The needle vibrates at high frequency, repeatedly striking or agitating the bottom pins.

By adjusting the position and pressure of the tool, the operator seeks to achieve a resonant effect where the pins are repeatedly driven upward, increasing the likelihood that they will align at the shear line. This continuous motion distinguishes vibrational tools from manual pick guns and results in different patterns of wear and forensic evidence [160].

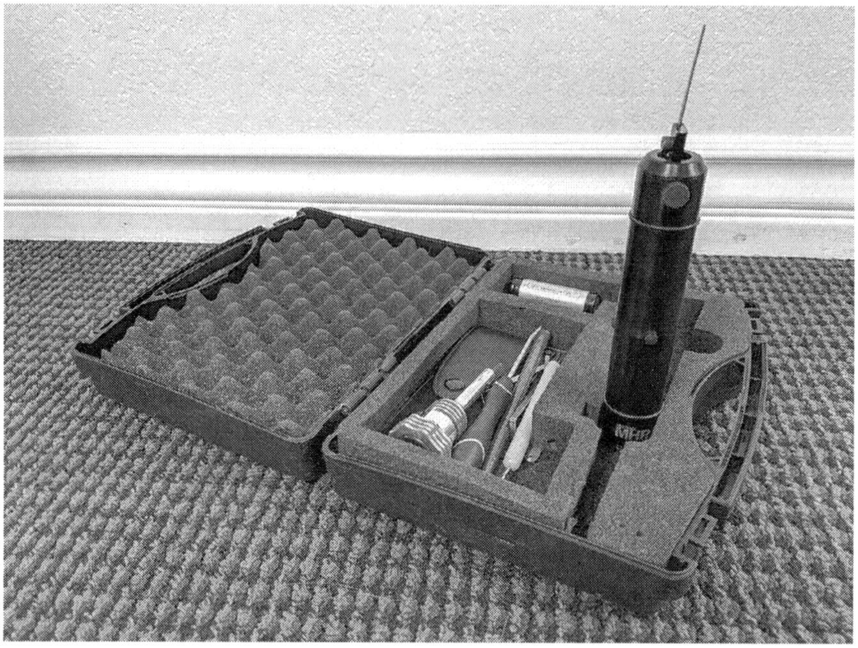

*Figure 58: Electric Pick Gun Toolkit. ASAP Locksmith, CC BY-SA 4.0, via Wikimedia Commons.*

From a forensic perspective, pick guns produce distinctive marks that differ from those created by traditional lockpicking. The primary evidence is typically found on the bottom surfaces of the pins, where the needle repeatedly strikes. Instead of the linear scratches associated with picking, pick guns create small impact dents or impressions. With repeated use, these marks can form patterns resembling radial spokes around the circumference of the pin, as slight rotation of the pins between impacts distributes the marks [160].

Additional evidence may be found within the plug itself. If the needle is not correctly positioned, it may strike or scrape against the plug walls, leaving gouges or scratches. The rear cam of the lock is also a common location for tool marks, particularly if the needle is inserted too deeply. These marks are often well-defined and may be numerous, allowing investigators to estimate how many times the pick gun was activated. As with other covert entry techniques, the tension tool may also leave characteristic marks at the point where torque was applied [160].

Electric and vibrational pick guns tend to produce more extensive internal damage due to their continuous motion. The constant oscillation of the needle can result in significant material removal from both the pins and the plug walls. This may produce a roughened surface texture on the pins, rather than discrete impact marks, and can generate visible metal particles within or exiting the keyway. The plug walls may exhibit numerous vertical or irregular scratches, sometimes with a staggered or "stuttering" appearance depending on the motion of the tool .

### Key Bumping Evidence

Key bumping is a covert entry technique used against pin-tumbler locks, where a specially prepared key is used to transfer kinetic energy through the pin stacks, momentarily lifting the top pins above the shear line so the plug can rotate. There are two primary methods, pull-out and minimal movement, but both rely on the same physical principle and tend to produce similar types of forensic evidence within the lock and on the key itself. In operation, key bumping is conceptually similar to the use of pick guns, as both techniques depend on rapid energy transfer rather than precise manipulation [160].

To perform key bumping, a key blank that matches the lock's keyway is first obtained. This key is then modified so that all cuts are made to their maximum depth (often referred to as a "999 key," where 9 represents the deepest cut). This ensures that the key does not fully lift the bottom pins during normal insertion. The key can be prepared manually using measuring tools such as a micrometre or key gauge, or cut using a key machine set to the lowest depth. In the pull-out method, the key is inserted fully and then withdrawn slightly, typically one pin position, before being struck. In the minimal movement method, additional material is removed from the key's shoulder and tip so it can be inserted fully while still allowing movement during impact [160].

Once the key is positioned, light rotational tension is applied, and the bow of the key is struck using a tool such as a bump hammer. This impact drives the key forward, transferring kinetic energy into the bottom pins, which in turn transmit the force to the top pins. If the timing and force are sufficient, the top pins will momentarily jump above the shear line, allowing the plug to rotate under applied tension. Because this process relies on repeated impacts rather than precision, multiple attempts are often required [160].

From a forensic perspective, key bumping produces distinct and recognisable forms of damage. The most significant evidence is typically found on the bottom pins, which absorb the direct impact from the key. These pins often exhibit pronounced dents, flattening, and impact marks rather than the fine scratches associated with traditional lockpicking. In some cases, particularly where harder key materials are used, additional linear scratches may appear as the key acts momentarily like a file during impact [160].

Repeated bumping can result in multiple impact marks distributed across the surface of the pins. Under appropriate lighting, these marks may be clearly visible and, in some cases, can even be counted to estimate the number of attempts made. Secondary effects may also be observed on the top pins, particularly when bumping attempts are unsuccessful. In these situations, misalignment can cause the top pins to bind and experience light shearing or deformation at their contact surfaces [160].

The pin chambers themselves may also show signs of damage. When energy transfer is incomplete or uneven, the pin stacks can be forced against the chamber walls, leading to distortion or elongation over time. Additionally, visible damage is often present on the face of the plug, particularly above and below the keyway, where the shoulder of the key repeatedly strikes during the bumping process. Modified keys used in the minimal movement method may reduce some of this external damage but can introduce wear or deformation within the keyway due to altered geometry [160].

In some cases, attackers may attempt to use bumping techniques on locks that are not susceptible to this method, such as high-security or non-pin-based systems. These attempts often result in significant internal damage, including broken components, deformed elements, or complete lock failure. Such damage is typically inconsistent with normal operation or covert entry and can provide clear evidence of an attempted but unsuccessful attack [160].

Overall, the forensic signature of key bumping is characterised by impact-related damage, particularly on the bottom pins and surrounding components. These patterns differ markedly from those produced by picking or other covert techniques, allowing trained locksmiths and investigators to identify bumping as the likely method of entry and assess the nature and extent of the attack [160].

**Impressioning Evidence**

Impressioning is a covert entry technique used to create a working key for a specific lock without having an original key. It generally exists in two forms: copying, which involves taking a mould or duplicate of an existing key, and manipulation, which involves using a blank key to interact directly with the lock's internal components. In this context, the focus is on manipulation-based impressioning, which is widely recognised in forensic locksmithing [160].

The process begins with a key blank that matches the lock's keyway profile. The locksmith or attacker inserts this blank into the lock and applies significant rotational force (torque). This force causes the internal components, typically the bottom pins in a pin-tumbler lock, to bind against the plug. The blank key is then moved slightly within the lock, allowing the bound components to press into the softer brass surface of the key. These contact points create small marks on the key blank, indicating where material must be removed [160].

The underlying principle is that incorrectly positioned components will bind and leave marks, while correctly positioned components will no longer bind and therefore stop marking the key. After each marking cycle, the key is removed and carefully filed at the marked locations. This process is repeated multiple times, gradually refining the key until all components align at the shear line and the lock can be operated. While this explanation is based on pin-tumbler systems, the same general concept can be adapted to other lock types, with variations in technique depending on the mechanism [160].

Because manipulation-based impressioning places considerable stress on both the key and the lock, it often leaves identifiable forensic evidence. The forced binding of components, particularly bottom pins being pushed to or above the shear line, can result in distinct marks on the pins themselves. Additionally, if the person performing the impressioning does not clean the key properly between filing stages, traces of metal filings may be transferred into the lock [160].

Some variations of impressioning use pressure-sensitive materials such as lead, tape, plastic, or even ultraviolet (UV) ink to enhance the visibility of marks. In these cases, traces of these materials may be transferred to internal components such as pins, plug walls, or the face of the lock. For example, UV ink techniques can leave visible residue patterns under appropriate lighting, providing strong evidence of manipulation-based impressioning [160].

From a forensic perspective, impressioning typically produces linear or shearing marks on the pins where they have been forced against the plug under torque. In some instances, these marks are sufficiently distinct to allow investigators to estimate how many impressioning cycles were performed. The spacing and pattern of marks may also indicate the skill level of the individual, particularly if consistent depth increments suggest a systematic approach [160].

## Decoding Evidence

Decoding is a broad term used to describe a range of covert and surreptitious entry methods aimed at determining the correct positions of components within a lock. Rather than directly forcing or manipulating the lock to open, decoding focuses on extracting information, either from the key, the lock's internal structure, or user interaction, to understand how the locking system operates. It is one of the most diverse and ambiguous categories of compromise techniques, encompassing approaches such as key analysis, manipulative decoding, disassembly, visual or optical methods, as well as more advanced techniques like thermal and radiological imaging. Among these, key analysis, manipulative, and visual decoding are the most commonly encountered in practice, with visual decoding being particularly prevalent due to its simplicity and low risk of detection [160].

Unlike impressioning, which aims to create a working key, or lockpicking, which aims to open the lock directly, decoding is primarily about gathering information. This information can later be used to produce a functional key or to facilitate access at a later time. One of the key advantages of decoding is that many of its methods are surreptitious, meaning they leave little to no forensic evidence, making detection extremely difficult .

One of the simplest forms of decoding involves direct key analysis. In this process, the bitting pattern of a key is examined either by reading stamped code numbers or by physically measuring each cut using tools such as calipers or micrometers. These measurements can then be translated into a manufacturer's key code, allowing an identical key to be produced. Modern locksmithing tools can automate this process by analysing both the keyway profile and the cut depths. However, this method becomes more complex when dealing with high-security keys that incorporate additional features such as sidebars, angled cuts, moving elements, or magnetic and electronic components [160].

Another common approach is manipulative or invasive decoding, which involves using specialised tools to interact with the internal components of the lock. These tools are often designed for specific lock types and may combine traditional picking techniques with measurement capabilities. The goal is to assess characteristics such as the position, movement, spacing, or resistance of internal components. In some cases, partial picking of the lock is required before accurate decoding can occur. Tools such as tubular lock picks or dedicated decoding devices are commonly used in this category, and some are capable of both opening and decoding the lock simultaneously [160].

Disassembly is a more direct but intrusive form of decoding, where the lock is physically removed and taken apart to allow full inspection of its internal components. This method provides the most accurate information but typically requires prior access to the secured area. It is often only feasible in situations where security is already compromised or where access controls are weak. Proper reassembly is critical, as incorrect installation can render the lock inoperable or alter its performance [160].

Visual and optical decoding methods rely on observation rather than physical interaction. For example, a clear photograph of a key can often provide enough detail to determine its bitting pattern. Similarly, tools such as borescopes or otoscopes can be used to inspect the internal components of a lock through the keyway. In some cases, surveillance techniques are used to observe users entering combinations on keypad locks. These methods are particularly effective because they do not require direct manipulation of the lock and typically leave no trace [160].

More advanced forms of decoding include radiological and thermal techniques. Radiological imaging uses penetrating radiation, such as X-rays, to reveal the internal structure of a lock or safe, including the position of components like combination wheels. Thermal imaging, on the other hand, detects heat signatures left on recently used keypads, potentially revealing which buttons were pressed. While highly effective, these methods require specialised equipment and are generally limited to higher-skill or well-resourced attackers [160].

From a forensic perspective, the evidence associated with decoding varies significantly depending on the method used. Manipulative decoding may leave marks similar to those produced by lockpicking, while key analysis and disassembly may leave traces depending on the tools and handling involved. However, visual, optical, thermal, and radiological methods are typically non-invasive and leave no identifiable marks on the lock itself, making them particularly challenging to detect [160].

Certain indicators may suggest that decoding has occurred. For example, coloured pins, often used in rekeying kits, can make optical decoding easier by visually indicating component sizes. Low-security wafer locks may also be decoded visually by observing wafer positions. In keypad systems, wear patterns on frequently used buttons can reduce the number of possible combinations, providing valuable clues to an observer. These subtle signs highlight the importance of considering not only mechanical security but also information exposure when assessing lock systems [160].

### Bypass Evidence

Bypass is a form of covert entry that seeks to circumvent the security of a lock by directly manipulating internal operating components such as the cam, bolt, or actuator. Unlike lockpicking, which involves aligning internal components (such as pins or wafers) to meet the lock's intended operating conditions, bypass techniques aim to operate the locking mechanism without properly engaging those security features. In many cases, this allows the lock to be opened without damaging or altering its core components, making bypass a predominantly non-destructive method, although some variations may involve minor damage [160].

The underlying principle of bypass is to exploit design vulnerabilities or access points that allow direct interaction with the locking mechanism. One common approach involves targeting the cam or actuator. In certain poorly designed locks, these components can be manipulated independently of the locking system, allowing the bolt to retract without the need for a correct key or alignment of internal elements. While such vulnerabilities are relatively uncommon in modern high-security locks, they can be highly effective when present. Because tools used in these attacks must apply torque and physically interact with internal parts, they often leave identifiable tool marks on the cam or surrounding components [160].

Spring-loaded bolts and latches are particularly susceptible to bypass techniques such as shimming. This method involves inserting a thin object, such as a shim or wedge, between the bolt and its housing to disengage the locking mechanism. A well-known example is the "credit card" technique used to open simple latch-based doors. Similar principles apply to low-security padlocks, where shims can be used to release the shackle. When applied to doors, this technique is often referred to as loiding. These methods are effective because they exploit the mechanical behaviour of spring-loaded components rather than defeating the lock's internal security features [160].

Another form of bypass targets internal thumb-turns or lever handles. In this scenario, a tool is inserted under or through gaps around the door to physically engage and operate the internal handle. By catching and turning the thumb-turn or lever, the attacker can unlock the door without interacting with the cylinder at all. The presence and extent of forensic evidence in these cases depend on factors such as the materials involved, the type of tool used, and the number of attempts required [160].

In automotive contexts, bypass techniques often involve the use of an air wedge. A wedge is first inserted to create a small gap between the door and frame, after which an inflatable air wedge is used to widen the opening. This allows a tool to be inserted into the vehicle to manipulate internal locking controls. While commonly used by locksmiths in legitimate lockout situations, the same technique can be exploited for unauthorised entry. Evidence may include deformation of door seals or minor marks around the frame [160].

From a forensic perspective, bypass attacks often leave distinct evidence, particularly on components such as cams, actuators, and internal linkages. For example, certain older padlock models, such as the American 700 series, were known to be vulnerable to cam manipulation. In such cases, tool marks on the cam and back plate can clearly indicate bypass as the method of entry. Even when manufacturers introduce design modifications or hardware patches to address these vulnerabilities, attempted bypasses may still leave visible marks on modified components [160].

Electromechanical locks can also be vulnerable to bypass techniques that target their actuators. In systems such as the CodeLock 4000 series, attacks may involve inserting a thin tool to manipulate the bolt actuator directly. This can result in material transfer, debris within the lock, and distinct tool marks on the actuator surface. These marks are typically irregular in angle and depth and are inconsistent with normal operation, making them valuable indicators in forensic analysis [160].

### Lock Disassembly

Evidence handling and examination procedures can vary between jurisdictions; however, the techniques selected in forensic locksmithing are consistently chosen for their minimal impact on evidence integrity and the investigative process. The primary objective is to preserve all physical and trace evidence while enabling a thorough and accurate analysis of the lock and its components [160].

Before any laboratory examination can take place, the lock must be carefully removed and disassembled. This process can range from straightforward to highly complex, depending on the lock type and its installation. In many cases, the forensic locksmith is not provided with a working key, meaning the lock must be opened using alternative methods that do not compromise evidence. Importantly, techniques such as picking or bumping are avoided, as they can introduce misleading marks or destroy existing evidence. During disassembly, all components must be maintained in their original position, sequence, and, where possible, alignment. Pinning trays are typically used to organise pin-tumbler components, while wafer and lever lock components are laid out in order to reflect their original configuration [160].

It is essential that the lock, door, and surrounding area are thoroughly examined prior to disassembly. Many indicators of entry method, tool use, and attack direction are more clearly visible when the lock remains installed. Observations made at this stage provide critical context that may be lost once the lock is removed and taken apart [160].

Non-destructive disassembly is always the preferred approach. This involves removing the lock from the door and accessing internal components without causing damage. In pin-tumbler cylinders, this may include removing retaining screws, housings, or caps. For lever, warded, or combination locks, disassembly often involves removing external casings. While many lock designs, particularly American models, facilitate this process, others, such as certain European profile cylinders, may lack accessible fasteners, making non-destructive disassembly more challenging [160].

Rapping is one technique that may be used to open a pin-tumbler cylinder without introducing forensic contamination. This method involves applying light rotational tension to the plug while gently striking the cylinder body with a soft mallet. The impact can cause the top pins to momentarily separate at the shear line, allowing the plug to rotate. Although similar in principle to bumping, rapping does not involve a key and therefore typically leaves no identifiable forensic marks. Its effectiveness is limited, but it is often attempted because of its low risk to evidence [160].

Shimming is another non-destructive or minimally invasive method. It involves inserting a thin piece of metal between the plug and cylinder to separate the pin pairs at the shear line. A blank key is often used in conjunction with the shim, which is inserted from the rear of the cylinder. As the key is gradually withdrawn, the shim captures the top pins above the shear line, eventually allowing the plug to rotate. While shimming may leave minor traces, these are generally minimal and do not significantly interfere with forensic analysis [160].

In some cases, destructive methods are required to access internal components, particularly when non-destructive options are not feasible. These methods must be applied with extreme care and precision, ensuring that the lock is only damaged to the extent necessary for examination. A thorough understanding of the lock's security features, such as anti-drill inserts or hardened components, is essential to avoid unnecessary damage and ensure investigator safety. Throughout the process, detailed documentation, including photographs and notes, must be maintained to preserve the evidentiary value and demonstrate proper procedure [160].

One specific destructive technique used in forensic analysis is the controlled cutting, or "vivisection," of lock components. This is particularly useful in pin-tumbler and wafer cylinders, where internal surfaces may contain valuable evidence. Rather than cutting the plug in half, it is typically sectioned into thirds. This approach preserves the upper portion of the keyway, which is normally untouched by keys or components and may contain undisturbed forensic evidence. Fine-toothed saws, such as jeweller's saws, are used to ensure precision and minimise disruption to evidence [160].

The use of a working key during disassembly is generally avoided. Since the key represents the last object to interact with the lock, inserting it may alter or destroy existing evidence or introduce new marks that could mislead the investigation. A key should only be used when no other method is available and when non-destructive or controlled destructive techniques are not permitted [160].

## Key Analysis

While the examination of locks is an important aspect of forensic investigation, it is often the keying system itself that has been compromised. Much like in the field of cryptography, security failures are rarely due to complex or sophisticated flaws; instead, they frequently result from the simple acquisition or misuse of legitimate keys. Keys can provide as much, if not more, information than the lock, particularly because they are handled directly and may carry trace evidence such as hair, fibres, or fingerprints. As a result, while lock analysis is valuable for determining the method of entry, key analysis is often even more valuable for identifying suspects and understanding how access was gained.

A fundamental aspect of key examination involves analysing the cuts, or bitting, along the blade of the key. These cuts interact with the internal components of a lock and determine whether it will operate. By examining the depth, spacing, and pattern of these cuts, an investigator can determine whether a specific key could have operated a particular lock. In some cases, the key may also be assessed for its ability to be partially inserted or manipulated (for example, jiggled) to achieve entry, which can indicate vulnerabilities in the system.

The keyway profile is another critical feature. This refers to the shape of the key and its compatibility with the internal warding of the lock. Many keys have a keyway code stamped on the bow, while others require comparison with reference databases or codebooks for identification. Modern key-cutting machines may even identify keyways using optical or laser systems. Establishing whether a key matches a lock's keyway is essential in determining whether it could physically engage the locking mechanism.

Codes stamped on keys can also provide valuable forensic information. These may include direct bitting codes, which indicate the exact cut depths, or indirect codes that can be traced by manufacturers to specific systems, locations, or authorised users. Additional markings, such as brand names or locksmith identifiers, may further assist in tracing the origin of the key or linking it to a particular supplier or individual.

The material composition of a key can also be informative. Different manufacturers use specific alloys, and in some cases, these materials can be traced back to their source. The plating of a key is another indicator—factory-original keys are typically plated after cutting, whereas duplicated keys often show exposed base metal where the plating has been removed during cutting. This distinction can help determine whether a key is original or a duplicate.

Identifying whether a key has been duplicated is a key function of forensic locksmithing. Factory-original keys typically exhibit clean, precise cuts with consistent angles and smooth valleys. In contrast, duplicated keys may show slight variations in cut geometry, including uneven slopes or irregular valleys. Stylus-based duplication machines may also leave distinctive marks on the original key, such as polished lines along the blade. Additionally, the texture of the cuts can reveal the type and speed of the machine used, which may help link a key to specific equipment.

Handmade keys present another important area of analysis. These keys are often identifiable by irregular spacing, inconsistent angles, and visible tool marks from files or rotary tools. For example, a key shaped with a file may display straight, linear scratch patterns, while one cut with a rotary tool may show uneven material removal and rounded edges. Such characteristics can indicate deliberate attempts to create a working key without proper equipment and may be linked to tools found in a suspect's possession.

Tool mark identification on keys can provide further insights into their use. Marks left by clamps, pliers, or other tools may indicate duplication, impressioning, or other forms of manipulation. In some cases, distinctive patterns—such as those left by vice grips—can be matched to specific tools. Keys used in covert entry techniques may also show deformation or unusual wear, such as compressed shoulders from repeated impacts in bumping.

Material transfer is another valuable source of forensic evidence. Keys may carry residues such as wax, clay, or ink, which can indicate impressioning techniques. These materials may be visible under normal lighting or require alternative light sources, such as ultraviolet illumination, to detect. In addition, microscopic traces such as fibres or particles can be analysed to establish links between the key, the environment, and potential suspects.

Overall, key analysis provides a comprehensive view of how access may have been achieved. By examining physical characteristics, tool marks, and trace evidence, forensic locksmiths can reconstruct the history and use of a key, making it a critical component of any investigation involving locks and access control systems.

### Evidence Preservation in Locksmithing Contexts

Evidence preservation is equally critical and refers to the careful handling and protection of physical evidence to ensure it remains uncontaminated and admissible for investigation or legal proceedings. In locksmithing, this often involves preserving the condition of locks, cylinders, doors, and surrounding areas after a suspected break-in.

The first priority is to avoid contamination or alteration of evidence. Locksmiths should minimise handling of the lock and surrounding surfaces, ideally using gloves to prevent transferring fingerprints or oils. Any unnecessary operation of the lock—such as inserting keys or attempting to turn the mechanism—should be avoided unless required for safety, as this can destroy or obscure critical evidence such as internal tool marks or debris.

Accurate documentation is essential. This includes taking high-quality photographs from multiple angles, using scale references (such as rulers), and recording observations about the condition of the lock, door, and environment. Notes should include details such as visible damage, unusual wear patterns, and the position of components at the time of inspection. This documentation forms part of the evidentiary record and may be used in reports or court proceedings.

Where components need to be removed, for example, a cylinder or lock body, they should be carefully packaged and labelled. Individual items should be stored in separate, sealed containers to prevent cross-contamination. Maintaining a clear chain of custody is important, meaning that every person who handles the evidence is recorded, along with the time and reason for transfer.

### Impact on Locksmithing Practices

Tool mark analysis and evidence preservation significantly influence locksmithing practice, particularly in forensic and investigative contexts. Locksmiths are often among the first professionals to examine a compromised lock, placing them in a position of responsibility. Their ability to recognise tool marks can help distinguish between forced and non-forced entry, identify likely methods of attack, and support broader investigations.

Equally, poor evidence handling can compromise an entire investigation. If tool marks are damaged, contaminated, or undocumented, it may no longer be possible to determine how entry occurred or to link a suspect tool to the scene. This highlights the importance of a disciplined, methodical approach.

In modern locksmithing, these skills extend beyond traditional trade work. They require an understanding of forensic principles, attention to detail, and adherence to legal and procedural standards. Together, tool mark analysis and evidence preservation form a critical bridge between locksmithing, security assessment, and forensic investigation, reinforcing the profession's role in both prevention and post-incident analysis.

## Legal Responsibilities and Expert Witness Roles

Legal responsibilities and expert witness roles represent critical dimensions of professional practice across multiple disciplines, extending beyond traditional technical expertise to encompass formal legal obligations, ethical conduct, and accountability within judicial and investigative processes. While the provided text focuses on locksmithing, the underlying principles of professional liability, expert testimony, and legal accountability are applicable across numerous professions including healthcare, law, forensic science, and other regulated fields. This synthesis examines the foundational legal and ethical frameworks that govern professional conduct when practitioners assume roles with legal implications.

### Legal Framework and Professional Obligations

Professional practitioners operating within legal contexts must navigate a complex framework of statutory obligations, regulatory requirements, and ethical standards. The legal responsibilities of professionals are multifaceted, encompassing compliance with applicable legislation, adherence to professional standards, and accountability for their actions and conclusions [161-166].

The concept of professional liability extends across multiple dimensions. Healthcare professionals, for instance, face civil, criminal, administrative, and disciplinary liability [163-166]. Similarly, forensic experts and other professionals engaged in legal proceedings are subject to comparable frameworks of accountability [162]. These liability structures serve to establish minimum standards of care and competence, protecting both the public and the integrity of legal processes [161, 162, 164, 165].

Professional licensing and regulatory regimes form the foundation of legal responsibility. In regulated professions, practitioners must comply with statutory requirements governing their qualifications, conduct, and scope of practice [161, 163, 165, 166]. These regulatory frameworks establish baseline competency standards and provide mechanisms for disciplinary action when professionals fail to meet established norms [162, 165, 166].

### Evidence Preservation and Documentation Standards

A fundamental legal responsibility for professionals engaged in investigative or forensic work is the preservation of evidence and maintenance of accurate documentation. The integrity of evidence is paramount in legal proceedings, and professionals must ensure that their actions do not compromise investigative outcomes [162-164].

Documentation serves multiple critical functions within professional practice. Accurate, objective, and detailed records establish a contemporaneous account of observations, methodologies, and conclusions [162-164]. Such documentation must be based on observable evidence rather than speculation, with clear delineation between facts, interpretations, and opinions [162, 163]. The quality and completeness of documentation directly impacts the credibility and defensibility of professional conclusions in legal contexts [162-164].

Professional documentation standards require that all findings be recorded with precision, including measurements, photographs, diagrams, and written descriptions Mulyk, 2021; Wibowo, 2026; . Limitations and uncertainties must be explicitly stated, ensuring transparency regarding the scope and reliability of conclusions [162, 163]. This approach supports both accountability and the ability to withstand scrutiny in legal or insurance proceedings [162-164].

### Confidentiality and Privacy Obligations

Confidentiality represents a cornerstone of professional responsibility, particularly for practitioners who access sensitive information during the course of their work [161, 166-168]. The handling of confidential information is governed by both statutory privacy laws and professional ethical codes [161, 167, 168].

The duty of confidentiality extends to information regarding security vulnerabilities, access control systems, client details, and other sensitive data [161, 167, 168]. Unauthorized disclosure of such information can result in serious legal consequences, including civil liability, criminal prosecution, and professional disciplinary action [166-168]. The breach of confidentiality may also compromise the security of individuals or organizations, creating secondary harms [161, 167, 168].

Professional ethics codes establish that confidentiality is fundamental to maintaining trust and enabling effective professional relationships [161, 167, 168]. However, confidentiality is not absolute; statutory obligations may require disclosure in specific circumstances, such as threats to public safety or mandatory reporting requirements [161, 168]. Professionals must understand the boundaries of confidentiality obligations and the circumstances under which disclosure is legally required or ethically justified [161, 167, 168].

## The Expert Witness Role and Responsibilities

When professionals assume the role of expert witness, their responsibilities expand significantly beyond technical expertise to encompass formal legal obligations and duties to the court [161, 162] . The expert witness role is distinct from advocacy; the primary duty of the expert witness is to the court or tribunal, not to the party who engaged them [161, 162].

### Independence and Impartiality

Expert witnesses must maintain independence and provide impartial, evidence-based opinions [161, 162]. This requirement reflects the fundamental principle that expert testimony should assist the court in understanding technical or specialized matters, not advance the interests of either party [161, 162]. The expert's duty to the court supersedes any obligation to the client who retained them [161, 162] .

Maintaining impartiality requires that experts avoid bias, whether conscious or unconscious, in their analysis and conclusions [161, 162]. Experts must be prepared to acknowledge evidence that contradicts their initial assumptions or conclusions, and they must present their findings objectively regardless of whether those findings favour the party who engaged them [161, 162].

### Expert Reports and Testimony

Expert witness work typically involves the preparation of formal reports that document the scope of investigation, methodologies employed, evidence examined, and conclusions reached [161, 162]. These reports must be clear, logical, and supported by evidence, with technical concepts explained in language accessible to non-specialists such as judges or jurors [161, 162].

The expert report serves as the foundation for expert testimony and must withstand scrutiny during cross-examination [161, 162]. Reports should distinguish clearly between facts observed, interpretations of those facts, and opinions based on professional expertise [161, 162]. Any limitations in the investigation, analysis, or conclusions should be explicitly stated [161, 162].

Oral testimony in court requires the expert to explain not only what was observed but what those observations mean within the relevant professional context [161, 162]. The expert must be able to justify conclusions by referencing physical evidence, established professional principles, and accepted methodologies [161, 162]. Speculation or unsupported opinions undermine credibility and are vulnerable to challenge during cross-examination [161, 162].

### Professional Competence and Knowledge

Maintaining current professional knowledge is essential for credible expert testimony [161, 162]. Experts must be familiar with modern technologies, contemporary forensic or investigative techniques, and established professional standards within their field [161, 162]. Continuing professional development and awareness of emerging practices contribute to the reliability and defensibility of expert conclusions [161, 162].

Expert competence encompasses not only technical knowledge but also understanding of the legal context in which expertise is applied [161, 162]. Professionals functioning as expert witnesses benefit from formal legal training or at minimum a working knowledge of relevant legal principles, court procedures, and the standards governing expert testimony [161, 162].

**Legal Risks and Professional Consequences**

Professional practice within legal contexts carries significant legal risks. Providing inaccurate, misleading, or biased evidence can result in multiple adverse consequences, including reputational damage, professional disciplinary action, and legal liability [161-166].

Types of Liability

Professionals may face civil liability when their actions or omissions cause harm to others [161-166]. Civil liability typically requires proof of duty, breach of that duty, causation, and damages [161-166]. In professional contexts, the standard of care is generally defined by reference to what a reasonably prudent professional in similar circumstances would do [161, 162, 164, 165].

Criminal liability may apply when professional conduct violates criminal statutes, such as falsification of documents, fraud, or negligence resulting in serious harm [161-166]. Criminal liability typically requires proof of intent or gross negligence, depending on the specific offense [161-166].

Administrative and disciplinary liability are imposed through professional regulatory bodies and may result in sanctions ranging from warnings to suspension or revocation of professional credentials [161-166]. Disciplinary proceedings focus on violations of professional standards and ethical codes rather than criminal intent [161-166].

Standard of Care and Professional Negligence

Professional negligence occurs when a practitioner fails to exercise the standard of care expected of a reasonably competent professional in similar circumstances [161-166]. The standard of care is typically established by reference to professional guidelines, accepted practices, and expert testimony regarding customary standards [161, 162, 164-166].

Importantly, the legal standard of care may diverge from the scientific or evidence-based standard [169]. Courts adjudicate liability based on what most practitioners are doing, not necessarily on what scientific evidence suggests they should be doing [169]. This can create a "standard-of-care sprawl" where practices undertaken primarily to avoid liability become incorporated into the accepted standard of care, even when evidence does not support their necessity [169].

## Ethical Dimensions of Professional Practice

Professional ethics and legal obligations are closely intertwined, though they are not identical [161, 164, 166, 170]. Ethical standards often precede and exceed legal requirements, establishing professional expectations that go beyond minimum legal compliance [161, 164, 166, 170]. However, ethical guidelines cannot substitute for legal accountability [161, 164, 166, 170].

Conflicts may arise between ethical principles and legal requirements, particularly regarding confidentiality, informed consent, and reporting obligations [161, 166, 168]. Professionals must navigate these conflicts carefully, understanding both their legal obligations and their ethical responsibilities [161, 166, 168]. In some cases, clinical or professional judgment may permit flexibility in applying rigid legal requirements when doing so serves the interests of clients or the public [168].

The relationship between ethics and law reflects broader societal values and evolves over time [161, 166, 170, 171]. Changes in legal frameworks often reflect evolving ethical perspectives regarding patient autonomy, professional accountability, and the balance between individual rights and public protection [161, 166, 170, 171].

## Professional Development and Continuing Competence

Maintaining professional competence is both a legal obligation and an ethical imperative [161-166, 172]. Professionals must ensure that their knowledge remains current and that they are familiar with developments in their field, including new technologies, methodologies, and standards [161-166, 172].

Continuing professional development serves multiple functions: it ensures that practitioners maintain competence, it demonstrates commitment to professional standards, and it provides evidence of diligence in the event of legal challenges [161-166, 172]. Professional organizations and regulatory bodies often establish requirements for continuing education as a condition of maintaining licensure or professional credentials [161-166, 172].

Legal responsibilities and expert witness roles represent essential dimensions of professional practice in contemporary society. Professionals engaged in forensic investigations, insurance assessments, security consultancy, or other legally significant work must understand and comply with applicable legal frameworks, maintain rigorous standards of documentation and evidence preservation, and uphold confidentiality obligations. When assuming the role of expert witness, professionals must recognize that their primary duty is to the court, not to the party who engaged them, and must provide impartial, evidence-based opinions grounded in current professional knowledge and accepted methodologies.

The integration of legal, ethical, and professional standards creates a framework within which practitioners can contribute effectively to judicial and investigative processes while maintaining integrity and accountability. Understanding these responsibilities ensures that professionals can navigate complex legal contexts, protect themselves from liability, and uphold the credibility of their profession. As professional practice continues to evolve, particularly with the introduction of new technologies and changing legal frameworks, the importance of maintaining current knowledge, adhering to established standards, and demonstrating professional competence becomes increasingly critical.

### Impact on Locksmithing Practices

Legal responsibilities and expert witness roles are highly relevant to anyone entering locksmithing because they fundamentally shape how the profession is practiced, evaluated, and trusted. Locksmithing is not just a technical trade involving locks and keys—it operates within legal, insurance, and investigative frameworks where the locksmith's actions and opinions can have significant legal consequences. This means that every task, from routine lock installation to forensic analysis after a break-in, must be carried out with an awareness of legal obligations, professional standards, and potential scrutiny.

One of the most important influences on locksmithing practice is accountability. Locksmiths are often involved in situations where their work may later be examined in legal or insurance contexts, such as after a burglary or disputed entry. This requires locksmiths to follow recognised standards of care, ensuring that their work is competent, documented, and defensible. Poor workmanship, incorrect conclusions, or failure to follow accepted procedures can lead to civil liability, professional complaints, or loss of credibility. As a result, locksmiths must approach their work systematically and professionally, rather than relying solely on practical skill or experience.

Evidence preservation and documentation are particularly critical in forensic or investigative locksmithing. When attending a scene, a locksmith may be responsible for identifying whether entry was forced or non-forced, and their findings can directly influence insurance outcomes or legal proceedings. This means observations must be carefully recorded using photographs, notes, and measurements, with a clear distinction between fact and interpretation. The influence on practice is significant: locksmiths must develop investigative discipline, not just technical ability, ensuring that their work can withstand scrutiny in court or by insurers.

Confidentiality also plays a major role in shaping locksmithing behaviour. Locksmiths routinely handle sensitive information, such as access systems, key control arrangements, and security vulnerabilities. Improper disclosure of this information can compromise client security and result in legal consequences. This reinforces the need for ethical conduct, discretion, and an understanding of privacy obligations. In practice, this means locksmiths must be cautious about what they share, how they store information, and who they provide access to.

The expert witness role represents an advanced but increasingly important aspect of locksmithing. In this capacity, a locksmith may be called upon to provide an independent, evidence-based opinion in legal proceedings. This requires a shift in mindset: the locksmith's duty is not to the client, but to the court. Opinions must be impartial, supported by evidence, and clearly explained in a way that non-specialists can understand. This influences locksmithing practice by encouraging objectivity, critical thinking, and the ability to justify conclusions using recognised principles and methodologies.

Professional competence and ongoing learning are also directly influenced by these legal responsibilities. Locksmiths must stay up to date with evolving technologies, including high-security mechanical systems, electronic access control, and hybrid locking solutions. In legal contexts, outdated knowledge or incorrect assumptions can undermine credibility and expose the locksmith to challenge. Therefore, continuous professional development becomes not just a career advantage, but a legal and ethical necessity.

Legal risk further reinforces the importance of accuracy and professionalism. Providing incorrect advice, misidentifying entry methods, or producing biased reports can result in serious consequences, including legal action or disciplinary measures. This encourages locksmiths to adopt a cautious, evidence-based approach, ensuring that conclusions are well supported and that limitations are clearly acknowledged. It also highlights the importance of understanding the difference between what is known, what is likely, and what cannot be determined.

## Case Studies of Staged Break-Ins

The following are hypothetical case studies that illustrate how staged break-ins are identified in forensic locksmithing and crime analysis. These examples highlight the mismatch between claimed events and physical evidence, which is the key indicator of staging.

### Case Study 1: Simulated Forced Entry for Insurance Claim

A homeowner reports a burglary, claiming that entry was forced through the front door using a crowbar. The door frame shows splintering near the strike plate, and the lock face has heavy gouging around the keyway.

On closer forensic examination, several inconsistencies are identified. The damage to the door frame is superficial and does not align with the force required to actually defeat the lock. The strike plate is bent outward, suggesting force applied from inside rather than outside. Additionally, the lock remains fully functional, and the internal components show no signs of forced manipulation or failure.

Inside the property, there is minimal disturbance. High-value items are reportedly missing, but there is no evidence of a typical search pattern or hurried activity.

The conclusion is that the damage was staged after entry was gained using a key. The tool marks on the lock face are consistent with a screwdriver being inserted and twisted, rather than a genuine forced entry attempt. This case demonstrates how superficial damage, incorrect force direction, and lack of internal disruption can indicate fraud.

### Case Study 2: False Claim of Lock Picking

A commercial tenant reports that their premises were accessed by a "professional lockpicker," claiming no signs of forced entry but suspecting covert entry.

The lock is removed and examined. The keyway shows heavy gouging and material removal at the entrance, with deep, irregular scratches extending into the first pin chamber. Internal inspection reveals significant deformation of the bottom pins, with large gouges inconsistent with normal picking techniques.

True lockpicking typically leaves fine, controlled scratches, often limited to specific areas such as pin contact points or plug walls. In this case, the damage is excessive and inconsistent with covert methods. There are no corresponding marks deeper in the lock, suggesting that the tool did not penetrate fully.

Further investigation reveals that the tenant had recently lost a key and had not implemented any key control measures.

The conclusion is that entry was likely achieved using a legitimate or duplicated key, and the visible damage was staged to simulate lockpicking. This case highlights the importance of distinguishing between genuine covert entry and crude attempts to mimic it.

**Case Study 3: Staged Padlock Destruction**

A storage facility reports that a unit was broken into, with the padlock found cut. The shackle shows clear signs of cutting, with deformation consistent with bolt cutters.

However, further analysis reveals that the cut surface is clean and uniform, with no secondary marks or slippage that would be expected in a real attack under time pressure. The surrounding gate and latch show no signs of force or disturbance.

Additionally, the lock appears to have been cut in a controlled position, not while under tension. In genuine forced entry, the lock is typically under load, resulting in irregular or angled cuts.

The tenant claims high-value items were stolen, but there is no evidence of forced access to the unit itself beyond the lock.

The conclusion is that the padlock was likely removed, cut elsewhere using proper tools, and then reattached or presented as evidence of forced entry. This case demonstrates how tool mark characteristics and context (e.g., lack of surrounding damage) are critical in identifying staged events.

**Case Study 4: Insider Access Disguised as Break-In**

An office reports a break-in overnight, with sensitive equipment missing. There are no visible signs of forced entry, and the locks appear intact and fully operational.

Electronic access logs are reviewed and show that the door was opened using a valid credential during the reported time of the incident. The access corresponds to an employee who claims not to have been present.

Further investigation reveals that the employee had shared their access card with a colleague, violating company policy. There are no signs of forced entry, lock manipulation, or damage.

The conclusion is that the incident involved authorised access rather than a break-in. The attempt to frame it as a forced entry was likely intended to avoid accountability. This case highlights the importance of integrating physical and electronic evidence in modern forensic locksmithing.

### Case Study 5: Attempted Staging of Cylinder Snapping

A property owner reports that their euro-profile cylinder was snapped during a burglary. The cylinder is indeed broken at the central fixing point, a common weak spot.

However, forensic analysis shows that the break direction is inconsistent with an external attack. The fracture pattern suggests force applied from an angle that would only be possible if the door were open. Additionally, there are no tool marks on the exposed portion of the cylinder that would indicate gripping with pliers or a wrench.

The cam inside the cylinder shows deformation inconsistent with normal snapping attacks, further supporting the conclusion that the break occurred under non-operational conditions.

The conclusion is that the cylinder was intentionally broken after removal or with the door open, then presented as evidence of forced entry. This case illustrates how mechanical analysis of fracture patterns can reveal staging.

### Key Takeaways for Locksmithing and Forensics

Across the examined case studies, several consistent indicators of staged break-ins become apparent. One of the most significant is a mismatch between the visible damage and the actual function of the lock. For example, a lock may display apparent signs of forced entry, yet remain fully operational, suggesting that the damage was applied after access had already been gained.

Another common indicator is the presence of tool marks that are inconsistent with recognised methods of entry. Genuine forced or covert entry techniques tend to leave characteristic and predictable marks, whereas staged damage is often crude, excessive, or poorly aligned with known attack patterns. This inconsistency can provide strong evidence that the damage was artificially created.

The location and direction of damage also play a critical role in analysis. In many staged scenarios, the force applied does not correspond with how an intruder would realistically approach or attack the lock. For instance, marks may indicate force applied from the wrong direction or at an angle that would not be possible during an actual break-in.

A further indicator is the absence of supporting environmental evidence. Genuine break-ins typically involve a range of associated signs, such as disturbance within the premises, secondary damage, or evidence of hurried activity. When such indicators are missing, despite claims of forced entry, the likelihood of staging increases.

Additionally, evidence of authorised access or failures in key control frequently emerges in these cases. The use of legitimate or duplicated keys often explains entry more convincingly than the alleged forced methods, particularly when no credible forensic evidence supports the latter.

For locksmiths, these findings reinforce the importance of objective, evidence-based analysis. The ability to identify staged break-ins is essential not only for accurate technical reporting, but also for supporting insurance investigations, legal proceedings, and maintaining professional integrity and credibility within the field.

### Impact on Locksmithing Practices

These case studies are highly relevant to anyone interested in locksmithing because they demonstrate that the profession extends well beyond installing and opening locks. Modern locksmithing increasingly intersects with forensic analysis, legal processes, and investigative work. A locksmith is often called upon not only to fix a problem, but to explain *how* and *why* a lock was compromised. Understanding staged break-ins equips a locksmith with the ability to distinguish between genuine attacks and fabricated scenarios, which is critical in real-world practice.

From a practical perspective, these examples reinforce the importance of evidence-based assessment. Locksmiths must learn to interpret tool marks, damage patterns, and lock functionality in a systematic and objective way. Rather than relying on assumptions or client statements, a competent locksmith evaluates physical evidence—such as direction of force, internal component condition, and consistency with known entry methods—to form defensible conclusions. This analytical mindset directly influences how locksmiths approach inspections, reporting, and decision-making on site.

The case studies also highlight the growing role of locksmiths in insurance and legal contexts. In many situations, locksmiths are asked to provide reports that may be used in insurance claims or court proceedings. Identifying inconsistencies—such as damage that does not match the claimed method of entry or evidence of key-based access—can prevent fraudulent claims and ensure accurate outcomes. As a result, locksmiths must maintain high standards of documentation, impartiality, and professional integrity, knowing their findings may be scrutinised in formal settings.

Another key influence on locksmithing practice is the emphasis on understanding real attack methods versus simulated damage. These scenarios teach locksmiths what genuine forced entry, covert entry, and bypass techniques actually look like in practice. This knowledge improves both forensic capability and general trade competence. For example, recognising the difference between authentic lockpicking marks and crude gouging helps avoid misdiagnosis and ensures that appropriate security recommendations are made.

The importance of key control and access management is also reinforced. Several case studies show that alleged break-ins were actually the result of authorised or duplicated key use. This shifts the locksmith's role toward advising clients on proper key management systems, restricted keyways, and access control solutions. It highlights that security is not just about hardware strength, but also about controlling who has legitimate access.

In addition, these examples influence how locksmiths approach damage interpretation and reporting. Locksmiths must be cautious not to overstate conclusions or misinterpret evidence. Instead, they are expected to clearly differentiate between observations, interpretations, and professional opinions. This aligns with broader forensic principles and ensures that reports remain credible, especially if challenged.

Finally, these case studies reinforce the importance of professional ethics and accountability. A locksmith may encounter situations where clients attempt to mislead or manipulate outcomes for financial or legal gain. The ability to recognise staged break-ins—and the willingness to report findings honestly—protects the locksmith's reputation and the integrity of the profession as a whole.

## Chain of Custody in Locksmith Investigations

Chain of custody represents a critical procedural framework within forensic investigations, extending beyond digital and biological evidence to encompass physical evidence such as locks, keys, and access control devices. This comprehensive examination synthesizes current research and best practices to establish the foundational principles, implementation procedures, and consequences of chain of custody maintenance in forensic locksmithing contexts.

### Defining Chain of Custody and Its Fundamental Purpose

Chain of custody is fundamentally defined as the chronological documentation and accountability of the custody, control, transfer, analysis, and disposition of evidence [173]. In forensic contexts, it constitutes a documented, unbroken control mechanism that tracks evidence from initial identification through final presentation in legal or insurance proceedings [173]. The primary purpose of maintaining chain of custody is to ensure that evidence remains authentic, untampered, and legally admissible [174, 175].

The concept extends across multiple forensic disciplines. While extensively researched in digital forensics [176-181], the principles apply equally to physical evidence handling in locksmith investigations [175]. The chain of custody serves as a transparent timeline demonstrating evidence integrity, enabling investigators, insurers, and courts to trust that findings are reliable and scientifically sound [174, 175].

### Core Principles of Evidence Integrity

The integrity of evidence depends upon strict adherence to established protocols and procedures [182]. Every person who handles evidence must be identified, and every transfer of possession must be recorded [174, 175]. This creates an auditable trail that establishes the authenticity of evidence and prevents questions regarding its reliability [177].

In locksmith investigations specifically, evidence integrity begins at the crime scene. When a lock, cylinder, key, or associated hardware is identified as potential evidence, it must be carefully documented in situ before removal [174]. This documentation includes photographs, detailed notes on position and condition, and observations of visible damage or tool marks [174]. The locksmith must exercise particular care during removal to avoid altering the evidence, for instance, by not using a working key or applying force that could introduce new marks [174].

The importance of maintaining evidence integrity is underscored by research demonstrating that failure to preserve chain of custody can result in evidence being challenged or deemed inadmissible in legal proceedings [174, 175]. In insurance contexts, broken chain of custody can result in claims being denied or disputed [174]. For the locksmith professional, such failures damage credibility and raise questions about the validity of findings [174].

### Documentation and Labelling Procedures

Proper documentation forms the backbone of chain of custody maintenance [174, 175]. Each piece of evidence must be labelled and logged with critical details including [174]:

- Description of the item (e.g., "front door euro cylinder")
- Location where it was found
- Date and time of collection
- Name and signature of the person collecting it
- Unique reference or case number

Research on forensic evidence management emphasizes that documentation must be contemporaneous—recorded at the time of collection rather than retrospectively [174]. This contemporaneous documentation creates a reliable record that cannot be questioned regarding accuracy or timing [174].

Beyond initial collection documentation, any subsequent transfer of evidence must be recorded with specificity [174]. This includes documentation of who transferred the item, who received it, when the transfer occurred, and the condition of the evidence at that time [174]. Even temporary handling, such as examination under a microscope or analysis in a laboratory setting, should be documented [174] .

The documentation process has evolved with technological advancement. While traditional paper-based documentation remains valid, emerging approaches utilize XML-based systems [178, 181], digital evidence management systems [183], and blockchain-based solutions [173, 177, 184-187] to enhance the reliability and accessibility of chain of custody records. However, the fundamental principle remains constant: comprehensive, contemporaneous documentation of all evidence handling [174, 175].

### Evidence Packaging and Storage Requirements

Proper packaging and storage are essential components of chain of custody maintenance [174, 175]. Once removed from the scene, evidence must be packaged appropriately to prevent contamination or loss [174]. In locksmith investigations, this typically involves sealed evidence bags or labelled containers designed to protect the integrity of locks, keys, and associated hardware [174].

Secure storage requires a clean, designated environment where evidence cannot be tampered with, mixed with other items, or exposed to contaminants [174]. Access must be restricted to authorized personnel only, and any access must be logged [174]. This controlled environment approach is particularly important for physical evidence, which can be subject to environmental degradation, contamination, or deliberate tampering [174, 175].

Research on forensic evidence management emphasizes that storage facilities must meet specific standards [174, 175]. The environment should be monitored to prevent contamination from external sources [182]. For certain types of evidence, such as biological materials, specialized storage conditions may be required [182]. In locksmith investigations, storage should protect evidence from moisture, temperature fluctuations, and physical damage that could alter tool marks or other forensic characteristics [174].

### Consequences of Chain of Custody Failure

The consequences of failing to maintain proper chain of custody are severe and well-documented across forensic disciplines [174, 175, 179, 182, 188]. If gaps, inconsistencies, or unexplained handling occur, evidence may be challenged or deemed inadmissible in legal proceedings [174]. Research demonstrates that broken chain of custody can result in evidence being excluded from trial or even lead to mistrial declarations [182].

In insurance contexts, failure to maintain chain of custody can result in claims being denied or disputed [174] . For the locksmith professional, such failures damage professional credibility and raise fundamental questions about the validity of their findings [174]. Beyond professional consequences, broken chain of custody can compromise justice by rendering otherwise probative evidence inadmissible [17, 18, 31.

The consequences extend to the broader investigative process. Any interruption in chain of custody can lead to disqualification of evidence and, consequently, potential annulment of the legal process [174]. This underscores the critical importance of maintaining unbroken documentation and control throughout the evidence lifecycle [174, 175].

### Practical Implementation in Locksmith Investigations

Chain of custody directly influences how locksmiths conduct their forensic work [174]. The requirement to maintain chain of custody reinforces the need for careful, non-destructive handling of locks and keys [174]. This means avoiding techniques that could introduce new marks, alter existing damage patterns, or compromise the forensic value of the evidence [174].

Detailed and contemporaneous documentation becomes essential practice [174]. Locksmiths must record observations at the time of examination, noting the condition of evidence before and after any handling [174]. This documentation should be sufficiently detailed that another qualified professional could understand the evidence's condition and the basis for any conclusions [174].

The use of clean tools and controlled environments is mandatory [174]. Contamination from previous evidence or environmental sources can compromise the forensic value of locksmith evidence [174]. Locksmiths must maintain separate tools and work areas to prevent cross-contamination [174].

Clear separation between observation and intervention is critical [174]. Locksmiths must distinguish between what they observed at the scene and what they determined through subsequent examination [174]. This distinction is essential for establishing the reliability of their findings and maintaining the integrity of the chain of custody [174].

### Professional Accountability and Forensic Standards

Chain of custody transforms locksmithing from a purely technical trade into a forensically accountable discipline [174]. This transformation requires locksmiths to understand and implement forensic principles that extend beyond traditional locksmith knowledge [174]. Professional accountability means accepting responsibility for maintaining evidence integrity and being prepared to defend the chain of custody in legal or insurance proceedings [174].

Research on forensic evidence management emphasizes the importance of professional training and adherence to established standards [174, 175, 188]. Locksmiths engaged in forensic work should be familiar with relevant forensic standards and best practices [174]. This includes understanding the legal requirements for evidence admissibility and the procedures necessary to meet those requirements [174]

The role of the locksmith in forensic investigations requires not only technical expertise but also meticulous attention to procedural requirements [174]. Any conclusions drawn about entry methods, tampering, or security breaches must be supported by evidence that can withstand scrutiny in both investigative and legal settings [174]. This requires that the chain of custody be maintained throughout the investigation process [174].

### Emerging Technologies and Chain of Custody Enhancement

While traditional chain of custody procedures remain valid and necessary, emerging technologies offer potential enhancements to evidence management [173, 177, 184-187]. Blockchain technology has been proposed as a solution for creating immutable, decentralized records of evidence handling [173, 177, 184-187]. These systems can provide tamper-proof documentation of evidence transfers and maintain transparent audit trails [173, 177, 184, 185].

Digital evidence management systems offer centralized platforms for documenting and tracking evidence throughout the investigation process [183]. These systems can incorporate secure user authentication, role-based access control, file integrity verification using hash functions, and comprehensive audit logging [183]. Such technological approaches can enhance the efficiency and reliability of chain of custody maintenance [183].

However, it is important to note that technological solutions supplement rather than replace fundamental chain of custody principles [173-175]. The core requirement—documented, unbroken control of evidence with identification of all handlers and recording of all transfers—remains essential regardless of the technology employed [174, 175].

Chain of custody is a fundamental concept in forensic locksmithing that ensures evidence authenticity, integrity, and legal admissibility [174]. The process begins at the scene with careful documentation and non-destructive removal, continues through appropriate packaging and secure storage, and extends through all transfers and examinations until final presentation [174]. Every person who handles evidence must be identified, and every transfer must be recorded, creating a transparent timeline that demonstrates evidence integrity [174, 175].

Failure to maintain proper chain of custody has serious consequences, potentially rendering evidence inadmissible in legal proceedings and damaging professional credibility [174]. For locksmiths engaged in forensic work, chain of custody transforms their practice into a forensically accountable discipline that requires careful, non-destructive handling, detailed documentation, use of clean tools and controlled environments, and clear separation between observation and intervention [174].

The principles of chain of custody are well-established across forensic disciplines and apply directly to locksmith investigations [174, 175]. By maintaining rigorous chain of custody procedures, locksmiths ensure that their findings are reliable, defensible, and capable of withstanding scrutiny in investigative, insurance, and legal contexts [174].

### Impact on Locksmithing Practices

Chain of custody is highly relevant to anyone interested in locksmithing because it fundamentally changes how the profession is understood and practiced. While traditional locksmithing focuses on installing, repairing, and defeating locks, forensic locksmithing requires a shift toward evidence handling, legal accountability, and procedural discipline. This means that a locksmith is no longer just a technician, but also a potential investigator whose work may directly influence insurance outcomes, criminal cases, and legal decisions.

From a practical perspective, chain of custody introduces a strong emphasis on how locks and keys are handled, not just what is done to them. A locksmith must approach every job—particularly those involving suspected break-ins—with the mindset that the lock or key could be evidence. This influences behaviour at the scene, requiring careful observation before any intervention. For example, instead of immediately opening or replacing a lock, a forensic-aware locksmith will document its condition, photograph damage, and consider whether their actions could alter or destroy critical evidence.

It also directly influences the methods and techniques used in locksmithing practice. Non-destructive techniques become preferred, not just for preserving hardware, but for preserving forensic value. Actions such as using a working key, forcing a mechanism, or cleaning components prematurely can compromise evidence and break the chain of custody. As a result, locksmiths must adopt more controlled, methodical approaches, often separating routine service work from investigative work entirely.

Documentation becomes a central professional skill rather than an administrative afterthought. Chain of custody requires locksmiths to produce clear, contemporaneous, and defensible records of what was observed, what actions were taken, and who handled the evidence. This elevates the importance of written reports, photography, labelling, and evidence logs. In practice, this means locksmiths must develop competencies similar to those in forensic science—accurate note-taking, objective language, and structured reporting.

The concept also reinforces the importance of professional integrity and credibility. Because locksmith findings may be scrutinised in court or by insurers, maintaining an unbroken chain of custody ensures that conclusions are trusted. A locksmith who cannot account for how evidence was handled risks having their work dismissed, regardless of technical accuracy. This has a direct impact on reputation, liability, and professional standing within the industry.

Chain of custody further influences the design of workspaces and tools used by locksmiths engaged in forensic work. Clean, controlled environments, dedicated tools, and contamination prevention become essential. This may lead to the separation of general workshop activities from forensic examinations, the use of non-marking tools (e.g., plastic tweezers, soft followers), and stricter handling protocols to avoid cross-contamination or accidental damage.

In a broader sense, chain of custody expands the role of the locksmith into security consultancy and investigative support. Understanding how evidence is preserved and interpreted allows locksmiths to contribute meaningfully to crime analysis, insurance assessments, and legal proceedings. It also improves their ability to identify issues such as staged break-ins, key control failures, or covert entry techniques, all of which rely on reliable evidence handling.

Chain of custody shapes the mindset of the locksmith by reinforcing a clear distinction between observation and intervention. This is critical in forensic contexts. A locksmith must be able to explain not only what they found, but also demonstrate that their findings were not influenced or altered by their own actions. This disciplined approach ensures that conclusions about entry methods, tampering, or system compromise are based on preserved evidence rather than assumptions.

## Chapter 5 Review Questions

### Question 1

A shop owner claims a burglar forced entry through the front door. When the locksmith examines the door, the lock cylinder, door frame, and strike plate show no visible damage.

What is the most likely explanation?

A. The lock was defective
B. The entry may have occurred using a legitimate or duplicated key

C. The door automatically unlocked overnight
D. The lock repaired itself after the attack

**Question 2**

A locksmith inspecting a break-in finds drill marks and metal shavings around the cylinder. The keyway is distorted and internal components are damaged.

What type of entry does this evidence indicate?

A. Non-forced entry
B. Electronic access
C. Forced entry
D. Administrative access

**Question 3**

During an investigation, a locksmith finds a snapped euro-profile cylinder at the fixing point.

What does this damage most strongly suggest?

A. A decoding attack
B. A cylinder snapping attack involving physical force
C. A duplicated key was used
D. Electronic access was granted

**Question 4**

A warehouse reports a break-in, but the locksmith notices that tool marks on the lock appear crude and inconsistent with recognised burglary techniques.

What should the locksmith consider?

A. The marks were made by advanced lock-picking tools
B. The damage may have been staged
C. The lock manufacturer is at fault
D. The lock cannot be analysed

**Question 5**

A property owner claims a door was forced open, but the surrounding environment shows no disturbance, no broken items, and no signs of hurried activity.

What does this lack of environmental evidence suggest?

A. A professional burglar left no traces
B. The break-in may have been staged
C. The lock automatically opened
D. The locksmith misidentified the lock

**Question 6**

A security investigation reveals that a door was opened using an employee's valid access credential at the time of the reported "break-in."

What type of entry occurred?

A. Forced entry
B. Covert entry using authorised access
C. Mechanical failure
D. Lock bypass through drilling

**Question 7**

A locksmith observes pry marks on the door frame and a bent strike plate after a burglary.

What type of attack likely occurred?

A. Bypass entry
B. Key decoding
C. Levering or prying forced entry
D. Credential misuse

**Question 8**

A locksmith discovers that a padlock was cut cleanly with bolt cutters, but there are no marks on the surrounding gate or structure where the lock was supposedly attacked.

What should the locksmith suspect?

A. The lock failed spontaneously
B. The lock was cut elsewhere and staged
C. The gate material absorbed the force
D. The attack was performed using heat

**Question 9**

A lock appears to have damage on the outside, but the internal mechanism functions normally and shows no signs of internal force.

What forensic conclusion might be drawn?

A. The lock was damaged after it was opened
B. The lock is defective
C. The lock is immune to forced entry
D. The key was incorrectly cut

**Question 10**

During a forensic inspection, a locksmith documents tool marks, photographs damage, and records observations before removing the lock.

Why is this process important?

A. To avoid replacing the lock unnecessarily
B. To maintain accurate evidence for investigation or legal proceedings
C. To reduce installation time
D. To avoid speaking with the client

# Chapter 6

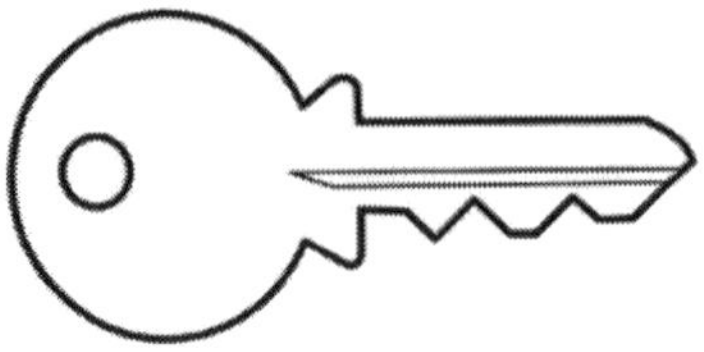

# The Science of Lock Manipulation (Ethical Perspective)

This chapter introduces lock manipulation as a *sensory-mechanical* discipline rather than a "bag of tricks." In ethical locksmithing, the point is not simply gaining access—it is understanding how a lock behaves internally when torque and tool contact are applied, and using that understanding to work with control, restraint, and accountability. This chapter frames manipulation as a process of *reading* a concealed mechanism through feedback, where the practitioner's judgement and touch matter more than brute force or speed.

At the centre of this science is manipulation feedback: the combined signals of resistance, micro-movement, sound, and vibration produced by friction, tolerances, spring tension, and binding. Because no lock is manufactured perfectly, internal components bind in a sequence and "announce" their state through subtle cues. Ethical practice depends on learning to interpret those cues correctly—using minimal force to preserve the lock, reduce unnecessary wear, and maintain professional integrity. In real locksmithing work, this supports non-destructive entry, accurate fault diagnosis, and a deeper understanding of how different lock types (pin-tumbler, wafer, lever, disc detainer) "communicate" differently.

The chapter then builds skill in a structured way: it explains how tactile feedback is developed and refined, how auditory and vibrational cues confirm (or contradict) what the fingers are reporting, and how indirect visual indicators, such as slight plug rotation or actuator movement, can reinforce interpretation. It also connects technique to consequence: how controlled manipulation differs from tool-assisted or force-based methods, what kinds of marks or disturbances different approaches can leave, and why this matters in professional reporting, security assessment, and forensic contexts.

The chapter concludes by profiling a range of lock picking techniques for non-destructive access. Overall, Chapter 6 positions manipulation as a disciplined craft grounded in mechanics, perception, and ethics, where competence is measured not only by outcomes, but by method, care, and defensibility.

## Understanding Manipulation Feedback

Manipulation feedback is the fundamental mechanism that allows a locksmith or attacker to interpret what is happening inside a lock without direct visual access. In essence, it is the process of "reading" the lock through touch, sound, and movement. Every lock, particularly mechanical systems such as pin-tumbler, wafer, and lever locks, provides subtle physical signals when its internal components are engaged, resisted, or correctly aligned.

At its core, manipulation feedback is governed by basic mechanical principles: friction, binding, spring tension, and tolerances. When torque is applied to a lock (via a tension tool or key), internal components such as pins or wafers begin to bind against the plug or housing. This binding creates resistance, which is transmitted back through the tool to the operator. By carefully interpreting these signals, the operator can determine which components are out of position and require adjustment.

**Types of Manipulation Feedback**

**1. Tactile Feedback (Touch)**

This is the most important and commonly used form of feedback. Through the fingers, the operator can feel:

- **Binding pins**: Pins that resist movement indicate incorrect positioning and must be lifted or adjusted
- **Set pins**: A slight "click" or change in resistance signals that a pin has reached the shear line
- **Spring tension differences**: Variations in resistance can indicate security features such as spool or serrated pins

Experienced operators develop a highly refined sense of touch, allowing them to distinguish between normal resistance and meaningful feedback.

The ability to interpret tactile feedback in lock manipulation—such as feeling binding pins, recognising when pins set, and detecting subtle variations in spring tension—is not an innate talent but a learned sensory skill. It is developed through structured practice, careful technique, and repeated exposure to different lock behaviours. For anyone serious about locksmithing or the science of lock manipulation, refining this sense of touch is fundamental to progressing beyond basic mechanical operation to true technical proficiency.

Skill development should begin with controlled, progressive practice using simple pin-tumbler locks that contain no security features. These locks provide clear and predictable feedback, allowing the learner to focus on understanding fundamental sensations without interference. At this stage, the objective is not speed or even successful opening, but awareness. The learner should apply very light tension, lift each pin slowly and deliberately, and pause to observe subtle changes in resistance. This process builds an initial understanding of how internal components behave under controlled conditions.

A critical aspect of developing tactile feedback is mastering tension control. The quality of feedback is directly influenced by how much force is applied. Excessive tension can mask feedback entirely, while insufficient tension prevents the binding necessary to generate meaningful signals. Effective practice involves using the minimum tension required to create binding, then experimenting with slight variations to observe how feedback clarity changes. A useful exercise is to intentionally apply too much tension and then gradually reduce it until feedback becomes perceptible, reinforcing the relationship between force and sensitivity.

As skill improves, attention should shift to identifying the binding order within the lock. Due to manufacturing tolerances, each lock develops a natural sequence in which pins bind. Developing tactile sensitivity involves learning to detect which pin offers the greatest resistance and prioritising it. By probing each pin lightly, focusing on the stiffest pin, and repeating this process sequentially, the operator trains their fingers to recognise meaningful signals rather than reacting randomly.

Equally important is recognising the sensation of a pin setting at the shear line. This is typically experienced as a subtle click, a slight rotation of the plug, or a reduction in resistance. Developing this skill requires slow, deliberate lifting of pins and stopping at the first indication of movement. By repeatedly resetting the lock and comparing the feel of pins before and after setting, the learner builds both sensory recognition and muscle memory.

Once the fundamentals are established, practice should introduce variability and complexity. Locks containing security features such as spool or serrated pins generate misleading feedback, including false sets and counter-rotation. Working with these locks develops the ability to distinguish between genuine and deceptive signals, an essential capability for advanced manipulation. At this stage, the operator learns to interpret more complex feedback patterns rather than relying on simple, predictable cues.

It is important to emphasise that skill development should prioritise slow, deliberate practice over speed. Attempting to open locks quickly often leads to excessive force and reduced sensitivity, undermining learning. Instead, practitioners should focus on what they feel rather than the outcome, repeating the same lock multiple times to build consistency. Experienced operators often spend more time interpreting feedback than actively manipulating components.

Tactile skill is also influenced by physical factors such as finger sensitivity and motor control. Consistent practice using the same tools helps build familiarity, while maintaining a relaxed grip reduces unnecessary tension in the hands. Practising with both dominant and non-dominant hands further enhances control and adaptability. Over time, the fingers become increasingly responsive to very small changes in resistance, enabling more precise interpretation of feedback.

To build real-world capability, training should extend beyond ideal conditions. Practising in low-light environments encourages reliance on touch rather than vision, while working with locks in different orientations develops spatial awareness. Introducing mild distractions can also help simulate real scenarios, strengthening focus and consistency under less controlled conditions.

Finally, developing tactile feedback has important forensic implications. Through controlled manipulation, locksmiths gain an understanding of what normal tool interaction looks and feels like, how excessive force alters internal components, and how different techniques leave distinct marks. This knowledge enhances both practical skill and the ability to interpret evidence in investigative contexts.

In summary, developing tactile feedback is a disciplined process of training the hands to detect and interpret subtle mechanical signals. It requires structured progression, precise control of tension, repetition, and exposure to increasing complexity. With time and practice, the operator moves beyond simply using tools to actively understanding and responding to the internal behaviour of the lock, which is the defining characteristic of true expertise in lock manipulation.

**2. Auditory Feedback (Sound)**

Although subtle, sound can provide useful cues:

- Clicks or ticks when pins set into position
- Scraping sounds indicating tool contact with the keyway or internal components
- Dull or muted sounds suggesting improper technique or excessive force

In quiet environments, these sounds can reinforce tactile information and improve accuracy.

Auditory feedback in lock manipulation is a subtle but valuable source of information that complements tactile sensing. While the primary skill lies in what is felt through the fingers, the sounds produced inside the lock, such as faint clicks, ticks, or scraping noises, can reinforce or clarify what is happening mechanically. Developing this skill requires deliberate practice in controlled conditions, where the operator can isolate and recognise these small acoustic cues without distraction.

The development of auditory sensitivity should begin in a quiet environment. Background noise can easily mask the very faint sounds produced by pins setting or tools contacting internal components. Early practice should therefore be conducted in near-silent conditions, allowing the learner to focus on identifying distinct sounds such as the soft "click" of a pin reaching the shear line. By pairing these sounds with the corresponding tactile sensations, the operator begins to build a multi-sensory understanding of lock behaviour.

A useful method for developing auditory awareness is to work slowly and deliberately while actively listening for each interaction inside the lock. As each pin is lifted, the learner should pause and concentrate on whether a sound accompanies the movement. The goal is to associate specific sounds with specific mechanical events—for example, recognising that a clean, sharp click often corresponds with a successful pin set, while a dull or muted sound may indicate that the pin has not reached the correct position or that excessive force is being applied.

Practising with varying levels of tension is also essential for refining auditory feedback. Light, controlled tension tends to produce clearer and more distinct sounds, while excessive tension can dampen or distort them. By intentionally varying tension during practice, the learner can observe how sound quality changes and begin to understand how proper technique enhances both tactile and auditory feedback. This reinforces the importance of precision and control in manipulation.

Another important aspect of training is learning to distinguish between meaningful sounds and irrelevant noise. Scraping sounds, for instance, often indicate that the tool is contacting the keyway walls rather than the pins, which may suggest poor tool positioning or technique. By consciously identifying and correcting these sounds during practice, the operator can improve tool control and reduce unnecessary contact that could interfere with accurate feedback or leave unwanted marks.

As skill develops, more complex locks with security features can be introduced. These locks may produce misleading or less distinct sounds, requiring the operator to rely on a combination of auditory and tactile cues. For example, spool pins may produce subtle clicks associated with false sets or counter-rotation. Learning to interpret these more complex acoustic signals helps build a deeper understanding of internal lock dynamics and improves overall manipulation accuracy.

Repetition plays a critical role in developing auditory skill. Working with the same lock multiple times allows the operator to become familiar with its characteristic sounds, making it easier to detect variations or anomalies. Over time, this familiarity extends to different lock types, enabling the practitioner to recognise patterns and differences in sound across various mechanisms.

Finally, auditory training should be integrated with tactile practice rather than treated as a separate skill. The most effective operators do not rely solely on sound but use it to confirm or refine what they feel through their tools. By combining these sensory inputs, the locksmith develops a more complete and accurate interpretation of the lock's internal state.

Developing auditory feedback requires quiet, focused practice, careful attention to sound-tactile relationships, and consistent repetition. By learning to recognise and interpret subtle acoustic cues, the operator enhances their ability to manipulate locks with precision and gains a deeper, more intuitive understanding of the mechanisms at work.

**3. Visual Feedback (Indirect Observation)**

While the internal components are not visible, indirect visual cues may be observed:

- Slight rotation of the plug when a pin sets
- Movement of external components (e.g., cam or actuator)
- Changes in tool position or alignment

In some cases (e.g., transparent training locks or optical decoding tools), visual feedback can be directly observed for learning purposes.

Visual feedback in lock manipulation is typically indirect, as the internal components of most locks are concealed. However, skilled operators learn to observe subtle external cues—such as slight plug rotation, tool movement, or changes in alignment—that indicate what is happening inside the mechanism. Developing this skill involves training the eye to detect very small changes and linking those observations to tactile and mechanical events within the lock.

A strong starting point for developing visual feedback is the use of transparent or cutaway training locks. These allow the learner to directly observe pin movement, binding, and setting while manipulating the lock. By watching how pins respond to tension and picking, the learner can clearly see the relationship between cause (tool input) and effect (pin movement and plug rotation). This visual grounding is extremely valuable in the early stages, as it builds a mental model of how the lock functions internally.

Once the basic principles are understood, practice should shift toward standard (non-transparent) locks while maintaining a strong focus on indirect visual cues. One of the most important of these is plug rotation.

Even a very slight movement of the plug—often barely perceptible—can indicate that a pin has set correctly. Learners should practise applying light tension and carefully watching the plug for these micro-movements, reinforcing what they feel with what they see.

Tool position and alignment also provide important visual information. As pins are lifted and set, the angle and depth of the pick may change slightly. Observing how the tool behaves—whether it shifts, binds, or moves freely—can help confirm which pins are being engaged and whether the technique is correct. Developing this awareness requires slow, deliberate manipulation and conscious observation of the tools at all times.

Another useful method is to practise in well-lit conditions with a stable visual focus on the lock and tools. Good lighting enhances the ability to detect subtle movements, such as minute rotations of the plug or slight deflections in the tension tool. Using a consistent viewpoint and minimising unnecessary movement helps train the eyes to pick up small changes that would otherwise go unnoticed.

As skill improves, learners can introduce more complex scenarios, such as locks with security pins or varying tolerances. These locks often produce less obvious visual cues, requiring greater attention and integration with tactile feedback. For example, a false set may produce a noticeable plug rotation that differs from a true set, and learning to recognise this visually can improve accuracy in diagnosing lock state.

Repetition is essential in developing visual feedback skills. By working with the same lock repeatedly, the operator becomes familiar with its normal behaviour and can more easily detect subtle deviations. Over time, this experience transfers to other locks, allowing the practitioner to quickly interpret visual cues across different mechanisms and designs.

It is also beneficial to occasionally combine visual training with other techniques such as slow-motion observation or recording practice sessions. Reviewing footage can help identify small movements or habits that were not noticed in real time, further refining the operator's ability to see and interpret visual feedback.

Visual feedback is not a standalone skill but part of a broader sensory system that includes touch and, to a lesser extent, sound. The goal is to integrate these inputs so that visual cues confirm and enhance tactile understanding. With consistent, focused practice, the operator develops the ability to "read" the lock externally, gaining insight into internal behaviour without direct visibility—an essential capability in advanced lock manipulation.

## Feedback in Different Lock Types

Different lock types produce different forms of tactile feedback because each design uses unique internal components, tolerances, and operating principles. As a result, the sensations felt through tools—such as resistance, movement, clicks, or counter-rotation—vary significantly between lock mechanisms. For a locksmith or operator, understanding these differences is essential, as applying the same expectations or techniques across all lock types can lead to misinterpretation of feedback and unsuccessful manipulation.

In pin-tumbler locks, which are the most common, feedback is typically clear and structured. Binding pins create noticeable resistance, and correctly set pins often produce a subtle click accompanied by slight plug movement. Variations in feedback occur when security pins such as spool or serrated pins are present, introducing false sets and counter-rotation. This type of lock is often used as the foundation for learning tactile feedback because its signals are relatively consistent and predictable.

In contrast, wafer locks tend to provide much lighter and less distinct feedback. The wafers are broader and flatter than pins, and the springs are often weaker, resulting in a softer, less defined sensation. Instead of sharp clicks, the operator may feel more of a gradual shift or smoothing of resistance. Because of this, wafer locks can be more difficult for beginners to interpret, as the feedback lacks the crisp indicators found in pin-tumbler systems.

Disc detainer locks present a completely different feedback profile. Rather than lifting components vertically, the operator rotates discs to align gates. Feedback is rotational rather than vertical, and the operator must detect binding through slight resistance in turning individual discs. There are no clicks in the traditional sense; instead, the feedback is subtle changes in rotational tension and alignment. This requires a different skill set focused on torque control and rotational sensitivity.

Lever locks also differ significantly, producing feedback through the lifting and positioning of levers to align with a gate. The operator feels resistance as each lever is lifted, and a correct position may result in a faint give or reduction in pressure. However, feedback can be dampened by the weight of the bolt stump or false gates, making it more ambiguous. Skilled operators learn to distinguish between true gates and false feedback through careful, incremental movement.

In combination locks, particularly rotary types, tactile feedback is minimal and often indirect. Instead of feeling individual components set, the operator relies on subtle changes in resistance as wheels align internally. Feedback may be detected as slight drops, changes in dial resistance, or contact points during manipulation. This form of feedback is far less intuitive and requires patience, consistency, and a highly refined sense of touch.

Modern high-security locks further complicate feedback by incorporating features designed specifically to obscure or distort it. Sidebars, magnetic elements, rotating pins, and electronic components can either reduce tactile feedback or introduce misleading signals. In these systems, traditional feedback cues may be absent or intentionally deceptive, requiring advanced knowledge and alternative approaches to interpretation.

**Pin-Tumbler Locks:** Feedback is primarily derived from the binding order of pins. Due to manufacturing tolerances, one pin stack will bind before others. The operator uses this feedback to sequentially set each pin until all reach the shear line.

**Wafer Locks:** Feedback tends to be less distinct due to simpler construction. However, resistance differences still indicate which wafers are obstructing rotation.

**Lever Locks:** Feedback involves lifting levers to precise heights. A correctly set lever produces a subtle "gate" feeling, allowing partial movement of the bolt stump.

**Combination Locks:** Feedback is often auditory and tactile, involving detection of contact points, wheel gates, or resistance changes as components align.

Recognising that each lock type "communicates" differently is fundamental to developing proficiency in lock manipulation. A skilled practitioner does not rely on a single set of sensations but adapts their expectations and technique to the specific mechanism being encountered. This adaptability is what allows the operator to accurately interpret feedback, avoid false assumptions, and successfully interact with a wide range of locking systems.

### The Role of Tolerances and Binding

Manipulation feedback exists because no lock is manufactured to absolute perfection. Even in high-quality locks, there are microscopic variations in component dimensions, alignment, surface finish, and wear. These small inconsistencies create differences in how individual components interact under tension, resulting in a predictable binding order within the mechanism.

This binding order is fundamental to lock manipulation. When tension is applied, one component—typically the one with the greatest misalignment—will bind first and produce the strongest and most noticeable feedback. This allows the operator to identify which component requires adjustment.

Once that component is correctly positioned, it no longer binds, and the next component in the sequence begins to resist movement. The process then repeats, with each component providing feedback in turn. This sequential binding continues until all components are aligned at the shear line, allowing the lock to operate.

Without these inherent imperfections, all components would move freely and simultaneously under tension, providing no distinguishable feedback. In such a perfectly manufactured system, manipulation would be significantly more difficult, if not impossible, as there would be no way to identify or prioritise individual components.

### Interpretation vs Skill

Understanding manipulation feedback is not simply about sensing mechanical signals—it is about accurately interpreting what those signals mean. Beginners commonly struggle at this stage. They often apply excessive torque, which masks the subtle feedback needed to identify binding components. They may also mistake normal friction for meaningful resistance, leading to incorrect adjustments, or fail to recognise when a component has reached its correct position. In contrast, skilled practitioners apply minimal, controlled force, allowing the lock to "communicate" clearly through consistent and interpretable feedback.

From a forensic locksmithing perspective, manipulation feedback has direct relevance to the identification of covert entry techniques. Skilled manipulation typically leaves very fine, controlled scratches within the keyway and on internal components, reflecting precise and deliberate tool use. In many cases, there is minimal or no visible damage, indicating that the operator relied on feedback rather than force. However, repeated manipulation can still leave consistent micro-mark patterns on pins and plug surfaces. By understanding how feedback is used during entry, the forensic locksmith is better equipped to distinguish between genuine covert techniques and crude or staged damage.

Manipulation feedback has also had a significant influence on modern lock design and security engineering. High-security locks are deliberately designed to disrupt or distort feedback, making it difficult for an operator to interpret what is happening inside the mechanism. Features such as spool, serrated, and mushroom pins introduce false feedback, while tighter manufacturing tolerances reduce predictable binding order. More advanced systems incorporate sidebar mechanisms that require multiple conditions to be met simultaneously, and false gates in combination or lever locks further complicate interpretation. These design innovations are intended to confuse the operator, making feedback unreliable or misleading and significantly increasing resistance to manipulation.

Manipulation feedback is the "language" of a lock. It transforms a closed, invisible mechanism into a system that can be understood through touch, sound, and movement. Mastery of this feedback allows precise, non-destructive entry, while misunderstanding it leads to failure or unnecessary damage. For locksmiths, it represents both a practical skill and a critical concept in understanding security, vulnerability, and forensic evidence.

**Impact on Locksmithing Practices**

Manipulation feedback is highly relevant to anyone interested in locksmithing because it represents the core mechanism through which locks are understood, serviced, and, where authorised, opened without damage. At a practical level, locksmithing is not simply about tools or procedures—it is about interpreting the internal state of a concealed mechanical system. Manipulation feedback provides the only real-time insight into that system. Without an understanding of how locks "communicate" through touch, sound, and movement, a practitioner is limited to force-based or trial-and-error approaches, which are inefficient, potentially damaging, and professionally unacceptable in many contexts.

This concept fundamentally shapes how locksmiths develop their technical skills. Training is not just about learning techniques, but about learning to interpret subtle signals such as resistance, clicks, and micro-movements. As a result, locksmithing practice places strong emphasis on controlled technique, minimal force, and deliberate, sensory-based interaction with locks. Skilled practitioners rely on feedback to guide every action, allowing for precise, non-destructive entry. This directly influences day-to-day work, particularly in scenarios such as lockouts, safe opening, and maintenance, where preserving the integrity of the lock is critical.

Manipulation feedback also drives the progression from basic to advanced competence. Beginners often focus on tool movement, while experienced locksmiths focus on what the lock is telling them. This shift in mindset—from doing to interpreting—is what distinguishes a trained professional from an unskilled operator. As such, locksmithing practices increasingly incorporate structured training methods, including progressive lock difficulty, controlled environments, and repetition, all designed to refine sensory interpretation rather than brute technique.

From a forensic perspective, understanding manipulation feedback is equally important. The way a lock is manipulated leaves characteristic evidence, such as fine scratches, consistent micro-mark patterns, or minimal internal damage. A locksmith trained in feedback interpretation is better equipped to recognise these patterns and distinguish between skilled covert entry, forced entry, and staged damage. This has direct implications for insurance investigations, legal proceedings, and security assessments, where accurate interpretation of evidence is essential.

The concept also influences how locksmiths approach security and risk. By understanding how feedback enables manipulation, locksmiths gain insight into how locks can be compromised. This knowledge informs recommendations for improved security, such as the use of high-security cylinders, restricted key systems, or locks with features designed to disrupt feedback. In this way, manipulation feedback is not just a tool for entry, but a foundation for evaluating and improving security systems.

Importantly, manipulation feedback has shaped the evolution of modern lock design. Manufacturers actively design locks to reduce, distort, or mislead feedback through features such as security pins, sidebars, and false gates. Locksmiths must therefore continually update their knowledge and adapt their techniques to account for these changes. This creates a dynamic relationship between locksmithing practice and lock engineering, where understanding feedback is essential to both overcoming and designing security measures.

Finally, manipulation feedback reinforces professional standards within locksmithing. The ability to perform non-destructive entry, interpret evidence accurately, and apply minimal force reflects a high level of skill, care, and professionalism. It supports ethical practice by reducing unnecessary damage and ensuring that work is carried out with precision and accountability. For those entering the field, mastering manipulation feedback is not optional—it is central to competent, credible, and modern locksmithing practice.

## Sound, Vibration, and Tactile Sensitivity

The concepts of sound, vibration, and tactile sensitivity form a combined sensory system that allows a practitioner to interpret the internal state of a lock without direct visual access. These are not separate skills, but interconnected channels of feedback that, when developed together, enable precise, controlled, and non-destructive manipulation.

Tactile sensitivity is the primary and most critical of these senses. It involves the ability to feel minute changes in resistance, pressure, and movement through the fingertips via the tension tool and pick. When torque is applied, internal components such as pins or wafers bind due to slight manufacturing tolerances. This binding creates resistance that is transmitted back through the tool. A skilled operator can detect which component is binding, when it begins to move, and when it reaches the correct position (such as the shear line). Subtle cues—like a slight reduction in resistance, a faint "give," or controlled counter-rotation—provide essential information. In ethical lock picking, this sensitivity allows the practitioner to work with minimal force, preserving the lock and avoiding unnecessary damage.

Closely linked to tactile input is vibration sensitivity, which can be understood as a finer extension of touch. Vibrations are micro-signals generated by internal interactions—such as pins contacting the plug, springs compressing, or components snapping into place. These vibrations travel through the tools and into the fingers. While often imperceptible to beginners, experienced practitioners learn to detect these faint oscillations and use them to confirm what they feel. For example, a correctly set pin may produce not only a slight click but also a subtle vibration that distinguishes it from normal friction. In techniques involving rapid movement—such as raking or the use of pick guns—vibration becomes even more significant, as it reflects how energy is transferred through the lock. Developing sensitivity to vibration enhances precision and reduces reliance on excessive force.

Sound provides a complementary layer of feedback that reinforces tactile and vibrational information. Inside a lock, mechanical interactions generate faint acoustic signals—clicks, ticks, scrapes, and dull impacts. A clean, sharp click often indicates that a component has set correctly, while scraping sounds may suggest improper tool placement or contact with the keyway walls. Dull or muted sounds can indicate excessive tension or poor technique, where components are being forced rather than guided into position. In quiet environments, these sounds can significantly improve accuracy by confirming what is being felt through the tools. Ethical practitioners use sound not as the primary guide, but as a secondary confirmation that supports careful, controlled manipulation.

From a scientific perspective, all three forms of feedback, tactile, vibrational, and auditory, are governed by the same underlying mechanical principles: friction, binding, elasticity, and energy transfer. When a lock is manipulated, these principles produce measurable physical signals. The practitioner's role is to detect and interpret these signals correctly. This requires not only sensory acuity but also an understanding of how different lock designs influence feedback. For example, a standard pin-tumbler lock may produce clear clicks and resistance changes, while a high-security lock with spool pins may generate misleading feedback such as false sets and counter-rotation.

In ethical lock picking, the development of these sensory skills is directly tied to professional responsibility. The goal is not simply to open a lock, but to do so in a controlled, non-destructive manner that respects the integrity of the mechanism. By relying on sound, vibration, and tactile sensitivity rather than force, the practitioner minimises damage, preserves evidence (in forensic contexts), and demonstrates technical competence. This approach aligns with the broader ethical framework of locksmithing, where skill, precision, and accountability are prioritised over speed or brute methods.

## Impact on Locksmithing Practices

Understanding sound, vibration, and tactile sensitivity is highly relevant to anyone interested in locksmithing because these sensory skills underpin the transition from basic mechanical operation to true technical expertise. Locksmithing is not simply about using tools, it is about interpreting the internal behaviour of a lock without direct visibility. This sensory system becomes the primary interface between the practitioner and the mechanism, allowing informed, precise interaction rather than guesswork or force.

From a practical perspective, these skills directly influence how a locksmith approaches everyday tasks such as non-destructive entry, servicing, and diagnostics. A practitioner who can accurately interpret tactile resistance, subtle vibrations, and faint acoustic cues is able to identify binding components, detect faults, and manipulate mechanisms with a high degree of control. This reduces reliance on destructive methods, preserves the integrity of the lock, and improves efficiency. In contrast, a lack of sensitivity often leads to excessive force, increased wear, and potential damage to both the lock and surrounding hardware.

These concepts also shape professional standards and expectations within locksmithing practice. Ethical locksmithing places strong emphasis on non-destructive techniques wherever possible. The ability to "read" a lock through sound, vibration, and touch enables the practitioner to meet this expectation by working with the mechanism rather than against it. This not only protects client property but also aligns with industry responsibilities related to minimising damage, maintaining security, and demonstrating competence.

In a forensic context, the relevance becomes even more significant. Skilled manipulation leaves distinct but minimal evidence, whereas forceful or unskilled techniques leave obvious damage. A locksmith who understands how feedback is generated and interpreted is better equipped to recognise these differences during investigations. This allows them to distinguish between covert entry, forced entry, and staged damage, which is critical in insurance assessments, legal proceedings, and security evaluations. The same sensory knowledge used to open a lock is therefore directly applied to analysing how a lock was opened.

The influence of these principles extends into lock design and security awareness. Modern high-security locks are engineered specifically to disrupt or distort feedback, using features such as security pins, sidebars, and false gates. A locksmith who understands sound, vibration, and tactile sensitivity can better appreciate how and why these features work, and can make informed recommendations regarding appropriate security solutions. This knowledge supports better system design, product selection, and client advice.

These skills also impact training and professional development within the trade. Developing sensitivity to feedback encourages disciplined, methodical practice rather than speed-focused or force-based approaches. It reinforces the importance of tool control, tension management, and patience. Over time, this leads to more consistent outcomes, higher-quality workmanship, and greater adaptability across different lock types and technologies.

Ultimately, sound, vibration, and tactile sensitivity are not just technical concepts, they shape how locksmithing is performed at a fundamental level. They influence technique, ethics, forensic capability, and security understanding. For anyone entering or advancing in the field, mastering these sensory skills is essential for achieving precision, maintaining professional integrity, and operating effectively across both practical and investigative aspects of locksmithing.

## Human Skill Vs Tool-Assisted Entry

Human skill and tool-assisted entry represent two fundamentally different approaches to interacting with locks, and understanding the distinction is essential for anyone studying locksmithing, security, or forensic analysis. While both approaches aim to achieve access, they differ significantly in method, reliance on feedback, level of control, and the type of evidence they leave behind.

From the perspective of human skill, entry is achieved through the operator's ability to interpret manipulation feedback—primarily tactile, but also vibrational and auditory. The practitioner uses simple tools such as picks and tension wrenches, but the outcome depends almost entirely on their ability to "read" the lock. This involves identifying binding components, recognising when elements are correctly positioned, and applying precise, minimal force. Skilled operators work slowly and deliberately, allowing the lock to communicate its internal state. As a result, entry can often be achieved non-destructively, with little to no visible damage. The defining characteristic of human skill-based entry is that the tool is passive—the intelligence and decision-making come from the operator.

In contrast, tool-assisted entry relies more heavily on the design and function of specialised tools to achieve access, often reducing the need for advanced sensory interpretation. These tools may include pick guns, bump keys, decoding devices, or lock-specific bypass tools. Instead of carefully interpreting feedback, the operator uses tools that apply kinetic energy, automate movement, or mechanically determine internal positions. For example, a pick gun rapidly transfers energy to pin stacks to momentarily separate them, while a decoding tool may measure component positions directly. In these cases, the tool performs much of the "work," and the operator's role shifts toward correct positioning and activation rather than continuous interpretation.

The difference between these approaches has important implications for control and precision. Human skill-based manipulation allows for extremely fine control, making it suitable for delicate work, high-security systems, and situations where preserving the lock is critical. Tool-assisted methods, while often faster, can be less precise and may introduce additional wear or disturbance to internal components. Some tools, particularly those that rely on impact or vibration, can generate more noticeable forensic evidence due to repeated or forceful interactions.

From a forensic locksmithing perspective, the distinction is highly significant. Skilled manual manipulation typically leaves very fine, consistent micro-markings—such as light scratches on pins or plug surfaces—reflecting controlled tool movement. In contrast, tool-assisted entry often produces more distinctive patterns. For example, pick guns may leave impact marks on pins, bump keys can create characteristic dents and wear patterns, and bypass tools may leave marks on cams or actuators.

By analysing these patterns, a forensic locksmith can often determine not only that entry occurred, but how it was achieved and the level of skill involved.

The distinction also influences training and professional development in locksmithing. Developing human skill requires time, repetition, and disciplined practice to refine sensory awareness and technique. It builds a deep understanding of lock mechanics and behaviour. Tool-assisted methods, while still requiring knowledge and proper use, can be learned more quickly but may not provide the same depth of understanding. For this reason, strong foundational skills in manual manipulation are generally considered essential before relying on advanced tools.

In terms of security and lock design, both approaches have driven innovation. High-security locks are often designed to resist skilled manipulation by introducing false feedback, tight tolerances, and multiple alignment requirements. At the same time, they are also engineered to resist tool-assisted attacks by strengthening components, shielding critical areas, and eliminating known bypass vulnerabilities. Understanding both human and tool-assisted entry methods allows locksmiths to better evaluate the strengths and weaknesses of different locking systems.

Finally, from an ethical standpoint, the distinction reinforces the principles of responsible locksmithing. Ethical practitioners prioritise non-destructive, controlled methods and use tools appropriately within legal and professional boundaries. The goal is not simply to gain entry, but to do so in a way that preserves security, minimises damage, and maintains trust.

### Impact on Locksmithing Practices

Understanding the distinction between human skill and tool-assisted entry is highly relevant to anyone interested in locksmithing because it defines not only *how locks are opened*, but also *what it means to be a competent, ethical, and professionally credible locksmith*. This distinction directly shapes training pathways, day-to-day work practices, forensic awareness, and even legal responsibility.

From a skills development perspective, this concept establishes the foundation of locksmithing as a craft based on sensory interpretation and mechanical understanding, rather than simply tool usage. Human skill-based manipulation requires the practitioner to interpret tactile, vibrational, and auditory feedback to understand the internal state of the lock. This develops a deep, transferable knowledge of lock behaviour, tolerances, and failure points. As a result, locksmiths who prioritise manual skill tend to have a far greater ability to diagnose faults, select appropriate entry methods, and work across a wide range of lock types. In contrast, over-reliance on tool-assisted methods can limit this deeper understanding, as the tool performs much of the work that would otherwise build technical insight.

This distinction also has a direct influence on professional locksmithing practices, particularly in relation to non-destructive entry. Industry best practice strongly favours techniques that preserve the integrity of the lock wherever possible. Human skill-based manipulation aligns with this objective, as it allows controlled, precise entry with minimal or no damage. This is especially important in residential, commercial, and automotive contexts where maintaining the functionality of the lock is critical. Tool-assisted methods, while often faster, may introduce unnecessary wear, internal disturbance, or visible damage, which can increase costs, reduce security, and negatively impact client trust. As such, skilled locksmiths typically treat tool-assisted entry as a secondary option, used only when appropriate.

The relevance becomes even more significant in forensic locksmithing and investigative work. The ability to distinguish between human skill-based entry and tool-assisted entry is essential when analysing evidence. Each method leaves different types of marks and patterns, and without an understanding of both, a locksmith cannot reliably interpret how entry was achieved. This has implications for insurance investigations, legal disputes, and criminal cases. A locksmith with strong manual manipulation skills is better equipped to recognise subtle micro-markings, differentiate between genuine covert entry and staged damage, and provide defensible, evidence-based conclusions. This elevates locksmithing from a trade to a forensically accountable profession.

There is also a strong connection to legal and ethical practice. As outlined in global legislation, the legality of lock picking tools is often determined by intent. A locksmith who can demonstrate that their work is based on controlled, professional, and non-destructive techniques is better positioned to justify their actions as legitimate. Ethical principles—such as only working on locks with permission and avoiding locks in active use—are reinforced by a skill-based approach that prioritises care, precision, and accountability. Tool misuse or reliance on forceful methods can raise questions about intent, professionalism, and compliance with legal standards.

In terms of training and professional development, this distinction influences how locksmiths are educated and assessed. High-quality training programs emphasise the development of tactile sensitivity, feedback interpretation, and mechanical reasoning before introducing advanced tools. This ensures that practitioners understand *why* a lock opens, not just *how* to open it. Such an approach produces more adaptable, competent professionals who can respond effectively to unfamiliar systems, high-security locks, and complex access scenarios.

The distinction has a broader impact on security awareness and lock design evaluation. By understanding both human skill-based and tool-assisted entry methods, locksmiths can better assess the vulnerabilities of different locking systems. This knowledge informs recommendations to clients, supports the selection of appropriate security hardware, and contributes to improving overall security standards. It also helps locksmiths stay aligned with evolving threats, as modern lock design increasingly aims to resist both skilled manipulation and specialised tool attacks.

## Training the "Feel" of Locks

Training the "feel" of locks is the process of developing the sensory and cognitive ability to interpret what is happening inside a lock through touch, movement, and subtle resistance. In the science of lock manipulation, this "feel" is not instinctive, it is a learned skill that combines tactile sensitivity, mechanical understanding, and controlled technique. It allows a locksmith to move beyond simply using tools and instead interact with the lock as a responsive mechanical system.

At its core, training this skill begins with developing awareness of resistance and movement. When torque is applied to a lock, internal components such as pins, wafers, or levers begin to bind due to slight manufacturing tolerances. This binding creates feedback that is transmitted through the tension tool and pick. A learner must train their fingers to detect these very small differences in resistance—distinguishing between a free-moving component, a binding component, and one that has reached its correct position. This requires slow, deliberate manipulation rather than force or speed, as excessive pressure masks the very signals the operator is trying to learn.

*Figure 59: A transparent padlock for lock-picking practice, with the key inserted and being turned. Simon Speed, CC0, via Wikimedia Commons.*

A critical part of this training is mastering tension control, as tension directly influences the clarity of feedback. Too much tension will lock components in place and eliminate subtle signals, while too little tension will prevent binding from occurring at all. Effective training involves experimenting with minimal tension—just enough to create binding—while observing how slight changes in pressure affect what can be felt. Over time, the operator learns to apply consistent, controlled torque that allows the lock to "communicate" clearly.

Another essential element is learning to identify and follow the binding sequence. Because no lock is perfectly manufactured, components will bind in a predictable order. Training the "feel" involves probing each component, identifying which one offers the most resistance, and focusing on it until it sets. Once set, the next component begins to bind, and the process repeats. This teaches the operator to prioritise feedback rather than act randomly, building a structured approach to manipulation.

Equally important is recognising the sensation of a correct set. This may be experienced as a subtle click, a slight movement of the plug, or a reduction in resistance. These signals are often very faint, especially in higher-quality locks, and require repeated practice to recognise consistently. By resetting the lock and repeating the same actions, the operator develops muscle memory and a mental reference for what a "correct" set feels like compared to incorrect movement or oversetting.

As skill develops, training expands to include variation and complexity. Different lock types and security features produce different forms of feedback. For example, spool or serrated pins create false sets and counter-rotation, which can mislead the operator. Learning to interpret these deceptive signals is a key part of refining the "feel," as it teaches the practitioner to differentiate between genuine progress and false feedback. This stage moves the learner from basic sensation recognition to deeper mechanical interpretation.

Consistency and repetition are fundamental to the process. Training the "feel" is similar to developing fine motor skills in any technical discipline—it requires repeated exposure under controlled conditions. Working with the same lock multiple times allows the operator to become familiar with its behaviour, making it easier to detect subtle changes. Over time, this familiarity extends to new locks, enabling quicker and more accurate interpretation of feedback across different mechanisms.

There is also a physical component to this skill. Finger sensitivity, grip control, and relaxation all influence how well feedback is perceived. A tense grip or rigid hand can dampen sensitivity, while a relaxed, controlled hold allows finer detection of movement and resistance. Practising with consistent tools and maintaining proper hand positioning helps build stability and precision, which are essential for reliable feedback interpretation.

Finally, training the "feel" of locks has important implications beyond manipulation itself. It enhances a locksmith's ability to diagnose faults, understand lock design, and recognise signs of wear or tampering. In forensic contexts, it also supports the interpretation of tool marks and manipulation evidence, as the practitioner understands how different techniques interact with internal components.

Training the "feel" of locks is a disciplined process of developing tactile awareness, tension control, and mechanical insight. Through slow, deliberate practice and repeated exposure, the operator learns to interpret subtle signals and respond with precision. This skill represents a fundamental transition from simply using tools to truly understanding and interacting with the internal behaviour of locking systems.

## Impact on Locksmithing Practices

Training the “feel” of locks is highly relevant to anyone interested in locksmithing because it represents the transition from basic tool use to true technical mastery. Locksmithing is not simply about inserting tools into a lock—it is about understanding and interpreting the internal mechanical behaviour of that lock without direct visibility. Developing this sensory skill allows a practitioner to “read” the lock, making informed decisions rather than relying on guesswork or force. This distinction is what separates a novice from a skilled locksmith.

From a practical perspective, this skill directly influences how locksmiths approach entry techniques. A well-trained sense of feel enables controlled, non-destructive entry, which is a core expectation in professional locksmithing. Rather than forcing a lock or relying on invasive methods, the locksmith can manipulate components precisely, preserving the integrity of the lock and surrounding hardware. This is especially important in real-world situations where damage can result in additional costs, security risks, or liability for the practitioner.

The development of tactile awareness also shapes tool handling and technique. Locksmiths who have trained their sense of feel apply minimal, consistent tension and use deliberate, measured movements. This reduces wear on both tools and locks, improves success rates, and enhances efficiency over time. In contrast, a lack of tactile skill often leads to excessive force, poor control, and increased likelihood of damaging internal components. As a result, training the “feel” of locks fundamentally changes how tools are used—from blunt instruments to precise extensions of the operator’s senses.

This skill also has a strong influence on diagnostic capability. Locksmiths frequently encounter locks that are worn, damaged, or malfunctioning. A refined sense of feel allows them to detect irregularities such as binding issues, broken springs, misaligned components, or foreign debris inside the mechanism. Instead of relying solely on visual inspection or disassembly, the locksmith can often diagnose problems through manipulation alone. This improves efficiency and supports more accurate repair or replacement decisions.

In terms of security assessment and lock selection, understanding how locks “feel” during manipulation provides insight into their strengths and weaknesses. Locksmiths who have developed this skill are better equipped to evaluate different lock types, identify vulnerabilities, and recommend appropriate security upgrades. For example, they can recognise how certain designs produce clear or misleading feedback, and therefore advise on high-security options that resist manipulation. This knowledge directly informs professional recommendations and enhances the quality of service provided to clients.

The relevance extends further into forensic locksmithing and investigative work. Training the “feel” of locks gives practitioners a deeper understanding of how different manipulation techniques interact with internal components. This allows them to interpret tool marks, distinguish between skilled and crude entry methods, and identify signs of covert access. Without this experiential knowledge, forensic conclusions may lack accuracy or credibility. Therefore, tactile skill not only supports entry techniques but also underpins evidence-based analysis.

There is also an important connection to professional standards and reputation. Clients expect locksmiths to work efficiently, minimise damage, and demonstrate expertise. A locksmith who can open a lock cleanly and precisely—without unnecessary force—builds trust and credibility. Conversely, repeated reliance on destructive methods or poor technique can damage both locks and professional standing. Training the "feel" of locks therefore contributes directly to the perception of competence and professionalism within the industry.

Finally, this skill influences long-term professional development. It encourages a disciplined, methodical approach to learning, where progress is based on sensitivity, control, and understanding rather than speed or shortcuts. This mindset carries over into other areas of locksmithing, including installation, maintenance, and security consulting. As a result, training the "feel" of locks is not just a technical exercise—it shapes how a locksmith thinks, works, and evolves within the profession.

## Ethical Boundaries and Legal Implications

Within the global locksport and ethical lock picking community, two fundamental rules are consistently recognised. The first is that a lock should never be picked unless it is owned by the individual or explicit permission has been granted by the owner [189]. This principle reinforces respect for property, legality, and professional ethics. The second rule is that a lock should never be picked while it is in active use [189].

The importance of the second rule lies in the nature of lock manipulation itself. Picking a lock involves exploiting mechanical tolerances and weaknesses within the design, and even when performed carefully, there is always a risk of causing damage. Internal components such as pins, springs, or wafers may be worn, distorted, or broken during manipulation. For this reason, practising on locks that are actively securing doors, vehicles, or other critical assets is strongly discouraged, as failure could result in loss of access or the need for replacement [189].

Safe and responsible practice requires the use of dedicated training locks or spare cylinders that are not installed in any operational setting. Under no circumstances should individuals practise on their own home entry locks, windows, vehicles, or any system that is relied upon for security. In situations such as accidental lockouts, the appropriate course of action is to contact a qualified locksmith rather than attempting to manipulate the lock [189].

In addition to these core principles, various locksport organisations publish broader codes of conduct that promote legality, safety, and ethical behaviour. Adhering to these guidelines helps ensure that lock picking remains a legitimate educational and technical activity, while maintaining public trust and protecting the reputation of the locksmithing profession [189].

Lock picking sits in a legally nuanced space worldwide. In most jurisdictions, the tools themselves are not outright illegal, but their possession, use, or intent can bring a person into conflict with criminal law.

## Beyond the Key

The key distinction across nearly all countries is intent, whether tools are held for legitimate purposes (e.g. locksmithing, locksport, training) or for unlawful entry.

The following is provided for general educational purposes only and does not constitute legal advice regarding the possession or use of lock picking tools. Laws relating to lock picks and lock manipulation can vary by jurisdiction, and individuals should ensure they understand and comply with all applicable legislation in their area.

### Australia

In Australia, lock picks are generally legal to own, but are classified as *"controlled" or "prohibited implements"* depending on the state or territory.

- Under laws such as the Crimes Act 1900 (NSW) and equivalent legislation in other states, possessing tools "for the purpose of committing an offence" is illegal.
- Police may charge a person if they cannot provide a reasonable lawful excuse for possession.
- Locksmiths, trainees, and hobbyists (locksport) typically rely on:
    - Employment or training justification
    - Membership of recognised associations
    - Demonstrable legitimate use

**Key point:** Ownership is not the issue—intent and circumstances determine legality.

### United States

In the U.S., laws vary by state, but a common legal concept is "possession of burglary tools."

- Many states (e.g., California, Texas, New York) allow possession of lock picks.
- However, under statutes such as New York Penal Law §140.35, it is illegal to possess such tools with intent to use them unlawfully.
- Some states (e.g., Nevada) require locksmith licensing.
- Locksport is widely accepted, and organisations like TOOOL (The Open Organisation of Lockpickers) promote ethical use.

**Key point:** Legal to own in most states, but intent to commit a crime makes possession illegal.

### United Kingdom

In the UK, lock picks are legal to own, but tightly regulated through intent-based laws.

- Under the Theft Act 1968 (Section 25):
    - It is an offence to have tools "for use in the course of burglary or theft."
- There is no licensing requirement to own picks, but:
    - Carrying them in public without a clear reason may lead to suspicion or arrest.

**Key point:** Legal possession, but very strict on inferred intent.

Germany

Germany is relatively liberal regarding lock picking.

- Lock picks are legal to own and purchase.
- Locksport is widely accepted, especially through groups like the Chaos Computer Club.
- However, using tools for unlawful entry is prosecuted under German Criminal Code (Strafgesetzbuch – StGB).

**Key point:** Legal and culturally accepted, but criminal use is strictly penalised.

New Zealand

In New Zealand, the law is similar to Australia.

- Under the Crimes Act 1961 (NZ):
  - Possession of tools is not illegal by itself.
  - It becomes an offence if tools are held without lawful excuse and with intent to commit burglary.
- Locksmiths and trainees can justify possession through legitimate purpose.

**Key point:** Legal possession, but burden of lawful excuse applies.

Canada

Canada follows a similar legal framework.

- Under the Criminal Code of Canada (Section 351):
  - It is illegal to possess tools for the purpose of breaking into a place.
- Lock picks are legal if there is no criminal intent.
- Locksmith licensing varies by province.

**Key point:** Legal tools, but intent defines legality.

Other Regions (General Trends)

**Europe (General)**

- Most EU countries follow the **intent-based model**.
- Lock picks are legal, but unlawful use is criminal.
- Some countries require locksmith licensing (e.g., France, Italy).

**Asia**

- Varies significantly:
    - **Japan**: Strict—possession of lock picks without justification is illegal under the *Act on Prohibition of Possession of Special Unlocking Tools*.
    - **Singapore**: Highly restrictive; possession may be illegal without clear reason.
    - **India**: Generally legal, but misuse prosecuted under criminal law.

**Middle East**

- Often stricter:
    - Possession may require justification or licensing.
    - Strong penalties for misuse.

Lock picking can be a legitimate technical skill, but it exists in a highly regulated legal environment. The difference between lawful practice and criminal behaviour is not the tool itself—but intent, context, and conduct.

## Impact on Locksmithing Practices

Understanding lock picking legality and ethical practice is fundamental for anyone interested in locksmithing because it defines the boundaries between legitimate professional activity and criminal behaviour. Locksmithing is not just a technical trade—it is a regulated, trust-based profession that operates within strict legal and ethical frameworks. Without a clear understanding of these rules, even technically skilled individuals can inadvertently place themselves at legal risk or damage their professional reputation.

At its core, the relevance lies in intent and accountability. Across most jurisdictions, lock picking tools are not inherently illegal; rather, it is the purpose for which they are possessed or used that determines legality. For a locksmith, this means that every action must be justifiable as lawful—such as performing authorised entry, training, or security assessment. This directly influences how locksmiths conduct their work, requiring them to maintain clear documentation, obtain client consent, and operate transparently at all times. The profession therefore demands not only technical competence but also a strong awareness of legal context.

The ethical rules commonly promoted within locksport—such as never picking a lock without permission and never picking a lock in use—translate directly into professional practice. These principles reinforce respect for property, client trust, and duty of care. In practical terms, locksmiths must ensure they are authorised before performing any work, and they must avoid actions that could compromise a client's security. For example, attempting to manipulate an in-service lock carries the risk of damage, which could result in lock failure, liability issues, or safety concerns. As a result, professional locksmiths often prioritise non-destructive methods but will also recognise when replacement or alternative techniques are more appropriate.

This legal and ethical framework also shapes training and skill development. Responsible practice requires the use of dedicated training locks rather than live security systems. This not only prevents accidental damage but also reinforces disciplined, controlled learning. For someone entering locksmithing, this establishes good habits early—such as working methodically, respecting equipment, and understanding the consequences of improper technique. It also aligns with broader industry expectations, particularly in regulated environments like Australia, where demonstrating a lawful excuse for tool possession is essential.

From a professional standpoint, adherence to legal and ethical standards directly affects credibility and employability. Locksmiths are often entrusted with access to secure environments, sensitive assets, and personal property. Clients, employers, and regulatory bodies expect high levels of integrity. Any association with unethical behaviour or questionable use of tools can undermine trust and lead to loss of licensing, contracts, or legal standing. Therefore, understanding and following these principles is not optional—it is central to maintaining a viable career in the field.

The influence on locksmithing practices is also evident in risk management and decision-making. Locksmiths must constantly assess whether a method is appropriate, lawful, and safe. For example, in a lockout situation, a professional will verify ownership or authority before proceeding. They will also choose techniques that minimise damage and preserve the lock where possible. In forensic or investigative contexts, ethical practice becomes even more critical, as improper handling or unauthorised actions can compromise evidence and legal outcomes.

Additionally, these principles contribute to the public perception and legitimacy of the profession. Locksmithing exists in a space that can easily be misunderstood due to its association with security bypass techniques. By adhering to clear ethical standards and legal requirements, locksmiths help distinguish their work from criminal activity. This is particularly important in maintaining industry reputation, supporting regulatory compliance, and ensuring that lock picking remains recognised as a legitimate technical and educational discipline rather than a suspicious or illicit activity.

Finally, the global variation in laws reinforces the need for jurisdictional awareness. A practice that is acceptable in one country may be restricted or illegal in another. For locksmiths who travel, train internationally, or engage with global locksport communities, understanding these differences is essential. It encourages a cautious, informed approach where practitioners actively verify local laws and adapt their behaviour accordingly.

## Lock Picking Techniques

### Pick a Pin Tumbler Lock

Pin tumbler locks are among the most common locking mechanisms used on residential doors, making them an important system for locksmiths to understand. While a detailed knowledge of their internal operation is not strictly required for basic interaction, a clear understanding of how they function provides valuable insight into their behaviour, vulnerabilities, and maintenance requirements.

The basic concept of the pin tumbler lock dates back thousands of years, with early forms used in ancient civilisations. The modern version, commonly found in door cylinders today, was refined in the 19th century and remains largely unchanged in principle. Despite advances in materials and security features, the fundamental design, relying on aligned pins to permit rotation, continues to underpin a vast proportion of locking systems in use worldwide.

A typical pin tumbler lock consists of an outer cylindrical housing and an inner rotating plug. The interface between these two components is known as the shear line, which is critical to the lock's operation. Within the plug are a series of vertical chambers containing pairs of pins: key pins (which contact the key) and driver pins (which sit above them), along with springs that apply downward pressure. In the locked state, these pins extend across the shear line, preventing the plug from turning.

When an incorrect key, or no key at all, is present, the differing lengths of the key pins cause the driver pins to obstruct the shear line. This blocks rotation of the plug and keeps the lock secure. When the correct key is inserted, each key pin is lifted to a precise height so that the gap between the key pin and driver pin aligns exactly with the shear line. Once all pin stacks are aligned in this way, the plug is free to rotate, allowing the lock to open.

From a conceptual standpoint, any method of manipulating a pin tumbler lock—whether through a key or other means—relies on achieving this same alignment at the shear line. The key provides a direct and precise method of doing so, while other approaches rely on interpreting mechanical feedback and positioning components individually. The underlying principle, however, remains identical: all pin stacks must be positioned so that nothing obstructs the rotation of the plug.

In professional locksmithing, specialised tools are used for legitimate purposes such as maintenance, servicing, and non-destructive access. These tools typically include devices for applying controlled rotational pressure and instruments designed to interact with internal components. The effectiveness of these tools depends less on their complexity and more on the operator's understanding of the lock's internal mechanics and their ability to apply controlled, precise movements.

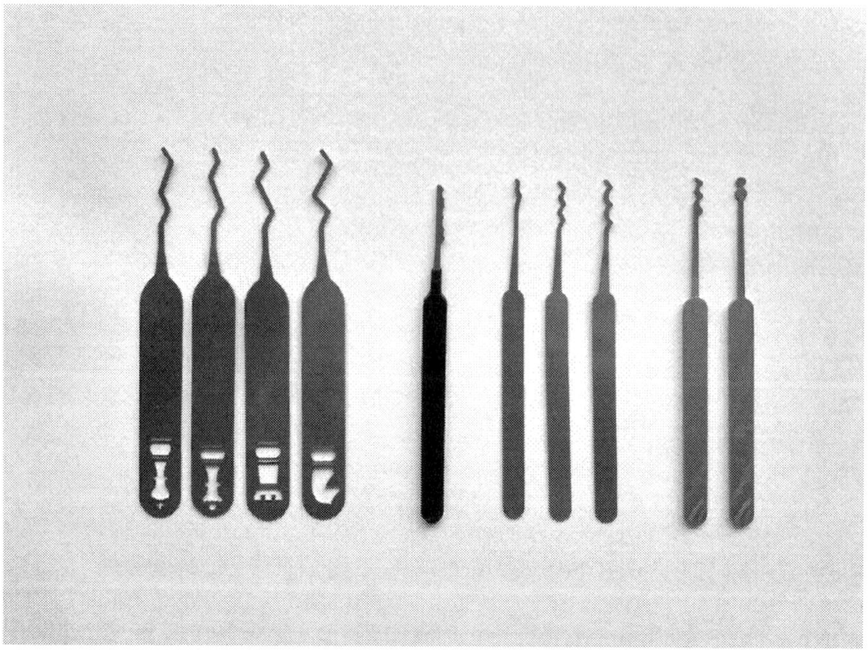

*Figure 60: Rake Lock Picks - From left to right: 1. king rake 2. queen rake 3. rook rake 4. horse rake 5. city rake 6. snake rake 7. worm rake 8. bogota rake 9. W rake 10. snowman rake. Lukkojaska, CC BY-SA 4.0, via Wikimedia Commons.*

Lock manipulation relies on two fundamental actions: applying controlled rotational pressure to the plug and precisely interacting with the internal components within the keyway. To achieve this, specialised tools are used, each designed to perform a specific function. While there are many variations and advanced tools available, a small number of core tools form the foundation of most locksmithing and lock manipulation work [190].

The first essential tool is the tension wrench. This tool is used to apply rotational pressure to the plug, simulating the turning force of a key. Tension tools are available in a range of shapes, sizes, and profiles, with the most common being the pry bar (typically used for top-of-keyway tension) and the L-shaped wrench (commonly used for bottom-of-keyway tension). Selecting the correct tension tool is critical—it must fit securely within the keyway without obstructing access to internal components [190]. Maintaining consistent, controlled tension is equally important, as any loss of tension can reset the internal alignment and undo previous progress.

Another commonly used tool is the rake. Rakes are designed to interact with multiple components simultaneously, using rapid or sweeping movements to create momentary alignment through chance and mechanical interaction. This approach is often used as a fast and simple method in suitable locks. Different rake profiles, such as the city rake or Bogota-style designs, are shaped to maximise contact with internal components and increase the likelihood of successful alignment through movement and vibration [190].

In contrast, hook-style tools are used for more controlled and deliberate interaction. Hooks come in a variety of shapes, including standard and offset designs, and are intended to engage individual components within the lock. The effectiveness of a hook depends on its ability to move freely within the keyway while providing sufficient reach and control. These tools are typically used when a more methodical approach is required, allowing the operator to work with individual components and respond to subtle mechanical feedback [190].

The selection and use of these tools are not just about availability, but about control, precision, and suitability for the specific lock being worked on. A well-chosen tool should allow clear access, minimal obstruction, and smooth movement within the keyway. Ultimately, while tools facilitate interaction with the lock, successful outcomes depend on the operator's understanding of the mechanism and their ability to apply controlled, consistent technique.

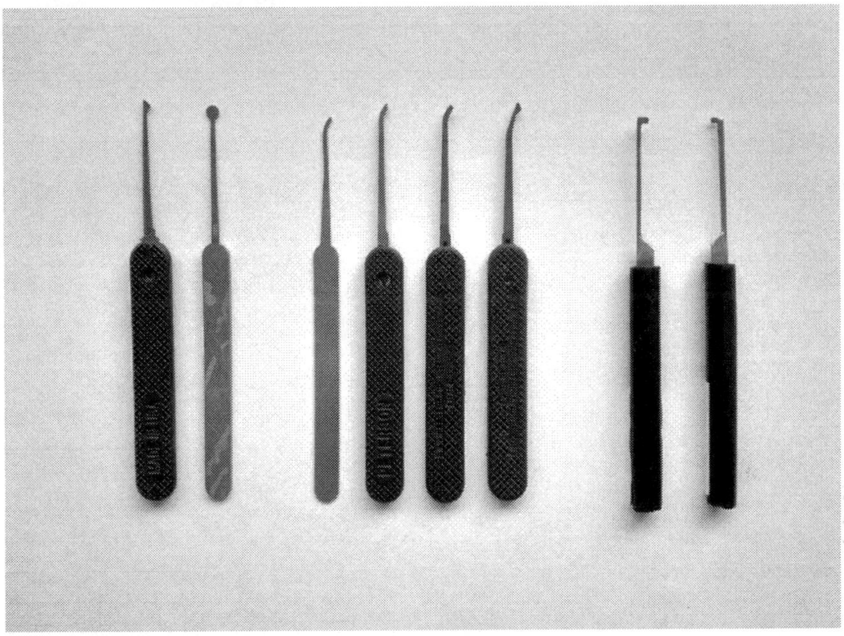

*Figure 61: Single-Pin Picking Lockpicks - From left to right: 1. half diamond 2. ball 3-4. short hooks 5. medium hook 6. deforest half diamond 7-8. flags. Lukkojaska, CC BY-SA 4.0, via Wikimedia Commons.*

The interaction with a lock is often described as requiring a developed "feel." This reflects the importance of interpreting subtle mechanical signals, such as resistance, movement, and changes in pressure, rather than relying on force. Each lock behaves slightly differently due to manufacturing tolerances and wear, and skilled practitioners learn to recognise these variations and respond accordingly. This is why locksmithing is often regarded as both a technical and sensory skill.

It is also important to recognise that not all locks respond identically. While many standard pin tumbler locks operate predictably, others incorporate additional security features that alter internal behaviour and feedback. In such cases, a more methodical and controlled approach is required, with greater emphasis on understanding the specific characteristics of the mechanism being worked on.

Overall, the operation of a pin tumbler lock is relatively simple in principle but highly nuanced in practice. By understanding the relationship between pins, the shear line, and plug rotation, a locksmith develops the foundation needed to work with these systems effectively, whether for installation, maintenance, security assessment, or forensic analysis.

### Tensioning a Lock

Lock picking begins with one fundamental principle: proper tensioning of the plug. Regardless of the technique used, the way tension is applied determines how the internal components behave and whether meaningful feedback can be detected. In simple terms, tension creates the binding condition that allows individual components to be identified and positioned correctly. Without controlled tension, effective manipulation is not possible [190].

**Lock Pick Tension Wrenches**

Effective lock picking begins with correct use of the tension wrench. The tool is inserted into the keyway, commonly at the bottom, and a very light, controlled amount of force is applied in the direction the key would normally turn to unlock the mechanism, typically clockwise. The amount of pressure should be minimal, comparable to the force required to press a key on a keyboard. This light tension is critical, as it creates the conditions for internal components to bind and produce usable feedback [191].

Once tension is applied, the next step is to identify the first binding component within the lock. This is typically the pin that offers the greatest resistance when probed. By gently lifting this binding pin, it can be moved to its correct position within the lock. When this occurs, the operator may feel a slight rotation of the plug or hear a faint "click," both of which indicate that the pin has been successfully set [191].

After setting the first pin, the binding condition transfers to another pin within the lock. The process is then repeated: locate the next pin that resists movement, lift it carefully, and allow it to set. Each correctly set pin contributes incrementally to the rotation of the plug. This sequence continues until all pins have been set to the correct height along the shear line [191].

When all internal components are properly aligned, the plug will rotate fully under the applied tension, and the lock will open. This process relies on maintaining consistent, minimal tension throughout, as excessive force can prevent pins from setting correctly or obscure the feedback needed to identify binding components [191].

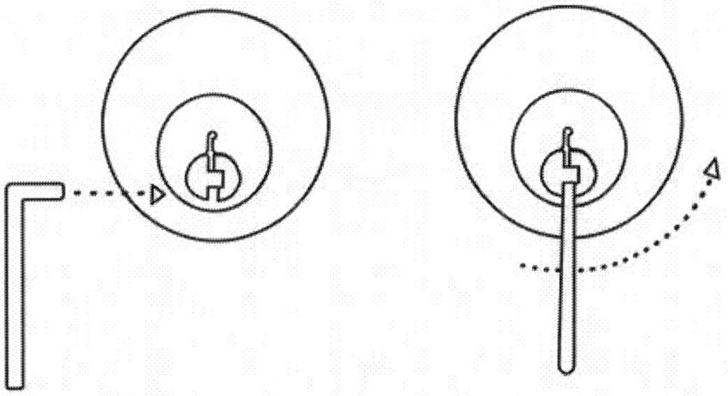

*Figure 62: Using lock pick wrenches.*

There are two primary approaches to applying tension within the keyway: top of the keyway (TOK) and bottom of the keyway (BOK). Top-of-the-keyway tension involves placing the tension tool at the upper portion of the keyway. This method is often preferred by experienced practitioners because it leaves more space within the keyway for tools to move freely, particularly in narrower or more complex profiles. TOK tools are also commonly double-ended, offering versatility and improved control in a range of lock types [190].

Bottom-of-the-keyway tension, on the other hand, is typically the starting point for beginners, as many basic tool sets include these types of tension wrenches. This method involves inserting the tool into the lower part of the keyway. While it can be effective, it may also restrict movement within the keyway and, if not carefully managed, can interfere with internal components. There is also a risk of the tool binding between the plug and the housing, which can reduce sensitivity and make manipulation more difficult [190].

Choosing between TOK and BOK is not simply a matter of preference but of suitability to the lock and the task. While beginners may initially rely on BOK due to accessibility, TOK often provides better control and clearer feedback once the operator develops familiarity with the technique. Understanding the advantages and limitations of each approach is essential for effective practice [190].

Beyond tool placement, the most critical factor in tensioning is the amount of pressure applied. Excessive tension will cause all internal components to bind too tightly, masking the subtle feedback needed for precise manipulation. Conversely, insufficient tension will fail to create the binding necessary for components to remain in position, preventing progress altogether. The goal is to apply the minimum amount of pressure required to produce controlled, detectable resistance [190].

Another common issue is applying tension in the wrong direction. Tension must always be applied in the same direction that the key would turn to operate the lock. If the plug is rotated the wrong way, the internal components may appear to align correctly, but the lock will not open. In such cases, the process must usually be reset and repeated in the correct direction [190].

A practical way to determine the correct direction of tension is to gently test both directions with the tension tool. Typically, the correct direction will feel slightly looser or offer more movement. This subtle difference indicates the natural operating direction of the lock. Applying tension in this direction improves feedback clarity and increases the likelihood of successful manipulation [190].

Ultimately, effective tensioning is a balance of control, sensitivity, and understanding. It is not about force, but about creating the precise conditions that allow the lock to communicate its internal state. Mastery of tension is often considered the most important skill in lock manipulation, as it directly influences every other aspect of the process [190].

### Simple Picking Process - Raking

The simplest approach to lock picking is a technique known as *raking*. This method uses a rake tool to rapidly manipulate multiple pins at once, rather than addressing each component individually. Unlike more advanced techniques, raking does not require a detailed, moment-by-moment understanding of what is happening inside the lock. However, it still requires a basic knowledge of the process, the sequence of actions, and how to adjust when the method is not producing results [190].

Raking works by introducing movement and variation across all pin stacks simultaneously. Instead of carefully setting each pin, the operator relies on repeated motion and probability to momentarily align the pins at the shear line. When combined with correct tension, this can allow the plug to rotate and the lock to open [190].

Apply light, controlled tension using the tension wrench in the direction the key would normally turn. Insert the rake into the keyway and slide it all the way to the back of the lock. Gently lift the rake upward so that it makes contact with the pins, raising them slightly [191].

With tension maintained, begin moving the rake in and out of the keyway using a quick, scrubbing motion, similar to brushing with a toothbrush. As you do this, vary the angle, height, and speed of the rake to create different contact patterns with the pins. This variation increases the likelihood of aligning multiple pins at the correct height [191].

If the lock does not open within approximately 10–20 seconds, release the tension completely to allow the pins to reset. Then reapply tension and repeat the process. It is common to reset and attempt the process multiple times before achieving success, as small adjustments in tension and movement can significantly affect the outcome [191].

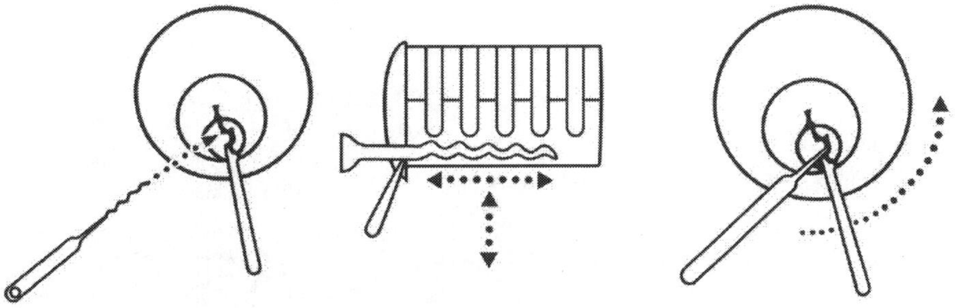

*Figure 63: Using a rake.*

The process begins by inserting a tension wrench into the keyway and applying light, controlled pressure in the correct direction, typically the same direction the key would turn. Once tension is applied, the rake is inserted fully into the keyway. The operator then moves the rake in and out while simultaneously lifting and lowering it, creating a scrubbing or sweeping motion across the pins. This action is repeated continuously for a short period, usually no more than a minute or two, while maintaining consistent tension [190].

During this process, the operator periodically tests the tension wrench to see if the plug has rotated and the lock has opened. If the lock does not open, the tension should be released, adjusted slightly—either increased or decreased—and the process repeated. Small changes in tension can significantly affect the outcome, as they influence how the pins bind and set [190].

Although raking is relatively simple and requires less precision than other methods, it is not without its limitations. One of the most common mistakes is applying incorrect tension, either too much, which prevents pins from moving, or too little, which stops them from setting. Another limitation is that not all locks will respond to raking. Locks with tighter tolerances or security features may require a more controlled, methodical approach, such as single pin picking [190].

In essence, raking is an efficient and accessible entry technique that relies on speed and probability rather than precision. While it can be highly effective on simpler locks, it also highlights the importance of understanding when to adapt technique and move to more advanced methods [190].

<u>Single pin picking (SPP)</u>

Single pin picking (SPP) is considered the more advanced and precise method of lock manipulation. Unlike raking, which relies on moving multiple pins at once, SPP involves manipulating each pin individually using a hook-style pick. This approach requires a deeper understanding of how the lock functions internally, as well as the ability to interpret subtle feedback during the process [190].

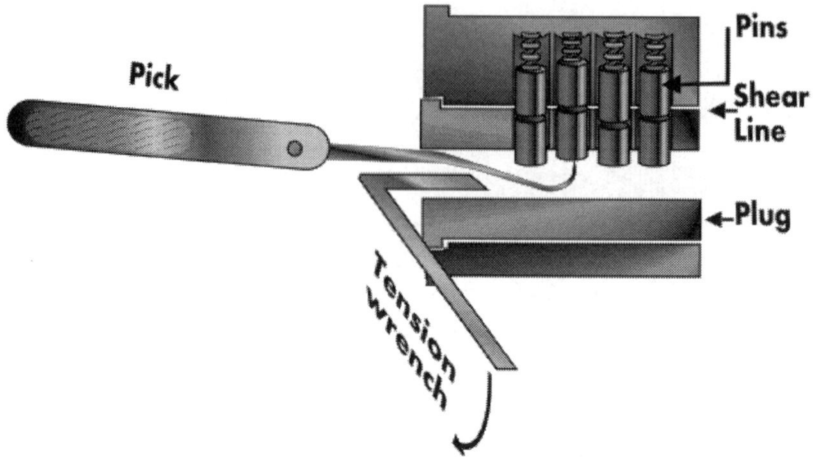


*Figure 64: Pin and tumbler lock picking. Oldie~commonswiki, CC BY-SA 3.0. via Wikimedia Commons.*

When performing single pin picking, the operator works through each pin stack one at a time, identifying and setting them in the correct sequence. This requires more than just a basic knowledge of lock components, it involves understanding how the lock behaves under tension at that exact moment. A critical part of this method is the ability to interpret feedback. The lock effectively "communicates" through resistance, movement, and changes in spring tension, and the operator must learn to recognise and respond to these signals [190].

The process typically begins by inserting a tension wrench into the keyway and applying light, controlled torque in the correct direction. A hook pick is then inserted, and the operator begins probing the pins, usually starting from the back of the lock and working forward (or vice versa, depending on preference). Each pin is gently tested to determine how it responds. Pins that feel loose or springy are generally ignored at first, as they are not yet binding [190].

*Figure 65: Using picking tools. Arkalonso, CC BY-SA 4.0, via Wikimedia Commons.*

Attention is then focused on the pin that offers the most resistance, this is the binding pin. By carefully lifting this pin while slightly adjusting tension, the operator attempts to raise it to the correct position at the shear line. When successful, this is often indicated by a subtle click, a slight rotation of the plug, or a noticeable reduction in resistance. Once a pin is set, it should no longer resist movement, and the next binding pin in the sequence will begin to stand out [190].

This process is repeated, pin by pin, following the natural binding order of the lock until all pins are correctly set and the plug is free to rotate. At that point, the tension wrench can be turned fully to open the lock. If the lock does not open, the operator must release tension, adjust it slightly, and begin again, often learning something new about the lock's behaviour with each attempt [190].

The most challenging aspect of single pin picking is not the physical movement of the tools, but the interpretation of feedback. Without the ability to feel and correctly interpret what is happening inside the lock, the method becomes ineffective. SPP requires a combination of theoretical understanding and practical sensitivity, where knowledge is translated into precise, controlled action [190].

Developing this skill takes time and consistent practice. Progress is not always linear, there may be moments of rapid improvement followed by periods of difficulty. However, by remaining patient, focusing on fundamentals such as tension control and feedback interpretation, and practising regularly, anyone can develop the ability to pick locks using this method with a high degree of control and understanding [190].

Picking Locks with Security Pins

Locks that contain security pins introduce an additional level of complexity to lock manipulation. Unlike standard pins, security pins are specifically designed to disrupt normal feedback and make picking more difficult. As a result, it is important to recognise when a lock is not opening due to the presence of these features. One of the key indicators is the sensation of a "false set," where the lock appears to be partially picked, but the plug will not fully rotate [190].

A false set occurs because security pins can catch between the plug and the housing (bible), rather than clearing the shear line completely. To the operator, this can feel very similar to a correctly set pin, as the usual spring tension may seem to disappear. However, the lock remains jammed internally. This misleading feedback is what makes security pins effective, they create the illusion of progress while preventing the lock from opening [190].

The likelihood and intensity of false sets can be increased by additional design features such as threaded or counter-milled pin chambers. These modifications create extra ledges inside the lock, giving security pins more opportunities to catch. Although the lock often provides subtle clues about what is happening, these signals can be easily missed without experience or an understanding of how different security pins behave [190].

Spool pins are among the most common types of security pins. They are shaped with a narrow middle section and wider ends, resembling a spool. When a spool pin catches at the shear line, it allows the plug to rotate slightly, creating a noticeable false set. This rotation is often more pronounced than with standard pins. To set a spool pin correctly, the operator must reduce tension slightly and carefully lift the pin, allowing it to pass the ledge and reach the true shear line. This process often produces a sensation of counter-rotation, where the plug briefly moves in the opposite direction [190].

Mushroom pins function in a similar way to spool pins but have a more gradual shape. Instead of a sharply defined narrow section, they taper toward the top, creating a smoother transition. As a result, the feedback they produce is less abrupt, and the rotation of the plug tends to occur more gradually. Despite these differences, the approach to setting mushroom pins is essentially the same as with spool pins, lightening tension and carefully guiding the pin past the obstruction [190].

Serrated pins present a different challenge altogether. These pins have multiple grooves or ridges along their length, which can create several points where the pin catches. Unlike spool or mushroom pins, serrated pins may produce feedback that closely resembles a normal pin setting, making them difficult to identify. They can also create the impression that other pins are binding when they are not, leading to confusion in the picking process. When progress seems inconsistent or misleading, it is often a sign that serrated pins are present. In these cases, the operator must reduce tension and work the pins with greater precision [190].

Because of the deceptive feedback created by security pins, techniques such as raking are often less effective. While raking may bring the lock into a false set, it rarely completes the process. At that point, single pin picking becomes necessary, allowing each pin to be addressed individually. This highlights the importance of controlled, deliberate manipulation when dealing with higher-security locks [190].

When working with security pins, careful tension control is critical. Tension should never be completely released, as this will reset the lock and undo progress. Instead, the operator must ease off tension just enough to allow the pin to move past the obstruction while still maintaining control. Reducing tension too much can cause previously set pins to drop, requiring the process to be repeated. Success with security pins therefore depends on patience, precision, and a clear understanding of how these components alter feedback within the lock [190].

Other Methods for Picking Locks

Lock picking can be broadly defined as the manipulation of a lock's internal components without the use of the correct key. While this definition captures the core idea, the reality is far more complex. Beyond basic methods such as raking or single pin picking, there are additional techniques that rely on different principles, tools, and levels of understanding. These methods highlight that lock manipulation is not a single skill, but a spectrum of approaches with varying degrees of precision, control, and effectiveness [190].

One such method involves the use of pick guns. These tools, whether manual or electric, operate on a principle similar to raking. A tension wrench is still required, as torque is essential for creating the binding conditions needed for pins to set. The pick gun rapidly strikes or "bumps" the pins upward while the tool is moved in and out of the keyway. This semi-automated action aims to set multiple pins quickly, increasing the chance of opening the lock. However, the effectiveness of pick guns depends on correct adjustment—if the pins are not lifted high enough, they will not set, and if they are lifted too far, they may overset. Because this method treats all pins simultaneously rather than individually, it is generally less effective on locks that incorporate security pins [190].

Bump keys represent another technique that relies on kinetic energy rather than precise manipulation. A bump key is specially cut so that all cuts are at the lowest possible depth for a specific keyway. When inserted into the lock and struck (typically with a light impact), the force transfers through the key to the pins, momentarily separating the driver pins from the key pins. If torque is applied at the same time, the plug can rotate during this brief window when the shear line is clear. Although this method can be effective on standard pin-tumbler locks, it requires a correctly matched bump key for the specific lock. Additionally, many modern locks are designed to resist bumping, and the technique is ineffective on non–pin-tumbler systems [190].

*Figure 66: 16 piece bumpkey set plus bonus weighted hammer. Cripton911, CC0, via Wikimedia Commons.*

Bypass techniques differ fundamentally from traditional lock picking. Rather than manipulating the components that interact with a key, bypass methods aim to circumvent the lock mechanism entirely by directly engaging the unlocking components, such as the cam, actuator, or bolt. These techniques vary widely depending on the lock design and often require specific knowledge of vulnerabilities within a particular model. In some cases, bypass methods can render even high-security features irrelevant, as they avoid the need to interact with pins or other security elements altogether [190].

The existence of these alternative methods demonstrates that lock picking cannot be reduced to a small set of simple procedures. Each approach, whether based on manual skill, kinetic force, or mechanical bypass, introduces different challenges and considerations. Understanding these methods is important not only for developing technical capability, but also for appreciating the broader complexity of lock design, security vulnerabilities, and the ongoing evolution of both attack and defence in locksmithing [190].

<u>Lock Style and Orientation</u>

The way a lock is installed or positioned has a significant impact on how it is approached during manipulation. Although padlocks, door cylinders, and deadbolts may all use similar internal mechanisms—such as pin tumbler systems—their external configuration and mounting conditions introduce different practical challenges. Understanding these differences is important for developing adaptable locksmithing skills [190].

**Padlocks:** Padlocks are often free-hanging, typically secured on chains or hasps, which means they are usually picked while being held in the hand. This introduces an additional layer of difficulty, as the operator must stabilise both the lock and the tools simultaneously. Developing a comfortable grip and consistent tool control takes practice. Unlike fixed locks, padlocks may move or rotate during manipulation, reducing feedback clarity. In many cases, padlocks cannot be easily disassembled, limiting the ability to access or isolate the core, and without a vice or mounting device, maintaining stability can be challenging [190].

**Exposed Core:** Some practitioners choose to remove the lock core from its housing, particularly in padlocks or certain deadbolts. This exposes the plug and bible without the surrounding hardware, allowing a clearer focus on the internal mechanics. Working with an exposed core can be useful for training, as it reduces external variables and allows the learner to concentrate on feedback, tension control, and pin interaction. However, this approach does not fully replicate real-world conditions, where restricted access, keyway orientation, and housing constraints affect technique. As a result, exposed cores are best used as a stepping stone before progressing to fully assembled locks [190].

**Mounted Locks:** Locks that are fixed in place—such as door knobs and deadbolts—offer greater stability during manipulation. This can make the process more controlled, as the lock effectively acts as though it is secured in a vice. For training purposes, mounting practice locks in timber boards or simulated door setups provides a realistic working environment. However, orientation becomes an important factor. Locks installed upside down or at unusual angles can disrupt familiar techniques, particularly when applying top-of-keyway (TOK) or bottom-of-keyway (BOK) tension. Adapting to different orientations is an essential skill, as real-world locksmithing rarely presents ideal conditions [190].

Overall, lock style and orientation influence stability, access, and feedback. Developing competence across different configurations ensures that a locksmith can apply consistent techniques regardless of how or where a lock is installed.

### Lever Locks

Lever locks are often considered the next step for those progressing beyond pin tumbler locks, both because of their widespread use and the additional level of skill they require. After pin cylinders, lever locks are among the most common locking mechanisms, particularly in residential doors and certain types of padlocks. Their origins can be traced back to 1778, when Robert Barron introduced the "double lever lock" [192]. Over time, manufacturers such as Chubb refined and popularised the design, to the point where many people now refer to lever locks simply as "Chubb locks" [192].

The operating principle of a lever lock is based on a series of internal levers, each containing a cut-out section known as a gate. For the lock to open, each lever must be lifted to a precise height so that all gates align. When the correct key is inserted and turned, its varying cut depths raise each lever to the required position. Once aligned, these gates create a clear path for the bolt stump to move. As the key continues to turn, a final section of the key, often referred to as the bolt thrower, drives the bolt. With the gates aligned, the bolt retracts smoothly into the lock body, allowing the door to open [192].

This principle remains consistent regardless of how many levers are present. Simpler internal locks may contain two or three levers, while external door locks, such as mortice locks, commonly contain five or more for increased security. Lever padlocks also follow the same fundamental design, though their construction may vary depending on application and security level [192].

In terms of manipulation, basic lever locks can sometimes be operated using simple tools, such as bent wires. One tool is used to apply tension to the bolt stump, while another is used to lift the levers individually. Applying tension creates a slight internal ledge, similar in concept to the shear line in pin tumbler locks, which allows each correctly positioned lever to hold its place. The operator then lifts each lever until it rests on this ledge, repeating the process until all levers are aligned and the lock opens [192].

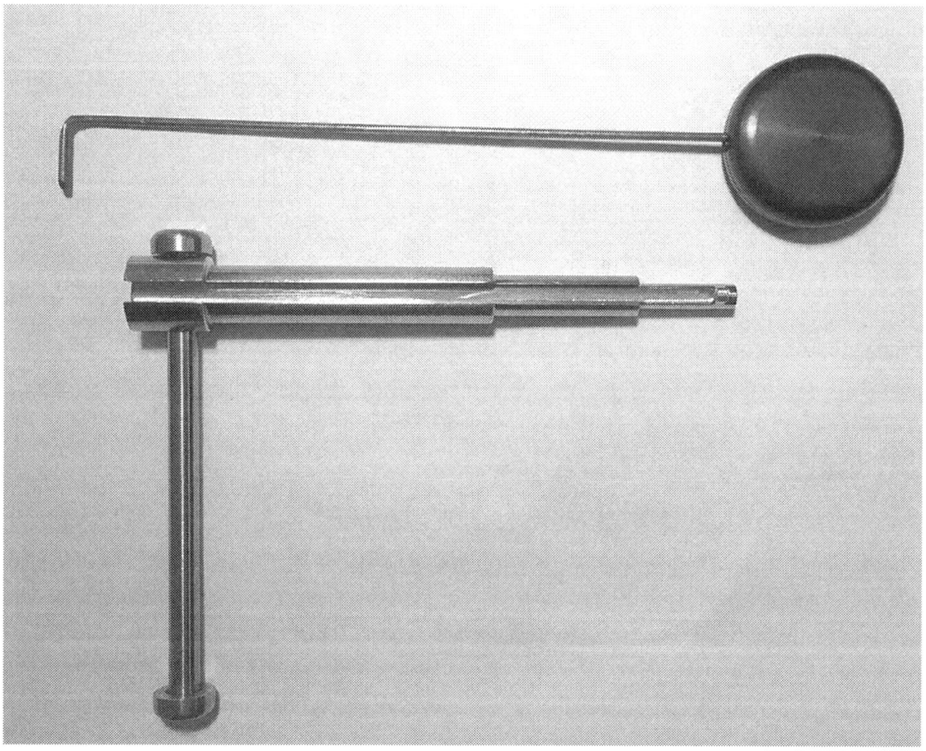

*Figure 67: Lever Lock Pick Tool. Willh26, CC BY-SA 4.0, via Wikimedia Commons.*

However, modern lever locks often include additional security features that make this process more complex. For example, alternating high and low gates can mislead the operator, and many mortice locks incorporate a component known as a curtain. This rotating disc restricts direct access to the levers, preventing the use of simple tools. To work around this, specialised tools such as curtain picks are used. These tools are designed to navigate past the curtain, apply tension to the bolt stump, and allow individual manipulation of the levers [192].

For lever locks without such protective features, tools known as "2-in-1 picks" are commonly used. These combine tensioning and lever manipulation into a single instrument and are available in different gauges to match the dimensions of specific locks. More advanced lever locks and high-security padlocks often require purpose-built tools designed for particular models, reflecting the increased complexity and precision of these mechanisms [192].

In practice, learning to work with lever locks ranges from basic experimentation with simple mechanisms to the use of specialised tools for more secure designs. Improvised tools may occasionally be effective, but they carry a higher risk of damaging the lock or the tool itself. For this reason, professional locksmithing typically relies on purpose-designed equipment that provides greater control, precision, and consistency when interacting with lever-based locking systems [192].

### Wafer Locks

Wafer locks, while not as widely recognised as pin tumbler or lever locks, are extremely important due to their widespread use—particularly in the automotive industry. Millions of vehicles rely on wafer lock mechanisms, and they are also commonly found in everyday applications such as drawers, lockers, cabinets, and some padlocks. For anyone involved in locksmithing, understanding wafer locks is essential because of their prevalence and practical significance [192].

The wafer lock was first introduced in 1868 through a patent by Philo Felter. His design featured a double-bitted key system, meaning the key had cuts on both edges and operated two sets of wafers within the lock. Shortly after, in 1871, Hiram S. Shepardson developed a single-sided wafer lock. This simplified design used only one set of wafers and proved to be more cost-effective and durable, leading to widespread commercial adoption [192]. The single-sided wafer lock remains the most common type found in applications such as lockers and cabinets [192].

Functionally, wafer locks operate on principles similar to pin tumbler locks. Instead of pins, however, they use flat metal components called wafers. These wafers are spring-loaded and extend into the housing of the lock, preventing the plug from rotating. When the correct key is inserted, it lifts each wafer to the correct height so that none protrude into the housing. At this point, the wafers align with the shear line, allowing the plug to rotate freely and the lock to open. If a wafer is lifted too high or not enough, it will still obstruct the plug, preventing operation [192].

Because of their structural similarity to pin tumbler locks, wafer locks can be manipulated using many of the same tools and techniques. A tension wrench is used to apply rotational force, while a pick is used to lift each wafer individually into position. This process is often referred to as single wafer picking. In addition, wafer locks are generally more tolerant of raking techniques, where multiple wafers are manipulated simultaneously in a less precise but often effective manner [192].

There are also specialised tools designed specifically for wafer locks. These tools often combine tensioning and manipulation into a single instrument, allowing the operator to apply torque while simultaneously raking or jiggling the wafers. Common examples include lock jigglers and wafer rakes, which rely on rapid, varied movements to position the wafers correctly. As with all manipulation techniques, careful control of tension is critical—excessive force can prevent the wafers from aligning properly or reduce feedback [192].

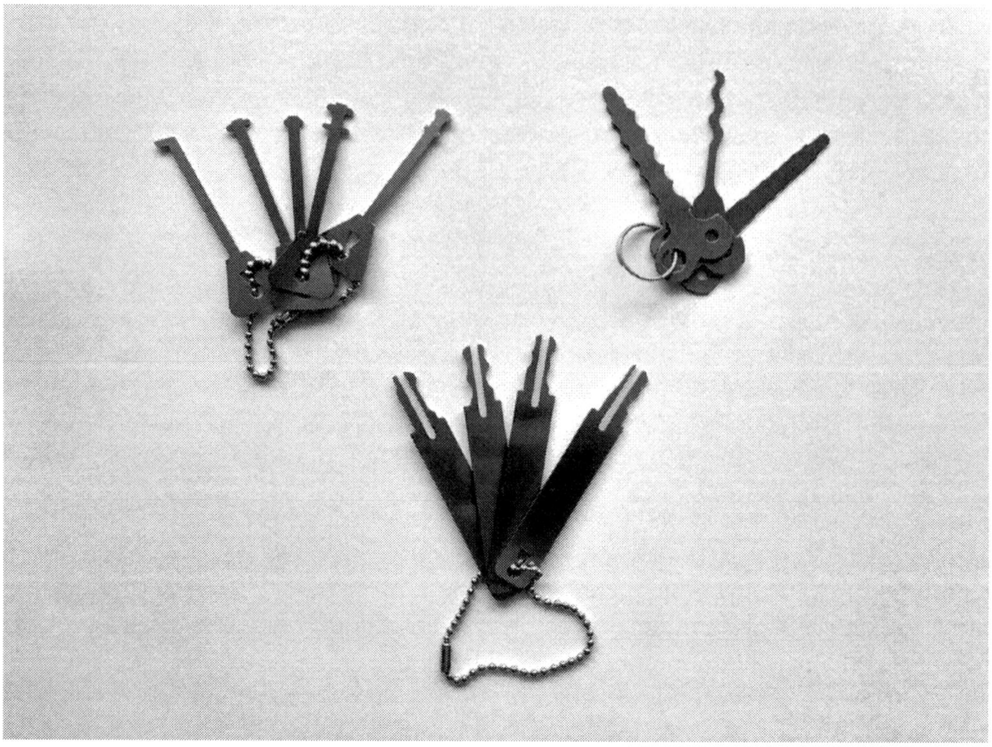

*Figure 68: Jiggler keys and tryout keys used to open weak locks when the matching keys are not available. The key is inserted into the lock and it is jiggled and turned in hopes of getting the lock to open. Since they are not actual keys there is no guarantee that the lock will open. Top left ones are used for warded locks, top right ones for car door wafer locks and bottom ones for filing cabinet wafer locks. Lukkojaska, CC BY-SA 4.0, via Wikimedia Commons.*

For more complex wafer systems, particularly those used in vehicles, highly specialised tools are available. One well-known example is the Lishi pick, which is designed for specific makes and models of locks. These tools incorporate guides that allow the operator to position the picking tip accurately on each wafer, enabling both manipulation and decoding of the lock. Another variation, inner groove picks, use the lock's warding as a guide and interact with multiple wafers simultaneously, functioning similarly to a rake while maintaining controlled tension [192].

Overall, wafer locks represent a practical and widely used locking mechanism that bridges simplicity and functionality. Their design makes them accessible for learning manipulation techniques, while their variations—especially in automotive applications—introduce complexity that requires specialised knowledge and tools. For locksmiths, mastering wafer locks is not only useful but essential for working across a broad range of real-world locking systems [192].

**Warded Locks**

## Beyond the Key

Warded locks are among the oldest known locking mechanisms, with origins dating back to ancient Rome. Despite their long history, they are not widely used in modern security applications. Today, they are most commonly found on heritage properties, antique furniture, or decorative padlocks, where maintaining a traditional appearance is more important than providing high levels of security. Their decline in widespread use is largely due to their relatively low resistance to manipulation [192].

The internal design of a warded lock is fundamentally different from pin tumbler or wafer locks. Rather than using moving components such as pins or wafers to block rotation, warded locks rely on fixed internal obstructions known as "wards." These wards are positioned within the lock to restrict access to the turning mechanism. The correct key is shaped in a way that allows it to pass around these obstructions and engage the mechanism at the back of the lock, enabling it to turn and open [192].

A key concept associated with warded locks is the original "skeleton key." While the term is often used more broadly today, it originally referred specifically to a simplified key designed for warded locks. By removing most of the key's blade and leaving only the portion required to turn the mechanism, the skeleton key can bypass the wards entirely. Because the wards only block incorrectly shaped keys rather than securing the mechanism itself, a stripped-down key or similar tool can often reach the back of the lock and rotate it without interference [192].

This highlights a fundamental limitation of warded locks: they do not secure the plug through internal alignment, but instead rely solely on restricting access. If a tool can navigate around or ignore the wards, the locking mechanism can often be operated directly. As a result, these locks are generally considered low-security by modern standards [192].

In practical terms, warded locks can often be manipulated using simple tools rather than the more precise instruments required for other lock types. Bent wire or basic warded lock pick sets are commonly used to bypass the wards and engage the turning mechanism. These specialised pick sets are designed to fit around typical ward patterns and are particularly effective on warded padlocks, which remain one of the more common modern applications of this design [192].

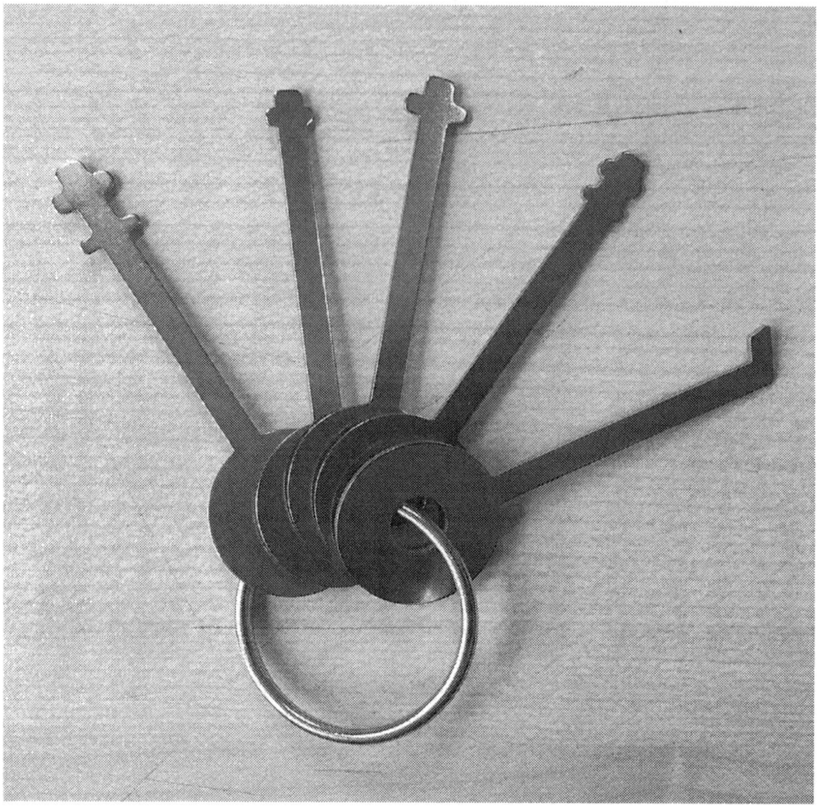

*Figure 69: A common set of "skeleton" keys used to open most types of warded padlocks, through a type of lock bypass method. Willh26, CC BY-SA 4.0, via Wikimedia Commons.*

**Disc Detainer Locks**

Disc detainer locks were invented by Emil Henriksson in 1907 and later manufactured under the Abloy brand, the company he founded. Unlike traditional pin tumbler or wafer locks, disc detainer locks use a series of rotating metal discs, each with a slot cut into it. When the correct key is inserted, it rotates each disc to a precise position. Once all the slots are aligned, a sidebar is able to drop into place, allowing the lock to turn and open. This operating principle is similar to the alignment mechanisms found in high-security safes [192].

A key distinguishing feature of disc detainer locks is the absence of springs. Because there are no spring-loaded components, these locks are highly resistant to environmental factors such as dirt, moisture, salt, and debris. This makes them particularly suitable for harsh or outdoor conditions, where other lock types may fail due to corrosion or contamination. As a result, disc detainer locks are commonly used in applications such as padlocks, transport security, and industrial environments [192].

High-quality disc detainer locks, especially those manufactured by Abloy, are widely regarded as offering a high level of security. Their precision engineering, tight tolerances, and complex disc configurations make them significantly more resistant to manipulation compared to standard pin tumbler locks. However, many lower-cost copies exist on the market, and these often lack the same level of manufacturing precision. As a result, cheaper versions can sometimes be manipulated more easily using basic techniques and tools [192].

In lower-quality disc detainer locks, it is sometimes possible to apply principles similar to those used in pin tumbler manipulation, such as identifying binding components and working them sequentially. Basic picking tools or improvised methods may be effective in these cases. In contrast, genuine high-security disc detainer locks require specialised tools designed specifically for their mechanism. These tools must be robust and precisely engineered, as inferior tools may bend or fail when used against tightly manufactured components [192].

Overall, disc detainer locks represent a significant advancement in lock design, combining durability with enhanced security. Their design minimises common vulnerabilities found in spring-based systems, while their resistance to environmental damage makes them highly reliable in demanding conditions [192].

## Impact on Locksmithing Practices

Understanding pin tumbler locks is relevant to locksmithing because they are the "default" lock mechanism in everyday life: front doors, many padlocks, and a large share of basic commercial cylinders. If you can't confidently explain how a plug, pin stacks, and the shear line interact, you're missing the core mechanical model that underpins a huge proportion of service work—installation, rekeying, fault diagnosis, upgrades, and security assessments. In practice, this knowledge is what allows you to predict how a lock *should* behave, recognise when it's behaving abnormally, and choose the least invasive remedy.

It also directly shapes how locksmiths approach non-destructive entry ethically and professionally. The most important influence here is *mindset*: competent locksmithing prioritises controlled, minimal-force methods that preserve the client's hardware and security. When you understand what actually blocks rotation (and what "alignment" really means inside a cylinder), you stop thinking in terms of "forcing it open" and start thinking in terms of restoring normal operation, through correct technique, correct tool selection, and correct decision-making (or deciding that replacement is the safer option). That distinction is central to ethical practice, professional standards, and liability management.

The discussion of tensioning, feedback, and the "feel" of locks matters because it mirrors what locksmiths do daily: interpret tiny mechanical signals to diagnose and solve problems. In a service context, that sensory skill doesn't just relate to manipulation, it's the same sensitivity used to detect worn pin stacks, debris in a keyway, weak springs, plug drag, cam problems, and misalignment between latch/bolt and strike. In other words, training the "feel" improves both *entry capability* and *mechanical troubleshooting*, which is why experienced locksmiths often resolve issues faster and with less damage.

Pin tumbler fundamentals also influence locksmithing practice in security design and client advice. Once you understand the baseline mechanism, you can explain (in plain language) why certain upgrades work: tighter tolerances, security pins that create misleading feedback, restricted key systems, hardened components, anti-drill protections, or better cylinder/door hardware integration. That helps locksmiths move from being "the person who opens doors" to being a security professional who can justify recommendations, document risks, and select fit-for-purpose solutions.

From a forensic and investigative angle, this topic matters because claims about "forced entry" or "picked locks" are often evaluated against what a pin tumbler lock *normally* does under different conditions. A locksmith who understands shear-line behaviour, pin interaction, and the difference between controlled manipulation vs force-based interference is better placed to interpret evidence objectively—especially where insurance, disputes, or staged events are possible. The influence on practice is stronger reporting discipline: careful observation, conservative conclusions, and separating what is physically supported from what is merely alleged.

Finally, the legal/ethical framing you've included (permission, not picking locks in use, responsible training setups) is relevant because locksmithing is a trust profession. The same technical knowledge that enables legitimate work can also be misused, and that reality shapes real-world practice: identity checks, authorisation procedures, documentation, scope limits, and a strong bias toward non-destructive, accountable methods. In short, understanding pin tumbler locks influences not only *how* locksmiths work, but *how they justify, document, and govern* their work in a way that maintains public trust and professional credibility.

## Chapter 6 Review Questions

### Question 1

A locksmith trainee attempts to open a practice lock by applying heavy torque and rapidly moving the pick. The lock becomes jammed and the pins appear damaged.

What principle from lock manipulation training was ignored?

A. Using specialised electronic tools
B. Applying controlled tension and careful feedback interpretation
C. Replacing the cylinder immediately
D. Using destructive entry techniques

### Question 2

A locksmith is practicing manipulation on a customer's front door lock during quiet hours to "improve skills."

Why is this practice considered unethical?

A. Practice locks should always be replaced afterward
B. Locks in active use should not be manipulated for training
C. Manipulation always breaks the lock
D. Only electronic locks can be practiced on

**Question 3**

During lock manipulation training, a student learns to identify binding pins through subtle resistance transmitted through the tension tool.

What skill is the student developing?

A. Mechanical drilling technique
B. Tactile feedback interpretation
C. Cylinder replacement procedures
D. Digital access programming

**Question 4**

A locksmith is called to a lockout situation at a commercial office. The client requests entry but wants the lock to remain usable afterward.

Which method aligns with professional best practice?

A. Drilling the cylinder immediately
B. Using controlled, non-destructive manipulation
C. Breaking the door frame
D. Removing the entire door

**Question 5**

During forensic examination of a suspected break-in, a locksmith must determine whether the lock was opened through manipulation or by force.

Why is understanding lock manipulation techniques important?

A. It helps determine the manufacturer
B. It allows identification of entry methods and possible evidence
C. It prevents the need for photography
D. It eliminates legal responsibility

**Question 6**

A locksmith opens a customer's lock using a specialized bypass tool even though the lock could have been manipulated non-destructively.

What potential issue could arise from this approach?

A. Reduced professional reputation due to unnecessary damage
B. Faster service time
C. Improved client satisfaction
D. Lower equipment costs

**Question 7**

A trainee practices manipulation using transparent training locks before working on real locks.

Why are training locks recommended?

A. They reduce learning time to zero
B. They allow safe skill development without risking real security systems
C. They eliminate the need for tools
D. They automatically open

**Question 8**

A locksmith notices a lock repeatedly fails during service calls because excessive force was used during previous manipulation attempts.

What lesson does this situation illustrate?

A. Manipulation should be avoided entirely
B. Excessive force reduces lock lifespan and reliability
C. All locks should be replaced immediately
D. Force improves lock feedback

**Question 9**

A locksmith with strong tactile skills diagnoses a lock problem without disassembling it by detecting unusual resistance when turning the plug.

What professional capability does this demonstrate?

A. Digital access programming
B. Sensory-based mechanical diagnosis
C. Electronic hacking
D. Key duplication

**Question 10**

A locksmith documents authorisation before opening a locked property during a late-night emergency call.

Why is this important?

A. It reduces service time
B. It protects the locksmith legally and ensures the work is authorised
C. It guarantees the lock will open
D. It eliminates insurance requirements

# Part 4

# Human Factors and Security Failures

# Chapter 7

# Human Error as the Weakest Link

Real-world experience consistently demonstrates that the effectiveness of any locking system is not determined solely by its engineering, but by how it is used, managed, and maintained in everyday situations. The most advanced lock can be rendered ineffective not through technical defeat, but through simple, predictable human behaviours.

This chapter shifts the focus from the lock itself to the interaction between people and security systems. It explores how routine actions, such as poor key management, failure to properly engage locks, neglect of maintenance, and convenience-driven decisions, can unintentionally create vulnerabilities that undermine even well-designed systems. It also examines how these behaviours are not isolated incidents, but recurring patterns that must be anticipated and addressed as part of professional locksmithing practice.

In addition, the chapter extends beyond traditional mechanical security to consider modern challenges, including digital access systems, social engineering, insider threats, and organisational key management failures. As security systems become more integrated and complex, the human element remains the most variable—and often the weakest—component.

For the locksmith, this represents a fundamental shift in perspective. Security is no longer just about selecting and installing hardware; it is about designing systems that account for human behaviour, implementing controls that maintain accountability, and educating users to reduce risk. By understanding how and why people defeat their own security systems, locksmiths can move from reactive problem-solving to proactive risk management.

Ultimately, this chapter establishes a critical principle: a lock is only as effective as the way it is used. Recognising and addressing the human factors that influence security is essential to delivering reliable, resilient, and practical access control solutions in both residential and organisational environments

## Common User Behaviours that Defeat Locks

A common misconception in physical security is that stronger locks automatically result in better protection; however, in practice, many security breaches occur not because locks fail, but because they are used incorrectly or inconsistently. A high-quality locking system can be rendered ineffective by simple, everyday habits such as poor key control, incomplete locking, or neglect of maintenance. Understanding this gap between design performance and real-world use is essential, as it reveals that security is not just a function of engineering, but of human behaviour. It is within this interaction between user and system that many vulnerabilities emerge, making it critical to examine how routine actions can unintentionally defeat even the most robust locking mechanisms.

The effectiveness of physical security systems depends not solely on the mechanical design and engineering of locks, but significantly on how users interact with and maintain these systems. While locks are engineered to resist unauthorised access through mechanical precision and material strength, their real-world performance is heavily influenced by user behaviour patterns that often create vulnerabilities independent of hardware quality [24, 122, 193]. This analysis synthesises evidence regarding common user behaviours that undermine lock security and examines the implications for locksmithing practice and security design.

### Key Management and Storage Practices

One of the most critical vulnerabilities in physical security systems stems from improper key management. Users frequently engage in predictable behaviours that compromise key security, including the storage of spare keys in easily discoverable locations such as under doormats, inside plant pots, or above door frames [122]. These hiding places are well-known to opportunistic offenders and represent the first areas checked during attempted entry [122]. The casual sharing of keys with multiple individuals, such as friends, contractors, or temporary occupants, without proper tracking or rekeying creates uncontrolled key distribution and long-term security risks [122].

The problem extends beyond initial key distribution. Poor key management practices include the use of worn, duplicated, or low-quality keys, which can result in reduced accuracy and excessive wear on internal lock components such as pins and springs [122]. Repeated duplication of keys degrades their precision, shortening lock lifespan and increasing susceptibility to picking or malfunction [122]. Additionally, users may continue using damaged or bent keys, which compromise the alignment of the locking mechanism and increase the likelihood of failure [122].

Research on electronic locking systems has demonstrated that key management issues persist even in digital contexts. While electronic locks were designed to address some mechanical key vulnerabilities, such as the claim that electronic keys cannot be cloned as easily as mechanical counterparts, implementation flaws can still allow extraction of system-wide master secrets [24]. The fundamental principle remains that ownership of a key, whether mechanical or electronic, can be tracked through observable characteristics that may be duplicated [24].

### Failure to Engage Locking Mechanisms

A significant behavioural issue affecting lock security is the failure to fully engage locking mechanisms. Many users rely solely on spring latches by pulling doors closed without turning the key or engaging a deadbolt [122]. This leaves doors vulnerable to simple bypass methods such as shimming, slipping, or latch manipulation [122]. Even when deadlocks are installed, they are often left unused due to convenience, particularly during short absences, significantly reducing the intended level of security [122].

This behaviour is particularly problematic because it negates the security benefits of properly installed hardware. The distinction between mechanical engagement and passive closure is critical; spring latches alone provide minimal resistance to determined entry attempts [122]. Users frequently underestimate the security risk posed by incomplete engagement, particularly when they believe their absence will be brief [122].

### Maintenance Neglect and Environmental Degradation

Locks are mechanical devices requiring periodic inspection, cleaning, and lubrication, yet users often ignore early warning signs such as stiffness, grinding, or difficulty turning the key (Chesher, 2013). Over time, this neglect leads to internal wear, misalignment, or component failure, making locks easier to manipulate or bypass [122]. Environmental factors, such as dust, moisture, and corrosion, accelerate this deterioration if not properly managed [122].

The relationship between maintenance and security is direct and measurable. Mechanical locks depend on precise tolerances and alignment of internal components; degradation of these tolerances through wear or corrosion directly reduces security effectiveness [122]. Users often fail to recognise that maintenance is not merely a matter of convenience but a critical security requirement [122].

### Convenience-Driven Modifications

Users sometimes disable or alter locking mechanisms to facilitate access, such as taping over latches, removing strike plates, or leaving doors propped open [122]. Others may install incompatible hardware or attempt do-it-yourself repairs without understanding the mechanical tolerances required for proper operation [122]. These actions significantly weaken the structural integrity of the locking system [122].

Such modifications represent a deliberate trade-off between security and convenience, often made without full understanding of the security implications. The removal of strike plates, for example, eliminates a critical component that distributes force across the door frame, substantially reducing resistance to forced entry [122]. Similarly, taping latches to prevent automatic locking removes a passive security feature that operates independently of user action [122].

### Situational Awareness and Contextual Risk Assessment

User behaviour related to situational awareness significantly impacts lock security. Doors and windows are sometimes left unlocked during short trips, under the assumption that risk is minimal [122]. This creates opportunities for opportunistic entry, as many incidents occur when properties are briefly unattended [122]. Similarly, failure to check that doors are properly closed and aligned can result in incomplete engagement of the lock, leaving it ineffective [122].

The assumption that brief absences present minimal risk is frequently incorrect. Research on opportunistic crime indicates that many burglaries occur during short periods when properties are unattended, and offenders often target properties that appear to have minimal security measures [122]. Users frequently misjudge the probability and timing of criminal activity, leading to security decisions based on incomplete risk assessment [122].

### Digital-Era Vulnerabilities in Electronic and Smart Locks

Digital-era behaviours have introduced new vulnerabilities, particularly with electronic and smart locks. Users may choose weak PIN codes, reuse access credentials, or fail to update firmware and security settings [122]. Sharing access codes without proper control or failing to revoke access when no longer required can lead to unauthorised entry [122]. These behaviours mirror traditional key management issues but occur within a digital framework [122].

Research on IoT-based smart locks has identified that cloud-based access control rules present particular challenges, as lock-outs can occur if cloud services become unavailable [194]. Additionally, users may not understand the security implications of their digital choices. For example, weak PIN codes can be compromised through brute force attacks, and failure to update firmware leaves systems vulnerable to known exploits [122]. The convenience features of smart locks, such as remote access and temporary access codes, introduce new management requirements that users often fail to implement properly [122].

Electronic locking systems present additional complexity. While these systems were designed to provide more flexible access rights management and the ability to revoke physical keys, implementation flaws can compromise security [24]. Users may not understand that electronic keys, despite claims of resistance to cloning, can still be compromised through cryptographic attacks or power analysis [24]. The transition from mechanical to electronic locks has not eliminated key management challenges; it has merely transformed them into digital equivalents [24].

### Implications for Locksmithing Practice and Security Design

Understanding these common user behaviours is essential for locksmiths and security professionals, as it highlights that security is not solely dependent on hardware [122]. Effective locksmithing practice involves not only selecting and installing appropriate locking systems but also educating users on correct usage, maintenance, and key control [122]. By addressing human factors alongside mechanical design, locksmiths can significantly improve the overall effectiveness of physical security systems [122].

The evidence demonstrates that even well-designed and properly installed locks can be rendered ineffective through predictable human behaviours [24, 122]. This recognition has implications for security system design, suggesting that systems should incorporate features that reduce reliance on consistent user behaviour. For example, automatic deadbolts that engage without user action, or systems that provide feedback when locks are not fully engaged, can mitigate some behavioural vulnerabilities [122].

Furthermore, the convergence of physical and digital security in smart lock systems requires users to understand both mechanical and digital security principles [122]. This increased complexity may paradoxically increase vulnerability if users do not receive adequate education and support [122]. Security professionals must therefore consider not only the technical capabilities of locking systems but also the realistic behavioural patterns of users and design systems accordingly [122].

Common user behaviours play a significant role in reducing the effectiveness of locks, often undermining even well-designed and properly installed security systems [122]. These behaviours span multiple domains: improper key storage and management, failure to engage locking mechanisms, neglect of maintenance, convenience-driven modifications, poor situational awareness, and inadequate management of digital security features [122]. The evidence indicates that security failures frequently occur not because of flaws in the lock itself, but due to predictable human habits that create vulnerabilities [122].

Addressing these behavioural factors requires a multifaceted approach that combines improved lock design with user education and system features that reduce reliance on consistent user behaviour [122]. Locksmiths and security professionals must recognise that their role extends beyond hardware selection and installation to include user guidance on proper usage, maintenance, and key control [122]. By integrating human factors into security system design and practice, the overall effectiveness of physical security can be substantially improved [122].

### Locksmith's Role in Mitigation

Locksmiths play a critical role in mitigating the risks created by common user behaviours, not only through the selection and installation of appropriate hardware, but through education, system design, and ongoing support. Addressing human factors requires a proactive and holistic approach, where the locksmith anticipates predictable misuse and implements measures that either reduce reliance on user behaviour or guide it in safer directions.

One of the most effective strategies is improving key control and management practices. Locksmiths can recommend and supply restricted or patented key systems that prevent unauthorised duplication. These systems require proof of ownership or authorisation before additional keys can be cut, significantly reducing the risks associated with uncontrolled key sharing. In addition, locksmiths can advise clients on maintaining key registers, issuing logs, and implementing rekeying schedules when keys are lost, stolen, or when occupancy changes. For higher-risk environments, transitioning from traditional keys to electronic access systems can further enhance control and accountability.

To address the failure to fully engage locking mechanisms, locksmiths can install self-locking or automatically engaging hardware, such as deadlatching systems, electric strikes, or electronic locks that secure upon closure. Where manual locking is still required, clear user instruction and signage can reinforce correct operation. Door closers and properly aligned strike plates can also ensure that doors consistently latch and lock as intended, reducing reliance on user diligence.

In response to poor key usage and wear, locksmiths can promote the use of high-quality key blanks and precision duplication techniques, ensuring that keys maintain correct tolerances. They can also educate users about the risks of repeated copying and encourage the replacement of worn or damaged keys before they compromise the lock. Regular servicing schedules can be introduced, particularly in commercial environments, to inspect internal components and replace worn pins, springs, or cylinders before failure occurs.

Maintenance-related issues can be addressed through preventative maintenance programs. Locksmiths can offer scheduled inspections that include cleaning, lubrication with appropriate products, alignment checks, and functional testing. Providing clients with simple maintenance guidelines—such as what lubricants to use and how often to apply them—helps extend the life of the lock and maintain its security performance. Importantly, locksmiths can identify early signs of failure that users may overlook and rectify them before they develop into vulnerabilities.

To counter convenience-driven modifications and incorrect installations, locksmiths should emphasise the importance of professional installation and compliance with standards. Educating clients about the risks of DIY alterations and incompatible hardware is essential. Where misuse is likely, locksmiths can recommend more robust or tamper-resistant hardware, such as reinforced strike plates, security escutcheons, or anti-tamper fasteners. In some cases, redesigning the system to better suit user needs—such as installing keyless entry or access control systems—can reduce the temptation to bypass security features.

Improving situational awareness involves both education and system design. Locksmiths can conduct security audits to identify behavioural risks, such as habits of leaving doors unlocked or relying on inadequate locking points. Based on these assessments, they can recommend layered security solutions, including additional locking points, window locks, alarms, or surveillance systems. Simple measures such as visual reminders, checklists, or integrated locking systems can encourage consistent secure behaviour.

In the context of electronic and smart locks, locksmiths have an important role in guiding users toward secure digital practices. This includes advising on strong PIN selection, avoiding default codes, regularly updating firmware, and managing user access credentials. Locksmiths can configure systems to enforce security policies, such as automatic locking, audit trails, and timed access permissions. Training users on how to add, manage, and revoke access ensures that digital systems are used as intended and do not replicate the same vulnerabilities as traditional key systems.

Ultimately, locksmiths must adopt a user-centred approach to security, recognising that even the most advanced hardware can be compromised by poor usage. By combining technical expertise with client education, system design, and ongoing support, locksmiths can significantly reduce the impact of common user behaviours. This approach shifts the focus from simply installing locks to actively managing security as an integrated system, where human behaviour is anticipated, guided, and supported to achieve the intended level of protection.

**Impact on Locksmithing Practices**

This topic is highly relevant to anyone interested in locksmithing because it shifts the understanding of security from a purely mechanical or technical discipline to a human-centred practice. A developing locksmith may initially focus on how locks are designed, installed, and defeated from a technical perspective. However, this analysis demonstrates that, in real-world conditions, the effectiveness of any locking system is just as dependent—if not more so—on how users behave. This fundamentally reframes locksmithing as not only a trade of hardware expertise, but one of risk management, behavioural awareness, and system design.

For someone entering the locksmithing profession, this knowledge is critical because it explains why security failures often occur even when high-quality locks are installed correctly. Without understanding user behaviour, a locksmith may incorrectly attribute failures to poor hardware or installation, rather than recognising predictable human actions such as poor key control, failure to engage locks, or neglect of maintenance. This insight allows locksmiths to diagnose problems more accurately and provide more effective, long-term solutions rather than short-term fixes.

The influence on locksmithing practice is significant, as it expands the locksmith's role beyond installation and repair into that of a security advisor and educator. Locksmiths must be capable of explaining risks to clients in practical terms, such as why hiding a key under a mat defeats the purpose of a high-security cylinder, or why not engaging a deadbolt renders a door vulnerable despite having quality hardware fitted. This educational component becomes a routine part of service delivery, particularly in residential and small commercial contexts where user behaviour is often the weakest link.

It also directly influences how locksmiths select and specify locking systems. Rather than choosing hardware based solely on strength or resistance to attack, locksmiths must consider how real users will interact with the system. For example, if users are unlikely to consistently engage a deadbolt, the locksmith may specify an automatic locking system or electronic lock that removes reliance on user action. Similarly, where key control is a concern, restricted key systems or access-controlled environments may be recommended instead of standard keyed cylinders.

The analysis further highlights the importance of preventative and lifecycle-based locksmithing practices. Maintenance, servicing, and periodic reassessment become integral to ensuring ongoing security performance. Locksmiths are not just responding to failures but are actively managing the condition and effectiveness of locking systems over time. This is particularly relevant in commercial environments, where wear, duplication, and staff turnover can rapidly degrade security if not properly managed.

In modern locksmithing, the inclusion of electronic and smart locking systems means that practitioners must also understand digital security behaviours. The same human factors, such as poor credential management or complacency, apply in digital contexts, but with additional complexity. This requires locksmiths to develop competencies beyond traditional mechanical systems, including basic cybersecurity awareness, system configuration, and user training in digital access control.

Importantly, this topic reinforces the concept that effective security design must aim to reduce reliance on perfect user behaviour. This influences locksmithing practices toward designing systems that are more forgiving of human error, such as self-locking mechanisms, audit-enabled access systems, and tamper-resistant hardware. It encourages a proactive design philosophy where foreseeable misuse is anticipated and mitigated through engineering and system integration.

Ultimately, for someone interested in locksmithing, this knowledge is foundational because it defines the profession as one that operates at the intersection of engineering, human behaviour, and security risk. It elevates locksmithing from a purely technical trade to a professional discipline that requires critical thinking, client engagement, and system-level understanding. By recognising and addressing the human factors that defeat locks, locksmiths can deliver more effective, resilient, and context-appropriate security solutions.

## Social Engineering and Locksmith Vulnerabilities

Social engineering represents a critical vulnerability in security systems that extends beyond physical mechanisms to exploit human psychology and trust. In the context of locksmithing and access control, this threat is particularly significant because it circumvents the technical protections that locks provide by targeting the individuals who control access to secured premises. Understanding social engineering in locksmithing requires examination of how attackers manipulate trust, the specific vulnerabilities locksmiths face, and the verification procedures necessary to mitigate these risks.

Social engineering refers to the deliberate manipulation of individuals to gain unauthorised access to property, information, or systems by exploiting human trust rather than defeating physical security mechanisms [195]. In the context of locksmithing, this represents a significant and often underestimated vulnerability because it bypasses locks entirely by targeting the people who control access to them. Instead of picking, drilling, or bypassing a lock, an offender may simply convince someone, such as a property owner, tenant, receptionist, or even a locksmith, to grant access willingly [195].

The fundamental principle underlying social engineering attacks is that human psychology presents a more exploitable target than physical security systems. This is particularly relevant in access control contexts, where the verification of identity and authority becomes paramount [196]. The "never trust, always verify" principle, which underpins modern zero-trust security architectures, directly addresses the vulnerabilities that social engineering exploits [196].

### The Locksmith's Position of Trust and Vulnerability

For locksmiths, social engineering is particularly relevant because they are frequently placed in positions of trust and authority [195]. Clients rely on locksmiths to provide access to secured premises, rekey systems, or override existing security controls. This makes locksmiths potential targets for social engineering attacks where an individual falsely claims ownership, tenancy, or authority over a property [195]. For example, an offender may pose as a tenant locked out of their home, a contractor needing urgent access, or a new property owner requesting rekeying, all with the intention of gaining unlawful entry [195]. Without proper verification processes, the locksmith may unknowingly facilitate unauthorised access [195].

The vulnerability of locksmiths to social engineering is compounded by the nature of their work, which often requires rapid response to urgent requests. This time pressure can undermine careful verification procedures, creating opportunities for attackers to exploit the situation [195].

### Exploitation of Legitimacy Assumptions

One of the key vulnerabilities exploited through social engineering is the assumption of legitimacy [195]. People tend to trust individuals who appear confident, use industry terminology, wear high-visibility clothing, or present partial documentation [195]. This phenomenon reflects broader patterns in human psychology where external markers of authority and professionalism are often accepted at face value without rigorous verification [197].

Offenders may use urgency as a tactic, claiming emergencies such as a child locked inside, lost keys late at night, or critical business operations being disrupted [195]. This pressure can lead locksmiths or property managers to bypass standard verification procedures in favour of quick resolution, thereby compromising security [195]. The psychological impact of urgency on decision-making is well-documented, and attackers deliberately exploit this vulnerability to reduce the likelihood that their requests will be subjected to careful scrutiny [195].

### Information Gathering and Credibility Building

Another common tactic involves information gathering prior to engagement [195]. Offenders may observe a property, identify access points, learn occupant names, or gather details from social media to make their story more convincing [195]. When they approach a locksmith or occupant, this information can be used to build credibility. For instance, referencing a known tenant name, recent activity, or internal layout can make the request appear legitimate, increasing the likelihood of compliance [195].

This pre-engagement reconnaissance represents a sophisticated approach to social engineering, where attackers invest time in gathering intelligence to enhance the plausibility of their claims. The availability of personal information through social media and public records makes this tactic increasingly effective [195].

### Social Engineering in Access Control Systems

Social engineering vulnerabilities also extend to key systems and access control environments [195]. In organisations, individuals may request duplicate keys, access cards, or PIN resets by impersonating authorised personnel [195]. If procedures for issuing or duplicating credentials are not strictly controlled, this can result in unauthorised access without any physical breach (Jia et al., 2020). Similarly, in electronic systems, attackers may attempt to obtain access codes or administrative privileges by convincing users to disclose information or perform actions on their behalf [195].

The integration of identity and access management (IAM) systems with privileged access management (PAM) is increasingly recognised as essential for preventing such attacks [198]. These systems enforce continuous verification and control of access privileges, reducing the likelihood that social engineering attacks will succeed [198]. The principle of least privilege, granting users only the minimum access necessary to perform their functions, is a critical control that limits the damage that can result from successful social engineering attacks [198].

The principle of least privilege (PoLP) is a fundamental security concept that requires individuals, systems, or processes to be given only the minimum level of access necessary to perform their required tasks—no more, and no less. In the context of locksmithing and physical security, this means that access to keys, locks, areas, or systems should be strictly limited to what is operationally necessary at a given time. By reducing unnecessary access, the potential impact of misuse, loss, theft, or human error is significantly minimised.

This principle is highly relevant to locksmithing because many security failures arise not from forced entry, but from overexposure of access rights. For example, if multiple people hold master keys, if keys are freely duplicated, or if access codes are widely shared, the system becomes difficult to control and audit. The more individuals who have access, the greater the likelihood of compromise—whether intentional or accidental. Applying least privilege ensures that access remains controlled, traceable, and aligned with actual need.

In practical locksmithing applications, least privilege can be implemented through structured key systems. Master key systems can be designed with hierarchical access levels, where individuals only have keys that open the specific doors relevant to their role. For instance, a cleaner may have access to common areas only, while a manager may have broader access, and only a very limited number of authorised personnel hold master keys. Importantly, the distribution of high-level keys should be tightly controlled and documented.

Another key implementation strategy is the use of restricted key systems. These systems prevent unauthorised duplication by requiring proof of ownership or authorised signatures before keys can be copied. This ensures that access cannot be expanded without control, supporting the principle that privileges should not grow unchecked over time. Combined with key registers and issue logs, locksmiths can help organisations maintain clear accountability over who has access to what.

In electronic and smart locking environments, least privilege is applied through role-based access control (RBAC). Users are assigned access rights based on their role, with permissions configured to allow entry only to specific areas and at specific times. For example, a staff member may only be able to access a building during working hours, while a supervisor may have extended access. Temporary access credentials can be issued and automatically expire, ensuring that privileges are not retained longer than necessary.

The principle also requires regular review and revocation of access. Locksmiths and security managers should implement processes to remove or adjust access when roles change, staff leave, or keys are lost. Rekeying, deactivating credentials, or updating access permissions ensures that outdated privileges do not remain active. Without this step, systems gradually accumulate unnecessary access rights, weakening overall security.

From a system design perspective, least privilege encourages locksmiths to create solutions that are both functional and controlled. This may involve segmenting access across multiple locking zones, avoiding over-reliance on single master keys, and integrating audit capabilities where possible. In high-security environments, dual-control systems (requiring two authorised individuals) or logging mechanisms may be introduced to further restrict and monitor access.

The principle of least privilege reinforces the idea that security is strongest when access is limited, intentional, and accountable. For locksmiths, applying this principle enhances the integrity of both mechanical and electronic systems, reduces risk exposure, and supports best practice in modern security design. It transforms access control from a convenience-based approach into a disciplined, risk-managed system aligned with real operational needs.

**Verification-First Approach and Professional Practice**

The implications for locksmithing practice are significant [195]. Locksmiths must adopt a verification-first approach, ensuring that identity, authority, and right of access are confirmed before any work is carried out [195]. This may include requesting formal identification, proof of occupancy or ownership, written authorisation, or confirmation from property managers or landlords [195]. In some cases, locksmiths may need to refuse service if adequate verification cannot be obtained, even under pressure or in seemingly urgent situations [195].

Professional practice also requires the establishment of clear policies and procedures for handling access requests [195]. This includes documenting all work performed, recording client details, and maintaining job records that can be audited if required [195]. For commercial and high-security environments, locksmiths may implement restricted key systems, controlled authorisation processes, and multi-level approvals to reduce the risk of unauthorised duplication or access [195].

**Sample Policy: Handling Access Requests**

**1. Purpose**

This policy establishes the requirements and controls for verifying and responding to requests for access to secured premises, vehicles, safes, or restricted areas. It ensures that locksmith services are only provided to authorised individuals and that risks associated with unauthorised access and social engineering are minimised.

**2. Scope**

This policy applies to all employees, contractors, and authorised representatives of the organisation who are involved in providing locksmith services, including emergency lockouts, rekeying, key duplication, and access control system changes.

**3. Policy Statement**

The organisation is committed to ensuring that access is only granted to individuals who can demonstrate legitimate authority. No work is to be undertaken without appropriate verification of identity and authority, regardless of urgency or client pressure.

Failure to comply with this policy may result in disciplinary action and/or legal consequences.

**4. Principles**

- **Verification First:** Access must not be granted without confirming identity and authority.
- **No Assumptions:** Appearance, confidence, or urgency must not replace verification.
- **Document Everything:** All access-related work must be recorded.
- **Right to Refuse:** Staff are authorised to refuse service if verification is inadequate.
- **Compliance:** All actions must comply with relevant laws, regulations, and licensing requirements.

**Procedure: Handling Access Requests**

**Step 1: Initial Request Assessment**

Upon receiving a request for access:

- Identify the type of request (e.g. lockout, rekey, key duplication, electronic access reset).
- Determine the location and nature of the asset (residential, commercial, vehicle, safe).
- Assess whether the request presents potential risk indicators (e.g. urgency, lack of documentation, third-party request).

**Step 2: Identity Verification**

The requester must provide valid identification:

- Acceptable forms include driver licence, passport, or government-issued ID.
- The name on the ID must match supporting documentation or be verified through additional means.
- Record ID details in job documentation (in accordance with privacy requirements).

**Step 3: Authority Verification**

The requester must demonstrate legal right to access:

- **Residential:** Proof of occupancy or ownership (e.g. lease agreement, utility bill, title documents).
- **Commercial:** Written authorisation from management or verification through known contact within the organisation.
- **Vehicles:** Proof of ownership (registration papers) and ID matching registration.
- **Safes/Restricted Assets:** Proof of ownership and, where applicable, manufacturer or supplier verification protocols.

If authority cannot be confirmed, **do not proceed**.

**Step 4: Additional Verification (if required)**

Where risk is elevated:

- Contact property owner, landlord, or facility manager directly.
- Request written confirmation (email, SMS, or formal authorisation).

- Cross-check details against internal records or prior job history.

**Step 5: Risk Assessment**

Evaluate the request for potential social engineering indicators:

- Inconsistent or vague information
- Excessive urgency or pressure
- Reluctance to provide documentation
- Third-party requests without clear authority

If concerns remain unresolved, escalate or refuse service.

**Step 6: Approval to Proceed**

Work may only proceed when:

- Identity and authority are verified
- Risks have been assessed and deemed acceptable
- Required documentation is obtained

**Step 7: Service Delivery Controls**

During the job:

- Ensure work aligns strictly with the approved request
- Do not provide additional access (e.g. extra keys) without authorisation
- Use appropriate tools and methods in accordance with professional standards

**Step 8: Documentation and Record Keeping**

Record the following:

- Client name and contact details
- Identification type and verification method
- Proof of authority
- Description of work performed

- Date, time, and technician details

Maintain records securely in accordance with privacy and legal requirements.

**Step 9: Refusal of Service**

Service must be refused if:

- Identity cannot be verified
- Authority cannot be established
- The situation presents a high risk of unauthorised access

Where appropriate:

- Provide explanation to the requester
- Offer alternative steps (e.g. return with documentation, contact owner)
- Document the refusal

**Step 10: Incident Reporting**

Any suspected social engineering attempt or suspicious behaviour must be:

- Reported internally
- Documented in incident records
- Escalated where required (e.g. to management or authorities)

**Supporting Measures**

- Staff training in social engineering awareness
- Use of restricted key systems and controlled duplication
- Standardised forms for verification and authorisation
- Periodic audits of compliance with this procedure

**Summary**

This policy and procedure ensures that locksmith services are delivered in a controlled, professional, and legally compliant manner. By embedding verification, documentation, and risk assessment into every access request, locksmiths protect not only their clients, but also their own professional integrity and liability exposure.

The importance of formal verification procedures is underscored by research on identity verification in other contexts. In the financial services sector, for example, electronic Know Your Customer (e-KYC) systems have been shown to significantly reduce fraud and improve institutional transparency [199]. The implementation of robust verification procedures reduces identity-related fraud by approximately 28.9 percent and decreases customer onboarding time while lowering verification costs [199].

Know Your Customer (KYC), and its digital form, electronic KYC (e-KYC), refers to processes used to verify the identity of a client and confirm their legitimacy before providing a service. While traditionally associated with banking and financial services, the same principles are increasingly relevant in locksmithing and physical security, where granting access to property or systems carries significant legal and ethical responsibility. In this context, e-KYC provides a structured, auditable method of confirming that a person requesting access is who they claim to be and has the authority to do so.

At its core, an e-KYC system replaces or enhances manual identity checks with digital verification processes. This may include capturing identification documents (such as a driver licence or passport), verifying them against government or trusted databases, and confirming the individual's identity through biometric methods such as facial recognition or live video verification. Some systems also incorporate address verification, cross-referencing utility records or other data sources to confirm a person's connection to a property. The result is a higher level of confidence in identity verification, supported by digital records that can be stored and audited.

For locksmithing, the relevance of e-KYC lies in its ability to reduce the risk of social engineering and unauthorised access. Traditional verification methods—such as visually checking ID or relying on verbal claims—can be inconsistent and vulnerable to deception. An e-KYC system introduces standardisation and reduces reliance on subjective judgement by requiring clients to complete a formal verification process before services are provided. This is particularly valuable in high-risk situations such as emergency lockouts, rekeying requests, or access to commercial premises.

Implementation of e-KYC in locksmithing can be achieved through mobile and cloud-based platforms. For example, a locksmith business may use a secure application where clients upload identification documents prior to service. The system can automatically verify the authenticity of the document, match it to a live selfie, and record the verification outcome. For residential access, additional steps could include uploading proof of occupancy (such as a lease or utility bill), which can be reviewed and stored alongside the job record. For commercial clients, verification may involve confirming employment status or authorisation through a registered business contact.

Another practical approach is integrating e-KYC into job management and booking systems. When a client requests a service, they are prompted to complete identity verification before the job is confirmed. This ensures that verification occurs before the locksmith arrives on site, reducing risk and saving time. In higher-security environments, multi-factor verification can be implemented, combining identity checks with approval from a property owner, manager, or central authority.

e-KYC systems also support record keeping and compliance. Digital verification records—including timestamps, verification results, and supporting documents—can be securely stored and linked to each job. This provides an audit trail that protects both the locksmith and the client in the event of disputes, investigations, or legal proceedings. It demonstrates due diligence and adherence to professional standards, which is increasingly important in regulated or high-liability environments.

However, implementation must also consider privacy, data protection, and legal obligations. Locksmith businesses must ensure that any personal data collected is stored securely, used only for its intended purpose, and handled in accordance with applicable privacy laws (such as the Australian Privacy Principles). Clients should be informed about what data is collected, why it is required, and how it will be stored and protected. Systems used for e-KYC should be reputable, secure, and compliant with relevant standards.

In practice, e-KYC should be combined with procedural controls and professional judgement. While digital verification enhances reliability, it does not replace the need for situational awareness and risk assessment. Locksmiths should still be trained to identify suspicious behaviour, inconsistencies, or red flags that may not be captured by automated systems alone.

#### Education and Awareness

Education and awareness are equally important components of social engineering mitigation [195]. Locksmiths should inform clients about the risks of social engineering and encourage them to adopt similar verification practices [195]. For example, businesses can train staff not to grant access based solely on verbal requests or appearance, and to verify credentials before allowing entry or issuing keys [195]. Residential clients can be advised to be cautious when dealing with unknown individuals claiming to require access to their property [195].

The effectiveness of awareness training in reducing social engineering attacks is well-established in cybersecurity contexts [200]. Organisations that implement comprehensive security awareness programs, including training on phishing and impersonation attacks, demonstrate significantly improved resistance to social engineering attempts [200].

#### The Fundamental Principle of Access Control

Ultimately, social engineering highlights a fundamental principle in locksmithing and security: the strongest lock is ineffective if access is granted willingly to the wrong person [195]. By recognising and addressing these human-based vulnerabilities, locksmiths can better protect both their clients and their own professional integrity, ensuring that access is only provided to those who are legitimately entitled to it [195].

This principle aligns with contemporary security frameworks that emphasise continuous verification and identity-centric security [196]. The zero-trust architecture model, which requires continuous verification of all users and devices attempting to access resources, represents an evolution of security thinking that directly addresses the vulnerabilities that social engineering exploits [196].

Social engineering in locksmithing represents a multifaceted threat that requires comprehensive mitigation strategies encompassing verification procedures, professional standards, organisational policies, and user education. The vulnerability of locksmiths to social engineering attacks stems from their position of trust, the time pressure under which they often operate, and the psychological tactics that attackers employ to establish false legitimacy. By implementing verification-first approaches, maintaining clear documentation and audit trails, and educating both clients and staff about social engineering risks, locksmiths can significantly reduce the likelihood of unauthorised access resulting from human manipulation rather than technical bypass. The integration of these practices with broader identity and access management frameworks ensures that security is maintained across both physical and organisational dimensions.

## Impact on Locksmithing Practices

This topic is highly relevant to anyone interested in locksmithing because it fundamentally changes how security is understood and applied in real-world practice. While many entering the profession focus on the technical aspects of locks, such as installation, picking resistance, and mechanical strength, social engineering demonstrates that the greatest vulnerability in any security system is often the human element, not the hardware. A locksmith who does not understand this risk may unknowingly become the point of failure in an otherwise secure system.

Understanding social engineering is critical because locksmiths operate in a position of authority and trust, where their actions can directly grant or deny access to secured property. This places significant responsibility on the practitioner. Without awareness of social engineering tactics—such as impersonation, urgency, or manipulation—a locksmith may unintentionally facilitate unlawful entry. For someone entering the field, recognising these risks early is essential to developing sound professional judgement and ethical decision-making.

The influence on locksmithing practice is substantial, as it requires a shift from a purely task-based role to a verification-driven and risk-aware profession. Locksmiths must adopt structured processes to confirm identity and authority before performing any work. This includes checking identification, verifying occupancy or ownership, and refusing service where appropriate. As a result, locksmithing becomes not just about "opening doors," but about ensuring that doors are only opened for the right people under the right conditions.

This also leads to the development and implementation of formal policies and procedures, such as access request protocols, documentation requirements, and incident reporting systems. These procedures are not administrative burdens; they are essential safeguards that protect the locksmith, the client, and the broader community. For example, maintaining detailed job records and verification evidence creates an audit trail that supports accountability and legal defensibility.

The concept of least privilege further influences locksmithing practice by guiding how access systems are designed and managed. Locksmiths must think beyond individual locks and consider how access is distributed across a system. This includes designing master key systems with controlled hierarchies, limiting the number of high-level keys, and ensuring that access rights are aligned with actual needs. In electronic systems, this extends to configuring role-based access and ensuring that permissions are regularly reviewed and revoked when no longer required.

The integration of e-KYC and digital verification systems represents another significant shift in modern locksmithing practice. These tools allow locksmiths to standardise identity verification, reduce reliance on subjective judgement, and create secure, auditable records of access decisions. For a developing locksmith, this highlights the increasing convergence between physical security and digital identity management, requiring new skills and awareness beyond traditional mechanical knowledge.

Importantly, this topic reinforces the need for professional communication and client education. Locksmiths must be able to explain why verification is necessary, even in urgent situations, and help clients understand the risks of social engineering. This builds trust while also reinforcing secure behaviours in the broader community. It also positions the locksmith as a security professional rather than simply a tradesperson.

## Key Management Failures in Organisations

Key management represents a critical yet frequently overlooked component of organizational security infrastructure. While substantial resources are typically allocated to the selection and installation of high-quality locks and access control systems, the effectiveness of these physical security measures is fundamentally dependent upon the proper management of the keys and credentials that grant access to them [201]. When key management practices are inadequate, even the most sophisticated locking mechanisms become ineffective, as access control is compromised at the point of credential distribution and lifecycle management [201].

### Lack of Accountability and Tracking Systems

One of the most prevalent failures in organizational key management is the absence of comprehensive accountability and tracking mechanisms [201]. In many organizations, keys are distributed informally without maintaining proper documentation regarding key holders, issuance dates, or the specific areas to which keys provide access [201]. This informal approach inevitably leads to "key sprawl," a condition wherein multiple unauthorized copies of keys accumulate over time with minimal organizational oversight [201].

Key sprawl, the uncontrolled growth and poor visibility of keys and access credentials, can only be effectively prevented through a combination of system design, administrative control, and ongoing oversight. It is not a single issue with a single solution; rather, it reflects weaknesses in how access is issued, tracked, and revoked over time. Preventing it requires locksmiths and organisations to treat key control as a managed system, not a one-off activity.

The most important step is establishing a formal key management system. Every key issued should be recorded in a central key register that documents who holds the key, what it opens, when it was issued, and when it is due to be returned. This creates accountability and visibility. Without this foundation, key sprawl develops quickly because there is no reliable way to track or control distribution. In more advanced environments, electronic key management systems or asset tracking software can automate this process and provide real-time oversight.

A critical control measure is the use of restricted or patented key systems. Unlike standard keys, restricted systems prevent unauthorised duplication by requiring proof of authority before copies can be cut. This directly addresses one of the main drivers of key sprawl—uncontrolled copying. By ensuring that every additional key must be authorised and recorded, the organisation retains control over how access expands.

Effective access design using the principle of least privilege is also essential. Keys should only provide access to the specific areas required for a person's role, rather than broad or universal access. Well-designed master key systems should limit the number of high-level keys and avoid unnecessary overlap between access zones. By reducing the number of people who hold powerful keys, the risk of widespread compromise and uncontrolled distribution is significantly reduced.

Another key strategy is implementing strict issuance and return procedures. Keys should only be issued following formal authorisation, and recipients should acknowledge responsibility for their use and return. When staff leave, change roles, or no longer require access, keys must be recovered promptly. This process should be integrated with HR onboarding and offboarding procedures to ensure that access control is maintained throughout the employee lifecycle.

Regular audits and reconciliations are necessary to prevent gradual loss of control. Periodic checks should confirm that all issued keys are accounted for and held by the correct individuals. Any discrepancies—such as missing keys or unrecorded holders—should trigger immediate investigation and corrective action, such as rekeying affected locks. Audits also help identify patterns of over-issuance or poor practices that may lead to future sprawl.

Locksmiths and organisations should also address the physical handling of keys through secure storage and handling practices. Keys not in active use should be stored in controlled environments such as locked key cabinets or electronic key safes with access logging. Policies should prohibit key sharing, unattended storage, or informal handovers between staff, as these behaviours quickly lead to loss of control.

In environments where risks are higher or where key sprawl has already become unmanageable, transitioning to electronic access control systems can provide additional safeguards. Electronic systems allow for easier issuance, modification, and revocation of access without physically replacing locks. They also provide audit trails, enabling organisations to monitor who accessed specific areas and when. However, these systems must still be managed properly to avoid digital forms of sprawl, such as excessive or outdated permissions.

Finally, education and organisational culture play an important role. Staff must understand that keys are security assets, not convenience tools. Training should reinforce the importance of accountability, proper use, and immediate reporting of lost or stolen keys. When users understand the risks and their responsibilities, compliance with key control procedures improves significantly.

The consequences of inadequate tracking systems are particularly severe when keys are lost, stolen, or not recovered from departing employees [201]. Without a formal key register or systematic tracking mechanism, organizations lack the ability to accurately assess the scope and magnitude of security risks resulting from missing or unaccounted-for keys [201]. This absence of visibility prevents the implementation of timely corrective actions and undermines the organization's capacity to maintain effective access control [201].

Modern approaches to addressing tracking deficiencies have emerged through technological solutions. Smart key management systems utilizing RFID technology and microcontroller-based platforms can automatically log every access event, creating comprehensive audit trails ([202]. These systems record not only who accessed which areas, but also when access occurred, enabling organizations to maintain detailed accountability records [202]. Additionally, blockchain-based approaches have been proposed to create immutable access logs that cannot be tampered with, providing verifiable records of all access events across multiple sites [203].

### Uncontrolled Key Duplication

A fundamental vulnerability in traditional key management systems is the ease with which keys can be duplicated without authorization or knowledge [201]. Standard mechanical key systems allow keys to be copied at hardware stores or by unauthorized individuals, meaning that even when an organization carefully controls initial key distribution, additional copies may be created outside of official channels [201]. This uncontrolled duplication results in the number of individuals with access expanding beyond intended limits, thereby compromising the integrity of the entire access control system [201].

The problem of uncontrolled duplication is particularly acute when master keys or high-level access credentials are involved [201]. Master keys, by their nature, provide access to multiple areas or systems, and their unauthorized duplication can result in widespread security breaches affecting large portions of a facility [201]. To address this vulnerability, restricted key systems have been developed that employ patented key profiles and specialized manufacturing processes to prevent unauthorized duplication [201]. These systems require that keys be duplicated only through authorized channels, typically by the original key manufacturer or licensed locksmiths, thereby maintaining control over the total number of keys in circulation.

### Master Key System Vulnerabilities

While master key systems are designed to provide operational convenience and enable hierarchical access control, they introduce significant security risks when not properly designed and managed [201]. The fundamental problem arises when master keys are distributed widely throughout an organization for convenience rather than being restricted to essential personnel only [201]. This widespread distribution creates a single point of failure, if a master key is lost or duplicated, large portions of the facility become vulnerable to unauthorized access [201].

Research on hierarchical access control systems has demonstrated that effective key management for hierarchies requires careful design to balance security with operational efficiency [204]. Atallah et al. propose a solution wherein keys for descendant classes in an access hierarchy can be derived from parent class keys through efficient key derivation mechanisms, while maintaining security against collusion attacks (Atallah et al., 2009). This approach allows organizations to implement hierarchical access structures where users at higher levels can derive keys for lower-level areas, but the system is designed to prevent unauthorized key derivation and to enable efficient revocation of access rights [204].

Poorly planned master key systems often lack clear zoning or organizational structure, resulting in excessive access rights that do not align with actual job roles and responsibilities [201]. This misalignment between access rights and organizational roles creates unnecessary security exposure and violates the principle of least privilege [201].

### Failure to Revoke Access Upon Role Changes or Departure

A critical weakness in many organizational access control systems is the failure to revoke access when employees change roles, transition to different departments, or depart from the organization entirely [201]. When employees leave, organizations often neglect to recover keys or update access permissions, resulting in former staff and contractors retaining access long after their legitimate need has ended [201]. In mechanical key systems, revoking access typically requires rekeying locks, a process that is frequently delayed or avoided due to cost considerations or operational inconvenience [201].

The problem of access revocation is particularly significant in modern electronic access systems, where failure to deactivate credentials can have consequences equivalent to those in mechanical systems [201]. Research on certificate revocation in public key infrastructure systems demonstrates the complexity of implementing effective revocation mechanisms at scale [205]. Research proposes an Accountable and Transparent TLS Certificate Management system that makes certificate revocation publicly verifiable and introduces improved revocation policies to address vulnerabilities in conventional PKI systems [205].

> An Accountable and Transparent TLS Certificate Management system is an enhanced approach to managing digital certificates used in secure communications (such as HTTPS), designed to overcome weaknesses in traditional Public Key Infrastructure (PKI). In conventional PKI, Certificate Authorities (CAs) issue and revoke certificates, but the processes for issuance and especially revocation can lack visibility, timeliness, and independent verification. This creates risks where compromised or misissued certificates may remain trusted longer than they should, enabling impersonation or man-in-the-middle attacks.
>
> This improved system introduces two key concepts: transparency and accountability. Transparency is achieved through publicly accessible, append-only logs (such as Certificate Transparency logs) where all issued certificates—and potentially revocation events—are recorded. Accountability means that all actions taken by certificate authorities (issuing, renewing, revoking certificates) can be independently audited and verified by third parties. In this model, revocation is no longer a hidden or optional process; it becomes publicly visible, cryptographically verifiable, and tamper-evident.
>
> A core problem this system addresses is the weakness of traditional revocation mechanisms like CRLs (Certificate Revocation Lists) and OCSP (Online Certificate Status Protocol). These methods can be slow, unreliable, or ignored by clients. For example, browsers may “soft-fail” if they cannot

reach an OCSP server, meaning a revoked certificate might still be accepted. An accountable and transparent system replaces or supplements these mechanisms with public, verifiable revocation records, ensuring that all parties can confirm whether a certificate is still valid.

Implementation of such a system involves several components. First, organisations must integrate with Certificate Transparency (CT) logs, ensuring that every issued certificate is logged in a publicly auditable system. This is already standard practice for modern web certificates. To extend this to revocation, systems can adopt revocation transparency logs, where every revocation event is also recorded in an append-only structure. These logs are cryptographically secured using Merkle trees, allowing anyone to verify that entries have not been altered or removed.

Second, organisations need to implement automated certificate lifecycle management. This includes tracking certificate issuance, expiry, renewal, and revocation in a centralised system. Automation ensures that certificates are renewed before expiry and revoked immediately when compromise is suspected. Integration with monitoring tools allows detection of anomalies, such as unexpected certificates issued for a domain, which can then trigger investigation or revocation.

Third, improved revocation policies must be defined and enforced. This includes clearly specifying when certificates must be revoked (e.g. key compromise, misissuance, organisational changes), how quickly revocation must occur, and how revocation status is communicated. Policies should require immediate logging of revocation events to transparency systems and may include mechanisms such as short-lived certificates, which reduce reliance on revocation by expiring quickly.

Fourth, client-side systems (such as browsers, applications, or internal services) should be configured to enforce strict verification. This may include requiring proof that a certificate is present in a transparency log and checking revocation status through verifiable mechanisms. Emerging approaches, such as OCSP stapling or Must-Staple certificates, improve reliability by ensuring that revocation information is delivered directly during the TLS handshake.

From an organisational perspective, implementation also requires governance and auditing processes. Regular audits should verify that all certificates are properly logged, that revocation events are recorded and visible, and that no unauthorised certificates exist. Security teams should monitor transparency logs for certificates issued for their domains, enabling rapid detection of potential compromise.

Although this concept originates in cybersecurity, it has strong parallels to locksmithing and physical security. Just as key control systems require accountability, audit trails, and controlled issuance and revocation, digital certificate management systems must ensure that trust is continuously verifiable and not assumed. In both domains, the goal is to prevent unauthorised access not only by strengthening the “lock,” but by ensuring that the credentials used to access it are tightly controlled, monitored, and revocable in a transparent and trustworthy manner.

In cloud computing environments, the challenge of credential revocation becomes even more complex due to the distributed nature of systems and the need to maintain consistency across multiple nodes [206]. Research proposes a blockchain-based revocation framework that uses a federated time-bound approach to enable transparent revocation and maintain audit trails, allowing the system to deny decryption even if access policies technically match [206]. This approach demonstrates how blockchain technology can enforce revocation across distributed systems by maintaining a verifiable ledger of revoked credentials [206].

A blockchain-based revocation framework with a federated, time-bound approach is an advanced access-control model designed to solve a key limitation in traditional encryption and PKI systems: once access is granted (e.g. a key or credential is valid), it can be difficult to reliably and immediately revoke that access. This framework introduces a system where revocation decisions are distributed, transparent, time-limited, and enforceable, even after credentials have been issued—meaning a user can be prevented from decrypting data even if they technically still satisfy the original access policy.

At a conceptual level, the system combines three core elements. First, blockchain technology is used as a tamper-resistant, shared ledger to record access rights, revocation events, and policy updates. This ensures that all actions—granting access, modifying permissions, or revoking rights—are transparent, auditable, and cannot be altered retrospectively. Second, the federated model means that control is not held by a single authority; instead, multiple trusted entities (e.g. departments, organisations, or system nodes) collectively validate and record revocation decisions. Third, the time-bound approach ensures that access rights are inherently temporary, requiring periodic renewal, which reduces reliance on reactive revocation alone.

The key innovation is that revocation is actively enforced at the point of decryption, not just at the point of access approval. In traditional systems, if a user has a valid key, they may still be able to decrypt previously obtained data even after their access should have been revoked. In this framework, decryption requires checking the current state of permissions—stored on the blockchain—so if a revocation has been recorded, the system can deny decryption in real time, regardless of whether the user's original credentials still exist.

Implementation begins with establishing a permissioned blockchain network involving relevant stakeholders. In a federated environment, this might include multiple organisations, departments, or trusted authorities that jointly manage access control. Each participant operates a node that validates transactions, such as granting or revoking access. The blockchain maintains a secure, append-only record of all access-related events, ensuring transparency and auditability.

Next, access control must be integrated with encryption and key management systems, often using approaches such as attribute-based encryption (ABE) or policy-based encryption. In these systems, access to encrypted data is governed by policies (e.g. role, location, time). The blockchain is used to store and update these policies, including revocation conditions. When a user attempts to decrypt data, the system checks the blockchain to confirm that their access rights are still valid and have not been revoked or expired.

The time-bound component is implemented by issuing access credentials or decryption keys that are valid only for a defined period. These may require periodic revalidation or renewal, which ensures that access does not persist indefinitely. If a user's role changes or a security issue arises, access can simply expire or be explicitly revoked on the blockchain, preventing further decryption attempts without needing to recover or invalidate previously distributed keys.

A critical element is the design of revocation policies and smart contracts. Smart contracts—self-executing code on the blockchain—can automate revocation decisions based on predefined rules. For example, access may be revoked automatically if a user leaves an organisation, fails authentication checks, or if a federated authority submits a revocation request. These contracts ensure that revocation is applied consistently and immediately across the system.

To make the system practical, organisations must also implement client-side enforcement mechanisms. Applications or devices that perform decryption must be configured to check the blockchain (or a trusted interface to it) before allowing access to data. This ensures that revocation is not just recorded, but actually enforced. Without this step, the system would revert to a passive logging mechanism rather than an active control.

Governance and auditing are also essential. Because the system is federated, clear rules must define who can grant or revoke access, how consensus is achieved, and how disputes are handled. Regular audits of the blockchain ledger provide a complete history of access and revocation events, supporting compliance, forensic investigation, and accountability.

In practical terms, this framework is particularly useful in environments where data is highly sensitive and shared across multiple parties, such as government, defence, healthcare, or large enterprises. It ensures that access control is dynamic, verifiable, and resilient to insider threats or delayed revocation.

From a broader perspective, this approach reflects the same principles seen in advanced locksmithing and physical security: access should not be permanent, should be traceable, and must be revocable at any time, regardless of previous permissions. By combining blockchain transparency, federated control, and time-limited access, this framework creates a system where trust is continuously validated, rather than assumed.

#### Inadequate Policies and Procedures

The absence of clear, formal guidelines governing the issuance, use, storage, and return of keys contributes significantly to key management failures [201]. Without documented procedures, staff members may develop inconsistent or insecure practices, such as sharing keys between employees, leaving keys unattended, or storing keys in unsecured locations [201]. The lack of formal procedures also impedes the organization's ability to enforce accountability or respond effectively to security incidents such as lost or stolen keys [201].

Effective key management requires the establishment of comprehensive policies that address the entire lifecycle of keys and access credentials [201]. These policies should specify who is authorized to issue keys, under what circumstances keys may be issued, how keys should be stored when not in use, and what procedures must be followed when keys are lost or when access needs to be revoked [201]. Additionally, policies should address the handling of master keys and high-level access credentials, which require more stringent controls than standard keys [201].

### Disconnection Between Physical and Organizational Security Systems

A significant structural failure in many organizations is the treatment of key management as an isolated function rather than as an integrated component of a comprehensive access control strategy [201]. This fragmentation often results in disconnects between human resources processes—such as employee onboarding and offboarding—and physical access control systems [201]. For example, an employee's information technology access may be revoked immediately upon departure, while their physical access to facilities remains unchanged [201].

Effective access control requires integration between multiple organizational systems and processes [201]. When an employee is hired, the onboarding process should simultaneously provision both IT access and physical access credentials. Conversely, when an employee departs, the offboarding process should simultaneously revoke both IT and physical access [201]. This integration ensures that access control decisions are made consistently across all systems and that no gaps exist in the revocation process.

Modern identity management systems have begun to address this integration challenge through centralized identity governance platforms. Research examines the role of Privileged Access Management (PAM) systems as policy-driven orchestration layers for managing both human and machine identities across distributed environments [207]. These systems can serve as integration points between HR processes and access control systems, enabling automated provisioning and deprovisioning of access credentials [207].

Privileged Access Management (PAM) systems are specialised security platforms designed to control, monitor, and govern access to critical systems and sensitive resources. When described as policy-driven orchestration layers, PAM systems act as a central control point that enforces rules (policies) and coordinates how privileged access is granted, used, and revoked across complex, distributed environments. Rather than simply storing credentials, modern PAM solutions actively manage who can access what, under which conditions, for how long, and with what level of oversight—for both human users and machine identities (such as applications, services, and automated processes).

In traditional environments, privileged accounts—such as administrators, system operators, or service accounts—often have broad, persistent access. This creates significant risk, as these accounts can bypass normal controls. PAM systems address this by introducing controlled, temporary, and fully auditable access. As an orchestration layer, PAM integrates with identity systems, infrastructure, applications, and security tools to ensure that access is not static, but dynamically governed by policy. For example, a user may request elevated access, which is then approved, provisioned, monitored, and automatically revoked—all coordinated by the PAM system.

A key feature of PAM is its ability to manage both human identities and machine identities. Human users include administrators or support staff who require elevated privileges to perform tasks. Machine identities include scripts, APIs, containers, and services that need credentials to operate. In distributed environments—such as cloud platforms, hybrid infrastructure, and microservices architectures—the number of machine identities often exceeds human users, making automated credential management essential. PAM systems ensure that these identities are authenticated, authorised, and rotated securely, reducing the risk of credential exposure.

Implementation begins with defining access policies aligned to organisational roles and risk levels. These policies determine who can request privileged access, what approvals are required, what systems can be accessed, and under what conditions (e.g. time, location, device). This reflects the principle of least privilege, ensuring that users and systems only receive the access necessary to perform their functions. Policies also define session controls, such as recording activity, restricting commands, or requiring multi-factor authentication.

Next, organisations deploy a centralised credential vault, where sensitive credentials (passwords, keys, tokens) are securely stored and managed. Instead of users knowing or handling these credentials directly, the PAM system brokers access on their behalf. For example, when an administrator needs access to a server, the PAM system retrieves the credential, initiates the session, and logs all activity without exposing the password to the user. This reduces the risk of credential leakage and supports full auditability.

Another critical component is just-in-time (JIT) access provisioning. Rather than granting permanent privileges, PAM systems issue temporary access that expires automatically after a defined period. This significantly reduces the attack surface, as there are no standing privileges to exploit. Access can also be tied to specific tasks, ensuring that elevated permissions are only available when needed and only for the duration required.

For machine identities, PAM systems implement automated credential rotation and lifecycle management. Service accounts, API keys, and certificates are regularly updated without manual intervention, preventing long-lived credentials from becoming a vulnerability. Integration with DevOps pipelines and cloud platforms ensures that secrets are injected securely at runtime rather than hardcoded into applications.

Monitoring and auditing are central to PAM functionality. All privileged sessions are logged, recorded, and analysed, providing a complete audit trail of who accessed what, when, and what actions were performed. Advanced systems may include real-time monitoring and anomaly detection, enabling organisations to identify suspicious behaviour and respond quickly.

In distributed environments, PAM must integrate with a wide range of systems, including identity providers (e.g. Active Directory, IAM platforms), cloud services, container environments, and endpoint devices. This integration allows PAM to function as a unifying control layer, enforcing consistent policies across diverse infrastructure. APIs and automation tools enable PAM to scale and adapt to dynamic environments where resources are constantly being created and removed.

From a governance perspective, organisations must establish clear roles, responsibilities, and approval workflows. This includes defining who can grant access, how requests are reviewed, and how compliance is maintained. Regular audits and reviews ensure that access remains aligned with organisational needs and that unused or excessive privileges are removed.

In essence, PAM systems transform privileged access from a static, high-risk condition into a controlled, policy-driven process. They ensure that access is temporary, monitored, and aligned with defined rules, while also providing visibility and accountability across both human and machine interactions.

From a broader perspective, this mirrors principles seen in physical security and locksmithing. Just as restricted key systems, key registers, and controlled issuance prevent unauthorised access in physical environments, PAM systems provide the digital equivalent—ensuring that access to critical systems is limited, traceable, and revocable at all times.

## Electronic Access Systems and Digital Credential Management

While electronic access systems offer significant improvements in control and auditability compared to mechanical key systems, they remain vulnerable to poor management practices [201]. Weak PIN codes, shared access cards, failure to monitor access logs, and delayed revocation of credentials can all compromise the security of electronic systems [201]. If electronic systems are not properly configured or maintained, they may replicate the same weaknesses as traditional key systems in a digital form [201].

Smart key management systems utilizing RFID technology and electronic locks can address many of these vulnerabilities through automated logging and access control [202]. Research describes a smart key prototype management system that uses RFID technology, NodeMCU microcontrollers, and solenoid locks to authorize access while logging every activity in a database for security and monitoring purposes [202]. The system includes a web admin portal that enables administrators to manage user profiles, access schedules, and activity tracking, providing centralized control over access credentials [202].

Electronic access systems also enable more sophisticated access control policies than mechanical systems. For example, access can be restricted not only by location but also by time of day, day of week, or other contextual factors [202]. Additionally, electronic systems can implement role-based access control, wherein access rights are automatically adjusted based on an employee's current role and responsibilities [202].

However, electronic systems introduce new security challenges related to the management of digital credentials. Other research describes a secure file-sharing system that uses session key mechanisms to control access to sensitive files [208]. In their system, employees receive one-time session keys through email when attempting to download files, and access is permitted only through the correct key [208]. If an employee enters an incorrect key, their account is promptly blocked, preventing unauthorized access [208]. This approach demonstrates how electronic systems can implement more granular access control than traditional key systems, but it also highlights the need for careful management of digital credentials.

## Integration of Cryptographic and Access Control Technologies

Modern approaches to access control increasingly integrate cryptographic technologies with traditional access control mechanisms. Nagarajan et al. propose an identity-based encryption approach for cloud security wherein data owners generate encryption keys based on user identities, and these keys are distributed through secure channels [209]. This approach enables fine-grained access control wherein different users can be granted access to different data based on their identities and attributes [209].

Further research describes a cloud-based data sharing system that uses the Verifiable yet Accountable Keyword Searchable Encryption (VAKSE) scheme to enable secure data sharing while maintaining verifiability and accountability [210]. In this system, data owners upload files and generate keys to securely share data with intended clients, maintaining control over data at all times [210]. The system includes a verification mechanism wherein clients can verify the authenticity of data and detect invalid ciphertexts, enhancing the robustness of the access control system [210].

### Recommendations for Effective Key Management

For locksmiths and security professionals, addressing key management failures requires the implementation of structured, comprehensive key management systems [201]. These systems should incorporate several key components: restricted key systems to prevent unauthorized duplication, detailed key registers to maintain accountability, master key systems designed with clear hierarchy and minimal exposure, and formal procedures for issuing and recovering keys [201]. Regular audits and reviews are essential to ensure that access rights remain aligned with current organizational needs and that no unauthorized keys remain in circulation [201].

The implementation of smart key management systems utilizing RFID technology and electronic locks can significantly improve the effectiveness of key management by automating logging, enabling centralized control, and facilitating rapid revocation of access [202]. These systems should be integrated with organizational HR systems to ensure that access is automatically provisioned upon hire and revoked upon departure [202].

Additionally, organizations should implement comprehensive policies and procedures that govern all aspects of key management, from initial issuance through final recovery [201]. These policies should address the handling of master keys, the procedures for reporting lost or stolen keys, the requirements for key storage, and the processes for revoking access [201]. All staff members should receive training on these policies and procedures to ensure consistent implementation [201].

Key management failures represent a fundamental vulnerability in organizational security that is frequently overlooked in favour of more visible security measures such as locks and access systems [201]. Effective security requires that organizations recognize key management as a critical component of their overall access control strategy and implement comprehensive systems, policies, and procedures to ensure that keys and access credentials are properly issued, tracked, used, and recovered [201]. By addressing the failures identified in this analysis, including inadequate tracking, uncontrolled duplication, poor master key design, failure to revoke access, inadequate policies, disconnection from organizational systems, and poor management of electronic credentials. organizations can significantly improve the effectiveness and resilience of their access control systems [201].

The integration of modern technologies such as RFID-based smart key systems, blockchain-based audit logs, and centralized identity management platforms can substantially enhance key management effectiveness [202, 203]. However, technology alone is insufficient; organizations must also establish clear policies, procedures, and governance structures to ensure that key management practices are consistently implemented and regularly reviewed [201]. Only through this combination of technical solutions, organizational policies, and ongoing oversight can organizations achieve secure, accountable, and resilient access control systems [201].

## Impact on Locksmithing Practices

This topic is highly relevant to someone interested in locksmithing because it reveals that locks themselves are only one part of a much larger security system. A developing locksmith may initially focus on hardware, cylinders, mechanisms, installation techniques, but this analysis demonstrates that the real point of control is not the lock, but the management of the keys and credentials that operate it. Without effective key management, even the most advanced locking systems fail, meaning that locksmithing must be understood as a discipline of access control management, not just mechanical expertise.

For someone entering the profession, this knowledge is essential because it explains why security failures frequently occur despite correct installation of high-quality hardware. Issues such as key sprawl, uncontrolled duplication, or failure to revoke access are not technical faults in the lock—they are failures in system management. Recognising this allows locksmiths to diagnose problems more accurately and avoid the common mistake of repeatedly upgrading hardware when the underlying issue is poor key control.

The influence on locksmithing practice is significant, as it expands the role of the locksmith into that of a system designer and security consultant. Locksmiths are no longer simply supplying locks and cutting keys; they are responsible for designing structured key systems, implementing restricted keyways, and advising clients on how to maintain control over access. This includes developing key registers, issuing protocols, and audit processes that ensure accountability throughout the lifecycle of every key.

Key management failures such as lack of tracking and accountability directly influence how locksmiths structure their work. A professional locksmith must ensure that every key issued is documented, that systems are designed to support traceability, and that clients understand the importance of maintaining records. This shifts locksmithing practice toward a more formal, policy-driven approach, particularly in commercial and institutional environments.

The issue of uncontrolled key duplication has also driven the widespread adoption of restricted and patented key systems in modern locksmithing. Locksmiths must now consider not just how a key operates a lock, but how its duplication is controlled. This requires knowledge of proprietary systems, legal controls over key blanks, and client education on why restricted systems are critical for maintaining security integrity.

Failures in master key system design further influence locksmithing by highlighting the need for careful planning and adherence to the principle of least privilege. Locksmiths must design systems that balance convenience with security, ensuring that access is limited to what is necessary for each role. Poorly designed systems can create large-scale vulnerabilities, meaning that locksmiths must think in terms of hierarchies, zoning, and risk exposure, rather than simple key functionality.

The failure to revoke access—particularly when staff leave or roles change—emphasises the importance of lifecycle management in locksmithing practice. Locksmiths must advise on rekeying strategies, key recovery procedures, and integration with organisational processes such as HR onboarding and offboarding. This aligns locksmithing with broader security practices, where access is continuously reviewed and adjusted rather than statically assigned.

Modern developments, such as electronic access systems, blockchain audit trails, and Privileged Access Management (PAM) concepts, further extend locksmithing into the digital domain. These technologies introduce ideas such as auditability, transparency, and automated revocation, which have direct parallels in physical security. For locksmiths, this means developing an understanding of both mechanical and digital access control, as the two are increasingly integrated in modern security environments.

Importantly, this topic reinforces the idea that locksmithing is fundamentally about controlling trust and access, not just securing doors. Keys—whether physical or digital—are credentials that must be issued, monitored, and revoked with precision. The locksmith's role is to ensure that this process is structured, accountable, and resilient to failure.

## Insider Threats and Access Misuse

Insider threats and access misuse refer to security risks that originate from individuals who already have legitimate access to a system, facility, or information, but use that access inappropriately—either intentionally or unintentionally. In locksmithing and physical security contexts, this is one of the most significant risks because it bypasses traditional protective measures. Unlike external threats that must overcome locks, alarms, or barriers, insiders already possess the credentials—keys, access cards, codes, or permissions—needed to enter secured areas.

Insider threats can be broadly categorised into three types. Malicious insiders deliberately misuse their access for personal gain, sabotage, theft, or unauthorised disclosure. For example, an employee may use a master key to access restricted areas after hours. Negligent insiders do not intend harm but create vulnerabilities through poor practices, such as sharing keys, writing down access codes, or failing to secure doors. Compromised insiders are individuals whose access has been exploited by others, often through social engineering or coercion, allowing external actors to operate under legitimate credentials.

A key issue in insider threats is that access is often over-provisioned or poorly controlled. Individuals may hold keys or permissions that exceed their actual job requirements, violating the principle of least privilege. Over time, this leads to environments where many people have broad access, making it difficult to detect misuse or attribute responsibility. For example, if multiple staff members share a master key or access code, it becomes nearly impossible to determine who accessed a restricted area during a security incident.

Access misuse also occurs when there is no effective monitoring or audit trail. In traditional mechanical systems, once a key is issued, its use is largely invisible. There is no inherent record of when or where it was used. This lack of visibility creates opportunities for misuse without detection. Even in electronic systems, if access logs are not regularly reviewed, suspicious activity may go unnoticed.

Another contributing factor is failure to revoke or update access. Employees who change roles or leave an organisation may retain keys or credentials longer than they should. Contractors or temporary staff may be given access for convenience, but that access is not removed when their work is complete. This creates a growing pool of individuals with legitimate but outdated access rights, increasing the risk of misuse.

Insider threats are also closely linked to organisational culture and awareness. When security policies are unclear or poorly enforced, staff may develop informal practices such as sharing keys, bypassing locks, or leaving access points unsecured. Over time, these behaviours become normalised, significantly weakening the overall security posture.

For locksmiths and security professionals, addressing insider threats requires a shift toward controlled, accountable, and monitored access systems. This begins with designing key systems based on the principle of least privilege, ensuring that individuals only have access to the areas necessary for their role. Master keys and high-level access credentials should be strictly limited and carefully tracked.

The use of restricted key systems helps prevent unauthorised duplication, ensuring that access cannot be expanded without control. Key registers and issuance logs create accountability by recording who holds each key and when it was issued. Regular audits can identify discrepancies, such as missing keys or individuals holding inappropriate access.

In higher-risk environments, transitioning to electronic access control systems provides additional safeguards. These systems can log access events, enforce time-based restrictions, and allow rapid revocation of credentials. For example, an employee's access card can be deactivated immediately upon termination, eliminating the need for physical rekeying. Advanced systems may also include alerts for unusual access patterns, supporting early detection of misuse.

Policies and procedures are equally important. Organisations must establish clear rules for issuing, using, and returning keys, as well as reporting lost or stolen credentials. Training programs can reinforce awareness of insider risks and encourage responsible behaviour. Staff should understand that access is a privilege tied to their role, not a convenience to be shared or extended.

Ultimately, insider threats highlight a fundamental reality in locksmithing and security: the greatest risk often comes from those who already have access. Effective security is not just about keeping unauthorised people out, but about ensuring that authorised access is used appropriately, monitored continuously, and revoked when no longer needed. By combining system design, technology, policy, and education, locksmiths can help organisations reduce the risk of insider misuse and maintain stronger, more resilient access control systems.

### Impact on Locksmithing Practices

Understanding insider threats and access misuse is highly relevant to anyone interested in locksmithing because it fundamentally changes how security is perceived and delivered in real-world environments. A developing locksmith may initially focus on preventing unauthorised entry through strong hardware, such as high-security cylinders, reinforced doors, and advanced locking mechanisms. However, this topic demonstrates that many security failures do not occur because locks are defeated, but because access is misused by individuals who are already authorised. This shifts locksmithing from a purely mechanical discipline to one that must also address human behaviour, organisational processes, and access governance.

For someone entering the profession, this knowledge is critical because it explains why well-installed and high-quality locking systems can still fail. Without an understanding of insider threats, a locksmith may incorrectly assume that improving hardware alone will improve security. In reality, if keys are over-issued, shared, or not recovered, or if access is not properly controlled, even the most sophisticated system becomes ineffective. Recognising this allows locksmiths to diagnose security weaknesses more accurately and provide solutions that address the root cause rather than just the visible symptoms.

The influence on locksmithing practice is significant, as it expands the role of the locksmith beyond installation and repair into that of a security system designer and advisor. Locksmiths must consider not only how a lock functions, but who will use it, how access will be managed, and how misuse can occur. This directly impacts how master key systems are designed, encouraging the use of structured hierarchies, limited distribution of master keys, and strict alignment with the principle of least privilege. It also reinforces the importance of avoiding convenience-driven decisions, such as issuing broad access to multiple users, which can introduce long-term vulnerabilities.

Insider threats also drive the need for accountability and traceability in locksmithing practice. Traditional mechanical systems provide limited visibility once keys are issued, which means locksmiths must compensate through strong key control measures such as registers, issue logs, and restricted key systems. In higher-risk environments, this has led to increased adoption of electronic access control systems that provide audit trails and real-time monitoring. As a result, modern locksmithing increasingly involves integrating mechanical and electronic solutions to achieve both physical security and accountability.

Another key influence is the emphasis on lifecycle management of access. Locksmiths must consider not just the installation of a system, but how access will be issued, monitored, and revoked over time. This includes aligning key control with organisational processes such as staff onboarding, role changes, and termination. Failure to manage this lifecycle results in accumulated access rights, which is a primary driver of insider risk. This perspective encourages locksmiths to design systems that are maintainable and adaptable, rather than static.

The topic also reinforces the importance of policies, procedures, and user education as part of locksmithing practice. Security is not achieved through hardware alone; it requires users to understand their responsibilities and follow consistent processes. Locksmiths therefore play a role in advising clients on best practices, such as not sharing keys, reporting lost credentials, and adhering to access control procedures. This educational component becomes a routine part of professional service delivery.

In modern contexts, insider threats highlight the growing need for locksmiths to understand electronic and digital access systems. As organisations move toward card-based, biometric, or smart lock systems, the risks associated with access misuse remain, but are expressed through digital credentials rather than physical keys. Locksmiths must therefore be capable of configuring, managing, and advising on these systems, including access levels, audit logs, and rapid revocation mechanisms.

## Designing Systems that Account for Human Error

### System Design Concepts

Designing systems that account for human error means building physical security and access control solutions that remain effective even when users are distracted, rushed, forgetful, undertrained, or acting for convenience. In locksmithing, this is a core practical reality: most "security failures" do not occur because a lock was technically defeated, but because the system was used inconsistently or incorrectly. Human error is predictable, so good security design assumes it will happen and reduces the likelihood that small mistakes become major vulnerabilities.

A fundamental principle is to reduce reliance on perfect behaviour. If a system depends on every user remembering to lock a deadbolt, never lending keys, always reporting losses immediately, and never propping a door open, it will eventually fail. The goal is to design systems where safe and secure outcomes occur by default. This involves selecting hardware and access methods that either automate critical security actions or make insecure behaviour difficult to maintain over time.

One of the most effective design strategies is using fail-secure defaults. In physical security, this means doors should default to being locked rather than unlocked, and secure closure should happen automatically where appropriate. Examples include deadlatching mechanisms that engage reliably when a door closes, self-locking locks where suitable, or electric strikes/maglocks that require deliberate, authorised input to release. The system should not require a separate step that is easy to forget; instead, normal use should naturally result in secure locking.

Another core approach is building in positive confirmation and feedback. Many user errors happen because people assume something is locked when it is not. Systems can reduce this through clear tactile and audible cues (a deadbolt that throws smoothly and firmly), visible indicators (locked/unlocked window displays), or electronic feedback (status lights, beeps, or app confirmations). In commercial settings, door position sensors and alarms can alert staff if a door is not fully closed, preventing the common error of "almost latched" doors that look shut but are not secured.

Security design should also account for misalignment and wear, which frequently cause human error indirectly. When doors sag, strike plates shift, or closers lose adjustment, users start "forcing" operation—slamming doors, leaving them ajar, or failing to lock because it feels stiff. Locksmiths can design for tolerance by selecting robust hardware, reinforcing frames, using adjustable strike plates, and ensuring closers and hinges are properly specified for door weight and usage. A system that remains smooth and reliable under real conditions reduces user workarounds.

Human error is strongly influenced by convenience, so systems should make the secure option the easiest option. This is known as secure usability. For example, if staff must carry multiple keys and repeatedly unlock different doors, they will tend to prop doors open. A better design may be a well-zoned master key system (least privilege), access cards, or time-limited credentials that reduce friction while maintaining control. If the secure workflow is quicker than the insecure workaround, compliance improves dramatically.

Another major design principle is limit the blast radius of mistakes. Because mistakes will happen, a system should be compartmentalised so that one lost key or one compromised credential does not expose the entire facility. This is achieved through zoning, layered security, and careful hierarchy design. For instance, rather than issuing a master key widely "for convenience," the system can provide role-based sub-masters and individual keys for specific areas. In electronic systems, access can be segmented by location and time, ensuring that errors or misuse affect only a limited area and only within defined conditions.

Designing for human error also requires strong key control and credential governance. People lose keys, share them, duplicate them, or fail to return them. Locksmiths can reduce this risk by specifying restricted key systems, requiring authorisation for duplication, and implementing key registers and issue/return logs. In higher-risk environments, electronic credential systems make revocation easier, but only if the organisation has disciplined processes for onboarding, offboarding, and role changes. The key point is that design must assume credentials will drift out of control unless lifecycle management is built in.

A practical method for handling human error is implementing time-bound access. In physical systems this can include temporary keys or controlled key issue processes; in electronic systems it is more direct, such as access that expires automatically after a shift ends or after a contractor's job is complete. Time-bound permissions reduce the harm caused when users forget to revoke access or return keys, because access is designed to decay automatically unless renewed.

Error-proofing (poka-yoke) is another useful concept from safety and manufacturing that can be applied to locksmithing. It involves designing systems that prevent incorrect actions or make them immediately obvious. Examples include keyways that prevent the wrong key being inserted, captive key systems that force a user to lock a door before removing the key, and door hardware that physically cannot remain in an insecure state without triggering an alarm or drawing attention. These approaches reduce reliance on memory and vigilance.

Training and signage still matter, but they are weakest when used alone. The best approach is design + training, where the system nudges correct behaviour and training reinforces it. Short, practical user instructions, such as "Close door firmly until latch clicks" or "Deadbolt must be engaged after hours", are more effective when supported by hardware that makes correct operation easy and incorrect operation noticeable. In workplaces, refresher training and toolbox talks can be aligned with observed failure points, such as propped doors or shared keys.

Maintenance is also a human-error control measure. Systems degrade, and users respond with unsafe workarounds. Designing for human error includes establishing planned maintenance schedules: inspection of alignment, lubrication, closer speed, strike plate integrity, key wear, and cylinder function. This keeps operation smooth and prevents the gradual transition from "secure and easy" to "stiff and annoying," which is when users start bypassing controls.

In electronic and smart lock environments, designing for human error includes protecting users from poor digital habits. Systems can enforce strong PIN rules, limit retries, require multi-factor authentication for sensitive areas, automatically revoke temporary codes, and provide audit logs and alerts for unusual access patterns. Importantly, systems should be designed so that losing a phone or forgetting a code does not force staff to disable security; robust fallback procedures (secure override protocols) prevent emergency workarounds becoming permanent vulnerabilities.

## Designing a Master Key System

Designing a master key system is a structured process that combines security planning, mechanical design, and controlled implementation. It is not simply about cutting keys that open multiple locks; it is about creating a hierarchical access control system that aligns with how a building or organisation operates, while maintaining strict control over who can access what.

### 1. Planning Phase (Understanding Requirements)

The planning stage is the most critical, as poor planning leads to long-term security weaknesses that are difficult to correct.

The planning phase is the foundation of any effective master key system and is widely regarded as the most critical stage in the entire process. Decisions made at this point determine not only how the system will function on day one, but also how resilient, scalable, and secure it will remain over time. Poor planning often leads to excessive key distribution, overlapping access, and the need for costly rekeying or complete system redesign. As such, this phase requires a structured, methodical approach that integrates physical layout, operational requirements, and human behaviour.

The process begins with a comprehensive site assessment, which involves physically inspecting and documenting every access point within the facility. This includes external doors, internal doors, gates, plant rooms, storage areas, safes, cabinets, and any other lockable assets. Each entry point must be recorded and, where possible, mapped onto a floor plan. The locksmith evaluates not only the number and type of locks, but also the function of each space. For example, a front entrance serving the public has a very different risk profile compared to a server room, records archive, or cash handling area. This stage also involves identifying vulnerabilities such as poorly secured secondary doors, shared access points, or areas where access control is currently inconsistent.

Once the physical layout is understood, the next step is to assess the security requirements of each space. This involves determining the sensitivity of the assets within each area and the potential consequences of unauthorised access. High-value or high-risk areas—such as IT infrastructure rooms, safes, or restricted storage—require tighter control and limited access. In contrast, low-risk areas such as general workspaces or common areas may allow broader access. This risk-based approach ensures that the level of security is proportionate to the importance of what is being protected, rather than applying a uniform level of control across all areas.

The locksmith then works closely with the client to define user groups and roles, which is essential for aligning access with actual operational needs. This involves identifying who requires access, why they require it, and when access is needed. Typical roles may include cleaners, administrative staff, supervisors, managers, contractors, and security personnel. Each role is analysed in terms of its responsibilities and movement within the facility. The principle of least privilege is applied at this stage, ensuring that each group is granted only the minimum level of access necessary to perform its function. This reduces unnecessary exposure and limits the impact of lost or misused keys.

Following the identification of roles, the site is structured into access zones. Zoning is a critical design concept that groups areas based on function, security level, and usage patterns. Rather than assigning access on a door-by-door basis, zoning allows the locksmith to create logical groupings that simplify both system design and ongoing management. For example, public areas may form one zone, staff work areas another, management offices a third, and high-security rooms a fourth. This approach enables the creation of sub-master keys for each zone, reducing the need for widespread distribution of higher-level keys and improving overall control.

In addition to physical and organisational considerations, the planning phase must account for operational requirements. These are the practical, real-world conditions under which the system will be used. Emergency access is a key consideration, ensuring that authorised personnel can gain rapid entry during incidents such as fire, medical emergencies, or security breaches. After-hours access must also be addressed, particularly in facilities that operate beyond standard business hours. Maintenance access is another important factor, as service personnel may require temporary or periodic access to plant rooms, equipment, or infrastructure. These requirements must be carefully balanced to provide necessary access without compromising security.

Future-proofing is another essential component of planning. Organisations are dynamic, with staff changes, role changes, and physical expansions occurring over time. A well-designed master key system anticipates these changes by allowing for scalability and flexibility. This may involve reserving unused key combinations, planning additional zones, or designing the hierarchy in a way that can accommodate expansion without disrupting the existing system. Without this foresight, even minor changes can require significant reconfiguration.

Ultimately, the planning phase is about translating a complex environment—comprising people, spaces, and risks—into a structured access control strategy. It requires not only technical knowledge of locks and key systems, but also an understanding of how organisations function and how people interact with security systems. When done correctly, it creates a solid foundation for all subsequent stages, ensuring that the master key system is logical, secure, and aligned with real-world use.

The process begins with a detailed assessment of the site. This includes identifying all doors, entry points, restricted areas, and critical assets. Each space must be evaluated based on its function, sensitivity, and required level of security. For example, public areas, staff-only zones, management offices, and high-security rooms (such as server rooms or safes) all require different access levels.

Next, the locksmith works with the client to define user groups and roles. These may include cleaners, general staff, supervisors, managers, contractors, and security personnel. Each group should only have access to the areas necessary for their role, applying the principle of least privilege.

From this, the site is divided into access zones. Zoning ensures that access is structured logically rather than randomly. For example:

- Zone A: Public/common areas
- Zone B: Staff work areas
- Zone C: Management offices

- Zone D: High-security or restricted areas

At this stage, the locksmith must also identify operational needs, such as:

- Emergency access requirements
- After-hours access
- Maintenance access
- Future expansion of the system

Good planning ensures the system remains scalable and does not require complete redesign when changes occur.

2. System Design Phase (Creating the Key Hierarchy)

Once the planning phase has established zones, user groups, and access requirements, the locksmith moves into the system design phase, where these requirements are translated into a structured key hierarchy. This hierarchy defines the controlled relationship between keys and locks, ensuring that access is granted in a logical, layered, and secure manner. Rather than issuing keys randomly, the locksmith creates a system where each key has a clearly defined scope of operation aligned with the user's role and the organisational structure.

A typical hierarchy may include:

- Change Keys (CK): Open only one lock
- Sub-Master Keys (SMK): Open a group of locks within a zone
- Master Key (MK): Opens all locks in the system
- Grand Master Key (GMK): (if required) Opens multiple master systems

At the lowest level of the hierarchy are change keys (CK), which are designed to operate a single lock only. These are typically issued to individual users such as office staff or occupants who require access to one specific area. Above this level are sub-master keys (SMK), which operate multiple locks within a defined zone, such as all doors within a department or floor. These are generally issued to supervisors or team leaders who require broader, but still limited, access. At a higher level is the master key (MK), which operates all locks within the system and is typically restricted to senior management or security personnel. In larger or more complex facilities, a grand master key (GMK) may be introduced to control multiple master systems, such as across separate buildings or divisions.

The development of this hierarchy requires careful consideration of both functionality and risk. Each additional level of access increases convenience but also increases the potential impact of loss or misuse. As a result, the locksmith must ensure that higher-level keys are limited in number, tightly controlled, and only issued where absolutely necessary. The hierarchy must reflect real operational needs, not convenience-based decisions that could compromise security.

To support this structure, the locksmith creates a keying schedule, which is a detailed document that maps the entire system. This schedule identifies every lock in the system and specifies which keys are authorised to operate it. It effectively serves as the blueprint of the master key system, showing the relationships between change keys, sub-masters, and master keys. The keying schedule must be clear, accurate, and securely stored, as it represents sensitive security information. Any errors at this stage can result in unintended access or system failure.

Alongside the keying schedule, the locksmith develops a pinning chart, which translates the hierarchy into the internal configuration of each lock cylinder. In pin tumbler systems, this involves assigning specific pin combinations to each lock so that different keys can align the pins at the correct shear line. In master key systems, this is achieved by introducing master wafers (or master pins), which create multiple shear lines within the same cylinder. This allows more than one key to operate the lock, each corresponding to a different level in the hierarchy.

From a mechanical perspective, this is where the complexity of the system is physically realised. However, this complexity must be carefully controlled. Each additional master wafer increases the number of shear lines, which can reduce the precision of the lock and potentially make it more susceptible to picking or manipulation. Therefore, the locksmith must balance the need for hierarchical access with the need to maintain mechanical integrity and security performance. Overly complex systems with too many master levels or combinations can weaken the overall system.

Several critical design decisions are made during this stage. One of the most important is limiting the number of master levels, as deeper hierarchies introduce more master wafers and increase vulnerability. Another is avoiding unnecessary cross-keying, where a key unintentionally operates locks outside its intended scope. Cross-keying can create confusion, reduce accountability, and compromise the principle of least privilege. Additionally, the locksmith must select appropriate key profiles, with restricted or patented systems preferred to prevent unauthorised duplication and maintain control over key distribution.

A well-designed key hierarchy results in a system that is both functional and secure. It provides a clear structure, where each key has a defined purpose and level of access. It ensures minimal overlap, so that users do not have access beyond what is required. Most importantly, it maintains strict control over high-level keys, recognising that these represent the greatest risk if lost, duplicated, or misused.

Ultimately, this phase represents the translation of organisational needs into a physical and mechanical system. It requires not only technical expertise in lock design and pinning, but also an understanding of risk management and human behaviour. When executed correctly, the key hierarchy becomes the backbone of the entire master key system, enabling controlled access while maintaining security integrity.

A well-designed system ensures:

- Clear hierarchy
- Minimal overlap
- Strong control over high-level keys

3. Hardware Selection

The hardware selection stage is critical because it determines how well the master key system will perform in real-world conditions. Even a perfectly designed key hierarchy can fail if the physical components are not suitable for the environment, usage level, or security requirements. At this stage, the locksmith translates the theoretical design into a practical system by selecting locks, cylinders, and key systems that will deliver durability, reliability, and controlled access over time.

The first consideration is the type of lock hardware to be used. Different applications require different lock types, such as mortice locks for high-strength door installations, cylindrical locks for standard commercial or residential doors, or padlocks for external gates and plant areas. Each type offers different levels of strength, resistance to attack, and suitability for integration into a master key system. The locksmith must ensure that all selected hardware can be keyed alike or incorporated into the same master key system without compromising functionality.

Equally important is the selection of the key system type, particularly whether the system will be restricted or unrestricted. In unrestricted systems, keys can be freely duplicated at hardware stores or by unauthorised individuals, which quickly leads to loss of control and key sprawl. In contrast, restricted key systems use patented or controlled key profiles that can only be duplicated through authorised channels, typically requiring proof of ownership or formal authorisation. For organisational environments, restricted systems are strongly recommended because they maintain control over key distribution, support accountability, and align with best practice in access management. Without this control, even the most carefully designed master key system can be undermined by uncontrolled duplication.

The locksmith must also consider the durability and security rating of the hardware. Locks are manufactured for different levels of use, ranging from light residential applications to heavy-duty commercial and high-security environments. In high-traffic areas, such as office entrances or shared facilities, locks must withstand frequent use without excessive wear. In higher-risk environments, such as storage areas, server rooms, or critical infrastructure, locks must also provide resistance to forced entry, picking, drilling, and other forms of attack. Selecting hardware that is under-rated for its intended use will lead to premature failure, increased maintenance, and reduced security performance.

Beyond the core components, several additional factors influence hardware selection. Compatibility with existing doors and infrastructure is essential, particularly in retrofit situations where new locks must integrate with existing door thicknesses, backsets, strike plates, and door materials. Incorrectly matched hardware can result in poor alignment, reduced strength, or operational issues such as binding or incomplete locking.

Environmental conditions must also be taken into account. External doors, coastal environments, industrial sites, or dusty and corrosive settings place additional stress on lock components. In such cases, materials and finishes must be selected to resist corrosion, moisture ingress, and contamination. Weather-resistant cylinders, protective covers, and appropriate lubricants may be required to ensure long-term reliability.

In modern systems, there is often a need to consider integration with electronic access control. Hybrid systems that combine mechanical locks with electronic components—such as card readers, keypads, or smart locks—require careful coordination. The locksmith must ensure that mechanical override functions, fail-safe or fail-secure configurations, and access control interfaces are compatible and aligned with the overall security design. This integration allows for greater flexibility, such as audit trails and rapid revocation of access, while still maintaining the reliability of mechanical systems.

> The effectiveness of the system depends heavily on the hardware chosen.
>
> The locksmith must select:
>
> - Lock types (e.g. mortice locks, cylinders, padlocks)
> - Key system type (restricted vs unrestricted)
> - Durability level (residential, commercial, high-security)
>
> For organisational environments, restricted key systems are strongly recommended. These prevent unauthorised duplication and ensure all keys remain under controlled management.
>
> Additional considerations include:
>
> - Compatibility with existing doors and hardware
> - Environmental conditions (weather, dust, corrosion)
> - Integration with electronic systems (if hybrid systems are used)

<u>4. Implementation Phase (Installation and Keying)</u>

The implementation phase is where the master key system transitions from design to reality. It involves the precise physical creation, assembly, and installation of all components defined in the keying schedule. At this stage, the theoretical hierarchy must be translated into exact mechanical configurations, and any deviation—no matter how minor—can affect the integrity of the entire system. This phase demands a high level of technical accuracy, attention to detail, and adherence to the design documentation.

The process begins with pinning the lock cylinders in accordance with the keying schedule and pinning chart. Each cylinder is disassembled, and the internal pins and springs are configured to match the required key combinations. In a master key system, this includes the careful placement of master wafers to create multiple shear lines, allowing different keys to operate the same lock at different hierarchy levels. The locksmith must ensure that each pin stack is assembled correctly, with the correct sizes and sequence, as even a single incorrect pin can result in a lock that either fails to operate or allows unintended access.

Following cylinder preparation, the locksmith proceeds with cutting the keys. Keys must be cut to precise depths and spacing as specified by the system design. This requires calibrated key cutting equipment and careful verification against the intended bitting codes. Inaccurate key cutting can lead to poor operation, increased wear, or failure to align the pins correctly within the cylinder. In master key systems, where multiple keys interact with the same lock, precision is even more critical to ensure that all authorised keys function correctly without compromising security.

## Beyond the Key

Once cylinders and keys are prepared, the next step is the installation of locks and hardware. This involves fitting the locks to doors, frames, or other access points, ensuring that all components are correctly aligned. Proper alignment between the lock, latch or bolt, and strike plate is essential for smooth operation and long-term durability. Misalignment can cause binding, incomplete locking, or excessive wear on internal components. The locksmith must also ensure that all fixings are secure and that the installation complies with relevant standards and manufacturer specifications.

After installation, a comprehensive testing process is carried out. Every lock must be tested with all relevant keys defined in the hierarchy. This includes verifying that each change key operates only its designated lock, that sub-master keys operate all intended locks within their zone, and that master or grand master keys function across the entire system as designed. Testing must also confirm that no unintended access occurs, meaning that no key should operate a lock outside its authorised scope. This step is critical in identifying cross-keying errors or pinning mistakes before the system is handed over.

In addition to access verification, the locksmith must check for smooth mechanical operation. Keys should insert easily, turn without excessive resistance, and withdraw cleanly. Any stiffness, grinding, or inconsistency may indicate issues such as incorrect pinning, poor alignment, or debris within the cylinder. These issues must be rectified immediately, as they can lead to premature wear or user frustration, which in turn may result in misuse or damage.

Accuracy throughout this phase is essential. Master key systems are highly interdependent, meaning that an error in one component can affect multiple locks and keys. A mis-pinned cylinder or incorrectly cut key can introduce security vulnerabilities, reduce system reliability, or require significant rework. For this reason, experienced locksmiths often follow strict quality control procedures, including double-checking pinning configurations, verifying key cuts, and systematically testing each component.

Implementation involves physically creating and installing the system.
This includes:

- Lock Pinning cylinders according to the keying schedule
- Cutting keys to precise specifications
- Installing locks and ensuring correct alignment
- Testing each lock with all relevant keys

Every lock must be tested to confirm:

- Correct operation with authorised keys
- No unintended access (no cross-keying errors)
- Smooth mechanical function
- Accuracy is critical.

Even a small error in pinning or key cutting can compromise the entire system.

5. Documentation and Key Control

Documentation and key control are the mechanisms that sustain the integrity of a master key system over time. While the physical hardware and key hierarchy establish how access should work, it is documentation and control processes that ensure access remains accurate, accountable, and secure in practice. Without these controls, even the most well-designed system will deteriorate as keys are lost, duplicated, or distributed without oversight. In this sense, key management is not a one-time task but an ongoing operational responsibility.

A central component of this process is the key register, which provides a complete record of all keys issued within the system. This register documents who holds each key, what access that key provides, when it was issued, and when it is expected to be returned. It creates accountability by linking individuals to specific access rights, making it possible to trace responsibility in the event of loss, misuse, or security incidents. In larger or higher-risk environments, this register may be maintained electronically, allowing real-time updates and integration with broader access control systems.

Below is a sample Key Register suitable for a master key system in a residential, commercial, or organisational environment. It is structured to support accountability, auditability, and compliance with best-practice locksmithing and security management.

*Table 4: Sample Key Register.*

| Key ID | Key Type | Access Level | Lock/Area Accessed | Issued To (Name & Role) | Issue Date | Return Date | Authorised By | Signature (Holder) | Status | Notes |
|---|---|---|---|---|---|---|---|---|---|---|
| K001 | CK | Single Door | Office 1 | John Smith (Admin) | 01/02/2026 | — | R. Sloan | ✓ | Active | — |
| K002 | CK | Single Door | Office 2 | Sarah Lee (Admin) | 01/02/2026 | — | R. Sloan | ✓ | Active | — |
| K010 | SMK-A | Zone A | All Staff Areas | Mark Brown (Supervisor) | 01/02/2026 | — | R. Sloan | ✓ | Active | Zone A access |
| K020 | SMK-B | Zone B | Management Offices | Lisa Green (Manager) | 01/02/2026 | — | R. Sloan | ✓ | Active | — |
| K100 | MK | Full System | All Doors | David White (Facility Manager) | 01/02/2026 | — | Director | ✓ | Active | Master key – restricted |

| Key ID | Key Type | Access Level | Lock/Area Accessed | Issued To (Name & Role) | Issue Date | Return Date | Authorised By | Signature (Holder) | Status | Notes |
|---|---|---|---|---|---|---|---|---|---|---|
| K101 | MK | Full System | All Doors | Spare (Secure Cabinet) | 01/02/2026 | — | Director | — | Stored | Cabinet #1 |
| K200 | CK | Single Door | Server Room | IT Dept (Controlled Issue) | 05/02/2026 | — | IT Manager | ✓ | Active | High security |
| K300 | SMK-C | Zone C | Maintenance Areas | Contractor (ABC Services) | 10/02/2026 | 15/02/2026 | Facilities | ✓ | Returned | Temporary issue |
| K400 | CK | Single Door | Storage Room | Emily Clark (Staff) | 12/02/2026 | — | Supervisor | ✓ | Lost | Reported 20/02/2026 |

**Key Register Field Explanation**

**Key ID:** A unique identifier assigned to each key. This should correspond to the stamped or coded marking on the key.

**Key Type:** Indicates hierarchy level (e.g. CK, SMK, MK, GMK).

**Access Level:** Defines scope of access (single door, zone, full system).

**Lock/Area Accessed:** Specifies which doors or zones the key operates.

**Issued To:** Full name and role of the key holder for accountability.

**Issue Date / Return Date:** Tracks lifecycle of the key.

**Authorised By:** Person approving issuance (e.g. manager, security officer).

**Signature (Holder):** Confirms responsibility for the key.

**Status:** Active / Returned / Lost / Stored / Revoked.

**Notes:** Additional information (e.g. temporary issue, restricted key, incident details).

**Best Practice Notes**

- Keys should never be issued without recording in the register
- se coded key IDs (do not label keys with door names)
- Store registers securely (physical or digital)
- Conduct regular audits (e.g. quarterly)

- Lost keys must trigger risk assessment and possible rekeying
- Integrate register with HR processes (onboarding/offboarding)

Equally important is the keying schedule, which serves as the technical blueprint of the system. This document outlines the relationship between locks and keys, including the hierarchy and access structure. Because it contains sensitive information that could be used to compromise the system, it must be stored securely and accessed only by authorised personnel. The keying schedule is essential for maintenance, troubleshooting, and future expansion, but its exposure represents a significant security risk.

The following provides a sample Keying Schedule suitable for a structured master key system. It reflects a simple commercial facility with multiple zones and a clear hierarchy (CK, SMK, MK).

*Table 5: Sample Keying Schedule (Master Key System).*

| Lock ID | Location / Door | Zone | Change Key (CK) | Sub-Master Key (SMK) | Master Key (MK) | Notes |
|---|---|---|---|---|---|---|
| L001 | Main Entrance | A | K001 | SMK-A | MK | High traffic |
| L002 | Reception Office | A | K002 | SMK-A | MK | — |
| L003 | Staff Room | A | K003 | SMK-A | MK | — |
| L004 | Meeting Room 1 | A | K004 | SMK-A | MK | Shared space |
| L005 | Open Office Area | B | K005 | SMK-B | MK | General staff |
| L006 | Office 1 | B | K006 | SMK-B | MK | — |
| L007 | Office 2 | B | K007 | SMK-B | MK | — |
| L008 | Records Room | B | K008 | SMK-B | MK | Restricted |
| L009 | Manager Office | C | K009 | SMK-C | MK | — |
| L010 | Director Office | C | K010 | SMK-C | MK | High priority |
| L011 | Finance Office | C | K011 | SMK-C | MK | Sensitive data |
| L012 | Server Room | D | K012 | — | MK | No sub-master (high security) |
| L013 | Safe Room | D | K013 | — | MK | Critical asset |
| L014 | Maintenance Room | E | K014 | SMK-E | MK | Contractors |
| L015 | Plant Room | E | K015 | SMK-E | MK | Limited access |

*Table 6: Key Hierarchy Overview.*

| Key Type | Key ID | Access Description |
|---|---|---|
| Change Keys (CK) | K001–K015 | Open individual locks only |
| Sub-Master A | SMK-A | Opens all Zone A locks |
| Sub-Master B | SMK-B | Opens all Zone B locks |
| Sub-Master C | SMK-C | Opens all Zone C locks |
| Sub-Master E | SMK-E | Opens all Zone E locks |
| Master Key | MK | Opens all locks (L001–L015) |

To support practical control, an identification system must be implemented. Keys are typically stamped or coded with identifiers that correspond to entries in the key register and keying schedule. These identifiers should not directly reveal the location or function of the lock, as doing so could aid unauthorised use if a key is lost. Instead, coded systems are used so that only authorised personnel can interpret the key's purpose. This approach balances usability with security by enabling tracking without exposing sensitive information.

To support practical control, an identification system must be implemented. Keys are typically stamped or coded with identifiers that correspond to entries in the key register and keying schedule. These identifiers should not directly reveal the location or function of the lock, as doing so could aid unauthorised use if a key is lost. Instead, coded systems are used so that only authorised personnel can interpret the key's purpose. This approach balances usability with security by enabling tracking without exposing sensitive information.

The following is a sample identification system you can adopt for a master key system. It is designed to support tracking and accountability while avoiding the common mistake of labelling keys in a way that reveals what they open.

**Sample Identification System for a Master Key System**
**1. Core Rules (How keys and locks will be identified)**
The identification system uses coded IDs that link to the key register and keying schedule, but do not reveal locations on the key itself. All keys are marked with a short, durable code (stamped or laser-marked). Doors/locks are also labelled with a corresponding internal ID (usually on the hinge side of the door, inside the frame, or on the lock body—never on the public-facing side).
Each key and lock ID contains:

- Site code
- System code
- Key/lock type
- Unique number
- Optional issue/revision **indicator**

**2. Key Coding Format**
**Key Marking Format (stamped on key bow)**
**SITE-SYS-LEVEL-####**
**Example:** HQ-01-CK-0047
**Meaning:**

- HQ = site identifier (Headquarters)
- 01 = system number (Master Key System 01)
- CK = key level/type (CK/SMK/MK/GMK)
- 0047 = unique serial number

**Key Level Codes (standardised)**

- CK = Change Key (single lock)
- SMK-A = Sub-master for Zone A
- SMK-B = Sub-master for Zone B
- SMK-C = Sub-master for Zone C
- MK = Master Key
- GMK = Grand Master Key (if used)

**Example zone keys:**

- HQ-01-SMK-A-0003
- HQ-01-SMK-B-0001

**3. Lock / Door Identification Format**
**Lock ID Format (internal label only)**
**SITE-SYS-L-####**
**Example:** HQ-01-L-0125
Where:

- L indicates "lock"
- The number matches the lock entry in the **keying schedule**

**Important:** This lock ID is not shown publicly. It is placed:

- inside the door edge, under the latch plate (common practice), or
- inside the frame rebate, hinge side, or
- on the rear of the lock case / mortice body.

**4. Key Register Linkage (How it ties together)**
In your **Key Register**, you record:

- Key Code: HQ-01-CK-0047
- Key Type: CK
- Issued To: (name/role)
- Access: Lock IDs it opens (e.g., HQ-01-L-0125)
- Issue/Return dates
- Status: Active / Stored / Returned / Lost / Revoked

In your **Keying Schedule**, you record:

- Lock ID: HQ-01-L-0125
- Door/Location (internal-only description)
- Zone
- CK code(s) that operate it
- SMK/MK relationships

This ensures that if a key is found or stolen, the code alone does not reveal which door it opens—only someone with access to the register/schedule can interpret it.

**5. Colour and Physical Tagging (Optional but useful)**
You can add **non-informative colour coding** for quick visual handling, without revealing locations. Example:

- CK = plain (no colour)
- SMK = coloured key caps by zone (internal convention)
  - Zone A = Blue cap
  - Zone B = Green cap
  - Zone C = Yellow cap
- MK = Red cap
- GMK = Black cap

**Important:** Colour indicates *level*, not *location*. Avoid tags like "Server Room" or "Finance".

**6. Duplicate & Authorisation Control Markings**
To support duplication control, include a suffix on the key code:

- -R = restricted (must be authorised to copy)
- -D# = duplicate number

**Example:**

- HQ-01-CK-0047-R (restricted)
- HQ-01-CK-0047-R-D2 (second authorised duplicate)

This helps prevent "silent growth" of keys and supports audit trails.

**7. Lost / Revoked Key Handling**
When a key is lost:

- Mark status as LOST in the register
- Record incident date/time
- Trigger risk assessment (higher-level keys may require rekeying)
- If rekeying occurs, increment the system revision:

**System revision example:**

- Current system: HQ-01
- After major rekey: HQ-02

Old keys are then invalid under the new system, and the ID structure remains clean.

**8. Sample Set (As it would look in practice)**
**Keys**

- HQ-01-CK-0001
- HQ-01-CK-0002
- HQ-01-SMK-A-0001
- HQ-01-SMK-B-0001
- HQ-01-MK-0001 (stored secure cabinet)
- HQ-01-MK-0002 (issued Facilities Manager)

**Locks**

- HQ-01-L-0001 (main entry)
- HQ-01-L-0002 (reception)
- HQ-01-L-0012 (IT room)

Beyond documentation, strict procedures must be established for the issuance and return of keys. Keys should only be issued following formal authorisation, with recipients acknowledging their responsibility for the key. This may include signing a key issue form or agreeing to conditions of use. When keys are no longer required—such as when an employee changes roles or leaves the organisation—they must be returned promptly and recorded in the register. Integrating this process with organisational workflows, such as HR onboarding and offboarding, helps ensure that access is consistently managed.

**Key Issuance and Return Procedure (Sample)**
**1. Purpose**
To ensure all keys are issued, controlled, and returned in a secure, accountable, and traceable manner, reducing the risk of unauthorised access, key loss, and key sprawl.

**2. Scope**
This procedure applies to all:

- Employees
- Contractors
- Visitors issued with keys
- Security personnel and authorised key custodians

It covers all key types including:

- Change keys
- Sub-master keys
- Master keys
- Restricted keys

**3. Responsibilities**
**3.1 Authorising Officer**

- Approves all key requests
- Ensures access aligns with job role (least privilege)
- Authorises duplication (if permitted)

**3.2 Key Custodian (e.g. Locksmith / Facilities Manager)**

- Maintains key register
- Issues and receives keys
- Verifies identity of recipients
- Conducts audits

**3.3 Key Holder**

- Safeguards issued keys
- Does not duplicate or share keys
- Reports lost or stolen keys immediately
- Returns keys when required

**4. Key Issuance Procedure**
**4.1 Request and Approval**

- A Key Request Form must be completed.
- The request must specify:
- Name of recipient

  - Role/position
- Required access areas
- Justification for access
- The request must be approved by the **Authorising Officer**.

**4.2 Verification Prior to Issue**
Before issuing a key, the Key Custodian must:

- Verify the identity of the recipient (e.g. ID badge)
- Confirm approval has been granted
- Check availability of the correct key

**4.3 Key Issue Process**

- Record the following in the Key Register:
  - Key ID / Code
  - Key type (CK, SMK, MK, etc.)
  - Areas accessed
  - Recipient name and ID
  - Date of issue
- Provide the key to the recipient.
- Obtain a signature acknowledging responsibility.
- Issue a Key Holder **Agreement** outlining obligations.

**4.4 Key Holder Conditions**
All key holders must:

- Use keys only for authorised purposes
- Not loan, share, or duplicate keys
- Keep keys secure at all times
- Not label keys with identifiable location details
- Immediately report:
  - Lost keys
  - Stolen keys
  - Damaged keys

**5. Key Return Procedure**
**5.1 Trigger Events for Return**
Keys must be returned when:

- Employment ends

- Role or access requirements change
- Contract work is completed
- Keys are replaced or reissued
- Requested by management

**5.2 Key Return Process**

- Key holder returns keys to the Key Custodian.
- Custodian verifies:
    - Key identity
    - Condition of key
- Update the Key Register:
    - Date of return
    - Condition (e.g. serviceable/damaged)
- Obtain signature confirming return.

**5.3 Failure to Return Keys**
If keys are not returned:

- Immediate notification to management
- Conduct risk assessment
- Consider rekeying affected locks
- Record incident in security log

**6. Lost or Stolen Keys**

- Must be reported immediately (no delay).
- Key Custodian must:
    - Record incident
    - Identify affected locks/zones
    - Assess risk level (CK vs MK criticality)
- Management must determine:
    - Rekeying requirements
    - Temporary security measures
- Incident must be documented and closed only after corrective action.

**7. Key Duplication Control**

- Duplication is strictly prohibited unless authorised
- Only approved locksmiths may cut keys
- Restricted key systems must be used where possible

- All duplicates must be recorded in the Key Register

**8. Auditing and Compliance**

- Key register must be audited:
    - Quarterly (minimum)
    - Or based on risk level
- Audit must verify:
    - All keys accounted for
    - Correct holders
    - No unauthorised duplication
- Discrepancies must trigger investigation

**9. Security Requirements**

- Keys must be stored securely when not in use:
    - Locked key cabinet
    - Electronic key safe (if applicable)
- Master and sub-master keys require higher security controls:
    - Restricted access
    - Separate storage
    - Additional logging

**10. Breach of Procedure**
Failure to comply may result in:

- Disciplinary action
- Removal of access privileges
- Financial liability (where applicable)

**11. Record Keeping**
The following documents must be maintained:

- Key Register
- Key Request Forms
- Key Holder Agreements
- Incident Reports
- Audit Reports

All records must be stored securely and retained in accordance with organisational policy.

**Summary (Operational Insight)**
This procedure ensures:

- Accountability (who has what key)
- Control (who is allowed access)
- Traceability (audit trail)
- Rapid response (when things go wrong)

Without strict issuance and return procedures, even a well-designed master key system will quickly degrade into uncontrolled access—effectively defeating the entire security framework.

Procedures must also address the handling of lost or stolen keys, as these events represent immediate security risks. The organisation must be able to assess the level of risk associated with the missing key, which depends on its access level within the hierarchy. For example, the loss of a change key may affect a single lock, while the loss of a master key may compromise an entire facility. Based on this assessment, appropriate actions—such as rekeying affected locks, updating records, or restricting access—must be implemented quickly to restore security.

Control over authorised duplication is another critical element. In unrestricted systems, keys can be copied without oversight, leading to uncontrolled growth in the number of keys in circulation. To prevent this, organisations should use restricted key systems and establish clear authorisation procedures for duplication. Only designated individuals should be permitted to request additional keys, and all duplicates must be recorded in the key register. This ensures that the total number of keys remains known and controlled at all times.

Without effective documentation and control, key sprawl becomes inevitable. Over time, keys are duplicated, shared, or lost without being recorded, resulting in a system where access is no longer understood or controlled. This undermines accountability, increases the risk of unauthorised access, and may ultimately require complete system rekeying to restore control. By contrast, a well-managed system maintains clear visibility over all keys and access rights, allowing issues to be identified and addressed before they escalate.

In practice, documentation and key control transform a master key system from a static installation into a managed security system. They ensure that access remains aligned with organisational needs, that responsibility is clearly defined, and that risks are actively controlled. For locksmiths, this step reinforces the importance of not only designing and installing systems, but also supporting clients in maintaining them effectively over time.

A master key system is only as secure as its management.

The locksmith must provide:

- A key register (who holds which key)
- A keying schedule (securely stored)
- Identification system (key stamping or coding)

Strict procedures should be established for:

- Key issuance and return
- Lost or stolen keys
- Authorised duplication

Without documentation and control, key sprawl will quickly undermine the system.

6. Commissioning and Handover

The commissioning and handover phase is the final step in delivering a master key system, and it is critical in ensuring that the system operates as intended once it is in the hands of the client. While the earlier stages focus on design and installation, this phase ensures that the system is validated, understood, and correctly managed in practice. A well-executed handover bridges the gap between technical completion and operational use, reducing the risk of misuse, misunderstanding, or early system failure.

Before any handover occurs, the locksmith must conduct a comprehensive final testing process. This goes beyond basic operational checks and involves systematically verifying that every lock and key functions exactly as defined in the keying schedule. Each change key must operate only its designated lock, sub-master keys must correctly operate all locks within their assigned zones, and master or grand master keys must function across the full system. At the same time, it is essential to confirm that no unintended access exists—ensuring there are no cross-keying errors or overlaps that could compromise security. This stage effectively acts as a final quality assurance process, confirming that the physical system aligns with the original design.

In addition to functional testing, the system must be verified against the design requirements established during the planning phase. This includes confirming that access levels align with user roles, that restricted areas remain appropriately controlled, and that any special operational requirements—such as emergency access or after-hours entry—have been correctly implemented. This verification ensures that the system is not only mechanically correct, but also operationally appropriate for the client's environment.

Once testing and verification are complete, the locksmith proceeds to demonstrate the system to the client. This demonstration is an essential part of the handover, as it provides a practical explanation of how the system works. The locksmith may walk the client through key areas, showing how different keys operate specific locks and how the hierarchy functions in real terms. This helps the client visualise the structure of the system and builds confidence in its operation.

A key component of this phase is user training, which ensures that those responsible for managing and using the system understand their roles and responsibilities. The locksmith must clearly explain how the key hierarchy operates, including the differences between change keys, sub-master keys, and master keys, and the implications of each level of access. Clients must understand that higher-level keys represent greater risk and therefore require stricter control.

Training also includes outlining the responsibilities of key holders. This involves reinforcing that keys are security assets, not convenience tools, and must not be shared, duplicated, or left unsecured. Key holders should understand their obligation to safeguard keys and to return them when no longer required. This is particularly important in organisational environments, where multiple users interact with the system.

Another critical element of training is establishing clear procedures for reporting lost or stolen keys. The client must understand the importance of immediate reporting, as delays can significantly increase security risk. The locksmith should explain the potential consequences of key loss—particularly for high-level keys—and outline the actions that may be required, such as rekeying affected locks or updating access control measures. Providing clear procedures ensures that incidents are managed quickly and effectively.

The commissioning and handover phase may also include the delivery of documentation, such as the key register, keying schedule (secured appropriately), and any policies or procedures related to key control. These documents support ongoing management and provide a reference for future use, maintenance, or expansion of the system.

Before handover, the system should be:

- Fully tested
- Verified against design requirements
- Demonstrated to the client

The locksmith should provide user training, including:

- How the hierarchy works
- Responsibilities for key holders
- Procedures for reporting lost keys

This step is critical in ensuring the system is used correctly.

7. Ongoing Management and Maintenance

Ongoing management and maintenance is the phase that determines whether a master key system stays secure for years or quietly collapses into disorder. A master key system is not a static product—it is a living access control arrangement that is constantly affected by staff turnover, changing roles, lost keys, wear-and-tear, and day-to-day convenience behaviours. Even a well-designed system will degrade if it is not actively managed, because the weakest point is rarely the lock itself; it is the gradual loss of visibility and control over who has which keys and what access those keys still legitimately represent.

A core activity is conducting periodic audits of key holders. This means routinely reconciling the key register against reality—confirming that each issued key is still held by the correct person, still required for their role, and still accounted for physically. Audits are where "key sprawl" is detected early: missing keys, unofficial duplicates, keys held by people who have changed jobs, or keys that were never returned after a contract ended. Audits should be scheduled (for example, quarterly or biannually depending on risk) and treated as a formal process, not a casual check. The outcome should be documented, discrepancies investigated, and corrective actions assigned with deadlines.

When keys are lost, stolen, or not returned, the organisation must be prepared to rekey promptly and proportionately. Rekeying is not simply a mechanical task; it is a risk management response. The impact depends heavily on the key level within the hierarchy. A lost change key may only require rekeying a single cylinder, while a lost master or sub-master key may require rekeying multiple locks within a zone—or, in high-risk contexts, redesigning part of the hierarchy to restore integrity. Delay is dangerous because an unaccounted-for key creates an unknown access risk. A well-managed system includes clear triggers for when rekeying is mandatory, who authorises it, and how costs and disruptions are managed.

Lock hardware itself requires regular maintenance to remain reliable and secure. Locks operate on tight mechanical tolerances, and over time they accumulate dust, wear, corrosion, and misalignment—especially in high-traffic doors or harsh environments. Maintenance includes cleaning cylinders, applying appropriate lubricants, checking and adjusting strike alignment, tightening fixings, inspecting door closers and hinges, and replacing worn keys before they damage the cylinder. Good maintenance reduces failures and prevents users from adopting insecure workarounds such as leaving doors unlocked "because the key sticks" or propping doors open "because the latch doesn't catch properly." In this sense, maintenance is not just about smooth operation—it directly protects the security integrity of the system.

A well-run master key system also requires ongoing updates as organisational needs change. Organisations are dynamic: departments shift, rooms change function, staffing levels fluctuate, contractors come and go, and facilities expand. Each change can create pressure to "just give someone a key," which is exactly how systems drift into uncontrolled overlap. Instead, changes should trigger a structured process: review the access requirement, decide the least-privilege solution, update the key register and keying schedule, issue keys through authorised channels, and—where necessary—reconfigure cylinders to reflect the new structure. This keeps the system aligned with actual operational reality rather than becoming a historical artefact that no longer matches how the building is used.

If ongoing management is neglected, the system predictably begins to fail in stages. First there is loss of control, where the organisation no longer has accurate knowledge of how many keys exist or who holds them. Then comes increased security risk, because access becomes over-provisioned and untraceable, making insider misuse or unauthorised entry easier and harder to investigate. Finally, the organisation reaches system failure, where rekeying single locks no longer fixes the problem and the only viable remedy is a major rekey or complete redesign—often at significantly higher cost and disruption than regular management would have required.

For locksmiths, this phase is where professional practice extends beyond installation into ongoing stewardship. Many locksmith businesses support clients through periodic audits, scheduled maintenance programs, incident-driven rekeying support, and system reviews after staffing or facility changes. This is also where locksmiths add long-term value: by keeping the system orderly, accountable, and fit for purpose, they help ensure the master key system remains what it was intended to be—an organised, controlled access structure—rather than becoming an unmanaged collection of keys that undermines the very security it was meant to provide.

A master key system is not "set and forget." It requires ongoing management.

Key activities include:

- Periodic audits of key holders
- Rekeying when keys are lost or staff leave
- Maintenance of locks (cleaning, lubrication, adjustment)
- Updating the system when organisational changes occur

Failure to maintain the system leads to:

- Loss of control
- Increased security risk
- Eventual system failure

**Designing a Restricted Key System**

Designing a restricted key system is a specialised process that combines mechanical design, legal control, and administrative procedures to ensure that keys cannot be duplicated or distributed without authorisation. Unlike standard (unrestricted) systems, a restricted system is not just about how locks are pinned, it is about maintaining ongoing control over the entire lifecycle of keys and access.

1. Understanding the Purpose of a Restricted Key System

Understanding the purpose of a restricted key system is fundamental to modern locksmithing because it reframes security from a purely mechanical function into a controlled access management process. In a conventional (unrestricted) system, the lock may be secure, but the key itself becomes the weakest point—anyone with temporary access to a key can potentially duplicate it without the owner's knowledge. A restricted key system is specifically designed to eliminate this vulnerability by ensuring that keys cannot be copied freely. This is achieved through legally protected key profiles, patented designs, and tightly controlled distribution of key blanks, meaning duplication can only occur through authorised channels under verified approval.

At a deeper level, the purpose of a restricted system is not simply to stop copying—it is to establish ongoing authority over access. The locksmith is no longer just installing hardware but is implementing a system where every key represents a controlled permission. This shifts the focus from "who has a key" to "who is authorised to have access," which is a critical distinction in organisational security. By controlling the production and duplication of keys, the system ensures that access cannot expand beyond what has been formally approved, preventing the gradual and often invisible growth of risk known as key sprawl.

For locksmiths, this introduces an important professional principle: security responsibility extends beyond installation. In a restricted system, control must be maintained throughout the entire lifecycle of each key. This includes the initial issuance of keys to authorised users, ensuring that each key is recorded and linked to a specific individual or role. It also includes controlling duplication, where no additional keys can be created without proper authorisation, thereby preserving the integrity of the system over time. Without these controls, even a restricted system can degrade into an uncontrolled one.

Key tracking is another essential component of this purpose. Every key in a restricted system should be traceable—there should be a clear record of who holds it, what it opens, and when it was issued. This creates accountability and allows organisations to respond effectively if a key is lost, stolen, or misused. Without tracking, the system loses visibility, and the benefits of restriction are significantly reduced. Tracking transforms keys from anonymous objects into accountable assets within a managed system.

Equally important is the ability to revoke access when necessary. In any environment, people change roles, leave organisations, or lose keys. A restricted key system must support controlled revocation, whether through recovering keys, rekeying affected locks, or updating access structures. The purpose here is to ensure that access is not permanent or uncontrolled, but instead remains aligned with current needs. This reinforces the principle that access is conditional and can be withdrawn when no longer appropriate.

Taken together, these elements demonstrate that a restricted key system is not just a collection of locks and keys, but a controlled ecosystem of access management. The locksmith's role evolves into that of a system designer and security advisor, responsible for ensuring that access is issued deliberately, monitored continuously, and adjusted as circumstances change. This approach significantly increases the effectiveness of physical security by addressing not only how locks resist attack, but how access to those locks is governed over time.

## 2. Selecting the Restricted Key Platform

Selecting the restricted key platform is one of the most important decisions in the design of a controlled key system, because it establishes the technical, legal, and operational foundation for the entire access control framework. Unlike standard lock selection, where components can often be replaced or upgraded with relative ease, a restricted key platform becomes embedded into the organisation's infrastructure. Once deployed across multiple doors, users, and access levels, changing to a different system can require widespread rekeying, replacement of cylinders, and redistribution of keys. For this reason, the selection process must be approached as a long-term strategic decision rather than a simple product choice.

At the core of a restricted key platform is the use of patented or legally protected key profiles. These profiles are specifically designed so that key blanks cannot be freely manufactured or obtained outside authorised channels. This legal protection is what prevents unauthorised duplication, as it restricts who can produce compatible key blanks and cutting tools. From a locksmithing perspective, this introduces a level of control that extends beyond mechanical design into intellectual property and supply chain management. However, it also means the locksmith must carefully consider the duration of patent protection. If a patent expires, the key profile may eventually become unrestricted, reducing the long-term effectiveness of the system.

> Patented or legally protected key profiles are a fundamental component of restricted key systems, designed to prevent unauthorised duplication of keys and maintain long-term control over access. In simple terms, a key profile refers to the shape and design of the keyway (the groove pattern along the blade of the key that fits into the lock). In unrestricted systems, these profiles are widely available, meaning key blanks can be easily purchased and copied by anyone with access to standard cutting equipment. In contrast, patented or legally protected profiles are controlled through intellectual

property laws and restricted manufacturing, ensuring that only authorised parties can produce and duplicate keys.

A patented key profile is protected under patent law, which gives the manufacturer exclusive rights to produce and distribute that specific key design for a defined period, typically up to 20 years. During this time, unauthorised production of matching key blanks is prohibited. This means that even if someone physically possesses a key, they cannot legally obtain or manufacture a compatible blank unless they go through an authorised provider. For locksmiths, this creates a controlled environment where duplication is not only technically restricted but also legally enforceable. It transforms key control from a purely mechanical issue into a regulated system backed by legal protection.

In addition to patents, some systems rely on design registrations, trademarks, or contractual controls to maintain restriction. While these may not offer the same level of protection as an active patent, they still limit how key blanks are distributed and who is authorised to cut them. Manufacturers often implement strict supply chain controls, ensuring that blanks are only available to licensed locksmiths who have agreed to specific conditions, such as verifying customer authorisation before duplication. This layered approach—combining legal protection with controlled distribution—strengthens the overall integrity of the system.

From a practical locksmithing perspective, patented profiles enable the creation of authorised duplication systems. Keys can only be copied when the requester provides proper authorisation, such as a signature, security card, or system registration details. This ensures that every additional key is accounted for and approved, preventing uncontrolled expansion of access. It also allows locksmiths to maintain accurate records of key issuance and duplication, supporting accountability and auditability within the client's security system.

Another important aspect is that patented key systems often incorporate advanced mechanical features beyond the profile itself. These may include sidebars, angled cuts, magnetic elements, or multi-dimensional key designs that increase resistance to picking, bumping, and unauthorised decoding. While the legal protection controls duplication, the mechanical complexity enhances physical security. Together, these features provide a comprehensive approach to access control, addressing both credential management and lock resistance.

However, it is important to recognise that patent protection is time-limited. Once a patent expires, the key profile may become available for general manufacture, potentially reducing its effectiveness as a restricted system. For this reason, locksmiths must consider the remaining life of a patent when selecting a system, particularly for long-term installations. Some manufacturers address this by continuously developing new profiles and transitioning clients to updated systems before older patents expire.

In locksmithing practice, the use of patented or legally protected key profiles significantly influences how systems are designed, installed, and managed. It allows locksmiths to offer clients a higher level of control over their keys, reduces the risk of unauthorised duplication, and supports structured key management processes. More importantly, it reinforces the principle that security is not just about the lock itself, but about controlling the credentials that operate it. By integrating legal protection with mechanical design, these systems provide a robust and enforceable framework for maintaining access control over time.

Equally important is the control of key blank distribution. In a restricted system, blanks are not available through general hardware suppliers but are instead distributed only to authorised locksmiths or service providers. This ensures that even if someone obtains a physical key, they cannot easily duplicate it without access to the correct blank and cutting authorisation. When selecting a platform, the locksmith must assess how tightly this distribution is controlled and whether the supply chain is secure and reliable. A system with weak distribution controls can undermine the very purpose of restriction.

Control of key blank distribution is a core mechanism within restricted key systems that ensures only authorised parties can obtain and duplicate keys, thereby maintaining strict control over who can create access credentials. A key blank is the uncut form of a key, and in unrestricted systems these blanks are widely available through hardware stores and online suppliers. This open availability allows anyone with access to a cutting machine to duplicate a key, regardless of whether they have permission. In contrast, restricted systems tightly control the manufacture, supply, and distribution of key blanks to prevent this loss of control.

In a controlled distribution model, key blanks are manufactured exclusively by the system provider and are only supplied to approved locksmiths or authorised service centres. These locksmiths typically enter into formal agreements with the manufacturer, which require them to follow strict procedures for storage, handling, and duplication of blanks. This means that even if an unauthorised person possesses a physical key, they cannot simply obtain a matching blank unless they go through an authorised channel. The supply chain itself becomes a security control, ensuring that access cannot expand beyond what has been formally approved.

A critical aspect of this control is the registration of the key system to a specific client or site. When a restricted system is installed, it is usually registered under the client's name, and duplication rights are tied to that registration. To obtain additional keys, a requester must provide proof of authorisation—such as a signature, security card, or written approval from the registered owner. The locksmith verifies this authority before releasing a key blank and performing duplication. This process ensures that every new key is both intentional and traceable, preventing informal or unauthorised copying.

From a practical perspective, controlled distribution also requires secure storage and inventory management of key blanks by the locksmith. Blanks must be kept in locked storage, with access limited to authorised personnel. Many professional operations maintain stock control records, tracking how many blanks are received, used, or remaining. This prevents internal misuse and ensures that blanks cannot be diverted or used without proper documentation. In higher-security environments, electronic inventory systems or audit logs may be used to further strengthen accountability.

Another important element is that controlled blank distribution supports enforcement of key control policies over time. Unlike standard systems where keys can proliferate without oversight, restricted systems ensure that every additional key passes through a controlled checkpoint. This allows organisations to maintain an accurate understanding of how many keys exist and who holds them. It also supports processes such as rekeying, auditing, and access reviews, because the system has not been compromised by unknown or unauthorised duplicates.

The influence on locksmithing practice is significant. Locksmiths are no longer simply cutting keys on request; they are acting as gatekeepers of access credentials. This requires adherence to verification procedures, record keeping, and professional judgement. It also elevates the locksmith's role from a technical tradesperson to a trusted security controller, responsible for maintaining the integrity of the entire key system.

The requirement for system registration adds another layer of control by formally linking the key system to a specific client, site, or organisation. This registration typically includes assigning a unique system number and establishing authorised signatories who can approve key duplication. From a design perspective, this creates a traceable and enforceable framework for managing access. The locksmith must ensure that the registration process is robust, well-documented, and aligned with the client's organisational structure. Without proper registration, the system loses accountability and becomes vulnerable to misuse.

Below is a sample system registration document suitable for a restricted key system.

**Restricted Key System Registration Form**
**1. System Details**
**System Name / Reference:** RK-2026-001
**Client / Organisation Name:** ABC Property Management Pty Ltd
**Site Address:** 123 Collins Street, Melbourne VIC 3000
**Date of Registration:** 02 March 2026
**Locksmith / Provider:** SecureKey Locksmiths Pty Ltd
**System Type:** Restricted Key System (Patented Keyway)
**Key Profile / Platform:** Everest Restricted Series (Example)

**Authorised System Owner**
The registered system owner is the individual or entity with legal authority over the key system.
**Name:** John Smith
**Position:** Facilities Manager
**Organisation:** ABC Property Management Pty Ltd
**Contact Number:** 0400 000 000

**3. Authorised Signatories**
Only the following individuals are authorised to request key duplication, system changes, or rekeying.

| Name | Position | Signature | Authority Level |
|---|---|---|---|
| John Smith | Facilities Manager | ____________ | Full Authority |
| Sarah Lee | Operations Manager | ____________ | Limited Authority |
| Mark Chen | Security Supervisor | ____________ | Key Requests Only |

**4. Key Control Agreement**
The client acknowledges and agrees to the following conditions:
All keys remain under **controlled distribution**
No unauthorised duplication is permitted
Keys may only be duplicated through **SecureKey Locksmiths Pty Ltd**
Written or verified authorisation is required for all duplication requests
Lost or stolen keys must be reported immediately
The locksmith reserves the right to refuse duplication without proper authorisation
**Client Signature:** ________________
**Date:** ________________

**5. Key System Scope**
**System Coverage:**

- Main building entry points
- Internal offices and staff areas
- Restricted management and server rooms

**Hierarchy Type:**

- Change Keys (CK)
- Sub-Master Keys (SMK)
- Master Key (MK)

**6. Security Controls**
The following controls apply to this system:

- Restricted key blanks supplied only to authorised locksmith
- Key register maintained by client
- Duplication requests recorded and logged
- Periodic audits recommended (every 6–12 months)

**7. Duplication Procedure**
To request additional keys:

- Submit written request (email or form)
- Provide authorised signature or approval
- Locksmith verifies identity and authority
- Keys are cut, recorded, and issued

**8. Record Retention**

The locksmith will retain:

- Registration details
- Key cutting records
- Authorisation logs

Records will be stored securely in accordance with privacy and legal requirements.

**9. Declaration**
This registration confirms that the restricted key system is established under controlled conditions, and all parties agree to comply with the procedures outlined above.
**Locksmith Representative:** ________________
**Client Representative:** ________________

When evaluating different platforms, the locksmith must also consider the level of security required. Not all restricted systems offer the same level of protection. Some are suitable for general commercial use, while others are designed for high-security environments and incorporate advanced features such as sidebars, multiple locking elements, and resistance to picking and decoding. The selection must align with the risk profile of the site. Over-specifying may lead to unnecessary cost and complexity, while under-specifying can leave critical assets inadequately protected.

Another key consideration is the availability of authorised cutting centres and ongoing support. A restricted system is only practical if keys can be duplicated (when authorised) and components can be serviced within a reasonable timeframe. If the system relies on a limited number of suppliers or requires long lead times for key cutting, it may create operational challenges for the client. The locksmith must ensure that the chosen platform has a reliable support network, including access to replacement parts, additional keys, and technical assistance.

Compatibility with master keying requirements is also critical. Not all restricted systems support complex key hierarchies equally well. The locksmith must ensure that the chosen platform can accommodate the required number of master levels, key combinations, and future expansion. Some systems are better suited to large, multi-level master key structures, while others are more limited in their capacity. Selecting a platform that cannot support the required hierarchy may lead to design compromises or security weaknesses.

Ultimately, selecting the restricted key platform is about balancing security, control, practicality, and longevity. It requires the locksmith to think beyond the immediate installation and consider how the system will perform over many years of use, change, and expansion. A well-chosen platform provides a stable foundation for controlled access, while a poor choice can result in operational difficulties, increased costs, and reduced security over time. This step therefore has a lasting influence on the effectiveness and sustainability of the entire key management system.

<u>3. System Registration and Ownership Control</u>

The system registration and ownership control stage is what transforms a restricted key system from a purely mechanical arrangement into a controlled security framework governed by authority, accountability, and traceability. At this point, the locksmith formally establishes who owns the system, who is permitted to authorise changes, and how the system will be identified and managed over its lifecycle. Without this step, even a patented key profile can be undermined by poor administrative control, as there would be no clear mechanism to validate who is entitled to request duplication or modification.

The first element is the appointment of a registered system owner, which may be an individual (such as a homeowner) or, more commonly in organisational settings, a designated role (such as a facilities manager or security officer). This owner is recognised as the ultimate authority over the key system. Their role is not only administrative but also legal, as they assume responsibility for how access is granted, controlled, and revoked. Establishing a clearly defined owner eliminates ambiguity and ensures that all decisions relating to the system can be traced back to a responsible authority.

Closely linked to ownership is the creation of an authorised signatory list. This is a controlled list of individuals who are permitted to approve key duplication, rekeying, or system changes. In practice, this prevents unauthorised or informal requests from being actioned, even if they appear legitimate. For example, a staff member requesting a key cannot simply rely on verbal authority; they must be approved by a recognised signatory. This introduces a formal approval process that protects both the client and the locksmith, ensuring that access is only expanded through controlled and verifiable channels. It also reduces the risk of social engineering, as the locksmith can rely on documented authority rather than judgement or assumption.

The assignment of a unique system reference or registration number is another critical control. This identifier links all aspects of the system, including key profiles, cutting records, authorisation forms, and maintenance history, into a single traceable framework. Every key issued, duplicated, or replaced can be recorded against this reference, creating a complete audit trail. In larger organisations or multi-site operations, this becomes particularly important, as it allows multiple systems to be managed independently without confusion or overlap.

From a security perspective, this step introduces procedural and legal protection that complements the mechanical restrictions of the key system. While patented key profiles prevent unauthorised physical duplication, registration ensures that even authorised locksmiths cannot produce keys without proper approval. This dual-layer approach, combining physical restriction with administrative control, significantly strengthens the overall integrity of the system. It ensures that access cannot be expanded quietly or informally, and that every change is documented and accountable.

This stage also has important implications for long-term system management. As organisations evolve, staff change roles, contractors come and go, and access requirements shift. With a registered system in place, these changes can be managed in a controlled manner, with all updates recorded against the system reference and approved by authorised personnel. This prevents the gradual loss of control that typically leads to key sprawl, unauthorised duplication, and eventual system failure.

For locksmithing practice, system registration represents a shift toward a more professional and risk-managed approach. The locksmith is no longer simply supplying hardware but is acting as a custodian of a controlled access system, responsible for enforcing authorisation protocols and maintaining system integrity. This requires clear procedures, secure record-keeping, and the discipline to refuse work when proper authorisation is not provided.

4. Designing the Key Hierarchy

Designing the key hierarchy is one of the most critical stages in a restricted key system because it determines how access is structured, controlled, and limited across the entire site. At its core, the hierarchy defines the relationship between keys and locks, ensuring that each user can access only the areas necessary for their role. In organisational environments, this structured approach replaces ad hoc key distribution with a clearly defined system that aligns access with operational needs and security requirements.

The restricted system must be integrated into a structured key hierarchy, particularly in organisational environments.

Typical structure:

- Change Keys (CK) – individual access
- Sub-Master Keys (SMK) – zone access
- Master Key (MK) – full system access

When designing this hierarchy, the locksmith must:

- Apply the principle of least privilege
- Minimise the number of high-level keys
- Avoid unnecessary overlap or cross-keying

Restricted systems are especially important here because:

- High-level keys (e.g. master keys) represent critical risk points

Duplication of such keys must be tightly controlled

The hierarchy typically begins with change keys (CK), which are the most restricted level of access. Each change key operates a single lock or a very limited number of locks, making it suitable for individual users such as staff members assigned to specific rooms or areas. These keys represent the lowest level of privilege and form the foundation of the system. By ensuring that most users are issued change keys rather than higher-level keys, the locksmith maintains tight control over access and reduces the risk of widespread compromise.

Above this level are sub-master keys (SMK), which provide access to a defined group of locks within a particular zone or functional area. For example, a supervisor or team leader may hold a sub-master key that opens all doors within their department. This allows operational efficiency without granting unrestricted access across the entire site. Zoning is essential at this level, as it ensures that access is logically grouped based on function rather than convenience. Poor zoning can lead to unnecessary overlap, where keys open areas that are unrelated to the user's role, increasing security risk.

At the top of the hierarchy is the master key (MK), which provides access to all locks within the system. This level of access is typically restricted to a very small number of authorised individuals, such as facility managers or security personnel. Because the master key represents a single point of control over the entire system, it also represents a significant vulnerability if lost, duplicated, or misused. For this reason, its distribution must be strictly limited, carefully documented, and supported by strong procedural controls.

When designing the hierarchy, the locksmith must apply the principle of least privilege, ensuring that no individual has more access than is necessary to perform their role. This requires careful analysis of how the site operates, who needs access to which areas, and under what conditions. The goal is to create a system where access is intentional and justified, rather than broad and convenient. This not only improves security but also enhances accountability, as it becomes easier to identify who had access to a particular area if an incident occurs.

A key design consideration is the need to minimise the number of high-level keys, particularly master and sub-master keys. The more high-level keys that exist, the greater the risk of loss, duplication, or misuse. Each additional master key increases the potential impact of a security breach. Therefore, the locksmith must balance operational convenience with risk, ensuring that higher-level access is only granted where absolutely necessary.

Another critical factor is the avoidance of unnecessary overlap and cross-keying. Cross-keying occurs when a key is designed to open multiple locks that are not part of a clear hierarchical relationship. While it may solve short-term access needs, it introduces complexity and can weaken the mechanical integrity of the system by increasing the number of shear lines within a cylinder. This can reduce pick resistance and make the system more difficult to manage and audit. A well-designed hierarchy avoids such overlaps by maintaining clear, logical relationships between keys and locks.

Restricted key systems play a particularly important role at this stage because they protect the most critical elements of the hierarchy—especially high-level keys. Since master and sub-master keys provide access to multiple areas, any unauthorised duplication would have serious consequences. By using restricted key profiles, controlled blanks, and authorised duplication processes, the locksmith ensures that these keys cannot be copied without proper approval. This adds a layer of administrative and legal control that complements the mechanical design of the hierarchy.

## 5. Key Profile and Cylinder Design

The key profile and cylinder design stage is where the mechanical security foundation of a restricted key system is established. While earlier steps define who should have access, this step determines how securely that access is enforced at a physical level. In restricted systems, the key profile and cylinder are not generic components—they are purpose-selected to prevent unauthorised duplication, resist manipulation, and support the complexity of the master key hierarchy.

A key profile refers to the unique shape and milling pattern of the key blade that fits into the keyway of the cylinder. In restricted systems, this profile is legally protected and controlled, meaning that only authorised locksmiths or manufacturers can supply compatible key blanks. This exclusivity is critical because it ensures that even if someone gains access to a key, they cannot easily obtain a blank to copy it. From a design perspective, the locksmith must select a profile that is not only restricted but also dedicated to the specific system or client, ensuring that it cannot be confused or interchanged with other systems.

The cylinder design works in conjunction with the key profile to provide both security and functionality. Advanced restricted systems often incorporate features such as sidebars, which add a secondary locking element that must align correctly before the plug can rotate. Unlike traditional pin tumbler systems that rely solely on vertical pin alignment, sidebar mechanisms introduce additional locking points, making picking and manipulation significantly more difficult. Some systems also include secondary locking mechanisms, such as rotating pins, sliders, or magnetic elements, which require precise interaction with the key beyond simple pin lifting.

Another important aspect is the use of complex pin configurations, particularly in master key systems. These configurations may involve multiple shear lines within a single chamber to allow different keys (e.g. change keys and master keys) to operate the same lock. While this is essential for hierarchy, it must be carefully controlled. Excessive complexity can increase the number of shear lines, which may reduce resistance to picking or create mechanical instability. The locksmith must therefore design pinning arrangements that support the hierarchy while maintaining security integrity.

Compatibility is a key consideration throughout this process. The selected cylinders must be fully compatible with the chosen restricted platform, including its key profile, blank availability, and duplication control systems. Additionally, the cylinders must support the required level of master keying without exceeding their mechanical limits. Not all cylinder designs are suitable for complex hierarchies, and selecting the wrong platform can result in limitations that are difficult or impossible to correct later.

Equally important is ensuring that the system remains reliable and serviceable over time. High-security features such as sidebars and complex pin stacks can improve resistance to attack, but they also introduce additional moving parts and tighter tolerances. If the design is too complex, it may lead to issues such as increased wear, sensitivity to dirt or debris, or difficulty in maintenance. The locksmith must strike a balance between security and practicality, ensuring that the system can be maintained, repaired, and rekeyed without excessive difficulty or cost.

From a broader perspective, this step represents the point where security design meets real-world usability. A well-designed restricted system must not only resist unauthorised access but also function smoothly in daily operation. Keys must insert and turn reliably, cylinders must withstand environmental conditions, and the system must remain adaptable to future changes. By carefully selecting the key profile and cylinder design, the locksmith ensures that the system is not only secure at installation but remains effective and manageable throughout its lifecycle.

<u>6. Controlled Key Duplication Process</u>

The controlled key duplication process is the core administrative control mechanism that gives a restricted key system its real-world effectiveness. While patented key profiles and specialised cylinders provide mechanical protection, it is the duplication control procedure that ensures those protections are enforced consistently over time. Without strict duplication controls, even a high-security system can quickly degrade, as unauthorised copies of keys begin to circulate beyond the control of the system owner.

At its foundation, this process ensures that keys can only be duplicated by authorised locksmiths who are recognised by the system provider or manufacturer. This restriction is achieved through controlled distribution of key blanks, which are not available through general retail channels. As a result, duplication cannot occur casually or informally; it must take place within a controlled environment where procedures are followed and records are maintained. This immediately eliminates one of the most common causes of security failure in standard systems—unregulated key copying.

Equally important is the requirement that all duplication requests must be approved by an authorised signatory. This introduces a formal decision-making layer into the process. Even if a person physically possesses a key, they cannot have it duplicated unless they are listed as an authorised approver or have obtained approval from one. This prevents unauthorised expansion of access and ensures that all additional keys are issued intentionally and with oversight. It also protects the locksmith, as they are acting on documented authority rather than verbal requests or assumptions.

To support this process, many systems require the presentation of identification credentials or key control cards. A key control card is typically issued at the time of system registration and serves as proof of ownership or authority. It may include details such as the system reference number, authorised contacts, and security features to prevent forgery. When requesting duplication, the client must present this card along with appropriate identification, ensuring that the request is both authenticated and traceable.

Additional controls, such as signature verification and system registration checks, further strengthen the process. The locksmith must verify that the request matches the authorised signatory list and that the system details correspond to the registered records. This may involve comparing signatures, confirming contact details, or cross-referencing the system database. These steps ensure that even if an individual attempts to impersonate an authorised person, the request can be identified and rejected.

From a design perspective, this process transforms key duplication from a simple mechanical task into a regulated and auditable procedure. Every key produced can be recorded, linked to an approval, and traced back to a specific request. This creates accountability and visibility, allowing the system owner to maintain control over how access expands over time. It also enables effective response to incidents, as the organisation can determine how many keys exist, who holds them, and when they were issued.

This level of control is particularly important for high-level keys, such as master or sub-master keys, where unauthorised duplication could compromise large portions of the system. By enforcing strict duplication procedures, the locksmith ensures that these critical access points remain protected not only by mechanical design but also by administrative governance.

## 7. Key Identification and Tracking

Key identification and tracking is the stage where a restricted key system becomes fully accountable and manageable over time, rather than simply secure at the point of installation. Every key issued within the system must be uniquely identifiable so that its purpose, holder, and level of access can be clearly understood at any moment. Without this level of traceability, the system quickly loses integrity, as keys become indistinguishable and their associated risks cannot be accurately assessed.

The first element of this process is the use of key stamping or coding. Each key is marked with a unique identifier—typically an alphanumeric code—that corresponds to internal system records. Importantly, this code must be non-descriptive to outsiders. It should not indicate the location, door, or level of access the key provides. For example, instead of marking a key "Server Room" or "Office 3," a neutral code such as "K-1042" is used. This ensures that if a key is lost or stolen, it does not provide immediate intelligence about what it opens.

These codes are then linked to a central key register, which acts as the authoritative record of the system. The register documents each key's unique identifier, the locks it operates, the level of access it represents (e.g. change key, sub-master, master), and the individual or role to whom it has been issued. This creates a clear and auditable relationship between keys, users, and access permissions. In well-managed systems, the register is treated as a controlled document, with restricted access and regular updates to reflect any changes in key distribution.

A critical aspect of this stage is maintaining a clear mapping between keys and access levels. The locksmith must ensure that each key code can be traced back to its position within the hierarchy and the zones or doors it operates. This mapping allows the organisation to understand not just who holds a key, but the extent of access that key provides. In complex systems, this is essential for managing risk, as higher-level keys (such as sub-masters or master keys) carry significantly greater exposure than individual change keys.

Security considerations are central to this process. Keys must never be labelled with obvious or descriptive information, as this creates a direct vulnerability if the key is lost. Similarly, the coding system must be designed so that it only has meaning when referenced against internal records. This separation between the physical key and its associated information ensures that possession of the key alone does not reveal its function.

The true value of key identification and tracking becomes evident when a key is lost, stolen, or unaccounted for. Because each key is uniquely coded and recorded, the organisation can quickly assess the level of risk by identifying exactly what the key opens. This allows for targeted corrective action, such as rekeying specific locks or zones, rather than undertaking costly and unnecessary system-wide changes. Without this information, the organisation may be forced to assume the worst-case scenario, leading to broader disruption and expense.

From a locksmithing perspective, this step reinforces the transition from simple key cutting to system-level security management. The locksmith is responsible for establishing a structure where every key is controlled, traceable, and accountable throughout its lifecycle. This not only improves security outcomes but also supports operational efficiency, as the system can be maintained, audited, and adapted with confidence.

Ultimately, effective key identification and tracking ensure that the restricted system remains visible, controlled, and responsive to change. It enables informed decision-making, reduces uncertainty during incidents, and preserves the long-term integrity of the access control system.

## 8. Integration with Organisational Procedures

Integration with organisational procedures is the step that ensures a restricted key system functions as a living, controlled security system rather than a static installation. While the mechanical and design elements of the system establish how access should work, it is the organisation's policies and processes that determine whether that control is maintained over time. Without this integration, even the most well-designed restricted system will gradually degrade due to inconsistent practices, poor accountability, and unmanaged changes.

At a practical level, the locksmith must ensure the client establishes formal key issuance procedures. This means that keys are only issued following documented approval, with clear records of who received the key, what it opens, and when it was issued. Each recipient should acknowledge responsibility for the key, reinforcing that access is controlled and accountable. This prevents informal or undocumented distribution, which is one of the primary causes of key sprawl and loss of control.

Equally important are key return procedures, which ensure that keys are recovered when no longer required. This typically occurs when employees leave, change roles, or complete temporary work. Without a structured return process, keys remain in circulation beyond their intended use, creating ongoing security risks. The return process should be linked to organisational workflows so that access is not overlooked during transitions.

A critical supporting control is the establishment of lost key reporting processes. Staff must know how and when to report a lost or stolen key, and the organisation must respond promptly by assessing the risk and implementing corrective actions, such as rekeying or restricting access. Delayed or informal reporting significantly increases the likelihood that lost keys will be misused before action is taken. Clear procedures ensure that incidents are managed consistently and effectively.

In addition, regular audits are essential to maintain system integrity. Periodic reviews of the key register and physical key holdings allow the organisation to confirm that all keys are accounted for and held by the correct individuals. Audits also help identify discrepancies, such as missing keys, unauthorised holders, or excessive access levels. These checks provide an opportunity to correct issues before they escalate into security failures.

For long-term effectiveness, the key system must be integrated with broader organisational functions, particularly human resources (HR) systems. During onboarding, access should be issued based on role requirements, ensuring alignment with the principle of least privilege. During offboarding, all keys must be returned and access revoked as part of the exit process. This integration ensures that access control is managed consistently throughout the employee lifecycle, rather than as a separate or secondary process.

Alignment with security policies is also essential. The key system should reflect the organisation's overall security framework, including rules around access control, incident reporting, and risk management. This ensures that physical security measures are consistent with other security controls, such as IT access or surveillance systems, creating a unified approach to risk.

Integration with facility management processes further strengthens control. Facility managers are often responsible for day-to-day operations, maintenance, and contractor access. By aligning key management with these functions, the organisation can ensure that access is granted appropriately for maintenance, repairs, or temporary work, and that it is revoked when no longer required. This prevents the accumulation of unmanaged access rights over time.

From a locksmithing perspective, this step represents a shift from delivering hardware to enabling sustainable security practices. The locksmith's role includes advising clients on how to embed the key system into their organisational processes, ensuring that it remains effective long after installation. This requires an understanding of how people, policies, and operations interact with physical security systems.

<u>9. Scalability and Future Expansion</u>

Scalability and future expansion is a critical design consideration that ensures a restricted key system remains functional, adaptable, and secure over its entire lifecycle, rather than becoming obsolete or compromised as organisational needs change. Unlike simple lock installations, restricted systems are long-term infrastructure investments. Buildings evolve, staff numbers change, departments are restructured, and new areas are added. If the system is not designed with these changes in mind, it can quickly become constrained, forcing inefficient workarounds or complete redesign.

At the core of scalability is the ability to accommodate additional locks and doors without disrupting the existing hierarchy. As new rooms, offices, or facilities are introduced, they must be integrated into the system in a way that maintains the established access structure. This requires the locksmith to design the system with spare capacity, ensuring that new locks can be assigned appropriate key combinations without conflicting with existing ones. Without this foresight, adding new locks may require rekeying large portions of the system or introducing inconsistent access arrangements.

Similarly, the system must support additional users and roles. Organisations rarely remain static—new employees, contractors, and departments will require access over time. A scalable system allows new keys to be issued within the existing hierarchy without expanding access beyond what is necessary. This is particularly important for maintaining the principle of least privilege, as poorly planned systems often result in new users being given overly broad access simply because the system cannot accommodate more precise control.

Another key factor is the ability to adapt to changes in organisational structure. Departments may merge, split, or relocate, and access requirements may shift accordingly. A well-designed system allows these changes to be implemented through reallocation of keys or minor adjustments, rather than requiring a complete redesign. This flexibility is essential for maintaining both operational efficiency and security integrity over time.

To achieve this, locksmiths incorporate reserved key ranges during the initial design. These are unused combinations or spaces within the keying system that are intentionally left available for future use. By reserving these ranges, the locksmith ensures that new keys and locks can be added without interfering with existing ones. This is a fundamental aspect of master key system design, as it preserves the internal logic and prevents conflicts that could compromise functionality or security.

The use of a flexible master key design is equally important. The hierarchy should be structured in a way that allows additional sub-master levels or zones to be introduced if needed. For example, a building may initially have a single department but later expand into multiple divisions, each requiring its own access grouping. A flexible design anticipates this possibility and provides the capacity to accommodate it without restructuring the entire system.

Forward planning during the initial design phase is therefore essential. The locksmith must consider not only the current requirements but also likely future scenarios, such as building extensions, staff growth, or changes in operational use. This requires consultation with the client to understand long-term plans and potential developments. Even if these changes are uncertain, designing with flexibility ensures that the system can adapt without compromising its integrity.

Failure to plan for scalability can have serious consequences. Systems that lack expansion capacity often reach a point where no additional keys or locks can be added without conflict, forcing organisations to adopt inefficient solutions such as issuing higher-level keys or introducing cross-keying. This not only increases security risk but also reduces accountability. In many cases, the only viable solution becomes a complete system redesign, which is costly, disruptive, and time-consuming.

In addition to cost, poor scalability can lead to security compromises. When a system cannot accommodate change, users may be given broader access than necessary, or informal practices may develop to work around limitations. Over time, this erodes the controlled nature of the system and increases the likelihood of misuse or unauthorised access.

From a locksmithing perspective, scalability represents a shift toward strategic system design. It requires thinking beyond immediate installation and considering how the system will perform over years or even decades. A well-planned restricted key system is not only secure at the time of installation but remains adaptable, efficient, and controlled as the organisation evolves.

## 10. Ongoing Management and Enforcement

Ongoing management and enforcement is the stage that determines whether a restricted key system remains secure and effective over time or gradually deteriorates into an uncontrolled environment. While earlier steps focus on design and implementation, this phase is about maintaining control, accountability, and alignment with organisational needs throughout the system's lifecycle. A restricted system is not self-sustaining—without consistent oversight, even the most robust design will be undermined by human behaviour, operational changes, and administrative gaps.

One of the most important activities in this phase is conducting regular audits of key holders. Audits involve verifying that all issued keys are accounted for, held by the correct individuals, and still appropriate for their roles. This process helps identify discrepancies such as missing keys, unauthorised holders, or excessive access levels. Regular audits also reinforce accountability, as key holders are aware that their responsibilities are actively monitored. Without audits, inaccuracies in the key register accumulate over time, leading to loss of visibility and control.

Closely related to this is the need to monitor duplication requests. Even within a restricted system, authorised duplication can lead to gradual expansion of access if not carefully controlled. Each duplication request should be reviewed to ensure it is justified, approved by the correct authority, and recorded accurately. Patterns of frequent or unnecessary duplication may indicate underlying issues, such as poor key management practices or inadequate access design. Monitoring this activity ensures that access grows in a controlled and intentional manner.

Another critical control is rekeying when keys are lost, stolen, or when staff leave the organisation. Rekeying is the primary method of revoking access in mechanical systems, and delays in carrying it out can leave the system exposed to unauthorised entry. When a key is unaccounted for, the organisation must assess the level of risk based on what that key opens and take appropriate action. Similarly, when employees or contractors depart, their access must be removed promptly. Failure to do so results in former personnel retaining access, which is a common and serious security vulnerability.

In addition, organisations must review access levels periodically to ensure they remain aligned with current roles and responsibilities. Over time, individuals may accumulate access beyond what they need due to role changes, temporary assignments, or historical practices. Periodic reviews allow the organisation to correct this by reducing unnecessary access and reinforcing the principle of least privilege. This helps maintain a clean, efficient hierarchy and reduces the risk of misuse.

At this stage, the locksmith's role evolves beyond technical installation into that of a long-term security advisor and system partner. Locksmiths can provide guidance on best practices for managing key systems, assist with the development of policies and procedures, and support clients in maintaining control. Their expertise is particularly valuable during audits, system reviews, and when addressing issues such as key sprawl or system inefficiencies.

Locksmiths may also play a key role in supporting upgrades and system improvements. As organisational needs change or new technologies become available, the system may need to be expanded, restructured, or integrated with electronic access control. The locksmith can assess the existing system, identify limitations, and recommend solutions that enhance both security and functionality while preserving as much of the original investment as possible.

Ensuring compliance with established procedures is another critical aspect of this phase. Policies relating to key issuance, duplication, storage, and return must be consistently followed. The locksmith can help reinforce these procedures by refusing unauthorised requests, maintaining accurate records, and educating clients on the importance of adherence. This consistency is essential for maintaining the integrity of the system and preventing gradual erosion of control.

**Design Application Example**

Scenario Title: "The Door That Was Always Locked... Except When It Wasn't"

A mid-sized commercial office building in Melbourne houses approximately 60 staff across administration, finance, IT, and management departments. The facility includes a reception area, open-plan offices, private offices, a server room, records storage, and a plant room. The organisation had previously installed a standard master key system using conventional cylinders and issued keys based on convenience rather than structured access control.

## Initial System and Emerging Problems

At first glance, the system appeared secure. All external doors were fitted with keyed locks, internal rooms were lockable, and a master key was held by several supervisors for operational efficiency. However, over time, a pattern of recurring security issues began to emerge.

Staff frequently relied on spring latches rather than engaging deadbolts, particularly during short absences. Cleaners and contractors were issued sub-master keys but often shared them informally to avoid delays. The server room door, which required frequent access by IT staff, was often left slightly ajar because the latch alignment had deteriorated and required force to close properly. In some cases, staff propped doors open for convenience during busy periods.

Several keys were lost but not reported immediately, as staff assumed the risk was low. Because the system relied heavily on user diligence—remembering to lock doors, return keys, and report issues—small lapses accumulated into significant vulnerabilities. Eventually, an incident occurred where unauthorised access to the records room was detected, but due to shared keys and lack of tracking, it was impossible to determine who had entered.

## Analysis of Failure

A review conducted by a locksmith and security consultant revealed that the failure was not due to the lock hardware itself, but due to system design that did not account for predictable human behaviour. The system assumed that users would consistently follow correct procedures, but in practice, users prioritised convenience, speed, and habit over strict compliance.

Key issues identified included:

- Over-reliance on manual locking (deadbolts not engaged)
- Excessive distribution of sub-master keys
- Lack of feedback when doors were not properly secured
- Poor tolerance to wear and misalignment
- No control over key duplication or tracking
- No containment of risk when keys were lost

The system was technically sound but behaviourally fragile.

Redesigned System (Human-Centred Approach)

The locksmith redesigned the system using principles that account for human error.

First, fail-secure hardware was introduced. External doors were fitted with deadlatching mechanisms and self-locking electronic strikes, ensuring that doors automatically secured upon closing without requiring a separate action. This removed reliance on users remembering to engage locks.

Second, positive feedback mechanisms were added. Door position sensors were installed on critical doors such as the server room and records storage. If a door was not fully closed or latched, an alert was triggered at reception. Locks were also selected to provide clear tactile feedback when engaged.

Third, the system was re-zoned and simplified. Instead of issuing multiple sub-master keys, access was restructured using the principle of least privilege. IT staff were given access only to relevant areas, and high-risk areas such as the server room were isolated with no sub-master access—only authorised change keys and controlled electronic credentials.

Fourth, a restricted key system was implemented. This prevented unauthorised duplication and ensured all keys were issued, tracked, and approved. A key register was introduced, and all keys were coded rather than labelled.

Fifth, the system was designed to limit the blast radius of mistakes. If a key was lost, it would only affect a single area or zone, rather than the entire facility. High-level keys were reduced and stored securely.

Sixth, time-bound electronic access was introduced for contractors and cleaners. Instead of issuing physical keys, temporary access cards were programmed to work only during specific hours and automatically expired after the contract period.

Seventh, maintenance and usability improvements were implemented. Door alignment was corrected, closers were adjusted, and hardware was upgraded to ensure smooth operation. This removed the need for users to "force" doors or bypass locks due to frustration.

Outcome

Following implementation, the organisation experienced a significant improvement in security and operational efficiency. Doors consistently secured without relying on user memory, key control was restored, and access became traceable and accountable.

Importantly, user behaviour improved—not because staff became more disciplined, but because the system made the secure action the easiest and most natural action. Staff no longer needed to remember multiple steps, and insecure workarounds became unnecessary.

When a key was later reported lost, the organisation was able to quickly identify the affected area and rekey only a single cylinder, avoiding widespread disruption.

Key Lessons for Locksmithing Practice

This case study demonstrates that effective locksmithing is not just about installing strong locks, but about designing systems that anticipate human behaviour. The original system failed because it depended on perfect use; the redesigned system succeeded because it assumed imperfection.

For locksmiths, the key insights are:

- Security must be designed around real user behaviour, not ideal behaviour
- Systems should default to secure states automatically
- Feedback and visibility reduce user error
- Convenience drives behaviour, so secure options must be easy
- Risk must be contained, not eliminated entirely
- Ongoing management and maintenance are essential

Ultimately, designing for human error transforms locksmithing from a purely mechanical trade into a system-level security discipline—one that integrates engineering, behaviour, and risk management to create resilient and practical security solutions.

**Impact on Locksmithing Practices**

Understanding system design concepts that account for human error is highly relevant to anyone interested in locksmithing because it fundamentally changes how security is viewed, designed, and delivered. Traditionally, locksmithing has been associated with the mechanical strength of locks—how well a lock resists picking, drilling, or forced entry. However, in real-world environments, most security failures do not occur because a lock is technically defeated, but because it is used incorrectly, inconsistently, or bypassed for convenience. This means that a locksmith who focuses only on hardware without considering user behaviour will often deliver systems that are technically sound but practically ineffective.

For someone entering or developing within the locksmithing profession, this concept introduces a critical shift from product-focused thinking to system-focused thinking. It highlights that security is not just about installing locks, but about designing complete access control systems that function reliably under real human conditions. People forget to lock doors, share keys, ignore procedures, and take shortcuts—these behaviours are predictable, not exceptional. A skilled locksmith must anticipate these behaviours and design systems that either prevent errors or minimise their impact. This elevates locksmithing from a reactive trade—fixing problems after they occur—to a proactive discipline focused on risk prevention.

The influence on locksmithing practice is significant, particularly in how systems are specified and installed. Instead of relying on users to perform critical actions, locksmiths are encouraged to select fail-secure hardware and automated solutions that make secure behaviour the default. For example, choosing deadlatching locks, self-locking mechanisms, or electronic access systems reduces dependence on users remembering to lock doors. This directly improves real-world security outcomes and reduces the likelihood of common failures such as unlocked doors or partially latched entry points.

Another major impact is the increased emphasis on usability and convenience as security factors. Locksmiths must recognise that if a system is difficult, time-consuming, or frustrating to use, people will find ways around it. This leads to insecure practices such as propping doors open or sharing keys. As a result, modern locksmithing involves designing systems where the secure option is also the easiest option. This may include well-structured master key systems, access cards, or simplified zoning that reduces the need for multiple keys. By aligning security with user behaviour, locksmiths can significantly improve compliance without relying on constant supervision or enforcement.

The concept also reinforces the importance of risk containment rather than risk elimination. Since human error cannot be completely prevented, systems must be designed so that mistakes have limited consequences. This influences how locksmiths design key hierarchies, encouraging the use of zoning, restricted access levels, and minimal distribution of high-level keys. For example, instead of issuing a widely used master key, a locksmith may design a system where access is compartmentalised, ensuring that a lost key only affects a small part of the facility. This approach improves resilience and reduces the impact of inevitable errors.

In addition, these concepts expand the locksmith's role into education, advisory, and system management functions. Locksmiths are no longer just installers of hardware; they are responsible for guiding clients on best practices, such as key control, maintenance, and access procedures. They must be able to explain why certain design decisions are made, such as restricting key duplication or implementing time-bound access. This advisory role is particularly important in organisational environments, where security depends on both technical systems and administrative processes.

The integration of mechanical and electronic systems further strengthens the relevance of designing for human error. In modern locksmithing, access control increasingly involves digital credentials, audit trails, and automated controls. These systems can enforce rules such as limited access times, automatic locking, and real-time alerts, reducing reliance on human memory and judgement. However, they also introduce new risks if users choose weak codes or fail to manage credentials properly. As a result, locksmiths must understand both physical and digital behaviours and design systems that address vulnerabilities in both domains.

Finally, this concept influences long-term locksmithing practices by emphasising maintenance and lifecycle management. Systems degrade over time, and as they become harder to use, users are more likely to bypass them. Designing for human error includes ensuring that systems remain easy to operate through regular maintenance, adjustment, and servicing. This prevents the gradual shift from secure operation to insecure workarounds, which is a common cause of system failure.

In summary, designing systems that account for human error is essential to modern locksmithing because it aligns security with real-world conditions. It transforms locksmithing from a purely mechanical trade into a holistic security discipline that integrates engineering, human behaviour, and risk management. For practitioners, this means designing systems that are not only strong, but also practical, intuitive, and resilient—ensuring that security is maintained even when users are imperfect.

## Chapter 7 Review Questions

### Question 1

A facilities manager requests several additional copies of a restricted key for an office building. When the locksmith checks the system records, the manager's name is not listed as an authorised signatory.

What should the locksmith do?

A. Cut the keys immediately because the request came from management
B. Refuse the request until proper authorisation is verified
C. Cut the keys but do not record them
D. Ask another employee to confirm the request

### Question 2

A locksmith installs a master key system in a commercial building but fails to keep documentation of which keys were issued to each staff member.

What problem is most likely to occur?

A. The locks will stop functioning
B. The master key will no longer operate
C. Loss of accountability and inability to track access
D. Doors will open automatically

### Question 3

During a security audit, the locksmith discovers that several employees have duplicated keys without authorisation.

Which system feature is designed specifically to prevent this problem?

A. Standard key blanks available at hardware stores
B. Restricted key blanks supplied only to authorised locksmiths

C. Colour-coded key caps
D. Electronic door closers

**Question 4**

A key coded HQ-01-CK-0047-R-D2 appears in the key register.

What does the "D2" designation most likely indicate?

A. The second door the key opens
B. The second authorised duplicate of that key
C. The second master key level
D. The second locksmith who cut the key

**Question 5**

An employee leaves a company but forgets to return their issued office key.

According to good key control practice, what should happen first?

A. Ignore the issue
B. Immediately change every lock in the building
C. Notify management and conduct a risk assessment
D. Destroy the key register

**Question 6**

A locksmith designs a master key hierarchy including Change Keys (CK), Sub-Master Keys (SMK), and a Master Key (MK).

What is the purpose of this hierarchy?

A. To simplify manufacturing of locks
B. To organise access levels across different areas
C. To reduce the number of doors in a building
D. To eliminate the need for documentation

**Question 7**

A company stores restricted key blanks in an unlocked cabinet where multiple employees can access them.

Why is this a security concern?

A. The keys will wear out faster
B. Uncontrolled access may allow unauthorized key duplication
C. Locks will jam
D. Key codes will automatically change

**Question 8**

A locksmith uses a key register to track issued keys.

Which information would most likely appear in this register?

A. The manufacturer's advertising material
B. Who holds the key and when it was issued
C. The colour of the building walls
D. The type of door hinges installed

**Question 9**

A key with the code HQ-01-MK-0001 is stored in a secure cabinet and labelled as "Master Key."

Why is this key usually stored under strict control?

A. It opens only one door
B. It opens multiple locks across the system
C. It cannot be duplicated
D. It is used only during emergencies

**Question 10**

During a periodic audit, the locksmith compares the key register with the keying schedule.

Why is this comparison important?

A. To ensure key codes are aesthetically pleasing
B. To verify that keys and locks match the intended access hierarchy
C. To reduce locksmith workload
D. To eliminate master keys entirely

# Chapter 8

# Security Psychology and Behavioural Design

This chapter marks a pivotal transition in the study of locksmithing, moving beyond the traditional focus on hardware and mechanical security into the deeper, often overlooked domain of human behaviour and decision-making. While earlier chapters have explored how locks are designed, installed, and managed, this chapter examines a more fundamental question: why security systems fail in real-world environments—even when they are technically sound.

At the centre of this discussion is a critical insight: security is not determined solely by the strength of a lock, but by how people interact with it. Human behaviour, shaped by convenience, habit, time pressure, and perception of risk, often overrides formal procedures and technical design. As a result, many security failures are not caused by weaknesses in locks themselves, but by predictable patterns in how users behave in everyday situations.

This chapter introduces the principles of security psychology and behavioural design, providing a framework for understanding how and why people ignore security advice, how habits are formed, and how environmental cues influence decision-making. It challenges the assumption that users are the weak link, instead reframing security as a design problem, one that must account for real human behaviour rather than idealised compliance.

Through practical examples, case studies, and applied design strategies, the chapter demonstrates how locksmiths can move toward human-centred security solutions. Concepts such as nudging behaviour through design, making secure actions automatic, reducing complexity, and providing clear feedback are explored as essential tools for improving real-world security outcomes.

## Why People Ignore Security Advice

People rarely ignore security advice because they do not understand it; more often, they ignore it because it conflicts with how they actually live and work. Research in human behaviour and security consistently shows that users act rationally within the constraints of their environment, prioritising efficiency, time, and task completion over abstract security requirements. In locksmithing and physical security contexts, this is a critical insight. Users generally intend to behave securely, but their actions are shaped by convenience, habit, time pressure, and perceived risk rather than formal procedures [211, 212].

One of the most significant factors is the dominance of convenience over security. Individuals naturally gravitate toward the simplest and fastest way to complete routine tasks. If securing a door requires multiple deliberate actions, such as checking alignment and engaging secondary locks, users are likely to omit steps, particularly during frequent or short-duration use. Over time, these shortcuts become embedded behaviours. This aligns with the concept of the "compliance budget," where users allocate limited effort toward security tasks and will bypass controls that exceed that effort threshold [213]. For locksmiths, this demonstrates that systems relying on consistent user diligence are inherently vulnerable.

Risk perception also plays a central role. People routinely underestimate the likelihood of security incidents, particularly in familiar environments. Behavioural research shows that individuals assess risk based on experience rather than statistical probability, leading to statements such as "it won't happen here" or "just for a minute." This aligns with cognitive bias frameworks, where intuitive (fast) thinking overrides analytical risk assessment [214]. As a result, users may leave doors unlocked, delay reporting lost keys, or adopt insecure practices because the perceived risk does not justify the perceived inconvenience.

Habit and routine further reinforce insecure behaviour. Once a pattern is established, it becomes automatic and resistant to change. Human factors research demonstrates that behaviour is often driven by learned routines rather than conscious decision-making [215]. Even when users receive training or updated instructions, established habits tend to persist, especially under time pressure or stress. For locksmithing practice, this highlights the importance of designing systems that align with natural human behaviour rather than relying solely on procedural compliance.

Another contributing factor is the absence of immediate consequences. When insecure behaviour does not result in visible negative outcomes, it becomes reinforced over time. Users interpret the absence of incidents as evidence that their behaviour is acceptable, creating a false sense of security. This phenomenon is well documented in safety and human error research, where latent conditions remain undetected until a failure occurs [215]. In physical security, this means that repeated non-compliance can continue unchecked until a breach exposes the underlying vulnerability.

Complexity and poor system design also reduce compliance. Systems that are difficult to understand or operate increase the likelihood of user error or avoidance. From a design perspective, usability is directly linked to security effectiveness; poorly designed systems encourage workarounds that bypass intended controls [216]. In locksmithing, this may include confusing key hierarchies, unclear labelling, or overly complex access procedures, all of which lead users to simplify processes in ways that reduce security.

Social and organisational influences are equally powerful. Behaviour within workplaces is often shaped more by observed norms than by formal policies. If employees see others bypassing security measures without consequence, those behaviours become normalised. Over time, informal practices can override official procedures, particularly in environments where productivity is prioritised over compliance. This reflects broader organisational behaviour patterns where culture determines actual practice more than documented rules.

A further issue is the lack of personal accountability. In many organisational settings, users do not feel directly responsible for security outcomes. When responsibility is diffused, motivation to follow procedures decreases. This is particularly evident in large systems where the consequences of individual actions are not immediately visible or personally experienced.

In digital and hybrid systems, similar patterns emerge. Users select weak PINs, reuse credentials, or avoid updating access permissions because these behaviours are easier and more familiar. The intangible nature of digital risk further reduces perceived urgency, even though the underlying principles mirror those seen in physical security.

For locksmithing practice, these insights are highly significant. They demonstrate that security systems must be designed with human behaviour in mind. Effective systems reduce reliance on perfect user compliance by making secure actions simple, intuitive, and automatic wherever possible. Education and policy remain important, but they must be supported by design strategies that anticipate real-world behaviour.

People ignore security advice not because they are careless, but because security competes with convenience, habit, and perception. Effective locksmithing practice recognises this reality and focuses on designing systems that remain secure even when users behave imperfectly.

Locksmiths address this problem most effectively when they stop treating non-compliance as a user failure and instead treat it as a design problem. The core insight is that people are behaving rationally within their environment—so the system must be designed to work with that behaviour, not against it. This shifts locksmithing from a purely technical trade into a discipline that combines engineering, human factors, and risk management.

**Designing for Convenience (Making Security the Default):** The most effective strategy is to make the secure action the easiest action. If users naturally prioritise speed and simplicity, then systems must remove friction from secure behaviour. This can be achieved through hardware and system design such as self-locking mechanisms, automatic deadbolts, or electronic locks that engage upon closure.

When a door locks automatically, the user no longer has to remember to engage a deadbolt. This eliminates reliance on memory and effort, which are the primary failure points. In this way, locksmiths reduce the "compliance budget" required from users—security becomes passive rather than active.

**Reducing Steps and Complexity:** Complex systems are consistently bypassed. Locksmiths can address this by simplifying how users interact with locks and access systems. This may involve reducing the number of keys a person carries, simplifying master key hierarchies, or replacing multi-step locking processes with single-action systems.

Clear, intuitive design is critical. If a user has to think about which key to use or how a system works, they are more likely to make errors or develop shortcuts. Good locksmithing design ensures that correct use is obvious and requires minimal decision-making.

**Designing for Habit and Routine:** Because behaviour is driven by habit, locksmiths should design systems that align with natural routines. For example, doors that automatically close and lock reinforce secure behaviour without requiring conscious effort. Similarly, access systems that integrate with daily workflows—such as swipe access at normal entry points—encourage consistent use.

Where habits already exist, systems should adapt to them rather than attempt to overwrite them entirely. Trying to change behaviour through instruction alone is far less effective than embedding security into routine actions.

**Providing Feedback and Visibility:** One reason users ignore security is the lack of immediate consequences. Locksmiths can address this by introducing clear feedback mechanisms. This might include locks that provide audible or visual confirmation when engaged, or electronic systems that notify users when a door is left unsecured.

Feedback makes security visible. When users can see or hear that a door is not properly secured, they are more likely to correct their behaviour. This bridges the gap between action and consequence, which is often missing in physical security systems.

**Controlling Risk Through System Design:** Locksmiths must assume that users will occasionally act insecurely and design systems that limit the impact of those actions. This includes implementing the principle of least privilege, ensuring that individuals only have access to what they need.

For example, if a key is lost, a well-designed system ensures that only a small, defined area is affected. Similarly, restricted key systems prevent unauthorised duplication, limiting how far a single behavioural failure can spread. This approach accepts human error but prevents it from becoming a system-wide vulnerability.

**Strengthening Accountability Without Friction:** Rather than relying on strict enforcement alone, locksmiths can build accountability into the system in a way that does not increase user burden. Key registers, controlled issuance, and identification systems ensure that responsibility is clear without making daily use more difficult.

In electronic systems, audit trails can provide visibility of access without requiring additional effort from the user. When people know that access is traceable, behaviour often improves naturally, even without direct supervision.

**Integrating Security into Organisational Processes:** Security failures often occur because systems are disconnected from organisational processes. Locksmiths can address this by ensuring integration with workflows such as onboarding and offboarding. When access is automatically issued and revoked as part of employment changes, the system no longer relies on individuals remembering to act.

This reduces the risk created by time pressure, oversight, or competing priorities, which are common causes of non-compliance in real environments.

**Supporting Users Through Education (But Not Relying on It):** Education remains important, but it must be practical and relevant. Locksmiths should focus on explaining why certain behaviours matter in real-world terms, rather than relying on abstract rules. However, education should support system design—not replace it.

The key principle is that training alone cannot overcome poor design. Even well-informed users will revert to convenient behaviour if the system makes secure actions difficult.

**Moving Toward Hybrid and Adaptive Systems:** Modern locksmithing increasingly involves hybrid systems that combine mechanical and electronic controls. These systems allow for adaptive security measures such as time-based access, automatic locking, and rapid credential revocation.

By using these technologies, locksmiths can create systems that adjust to user behaviour rather than expecting users to adapt perfectly to the system.

## Impact on Locksmithing Practices

This concept is highly relevant to anyone interested in locksmithing because it fundamentally changes how security is understood, designed, and delivered in real-world environments. A new or developing locksmith may initially focus on the technical strength of locks—materials, pick resistance, installation quality, and compliance with standards. However, this analysis makes it clear that even the most secure hardware can be rendered ineffective if it does not align with how people actually behave. In practice, security failures are rarely caused by lock failure; they are caused by predictable human behaviour. Understanding this shifts locksmithing from a purely mechanical trade into a discipline that must account for human factors, behavioural patterns, and environmental constraints.

For someone entering the profession, this insight is critical because it explains why "correct" solutions often fail in real-world settings. A locksmith might install a high-security deadlock, but if the user does not consistently engage it due to inconvenience, the intended level of protection is never achieved. Without understanding user behaviour, a locksmith may incorrectly assume that the solution was effective, when in reality it was never properly used. This knowledge allows locksmiths to diagnose security weaknesses more accurately, recognising that the issue may lie in system design and usability rather than hardware quality.

The influence on locksmithing practice is significant, as it expands the role of the locksmith beyond installation and repair into system design and behavioural risk management. Locksmiths must consider how users interact with locks on a daily basis, under time pressure, habit, and routine, not just how locks perform under ideal conditions. This leads to a design philosophy where systems are built to function effectively even when users are inconsistent, distracted, or prioritising convenience. In practical terms, this means selecting and recommending solutions such as self-locking mechanisms, simplified key systems, and intuitive access controls that reduce reliance on user effort.

It also directly impacts how locksmiths approach master key systems and key management. Rather than simply creating hierarchical access structures, locksmiths must ensure that these systems are easy to understand and use. Overly complex key hierarchies or unclear labelling can lead to misuse, workarounds, or loss of control. By designing systems with clarity and simplicity, locksmiths reduce the likelihood of user error and improve overall security outcomes. This reflects a shift from designing for theoretical control to designing for practical usability.

Another important influence is the emphasis on limiting the consequences of human error. Since it is unrealistic to expect perfect compliance, locksmiths must design systems that contain and minimise risk when mistakes occur. This includes applying the principle of least privilege, restricting key duplication through controlled systems, and ensuring that the loss or misuse of a single key does not compromise an entire facility. This approach treats human error as an expected condition rather than an exception, and designs systems to remain resilient under those conditions.

The concept also reinforces the importance of integrating locksmithing with organisational processes. Security cannot rely solely on individual behaviour; it must be supported by structured procedures such as key registers, controlled issuance, and integration with HR processes. For example, ensuring that keys are issued and revoked as part of onboarding and offboarding reduces reliance on memory and manual action. This reflects a broader systems-based approach, where security is embedded into workflows rather than imposed as an additional task.

In modern locksmithing, this understanding extends into electronic and hybrid systems. The same behavioural patterns—convenience, habit, and risk perception—apply to digital credentials as they do to physical keys. As a result, locksmiths must design and configure systems that enforce good security practices automatically, such as time-based access, automatic locking, and audit trails. This requires an expanded skill set that includes not only mechanical expertise but also an understanding of access control technologies and user interaction.

## Fear, Convenience, and Habit in Lock Usage

### Fear as a Driver of Security Behaviour

Fear is one of the most powerful motivators influencing how and when people use locks. When individuals perceive a high level of risk, such as in unfamiliar environments, high-crime areas, or after a recent incident, they are far more likely to engage all available security measures. This may include double-locking doors, checking locks repeatedly, or upgrading to higher-security systems. In these situations, behaviour aligns closely with recommended security practices because the perceived consequences of failure are immediate and tangible.

However, fear is inconsistent and often short-lived. In familiar environments such as homes or workplaces, perceived risk decreases over time. As a result, users become less vigilant, even if the actual risk remains unchanged. This creates a pattern where security behaviour fluctuates depending on recent experiences rather than objective risk. For locksmiths, this highlights that systems cannot rely on fear to ensure compliance, as it is not a stable or predictable driver of behaviour.

### Convenience as the Dominant Influence

Convenience is often the strongest and most consistent factor influencing lock usage. People naturally prioritise speed, ease, and efficiency, especially during routine tasks. If a locking process is perceived as inconvenient, requiring extra steps, effort, or time, users are likely to bypass or simplify it. For example, they may rely on a latch instead of engaging a deadbolt, leave doors unlocked during short absences, or avoid using secondary locking mechanisms altogether.

Over time, convenience-driven shortcuts become normalised behaviour. This is particularly evident in environments where locks are used frequently, such as workplaces or shared facilities. From a locksmithing perspective, this demonstrates that systems requiring repeated user effort are inherently vulnerable. If secure behaviour is not the easiest option, it will not be the most common option. Therefore, effective system design must reduce friction by making security automatic or effortless wherever possible.

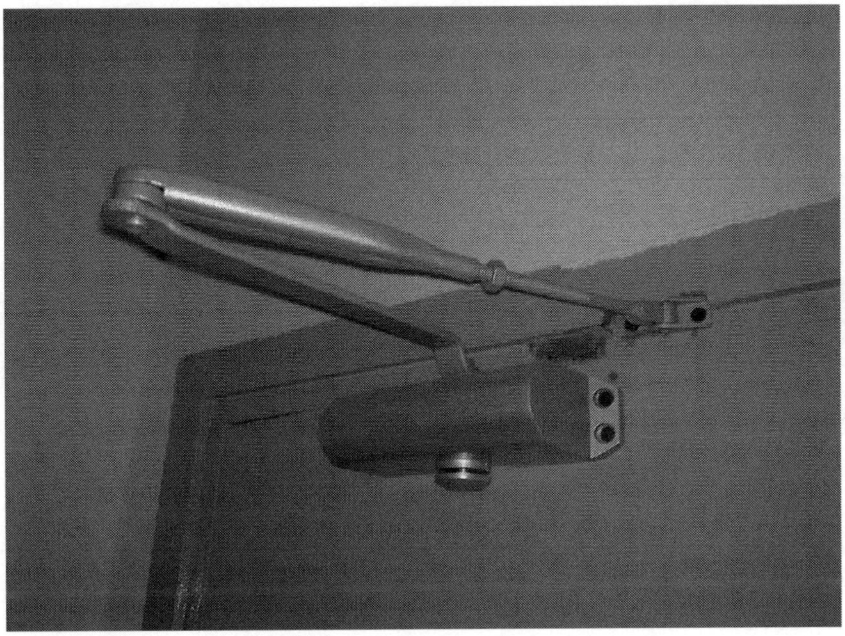

*Figure 70: Automatic door closer to self-close door which may be incorporated with self-locking. Irvin calicut, CC BY-SA 3.0, via Wikimedia Commons.*

**Habit as the Reinforcing Mechanism**

Habit plays a critical role in sustaining both secure and insecure behaviours. Once a pattern of lock usage is established, it becomes automatic and is performed with little conscious thought. This means that initial behaviour, whether correct or incorrect, tends to persist over time. For example, if a user develops the habit of locking a door fully every time they leave, that behaviour becomes reliable. Conversely, if they habitually skip certain steps, such as engaging a deadbolt, that insecurity becomes embedded.

Habits are particularly resistant to change, even when users are given new instructions or training. Under time pressure or stress, people revert to their established routines rather than consciously applying new procedures. For locksmiths, this reinforces the importance of designing systems that encourage correct habits from the outset. Systems should support consistent behaviour through simplicity, feedback, and alignment with natural routines.

**Interaction Between Fear, Convenience, and Habit**

These three factors do not operate independently—they interact to shape overall behaviour. Fear may temporarily increase compliance, but convenience and habit tend to dominate over time. For example, after a security incident, a user may be highly diligent for a period, but as the perceived threat fades, they gradually return to more convenient habits. This cycle explains why security behaviour often degrades over time, even when users are aware of best practices.

From a locksmithing perspective, this interaction highlights a key principle: security systems must be designed for long-term behaviour, not short-term reactions. Systems that rely on users maintaining high levels of vigilance or effort are unlikely to remain effective.

**Impact on Locksmithing Practices**

For locksmiths, recognising the roles of fear, convenience, and habit fundamentally changes how security solutions are designed and recommended. It shifts the focus from simply installing strong locks to creating systems that are usable, intuitive, and aligned with real human behaviour.

In practice, this means:

- Designing systems where the secure action is automatic or requires minimal effort
- Reducing complexity so that correct use is obvious and consistent
- Supporting the development of good habits through intuitive operation
- Avoiding reliance on user vigilance or memory

For example, self-locking doors, simplified key systems, and clearly structured access controls all help ensure that security is maintained without requiring constant user attention.

## Designing Locks People Will Actually Use Correctly

Designing locks that people will actually use correctly requires a shift from purely mechanical thinking to human-centred security design. The core principle is simple but often overlooked: a lock is only effective if it is used properly, every time. Because users naturally prioritise convenience, speed, and routine, locksmiths must design systems where the secure behaviour is the easiest and most natural behaviour. This perspective is echoed across studies and engineering discussions that connect usability, human factors, and security in physical access systems. Multiple sources converge on the idea that default secure states, reduced cognitive workload, alignment with natural behaviour, clear feedback, and error-tolerant design are central to achieving real-world secure usage of locks and access systems [217-222]. In syntheses of usability-centred security, the overarching claim is that security is most effective when embedded into normal operation, rather than requiring deliberate, memory-heavy actions by users [217, 218] These themes are reinforced by examinations of door and latch design, where automatic or self-locking behaviour, intuitive cues, and streamlined access management reduce improper use and improve overall security outcomes [217, 219, 220].

**Make the Secure Action the Default:** A core design prescription is to default to security and minimize reliance on user memory or extra steps. Locks that self-latch, self-lock, or automatically secure when the door closes illustrate this principle by embedding security into the door's normal operation, thereby reducing the likelihood of forgetfulness or skipped steps. The literature notes that automatic locking and self-locking mechanisms can significantly reduce the common failure mode of users omitting a deliberate locking action [217, 219, 220]. In practice, electronic and automatic locking designs that engage as doors close are highlighted as effective embodiments of "make the secure action the default" [217, 219, 220].

**Minimise Steps and Cognitive:** Effort Reducing the number of user decisions and steps in the locking process lowers the probability of non-compliance and incorrect use. The literature consistently argues that concentrating interactions into a single, obvious action—where possible—improves correct usage by making security behaviour automatic and intuitive [217-221]. Consolidating locking functions and employing intuitive mechanisms helps avoid reliance on memory or complex sequences, which aligns with evidence about human factors in locking and access design [217-221].

**Align with Natural Behaviour and Movement:** Lock design should fit the way people move and behave in real environments. Examples include doors that automatically close and latch as part of a normal routine, rather than requiring manual closing or multiple steps to secure, which can introduce errors [217-221]. Access systems should be placed and operate in ways that align with typical workflows, increasing compliance without extra effort. The alignment between security actions and everyday movement is repeatedly identified as a determinant of effective usage [217-221].

**Provide Clear Feedback to the User:** Users frequently misuse locks when they cannot tell whether a lock is engaged. Immediate and unambiguous feedback, tactile, visual, or audible, helps reinforce correct behaviour and enables real-time error detection. The literature emphasizes that feedback reduces misperceptions about lock status and helps users recognise when engagement has failed, thereby improving correct usage [217-221].

**Design for Error Tolerance:** Good security design assumes human error and builds safeguards to limit its impact. Examples include lock mechanisms that resist bypass if a door is not fully closed and that provide partial engagement with some security level, or that generate alarms when a door is left unsecured. This principle accepts inevitability of mistakes while focusing on preventing errors from becoming security failures [217-221].

**Reduce the Consequences of Non-Compliance:** To mitigate imperfect user behaviour, designs should limit the reach of mistakes. Principles such as least privilege in key systems, restricted key systems to prevent duplication, and compartmentalised access aim to ensure that a fault in one area does not compromise the entire system. This resilience-oriented approach is consistent with security design thinking that prioritises containment and controlled access [217, 221].

**Simplify Key and Access Management:** Clear key hierarchies, intuitive identification, and limiting the number of keys per user reduce misuse. In electronic systems, role-based access can further simplify permissions. The literature notes that complex key systems are a major source of confusion and misuse, and simplification, along with proper access management, contributes to correct usage [217, 221].

**Integrate Security into Everyday Processes:** Locks should be part of broader processes and workflows (e.g., onboarding/offboarding and automatic access updates) rather than standalone elements. When security is embedded in routine procedures, consistent and correct usage is more likely, reducing reliance on individuals to remember procedures. This integration perspective is repeatedly highlighted as crucial for achieving sustained secure usage [217, 218, 221].

**Support Behaviour with Education—But Don't Depend on It:** Education can illuminate why security matters, but good design should do most of the work. Training should reinforce how to use the design, while the system itself should steer users toward correct behaviour. The consensus is that education alone cannot compensate for poor design and that simple, practical guidance complements secure-by-default design, not replaces it [217, 218, 221].

**Leverage Technology Where Appropriate:** Modern technologies (electronic locks, access cards, mobile credentials) can enhance usability and compliance by enabling automatic locking, time-based access, and audit trails. However, these technologies must be designed with usability in mind to avoid introducing new complexities and user errors. The literature emphasizes using technology to simplify interaction and support secure use rather than to complicate it [217, 221].

Across these sources, the synthesis converges on a human-centered view of lock design: security is most effective when it operates automatically, is simple and intuitive, fits with real-world behaviour, provides clear feedback, tolerates human error, minimizes the consequences of non-compliance, simplifies access management, and is integrated with everyday processes. By these standards, locksmithing can move beyond purely mechanical thinking toward secure-by-default designs that people will use correctly, every time. The body of work supports the claim that embedding security into normal use, rather than relying on user discipline, yields stronger practical security outcomes [217-222].

### Practical Example: Office Side Entrance – From Poor Design to Human-Centred Design

Scenario Overview

A medium-sized office building has a side entrance used daily by staff. The door is fitted with a standard mortice lock and a separate deadbolt. Staff enter using a key and are expected to manually lock the door behind them by turning the key and engaging the deadbolt.

Before: Traditional (Mechanically Sound but Behaviourally Weak)

In practice, the system fails—not because the lock is poor, but because people don't use it correctly:

- Staff often pull the door shut without turning the key, relying only on the latch

- During busy periods, people skip locking steps to save time
- Some assume "someone else will lock it" (diffused responsibility)
- There is no feedback confirming whether the door is secure

**Result:** The door is frequently left unsecured, creating a real vulnerability despite having a high-quality lock installed.

<u>After: Human-Centred Lock Design</u>

The locksmith redesigns the system based on human behaviour principles:

1. Secure Action is Automatic

- The door is fitted with a self-locking mechanism (deadlatch or electronic lock)
- When the door closes, it automatically locks
- No need for staff to remember or perform extra steps

2. Steps are Reduced

- Replace key entry with access card or keypad
- Entry and exit become single-action processes

3. Aligned with Natural Behaviour

- Staff simply walk through the door as normal
- The door closer ensures it shuts fully every time

4. Clear Feedback

- Lock provides:
    - Audible click when locked
    - LED indicator showing secure status
- Users instantly know if the door is secure

5. Error Tolerance

- If the door is not fully closed:
    - The system triggers a "door ajar" alert
- Prevents silent failure

6. Reduced Consequences of Mistakes

- Access is controlled via role-based permissions

- If a credential is lost:
    - It can be immediately revoked
    - No need to rekey the entire system

7. Simplified Management

- No multiple keys
- No confusion about which key to use
- Access is tied to individual identity

*Table 7: Outcome Comparison.*

| Aspect | Before | After |
|---|---|---|
| User effort | High (manual locking) | Low (automatic) |
| Reliance on memory | High | None |
| Error rate | Frequent | Minimal |
| Feedback | None | Immediate |
| Security outcome | Inconsistent | Reliable |

The example highlights a fundamental shift in locksmithing practice, moving away from a purely hardware-focused mindset toward a human-centred approach to security design. Traditionally, locksmithing relied on the assumption that installing a strong, well-engineered lock and providing instructions would be sufficient to ensure security. However, real-world experience demonstrates that this approach is limited. Modern locksmithing recognises that security systems must be designed so that users cannot easily operate them incorrectly. In this context, the locksmith's role extends beyond installation to actively shaping user behaviour through thoughtful system design.

This shift is important because it reflects how people actually behave in everyday environments. Users are often busy, working under time pressure, and focused on completing tasks efficiently. As a result, they may forget to follow procedures, skip steps, or take shortcuts that compromise security. These behaviours are not necessarily due to carelessness, but rather a natural response to competing priorities. A well-designed lock system anticipates these realities and removes reliance on perfect user behaviour by making secure actions automatic, simple, and intuitive.

The key takeaway is that the effectiveness of a lock is not determined solely by its mechanical strength, but by how consistently it is used correctly. The most effective systems are those that require minimal effort, align with natural human behaviour, provide clear and immediate feedback, and maintain security even when users make mistakes. This approach defines human-centred locksmithing, where the focus is on creating systems that work reliably in real-world conditions. By embedding security into normal use rather than relying on user discipline, locksmiths can transform theoretical security into practical, dependable protection.

## Impact on Locksmithing Practices

This concept is highly relevant to anyone interested in locksmithing because it fundamentally redefines what "good security" actually means in practice. A new or developing locksmith might initially focus on technical strength—stronger cylinders, better materials, higher pick resistance. However, this analysis demonstrates that even the most advanced lock is ineffective if it is not used correctly and consistently. This shifts the focus of locksmithing from purely mechanical performance to real-world usability, where human behaviour becomes just as important as engineering design.

For someone entering the profession, this is a critical insight because it explains why security failures so often occur despite high-quality installations. It is not usually the lock that fails, it is the interaction between the user and the system. Understanding this helps locksmiths move beyond simply installing hardware and toward diagnosing the true cause of vulnerabilities. For example, a door repeatedly found unlocked is not necessarily a hardware problem, but a design problem where the system does not align with how people actually behave. Recognising this distinction allows locksmiths to provide more effective, long-term solutions rather than repeatedly addressing symptoms.

The influence on locksmithing practice is significant, as it expands the role of the locksmith into that of a system designer. Locksmiths must consider how users move, how often doors are used, what level of effort users are willing to expend, and how decisions are made under time pressure. This leads to the selection of hardware and systems that reduce reliance on memory and discipline, such as self-locking mechanisms, automatic deadbolts, or electronic access systems. In this way, locksmiths begin to design environments where secure behaviour happens by default, rather than requiring constant user intervention.

It also directly affects how master key systems and access control systems are designed. Simplification becomes a priority—reducing the number of keys, clarifying hierarchy, and aligning access with actual roles. Complex systems that require users to think, remember, or interpret instructions are more likely to fail in practice. By contrast, intuitive systems that match natural behaviour patterns improve compliance without additional effort. This reinforces the importance of applying principles such as least privilege, clear zoning, and controlled key distribution in a way that is easy for users to understand and follow.

Another major influence is the integration of feedback and error tolerance into locksmithing solutions. Traditional mechanical systems often provide little or no indication of whether a lock is properly engaged. Human-centred design introduces feedback mechanisms, such as audible clicks, visual indicators, or electronic alerts, that allow users to immediately recognise whether an action has been completed correctly. This reduces uncertainty and helps prevent small errors from becoming security failures.

The concept also highlights the importance of designing systems that limit the consequences of inevitable human error. Locksmiths must assume that users will occasionally forget, make mistakes, or take shortcuts. As a result, systems should be compartmentalised and resilient. For example, restricted key systems prevent uncontrolled duplication, and role-based electronic access ensures that a single lost credential does not compromise an entire facility. This approach transforms security from being fragile and dependent on perfect behaviour into something robust and fault-tolerant.

In modern locksmithing, this perspective also drives the adoption of hybrid and electronic systems. Technologies such as access cards, mobile credentials, and time-based permissions allow locksmiths to automate security, reduce user effort, and improve control. However, the key lesson is that technology must simplify, not complicate. Poorly designed electronic systems can introduce new usability problems and recreate the same behavioural failures seen in mechanical systems. Therefore, locksmiths must apply the same human-centred principles regardless of the technology used.

## Nudging Behaviour Through Design

Nudging behaviour through design involves subtly shaping how people act by structuring environments so that the desired behaviour becomes the easiest, most natural, or most obvious choice. Rather than relying on strict rules, enforcement, or perpetual reminders, nudging works by influencing decision-making at the point of action. In locksmithing and physical security, this translates to designing locks, doors, and access systems so that users are gently guided toward secure behaviour without needing to consciously think about it. This understanding rests on the insight that decision-making is often heuristic, habit-driven, and guided by environmental cues rather than careful, deliberative analysis [223-225]. When systems are well designed, these natural tendencies can be harnessed to support security rather than undermine it, for example through doors that automatically close and lock, thereby nudging users toward secure conduct by default rather than by enforcement alone [224, 226].

At the core of nudging is the recognition that people frequently rely on habits and environmental cues rather than performing exhaustive analyses before acting. If a system is crafted to align with these tendencies, security can be reinforced as the natural outcome of ordinary use. A door that automatically closes and locks exemplifies this logic: secure action becomes the default because the environment removes the need to remember to lock manually. Research and theoretical treatments of security design emphasize that nudging aims to embed secure choices into routine operation, rather than forcing behaviour through coercion or constant oversight [223-225].

A central, practical nudging strategy in locksmithing is to make the secure state automatic. Self-locking mechanisms, deadlatches, and electronic locks that engage upon closure ensure that security does not depend on user memory or effort. This directly addresses a pervasive failure mode in real-world security: people forgetting or choosing not to lock doors. By embedding the secure action into normal operation, the system eliminates opportunities for non-compliance and aligns user experience with security goals [223-225]. Such automatic-secure designs are widely regarded as effective nudges because they transform secure behaviour from a deliberate act into a default consequence of using the hardware and systems [223-225].

Another impactful nudging approach is reducing friction for secure actions while increasing friction for insecure ones. An effective access-control system can permit easy entry via a card or keypad while making it harder to prop doors open or bypass the lock. Proper door closers and well-aligned strike plates encourage proper closure, whereas misalignment, heavy doors, or other poor configurations tend to drive users toward insecure outcomes such as doors left ajar. By shaping the physical experience, locksmiths can influence behaviour without explicit instructions, guiding users toward secure conduct through what feels natural and effortless [223-225].

Feedback mechanisms are another potent nudge. Immediate confirmation, such as a distinct lock click, a visual indicator, or an audible signal, helps users recognize that an action has succeeded and that the door is secure. Without such feedback, users may incorrectly assume security is established when it is not. Clear, real-time feedback reinforces correct behaviour and enables users to detect and correct mistakes promptly. This principle, making outcomes perceptible and immediate, has been identified as a robust driver of secure user action in physical security design [223-225].

Design that aligns with natural movement and routine can facilitate compliance with security measures. People tend to follow established paths and habitual workflows, particularly in workplaces. Positioning access readers at logical stopping points near doors encourages use, while poorly placed systems may be ignored or bypassed. By understanding human movement patterns and daily routines, locksmiths can situate access points to support voluntary compliance with security protocols, turning secure behaviour into an intuitive part of the daily workflow [223-225].

Visibility and subtle accountability are human-focused nudges that influence behaviour without overt enforcement. Systems that make actions observable, such as access logs, key registers, or the underlying knowledge that access is monitored, can encourage proper procedures even in the absence of explicit coercion. The goal is to create an environment where responsible behaviour is the norm rather than the exception, not to rely on surveillance per se but to leverage social and epistemic cues that promote secure practices [223-225].

Nudging does not imply the elimination of rules or training; rather, it reduces reliance on them by embedding secure choices into the environment. Traditional approaches that emphasize explicit rules and training can be unreliable due to human variability and non-compliance. In locksmithing, this means combining high-quality hardware with thoughtful system design so that the environment itself supports secure behaviour rather than depending solely on user discipline or memory. This holistic view integrates hardware design with behavioural insights to produce reliable security in everyday use [223-225].

From a practical standpoint, nudging behaviour through design transforms locksmithing into a more holistic discipline. The locksmith is not merely installing locks but actively shaping how people interact with security systems. By integrating an understanding of human behaviour with design principles, locksmiths can create solutions that are secure in theory and effective in practice, reducing the incidence of security failures due to human error and enhancing reliability across operational contexts [223-225].

Working with Human Nature to Improve Security Outcomes In sum, nudging seeks to work with human nature by making secure behaviour easy, intuitive, and automatic. When security is seamlessly integrated into the everyday use of locks, doors, and access systems, reliability improves and the likelihood of human-caused security failures declines. The combined emphasis on automatic secure states, friction management, feedback, alignment with routine movement, visibility, and prudent use of training supports a comprehensive approach to modern locksmithing that aligns design with behavioural science to achieve durable security outcomes [223-225].

## Impact on Locksmithing Practices

This concept is highly relevant to someone interested in locksmithing because it fundamentally changes how security problems are understood and solved. A beginner may assume that improving security is primarily about stronger locks, better materials, or resistance to attack. However, nudging behaviour through design demonstrates that many real-world security failures occur not because locks are weak, but because people use them inconsistently or incorrectly. This insight shifts locksmithing from a purely technical discipline into one that integrates human behaviour, environmental design, and risk management.

For an aspiring locksmith, this is critical because it explains why traditional approaches—installing high-quality hardware and providing instructions—often fail in practice. People do not consistently follow procedures, not due to ignorance, but because their behaviour is shaped by convenience, habit, and environmental cues. Understanding nudging allows locksmiths to recognise that users are not the problem to be fixed; rather, the system must be designed to guide behaviour. This leads to more effective problem diagnosis and more durable solutions.

The influence on locksmithing practice is significant, as it expands the role of the locksmith into that of a behavioural designer. Locksmiths must consider how users approach doors, how quickly they move, what they prioritise, and how they respond to feedback. This directly affects hardware selection and system configuration. For example, instead of relying on users to manually lock a door, a locksmith may specify a self-locking mechanism or an electronic lock that secures automatically. In doing so, the locksmith removes the need for conscious compliance and embeds security into normal use.

Nudging also influences how locksmiths design access control systems and key management structures. Systems must be intuitive, with minimal decision points, so users do not have to think about what to do. This reduces errors and prevents the development of insecure shortcuts. Clear zoning, simplified key hierarchies, and well-placed access points all reflect this principle. When systems align with natural movement and workflow, compliance improves without additional effort or enforcement.

Another important impact is the incorporation of feedback and visibility into locksmithing solutions. By providing immediate confirmation—such as a clear locking sound, visual indicators, or alerts—locksmiths help users understand whether an action has been completed correctly. This reduces uncertainty and reinforces correct behaviour. Similarly, introducing subtle accountability through audit trails or controlled key systems encourages responsible use without creating friction.

The concept also drives a more resilient approach to security design. Rather than assuming perfect user behaviour, locksmiths design systems that tolerate error and limit its consequences. For example, restricted key systems prevent uncontrolled duplication, while role-based access ensures that mistakes or misuse are contained within defined boundaries. This reflects a shift toward designing systems that remain secure even when users act imperfectly.

In modern practice, nudging is closely linked to the integration of mechanical and electronic systems. Technologies such as automatic locking, time-based access, and monitored entry points allow locksmiths to implement behavioural design principles more effectively. However, the key lesson is that technology must support simplicity and usability. Poorly designed systems can create confusion and new vulnerabilities, so the same human-centred principles must be applied regardless of the technology used.

## Case Studies in Behavioural Security Failure

Below are practical case study scenarios that demonstrate how behavioural factors, not mechanical weaknesses, lead to security failures in locksmithing and physical access systems.

### Case Study 1: The Unlocked Office Door

A commercial office installs high-quality mortice locks with deadbolts on all external doors. Staff are trained to lock doors when leaving, even for short periods.

In practice, employees frequently pull the door shut without engaging the deadbolt. During busy periods, they prioritise speed and assume someone else will secure the door. Because nothing immediately goes wrong, this behaviour becomes normal.

Eventually, an opportunistic intruder enters through an unlocked door and steals equipment.

**Failure Cause:** Not a lock failure—but reliance on user memory and effort.

**Lesson for Locksmithing:** Security must not depend on deliberate user action. A self-locking system or electronic access control would have eliminated this vulnerability.

### Case Study 2: The Hidden Spare Key

A homeowner hides a spare key under a doormat for convenience. They believe the risk is low because the neighbourhood is quiet.

An intruder checks common hiding locations and gains entry without forced access.

**Failure Cause:** Risk perception bias—user assumes "it won't happen here."

**Lesson for Locksmithing:** Users prioritise convenience over abstract risk. Locksmiths should recommend key safes or controlled access solutions instead of relying on user judgement.

**Case Study 3: The Propped-Open Fire Door**

In a warehouse, a secure fire exit door is frequently propped open with a wedge to allow airflow and easier movement of goods.

Staff know this violates policy, but it becomes common practice because it improves workflow efficiency.

An unauthorised person enters through the open door and accesses restricted stock areas.

**Failure Cause:** Conflict between productivity and security.

**Lesson for Locksmithing:** If security interferes with workflow, it will be bypassed. Solutions may include alarmed doors, delayed egress systems, or redesigning access routes.

**Case Study 4: Master Key Misuse**

A facility uses a master key system for convenience. Several staff members are issued master keys to reduce operational delays.

Over time:

- Keys are shared informally
- One is lost but not reported immediately
- No tracking system exists

Eventually, unauthorised access occurs across multiple areas.

**Failure Cause:** Lack of accountability and over-allocation of high-level access.

**Lesson for Locksmithing:** Apply least privilege and strict key control. Master keys represent high-risk assets and must be tightly managed.

**Case Study 5: Confusing Key System**

A building has a complex key system with multiple similar-looking keys and unclear labelling.

Users:

- Frequently try the wrong keys
- Leave doors unlocked to avoid hassle
- Share keys to simplify access

**Failure Cause:** System complexity leading to user workarounds.

**Lesson for Locksmithing:** If users have to think too much, they will bypass the system. Simplify key hierarchies and improve usability.

**Case Study 6: No Feedback from Lock**

A door appears closed but is not fully latched due to misalignment. There is no clear feedback to the user.

Users assume the door is secure and walk away.

The door can be pushed open without a key.

**Failure Cause:** Lack of feedback and poor door/lock alignment.

**Lesson for Locksmithing:** Security requires clear confirmation. Locks should provide tactile/audible feedback, and installation quality is critical.

**Case Study 7: PIN Code Sharing**

A keypad lock is installed to eliminate physical keys. Over time:

- Staff share PIN codes
- Codes are not updated
- Frequently used digits become visible

Unauthorised access occurs.

**Failure Cause:** Convenience and social behaviour override policy.

**Lesson for Locksmithing:** Digital systems must include:

- Unique user credentials
- Audit trails
- Regular code changes

Across all of these case studies, a clear and consistent pattern emerges: most security failures are behavioural rather than mechanical. The locks themselves are often functioning exactly as designed, yet the system still fails because of how people interact with it. Users are not acting irrationally or carelessly; instead, they are behaving in ways that make sense within their environment. Convenience, habit, and the need to complete tasks efficiently tend to dominate behaviour, often outweighing formal security procedures.

**Impact on Locksmithing Practices**

This has significant implications for locksmithing practice. It highlights the need for a shift away from designs that depend on perfect user behaviour, memory, or discipline. Instead, modern locksmithing must focus on creating systems where secure behaviour occurs naturally. This means making the secure action automatic wherever possible, reducing the number of steps required to secure a door, and minimising the effort required from the user. At the same time, systems should provide clear and immediate feedback so users can easily confirm that security has been achieved, while also limiting the impact of inevitable mistakes through thoughtful design.

These insights reinforce a critical principle: the strongest lock in the world can still fail if it is not used correctly. Effective security is not defined solely by mechanical strength, but by how reliably the system performs in real-world conditions. The most successful locksmithing solutions are those that anticipate human behaviour, remove opportunities for error, and integrate security seamlessly into everyday use. This approach forms the foundation of behavioural security and human-centred locksmithing, where the focus is on creating systems that remain secure even when users are not perfect.

## Chapter 8 Review Questions

**Question 1**

A locksmith installs a high-security deadbolt on an office door. Employees must manually turn the key each time they leave. Over time, staff begin simply pulling the door shut without engaging the deadbolt.

What is the most likely reason for this behaviour?

A. Convenience and routine outweigh security procedures
B. The lock mechanism is defective
C. The lock is too strong
D. Staff are intentionally ignoring security rules

**Question 2**

A warehouse fire exit door is frequently wedged open so workers can move equipment more easily.

What behavioural factor most strongly explains this security failure?

A. Mechanical failure of the lock
B. Lack of door hinges
C. Excessive lock strength
D. Conflict between productivity and security procedures

**Question 3**

A locksmith installs a self-locking electronic access control system that automatically locks the door when it closes.

Which behavioural security principle does this system demonstrate?

A. Eliminating access control
B. Increasing user workload
C. Making the secure action the default
D. Security through obscurity

**Question 4**

A homeowner hides a spare key under a doormat. The house is later burglarized after the intruder checks common hiding places.

Which behavioural factor contributed most to this failure?

A. Improper installation
B. Overestimation of lock strength
C. Poor lock manufacturing
D. Risk perception bias

**Question 5**

In an office building, a keypad lock is installed and staff share the same PIN code for convenience.

Why does this create a security vulnerability?

A. Keypads cannot secure doors
B. PIN codes automatically expire
C. Locks stop functioning when codes are shared
D. Shared credentials remove accountability and traceability

**Question 6**

A locksmith notices that staff frequently leave a door unlocked because they are unsure whether the lock engaged.

What design improvement would most likely solve this issue?

A. Increasing lock size
B. Using heavier door materials
C. Removing the lock entirely
D. Adding clear audible or visual feedback when the lock engages

**Question 7**

A facility installs a complex master key system with poorly labelled keys. Employees often try multiple keys before finding the correct one and eventually begin leaving doors unlocked to avoid the inconvenience.

What was the main design problem?

A. Excessive user effort and system complexity
B. Insufficient lock strength
C. Poor lighting in the building
D. Incorrect door material

**Question 8**

A company introduces access cards that automatically log every door entry.

What behavioural security concept does this most strongly support?

A. Increasing physical door strength
B. Eliminating security policies
C. Preventing door closures
D. Improving accountability through audit trails

**Question 9**

After a burglary in a neighbourhood, residents temporarily become much more careful about locking doors. Several months later, most return to their previous habits.

What behavioural concept explains this pattern?

A. Mechanical fatigue
B. Lock design failure
C. Short-term fear followed by habit-driven behaviour
D. Incorrect key duplication

**Question 10**

A locksmith redesigns a building entrance so that the door automatically closes, locks, and gives a confirmation sound.

Which locksmithing design philosophy does this represent?

A. Decorative lock design
B. Randomised access control
C. Mechanical-only security
D. Human-centred security design

# Part 5

# Digital, Smart, and Connected Lock Systems

# Chapter 9

# Smart Locks and IoT Vulnerabilities

This chapter moves beyond the traditional boundaries of locksmithing and into the complex, evolving landscape of smart locks and connected access systems. As locking technology becomes increasingly integrated with software, networks, and cloud platforms, the nature of security is no longer confined to physical components alone. Instead, it is shaped by the interaction between hardware, firmware, communication protocols, and user behaviour, where weaknesses in any layer can compromise the entire system.

Throughout this chapter, real-world scenarios illustrate how modern smart locks can be defeated not only through physical manipulation, but also through vulnerabilities in firmware, communication, credential management, and system configuration. These examples demonstrate that many security failures do not arise from the lock itself, but from how it is designed, installed, maintained, and used within a broader digital environment. From insecure update processes and unpatched software to poor access control and network weaknesses, the risks are often subtle, systemic, and difficult to detect without a deeper understanding of integrated security systems.

For the modern locksmith, this represents a fundamental shift in practice. The role is no longer limited to mechanical installation and key control, but extends to system integration, cybersecurity awareness, and lifecycle management. Evaluating a lock now requires consideration of authentication methods, firmware integrity, communication security, and user behaviour, alongside traditional concerns such as build quality and resistance to physical attack.

This chapter therefore serves as a bridge between traditional locksmithing and contemporary access control systems. It highlights the need for a layered, holistic approach to security—one that recognises that the effectiveness of any lock is determined not by a single component, but by the strength of the entire system that supports it.

## Architecture of Smart Locking Systems

The architecture of smart locking systems represents a significant evolution from traditional mechanical locks, integrating electronics, software, and networked communication to create dynamic, controllable access systems. Rather than relying solely on physical keys and mechanical interaction, smart locks operate as part of a broader ecosystem that combines hardware, identity management, communication protocols, and control platforms. Understanding this architecture is essential for locksmiths, as it shifts the role from purely mechanical installation to system integration, configuration, and lifecycle management.

**Core Hardware Layer (Physical Lock Components)**

At the foundation of any smart locking system is the physical hardware installed on the door. This hardware includes the core lock mechanism, such as a deadbolt, latch, or mortice system, along with an actuator, sensors, and a power source. The actuator, typically a motor or solenoid, is responsible for physically moving the bolt, replacing the need for manual key operation. Sensors are integrated to monitor conditions such as door position, lock status, and potential tampering, while the system is powered either by batteries or, in some cases, a wired electrical supply.

The actuator is a key component that enables electronic control of the lock. Instead of turning a key, the system activates the motor or solenoid to engage or disengage the locking mechanism. At the same time, sensors continuously provide feedback to the system, confirming whether the door is open or closed, whether the lock is engaged, and whether any tampering has occurred. This real-time feedback is essential for ensuring that the system operates reliably and that users or administrators are aware of the lock's status.

From a locksmithing perspective, this hardware layer still relies heavily on traditional mechanical skills. Proper installation, precise alignment, and overall mechanical integrity remain critical to system performance. Even the most advanced smart locking system can fail if the door is misaligned, the lock is poorly fitted, or the mechanical components do not operate smoothly. As a result, foundational locksmithing practices continue to play a vital role in ensuring that smart locks function effectively in real-world conditions.

In the residential and light commercial space, products such as the Yale Assure Lock 2 and the Schlage Encode Smart WiFi Deadbolt are widely used examples of electronic deadbolt systems. These locks retain a conventional deadbolt mechanism but incorporate an internal motor-driven actuator that extends and retracts the bolt electronically. When a user enters a PIN, uses a mobile app, or triggers remote access, the actuator replaces the manual turning of a key. These systems are typically battery-powered and include sensors that detect whether the door is locked or unlocked. Some models also provide alerts if the door is left unsecured, demonstrating how feedback is built into the hardware layer.

In more advanced residential and multi-residential applications, the August Wi-Fi Smart Lock illustrates a retrofit approach. This device attaches to the internal side of an existing deadbolt and physically turns the thumbturn using a motorised actuator. While the external hardware remains unchanged, the internal actuator and sensors enable remote locking, unlocking, and status monitoring. This highlights how the actuator's role is purely mechanical in function but digitally controlled, bridging traditional locksmithing hardware with modern smart systems.

For commercial environments, systems such as the Salto XS4 Electronic Lock and dormakaba Saflok Quantum RFID Lock demonstrate integrated mortice lock platforms. These locks combine a full mortice lock body with an embedded electronic actuator and multiple sensors. They commonly include door position sensors, latch monitoring, and tamper detection. Access is granted via RFID cards, mobile credentials, or PIN codes, and the actuator engages the locking mechanism accordingly. These systems may be battery-powered but are often supported by networked infrastructure for audit trails and centralised control.

## Beyond the Key

At a higher-security level, the ASSA ABLOY Aperio E100 represents a wireless electronic cylinder solution used in enterprise environments. This system replaces a traditional mechanical cylinder with an electronic one, where the actuator controls the cam rotation inside the cylinder. It integrates with access control systems and includes encrypted communication, audit logging, and real-time status updates. Sensors within the system provide feedback on lock state and attempted access, reinforcing the importance of continuous monitoring.

Across all these examples, the same architectural principles apply. The actuator physically drives the locking mechanism, replacing manual key operation. Sensors provide continuous feedback on door and lock status, enabling visibility and control. Power is typically supplied by batteries, requiring consideration of maintenance cycles and fail-safe behaviour.

From a locksmithing perspective, these examples reinforce that despite the added electronics, the foundation remains mechanical. A Schlage Encode Smart WiFi Deadbolt will still fail if the door is misaligned or the strike is incorrectly positioned. Similarly, a Salto XS4 Electronic Lock requires precise mortice installation to ensure correct latch and bolt operation. The integration of actuators and sensors does not eliminate the need for correct door preparation, alignment, and hardware selection—it makes these factors even more critical.

These commercial systems clearly show that smart locking technology is not a replacement for traditional locksmithing skills, but an extension of them. The effectiveness of the entire system still depends on the quality of the physical installation, the reliability of the mechanical components, and the locksmith's ability to ensure that the hardware operates smoothly under real-world conditions.

### Identity and Credential Layer

Instead of relying on traditional physical keys, smart locking systems use digital credentials to grant and control access. These credentials can take several forms, including PIN codes entered via a keypad, RFID or proximity cards, mobile credentials delivered through Bluetooth, NFC, or smartphone applications, and biometric identifiers such as fingerprints or facial recognition. Each method provides a different balance of convenience, security, and suitability depending on the environment in which it is used.

*Figure 71: dormakaba RFID key fob. Paul2520, CC BY-SA 4.0, via Wikimedia Commons.*

A key advantage of digital credentials is that they are linked to individual user identities and specific access permissions. This enables fine-grained control over access, allowing organisations to define exactly who can enter certain areas and at what times. For example, a staff member may have access to a building only during working hours, while a manager may have broader or unrestricted access. This level of control is not easily achievable with traditional key systems, where access is tied to possession of a physical key rather than a defined identity.

For locksmiths, this represents a significant shift in practice. The role moves beyond cutting and issuing physical keys to managing identity-based access systems. Locksmiths must now be involved in provisioning credentials, modifying access permissions as roles change, and revoking access when it is no longer required. This requires an understanding of both physical security hardware and digital access management, as well as the ability to integrate these systems into broader organisational processes.

Specific commercial products clearly illustrate how smart locking systems use different types of digital credentials and how these are applied in real-world environments.

For PIN code access, a common example is the Lockwood 001Touch Plus Smart Deadlatch, widely used in Australia for residential and light commercial applications. It allows users to enter a PIN on a touchscreen keypad, eliminating the need for a physical key. In a more commercial setting, the Kaba Simplex 5000 Series provides a fully mechanical push-button PIN system often used in offices, schools, and plant rooms where electronic infrastructure is not required but controlled access is still needed.

For RFID and proximity card systems, large-scale commercial environments commonly use platforms such as HID Global Signo Readers integrated with access control systems. These allow users to tap a proximity card or fob to gain entry. Similarly, Salto XS4 Electronic Lock systems are widely used in hotels, universities, and corporate buildings, where staff or guests are issued RFID cards programmed with specific access rights, including time-based permissions.

Mobile credentials are increasingly common in modern smart buildings. Systems such as ASSA ABLOY CLIQ Connect and SALTO KS (Keys as a Service) allow users to unlock doors using smartphone apps via Bluetooth or NFC. In office environments, platforms like HID Mobile Access enable employees to use their phones instead of cards, with credentials issued and revoked remotely. This is particularly useful for managing contractors, temporary staff, or remote access without needing to physically hand over keys.

For biometric access, products such as the Samsung SHP-DP609 Digital Door Lock and ZKTeco Biometric Access Control Systems use fingerprint recognition to grant entry. These are commonly used in high-security areas, server rooms, laboratories, and executive offices where identity verification must be tightly controlled. Some systems also combine biometrics with PIN or card access for multi-factor authentication.

These commercial examples demonstrate how digital credentials are directly linked to user identities and access permissions. For instance, in a system like Salto or HID, an employee's card or phone can be programmed to allow access only to specific doors and only during certain hours. If that employee changes roles or leaves the organisation, their access can be modified or revoked instantly without changing any physical locks. This is a major advantage over traditional key systems, where access is difficult to track and revoke.

For locksmiths, these systems fundamentally change the nature of the work. Instead of simply cutting keys, locksmiths must now configure access control software, enrol users, assign permissions, and manage credential lifecycles. They may also need to integrate these systems with building management systems, HR processes, or cloud-based platforms. In practice, this means the locksmith's role expands into areas such as system programming, user management, and ongoing security administration—bridging the gap between physical security and digital identity management.

## Communication Layer

Smart locks rely on communication technologies to connect with other devices, systems, and networks, enabling advanced functionality beyond basic locking and unlocking. Common communication methods include Bluetooth, Wi-Fi, Zigbee or Z-Wave, and Near Field Communication (NFC). Bluetooth is typically used for short-range, direct interaction between a user's smartphone and the lock, making it suitable for local access without requiring an internet connection. Wi-Fi enables remote access, allowing users or administrators to control and monitor locks from anywhere via the internet. Zigbee and Z-Wave are low-power communication protocols commonly used in smart home and building automation systems, allowing locks to integrate with other connected devices such as lighting, alarms, and sensors. NFC provides a simple tap-based interaction, often used with smartphones or access cards for quick and intuitive entry.

This communication layer enables a range of practical features that enhance both convenience and control. Users can lock or unlock doors remotely, check the real-time status of a door (such as whether it is open, closed, or secured), and receive notifications about access events. In more advanced systems, smart locks can be integrated with alarm systems, surveillance platforms, or broader building management systems, creating a coordinated security environment. For example, unlocking a door could automatically disarm an alarm or log an access event in a central system.

However, this connectivity also introduces new security considerations. Unlike purely mechanical locks, smart locks are exposed to cyber risks such as unauthorised access, interception of communication, or network-based attacks. As a result, secure system design becomes essential. This includes implementing strong encryption to protect data transmission, ensuring secure pairing between devices to prevent unauthorised connections, and maintaining robust network security practices. For locksmiths, this means that understanding communication protocols and cybersecurity fundamentals is increasingly important, as protecting access now extends beyond physical hardware to include digital and networked components.

For Bluetooth-based access, a well-known example is the August Wi-Fi Smart Lock (4th Gen), which uses Bluetooth for local smartphone control when the user is near the door. Similarly, the SALTO KS Smart Lock uses Bluetooth Low Energy (BLE) to allow mobile key access in commercial environments such as offices and co-working spaces. In both cases, Bluetooth enables secure, short-range communication without relying on constant internet connectivity.

For Wi-Fi-enabled systems, products such as the Yale Assure Lock SL with Wi-Fi and Schlage Encode Smart WiFi Deadbolt provide full remote control via cloud-connected apps. These locks allow users to lock or unlock doors from anywhere, receive notifications, and monitor activity in real time. In commercial settings, platforms like Brivo Access Control System use Wi-Fi and cloud infrastructure to manage multiple doors across sites, providing centralised control and audit logging.

For Zigbee and Z-Wave integration, smart home ecosystems often use locks such as the Yale Real Living Z-Wave Plus Deadbolt or Kwikset SmartCode 916 Z-Wave. These locks connect to home automation hubs (such as SmartThings or Hubitat), allowing integration with lighting, alarms, and sensors. For example, locking the door can automatically arm a security system or turn off lights, demonstrating how these protocols support coordinated building automation.

For NFC-based access, commercial and institutional systems frequently use solutions such as HID Mobile Access and SALTO XS4 NFC Lock. These allow users to tap a smartphone or card against a reader to gain entry. NFC is widely used in offices, universities, and hotels because it provides fast, intuitive, and contactless access while supporting secure credential management.

These technologies enable practical features such as remote locking and unlocking, real-time monitoring, and system integration. For example, in a system like Brivo or SALTO KS, an administrator can see exactly when a door was accessed, by whom, and can instantly revoke access if needed. In a smart home environment, unlocking a door via a Yale or August lock can automatically disarm an alarm system or trigger cameras, illustrating how locks now function as part of a broader security ecosystem.

However, these same communication capabilities introduce cybersecurity risks. Commercial systems address this through features such as encrypted communication (e.g. AES encryption in Z-Wave devices), secure pairing processes, and cloud-based authentication. For instance, HID and SALTO systems use encrypted credential exchange and managed identity platforms to prevent unauthorised access. Locksmiths working with these systems must therefore understand not only the physical installation, but also how devices are paired, how credentials are managed, and how networks are secured.

These examples highlight a major shift in locksmithing practice. Locks are no longer standalone mechanical devices, they are connected endpoints within a wider digital system. As a result, locksmiths must increasingly engage with networking concepts, mobile credential management, and cybersecurity principles to ensure that both physical and digital access remain secure.

### Control and Management Layer (Software Platform)

At the centre of any smart locking system is the control layer, often described as the "brain" of the system. This component is responsible for managing how the lock operates, who can access it, and under what conditions. Depending on the scale and complexity of the installation, this control layer may take the form of a mobile application, a cloud-based management platform, or a locally hosted access control server. In residential settings, this is commonly a smartphone app, while in commercial or enterprise environments, it is typically a centralised software platform that manages multiple doors and users across a site or organisation.

This control system performs a range of critical functions. It allows administrators to create and manage users, assigning each individual a unique identity within the system. Access rights can then be configured to define exactly which doors a user can open and at what times. For example, a staff member may be granted access only during working hours, while a supervisor may have extended or unrestricted access. The system also enables scheduling, allowing access permissions to be automatically activated or restricted based on time, day, or specific conditions.

In addition to access control, the system provides monitoring and audit capabilities. Every access event, such as a door being unlocked, a failed attempt, or a credential being used, is recorded in an audit log. This creates a detailed history of activity that can be reviewed for security, compliance, or investigation purposes. One of the most significant advantages of this digital control layer is the ability to revoke credentials instantly. If a key card is lost or an employee leaves, access can be removed immediately without the need to physically change locks.

For locksmiths, this layer represents a major expansion of their traditional role. The focus shifts from purely mechanical tasks to include software configuration, user management, and policy enforcement. Locksmiths must understand how to set up and manage these systems, define access rules, troubleshoot configuration issues, and ensure that security policies are correctly implemented. This evolution transforms locksmithing into a hybrid discipline that combines physical security expertise with digital system management, reflecting the increasing integration of technology in modern access control systems.

In residential and small business environments, the control layer of a smart locking system is typically delivered through a mobile application. These systems are designed for ease of use, allowing users to manage access directly from their smartphones. For example, platforms such as August Smart Lock enable homeowners to create virtual keys for family members or guests, set temporary or recurring access schedules, view activity logs showing who accessed the door and when, and instantly revoke access if required. Similarly, Yale Assure Lock systems provide additional functionality such as remote locking and unlocking via Wi-Fi, the ability to create guest PIN codes, and integration with broader smart home ecosystems. Schlage Encode systems focus on simplicity and reliability, offering user code management, basic scheduling, and straightforward audit trails.

From a locksmithing perspective, even at this entry level, the role extends beyond physical installation. Locksmiths must be capable of configuring mobile applications, setting up user accounts, assisting clients with onboarding and troubleshooting, and explaining how access permissions and security settings operate. This represents an early shift from purely mechanical work toward user support and digital system configuration.

In commercial environments, the control layer becomes more complex and is typically delivered through cloud-based access control platforms. These systems provide centralised management of multiple doors, users, and sites. Platforms such as Brivo Access are widely used in offices, gyms, and multi-tenant buildings, offering centralised user management, role-based access control, real-time event monitoring, and the ability to revoke credentials remotely. SALTO Systems, through platforms such as SALTO KS and SALTO Space, provide enterprise-grade functionality including time-based access control, integration of mobile keys and RFID credentials, detailed audit trails, and compatibility with hospitality and facility management systems. Systems like Kisi are common in modern offices and co-working spaces, enabling smartphone-based access, remote unlocking, and integration with other business platforms through APIs. Openpath, now part of Motorola Solutions, focuses on mobile-first access, offering touchless entry, cloud-based credential management, and real-time monitoring and alerts.

At this level, the locksmith's role expands significantly. Locksmiths are required to configure cloud platforms, define and manage user permissions, design access roles and schedules, integrate locking systems with other building or business systems, and support clients with ongoing audits and compliance requirements. The work becomes increasingly systems-oriented, requiring both technical and administrative competence.

In large organisations and high-security environments, the control layer is often implemented through on-premise, server-based access control systems. These systems provide a high level of control, security, and integration across entire facilities or multiple sites. Solutions such as HID Global enterprise systems are used in government, healthcare, and critical infrastructure, offering centralised identity management, integration with biometric and credential technologies, and advanced audit and compliance capabilities. LenelS2 OnGuard provides a comprehensive enterprise platform with advanced role-based access control, integration with CCTV and alarm systems, and detailed reporting and analytics. Gallagher Command Centre is widely used in high-security environments, delivering site-wide control of access, alarms, and perimeter security, along with real-time monitoring and automation. In Australia, Inner Range Integriti is a prominent system used in commercial and government settings, combining access control with intrusion detection, flexible permission structures, and reliable local server operation.

At this level, locksmiths transition into the role of security system specialists. Their responsibilities include understanding networked systems and server architecture, designing access policies aligned with organisational requirements, integrating access control with HR and broader security systems, and providing ongoing maintenance, updates, and system optimisation. The profession evolves from a trade focused on hardware to a hybrid discipline that combines physical security, information technology, and organisational risk management.

### Policy and Security Layer

At the highest level, smart locking systems are governed by a framework of rules and policies that determine how access is controlled and enforced. These policies define who is allowed access, when that access is permitted, and under what specific conditions it can occur. Rather than relying solely on the physical act of locking and unlocking, the system applies structured logic to ensure that access aligns with organisational requirements and security objectives.

This policy layer enables the implementation of key security principles. One of the most important is the concept of least privilege, where individuals are only granted the minimum level of access required to perform their role. This reduces unnecessary exposure and limits the potential impact of misuse or compromise. Time-based access is another critical feature, allowing permissions to be restricted to certain hours or days—for example, ensuring that contractors can only enter a site during approved working periods. In addition, the system supports audit and accountability by recording access events, linking actions to specific users, and enabling traceability across the entire system.

It is this layer that elevates a lock from being a simple physical barrier to becoming part of an integrated security system. By combining access rules, user identity, and real-time monitoring, the system moves beyond passive protection and becomes an active tool for managing risk, enforcing policy, and maintaining control over secure environments.

At the policy and rules layer of smart locking systems, a number of well-established commercial platforms demonstrate how access control moves beyond simple locking into structured, enforceable security management.

A widely used example is Brivo Access, a cloud-based access control platform commonly deployed in offices, gyms, and multi-tenant buildings. In Brivo systems, administrators define user roles such as "staff," "cleaner," or "manager," and assign access rights accordingly. For instance, cleaning contractors may only be granted access between 6:00 pm and 10:00 pm, while management retains 24/7 access. This is a direct implementation of both least privilege and time-based access. Every entry attempt is logged, creating a full audit trail that links each action to a specific credential and user.

Another strong example is SALTO Systems, particularly their SALTO KS and SALTO Space platforms. These systems are widely used in hotels, universities, and large facilities. SALTO enables highly granular policy control, such as restricting access to specific rooms, floors, or zones, and automatically expiring credentials after a defined period. For example, a hotel guest's mobile key or RFID card is valid only for the duration of their stay, while staff access is segmented by department (housekeeping, maintenance, management). The system also records detailed audit logs, supporting accountability and compliance.

Kisi provides another modern, cloud-based example often used in co-working spaces and technology offices. Kisi focuses heavily on policy-driven access through role-based permissions and automation. Administrators can create rules such as allowing employees access during business hours while restricting server room access to authorised IT personnel only. Kisi also integrates with business tools like Slack or HR systems, automatically updating access when staff join or leave the organisation. This demonstrates how policy layers can be embedded into organisational workflows, ensuring access control remains accurate and up to date.

In higher-security environments, enterprise platforms such as HID Global and Gallagher Command Centre illustrate the full capability of policy-driven security. These systems are used in government, healthcare, and critical infrastructure. They support advanced rules such as multi-factor authentication (e.g. card + PIN or biometric), anti-passback (preventing credential sharing), and conditional access based on security status (e.g. lockdown scenarios). Access policies can be dynamically adjusted in real time, and every interaction is logged for investigation and compliance purposes.

An Australian-focused example is Inner Range Integriti, widely deployed across commercial and government sites. Integriti combines access control with intrusion detection and automation, allowing policies to extend beyond doors. For example, access to a secure area may automatically disarm an alarm for authorised users while triggering alerts if accessed outside permitted times. Permissions are structured around roles, time schedules, and site zones, reinforcing least privilege and controlled access.

Across all these systems, the common feature is that access is no longer determined simply by possession of a key, but by a defined set of rules enforced by software. Locksmiths working with these platforms must understand how to configure roles, apply time-based restrictions, manage user identities, and interpret audit logs. This represents a significant evolution in practice, where the locksmith is not just installing hardware but actively implementing and managing security policy through technology.

**Integration Layer (Ecosystem Connectivity)**

Modern smart locking systems are rarely standalone devices; instead, they form part of a broader, interconnected security ecosystem. In contemporary installations, locks are commonly integrated with alarm systems, CCTV and surveillance platforms, building management systems (BMS), and smart home technologies. This integration allows multiple systems to communicate and respond to events in a coordinated manner, significantly enhancing overall security and operational efficiency.

In practical terms, this means that actions at a door can trigger responses across other systems. For example, when an authorised user unlocks a door, the system may automatically disarm the alarm for that area, eliminating the need for a separate action. Conversely, if a forced entry or unauthorised access attempt is detected, the system can trigger alarms, activate nearby cameras, and send real-time alerts to security personnel or building managers. These automated responses ensure that security events are detected and addressed immediately, rather than relying on manual intervention.

Integration with building management systems further extends functionality by linking access control with lighting, HVAC, and energy management. For instance, unlocking a door after hours could activate lighting in a specific area, while also logging the event for audit purposes. In residential environments, integration with smart home platforms enables similar automation, such as unlocking a door triggering lights, adjusting climate settings, or notifying the homeowner via a mobile app.

This interconnected approach transforms security from a collection of isolated components into a cohesive, intelligent system. For locksmiths, this represents an important shift in practice. They must understand not only the operation of individual locks, but also how those locks interact with other technologies. The role expands to include system integration, coordination with other trades (such as security technicians and IT professionals), and ensuring that all components work together reliably to create a comprehensive and responsive security environment.

In commercial environments, platforms such as Gallagher Security (Command Centre) provide a clear example of deep integration. In a typical installation, electronic locks on doors are connected to the Gallagher system, which also manages intrusion alarms, perimeter sensors, and CCTV. When an authorised employee presents a credential and unlocks a door, the system can automatically disarm the alarm zone for that area. If a door is forced open, the system immediately triggers an alarm, activates nearby cameras, and sends alerts to security staff. This creates a unified response where access control, alarms, and surveillance operate as a single system rather than separate layers.

Another strong example is LenelS2 (OnGuard platform), widely used in large commercial buildings and critical infrastructure. OnGuard integrates door hardware with CCTV systems such as Axis Communications cameras and video management systems like Genetec. When a door event occurs—such as a denied access attempt—the system can automatically display live camera footage of that door to security operators. This allows immediate visual verification of incidents. Additionally, OnGuard integrates with building management systems, enabling actions such as unlocking emergency exits during fire alarms or coordinating with HVAC systems during lockdown scenarios.

Cloud-based systems such as Brivo also demonstrate integration in multi-tenant and commercial office environments. Brivo Access can connect with video surveillance platforms and alarm systems, allowing administrators to view access events alongside video footage in real time. For example, when a user unlocks a door, the system logs the event and can simultaneously display a corresponding video clip. If suspicious activity is detected, alerts can be sent instantly to mobile devices, enabling rapid response even when managers are off-site.

In modern workplaces and co-working environments, Kisi provides a more software-driven integration model. Kisi integrates with smart building platforms, communication tools, and IoT devices. Unlocking a door via smartphone can trigger lighting systems, log attendance, or integrate with workplace management software. Kisi also connects with surveillance and alarm systems, enabling automated responses such as notifications when doors are left open or accessed outside authorised hours.

In the Australian context, Inner Range (Integriti system) is a widely used solution that combines access control, intrusion detection, and building automation. Integriti can link door access events with alarms, CCTV, and building services. For example, after-hours access can automatically turn on lighting in specific zones, while unauthorised access attempts can trigger both alarms and camera recording. The system also supports integration with fire systems and emergency protocols, ensuring coordinated building-wide responses.

Residential smart ecosystems provide similar integration on a smaller scale. Platforms like Amazon (Alexa-enabled smart home systems) and Google (Google Home) integrate smart locks such as Yale or Schlage with cameras (e.g. Ring), lighting, and alarms. Unlocking a front door can automatically disarm a home alarm, turn on lights, and send a notification to the homeowner. If motion is detected or a door is opened unexpectedly, cameras activate and alerts are sent instantly.

These examples illustrate how integration transforms security into a responsive, intelligent system. For locksmiths, this has a direct impact on practice. It is no longer sufficient to install a lock in isolation; the locksmith must understand how that lock interacts with access control platforms, surveillance systems, alarms, and building automation. This requires coordination with IT networks, security integrators, and facility managers, as well as an understanding of system interoperability and configuration.

### Power and Reliability Considerations

Unlike traditional mechanical locks, smart locking systems rely on a continuous power source to function effectively. In most cases, this is provided by batteries, although some systems may use wired power. As a result, battery life must be actively monitored and managed to prevent unexpected failure. If power is depleted, the lock may become inoperable or default to a predefined state, which can have significant security or safety implications.

A critical design consideration in smart locking systems is whether the lock operates in a fail-safe or fail-secure mode. Fail-safe systems are designed to unlock when power is lost, prioritising safety and emergency egress, which is common in fire exits or life-safety applications. In contrast, fail-secure systems remain locked during a power failure, maintaining security but potentially restricting access. The appropriate choice depends on the environment, risk profile, and regulatory requirements.

Because of these dependencies, backup access methods are essential. Most smart locks incorporate a physical key override, emergency power contacts, or alternative entry methods to ensure access can still be gained if the electronic system fails. These redundancies are critical in preventing lockouts and maintaining operational continuity.

From a locksmithing perspective, ensuring system reliability under adverse conditions is a key responsibility. This includes planning for scenarios such as power loss, network failure, or component malfunction. Locksmiths must design and install systems that remain functional—or safely fail—in these situations, while also advising clients on maintenance practices, such as regular battery replacement and system checks, to ensure long-term reliability and security.

### Key Architectural Insight

Smart locking systems should be understood as layered systems rather than standalone devices. Their effectiveness does not depend on a single component, but on how multiple layers interact and function together as an integrated whole. Each layer contributes to overall security, usability, and reliability, and a weakness in any one area can undermine the entire system.

At the foundation is mechanical reliability. Regardless of how advanced the electronics or software may be, the physical lock must operate smoothly and consistently. Poor installation, misalignment, or mechanical wear can compromise performance and lead to failures that no amount of digital control can compensate for.

Above this sits secure credential management, which governs how users are identified and granted access. This includes the creation, assignment, modification, and revocation of credentials such as PIN codes, cards, or mobile access. Effective management ensures that only authorised individuals can gain entry and that access rights remain appropriate over time.

Robust communication is another critical layer. Smart locks rely on technologies such as Bluetooth, Wi-Fi, or other protocols to exchange information with devices and systems. These communications must be reliable and secure, with appropriate encryption and protection against unauthorised access or interference.

Well-defined policies provide the rules that govern system behaviour. These policies determine who can access specific areas, when access is permitted, and under what conditions. They embed principles such as least privilege, time-based access, and accountability into the system, transforming it from a simple locking mechanism into an active security control.

Finally, ongoing management is essential to maintain system effectiveness. This includes monitoring system performance, updating access permissions, maintaining hardware, and responding to changes such as staff turnover or evolving security requirements. Without active management, even a well-designed system will degrade over time and become less secure.

Together, these layers form a cohesive system where security is achieved through integration, not isolation.

### Impact on Locksmithing Practices

For someone working in locksmithing, the architecture of smart locking systems represents a fundamental shift in the nature of the profession. Traditional locksmithing focused primarily on cutting keys and installing mechanical locks. In modern environments, this role expands into managing digital credentials, where access is tied to user identity rather than possession of a physical key. This transition requires locksmiths to think in terms of access rights, permissions, and lifecycle management rather than simply issuing hardware.

The role also evolves from basic lock installation to full system integration. Locksmiths are no longer dealing with isolated devices but with interconnected systems that must work seamlessly with access control platforms, alarms, surveillance, and building management systems. This requires an understanding of how different technologies interact and how to configure them to operate as a cohesive security solution.

In practice, locksmithing is becoming a hybrid discipline that combines physical and digital security. While mechanical reliability remains essential, it is now only one part of a broader system that includes software, communication networks, and policy-driven control. A failure in any of these areas can compromise the overall system, so the locksmith must be capable of working across both domains.

To operate effectively in this environment, locksmiths must develop new knowledge and skills. This includes a working understanding of electronics and basic networking concepts, such as how devices communicate and how systems are connected. They must also understand access control principles, including user roles, permissions, and the principle of least privilege. Cybersecurity considerations are increasingly important, particularly in protecting communication channels and preventing unauthorised digital access. In addition, an understanding of human behaviour and usability is essential, as systems must be designed in a way that users will operate correctly and consistently.

## Wireless Protocols (Bluetooth, Zigbee, Wi-Fi)

Wireless protocols are the communication backbone of modern smart locking systems. They allow locks, access devices, and control platforms to exchange information without physical wiring, enabling features such as remote access, real-time monitoring, and system integration. In locksmithing, understanding these protocols is essential because they directly affect how a system performs, how secure it is, and how it should be installed and maintained.

### Bluetooth (Short-Range, Direct Control)

Bluetooth is a short-range wireless protocol designed for direct communication between devices, typically within 5–10 metres. In smart locks, it is most commonly used to connect a user's smartphone directly to the lock without requiring an internet connection.

*Figure 72: Unlocking a smart lock via Bluetooth. Gettapkey, CC BY-SA 4.0, via Wikimedia Commons.*

From a functional perspective, Bluetooth enables local control. A user approaches the door, opens an app (or uses automatic proximity detection), and the lock communicates directly with the phone to authenticate and unlock. This makes Bluetooth systems fast, energy-efficient, and relatively simple to deploy.

In locksmithing practice, Bluetooth is often used in residential and small commercial applications where remote access is not essential. It reduces dependency on network infrastructure and is less exposed to external cyber threats compared to internet-connected systems. However, its limitation is range, users must be physically near the lock, and remote management typically requires an additional gateway device.

### Wi-Fi (Networked, Remote Access)

Wi-Fi enables smart locks to connect directly to a local network and, through it, to the internet. This allows users and administrators to control and monitor locks remotely from anywhere using a mobile app or cloud platform.

This protocol is central to features such as remote locking and unlocking, real-time notifications, and integration with cloud-based access control systems. For example, a user can unlock a door for a visitor while away from the property or receive alerts if a door is left open.

From a locksmithing perspective, Wi-Fi introduces both powerful capabilities and additional responsibilities. While it enables advanced functionality, it also exposes the lock to network-based risks. Proper configuration becomes critical, including secure passwords, encrypted communication, and network protection. Wi-Fi locks also consume more power than other protocols, making battery management an important consideration.

Wi-Fi is commonly used in residential systems and small-to-medium commercial installations where remote access and monitoring are key requirements.

**Zigbee (Low-Power Mesh Networking)**

Zigbee is a low-power wireless protocol designed for device-to-device communication within a mesh network. In this setup, devices such as locks, sensors, and lights communicate with each other and relay signals across the network, rather than connecting directly to the internet.

A key advantage of Zigbee is its efficiency and scalability. Because it uses minimal power, it is well suited to battery-operated devices like smart locks. The mesh network structure also improves reliability, as signals can travel through multiple devices to reach their destination.

*Figure 73: Telegesis ETRX357 2.4GHz ZigBee module with ceramic chip antenna. AutolycusQ, CC BY-SA 3.0, via Wikimedia Commons.*

In locksmithing and security system design, Zigbee is commonly used in smart homes and building automation systems where multiple devices must work together. However, Zigbee typically requires a central hub or gateway to connect the network to the internet or a control platform.

For locksmiths, this means understanding not just the lock itself, but the broader ecosystem it operates within. Installation involves ensuring proper network coverage, device pairing, and compatibility with the chosen hub or automation system.

**Key Differences and Practical Implications**

Each protocol serves a different purpose, and the choice depends on the application:

- Bluetooth → Best for simple, local access with minimal infrastructure
- Wi-Fi → Best for remote control and cloud-based management
- Zigbee → Best for integrated systems with multiple connected devices

In practice, many modern systems combine these protocols. For example, a lock may use Bluetooth for local access and Wi-Fi (via a gateway) for remote control, or Zigbee for integration into a larger building system.

Below are commercially available examples of smart locking systems that use Bluetooth, Wi-Fi, and Zigbee. These examples help connect the theory of wireless protocols to actual products a locksmith will encounter in residential, commercial, and enterprise environments.

Bluetooth-based smart locks are widely used in residential and light commercial settings where local, phone-based access is the primary requirement.

Commercial examples include:

- August Smart Lock (4th Gen): Uses Bluetooth for direct smartphone control. Users unlock the door via the August app when in proximity. Optional Wi-Fi bridge enables remote access.
- Danalock V3 (Bluetooth version): A compact retrofit smart lock that communicates via Bluetooth with smartphones. Common in apartments and small offices.
- Ultraloq U-Bolt Pro: Combines Bluetooth with keypad and fingerprint access. Bluetooth enables app-based control and user management at close range.
- Level Lock / Level Bolt: Uses Bluetooth for discreet smart access while maintaining a traditional lock appearance—popular in high-end residential installs.

These systems are simple to install and configure but require understanding of app pairing, device calibration, and user onboarding. They are ideal where low infrastructure and minimal networking are priorities.

Wi-Fi smart locks are designed for remote access and cloud-based control, making them popular in both residential and small commercial environments.

Commercial examples include:

- Schlage Encode Smart WiFi Deadbolt: Connects directly to Wi-Fi (no hub required). Users can manage codes, schedules, and receive notifications remotely.
- Yale Assure Lock 2 (Wi-Fi version): Offers remote locking/unlocking, guest PIN codes, and integration with smart home systems (e.g. Google Home, Apple HomeKit).
- Lockly Secure Pro: Combines Wi-Fi with biometric and keypad access. Provides real-time alerts and remote management via app.
- Eufy Smart Lock (Wi-Fi models): Focuses on integrated ecosystems with video doorbells and security cameras, enabling combined monitoring and control.

These systems require knowledge of network setup, Wi-Fi reliability, cybersecurity basics, and battery management. Locksmiths must often assist with router connectivity, app configuration, and troubleshooting remote access issues.

Zigbee-based locks are typically used in integrated smart home or building automation systems, where multiple devices communicate through a central hub.

Commercial examples include:

- Yale Assure Lock (Zigbee module version): Integrates with platforms like Samsung SmartThings. Allows automation such as locking doors when alarms are armed.
- Kwikset SmartCode 914 Zigbee: Designed for home automation ecosystems. Communicates via Zigbee hub rather than directly to Wi-Fi.
- Schlage Connect Zigbee Deadbolt: Common in smart home installations using SmartThings or similar hubs. Supports automation routines and event triggers.
- SALTO KS (Zigbee/BLE hybrid in some deployments): In commercial environments, SALTO uses low-power wireless communication (including Zigbee-like mesh concepts) for networked access control across doors.

These systems require understanding of hubs, device pairing, mesh networks, and ecosystem compatibility. Installation involves more than the lock—it includes ensuring reliable network coverage and correct integration with automation platforms.

In modern smart locking systems, it is increasingly common for devices to combine multiple wireless communication protocols in order to maximise functionality, flexibility, and user convenience. Rather than relying on a single method of communication, manufacturers design systems that leverage the strengths of different protocols to support both local and remote access, as well as integration with broader security and automation platforms.

For example, the August Smart Lock paired with a Wi-Fi bridge uses Bluetooth for direct, short-range communication between the lock and a user's smartphone, while the Wi-Fi bridge enables remote access via the internet. Similarly, the Yale Assure Lock can incorporate multiple communication modules, such as Bluetooth for local access and Zigbee or Z-Wave for integration with smart home systems, with optional Wi-Fi capability for remote control. In commercial environments, systems such as SALTO and Kisi combine Bluetooth Low Energy (BLE) for mobile credential access with cloud-based platforms that operate over Wi-Fi or internet connections, enabling centralised management and real-time monitoring. The Schlage Encode

Plus integrates Wi-Fi for remote access with NFC technology, allowing features such as Apple Home Key for tap-based entry.

These examples reflect a broader shift in real-world practice. Smart locks are no longer single-protocol devices but are instead multi-layered communication systems. This layered approach allows them to operate across different environments and use cases, providing local convenience, remote accessibility, and system-wide integration within a single solution.

## Impact on Locksmithing Practices

Understanding wireless protocols such as Bluetooth, Wi-Fi, and Zigbee is highly relevant to anyone interested in locksmithing because it reflects a fundamental shift in how access control systems operate. Traditional locksmithing focused almost entirely on mechanical components—locks, keys, and physical installation. However, modern smart locking systems rely on communication between devices, meaning that the effectiveness of a lock is no longer determined solely by its physical strength, but also by how well it communicates, integrates, and is secured within a networked environment. For a locksmith, this expands the role from purely mechanical work to a hybrid discipline that includes elements of electronics, networking, and cybersecurity.

From a practical perspective, each wireless protocol directly influences how a system should be designed, installed, and maintained. Bluetooth-based systems, for example, are typically simple, localised, and energy-efficient, making them well suited to residential or small-scale applications. A locksmith working with these systems must understand device pairing, app configuration, and user onboarding, ensuring that the lock communicates reliably with the user's smartphone. The limitation of Bluetooth—its short range—also means the locksmith must consider whether the client requires remote access and, if so, whether additional components such as gateways are needed.

Wi-Fi systems introduce a different set of considerations and responsibilities. While they enable powerful features such as remote access, real-time monitoring, and cloud-based management, they also expose the lock to network-related risks. This means that locksmiths must understand basic cybersecurity principles, including secure password practices, encryption, and network configuration. Installation is no longer limited to fitting hardware; it may involve connecting the lock to a router, troubleshooting connectivity issues, and ensuring reliable operation in environments where signal strength and interference can affect performance. Battery management also becomes more critical, as Wi-Fi communication typically consumes more power than other protocols.

Zigbee and similar mesh networking protocols further expand the scope of locksmithing practice by introducing system-wide integration. These systems are commonly used in smart homes and building automation environments, where locks operate as part of a larger ecosystem that may include lighting, alarms, and sensors. For locksmiths, this means understanding how devices communicate within a mesh network, how hubs or gateways function, and how to ensure compatibility across different platforms.

Installation becomes more complex, requiring attention not only to the lock itself but also to network coverage, device placement, and system integration.

The relevance of these protocols is even more apparent when considering that many modern systems combine multiple communication methods. A single lock may use Bluetooth for local access, Wi-Fi for remote control, and Zigbee for integration with other devices. This layered approach means that locksmiths must be able to work across multiple technologies simultaneously, understanding how they interact and how failures in one layer—such as network connectivity—can affect overall system performance.

The influence on locksmithing practice is significant. The profession is no longer limited to cutting keys and installing locks; it now involves configuring systems, managing digital access, troubleshooting connectivity issues, and advising clients on both physical and digital security. Locksmiths must be able to assess which protocol is most appropriate for a given application, balancing factors such as convenience, security, reliability, and infrastructure requirements. They must also be prepared to support clients after installation, as issues may arise not from the lock itself, but from the communication layer or the surrounding network.

## Common Vulnerabilities and Exploits

Smart locks and IoT-based access systems introduce significant advancements in convenience and control, but they also expand the attack surface beyond traditional mechanical vulnerabilities. In contrast to conventional locks, where threats are primarily physical (e.g. picking, bumping, or forced entry), smart locks are exposed to both physical and digital risks. These vulnerabilities arise from weaknesses in communication protocols, software design, user behaviour, and system configuration. For anyone involved in locksmithing, understanding these risks is essential because security is no longer confined to the lock itself; it extends across networks, devices, and user practices.

One of the most common vulnerabilities is weak authentication and credential management. Many smart locks rely on PIN codes, mobile apps, or digital keys. If users select weak PINs, reuse credentials, or fail to revoke access when it is no longer needed, the system becomes vulnerable to unauthorised entry. In some cases, default credentials are left unchanged, creating an immediate security gap. Unlike mechanical keys, which require physical possession, digital credentials can be guessed, shared, or intercepted, making poor credential management a critical weakness.

Many smart-lock and IoT devices depend on single-factor authentication, such as PINs or passwords, which are inherently vulnerable to guessing, interception, or reuse. The broader IoT authentication literature documents widespread design flaws in shared IoT devices, where weak authentication and insufficient access-control mechanisms enable unauthorized entry or control after credential modification or revocation [227, 228]. In line with this, studies of door access systems with mobile alerts demonstrate the added risk when only password-based or biometric methods are used in isolation, highlighting the need for multi-factor approaches to resist credential-based attacks [227, 229, 230]. These findings collectively support the assertion that single-layer credential schemes in physical-access contexts increase risk of unauthorized access, particularly when credentials can be guessed, intercepted, or revoked improperly [227, 229, 230].

Token-based authentication (e.g., JWTs) is repeatedly presented as a design choice that, when implemented with proper storage and expiration controls, can strengthen authentication in web services and APIs and reduce the window for credential abuse. Several sources emphasize JWT-based authentication and proper token lifecycle management (expiration, secure storage, refresh flows) as central to mitigating unauthorized access risks in modern systems [231-237]. The literature also notes that secure handling of tokens, secret key management, and adherence to OAuth2 flows are critical to preventing token theft and misuse in credential exchanges [231-233, 237]. In the context of smart locks or access-control ecosystems, adopting token-based schemes alongside MFA can limit the impact of compromised credentials and reduce opportunities for unauthorized entry [227, 230-233].

MFA is consistently framed as a crucial reinforcement against credential-based breaches. A body of work discusses MFA enhancements through OAuth, short-lived access tokens, and granular scopes to minimize exposure if a credential is compromised [232, 234]. Practical deployments in security-conscious domains (such as EHRs and data centres) exemplify MFA's effectiveness in limiting credential misuse and improving access control granularity [235, 238, 239]. While some sources focus on specialized domains (e.g., healthcare, data centres, CubeSats), the underlying principle, adding independent factors (something you know, have, or are) to authentication, translates to residential or small-business smart-lock contexts as a means to mitigate weak credential risks [235, 238, 239]. Some discussions also acknowledge adoption challenges, such as integrating OAuth with existing systems or managing refresh tokens securely, which are relevant to real-world smart-lock ecosystems that must interoperate with mobile apps and cloud services [232, 234].

Several studies explicitly identify problems with credential management practices that enable unauthorized access when defaults remain unchanged, passwords are reused, or revocation is not promptly enforced. Shared IoT devices and consumer security camera ecosystems are shown to suffer from credentials modification and revocation issues that permit continued access by an adversary [227]. In the broader literature on credential cloning and automatic login mechanisms, weaknesses in credential handling and token-based trust models are documented as sources of breach when users or devices fail to revoke or rotate credentials appropriately [17]. Such findings reinforce the claim that default or weak credentials, plus poor revocation and rotation policies, create immediate security gaps in smart-lock and IoT access systems [17, 227].

The literature explores biometric and multi-modal authentication (e.g., fingerprint, facial recognition, and PIN) as means to raise resilience against credential guessing and interception. Several studies present multi-factor or biometric-inclusive door-access designs with mobile alerts and biometrics to augment security beyond single-factor passwords or PINs [229, 234, 240]. Deep-learning–based facial recognition and biometric-centric encryption frameworks are discussed as advanced approaches to reduce reliance on static credentials while maintaining usability [241, 242]. While these sources emphasize security benefits, they also acknowledge implementation constraints, such as privacy, processing requirements, and integration with existing systems, which are important considerations for smart-lock deployments [241, 242].

To mitigate credential vulnerabilities in smart-lock ecosystems, the literature converges on several practical recommendations: adopt MFA with token-based authentication and short-lived tokens; implement secure key/secret management (secret rotation, secure storage, and minimized exposure); enforce robust password policies and discourage reuse; ensure prompt credential revocation and revocation propagation across devices and services; and consider biometric or graphical password approaches as supplementary layers where appropriate [230-233, 235, 238, 239, 241]. Several works also stress the need for secure configuration, rate limiting, input validation, HTTPS, and proper CORS policies as part of a defence-in-depth strategy for API-facing components in smart-lock ecosystems [231, 233, 237, 243].

A practical example of weak authentication and poor credential management can be seen in a small office environment using a keypad-enabled smart lock. The office installs a smart lock on its main entrance, allowing staff to enter using a shared PIN code instead of physical keys. For convenience, management sets a simple code such as "1234" and distributes it to all employees, including temporary contractors. Over time, the code is also shared informally with cleaning staff, delivery personnel, and former employees who were never removed from the system.

Several vulnerabilities quickly emerge. First, the PIN itself is weak and easily guessable, making it susceptible to brute-force attempts. Second, because the same code is shared among multiple users, there is no accountability, if unauthorised access occurs, it is impossible to determine who used the code. Third, when a staff member leaves the company, the PIN is not changed, meaning that former employees can still access the premises.

The situation escalates when a former contractor, who still knows the PIN, enters the building after hours and removes equipment. Because the system only logs that the correct PIN was used (not who used it), there is no clear audit trail. The breach occurs not because the lock failed mechanically, but because credential management was poorly implemented.

A locksmith addressing this scenario would recommend several improvements. Each user should be assigned a unique PIN or digital credential linked to their identity, rather than sharing a common code. Stronger PIN policies (e.g. longer, non-sequential codes) should be enforced. Access should be role-based and time-limited, ensuring that contractors can only enter during authorised hours. Most importantly, credentials must be revoked immediately when a user no longer requires access.

This example highlights how the security of a smart lock is heavily dependent on how credentials are managed. Even a high-quality system can be compromised if authentication practices are weak, reinforcing the importance of proper configuration, user management, and ongoing oversight in modern locksmithing.

Another major area of concern is insecure communication protocols. Smart locks often use Bluetooth, Wi-Fi, or other wireless technologies to communicate. If these communications are not properly encrypted, they can be intercepted or manipulated. For example, attackers may perform "man-in-the-middle" attacks, capturing data exchanged between a smartphone and a lock, or replay attacks, where previously captured signals are reused to gain access. Bluetooth systems, in particular, may be vulnerable if pairing processes are not secure, while Wi-Fi systems depend heavily on the security of the surrounding network.

A practical example of insecure communication protocols can be seen in a residential smart lock that relies on Bluetooth for smartphone access. A homeowner installs a Bluetooth-enabled smart lock that allows the door to be unlocked via a mobile app. During setup, the lock uses a basic pairing process without strong encryption or mutual authentication. The system prioritises ease of use, automatically connecting to any previously paired device when it comes within range.

An attacker observes the homeowner regularly entering the property and uses a nearby device to capture the Bluetooth communication between the phone and the lock. Because the communication is not properly encrypted, the attacker is able to record the authentication signal transmitted during the unlocking process.

Later, the attacker returns and performs a replay attack. Using the captured signal, they transmit the same data back to the lock, which recognises it as valid and unlocks the door. In this case, the attacker does not need the homeowner's phone or credentials—only the previously captured communication.

In a Wi-Fi-based variation of this scenario, a poorly secured home network (e.g. weak router password or outdated encryption) allows an attacker to gain access to the network. Once inside, they intercept communication between the smart lock and its cloud service, potentially capturing credentials or issuing unauthorised commands if proper encryption and authentication are not enforced.

From a locksmithing perspective, this example highlights that security is no longer limited to the physical lock. The communication layer becomes a critical point of vulnerability. To mitigate these risks, locksmiths must ensure that smart locks use strong, end-to-end encryption, secure pairing processes, and regularly updated firmware. They should also advise clients on securing their Wi-Fi networks, using strong passwords, and enabling network protections such as WPA3 encryption.

This example demonstrates that even when the hardware is robust, weaknesses in communication protocols can allow attackers to bypass the lock entirely, reinforcing the need for locksmiths to understand both physical and digital security principles.

Software and firmware vulnerabilities are also a significant risk. Smart locks rely on embedded software to manage authentication, communication, and control functions. If this software contains bugs or is not regularly updated, it may be exploited by attackers. Vulnerabilities can allow unauthorised access, privilege escalation, or complete system compromise. In many IoT devices, manufacturers may not provide timely updates, or users may fail to install them, leaving known vulnerabilities unpatched.

A practical example of software and firmware vulnerabilities can be seen in a short-term rental property using a Wi-Fi–enabled smart lock managed through a mobile app.

The property owner installs the lock to allow guests to receive temporary access codes. The system works well initially, but the lock's firmware contains a known vulnerability that allows an attacker to bypass normal authentication checks by sending a specially crafted command over the network. The manufacturer releases a firmware update to fix the issue, but the owner is unaware of the update and never installs it.

An attacker scans for vulnerable devices on the local network or via exposed internet services and identifies the smart lock model. Using publicly available information about the vulnerability, the attacker sends the exploit command to the lock. Because the firmware has not been updated, the lock accepts the command and unlocks without requiring valid credentials.

In a more advanced scenario, the attacker may exploit the vulnerability to gain administrative control over the lock. This could allow them to create new access codes, disable existing ones, or prevent legitimate users from locking the door. In some cases, the attacker could maintain persistent access, meaning the system remains compromised even after normal use resumes.

From a locksmithing perspective, this example highlights that the security of a smart lock depends not only on its physical construction but also on the integrity of its software. Even a well-installed and mechanically sound lock can be compromised if its firmware is outdated or vulnerable.

To address this risk, locksmiths must ensure that devices are sourced from reputable manufacturers that provide ongoing security updates. They should educate clients on the importance of regularly updating firmware and, where possible, configure systems to apply updates automatically. During maintenance visits, checking firmware status should become part of routine service, alongside traditional mechanical inspections.

This example demonstrates that in modern locksmithing, maintaining security includes managing the lifecycle of software as well as hardware. Unpatched vulnerabilities can effectively bypass all physical protections, making firmware management a critical component of overall system security.

A further issue is poor access control and over-permissioning, which mirrors problems seen in traditional key systems. Users may be granted broader access than necessary, or temporary access may not be revoked. In digital systems, this can be even more problematic because credentials can be duplicated or distributed instantly. Without proper implementation of the principle of least privilege, a single compromised account can provide access to multiple areas or systems.

A practical example of poor access control and over-permissioning can be seen in a medium-sized commercial office using a cloud-based smart access control system. The organisation implements a smart locking system where employees use mobile credentials to access different areas of the building. To simplify setup, the administrator assigns broad access permissions to most staff, allowing them to enter not only general work areas but also storage rooms, meeting rooms, and certain restricted zones. Temporary contractors are also given similar access "for convenience," rather than being restricted to specific areas and timeframes.

Over time, several issues develop. A contractor completes their work but their access is never revoked. Because their credential remains active, they are still able to enter the building after hours. Additionally, one employee's account becomes compromised—either through a weak password or phishing attack. Because that account has broad permissions, the attacker is able to move freely through multiple areas, including spaces that should have been restricted.

The situation escalates when the former contractor uses their still-active credential to enter the building outside normal hours and access a storage area containing valuable equipment. Since the system logs only show that a valid credential was used, the access initially appears legitimate, delaying detection.

From a locksmithing perspective, this example highlights how digital systems can amplify traditional key management problems. In a mechanical system, issuing a master key to too many people is already a risk; in a digital system, over-permissioning can be even more dangerous because access can be distributed instantly and remotely, often without physical control or visibility.

To address this, locksmiths and security professionals must implement the principle of least privilege. Each user should only have access to the specific areas required for their role. Temporary access should be time-limited and automatically expire. Regular audits of user permissions should be conducted to ensure that access remains appropriate. When staff leave or roles change, credentials must be updated or revoked immediately.

This example demonstrates that the strength of a smart locking system is not just in its technology, but in how access is managed. Without proper control and discipline, over-permissioning can turn a secure system into a wide-open environment, where a single compromised credential can expose the entire site

Physical tampering and hardware attacks remain relevant even in smart systems. Attackers may attempt to bypass the electronic components by physically manipulating the lock, accessing internal wiring, or exploiting weaknesses in the actuator mechanism. Some lower-quality smart locks have been shown to expose internal circuits that can be triggered to unlock the device. This highlights that digital security does not replace the need for strong mechanical design, it must complement it.

A practical example of physical tampering and hardware exploitation can be seen in a poorly designed smart lock installed on a small retail shop's rear service door.

The lock is a budget smart deadbolt that uses a motorised actuator to retract the bolt when a valid credential is presented. While the electronic features—such as keypad entry and mobile app control—appear functional, the physical design has a critical weakness: the interior housing is made of thin plastic, and the wiring between the control board and the actuator is easily accessible once the cover is removed.

An attacker approaches the rear door after hours and removes the external escutcheon using basic hand tools. Once the casing is exposed, they identify two low-voltage wires connected to the actuator. By briefly applying power from a small battery, they trigger the actuator directly, causing the bolt to retract without needing any valid credential. The lock opens as if it had received an authorised command.

In another variation of this scenario, the attacker does not even need to apply external power. Instead, they short specific contacts on the circuit board, exploiting a design flaw where the actuator can be triggered without authentication if certain pins are bridged. This type of vulnerability has been demonstrated in some low-cost IoT locks where security was focused on software features rather than hardware protection.

From a locksmithing perspective, this example highlights a critical lesson: electronic security does not replace mechanical security—it must be built on top of it. If the physical construction of the lock is weak, attackers can bypass the entire digital system by directly manipulating the hardware. In effect, the "smart" features become irrelevant because the lock can be defeated at a more basic level.

To mitigate these risks, locksmiths must evaluate not only the digital capabilities of a smart lock but also its physical integrity. This includes selecting products with robust housings, tamper-resistant screws, shielded wiring, and protected internal components. High-quality locks often include tamper detection sensors that trigger alerts if the device is opened or interfered with.

Additionally, proper installation plays a key role. Ensuring that mounting hardware is secure, internal components are not exposed, and access to critical parts is restricted can significantly reduce the risk of tampering. In higher-risk environments, combining smart locks with reinforced doors, security plates, and alarm integration provides layered protection.

This example demonstrates that smart locks are still physical devices subject to physical attack. Effective locksmithing practice requires treating electronic and mechanical security as complementary layers, ensuring that weaknesses in one cannot be exploited to defeat the entire system.

Power-related vulnerabilities are unique to electronic systems. Smart locks depend on batteries or electrical supply, and failure to manage power can lead to denial of access or insecure states. For example, a lock configured as "fail-safe" may unlock during a power loss, while a "fail-secure" lock may prevent legitimate access. Attackers may attempt to exploit these behaviours, or simply take advantage of poorly maintained systems where batteries are depleted.

A practical example of power-related vulnerabilities can be seen in a commercial office building that uses electronic access-controlled doors with magnetic locks on emergency exits.

In this installation, the doors are configured as fail-safe, meaning they automatically unlock when power is lost. This is appropriate for life safety compliance, as it ensures occupants can exit during an emergency such as a fire. However, the system is not supported by adequate backup power or monitoring.

An attacker identifies this weakness and accesses an external electrical panel that supplies power to the door. By cutting or interrupting the power feed, the magnetic lock immediately releases, unlocking the door without triggering any alarm. Because the system lacks battery backup and does not generate a real-time alert, the attacker is able to enter the building undetected.

In a different scenario, a residential property uses a battery-powered smart deadbolt configured as fail-secure, meaning it remains locked when power is lost. Over time, the batteries are not replaced, and the system does not provide clear low-battery warnings. One evening, the batteries fully deplete, and the homeowner is unable to unlock the door using the keypad or mobile app. With no immediate backup method available, access is denied until a physical override key is used or external power is temporarily applied.

A more subtle variation occurs in a small business using Wi-Fi-enabled smart locks. The locks rely on batteries for operation and network connectivity for remote management. As batteries weaken, the lock becomes intermittent—sometimes responding slowly or failing to lock completely. Staff assume the door is secure, but due to insufficient power, the actuator does not fully engage the bolt. This creates a hidden vulnerability where the door appears locked but can be opened with minimal force.

From a locksmithing perspective, these examples highlight that power is a critical dependency in electronic locking systems. Unlike mechanical locks, which function independently of external energy sources, smart locks require ongoing power management to maintain both security and usability.

To address these risks, locksmiths must ensure that power considerations are built into system design and maintenance. This includes selecting the appropriate fail-safe or fail-secure configuration based on the environment, installing backup power supplies such as batteries or UPS systems where required, and ensuring that low-battery warnings are enabled and clearly communicated to users. Regular maintenance schedules should include battery replacement and system checks to confirm reliable operation.

Backup access methods are also essential. Mechanical key overrides, external power contacts, or emergency access procedures ensure that loss of power does not result in total system failure. In higher-security environments, power systems should be monitored so that any interruption triggers an immediate alert.

This example demonstrates that power is both an operational requirement and a potential vulnerability. Effective locksmithing practice must account for how locks behave under power loss conditions and ensure that systems remain secure, functional, and predictable even when power is disrupted.

Another critical vulnerability lies in cloud and backend system security. Many smart locks rely on cloud platforms for remote access and management. If these platforms are compromised, attackers may gain control over multiple locks simultaneously. Weak API security, poor authentication mechanisms, or data breaches can expose user credentials and access permissions on a large scale. This shifts part of the security responsibility away from the physical device to the broader digital infrastructure.

A practical example of cloud and backend system vulnerability can be seen in a multi-site commercial property management company that uses a cloud-based access control platform to manage smart locks across several office buildings.

In this scenario, all locks are connected to a central cloud service that allows administrators to remotely manage users, assign access permissions, and unlock doors. Staff, contractors, and tenants are issued mobile credentials or PIN codes, all controlled through the platform's web interface and APIs.

However, the system is configured with weak backend security controls. The administrative account uses a simple password without multi-factor authentication, and the platform exposes an API that is not properly secured against unauthorised requests. An attacker obtains the administrator's login credentials through a phishing email, gaining full access to the cloud dashboard.

Once inside the system, the attacker is able to:

- View all registered users and their access permissions
- Unlock doors remotely across multiple sites
- Create new user accounts with full access privileges
- Disable or modify existing access controls without detection

Because the locks rely on the cloud platform for command and control, the attacker does not need to physically interact with any door. Instead, they exploit the backend system to control the locks at scale. In effect, a single compromised account results in a system-wide breach.

In another variation, a vulnerability in the platform's API allows attackers to send unauthorised commands directly to the system. By exploiting this flaw, they can issue unlock commands or retrieve access data without valid authentication. This type of attack highlights how weaknesses in software design can bypass even well-configured user accounts.

From a locksmithing perspective, this example demonstrates a fundamental shift in where risk resides. The security of the lock is no longer determined solely by its mechanical strength or installation quality, but also by the integrity of the cloud infrastructure that manages it. Even a high-quality, properly installed smart lock can be compromised if the backend system is insecure.

To mitigate these risks, locksmiths and security professionals must consider cloud security as part of the overall system design. This includes ensuring that:

- Strong authentication methods (such as multi-factor authentication) are enabled
- Access to administrative accounts is tightly controlled
- API endpoints are secured and regularly tested
- Systems are kept up to date with security patches
- Audit logs are monitored for unusual activity

In addition, clients should be educated about the importance of protecting credentials and recognising phishing attempts, as human factors often play a role in cloud-based breaches.

This example highlights that smart lock security extends beyond the door itself. The lock is only one component in a larger, interconnected system, and vulnerabilities in cloud platforms or backend services can undermine the entire access control environment. Effective locksmithing practice must therefore account for both physical and digital security layers to ensure comprehensive protection.

Lack of monitoring and audit utilisation also contributes to exploitation. While smart systems often generate detailed logs, these are only useful if they are actively monitored. Unusual access patterns, repeated failed attempts, or access outside normal hours may indicate a security issue, but if no one reviews the logs, these warning signs go unnoticed. This parallels traditional locksmithing issues where lack of key tracking leads to undetected misuse.

A practical example of this vulnerability can be seen in a medium-sized warehouse that has implemented an electronic access control system across multiple entry points, including staff entrances, storage areas, and loading docks.

The system is technically well-configured. Each employee has an individual access card, permissions are assigned based on role, and every access event is automatically recorded in a central log. The system captures detailed information, including the user ID, door location, date, and time of each access attempt, as well as failed entry attempts.

However, despite this capability, the organisation does not actively monitor or review the audit logs. There is no routine process for checking access activity, no alerts configured for unusual behaviour, and no designated person responsible for oversight.

Over time, a staff member begins accessing a restricted storage area outside of normal working hours. This access is technically permitted due to overly broad permissions, but the pattern is unusual. The employee repeatedly enters the area late at night and on weekends, removing small quantities of inventory. Each entry is logged by the system, and there are also several failed access attempts at other doors, indicating probing behaviour.

Because no one is reviewing the logs, these warning signs go unnoticed for weeks. The activity only comes to light when a stocktake reveals missing inventory. When the logs are finally examined, they clearly show the pattern of after-hours access and the identity of the individual responsible. By this point, the loss has already occurred and could have been prevented with earlier detection.

A similar situation can occur in an office environment using smart locks with mobile credentials. An attacker gains access to a valid user account and begins entering the building at unusual times. The system records each entry, but without active monitoring or automated alerts, the behaviour is not flagged. The attacker is able to operate undetected, despite the system having all the data required to identify the breach.

From a locksmithing and security perspective, this example highlights that audit logs alone do not provide security, they provide potential for security. Their value depends entirely on how they are used. Without monitoring, analysis, and response, logs become passive records rather than active controls.

To address this, locksmiths and security professionals should ensure that systems are configured not only to record events, but to actively support monitoring. This may include setting up alerts for abnormal activity (such as access outside defined hours), assigning responsibility for regular log reviews, and integrating access control data with broader security systems.

This example reinforces a key principle: visibility without action does not improve security. Effective access control requires not just the ability to record events, but the discipline and processes to interpret and respond to them.

User behaviour continues to play a central role through social engineering and human error. Attackers may trick users into sharing access codes, approving fraudulent requests, or installing malicious applications. In IoT environments, users may also connect devices to insecure networks or fail to follow basic security practices. These behavioural vulnerabilities often bypass even well-designed technical controls.

A practical example of this vulnerability can be seen in a co-working office that uses a smartphone-based smart lock system for entry. Staff and members access the building using a mobile app, and administrators can remotely grant or revoke access.

An attacker targets the system not by attempting to break the lock, but by exploiting user behaviour. They send a convincing phishing message to several members, posing as building management. The message states that a "system upgrade" requires users to log in and confirm their access credentials. The link directs users to a fake website that closely resembles the legitimate access control platform.

One user, assuming the request is genuine, enters their login details. The attacker captures these credentials and gains access to the user's account. Because the system trusts authenticated users, the attacker can now unlock doors remotely, access restricted areas, or even invite additional "users" depending on permission levels.

In another variation, the attacker physically approaches the building and waits near the entrance. When a legitimate user arrives, the attacker claims they have forgotten their phone or access credentials and asks the user to unlock the door for them. Wanting to be helpful, the user grants access without verification. The attacker then enters the building without ever interacting with the lock system directly.

A further example involves a residential smart lock connected to a home Wi-Fi network. The homeowner installs the lock but connects it to an unsecured or poorly configured network with a weak password. The attacker gains access to the network and is then able to scan for connected devices, including the smart lock, potentially exploiting vulnerabilities or intercepting communications.

From a locksmithing perspective, these examples demonstrate that the strongest hardware and well-designed systems can still be bypassed through human behaviour. The attack does not target the lock itself, but the trust placed in users and their interactions with the system.

To address this, locksmiths and security professionals must consider user behaviour as part of the overall security design. This includes advising clients on secure practices, such as not sharing credentials, recognising phishing attempts, and verifying identities before granting access. Systems can also be designed to reduce reliance on user judgement, for example by limiting permissions, requiring multi-factor authentication, or restricting the ability to share access without approval.

This example highlights a key reality of modern security: technical controls alone are not sufficient. Human behaviour is often the weakest link, and effective security must account for how people think, act, and respond in real-world situations.

From a locksmithing perspective, these vulnerabilities fundamentally influence practice. Locksmiths must now consider not only the strength of the lock, but also the security of communication channels, the integrity of software, and the behaviour of users. This requires a broader skill set that includes basic cybersecurity knowledge, system configuration, and risk assessment. Installation must include secure setup procedures, such as changing default credentials, enabling encryption, and configuring access controls correctly.

Mitigation strategies are equally important. These include enforcing strong authentication practices, ensuring regular firmware updates, using encrypted communication protocols, and implementing the principle of least privilege. Physical security must still be maintained through robust hardware and proper installation. In addition, users should be educated about secure practices, and systems should be designed to minimise reliance on user behaviour wherever possible.

**Impact on Locksmithing Practices**

Understanding vulnerabilities and exploits in smart locks and IoT systems is highly relevant to anyone involved in locksmithing because it fundamentally redefines what "security" means in practice. Traditionally, locksmithing focused on resisting physical attacks such as picking, bumping, or forced entry. In modern systems, however, a lock can be perfectly secure mechanically and still be compromised through weak credentials, insecure networks, outdated firmware, or user error. This expands the locksmith's responsibility from protecting a physical point of entry to managing a complete, interconnected security ecosystem.

This shift has a direct influence on locksmithing practice by changing the nature of risk assessment. A locksmith can no longer evaluate a lock based solely on its mechanical strength or build quality. Instead, they must consider how the system authenticates users, how data is transmitted, where it is stored, and how it is managed over time. For example, a high-quality smart lock installed on a door offers little real security if it is configured with default credentials, connected to an insecure Wi-Fi network, or managed through a compromised cloud platform. The weakest point in the system determines the overall level of security.

Credential management becomes a central responsibility. In traditional systems, issuing a key was a relatively static process. In smart systems, credentials are dynamic, easily shared, and potentially exposed. This means locksmiths must guide clients in implementing strong authentication practices, such as unique user credentials, multi-factor authentication, and timely revocation of access. The focus shifts from "who has the key" to "who has access, under what conditions, and for how long," requiring ongoing management rather than one-time installation.

Communication security also becomes a critical consideration. Wireless protocols such as Bluetooth and Wi-Fi introduce risks that did not exist in purely mechanical systems. Locksmiths must understand how these technologies work, how they can be attacked (e.g. interception or replay), and how to configure them securely. This includes ensuring encryption is enabled, pairing processes are secure, and networks are properly protected. In practice, this often means the locksmith must engage with aspects of IT infrastructure, such as routers, gateways, and mobile device configuration.

Software and firmware management introduces another layer of responsibility. Unlike traditional locks, smart locks can become insecure over time if vulnerabilities are discovered and not patched. This creates a lifecycle responsibility for locksmiths, where maintenance includes checking for updates, verifying firmware versions, and ensuring systems remain secure as threats evolve. The role extends beyond installation into ongoing support and risk management.

The principle of least privilege and proper access control significantly influence system design. Locksmiths must now think in terms of roles, permissions, and time-based access rather than simply issuing keys. Poorly managed permissions can create large-scale vulnerabilities, where a single compromised account grants access to multiple areas. This requires a more structured and policy-driven approach to access design, aligning security with organisational needs and workflows.

Importantly, these vulnerabilities reinforce that physical security has not become less important—it has become one layer within a broader system. Physical weaknesses, such as exposed wiring or poorly constructed hardware, can still allow attackers to bypass electronic controls entirely. As a result, locksmiths must ensure that mechanical integrity, installation quality, and tamper resistance remain high, while also addressing digital risks.

User behaviour remains a decisive factor, and this has a strong influence on locksmithing practice. Many vulnerabilities arise not from technical failure, but from how people interact with the system—choosing weak passwords, sharing access, ignoring updates, or falling victim to social engineering. This means locksmiths must take on an advisory role, educating clients and designing systems that minimise reliance on correct user behaviour. Concepts such as automation, default-secure configurations, and limited permissions become essential tools in reducing human error.

Finally, these developments transform locksmithing into a hybrid discipline that combines physical security, digital systems, and human factors. The locksmith is no longer just a tradesperson installing hardware, but a security professional responsible for designing, configuring, and maintaining integrated access systems. This includes understanding cybersecurity principles, system architecture, and behavioural risks, alongside traditional mechanical skills.

## Firmware Attacks and Update Risks

Firmware attacks and update risks are among the most critical, and often least understood, security issues in smart locks and IoT-based access systems. Firmware is the embedded software that controls how a smart lock operates, including authentication, communication, motor control, and sensor feedback. Unlike mobile apps or cloud platforms, firmware runs directly on the device itself, meaning that if it is compromised, the attacker can potentially gain full control over the lock. The security of firmware updates is a fundamental aspect of protecting modern embedded devices, as updates are a frequent and highly sensitive vector for introducing or remedying vulnerabilities that could be exploited to bypass access controls or exfiltrate data [244, 245].

A firmware attack typically targets weaknesses in how this embedded software is written, stored, or updated. If an attacker is able to modify or replace the firmware, they can bypass normal security controls entirely, for example by installing malicious firmware that ignores authentication checks or secretly transmits access data to an external device, thereby making a lock appear functional while operating under an attacker's control [244]. Such compromises operate at a low level and may evade visible indicators in logs or user interfaces, underscoring the importance of hardware-anchored and update-time protections in addition to application-layer defences [244]

One of the most common entry points for firmware attacks is the update process itself. Smart locks and other IoT devices rely on firmware updates to patch vulnerabilities, fix bugs, and improve performance; when the update mechanism is poorly secured, it becomes a prime attack surface. Updates that are not digitally signed and verified can be spoofed by adversaries delivering malicious firmware packages that the device accepts as legitimate, and updates transmitted without encryption can be intercepted and altered, enabling a man-in-the-middle (MITM) attack on the update process. The literature repeatedly emphasizes the need for cryptographic authentication, integrity verification, and secure transmission channels to defend the update pipeline against such threats, including the use of digital signatures, secure boot, and trusted hardware modules to thwart MITM and tampering during firmware delivery [244, 246-249].

Insecure or poorly managed update practices further exacerbate risk: devices may not receive timely updates or users may neglect installation, leaving known vulnerabilities unpatched or devices to reach end-of-life without security updates, creating a widening security gap between device functionality and the evolving threat landscape [244, 245, 249]. Firmware rollback and downgrade attacks are a particular concern when version control or rollback protections are weak, enabling an attacker to force installation of older, vulnerable firmware even after prior updates have been applied [244-246].

Security mechanisms designed to protect the firmware lifecycle are diverse and often complementary. Secure boot and rollback protection are central to preventing unauthorized firmware from executing at boot or being reintroduced after updates; cryptographic authentication methods, including digital signatures and, in some contexts, post-quantum approaches, help ensure that only legitimate firmware is accepted and installed [244, 248, 250-253]. Hardware-assisted security primitives, such as cryptographic engines integrated into the device, secure enclaves, or hardware-based fingerprints, provide strong protection against tampering and impersonation during update delivery and verification [244]. In wireless FUOTA (Firmware Updates Over-The-Air) contexts, secure transmission methods and authenticated broadcast/multicast management become crucial due to resource constraints in IoT devices; the literature presents secure, authenticated, and efficient FUOTA designs that balance security with power and computational limitations while supporting anti-replay and secure update distribution across multiple devices [252-254]. For automotive and connected-vehicle ecosystems, secure OTA software updates are a distinct domain, with emphasis on authentication, encryption, integrity verification, rollback protection, and regulatory alignment to prevent unauthorized updates that could affect safety-critical systems [245, 255]. In parallel, several studies explore advanced cryptographic methodologies suitable for constrained devices, including hash-based signatures for post-quantum resilience and forward-secure schemes that limit exposure in the event of key compromise during updates [244, 251, 252].

Beyond cryptographic protections, robust update governance and physical security are essential. Insecure or exposed interfaces such as USB, serial, or debug ports can provide direct paths to extract or modify firmware if not adequately protected, highlighting the need for tamper resistance and properly sealed enclosures in lock hardware [244]. Update governance includes enforcing strict version control, authenticated update issuance, and ensuring that devices only install authorised, current firmware versions; where feasible, automation of secure updates and maintenance schedules helps ensure timely patching and reduce human error in the update process [244, 245]. Client education and installer best practices are also important: users should understand the importance of firmware updates, recognise legitimate update notifications, and avoid unofficial or third-party firmware that could introduce new risks; locksmiths and installers must evaluate firmware security as part of product selection and maintenance planning, prioritizing devices with secure update mechanisms, encryption, and protections such as secure boot and rollback prevention [244, 245].

A coherent security architecture for firmware and updates in smart lock and IoT access systems can be characterized by three interlocking layers: (i) secure boot and trusted execution environments that ensure only authenticated firmware runs on device startup and during updates; (ii) a secure, authenticated, and integrity-verified update pipeline with cryptographic protections (digital signatures, encryption, secure channels) and robust version control, including anti-rollback and certificate-based controls; and (iii) hardware-assisted security primitives and tamper resistance that protect the update process, storage of keys, and physical interfaces from exploitation. In addition, decentralization and transparency mechanisms such as blockchain-based update distribution or distributed ledger technologies can provide immutable provenance and verifiability for firmware updates, potentially reducing single points of failure and enabling verifiable update histories in complex ecosystems like connected vehicles and charging networks [253, 256, 257].

The literature also recognizes the potential role of DID and DLT-based approaches to secure OTA updates, including revocation and verifiable credentials, to improve authenticity and traceability of updates in modern automotive contexts [248].

Decentralised Identifiers (DIDs) and Distributed Ledger Technology (DLT) are increasingly being used to strengthen the security of over-the-air (OTA) updates in modern automotive systems. As vehicles become software-defined platforms, critical components—including door locking systems, immobilisers, and access control modules—are updated remotely. This creates a need for strong assurance that every update is authentic, authorised, and traceable. DID and DLT approaches address this by replacing traditional, centralised trust models with cryptographically verifiable, decentralised identities and tamper-evident records.

At the centre of this model is the DID, which acts as a secure digital identity for entities such as vehicle manufacturers, component suppliers, ECUs, and even individual vehicles. Each DID is linked to a set of cryptographic keys published on a distributed ledger. When an OTA update is issued—for example, a firmware update for a vehicle's electronic locking system—it is digitally signed using the private key associated with the issuer's DID. The vehicle can independently verify the update by resolving the DID on the ledger and confirming the corresponding public key. This ensures that the update genuinely originates from a trusted source and has not been altered in transit.

DLT complements this by providing an immutable and transparent record of update activity. Each update can be logged with metadata such as version number, issuing entity, timestamp, and affected components. This creates a permanent audit trail that supports traceability across the entire lifecycle of the system. In automotive contexts, this is particularly important for safety-critical components like central locking modules or keyless entry systems, where the ability to trace when and how software changes occurred is essential for diagnostics, compliance, and incident investigation.

Revocation is a critical feature of this architecture. If a signing key is compromised, a supplier is no longer authorised, or a vulnerability is discovered in a previously issued update, the associated DID or credential can be updated on the ledger to reflect its revoked status. Vehicles and connected systems can check the current state of the DID before accepting an update, ensuring that outdated or compromised credentials are rejected in real time. This dynamic revocation capability significantly improves response times compared to traditional certificate revocation mechanisms, which can be slow and inconsistently enforced.

Verifiable credentials extend this framework by enabling structured, cryptographically secure assertions about who is authorised to do what. For example, a third-party supplier responsible for developing firmware for a vehicle's smart locking system may be issued a verifiable credential by the manufacturer confirming their authority. When that supplier releases an update, the credential is presented alongside the update package. The receiving system can verify not only the signature but also the supplier's authorisation, ensuring that only approved entities can modify specific subsystems. This is particularly valuable in complex automotive supply chains, where multiple vendors contribute to different components.

The relevance of these technologies to locking systems is direct and increasingly significant. Modern vehicle locking systems are no longer purely mechanical; they are electronic and software-driven, often integrated with remote access, mobile credentials, and keyless entry technologies. OTA updates may modify how locks respond to credentials, how encryption is handled, or how anti-theft features operate. If these updates are not securely managed, they can introduce vulnerabilities that allow unauthorised access or disable security features entirely. DID and DLT-based approaches help ensure that only legitimate, verified updates can alter these critical systems.

For locksmiths, particularly those working with automotive or advanced electronic access systems, this represents a fundamental shift in practice. The role is expanding from mechanical key cutting and lock servicing to include an understanding of digital identity, firmware integrity, and secure update processes. Locksmiths may be required to verify that a vehicle's locking system is running authorised firmware, assist in diagnosing issues related to failed or compromised updates, and understand how digital credentials interact with physical locking mechanisms.

In addition, these technologies influence how locksmiths approach trust and security assurance. Instead of relying solely on physical inspection or manufacturer documentation, a locksmith may increasingly rely on cryptographic verification and system logs to confirm the integrity of a locking system. The ability to interpret update histories, verify digital signatures, and understand revocation status becomes part of ensuring that a system is secure and functioning as intended.

More broadly, the adoption of DID and DLT in OTA update ecosystems reflects a convergence between physical security and cybersecurity. Locksmithing, traditionally focused on tangible hardware, is evolving into a hybrid discipline where digital trust frameworks underpin physical access control. Understanding these systems enables locksmiths to remain relevant in environments where locks are controlled not just by keys, but by software, identities, and networked infrastructure.

Finally, emerging research on forward-secure and post-quantum-resilient schemes, such as hash-based signatures, offers potential long-term protections for firmware authentication under future quantum threats, particularly relevant for long-lived IoT deployments and critical infrastructure [251].

In practical terms for locksmithing practice, selecting products from reputable manufacturers that provide regular, securely delivered updates is essential, and installers should verify that devices support protections such as secure boot, encryption, and rollback prevention. During installation, systems should be configured to enable automatic updates where appropriate, or maintenance schedules should be established to ensure timely patching, with explicit guidance to clients on recognizing legitimate update channels and avoiding unofficial firmware. Education of end users remains critical to ensure that firmware integrity is maintained over the device's lifetime and that updates are treated as a standard security practice rather than optional maintenance.

Ultimately, firmware attacks and update risks illustrate a key reality of modern locksmithing: security is not static, and a lock's security posture must be continuously managed and validated over time through robust update governance, hardware-assisted protections, and user-centred education to address evolving threats in a persistent threat landscape [244, 245, 249, 258].

**Case Study Scenario: Compromised Firmware Update in a Smart Access System**

A mid-sized commercial office installs a networked smart locking system across multiple entry points, including the main entrance, server room, and staff-only areas. The system uses Wi-Fi–enabled locks managed through a cloud platform, allowing administrators to issue mobile credentials, monitor access, and receive firmware updates remotely. The manufacturer periodically releases firmware updates to improve performance and address security vulnerabilities, making the update process a routine part of system operation.

The organisation relies on automatic firmware updates to maintain security, but the implementation contains a critical weakness. The update process does not enforce strong cryptographic verification, meaning firmware packages downloaded over the internet are not properly validated before installation. In addition, the organisation's network lacks proper segmentation, allowing office devices and the smart locking system to operate on the same network, increasing exposure to potential threats.

An attacker gains initial access to the network through a phishing attack targeting an employee. Once inside, the attacker monitors network traffic and identifies communication between the smart locks and the update server. Recognising that the firmware update process lacks proper authentication, the attacker initiates a man-in-the-middle attack. When the next firmware update is requested, the attacker intercepts the communication and delivers a modified firmware package. Although it appears legitimate, the firmware has been altered to include a hidden backdoor. Because the system does not enforce robust digital signature verification, the locks accept and install the malicious update.

Following installation, the locks continue to operate normally from the perspective of users. Staff can still access the building using their credentials, and system logs appear unchanged. However, the compromised firmware introduces covert functionality. The lock can now accept a hidden "master" unlock command from the attacker, access events triggered by the attacker are not recorded in audit logs, and authentication checks can be selectively bypassed. These changes occur at a low level within the device, making them invisible to standard monitoring tools and user interfaces. As a result, the system appears secure while being fully compromised.

The attacker exploits this hidden access to enter the building after hours, targeting sensitive areas such as the server room. Because these actions are not logged, the breach remains undetected. Over time, the attacker is able to remove equipment and extract data without triggering alarms or raising suspicion. The situation escalates further when the organisation attempts to apply a legitimate firmware update. The malicious firmware prevents the installation of new updates by bypassing rollback protection mechanisms, effectively locking the system into a compromised state.

The breach is eventually discovered during a routine IT audit, when inconsistencies are identified between physical access events and recorded logs. A forensic investigation reveals that the firmware has been altered. In response, the organisation must replace or reflash affected locks, reset all user credentials and access permissions, and rebuild the network with appropriate segmentation and security controls. The financial cost and operational disruption are significant, far exceeding the original investment in the system.

From a locksmithing perspective, this scenario demonstrates that firmware and update security are critical components of modern access control systems. The failure did not result from a mechanical weakness, but from a compromised update process. It highlights that firmware integrity is as important as physical security, and that locks must only accept updates from trusted and authenticated sources. Secure update mechanisms, including encryption, digital signatures, and verification processes, are essential to prevent spoofed or malicious updates. It also reinforces the importance of lifecycle management, where ongoing monitoring, updating, and validation are required to maintain security over time.

If secure over-the-air update practices had been implemented, the attack could have been prevented. These practices include the use of digitally signed firmware verified with trusted cryptographic keys, secure boot processes to ensure only authorised firmware can run on the device, encrypted update delivery to prevent interception, and anti-rollback protections to block installation of unauthorised or outdated firmware. Hardware-backed key storage further strengthens protection by preventing tampering with critical security elements. In more advanced systems, decentralised identity and distributed ledger approaches can provide verifiable update provenance, ensuring that all updates are traceable, authentic, and capable of being revoked if compromised.

This case highlights a fundamental shift in locksmithing. The point of failure is no longer limited to the physical lock, but extends to the systems that control and update it. Firmware attacks operate below the surface and may not be visible through traditional inspection methods. For locksmiths, this underscores the need to evaluate not only the quality of hardware, but also the security of firmware, update mechanisms, and the broader digital environment in which modern locking systems operate.

## Impact on Locksmithing Practices

Firmware attacks and update risks are highly relevant to anyone involved in locksmithing because they fundamentally redefine what a “lock” is and where its vulnerabilities exist. Traditionally, locksmithing focused on mechanical integrity. resistance to picking, drilling, or forced entry. In smart locks and IoT-based systems, however, the critical point of control often lies in firmware, the embedded software that governs how the lock authenticates users, communicates with devices, and physically operates. If this firmware is compromised, the strongest mechanical components become irrelevant, as an attacker can bypass security controls entirely at the software level.

This shifts the locksmith’s role from purely mechanical expertise to a broader systems-based understanding of security. Firmware attacks demonstrate that access control is no longer confined to the physical interface of the lock. Instead, it extends into update mechanisms, communication pathways, and digital trust models. A locksmith must now recognise that a lock can be “picked” remotely through malicious code, without any physical interaction. This represents a significant evolution in threat models, where vulnerabilities may be invisible and operate below the level of traditional inspection.

The firmware update process itself becomes a critical area of concern and responsibility. In modern systems, updates are essential for maintaining security, but they also introduce risk if not properly secured. For locksmiths, this means evaluating whether a system uses secure update practices such as digital signature verification, encrypted delivery, and rollback protection. The case study clearly illustrates that an insecure update pipeline can be exploited to install malicious firmware, effectively handing control of the lock to an attacker while leaving no obvious signs of compromise. This highlights that installation alone is not sufficient, ongoing system integrity must be considered.

Another key influence on locksmithing practice is the concept of lifecycle security. Unlike traditional locks, which remain largely static once installed, smart locks require continuous management through updates, monitoring, and validation. Firmware vulnerabilities may emerge over time, and failure to update systems can leave them exposed. As a result, locksmiths must adopt a maintenance-oriented approach, where checking firmware status, ensuring updates are applied, and verifying system integrity become part of routine service. Security is no longer a one-time installation outcome but an ongoing process.

The emergence of advanced security frameworks such as secure boot, hardware-backed key storage, and cryptographic update verification further expands the technical expectations of the profession. Locksmiths must understand how these mechanisms protect against tampering and unauthorised firmware execution. In more advanced environments, concepts such as decentralised identifiers (DIDs), distributed ledger technology (DLT), and verifiable credentials introduce new ways of ensuring update authenticity and traceability. These technologies reinforce trust in the update process by enabling systems to verify not just the integrity of firmware, but also the identity and authority of the entity issuing it.

These developments also influence how locksmiths approach product selection and system design. It is no longer sufficient to choose a lock based solely on physical strength or brand reputation. Locksmiths must assess whether a device supports secure firmware practices, receives regular updates, and incorporates protections against rollback or tampering. Recommending appropriate systems becomes part of risk management, ensuring that clients are not exposed to avoidable vulnerabilities.

User education and organisational processes are also affected. Firmware security depends not only on design, but on how systems are managed in practice. Locksmiths must advise clients on recognising legitimate updates, avoiding unofficial firmware, and maintaining secure configurations. This includes ensuring that update mechanisms are enabled, credentials are protected, and systems are monitored for anomalies. In this sense, locksmithing increasingly overlaps with cybersecurity awareness and user behaviour management.

Firmware attacks highlight a fundamental transformation in locksmithing. The point of failure is no longer limited to physical components but extends into embedded software and digital infrastructure. A lock can appear secure while being fully compromised at the firmware level, making traditional indicators of security insufficient. For locksmiths, this means adopting a holistic view of security that integrates mechanical reliability, software integrity, secure update processes, and user practices.

## Securing Smart Locks in Residential and Commercial Settings

Securing smart locks in residential and commercial settings requires a layered approach that combines traditional locksmithing principles with modern cybersecurity practices. Unlike purely mechanical systems, smart locks operate within a broader ecosystem that includes wireless communication, user credentials, firmware, and cloud services. As a result, security must be considered across physical, digital, and operational domains.

### Residential Settings

In residential environments, smart locks are typically deployed for convenience, remote access, and integration with home automation systems. Security begins with correct installation and mechanical integrity. The lock must be properly aligned, securely mounted, and paired with a reinforced door and strike plate. A high-quality smart lock installed on a weak door still presents a vulnerability, so physical security remains the foundation.

Digital configuration is equally important. Home users often rely on PIN codes or mobile apps, which introduces risks if weak credentials are used. Strong, unique PINs should be enforced, and default passwords must be changed immediately during setup. Where possible, multi-factor authentication should be enabled to add an additional layer of protection beyond a single credential.

Wireless communication security plays a critical role. If the lock uses Wi-Fi, the home network must be secured using modern encryption (WPA2 or WPA3), strong router passwords, and regular firmware updates. For Bluetooth-based locks, secure pairing practices are essential to prevent unauthorised device connections. Where hubs or bridges are used, they should be positioned to ensure reliable coverage while minimising exposure.

Another key consideration in residential settings is user behaviour. Homeowners may unintentionally weaken security by sharing access codes, failing to remove old users, or ignoring firmware updates. Regular review of user access, audit logs, and device updates helps maintain system integrity over time. Battery management is also important, as depleted batteries can lead to lock failure or insecure states if not properly configured.

### Commercial Settings

In commercial environments, smart lock security becomes more complex due to scale, multiple users, and integration with broader access control systems. Security must be structured, managed, and auditable.

Access control policies are central. Users should be assigned individual credentials rather than shared codes, allowing precise tracking and accountability. Access should be based on roles, with permissions limited to specific areas and times. For example, cleaning staff may only have access outside business hours, while IT personnel may have restricted access to server rooms.

System integration introduces additional considerations. Smart locks may be connected to centralised management platforms, building automation systems, or cloud-based services. These systems must be secured through strong authentication, encrypted communication, and regular monitoring. Network segmentation is particularly important—smart lock systems should operate on a separate network from general office devices to reduce exposure to cyber threats.

Audit and monitoring capabilities are a major advantage in commercial settings, but only if they are actively used. Access logs should be reviewed regularly to detect unusual activity, such as repeated failed access attempts or entry outside authorised times. Alerts and automated notifications can support rapid response to potential incidents.

Firmware and update management are critical at scale. Organisations must ensure that all devices receive timely, authenticated updates to address vulnerabilities. Centralised update management can help maintain consistency, but it must be implemented securely to prevent compromised updates from affecting multiple devices simultaneously.

**Shared Security Principles**

Across both residential and commercial settings, several core principles apply:

- Physical and digital security must work together. A secure smart lock requires both robust hardware and well-configured software. Weakness in either area can compromise the entire system.
- Authentication and access management are fundamental. Strong credentials, individual user accounts, and timely revocation of access reduce the risk of unauthorised entry.
- Secure communication is essential. All data exchanged between locks, devices, and servers should be encrypted and protected against interception or manipulation.
- Lifecycle management is ongoing. Security does not end at installation—regular updates, maintenance, and monitoring are required to address evolving threats.
- User awareness is critical. Even the most secure system can be undermined by poor user practices, making education and clear procedures an important part of security.

**Smart Lock Installation**

Installing a smart lock requires both traditional locksmithing skills and an understanding of electronic components. The process must ensure correct mechanical alignment, reliable power, and proper digital configuration so the system operates securely and consistently.

<u>1. Pre-Installation Preparation</u>

Before installation begins, confirm that the door and hardware are suitable. Check that the door thickness, backset (typically 60 mm or 70 mm), and borehole sizes match the smart lock specifications. Ensure the door is correctly aligned, closes smoothly, and the latch or deadbolt operates without friction. Poor alignment will cause actuator strain and unreliable locking.

Disconnect any existing alarm or access system if required, and ensure you have all components from the smart lock kit, including mounting plates, screws, batteries, and any required modules (e.g. Wi-Fi bridge or Zigbee module).

<u>2. Remove Existing Hardware</u>

Remove the existing lockset carefully. Unscrew the internal handle or thumb turn, then remove the external cylinder or handle assembly. Extract the latch or deadbolt from the door edge.

Inspect the door preparation:

- Clean out debris from the borehole and latch cavity
- Check that the strike plate aligns with the latch or bolt
- Confirm the door frame is not causing binding

This step is critical because smart locks rely on smooth mechanical operation—unlike manual locks, they cannot "force" misaligned components.

3. Install the Mechanical Components

Insert the new latch or deadbolt into the door edge, ensuring:

- The bolt is oriented correctly (bevel facing the strike plate for latches)
- The mechanism sits flush with the door edge
- Screws are tightened securely but not overtightened

Attach the exterior assembly (keypad, reader, or escutcheon) through the borehole. Feed any connecting cables through the door to the interior side.

Mount the interior assembly, ensuring:

- The tailpiece or spindle aligns correctly with the lock mechanism
- Cables are not pinched or strained
- The unit sits flush against the door

Secure the assembly with the provided screws.

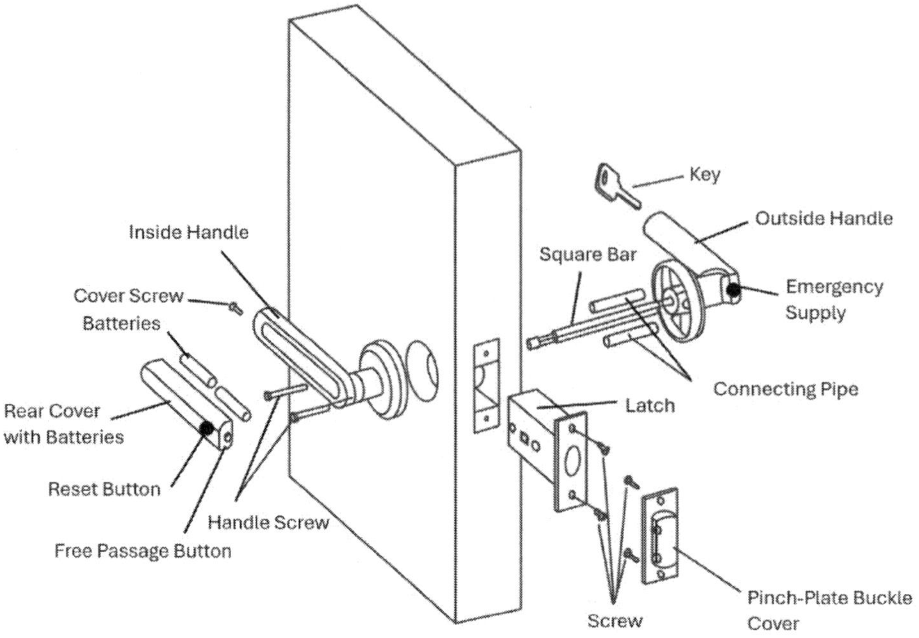


*Figure 74: Example diagram of door lock installation for a fingerprint lock.*

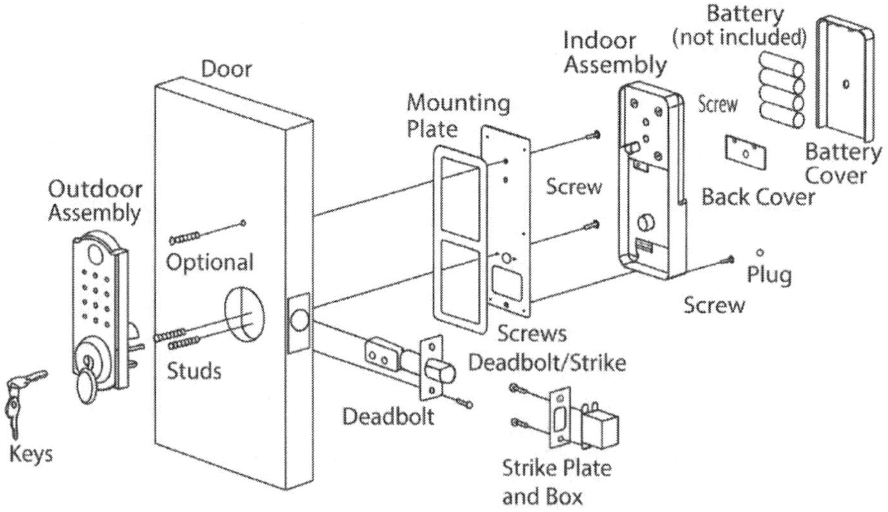


*Figure 75: Example diagram of door lock installation for a keypad and mechanical lock.*

4. Connect Electronics and Power

Connect the internal wiring according to the manufacturer's instructions. Most smart locks use plug-in connectors between the exterior and interior units.

Insert batteries (typically AA or lithium packs), ensuring correct polarity. Do not use mixed or low-quality batteries, as inconsistent power can cause erratic behaviour.

At this stage, the lock should power on. Many devices will indicate this with a light, sound, or startup sequence.

5. Mechanical Function Test (Critical Step)

Before configuring the digital system, test the lock mechanically:

- Operate the lock manually (thumb turn or key override)
- Confirm smooth extension and retraction of the bolt
- Close the door and test engagement with the strike plate
- Ensure the bolt fully extends without resistance

If the lock struggles, stops, or makes excessive noise, correct alignment issues before proceeding. Smart locks depend on low-resistance movement to avoid motor failure and battery drain.

6. Initial Digital Setup (Pairing and Configuration)

Follow the manufacturer's setup process, typically via a mobile app:

- Download the relevant app (e.g. Yale Home, Schlage Home, August App)
- Create or log into an account
- Pair the lock via Bluetooth or onboarding mode
- Assign an administrator account

Set up initial access credentials:

- Admin PIN code
- User PINs or mobile keys
- Optional biometric enrolment (if supported)

Ensure default credentials are changed immediately to secure values.

7. Network and Communication Setup

If the lock supports remote access:

- Connect to Wi-Fi (direct or via bridge)
- Pair with Zigbee/Z-Wave hub if applicable
- Test connectivity and signal strength

Verify:

- Remote lock/unlock functionality
- Real-time status updates
- Notification settings

Ensure the network is secure (strong Wi-Fi password, WPA2/WPA3 encryption).

8. Configure Access Control and Security Settings

Set up the system according to the environment:

- Assign individual users (avoid shared credentials)
- Configure access schedules (e.g. business hours)
- Enable audit logs and notifications
- Activate auto-lock or relock timers

Apply the principle of least privilege—users should only have access to what they need.

9. Test All Functions

Conduct a full operational test:

- Local unlocking (PIN, card, or app)
- Remote unlocking (if enabled)
- Auto-lock functionality
- Door closed / door ajar alerts (if sensors present)
- Backup access (physical key override or emergency power)

Test multiple times under normal use conditions to ensure reliability.

10. Finalise Installation and Handover

Secure all covers and ensure tamper-resistant screws are installed where applicable. Clean the installation area and confirm all components are firmly mounted.

Provide user instruction:

- How to lock/unlock the door
- How to manage users and credentials
- Battery replacement procedure
- Importance of firmware updates

### Impact on Locksmithing Practices

In residential settings, the relevance lies in the increasing demand for convenience-based security solutions such as app-controlled locks and remote access. Locksmiths are now expected not only to install locks correctly, but also to guide clients in setting strong credentials, securing home networks, and maintaining devices through updates and battery management. This shifts part of the role toward user education and ongoing support. A locksmith who cannot explain how to securely configure a smart lock is no longer providing a complete service.

In commercial environments, the influence is even more significant. Locksmithing becomes closely aligned with access control management and organisational security practices. Instead of issuing physical keys, locksmiths are involved in assigning digital credentials, configuring access levels, and ensuring that systems are auditable and compliant. This requires an understanding of concepts such as role-based access control, network segmentation, and system integration. The locksmith is no longer working in isolation but often alongside IT and security teams.

The installation process itself reflects this evolution. While mechanical alignment and correct fitting remain critical, they are only one part of the task. The locksmith must also manage pairing, network configuration, credential setup, and system testing. Errors in digital setup—such as leaving default passwords unchanged or failing to secure network connections—can undermine an otherwise perfect mechanical installation. This reinforces that installation is now both a physical and digital procedure.

Another key influence on locksmithing practice is the concept of lifecycle security. Unlike mechanical locks, which remain static once installed, smart locks require continuous management. Firmware updates, credential reviews, and monitoring of system logs are all part of maintaining security. Locksmiths may increasingly be expected to provide maintenance services, audits, or ongoing support agreements rather than one-off installations. This introduces a service-based model rather than a purely product-based one.

The emphasis on user behaviour also changes the locksmith's responsibilities. Many vulnerabilities arise not from the lock itself, but from how it is used—shared PINs, outdated access permissions, or ignored updates. This means locksmiths must act as advisors, helping clients understand risks and adopt secure practices. Communication and education become as important as technical skill.

## Chapter 9 Review Questions

### Question 1

A small business installs a keypad-enabled smart lock on its main entrance. All staff members share the same four-digit PIN code. Months later, a former contractor uses the code to enter the building after hours.

What was the primary security weakness?

A. The keypad was poorly manufactured
B. The lock used digital technology
C. Shared credentials prevented accountability and revocation
D. The door frame was incorrectly installed

### Question 2

A homeowner installs a Wi-Fi smart lock connected to their home network. The router password remains the default factory setting.

What is the most likely security risk?

A. Attackers may gain access through the unsecured network and target the lock
B. The lock will automatically unlock when Wi-Fi fails
C. Wi-Fi automatically encrypts all smart locks
D. The lock will disable manual keys

### Question 3

An attacker removes the outer casing of a poorly designed IoT smart lock and directly connects a small battery to exposed wires that control the actuator, causing the bolt to retract.

What vulnerability is being exploited?

A. Weak user authentication
B. Poor physical protection of internal electronics
C. Low battery levels
D. Weak Wi-Fi encryption

**Question 4**

A commercial building installs electronic access control using magnetic locks configured as **fail-safe**. An attacker cuts power to the door circuit and the doors unlock.

Why did this occur?

A. Fail-safe locks automatically release when power is lost
B. The doors were incorrectly aligned
C. The magnetic locks were mechanically damaged
D. The network connection failed

**Question 5**

A residential smart deadbolt is configured as **fail-secure** and powered by batteries. Over time the batteries completely drain, leaving the homeowner unable to unlock the door electronically.

What design feature could have prevented the lockout?

A. Removing electronic components
B. Eliminating authentication
C. Using only Wi-Fi control
D. Providing clear low-battery warnings and backup access methods

**Question 6**

A building uses smartphone-based access control. An attacker sends a fake email asking users to "confirm their account details," capturing login credentials.

What type of attack is this?

A. Mechanical lock bypass
B. Phishing-based social engineering
C. Key duplication attack
D. Magnetic lock manipulation

**Question 7**

An office installs cloud-based access control where administrators can instantly revoke user credentials when staff leave the organisation.

What security advantage does this system provide compared to traditional keys?

A. It eliminates the need for doors
B. It reduces mechanical wear on locks
C. Access can be revoked without physically changing locks
D. Users never need credentials

**Question 8**

A smart lock uses Bluetooth communication between the user's phone and the lock. An attacker intercepts and replays a previously captured signal to unlock the door.

What type of attack is this?

A. Replay attack exploiting wireless communication
B. Forced entry attack
C. Lock bumping attack
D. Key impressioning attack

**Question 9**

A company installs a smart lock system but fails to review access logs or audit records regularly.

What risk does this create?

A. Increased battery consumption
B. Security incidents may occur without detection
C. Locks will automatically unlock
D. Credentials will stop working

**Question 10**

A locksmith evaluating smart lock installations advises clients to implement multi-factor authentication (MFA) and strong credential management policies.

What is the main purpose of this recommendation?

A. To eliminate mechanical components
B. To reduce dependence on physical door hardware
C. To protect systems from weak authentication and credential theft
D. To simplify lock installation

# Chapter 10

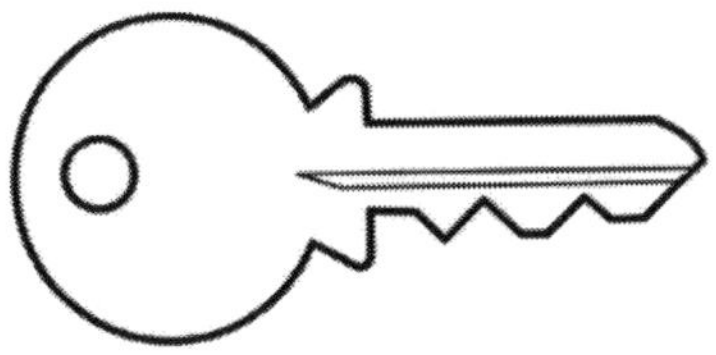

# Cybersecurity for Locksmiths

Modern locks are no longer isolated mechanical devices; they operate within complex digital ecosystems that involve wireless communication, cloud services, encryption, and identity management. As a result, protecting a door now requires an understanding of both physical security principles and the cybersecurity technologies that control and support them.

This chapter explores the growing intersection between locksmithing and cybersecurity. It introduces the concept of converged security—where physical access control systems interact with software, networks, and digital identities—and examines the technical mechanisms that make these systems secure. Topics such as encryption, secure communication protocols, access logging, privacy compliance, penetration testing, and collaboration with IT security professionals illustrate how the locksmith's role is evolving beyond mechanical installation to include system configuration, risk assessment, and digital security awareness.

For modern locksmiths and security professionals, understanding these concepts is essential. A lock may be mechanically strong, yet still vulnerable if its communication channels, credentials, or software systems are poorly protected. By examining how cybersecurity principles apply to access control systems, this chapter provides the knowledge needed to evaluate, install, and maintain modern locking technologies in a secure and responsible manner. In doing so, it highlights how the profession is expanding into a hybrid discipline that integrates traditional locksmithing skills with elements of information technology and cybersecurity.

## The Convergence of Physical and Digital Security

The convergence of physical and digital security describes a fundamental shift in how protection is designed, implemented, and managed. It reflects the move from traditional hardware-based protection—such as locks, doors, and barriers—to integrated, software-driven systems that connect networks, credentials, cloud platforms, and data. This integration blurs former boundaries between locksmithing and information technology, producing modern security ecosystems in which physical access is governed, monitored, and influenced by digital technologies.

At the heart of convergence is a shift from object-based access to identity-based security. Traditional locksmithing relied on physical keys and the possession of a tangible object to grant access. In contrast, modern systems tie access to digitally managed identities, user accounts, mobile credentials, or biometric profiles, whose verification occurs within software platforms or cloud services and is enforced physically by locking mechanisms. This transformation makes unlocking a door the culmination of a digital authentication process that determines access authorization, rather than a purely mechanical action [86, 259, 260]. In parallel, converged systems treat credentials as portable digital assets that enable conditional, auditable access across devices and locations, reinforcing the view that identity management sits at the core of physical security in the digital era [259, 261, 262].

The integration of physical and digital security hinges on communication technologies that link hardware to software, networks, and cloud platforms. Smart locks commonly employ wireless protocols such as Bluetooth, Wi-Fi, and Zigbee to communicate with smartphones, access control systems, and cloud services, enabling real-time control, remote provisioning, and centralized coordination. This connectivity supports remote unlocking, instantaneous grant or revocation of access, and comprehensive logging and analytics of entries and events. As a result, locks transition from passive devices to active elements within a broader, dynamic security system that responds to user actions and security events in real time [259, 260, 262, 263]. The role of these communication channels is widely discussed in the literature on converged security, where the reliability and security of the underlying networks directly affect physical access control outcomes [259, 262, 264].

Convergence gives rise to a layered security model in which multiple defensive strata interact to secure the end-to-end system. Physical strength alone is insufficient when software, credentials, and networks may be compromised. Security then depends on hardware integrity, credential management, secure communications, software reliability, and user behaviour. A weakness in any layer can compromise the entire system, illustrating the need for holistic design and implementation that addresses interdependencies across hardware, firmware, network protocols, authentication mechanisms, and user practices [259, 261, 262, 265]. This perspective aligns with discussions of converged security risk management, which emphasizes interrelated factors such as governance, people, processes, and technology, and acknowledges that organizational culture and training are critical to effective convergence [259].

A distinctive feature of converged security is the integration of otherwise discrete systems, locks, alarms, CCTV, building management systems, and IoT devices, into a coordinated security environment. In such ecosystems, actions in one domain trigger responses in others, enabling automated workflows: for example, an unlocked door can disarm an alarm, while a forced-entry attempt can prompt camera activation and alert generation. This level of integration enhances situational awareness and enables faster, automated responses to security events. The literature on converged security emphasizes these capabilities as central to modern security architectures, illustrating how physical access events become inputs to a broader security orchestration that spans multiple subsystems [259, 261, 262].

From a risk perspective, convergence broadens the attack surface by combining physical and digital threats. Traditional risks such as picking or forced entry now coexist with cyber threats such as credential theft, firmware manipulation, and network-based intrusions. Consequently, securing a door requires not only robust physical protections but also strong cybersecurity measures, encryption, secure authentication, regular software updates, and resilient firmware. The boundary between physical and digital risk becomes permeable, underscoring the need for integrated risk management that covers both hardware and software layers, as discussed in converged security frameworks and risk management literature [259, 261, 262].

The convergence has profound implications for locksmithing as a profession. The field is transitioning from a predominantly mechanical practice to a hybrid discipline that encompasses electronics, networking, and cybersecurity. Modern locksmiths must understand how digital credentials are managed, how systems communicate securely, how firmware is maintained, and how user behaviour shapes security outcomes. Installation tasks now include configuring software, securing networks, and advising clients on best practices for identity management and risk mitigation, reflecting a broader, interdisciplinary skill set that spans traditional locksmithing and digital security disciplines [259-262].

Ultimately, the convergence of physical and digital security signals a broader transformation in how access is controlled and managed. Locks are no longer standalone devices; they function as nodes within a connected system that harmonizes physical barriers with digital intelligence. For locksmiths and security professionals, understanding this convergence is essential to designing, installing, and maintaining security systems that are effective in the modern environment. The convergence narrative is reinforced across discussions of converged security management, IoT security, and secure update practices for networked devices, which collectively illustrate the practical, technical, and organizational dimensions of integrated security in contemporary settings [259, 261, 262, 265, 266].

## Practical Application

Modern access control systems tie physical entry to digitally managed identities by combining identity management, authentication, communication, and physical actuation into a single coordinated process. Rather than granting access based on possession of a physical key, these systems verify who the user is through software and then instruct the lock to open only if that identity is authorised. This approach can be seen in commercially available platforms such as Brivo, SALTO Systems, Kisi, and HID Global, all of which manage access through centrally controlled digital identities rather than traditional keys.

At the foundation is digital identity creation and management. Each user is assigned a unique identity within a system—this may exist in a mobile app, a cloud-based platform, or a local access control server. This identity is linked to credentials such as a PIN, smartphone, RFID card, or biometric profile. For example, in Brivo, administrators create user profiles in a cloud dashboard and assign mobile credentials or cards, while SALTO Systems links identities to mobile keys or RFID credentials within its access control ecosystem. Importantly, the identity is not just a label; it carries permissions, roles, and rules that define what the user is allowed to access, when, and under what conditions.

When a user attempts to unlock a door, the process begins with credential presentation and authentication. A user may enter a PIN on a keypad, tap an RFID card, use a smartphone via Bluetooth or NFC, or scan a fingerprint. In systems such as Kisi, users typically present a smartphone credential via Bluetooth, while HID Global solutions often use smart cards or mobile IDs. The system captures this input and checks it against stored data. In simpler devices, this verification may occur locally within the lock, such as in residential smart locks like Yale or Schlage products. In more advanced systems, the credential is validated through a connected platform—either a controller or a cloud service—which confirms whether the identity is valid and authorised.

Once the identity is verified, the system performs an authorisation check. This is where software plays a critical role. The platform evaluates rules such as whether the user is permitted to access that door, whether the request falls within allowed time periods, and whether any contextual restrictions apply. For instance,

SALTO Systems and Brivo both support time-based access and role-based permissions, allowing organisations to tightly control movement within a building. This separation of authentication (who you are) and authorisation (what you are allowed to do) is a defining feature of modern systems.

After authorisation, the system sends a command to the lock through its communication layer. Depending on the system, this may occur via Bluetooth (direct device-to-lock), Wi-Fi (via cloud infrastructure), or integrated building protocols. For example, Kisi uses cloud-based commands delivered over internet-connected controllers, while residential products such as Schlage Encode locks use Wi-Fi to receive remote instructions. The command activates the actuator within the lock, retracting the bolt or releasing the latch. At this point, the digital decision is physically enforced—the door unlocks.

Simultaneously, the system performs logging and feedback. Every access event is recorded and linked to a specific identity, time, and location. Platforms such as Brivo and HID Global provide detailed audit trails, enabling administrators to review access history, monitor activity, and investigate incidents. Users may also receive immediate feedback, such as app notifications or visual/audible confirmation from the lock. This level of accountability is a major advancement over traditional key systems, where access is anonymous.

In more advanced deployments, biometric and multi-factor authentication further strengthen the process. Systems from HID Global and enterprise platforms such as LenelS2 can combine credentials—for example, requiring both a card and a fingerprint, or a mobile credential and a PIN. This reduces the risk of unauthorised access by ensuring that multiple independent factors are verified.

Cloud-based platforms extend this model by enabling centralised control and real-time updates. Administrators can instantly grant or revoke access, adjust permissions, or monitor activity across multiple locations. For example, Kisi and Brivo allow administrators to manage access remotely through web dashboards, while SALTO Systems provides enterprise-level control across large facilities such as hotels, campuses, and commercial buildings. This means the physical lock becomes part of a dynamic, software-driven system.

From a locksmithing perspective, this represents a fundamental shift. The lock is no longer the primary decision-maker; it is the enforcement mechanism of decisions made by software platforms. Locksmiths must now understand how to configure systems like SALTO Systems or Brivo, manage digital credentials, troubleshoot connectivity, and ensure that access policies are correctly implemented. The role expands from mechanical installation to full system integration and lifecycle management.

In essence, modern systems tie access to identity by ensuring that every physical action at the door is the result of a verified digital decision. Through platforms such as Kisi, HID Global, and SALTO Systems, locks are transformed from passive barriers into active components of intelligent, identity-driven security systems.

**Impact on Locksmithing Practices**

In a converged environment the lock is only one component within a broader system that includes software platforms, digital identities, networks, and cloud services. This means that the effectiveness of a locking system is no longer determined solely by its physical strength, but by how well all these interconnected elements function together. For a locksmith, understanding this shift is essential to remaining competent and relevant in the industry.

One of the most significant influences on locksmithing practice is the transition from object-based to identity-based access control. Instead of managing physical keys, locksmiths are increasingly required to manage digital identities—user accounts, mobile credentials, and biometric profiles. This changes the nature of access control from something that is physically distributed to something that is digitally assigned, modified, and revoked. As a result, locksmiths must develop skills in configuring user permissions, setting access schedules, and managing credential lifecycles. The role expands from issuing keys to administering access systems, where decisions about who can enter a space are made in software and enforced by the lock.

The integration of communication technologies such as Wi-Fi, Bluetooth, and Zigbee also directly impacts locksmithing practices. Locks are now connected devices that rely on secure communication with smartphones, controllers, and cloud platforms. This means locksmiths must understand network fundamentals, including encryption, secure pairing, and signal reliability. Installation is no longer just about fitting hardware correctly; it also involves ensuring that the lock communicates securely and consistently within its environment. Troubleshooting may involve diagnosing connectivity issues, network misconfigurations, or interference—tasks that were previously outside the scope of traditional locksmithing.

Converged security also introduces a layered security approach that reshapes how locksmiths assess and mitigate risk. A lock can be mechanically strong but still vulnerable if its firmware is outdated, its credentials are poorly managed, or its network is insecure. Locksmiths must therefore evaluate security holistically, considering hardware, software, communication channels, and user behaviour as interconnected elements. This influences practices such as product selection, where preference is given to systems with secure update mechanisms, encryption, and robust authentication, as well as ongoing maintenance, where firmware updates and system audits become part of routine service.

Another key influence is the integration of locks with broader security ecosystems, including alarms, CCTV, and building management systems. In these environments, locksmiths are no longer working with standalone devices but with components of a coordinated system. This requires an understanding of how access events interact with other systems—for example, how unlocking a door may trigger disarming an alarm or logging an event in a central monitoring platform. Locksmiths may be involved in system integration, configuration, and testing, ensuring that all components work together as intended.

From a risk perspective, convergence expands the threat landscape and changes how locksmiths approach security. Traditional threats such as lock picking and forced entry still exist, but they are now accompanied by cyber risks such as credential theft, firmware attacks, and network intrusion. This dual threat environment requires locksmiths to adopt a broader security mindset, incorporating cybersecurity principles into their work. Practices such as advising clients on strong credential use, ensuring secure network configuration, and verifying firmware integrity become part of delivering a secure solution.

The convergence also significantly influences the skills and competencies required in the profession. Modern locksmiths must be comfortable working with software interfaces, mobile applications, and cloud platforms, in addition to mechanical hardware. They must understand concepts such as authentication, authorisation, encryption, and audit logging. This represents a shift toward a hybrid skill set that blends traditional trade skills with elements of IT and cybersecurity. Training and professional development must therefore evolve to reflect these new requirements.

Finally, this convergence changes the way locksmiths interact with clients and deliver services. The role increasingly includes education and advisory functions—helping clients understand how to manage digital credentials, maintain secure systems, and respond to emerging risks. Locksmiths are no longer just installers or repairers; they become security consultants who guide clients through the complexities of modern access control systems.

## Encryption Basics for Access Control

Encryption is a foundational concept in modern access control systems, ensuring that sensitive data, such as credentials, commands, and logs, cannot be read or altered by unauthorised parties. In traditional locksmithing, security relied on physical barriers and key control. In digital access systems, encryption performs a similar role by protecting information as it moves between devices and systems. Without encryption, smart locks and access control platforms would be vulnerable to interception, cloning, and manipulation.

Encryption is the process of converting readable data (plaintext) into an unreadable format (ciphertext) using a mathematical algorithm and a key. Only someone with the correct key can decrypt the data back into its original form. In access control, this ensures that when a credential, such as a mobile key, RFID card signal, or PIN verification request, is transmitted, it cannot be understood if intercepted by an attacker.

**Example: Encrypted Mobile Key Unlock**

Imagine a user approaching a smart lock on a commercial office door and unlocking it using a smartphone app.

When the user presses “unlock” in the app, the phone generates a message such as:

**Plaintext (readable data):**

“User ID 4582 requests access to Door A at 08:32”

Before this message is transmitted, the app encrypts it using a cryptographic key shared with the lock (or derived through a secure protocol). The message is transformed into something like:

**Ciphertext (unreadable data):**

“7F3A9C2D8E41B6A0F5C9E2...”

This encrypted data is then sent over Bluetooth or Wi-Fi to the lock.

If an attacker is nearby and intercepts the signal, they will only see the ciphertext, which appears as random data and cannot be understood or altered without the correct decryption key.

When the smart lock receives the message, it uses its stored key to decrypt the data. If the decrypted message is valid and the user is authorised, the lock proceeds to unlock the door.

**Key Point:**

Even though the communication is wireless and potentially exposed, encryption ensures that only the intended lock can interpret the message. Without the correct key, the intercepted data is useless, preventing attackers from copying, modifying, or replaying the credential.

This is the digital equivalent of a secure keyway in traditional locksmithing—only the correct "key" (in this case, a cryptographic key) can operate the lock.

There are two main types of encryption used in access control systems: symmetric and asymmetric encryption. Symmetric encryption uses a single shared key for both encryption and decryption. It is fast and efficient, making it suitable for communication between devices such as a smart lock and a controller. However, the challenge lies in securely sharing and storing that key. Asymmetric encryption, on the other hand, uses a pair of keys: a public key (which can be shared openly) and a private key (which is kept secret). This method is commonly used for secure authentication and key exchange, ensuring that devices can verify each other's identity before communicating.

**Example: Symmetric vs Asymmetric Encryption in a Smart Lock System**

Consider a modern commercial access control system where an employee uses a mobile credential to unlock a secure office door.

**Step 1: Asymmetric Encryption (Secure Authentication and Key Exchange)**

When the employee's smartphone first connects to the smart lock or access system, asymmetric encryption is used to establish trust.

The lock has a public key that is available to authorised devices. The smartphone uses this public key to send a secure, encrypted message to the lock, proving its identity without exposing sensitive data. The lock then uses its private key to decrypt the message and verify that the request is legitimate.

At the same time, the system may securely exchange a temporary session key. This process ensures that:

- The device communicating with the lock is genuine
- The lock itself is authentic (not a spoofed device)

This step is critical for preventing impersonation or "man-in-the-middle" attacks.

**Step 2: Symmetric Encryption (Fast Ongoing Communication)**

Once trust is established, the system switches to symmetric encryption for efficiency.

A shared session key (generated during the asymmetric exchange) is now used by both the smartphone and the lock to encrypt and decrypt messages. For example:

- The phone sends an encrypted "unlock request"
- The lock decrypts it using the same shared key
- The lock responds with an encrypted confirmation

Because symmetric encryption is faster and requires less processing power, it is ideal for real-time operations like unlocking doors.

**Practical Insight**

In this combined approach:

- Asymmetric encryption acts like an identity check and secure handshake at the start
- Symmetric encryption acts like a fast, secure communication channel once both sides trust each other

**Real-World Parallel**

This is similar to a locksmith verifying a client's identity before issuing a key (asymmetric stage), then allowing them to use that key repeatedly for quick access (symmetric stage).

**Key Takeaway**

Most modern smart lock systems use both methods together:

- Asymmetric encryption ensures secure identity verification and key exchange
- Symmetric encryption ensures fast, efficient day-to-day operation

This layered approach balances security and performance, which is essential in real-world access control systems.

Encryption is applied at multiple points within an access control system. One key area is communication security. When a smartphone sends an unlock command to a smart lock via Bluetooth or Wi-Fi, encryption ensures that the command cannot be intercepted or altered in transit. Protocols such as TLS (Transport Layer Security) are commonly used in cloud-connected systems to secure communication between devices and servers. In wireless credentials, such as RFID or NFC cards, encryption prevents cloning by ensuring that each interaction involves secure, often dynamic, data exchange rather than a fixed, easily copied code.

Transport Layer Security (TLS) is a cryptographic protocol used to secure communication between devices over a network. In access control systems, it ensures that data transmitted between components—such as smart locks, mobile applications, and cloud platforms—remains private, authentic, and protected from tampering. In practical terms, TLS prevents unauthorised parties from intercepting or altering sensitive information such as login credentials or unlock commands.

In modern access control environments, TLS is widely used whenever devices communicate over the internet or a local network. For example, when a user unlocks a door remotely using a smartphone app, the command is transmitted through the internet to a cloud server and then relayed to the lock or controller. TLS secures this entire communication pathway, ensuring that the data cannot be intercepted, read, or modified during transmission.

TLS operates through a process known as a handshake, followed by secure data exchange. During the handshake phase, trust is established between the communicating devices. The server presents a digital certificate to prove its identity, and the client verifies this certificate using trusted authorities. Once the identity of the server is confirmed, both parties agree on encryption methods and generate shared session keys. This process ensures that the user is communicating with a legitimate system and that a secure channel is established before any sensitive data is exchanged.

After the handshake is completed, encrypted communication begins. All data transmitted between the devices is encrypted using efficient symmetric encryption, ensuring that messages cannot be read or altered by third parties. In addition, integrity checks are applied to detect any tampering with the data during transmission. This combination of encryption and integrity verification ensures that communication remains secure and reliable.

TLS is critical in smart lock and access control systems because it protects key interactions between system components. These include communication between mobile apps and cloud platforms when sending unlock commands, between cloud platforms and locks or controllers when granting access, and between administrative interfaces and the system when managing users and permissions. Without TLS, attackers could intercept login credentials, capture unlock commands, or manipulate communications to gain unauthorised access. TLS mitigates these risks by encrypting and verifying all data in transit.

A practical example illustrates this process. A property manager uses a mobile app to unlock a door remotely. The app connects to the cloud server using TLS, and the server proves its identity through a digital certificate. The app then sends an encrypted unlock request, which the server verifies against access permissions. If authorised, the server sends an encrypted command to the lock, which executes the action. Throughout this process, the communication remains secure, and even if intercepted, the data cannot be understood or altered.

TLS provides several key security features that underpin its effectiveness. Encryption protects the confidentiality of data, authentication verifies the identity of communicating systems, and integrity mechanisms ensure that data has not been altered. The use of session keys further enhances security by limiting the impact of any potential key compromise.

For locksmiths, TLS is directly relevant because it affects the overall security of modern locking systems. A smart lock is only as secure as the communication that controls it, and weaknesses in TLS implementation can lead to remote unauthorised access. Installers must ensure that systems use current and secure versions of TLS, such as TLS 1.2 or 1.3, and that networks and applications are properly configured. Misconfigured systems or outdated protocols can undermine even the most robust physical hardware.

As a result, locksmiths must develop an understanding of how locks communicate with mobile applications and cloud platforms, verify that secure communication protocols are correctly implemented, and recognise the risks associated with insecure networks or outdated systems. In modern access control, TLS is not just a technical detail—it is a fundamental component of ensuring that digital commands are securely translated into physical access.

Another critical application is credential protection. Modern systems often use encrypted credentials stored on devices like smartphones or secure cards. For example, mobile access systems store digital keys in secure elements or encrypted storage areas, preventing extraction even if the device is compromised. Similarly, biometric data (such as fingerprints) is typically stored in encrypted form and processed locally to protect user privacy and prevent misuse.

**Example: Credential Protection**

A large commercial office deploys a mobile access system that allows employees to use their smartphones instead of physical keys or cards. Each employee is issued a digital credential through a secure app. Rather than storing this credential as plain data, the system places it within the phone's secure element—an isolated, hardware-protected area designed specifically for sensitive information. The credential is encrypted and cannot be accessed directly by other apps or the operating system.

When an employee approaches a door, they use their phone to unlock it via Bluetooth or NFC. The credential is not transmitted in raw form. Instead, the phone generates an encrypted authentication message using keys stored in the secure element. The lock verifies this encrypted message and, if valid, grants access. Even if an attacker intercepts the communication, the credential cannot be extracted or reused because it is protected by encryption and secure hardware.

In the same environment, a high-security area such as a server room requires biometric verification in addition to the mobile credential. Employees enrolled for access use a fingerprint scanner on their device or at the door. The fingerprint itself is never stored as an image; instead, it is converted into a mathematical template and stored in encrypted form within a secure enclave. When the user presents their fingerprint, the system compares it locally within this secure environment. The biometric data is never transmitted across the network, preventing interception or misuse.

This approach ensures that even if a smartphone is lost, stolen, or infected with malware, the attacker cannot extract the stored credential or biometric data. Access can also be remotely revoked by the administrator, rendering the digital key unusable. For locksmiths and security professionals, this example highlights how encryption and secure storage protect credentials at rest and in use, forming a critical layer of defence in modern access control systems.

Encryption also plays a vital role in authentication and trust. When a device or user attempts to gain access, the system must verify that the credential is genuine. Cryptographic techniques such as digital signatures and challenge-response mechanisms are used to confirm authenticity. For instance, a smart lock may issue a random challenge to a mobile device, which must respond with a correctly encrypted answer using its secret key. This ensures that even if communication is intercepted, it cannot be reused in a replay attack.

**Example: Authentication and Trust**

A corporate office deploys a smart access control system where employees use a mobile credential on their smartphone to unlock secure areas. Each authorised device is registered with the system and is assigned a unique cryptographic key stored securely on the phone.

When an employee approaches a restricted door, the smart lock does not simply accept a static credential. Instead, it initiates a challenge-response authentication process. The lock generates a random number (a "challenge") and sends it to the employee's mobile device via Bluetooth.

The mobile app receives this challenge and uses its securely stored private key to encrypt the challenge and generate a response. This response is then sent back to the lock. Because only the legitimate device possesses the correct secret key, only it can produce the correct encrypted answer.

The lock then verifies the response using the corresponding public key or shared secret. If the response is valid, the system confirms that the credential is genuine and has not been cloned or tampered with, and the lock actuates to allow entry.

This process provides strong protection against common attacks. Even if an attacker intercepts the communication between the phone and the lock, the captured data is useless. The challenge is different every time, so the response cannot be reused in a replay attack. Additionally, because the secret key never leaves the secure storage on the device, it cannot be extracted or copied.

In practice, similar authentication methods are used in modern systems such as mobile access platforms and high-security RFID cards, where cryptographic challenge-response ensures that access is granted only to verified, trusted devices. For locksmiths, this illustrates how encryption is not just protecting data—it is actively used to prove identity and enforce trust at the moment a door is unlocked.

Key management is one of the most important, and challenging, aspects of encryption in access control. Keys must be generated securely, stored safely (often in hardware-based secure elements), and rotated or revoked when necessary. Poor key management can undermine even strong encryption. In commercial systems, centralised platforms often manage keys and credentials, allowing administrators to revoke access instantly if a device is lost or a user leaves the organisation.

From a locksmithing perspective, understanding encryption is increasingly important because it directly affects how secure a system is in practice. A lock with strong physical construction can still be vulnerable if its communication is unencrypted or its credentials can be copied. Locksmiths must therefore evaluate not only the hardware, but also whether the system uses modern encryption standards, secure key storage, and protected communication channels.

In practical terms, this influences installation and maintenance practices. Locksmiths must ensure that devices are correctly configured to use secure protocols, that default settings (such as factory keys or passwords) are changed, and that firmware is kept up to date to address cryptographic vulnerabilities. They may also need to explain to clients why certain systems are more secure than others based on their encryption capabilities.

Encryption in access control serves the same purpose as a high-quality mechanical lock: it prevents unauthorised access. However, instead of resisting physical force or manipulation, it protects digital information and communication. For modern locksmiths, understanding encryption basics is essential to ensuring that access control systems are secure not only at the door, but across the entire digital pathway that controls it.

**Impact on Locksmithing Practices**

Encryption is highly relevant to anyone interested in locksmithing because it represents the digital equivalent of traditional key control, and it now underpins how modern locks actually function. In mechanical systems, security depends on physical barriers, keyways, pins, and restricted key profiles. In smart locks and access control systems, encryption performs the same role by ensuring that only authorised credentials can be understood, verified, and acted upon. Without encryption, wireless communication between devices could be intercepted, copied, or altered, effectively bypassing the lock without any physical interaction. This means that a locksmith who understands only the mechanical aspects of a lock is no longer fully equipped to assess or guarantee its security.

The influence on locksmithing practice is significant because the point of security has shifted from purely physical hardware to the entire communication pathway. When a user unlocks a door using a smartphone, card, or PIN, the decision to grant access is based on encrypted data being securely transmitted and verified. If encryption is weak, misconfigured, or absent, an attacker may be able to intercept credentials, replay signals, or manipulate commands. As a result, locksmiths must now evaluate not just the strength of the lock body, but also whether the system uses secure encryption protocols, protected communication channels (such as TLS), and proper authentication mechanisms like challenge-response.

Encryption also directly changes how locksmiths approach installation and system configuration. During installation, it is no longer sufficient to physically fit the lock correctly; the system must also be securely set up at a digital level. This includes ensuring that default passwords or keys are changed, secure pairing processes are followed, and devices are connected to protected networks. For Wi-Fi-enabled locks, this extends to advising on router security and encryption standards (e.g. WPA2/WPA3). For Bluetooth or RFID systems, it involves confirming that secure credential exchange and anti-cloning protections are in place.

Incorrect configuration at this stage can introduce vulnerabilities that negate the physical security of the lock.

Another major influence is in credential management and lifecycle control. In traditional locksmithing, managing security meant controlling physical keys—cutting, issuing, and recovering them. In encrypted systems, this responsibility shifts to managing digital credentials and cryptographic keys. Locksmiths must understand how credentials are stored (e.g. secure elements), how they are transmitted (encrypted), and how they can be revoked or updated. Poor key management, such as reused credentials, insecure storage, or failure to revoke access, can compromise an entire system, even if the encryption itself is strong. This expands the locksmith's role into areas traditionally associated with IT and cybersecurity.

Encryption also affects diagnostics, troubleshooting, and system evaluation. When a smart lock fails or behaves unexpectedly, the issue may not be mechanical, it may relate to failed authentication, expired certificates, or encryption mismatches. Locksmiths must be able to interpret these issues, understand how devices establish trust, and identify whether a fault lies in communication, credentials, or hardware. This requires a working knowledge of how encryption supports authentication, secure sessions, and data integrity, rather than relying solely on physical inspection.

From a risk perspective, encryption highlights the convergence of physical and digital threats. A door can now be compromised without being picked or forced—simply by exploiting weaknesses in encrypted communication or credential handling. This changes how risk assessments are performed. Locksmiths must consider threats such as interception, replay attacks, credential cloning, and cloud vulnerabilities alongside traditional risks like forced entry. Consequently, system selection and recommendation must account for encryption standards, update practices, and vendor security maturity, not just mechanical durability.

Finally, encryption reshapes the professional identity of locksmithing itself. The role is evolving from a trade centred on mechanical systems to a hybrid discipline combining physical security, electronics, and cybersecurity. Locksmiths are increasingly expected to advise clients on secure system design, explain the importance of encrypted communication, and ensure that digital controls are as robust as physical ones. In this context, encryption is not an abstract technical concept, it is a core component of how modern locks function and how security is maintained over time.

In essence, encryption influences locksmithing in the same way that restricted key systems once did: it defines who can access a lock and under what conditions. However, instead of controlling access through physical key duplication, it controls access through protected digital information. For modern locksmiths, understanding encryption is essential to ensuring that a lock is secure not only at the door, but across the entire system that governs access.

## Access Logs, Data Privacy, and Compliance

Modern access control systems inherently produce detailed records of activity known as access logs. These logs document events such as door unlock attempts, successful entries, failed authentication attempts, credential use, and administrative actions within the system, and each entry typically records information such as the user identity, time, location, and event type. This creates a verifiable audit trail that is absent in traditional mechanical locks, which provide no record of who entered or when. Access logs thereby enable organisations to maintain an auditable chronology of security-related activity and to support governance over physical access networks [267-269].

Access logs serve several important operational and security purposes. Administrators can monitor building access in near-real time, detect suspicious activity, and investigate incidents after they occur. Repeated failed access attempts may indicate attempts to guess a PIN or to misuse an unauthorised credential, while access outside normal hours may signal potential privilege misuse. Through regular log review, organisations can identify patterns, enforce policies, and respond swiftly to potential security threats, with logs also supporting internal audits and incident investigations in many commercial environments [267, 270, 271]. These functions rely on the logs' ability to capture a rich set of attributes per event, including user identity, timestamp, location, and event type, which collectively enable anomaly detection and forensic analysis [268, 269, 271].

While access logs provide security insights, they inherently involve the collection of personal or identifiable data such as employee names, identification numbers, and entry times. This necessitates strong data privacy protections in the design and operation of access control systems. Access logs should be viewable and analysable only by authorised personnel, stored securely with encryption and strict access controls, and governed by retention policies to avoid keeping data longer than necessary. The privacy implications require careful policy development and technical controls to safeguard individuals' information while enabling legitimate security and operational use of the logs [267, 269, 270]

Compliance with privacy and security regulations is a critical aspect of modern access control management. Many jurisdictions regulate how personal data must be collected, stored, and used. In Australia for example, organisations must comply with the Privacy Act 1988 and the Australian Privacy Principles (APPs) when handling personal information, requiring safeguards to protect personal data from misuse, loss, or unauthorised access.

The regulatory framework that governs the protection of personal information in access control systems varies between countries, but most developed jurisdictions have comparable laws requiring organisations to safeguard personal data. These laws influence how access logs, credentials, and identity information generated by smart locks and access control systems must be stored, processed, and protected.

**United States**

In the United States, there is no single nationwide equivalent to Australia's *Privacy Act 1988*. Instead, privacy regulation is based on a combination of federal sector-specific laws and state legislation. At the federal level, laws such as the Health Insurance Portability and Accountability Act (HIPAA) regulate the protection of health-related data, while the Gramm-Leach-Bliley Act (GLBA) governs personal financial information. However, the closest broad privacy protections for general personal data often come from state laws.

One of the most influential examples is the California Consumer Privacy Act (CCPA) and its expansion through the California Privacy Rights Act (CPRA). These laws give individuals rights over their personal data, including the right to know what information is collected, the right to request deletion, and the right to restrict certain uses of their data. For access control systems, this means organisations must carefully manage access logs and user identity data to ensure they are collected lawfully, stored securely, and accessible only to authorised personnel.

**Canada**

Canada regulates personal information through the Personal Information Protection and Electronic Documents Act (PIPEDA). This federal law governs how private-sector organisations collect, use, and disclose personal information during commercial activities. PIPEDA requires organisations to implement safeguards appropriate to the sensitivity of the information they hold and to limit data collection to legitimate purposes.

For access control systems, this means that logs containing identifiable information—such as employee entry times or credential use—must be stored securely and used only for legitimate operational or security purposes. Individuals also have the right to request access to personal information held about them and to challenge its accuracy.

Several Canadian provinces also have their own privacy legislation that operates alongside or in place of PIPEDA, such as Alberta's Personal Information Protection Act (PIPA), British Columbia's PIPA, and Quebec's privacy legislation under Law 25, which introduces stricter data protection requirements.

**United Kingdom**

In the United Kingdom, data protection is primarily governed by the UK General Data Protection Regulation (UK GDPR) and the Data Protection Act 2018. These laws are closely aligned with the European Union's GDPR framework and impose strict obligations on organisations that collect or process personal data.

Under the UK GDPR, organisations must ensure that personal data is processed lawfully, transparently, and for specific purposes. They must also implement appropriate technical and organisational measures to protect data from unauthorised access or loss. In the context of access control systems, this means that access logs, biometric data, and digital credentials must be securely stored, encrypted where appropriate, and accessible only to authorised personnel. Individuals also have rights such as access to their data, correction of inaccuracies, and, in some cases, the right to request deletion.

**Relevance to Access Control Systems**

Across all of these jurisdictions with examples provided, Australia, the United States, Canada, and the United Kingdom, the underlying principle is the same: organisations that collect personal data through access control systems must protect that data from misuse, loss, or unauthorised access. Access logs, user identities, and biometric records are considered personal information and must therefore be handled responsibly.

For locksmiths and security professionals installing modern smart locks and access control systems, understanding these regulatory frameworks is important. Systems must be configured in ways that support secure storage, restricted access to logs, encryption of sensitive information, and appropriate retention policies. Compliance with privacy regulations is therefore not just a legal obligation but also a critical component of responsible security system design and management.

Consequently, access logs must be stored securely, accessed only by authorised staff, and used strictly for legitimate security or operational purposes. Compliance considerations also shape how access control systems are implemented and maintained, including secure data storage, encrypted communications, appropriate user access controls, and clearly defined policies about log viewing, retention duration, and incident reporting for personal data incidents. Failure to manage this information responsibly can lead to legal consequences, reputational damage, and loss of trust [267, 268, 270, 272].

For locksmiths and security professionals, understanding access logs, privacy obligations, and compliance requirements is increasingly important. Modern locksmiths may install and configure systems that collect and manage access data, and they may advise clients on responsible use. This includes explaining the importance of secure configuration, proper credential management, and adherence to privacy regulations to ensure that clients gain the security benefits of access logs without compromising individuals' privacy or running afoul of regulatory requirements [267, 268, 270, 273].

### Methods and techniques for log analysis and interpretation

To derive security insights from access logs, organisations employ log analysis and anomaly detection techniques. Studies in forensic and security contexts demonstrate the value of graph clustering, anomaly detection, and visualization to identify unusual patterns in access control logs, such as anomalies in authentication events, illegitimate access attempts, and brute-force behaviours. Approaches include preprocessing of log data, clustering to reveal structure, and visualizations to aid investigators, with empirical results showing improved detection of security incidents in log data from operating systems and networked environments [268, 271, 272]. Beyond traditional statistical methods, digital forensics analyses contribute to identifying patterns linked to adversarial behaviour by examining security logs for failed login attempts and correlated threat indicators within cloud and enterprise environments [267, 270].

### Security, privacy, and governance integration

Effective access control governance integrates technical controls with policy frameworks. This requires secure data storage and encryption, controlled access to logs, retention management, and formal incident reporting procedures. In practice, these governance aspects are informed by regulatory guidance and frameworks, including privacy laws, industry best practices, and organisational risk management approaches. The literature indicates that audit trails founded on secure, tamper-resistant logging mechanisms, potentially augmented with distributed or blockchain-backed audit trails, enhance integrity and trust in access records, while also enabling auditability and accountability across complex environments [267, 274].

As access control systems evolve, research explores decentralised and verifiable audit trails, secure logging architectures, and user-centric authentication paradigms. Blockchain-based audit trail mechanisms are being investigated for their potential to provide tamper-evident and transparent logging capabilities, addressing attacks against log integrity and availability. While promising, these technologies also introduce complexity and deployment considerations that require careful evaluation in real-world settings [267, 275]. The ongoing evolution of authentication methods, including passwordless and public-key-based approaches, influences how logs capture credential usage and event provenance, with usability considerations also affecting operator and user behaviour and, consequently, the interpretability of log data [267, 269, 276].

**Impact on Locksmithing Practices**

Access logs are highly relevant to someone interested in locksmithing because they represent a fundamental shift in how physical security is monitored and governed. Traditional mechanical locks provide no record of who entered a space or when, meaning that security investigations relied on physical evidence or witness accounts. Modern access control systems, however, generate detailed, time-stamped audit trails that link entry events to specific digital identities. This transforms the lock from a passive barrier into an active data-generating component of a broader governance framework. For a locksmith, this means that installing a lock is no longer just about securing a door—it is about enabling accountable, traceable, and reviewable access within an organisation's security strategy.

The influence on locksmithing practices is significant because access logs introduce privacy, data protection, and compliance responsibilities. Logs contain personal information such as names, credential identifiers, timestamps, and location data, which are regulated in jurisdictions such as Australia, the United States, Canada, and the United Kingdom. Locksmiths who install and configure access control systems must therefore understand that they are working with systems that process regulated personal data. This affects how systems are configured, ensuring encrypted storage, restricted administrative access, appropriate retention periods, and secure communication channels. It also means locksmiths may need to advise clients on responsible log management, including who can access logs, how long they should be retained, and how to respond to data breaches or privacy requests.

Finally, access logs expand the professional scope of locksmithing into areas of governance, analytics, and risk management. Modern locksmiths may be involved in enabling log review functions, integrating systems with monitoring platforms, or supporting investigations through interpretation of access data. As technologies evolve, such as decentralised logging architectures or advanced anomaly detection, the locksmith's role increasingly intersects with IT security and compliance management. Understanding access logs is therefore not optional in contemporary practice; it is central to delivering secure, lawful, and professionally responsible access control solutions in a digitally regulated environment.

## Penetration Testing for Smart Locks

Penetration testing for smart locks is the process of deliberately evaluating a smart lock system to identify security weaknesses before they can be exploited by attackers. It involves simulating real-world attack methods against the lock, its communication channels, software, and supporting infrastructure. The purpose is not to damage the system but to discover vulnerabilities that could allow unauthorised access, credential theft, or system manipulation. In modern access control environments, penetration testing has become an important method for assessing the security of devices that combine physical hardware with wireless communication and software platforms.

Smart locks differ from traditional mechanical locks because they operate within a broader digital ecosystem that includes mobile applications, wireless networks, firmware, and cloud services. As a result, penetration testing examines several layers of the system rather than just the lock hardware. Testers analyse wireless communication protocols such as Bluetooth, Wi-Fi, or Zigbee to determine whether signals can be intercepted, replayed, or manipulated. They may attempt to capture encrypted traffic, perform replay attacks, or evaluate whether authentication processes prevent unauthorised devices from communicating with the lock.

Another important focus of penetration testing is credential security. Testers evaluate whether PIN codes, RFID credentials, mobile keys, or biometric authentication mechanisms can be bypassed or cloned. For example, they may assess whether RFID cards use secure encryption or whether a static credential could be copied. In mobile-based systems, penetration testers examine how digital keys are stored on devices and whether they can be extracted or duplicated if the smartphone is compromised.

Firmware and software security are also key components of penetration testing. Smart locks rely on embedded firmware to control their behaviour, and vulnerabilities in this software can allow attackers to manipulate lock functions or install malicious updates. Testers may analyse firmware update processes, check whether updates are cryptographically signed, and determine whether unauthorised firmware could be installed. They may also evaluate cloud platforms and application interfaces to ensure that administrative functions cannot be abused to unlock doors remotely.

Physical security testing remains part of the process as well. Even though smart locks are digitally controlled, the underlying hardware can still be attacked. Penetration testers may examine the lock casing, wiring, and actuator mechanisms to determine whether they can be bypassed or manipulated. For example, testers might attempt to access exposed circuitry, inject signals into the control electronics, or identify ways to trigger the unlocking mechanism directly.

Penetration testing also assesses system resilience against network-based attacks. This includes evaluating authentication systems, encryption protocols, and communication security such as TLS. Testers verify that devices properly authenticate each other, that credentials cannot be reused in replay attacks, and that encrypted communication cannot be easily intercepted or modified. Weak encryption, poor key management, or outdated protocols are common findings in poorly secured systems.

For locksmiths and security professionals, penetration testing highlights the importance of evaluating both physical and digital security. A lock may be mechanically robust yet vulnerable through its wireless communication or software configuration. Understanding penetration testing results helps locksmiths recommend more secure products, identify misconfigurations, and ensure that installations follow best security practices.

Following is a structured penetration testing procedure for smart lock systems designed to be systematic, controlled, and non-destructive, ensuring that vulnerabilities are identified without damaging equipment or disrupting legitimate operations.

**Procedure: Penetration Testing for Smart Locks**

**1. Authorisation and Scope Definition**

Before any penetration testing begins, formal authorisation must be obtained from the system owner or organisation responsible for the access control system. Testing without permission may violate legal or organisational policies. The scope of testing should clearly define:

- The specific smart lock models or systems to be tested
- Associated infrastructure (mobile apps, gateways, cloud platforms, controllers)
- Communication technologies involved (Bluetooth, Wi-Fi, Zigbee, RFID, etc.)
- Physical components and locations included in the assessment

Testing objectives should also be documented. These may include evaluating credential security, wireless communication integrity, firmware protection, or resistance to physical tampering.

**2. System Reconnaissance and Documentation**

The next step is to gather information about the smart lock system and its environment. This reconnaissance phase helps testers understand the architecture and identify potential entry points.

Tasks include:

- Identifying lock models, firmware versions, and hardware components
- Mapping communication pathways (lock → gateway → cloud server → mobile app)
- Identifying authentication mechanisms (PIN, RFID, mobile credentials, biometrics)
- Reviewing manufacturer documentation and configuration settings
- Identifying connected devices such as controllers, hubs, and management software

A network diagram or system architecture map should be created to visualise the components involved.

**3. Wireless Communication Testing**

Because smart locks often rely on wireless communication, testing should assess whether signals can be intercepted, analysed, or manipulated.

Key activities include:

- Monitoring wireless traffic (Bluetooth, Wi-Fi, Zigbee, NFC, or RFID)
- Attempting to capture authentication exchanges between the lock and devices
- Evaluating whether encrypted communication is properly implemented
- Testing for replay attack susceptibility by attempting to retransmit captured signals
- Assessing device pairing procedures to determine whether unauthorised devices can connect

The objective is to confirm that communication channels use secure encryption and authentication mechanisms.

**4. Credential Security Assessment**

Credential systems are a critical component of smart lock security. Testing should evaluate whether credentials can be copied, guessed, or bypassed.

Areas to test include:

- PIN code strength and rate-limiting protections
- RFID or NFC credential cloning resistance
- Mobile credential storage and protection on smartphones
- Biometric authentication protection and storage methods

Testers may attempt credential enumeration, brute-force attempts within permitted limits, or evaluation of how credentials are stored and transmitted.

**5. Firmware and Software Security Evaluation**

Smart locks operate using embedded firmware and software platforms. Vulnerabilities in these systems may allow attackers to manipulate lock behaviour.

Testing activities include:

- Verifying whether firmware updates require cryptographic signatures
- Checking whether firmware can be extracted or modified
- Evaluating whether unauthorised firmware updates can be installed
- Testing mobile apps and management platforms for authentication weaknesses

- Assessing API security and remote access functions

Special attention should be given to update mechanisms such as over-the-air (OTA) updates, as compromised update systems could affect multiple devices.

**6. Physical Hardware Assessment**

Despite digital controls, physical security remains an important element. Testing should examine whether the hardware can be manipulated to bypass the lock mechanism.

Physical testing may include:

- Inspecting casing for exposed screws or accessible electronics
- Evaluating tamper protection features
- Attempting to access internal wiring or actuator components
- Testing whether electronic signals can be injected into control circuits
- Checking emergency power ports or external interfaces for misuse

Testing should remain non-destructive and avoid permanent damage to the device.

**7. Network and Authentication Security Testing**

Penetration testing must also examine the broader network infrastructure supporting the smart lock system.

Tasks include:

- Evaluating authentication mechanisms between devices
- Confirming that encrypted protocols such as TLS are properly implemented
- Checking certificate validation processes
- Testing for man-in-the-middle attack vulnerabilities
- Assessing network segmentation and access control policies

These steps ensure that communication between locks, controllers, and cloud systems cannot be easily compromised.

**8. Log and Monitoring Evaluation**

Access control systems generate logs that record security events. Testing should confirm that logging functions correctly and cannot be manipulated.

Activities include:

- Verifying that authentication attempts are logged
- Confirming that administrative actions are recorded
- Checking whether logs can be modified or deleted without authorisation
- Evaluating alerting mechanisms for suspicious activity

Effective logging supports incident detection and forensic investigation.

**9. Vulnerability Analysis and Risk Assessment**

After testing is completed, findings must be analysed to determine the severity of identified vulnerabilities.

Each vulnerability should be evaluated based on:

- Likelihood of exploitation
- Potential impact (e.g. unauthorised entry, credential theft)
- Ease of attack execution
- Existing security controls

Vulnerabilities should then be categorised by risk level (e.g. low, medium, high, critical).

**10. Reporting and Remediation**

The final stage of penetration testing is documentation and remediation planning.

The final report should include:

- Summary of testing objectives and scope
- Methods used during the assessment
- Detailed description of vulnerabilities discovered
- Evidence supporting findings
- Risk assessment and severity ratings
- Recommended corrective actions

Recommendations may include firmware updates, improved encryption, stronger authentication policies, or hardware design improvements.

### Impact on Locksmithing Practices

Penetration testing of smart locks is highly relevant to locksmithing because modern security systems increasingly combine mechanical hardware with digital technologies. Traditional locksmithing focused primarily on mechanical vulnerabilities such as picking, bumping, drilling, or bypassing physical locking mechanisms. Smart locks, however, introduce additional layers of security including wireless communication, mobile applications, embedded firmware, and cloud-based management systems. Penetration testing provides a structured method for evaluating these digital components to identify weaknesses that could allow unauthorised access. For someone interested in locksmithing, understanding how penetration testing works broadens their knowledge beyond mechanical systems and helps them assess the overall security of modern access control devices.

Penetration testing also influences locksmithing practices by highlighting the importance of evaluating both physical and electronic attack surfaces. A smart lock may be mechanically strong but still vulnerable through weaknesses in wireless protocols, poor credential protection, or insecure firmware updates. For example, penetration testing may reveal that a Bluetooth communication channel can be intercepted or that RFID credentials can be cloned. These insights allow locksmiths to recommend stronger products, configure systems securely, and ensure that authentication mechanisms, encryption settings, and firmware updates are properly implemented. As locksmiths increasingly install smart locks and integrated access control systems, knowledge of these testing principles enables them to make informed decisions about system selection and deployment.

Finally, penetration testing influences the professional role of locksmiths by reinforcing the importance of security auditing, risk assessment, and responsible system management. Locksmiths may not always conduct full penetration tests themselves, but they benefit from understanding the results and implications of such assessments. This knowledge enables them to advise clients on system vulnerabilities, ensure that installations follow best security practices, and respond appropriately if security issues are discovered. In modern security environments, locksmiths are no longer only installers of hardware but also advisors on integrated security systems. Familiarity with penetration testing therefore supports the evolution of locksmithing from a purely mechanical trade into a broader security profession that incorporates both physical and digital protection.

## Working With IT Security Professionals

Working with IT security professionals has become an important aspect of modern locksmithing because many contemporary access control systems operate as part of a broader digital security infrastructure. Traditional locksmiths historically focused on the installation, repair, and maintenance of mechanical locks and physical security devices. However, smart locks, electronic access control systems, and networked security platforms now integrate with computer networks, mobile applications, cloud services, and identity management systems. As a result, locksmiths increasingly collaborate with IT security professionals who manage the cybersecurity aspects of these systems. This collaboration ensures that both the **physical** security of doors and entry points and the digital security of the systems controlling them are properly protected.

IT security professionals bring specialised expertise in areas such as network security, encryption, authentication protocols, and vulnerability management. When installing or maintaining smart locks or electronic access control systems, locksmiths may work alongside IT teams to configure secure communication between devices, integrate locks with building management systems, and ensure that access credentials are properly managed. For example, IT specialists may configure network segmentation, firewalls, and secure communication protocols such as TLS, while locksmiths ensure that the hardware components, locking mechanisms, and door installations function correctly. This cooperative approach reduces the risk that a system is mechanically secure but digitally vulnerable.

Effective collaboration also improves the management of user credentials and access permissions. Modern access control systems often rely on digital identities, such as employee credentials stored in central databases, mobile access keys, or RFID badges linked to identity management systems. IT security professionals typically manage these databases and authentication systems, while locksmiths ensure that physical access devices such as card readers, keypads, and smart locks operate reliably. By working together, both professions can ensure that access privileges are properly assigned, monitored, and revoked when required, reducing the risk of unauthorised access or credential misuse.

Another important aspect of collaboration involves incident response and system monitoring. If suspicious activity occurs within an access control system, IT security teams may analyse system logs, network activity, and authentication records to detect potential security incidents. Locksmiths may then assist in evaluating whether the issue relates to physical hardware, credential misuse, or system configuration. For example, if repeated failed entry attempts are detected, IT professionals may investigate digital logs while locksmiths inspect the physical lock or access point for signs of tampering. This coordinated approach enables organisations to respond quickly and effectively to security threats.

Working with IT security professionals also influences how locksmiths approach system design and installation. Secure installations now require consideration of factors such as encrypted communication, secure firmware updates, credential lifecycle management, and protection against network-based attacks. Locksmiths who understand these issues can ensure that smart locks are installed in ways that support cybersecurity best practices, such as limiting unnecessary network exposure, using secure authentication methods, and ensuring that firmware updates are properly managed.

**Impact on Locksmithing Practices**

Understanding how to work with IT security professionals is relevant to someone interested in locksmithing because modern access control systems increasingly combine physical locking hardware with digital technologies such as wireless communication, mobile applications, and cloud-based identity management systems. While traditional locksmithing focused mainly on mechanical security, many modern locks are now connected devices that rely on software, networks, and encrypted communications to function correctly. As a result, locksmiths must often collaborate with IT specialists who manage cybersecurity, networks, and authentication systems. This cooperation ensures that security systems are protected not only from physical attacks but also from digital threats such as credential theft, network intrusion, or system manipulation.

This collaboration influences locksmithing practices by expanding the role of locksmiths from purely mechanical specialists to professionals involved in integrated security system installation and management. Locksmiths must consider factors such as secure device configuration, credential management, firmware updates, and encrypted communication when installing smart locks and electronic access control systems. Working alongside IT professionals also helps ensure that locks are correctly integrated into broader security infrastructures such as building management systems and identity databases. As security technology continues to evolve, locksmiths who understand how physical security interacts with cybersecurity are better equipped to recommend secure products, design robust access control systems, and support organisations in maintaining effective protection against both physical and digital threats.

## Chapter 10 Review Questions

### Question 1

A corporate office replaces mechanical keys with a mobile credential system where employees unlock doors using a smartphone app linked to their company account.

What security principle does this system primarily rely on?

A. Identity-based access control
B. Mechanical key hierarchy
C. Physical lock complexity
D. Door frame reinforcement

### Question 2

A locksmith installs network-connected smart locks in a commercial building. Later, it is discovered that the locks are running outdated firmware containing a known vulnerability.

What cybersecurity practice would have reduced this risk?

A. Increasing door thickness
B. Regular firmware updates and patch management
C. Installing stronger strike plates
D. Using larger keyways

### Question 3

A facility manager notices that an attacker attempted to access several restricted doors using stolen login credentials obtained through a phishing email.

What aspect of modern security systems made this attack possible?

A. Physical door construction
B. Mechanical lock tolerances
C. Digital identity and credential management systems
D. Door hinge design

**Question 4**

A data centre uses a cloud-managed access control platform where administrators can manage permissions remotely from any location.

What is the main cybersecurity concern associated with this type of system?

A. The cloud system may allow remote management but also introduces potential network attack points
B. The doors may become heavier
C. Mechanical key duplication becomes impossible
D. The system eliminates the need for user authentication

**Question 5**

A building installs biometric fingerprint readers on sensitive doors. During a security review, administrators learn that biometric templates are stored on a central server.

What security risk must be considered?

A. Physical key cutting errors
B. Door alignment issues
C. Exposure of sensitive biometric data if the server is compromised
D. Reduced door durability

**Question 6**

A locksmith installs a smart lock that communicates with a mobile phone via Bluetooth. Security researchers demonstrate that intercepted communication signals can be replayed to unlock the door.

What type of vulnerability does this demonstrate?

A. Mechanical lock wear
B. Communication replay vulnerability in wireless systems
C. Door frame failure
D. Key cutting inaccuracies

### Question 7

A hospital installs a network-connected access system. During a power outage, doors remain secure but the network controller temporarily loses communication with the cloud server.

What security principle helps maintain protection in this situation?

A. Redundant mechanical bypass systems and local authentication control
B. Removal of all electronic components
C. Eliminating credential management
D. Using only physical keys

### Question 8

A security audit finds that several employees still have active electronic access permissions even though they left the company months earlier.

What failure does this represent?

A. Weak door hinges
B. Poor credential lifecycle management
C. Faulty strike plates
D. Incorrect key cutting

### Question 9

A locksmith recommends that a company enable multi-factor authentication (MFA) for administrators who manage their electronic access control system.

What risk does this primarily reduce?

A. Door frame damage
B. Mechanical key wear
C. Unauthorized administrative access to security systems
D. Lock cylinder misalignment

### Question 10

During a security consultation, a locksmith explains that physical access control systems should be integrated with cybersecurity policies and IT management.

Why is this integration important?

A. Physical security devices now rely on networked software systems
B. Mechanical locks require internet access
C. Doors cannot function without Wi-Fi
D. Key duplication machines require cloud accounts

# Part 6

# Advanced Access Control Systems

# Chapter 11

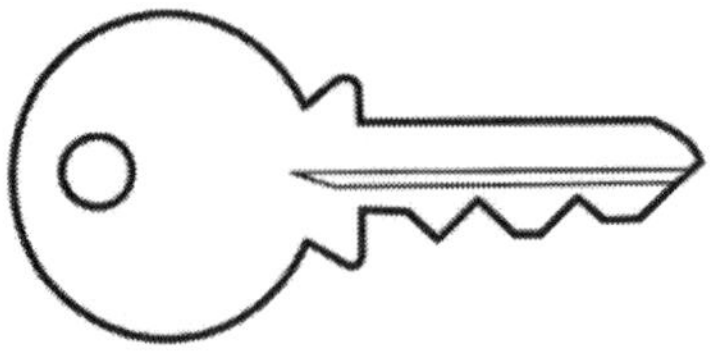

# Biometric Security Systems

While conventional locks have relied on keys, codes, or cards to control access, increasing attention is now being given to systems that identify individuals based on their biological characteristics. Biometric security systems represent a significant development in access control because they link identity verification directly to the person rather than to an external credential that can be lost, stolen, or duplicated.

This chapter examines the role of biometric technologies within contemporary security and locksmithing practices. It introduces the fundamental principles behind biometric authentication and explores three of the most widely used biometric methods in modern access control systems: fingerprint recognition, facial recognition, and iris scanning. Each of these technologies uses different biological features and technical processes to verify identity, and each has distinct advantages, limitations, and practical applications in residential, commercial, and high-security environments.

In addition to explaining how these systems operate, the chapter also explores the hardware components, operational processes, and commercially available devices used in biometric locking systems. Topics such as system accuracy, reliability, privacy protection, spoofing and bypass risks, and integration with traditional mechanical locks are examined to illustrate how biometric security technologies function in real-world installations. These issues are particularly important for locksmiths and security professionals who must install, configure, and maintain these systems.

By understanding how biometric authentication works and how it interacts with conventional locking hardware, security practitioners can better evaluate system performance, identify potential vulnerabilities, and implement effective access control solutions. This chapter therefore provides both a technical overview of biometric technologies and a practical perspective on their growing influence on modern locksmithing and security system design.

## Fingerprint, Facial Recognition, Iris Scanning

Biometric security systems use unique physical or behavioural characteristics of individuals to verify identity before granting access. Unlike traditional keys or passwords, biometric traits are inherently linked to a person and are therefore more difficult to lose, duplicate, or share. In access control systems, biometrics provide a method of authentication based on “who you are” rather than “what you have” or “what you know.” Modern biometric technologies are increasingly integrated into smart locks, mobile devices, and enterprise access control systems, where they may operate alone or as part of multi-factor authentication. Among the most widely used biometric methods in security systems are fingerprint recognition, facial recognition, and iris scanning, each of which uses different biological characteristics and technical processes to verify identity.

### Fingerprint Recognition

Fingerprint recognition is one of the most established biometric technologies and is widely used in smartphones, laptops, and access control devices. Fingerprint systems rely on the unique patterns formed by ridges and valleys on the surface of a person's fingertips. These patterns contain distinct features known as *minutiae points*, such as ridge endings and bifurcations, which differ from person to person, even between identical twins. When a fingerprint is enrolled into a system, a sensor scans the finger and converts the image into a mathematical template representing the positions of these key features. During authentication, the user places their finger on the sensor, and the system compares the scanned fingerprint with stored templates to determine whether a match exists.

Several sensor technologies may be used, including optical sensors that capture a visual image of the fingerprint, capacitive sensors that detect electrical differences between ridges and valleys, and ultrasonic sensors that create detailed three-dimensional images of the fingerprint structure. Fingerprint recognition is valued for its speed, reliability, and relatively low cost, making it common in residential smart locks and commercial access systems. However, systems must be carefully designed to prevent spoofing attempts using artificial fingerprints, which is why modern devices often include *liveness detection* techniques that verify characteristics such as skin conductivity, temperature, or blood flow.

*Figure 76: Fingerprint Scanning. panumas nikhomkhai, CC0, via Pexels.*

Fingerprint recognition locks use biometric identification technology to allow or deny access based on the unique patterns found on a person's fingertip. Instead of using a physical key or digital credential such as a card or PIN, the lock identifies the user by scanning and analysing their fingerprint. Each human fingerprint contains distinctive patterns of ridges and valleys that form identifiable characteristics known as *minutiae points*, including ridge endings, bifurcations, and ridge intersections. These patterns are unique to each individual and remain largely unchanged throughout a person's life. Because of this uniqueness, fingerprint recognition systems can reliably distinguish between authorised and unauthorised users when granting access.

The process begins with enrolment, which is the stage where a user's fingerprint is first registered with the lock system. During enrolment, the user places their finger on the fingerprint sensor, which captures a detailed image of the fingerprint pattern. The lock's internal processor then analyses this image and extracts the key features of the fingerprint, such as the position and orientation of minutiae points. Rather than storing the full fingerprint image, the system converts these features into a mathematical template, which is a compact digital representation of the fingerprint's unique characteristics. This template is stored securely in the lock's memory and associated with the user's identity or access permissions.

When the user later attempts to unlock the door, the system performs the authentication stage. The user places their finger on the sensor again, and the lock captures a new fingerprint scan. The processor converts this new scan into another template and compares it with the stored templates in the system's database. If the features match within an acceptable tolerance level, the system determines that the fingerprint belongs to an authorised user. The lock's control electronics then activate the actuator mechanism that retracts the latch or bolt, allowing the door to open. If the fingerprint does not match any stored template, access is denied and the lock remains secured.

Several types of fingerprint sensor technologies are used in fingerprint locks. Optical sensors capture a photographic image of the fingerprint using light and a small camera. Capacitive sensors, which are common in modern devices, detect the electrical differences between the ridges and valleys of the fingerprint using a grid of tiny capacitors. Ultrasonic sensors represent a newer technology that uses high-frequency sound waves to create a detailed three-dimensional image of the fingerprint surface. Each of these technologies converts the fingerprint pattern into digital data that can be analysed and compared by the system's software.

To operate effectively, fingerprint recognition locks require several key hardware components. The most important component is the fingerprint sensor module, which captures the fingerprint image. This module is connected to a microcontroller or embedded processor that performs the image processing, feature extraction, and template comparison operations. The lock also contains non-volatile memory for storing authorised fingerprint templates and user information. A locking actuator—such as a motor, solenoid, or electronic clutch—physically controls the locking mechanism once access is granted. Additional components typically include a power supply, which may consist of batteries or a wired power source, and an interface module such as a keypad, touchscreen, or wireless communication unit for configuration and administration.

Many modern fingerprint locks also include supporting electronic components that enhance security and usability. These may include liveness detection sensors that verify that the fingerprint comes from a living finger rather than an artificial copy, communication modules such as Bluetooth or Wi-Fi for remote management, and status indicators such as LEDs or displays that inform the user whether access has been granted or denied. Some systems also integrate fingerprint recognition with other authentication methods, such as PIN codes or smartphone credentials, to provide multi-factor authentication for higher levels of security.

Two widely available fingerprint recognition locking systems used in residential and commercial environments are the Ultraloq U-Bolt Pro WiFi Smart Lock and the Samsung SHP-DP609 Smart Door Lock. These systems illustrate how biometric authentication can be integrated into modern electronic access control hardware.

### Ultraloq U-Bolt Pro WiFi Smart Lock

The Ultraloq U-Bolt Pro WiFi Smart Lock is a commercially available biometric smart deadbolt designed primarily for residential properties but also suitable for small offices or rental properties. It integrates a fingerprint recognition sensor into the exterior lock housing, allowing authorised users to unlock the door by placing their finger on the scanner. The fingerprint sensor uses a capacitive biometric reader that scans the unique ridge patterns of the finger and compares them with stored templates inside the lock's internal processor. If the fingerprint matches an enrolled user, the lock activates a motorised deadbolt that retracts the locking bolt.

This system is designed as a multi-factor access control device. In addition to fingerprint access, the lock can be opened using a keypad PIN code, a smartphone application, Bluetooth or Wi-Fi connectivity, or a traditional mechanical key override. The lock supports mobile credential management through a web portal and smartphone app, allowing administrators to create temporary digital keys, monitor access logs, and control access remotely. The system's combination of biometric authentication and wireless connectivity provides convenience while maintaining a high level of security.

From a hardware perspective, the lock includes several integrated components. These typically include a capacitive fingerprint sensor, an embedded processor that runs the biometric matching algorithm, internal memory for storing fingerprint templates, a motorised deadbolt mechanism, and wireless communication modules for Bluetooth and Wi-Fi connectivity. Because it is designed for retrofit installation on standard doors, the unit replaces a traditional deadbolt and contains both exterior and interior assemblies powered by batteries.

### Samsung SHP-DP609 Smart Door Lock

The Samsung SHP-DP609 Smart Door Lock is a premium biometric access control system designed for modern homes, apartments, and commercial premises. This lock combines fingerprint recognition with several other authentication methods, including PIN codes, smartphone access, and mechanical keys. The fingerprint reader is positioned on the lock's exterior panel and allows rapid biometric identification. When the sensor scans a finger, the system compares the captured biometric data with stored templates and unlocks the door if a match is confirmed.

This lock is notable for its push-pull handle design, which allows the door to be opened by simply pushing or pulling the handle after authentication. It supports multiple unlocking methods such as fingerprint scanning, password entry, NFC or smartphone credentials, and emergency mechanical keys. Many installations also integrate the lock with mobile applications that allow remote management, temporary access codes, and monitoring of entry events. The system can store multiple fingerprint profiles and maintain logs of access activity for security management.

The hardware architecture of this lock includes a high-precision fingerprint sensor, a digital keypad interface, a microprocessor for biometric matching, secure memory storage for fingerprint templates, and a motorised mortise lock mechanism. Additional components include wireless communication modules for smartphone integration, a power supply (typically batteries), and electronic sensors that detect door status or tampering attempts. These components allow the lock to function as both a biometric identification system and a fully integrated electronic access control device.

**Facial Recognition**

Facial recognition is another widely used biometric method that identifies individuals based on the unique structure and proportions of their face. Facial recognition systems capture an image of the user's face using a camera and then analyse specific facial landmarks—such as the distance between the eyes, the shape of the jawline, the contour of the cheekbones, and the relationship between the nose, mouth, and forehead. These features are converted into a digital representation known as a *face template*. During authentication, the system compares a newly captured image with stored templates to determine whether the person is authorised. Modern facial recognition systems often use advanced machine learning algorithms and deep neural networks to improve accuracy and adapt to variations in lighting, facial expressions, or ageing. Some systems also incorporate 3D depth sensors or infrared imaging to distinguish real faces from photographs or video displays, helping prevent spoofing attacks. Facial recognition has the advantage of being a contactless authentication method, which makes it suitable for high-traffic environments such as offices, airports, and secure facilities. However, it requires careful management of privacy concerns and system accuracy, particularly in environments where lighting conditions, camera angles, or facial obstructions may affect recognition performance.

Facial recognition locks use biometric identification technology to grant or deny access based on the unique structure and features of a person's face. Instead of using a traditional key, card, or PIN, the system verifies a user's identity by analysing facial characteristics captured by a camera integrated into the lock or access control unit. Human faces contain distinctive features such as the spacing between the eyes, the shape of the nose, the contour of the cheekbones, and the outline of the jaw. These characteristics form a pattern that can be mathematically analysed and compared with stored biometric data. By using facial recognition, access control systems link entry to who a person is, rather than what they possess, making the system both convenient and difficult to duplicate.

The process begins with enrolment, where the authorised user's face is registered in the system. During this stage, the person stands in front of the camera, and the system captures one or more images of their face. Facial recognition software then identifies key landmarks on the face—often dozens or even hundreds of points—such as the distance between the eyes, the width of the nose, the depth of the eye sockets, and the shape of the chin. These measurements are converted into a mathematical representation known as a face template. Rather than storing a photographic image, the system stores this encrypted template in its internal memory or on a connected access control server. This template is then associated with the user's access permissions.

When the user later approaches the lock, the system performs the authentication stage. The camera captures a live image of the person's face and the software detects and isolates the face from the surrounding environment. Advanced facial recognition algorithms analyse the same landmarks used during enrolment and convert the captured image into a digital template. The system then compares this template with the stored templates of authorised users. If a match is found within the acceptable confidence threshold, the system confirms the user's identity and sends a signal to the lock mechanism to release the door. If no match is detected, the system denies access. Many systems also log the event, recording the time, identity, and location of the attempted access.

Modern facial recognition locks often incorporate anti-spoofing and liveness detection techniques to ensure that the system is recognising a real person rather than a photograph or video. These methods may include analysing depth information, detecting natural facial movements such as blinking, or using infrared imaging to identify heat signatures and facial contours. Some advanced systems use 3D facial mapping, which creates a three-dimensional model of the face to improve accuracy and prevent spoofing attempts. These security measures are important because facial recognition systems operate in open environments where attackers may attempt to trick the system using images or recordings.

Facial recognition locks require several key hardware components to function effectively. The most critical component is the camera module, which captures the facial image. This may be a standard high-resolution camera or a specialised sensor designed for biometric identification. Many systems also include infrared (IR) cameras or depth sensors to improve recognition accuracy in low light and to support liveness detection. The captured image is processed by an embedded processor or AI-enabled microcontroller, which runs the facial recognition algorithms that detect facial landmarks and compare them with stored templates.

The system also includes memory storage for biometric templates and system configuration data. A lock control module connects the recognition system to the physical locking mechanism. Once the system verifies an authorised identity, this module activates an electromechanical actuator, such as a motor or solenoid, which retracts the bolt or releases the latch. In addition, facial recognition locks typically require a power supply, which may be battery-based for residential units or connected to building power systems for commercial installations. Some systems also include wireless communication modules such as Wi-Fi, Ethernet, or Bluetooth to allow integration with access control platforms, mobile applications, or building management systems.

Additional hardware elements may include display screens, LED indicators, or touch panels that provide feedback to users and administrators. These interfaces allow users to enrol their faces, manage permissions, and receive confirmation when access is granted or denied. In enterprise environments, facial recognition devices may connect to centralised access control controllers and cloud platforms that manage identities and permissions across multiple doors or buildings.

Two commercially available facial recognition locking systems commonly used in residential and commercial access control are the Hikvision DS-K1T671M Face Recognition Terminal and the DDL-P6006S 3D Facial

Recognition Smart Lock. These systems illustrate how facial recognition technology is integrated into modern electronic locking and access control environments.

### Hikvision DS-K1T671M Face Recognition Access Terminal

The Hikvision DS-K1T671M is a professional facial recognition access control device designed for commercial environments such as offices, government buildings, schools, and secure facilities. The system uses a camera-based biometric identification process to recognise authorised users and unlock a connected electronic door lock or access control system.

The device includes a 7-inch LCD touchscreen interface and a 2-megapixel wide-angle camera that captures the user's face when they approach the door. Facial recognition software analyses facial features and compares them with stored biometric templates. If the system identifies a match, it triggers the door's electronic locking mechanism to unlock. The recognition process is very fast, typically completing identification in less than 0.2 seconds with an accuracy rate of approximately 99%.

The system can store thousands of authorised users, with capacity for up to around 6,000 facial profiles and tens of thousands of access events. It also supports additional authentication methods such as RFID cards or fingerprints, allowing organisations to implement multi-factor authentication.

From a hardware perspective, the terminal contains several integrated components, including a dual-lens camera, infrared lighting for low-light recognition, an embedded processor running facial recognition algorithms, internal memory for biometric templates, and network interfaces such as TCP/IP for integration with building access control systems. Once the user is verified, the terminal sends an electrical signal to an electronic strike or magnetic lock to release the door.

### DDL-P6006S 3D Facial Recognition Smart Door Lock

The DDL-P6006S is a consumer-grade smart lock that incorporates 3D facial recognition technology for residential or small commercial access control. Unlike traditional locks that require keys or PIN codes, this device allows authorised users to unlock the door simply by looking at the lock's camera.

The system uses facial recognition sensors to detect and analyse the user's face as they approach the door. The lock compares the captured facial image with stored templates of authorised users and grants access when a match is found. This allows contactless entry, making the system convenient and hygienic while still maintaining strong security.

The lock typically supports several authentication methods in addition to facial recognition, such as fingerprint recognition, passcodes, smartphone applications, and traditional mechanical key overrides. This multi-method design provides redundancy and allows the system to continue operating even if one authentication method is unavailable.

In terms of hardware, the lock includes a front-mounted camera module, infrared sensors for facial detection in low-light conditions, an embedded processor that performs facial recognition calculations, secure memory for storing facial templates, and a motorised electronic locking mechanism. The lock is powered by batteries and may connect to smartphone applications or smart-home systems through Wi-Fi or Bluetooth, allowing remote management of access permissions and monitoring of entry events.

### Iris Scanning

Iris scanning is considered one of the most accurate biometric identification technologies available. The iris is the coloured ring surrounding the pupil of the eye, and it contains extremely detailed patterns formed by fibres, furrows, and crypts during early development. These patterns are highly complex and remain stable throughout a person's lifetime, making them ideal for reliable identification.

Iris recognition systems use specialised cameras, often equipped with near-infrared illumination, to capture high-resolution images of the iris without being affected by normal variations in eye colour or lighting conditions. The captured image is processed to isolate the iris region, and distinctive pattern features are extracted and converted into a mathematical code. During authentication, the system compares this code with stored templates to determine whether there is a match. Iris recognition offers extremely low false-acceptance rates, making it suitable for high-security environments such as government facilities, border control, and data centres. Because the iris patterns are internal to the eye and difficult to replicate, iris scanning provides strong protection against forgery. However, the technology requires specialised hardware and precise image capture, which can increase system cost and complexity compared with other biometric methods.

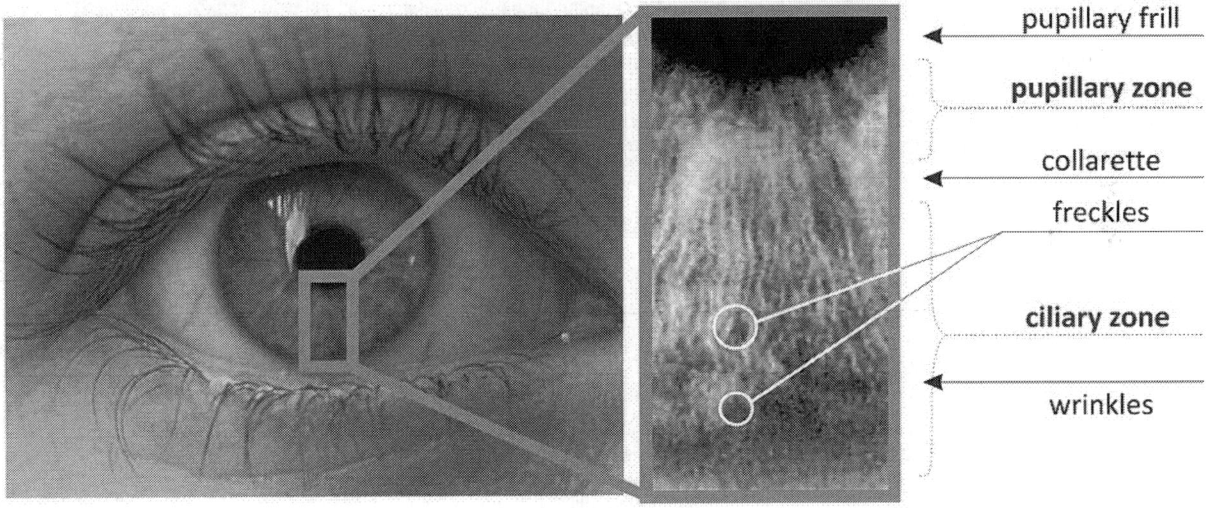


*Figure 77: Iris components. Aidan Boyd, Shivangi Yadav, Thomas Swearingen, Andrey Kuehlkamp, Mateusz Trokielewicz, Eric Benjamin, P. Maciejewicz, D. Chute, A. Ross, P. Flynn, K. Bowyer, A. Czajka, CC BY 4.0, via Wikimedia Commons.*

Iris scanning locks use biometric identification technology based on the unique patterns in the human iris, the coloured ring surrounding the pupil of the eye. The iris contains complex patterns formed by fibres, furrows, rings, and crypt-like structures that develop randomly before birth and remain stable throughout a person's lifetime. Because these patterns are extremely detailed and unique—even between identical twins—they provide a highly reliable method of identification. In iris recognition locking systems, access is granted only when the system confirms that the scanned iris pattern matches a stored template belonging to an authorised user. This allows doors to be unlocked based on biometric identity rather than physical keys, cards, or PIN codes.

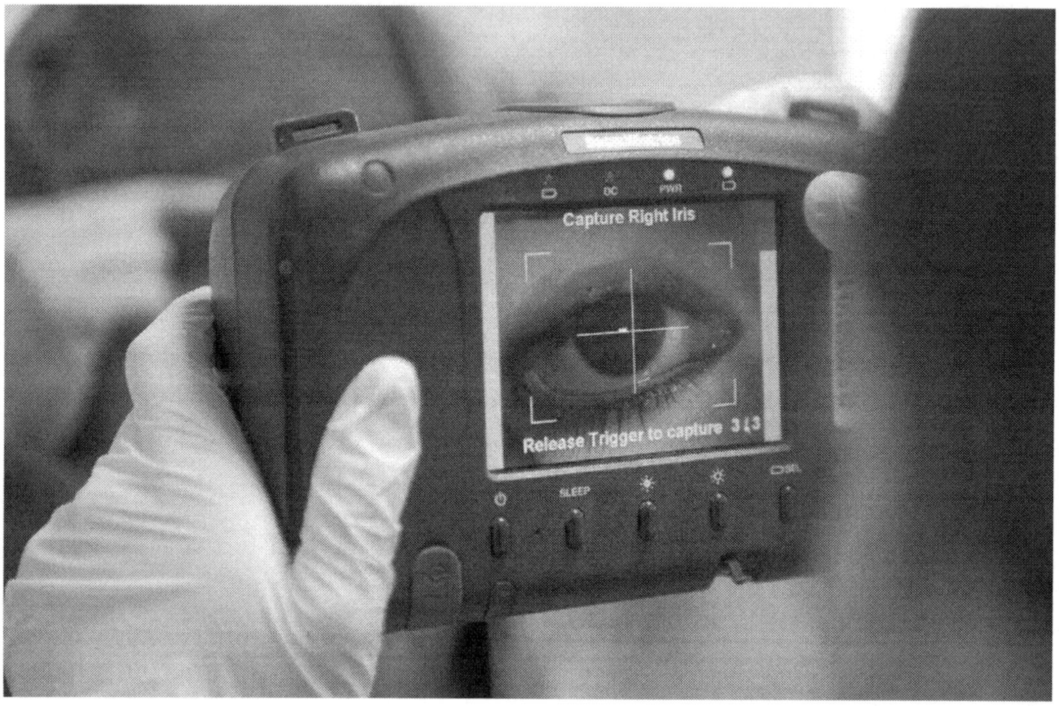


*Figure 78: Biometric eye scanner. Defense Visual Information Distribution Service, Public Domain, via NARA & DVIDS Public Domain Archive.*

The process begins with enrolment, where an authorised user's iris pattern is first registered in the system. During this stage, the user looks into a specialised camera that captures a detailed image of their eye. Most iris scanning systems use near-infrared illumination, which improves contrast and allows the camera to clearly capture iris patterns regardless of eye colour. The captured image is processed by software that isolates the iris region by locating the boundaries of the pupil and the outer iris. The system then analyses the iris texture and converts it into a mathematical representation called an iris template. This template encodes distinctive features of the iris pattern in a compact digital format and is stored in the lock's memory or within a connected access control system. Importantly, the system typically stores only this template rather than the original image to enhance privacy and security.

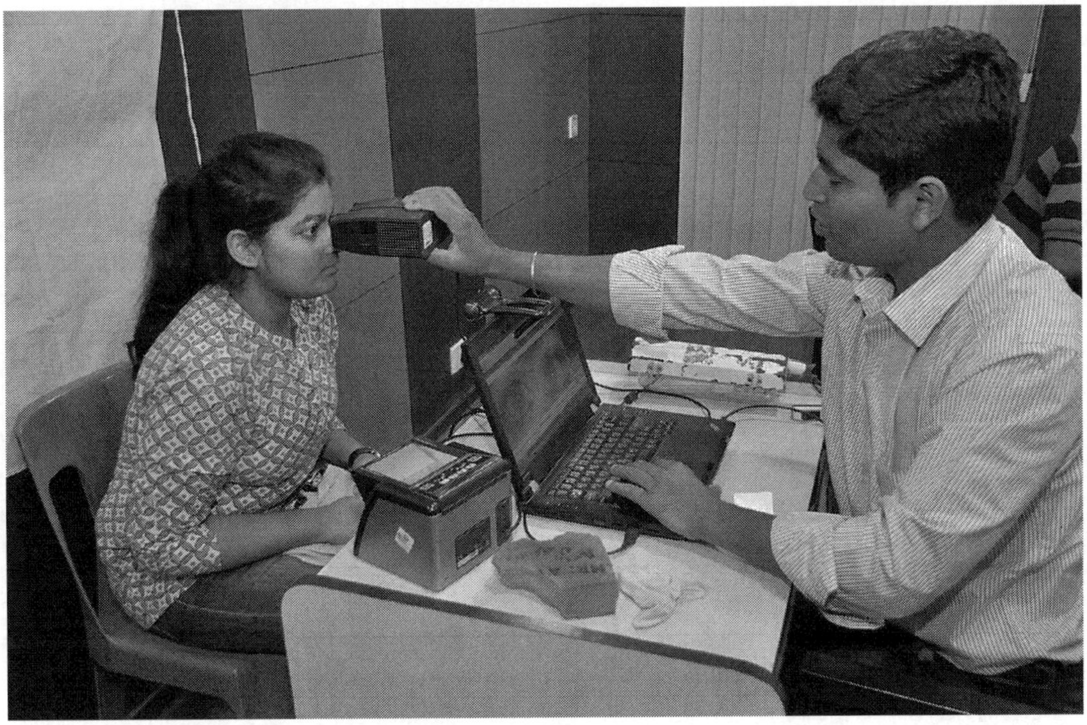

*Figure 79: Biometric Data Collection. Biswarup Ganguly, CC BY 3.0, via Wikimedia Commons.*

When the user attempts to unlock the door, the system performs the authentication stage. The person positions their eye in front of the scanner, and the camera captures a new image of the iris. The system again isolates the iris and extracts its unique pattern features. These features are converted into a new digital template, which is then compared against the stored templates of authorised users. If the comparison shows a sufficiently close match, the system verifies the user's identity and sends a command to the lock mechanism to release the door. If the iris pattern does not match any stored template, access is denied and the lock remains engaged. Many systems also record the event in system logs for auditing and security monitoring.

Modern iris recognition systems also include liveness detection mechanisms to prevent spoofing attempts using printed images or artificial eyes. These techniques may involve detecting natural pupil responses to light, analysing subtle eye movements, or using multiple infrared wavelengths to verify that the scanned eye is real and alive. Because the iris is protected behind the cornea and cannot easily be altered or replicated, iris recognition systems typically achieve very low false-acceptance rates, making them suitable for high-security environments such as research laboratories, government facilities, airports, and data centres.

To operate effectively, iris recognition locks require several specialised hardware components. The central component is the iris imaging camera, which captures high-resolution images of the user's eye. This camera is usually paired with near-infrared LED illumination, which provides consistent lighting and enhances the visibility of iris patterns without causing discomfort to the user. The captured images are processed by an embedded processor or biometric processing unit, which runs the algorithms responsible for iris detection, feature extraction, and template matching.

The system also includes secure memory storage for storing iris templates and user access data. A lock control module connects the biometric recognition unit to the physical locking mechanism. When the processor verifies a match, it sends an electrical signal to an electromechanical actuator, such as a motor or solenoid, which retracts the bolt or releases the door latch. Additional hardware components include a power supply (battery or wired), communication interfaces such as Ethernet, Wi-Fi, or RS-485 for integration with building access control systems, and user interface elements such as status LEDs or display panels that guide the user during the scanning process.

In summary, iris scanning locks combine advanced imaging technology, biometric pattern analysis, and electronic locking hardware to create a highly secure access control system. By linking physical entry to a unique biological feature that is extremely difficult to replicate, iris recognition provides one of the most reliable biometric authentication methods available. For locksmiths and security professionals, understanding how the camera sensors, infrared illumination, processing hardware, and locking mechanisms interact is essential when installing, configuring, or maintaining iris-based access control systems.

Two commercially available iris scanning locking systems commonly used in professional access control environments are the Iris ID iCAM 7S Series Iris Recognition Terminal and the Aratek IrisScan-based Access Control System. These systems demonstrate how iris recognition technology can be integrated into door access control systems to provide highly secure biometric authentication.

### Iris ID iCAM 7S Series Iris Recognition Access Control System

The Iris ID iCAM 7S Series is a widely used commercial iris recognition system designed for high-security environments such as airports, government facilities, data centres, and corporate buildings. The device functions as a biometric access control reader that scans the iris of a user and verifies their identity before unlocking a door or granting access to a restricted area.

The system uses a high-resolution camera and near-infrared illumination to capture detailed images of the iris. The coloured ring surrounding the pupil contains complex patterns such as furrows, rings, and crypts, which are unique to each individual. During enrolment, the system captures the iris image and converts it into a digital template that represents these patterns. When the user attempts to enter, the scanner captures a live iris image and compares it with stored templates in the database. If a match is detected, the access control system sends a signal to the electronic lock or door controller to release the door.

The iCAM 7S terminal is designed to integrate with existing access control infrastructure. It typically connects to door controllers, electronic strikes, or magnetic locks and can communicate with building security systems through network connections. The system includes components such as a dual iris camera, infrared LED illumination, embedded processor, biometric matching software, and network communication interface. These components allow the device to perform fast biometric authentication while maintaining a high level of accuracy and security.

<u>Aratek IrisScan Access Control System</u>

The Aratek IrisScan system is another commercially available iris recognition solution designed for identity verification and secure access control. This system is commonly used in secure buildings, border control environments, and enterprise facilities where reliable biometric authentication is required.

The device works by scanning the user's iris using a camera and infrared light to highlight the intricate patterns within the iris. These patterns are analysed by biometric algorithms and converted into a digital template that uniquely represents the user's eye. When a person attempts to access the door, the scanner captures the iris image again and compares it to the stored template. If the match is successful, the system triggers the door's electronic locking mechanism, allowing entry. Iris recognition is considered highly reliable because the iris pattern is extremely stable throughout a person's life and difficult to duplicate.

The hardware architecture of the system includes an iris scanning camera, infrared lighting module, embedded processor, biometric database storage, and communication interfaces for integration with building security systems. The device can be connected to electric door strikes, magnetic locks, or access control controllers to physically control door entry. In many installations, the iris scanner is combined with other authentication methods such as RFID cards or PIN codes to provide multi-factor authentication, increasing overall security.

Each of these biometric technologies contributes to modern access control systems in different ways. Fingerprint recognition offers a practical balance of cost, speed, and reliability, making it common in everyday devices. Facial recognition provides convenient, contactless authentication that can be used at a distance or in high-traffic areas. Iris scanning delivers extremely high accuracy and security for sensitive environments. In many modern systems, these biometric methods are combined with other authentication factors, such as mobile credentials, smart cards, or PIN codes, to create multi-factor authentication systems that significantly enhance security. For locksmiths and security professionals working with modern access control technologies, understanding how these biometric systems function is essential to selecting appropriate solutions, ensuring correct installation, and maintaining secure and reliable access management systems.

**Impact on Locksmithing Practices**

Biometric security systems are highly relevant to modern locksmithing because the profession is increasingly moving beyond purely mechanical locks into electronic and identity-based access control systems. Traditional locksmithing focused on physical keys, cylinders, and mechanical locking mechanisms, but modern security environments now incorporate biometric technologies such as fingerprint recognition, facial recognition, and iris scanning. These systems identify individuals based on unique biological characteristics rather than physical credentials. For locksmiths, understanding biometric authentication is essential because these technologies are now commonly integrated into smart locks, electronic access systems, and building security platforms that locksmiths may be required to install, configure, or maintain.

The growing use of biometric locks also influences the technical knowledge required in locksmithing practice. A locksmith working with biometric systems must understand how biometric sensors operate, how enrolment and authentication processes work, and how biometric templates are securely stored and processed within the lock's electronic hardware. This knowledge allows locksmiths to correctly install biometric devices, ensure sensors are properly positioned, configure user credentials, and troubleshoot issues such as recognition failures or system errors. Without this understanding, it would be difficult to diagnose problems related to sensors, firmware, or user enrolment that can affect the performance of biometric locks.

Biometric technologies also change how locksmiths approach security risk assessment and system design. Instead of focusing only on the strength of mechanical components such as cylinders and bolts, locksmiths must evaluate additional factors such as biometric spoofing risks, credential management, system integration, and privacy considerations. For example, fingerprint sensors must incorporate liveness detection to prevent fake fingerprints, facial recognition systems must account for lighting conditions and camera placement, and iris scanners require precise alignment for accurate image capture. Locksmiths therefore need to consider both physical security and biometric authentication reliability when recommending and installing modern locking systems.

Finally, the increasing adoption of biometric access control expands the professional role of locksmiths within integrated security environments. Many biometric locks connect to mobile applications, cloud services, or building access control platforms that manage identities and access permissions across multiple doors or facilities. Locksmiths are often responsible for installing the hardware, integrating it with door mechanisms, and ensuring that biometric authentication works correctly with the physical locking components. By understanding biometric technologies such as fingerprint, facial recognition, and iris scanning, locksmiths can provide more advanced security solutions, advise clients on appropriate systems, and remain relevant as the security industry continues to shift toward digital, identity-based access control systems.

## Accuracy Vs Reliability Vs Privacy

Biometric security systems are evaluated not only by how well they identify individuals but also by how consistently they operate and how responsibly they handle sensitive personal data. Three important considerations when assessing biometric systems are accuracy, reliability, and privacy. Although these concepts are related, they represent different aspects of system performance and security. Understanding the differences between them is important for locksmiths and security professionals who install or manage biometric access control systems.

Accuracy refers to how correctly a biometric system identifies authorised users and rejects unauthorised individuals. In biometric authentication, accuracy is often measured using metrics such as the False Acceptance Rate (FAR) and the False Rejection Rate (FRR). The False Acceptance Rate represents the likelihood that the system incorrectly grants access to an unauthorised person, while the False Rejection Rate represents the likelihood that the system denies access to an authorised user. A highly accurate biometric system minimises both of these errors. For example, iris recognition systems are frequently cited as having extremely low false acceptance rates because iris patterns are highly distinctive, and this characteristic underpins high accuracy in identification tasks [277-281].

Accuracy is influenced by the quality of the biometric sensor, the sophistication of the recognition algorithms, and the conditions in which the system operates, with literature repeatedly noting sensor quality, algorithmic design, and environmental conditions as determinants of accuracy [277, 281-285]. In practice, trade-offs between FAR and FRR are a central concern when configuring systems for specific security requirements, and several studies emphasize how accuracy must be evaluated in the context of template protection methods that may incur some performance degradation compared to unprotected baselines [278, 279, 286].

Reliability refers to the consistency and dependability of a biometric system during everyday operation. A system may be technically accurate under ideal conditions but still perform poorly if it fails frequently, struggles with environmental changes, or cannot consistently recognise users. Reliability therefore considers factors such as sensor durability, resistance to environmental conditions, processing speed, and the system's ability to function with variations in user presentation. For example, fingerprint scanners may fail to recognise a user when the finger is wet, dirty, or worn, while facial recognition systems may fail in poor lighting, illustrating how environmental and presentation variations can undermine reliable operation even when nominal accuracy is high [282, 284, 287-289]. A reliable biometric system maintains stable performance across different conditions and over long periods of use, underscoring the need for robust hardware, resilient matching algorithms, and well-managed operational workflows in access-control deployments [282, 287, 290].

Privacy concerns the protection and responsible handling of biometric data collected by the system. Unlike passwords or access cards, biometric identifiers such as fingerprints or iris patterns are permanent characteristics of a person, and once compromised they cannot be easily replaced. For this reason, modern biometric systems typically store biometric information as encrypted templates rather than raw images, reducing the risk of misuse.

Privacy considerations also involve how biometric data is collected, stored, transmitted, and accessed. Organisations must ensure that biometric systems comply with privacy laws and data protection regulations, and that access to biometric databases is strictly controlled. Privacy-preserving approaches for biometric systems are widely studied, including biometric template protection schemes (cancellable biometrics, biometric cryptosystems, and encryption-in-domain methods) and architectures that minimize data exposure during authentication, indexing, and identification tasks [277, 278, 280, 281, 285, 286, 291]. Recent work explicitly addresses the privacy implications of large-scale biometric indexing and identification, highlighting leakage risks in encrypted indexing and the need for post-quantum protected architectures to uphold privacy guarantees in real-world deployments [281, 289]. Collectively, the literature emphasizes irreversibility, unlinkability, and confidentiality as central privacy objectives, particularly when templates are stored or processed remotely [281, 283, 286, 288].

Balancing accuracy, reliability, and privacy is essential when designing or selecting biometric security systems. A highly accurate system that requires intrusive data collection may raise privacy concerns, while a system that prioritises privacy but lacks reliability may frustrate users and reduce security effectiveness. The literature consistently notes that effective biometric security solutions aim to achieve strong identification performance while maintaining dependable operation and protecting the personal data of users. In practical terms, this balance influences how locksmiths and security professionals implement biometric access control systems, from sensor selection and placement to the design of privacy-preserving template protection and the choice of cryptographic safeguards for storage and transmission [277-279, 281, 283, 285, 286]. Moreover, the integration of privacy-preserving techniques, such as homomorphic encryption, secure two-party computation, or cancellable templates, often involves trade-offs in computational efficiency and latency, which must be weighed against security and usability requirements in real-world installations [279, 280, 284, 286, 289]. When appropriately coordinated, these approaches can deliver robust identification performance while limiting privacy risks and maintaining operational reliability across diverse site conditions and long-term use [282, 287, 288, 290, 292].

A practical example of how accuracy, reliability, and privacy influence biometric security systems can be seen in a commercial office building that installs fingerprint-enabled smart locks for staff access.

In this scenario, the organisation installs fingerprint readers at the main entrance and several secure internal areas. Accuracy becomes important because the system must correctly identify authorised employees while preventing unauthorised individuals from gaining access. During configuration, the security administrator adjusts the system's matching threshold so that the False Acceptance Rate (FAR) is extremely low, ensuring that an unauthorised fingerprint cannot easily unlock the door. At the same time, the system must maintain a reasonable False Rejection Rate (FRR) so that legitimate employees are not repeatedly denied entry. If the system is configured too strictly, authorised staff may experience frequent access denials, while overly relaxed settings may increase the risk of unauthorised entry.

Reliability becomes evident during everyday use. Employees arrive at different times of day, sometimes with wet hands after rain or with slightly worn fingerprints from manual work. If the fingerprint reader cannot consistently recognise users under these conditions, the system becomes unreliable and causes delays at entry points. To maintain reliability, the organisation installs high-quality capacitive fingerprint sensors with protective housings and ensures that multiple fingerprint samples are enrolled for each employee. This allows the system to recognise users more consistently even if finger placement or environmental conditions vary.

Privacy considerations arise because the system collects and stores biometric information from employees. Instead of storing raw fingerprint images, the access control system converts each fingerprint into an encrypted biometric template stored in a secure database. Only authorised administrators can access the system, and all biometric data is encrypted during storage and transmission. The organisation also informs employees about how their biometric data is used and ensures compliance with relevant privacy legislation and workplace policies.

This example demonstrates how accuracy, reliability, and privacy must all be considered together when implementing biometric security systems. A fingerprint lock may be technically capable of identifying individuals, but it must also operate consistently in real-world conditions and protect the sensitive biometric data it collects. For locksmiths and security professionals, these considerations influence decisions about sensor selection, installation location, system configuration, and data protection measures when deploying biometric access control solutions.

### Impact on Locksmithing Practices

Understanding the relationship between accuracy, reliability, and privacy is highly relevant to someone interested in locksmithing because modern locksmiths increasingly work with biometric access control systems rather than purely mechanical locks. As fingerprint readers, facial recognition devices, and iris scanners become more common in residential and commercial security installations, locksmiths must understand how well these systems identify authorised users and how they behave in real-world conditions. A biometric system that is technically advanced but poorly configured can either allow unauthorised access or deny legitimate users entry. By understanding concepts such as False Acceptance Rate (FAR) and False Rejection Rate (FRR), locksmiths can better evaluate different biometric products and recommend systems that provide an appropriate balance between security and usability.

These concepts also influence how locksmiths approach system installation and configuration. For example, the placement of fingerprint sensors, cameras, or iris scanners can significantly affect system accuracy and reliability. Poor lighting, weather exposure, or incorrect sensor positioning can reduce recognition performance and cause repeated access failures. Locksmiths therefore need to consider environmental conditions, user behaviour, and sensor quality when installing biometric locks. In practice, this may involve selecting sensors suitable for outdoor conditions, enrolling multiple biometric samples for each user, or integrating alternative authentication methods such as PIN codes or key cards to maintain reliable access when biometric recognition fails.

Privacy considerations further expand the responsibilities of locksmiths working with biometric systems. Unlike mechanical keys, biometric identifiers such as fingerprints or iris patterns are permanent personal characteristics, and improper handling of this information can create serious privacy and legal risks. Locksmiths who install biometric systems must therefore understand how biometric templates are stored, encrypted, and transmitted within the system. They may also need to ensure that installations comply with relevant privacy legislation and organisational policies, particularly when biometric data is stored in networked databases or cloud-based access control systems.

Overall, the need to balance accuracy, reliability, and privacy has a direct influence on modern locksmithing practices and professional competencies. Locksmiths must now evaluate not only the mechanical strength of locks but also the performance of biometric sensors, the stability of electronic systems, and the protection of sensitive identity data. This shift reflects the broader transformation of locksmithing into a profession that integrates physical security, electronic access control, and digital identity management, requiring locksmiths to develop knowledge that extends beyond traditional mechanical locking systems.

## Spoofing and Bypass Techniques

Spoofing and bypass techniques refer to methods used to trick or circumvent security systems in order to gain unauthorised access. In the context of biometric security systems, spoofing involves presenting a false biometric sample, such as a fake fingerprint, photograph, or artificial eye, to fool the system into recognising it as a legitimate user. Bypass techniques, on the other hand, involve avoiding the biometric verification process entirely by exploiting weaknesses in the hardware, software, or installation of the locking system. Both types of attacks highlight potential vulnerabilities in biometric access control systems and are important considerations when evaluating their overall security.

Biometric spoofing targets the sensors used to capture biological data. Attackers attempt to create replicas of legitimate biometric traits to deceive the recognition system. For example, fingerprint systems can sometimes be fooled using artificial fingerprints created from materials such as silicone, gelatin, or latex if the sensor does not incorporate strong liveness detection. Facial recognition systems may be vulnerable to spoofing using photographs, video displays, or 3-dimensional masks that mimic a person's facial features. Similarly, iris recognition systems could potentially be targeted using high-resolution images of the eye if the scanner cannot distinguish between a live eye and a printed image. Modern biometric systems attempt to prevent these attacks using techniques such as liveness detection, infrared imaging, depth sensing, and behavioural analysis to confirm that the biometric sample comes from a real, living person.

Bypass techniques involve circumventing the biometric system rather than imitating a biometric characteristic. Attackers may exploit weaknesses in the lock's hardware, communication systems, or software controls. For instance, a smart lock connected to an electronic strike might be bypassed if exposed wiring allows the actuator to be triggered directly. In networked systems, attackers may attempt to intercept communication signals between the lock and its controller, manipulate software vulnerabilities, or exploit weak authentication protocols. Some attacks may also involve accessing maintenance ports, emergency override mechanisms, or poorly secured administrative interfaces.

These threats demonstrate that biometric systems must be designed and implemented with multiple layers of security. In addition to accurate biometric recognition, systems should incorporate encryption for communications, secure firmware updates, tamper detection mechanisms, and protected hardware components. Multi-factor authentication, combining biometrics with PIN codes, smart cards, or mobile credentials, can also reduce the risk that spoofing alone will grant access.

By means of biometric access system examples, two commercially available locking and biometric access control systems that incorporate protections against spoofing and bypass attacks are the Suprema BioStation 3 Facial Recognition Access Terminal and the ASSA ABLOY Aperio H100 Wireless Biometric Door Lock with Suprema BioEntry W2 Fingerprint Reader. These systems demonstrate how modern biometric security devices integrate anti-spoofing technologies, encrypted communications, and tamper protection to reduce the risk of unauthorised access.

<u>Suprema BioStation 3 Facial Recognition Access Control Terminal</u>

The Suprema BioStation 3 is an enterprise-grade biometric access control device commonly used in corporate buildings, government facilities, and data centres. The system primarily uses facial recognition technology, but it can also integrate with RFID credentials and mobile access systems to provide multi-factor authentication. The device uses an advanced camera and artificial intelligence-based recognition algorithms to identify authorised users before unlocking a connected electronic door lock or access control system.

To prevent spoofing attacks, BioStation 3 incorporates AI-based liveness detection and infrared imaging technology. These systems analyse depth, movement, and facial structure to determine whether the detected face belongs to a real person rather than a photograph, video display, or mask. The device uses infrared sensors to detect facial contours and heat signatures, allowing the system to distinguish between genuine human features and flat images. This helps prevent common spoofing methods that rely on photographs or digital screens.

The system also includes protections against bypass techniques. Communication between the device and the central access control system is encrypted, preventing attackers from intercepting authentication data or replaying signals. The device includes tamper detection sensors that trigger alarms or disable the system if the housing is opened or manipulated. Firmware updates are digitally signed to ensure that only authorised software can be installed, preventing attackers from introducing malicious firmware that could unlock the door remotely.

<u>ASSA ABLOY Aperio H100 Wireless Lock with Suprema BioEntry W2 Fingerprint Reader</u>

Another example of a commercially available biometric access control solution is the ASSA ABLOY Aperio H100 wireless door lock integrated with the Suprema BioEntry W2 fingerprint reader. This system is commonly used in commercial offices, healthcare facilities, and educational institutions where secure and convenient biometric access is required.

The BioEntry W2 fingerprint reader uses advanced fingerprint recognition technology combined with liveness detection features designed to prevent spoofing attempts. The sensor analyses characteristics of human skin such as electrical conductivity and subtle physiological properties that cannot easily be replicated by artificial materials. This allows the system to detect and reject fake fingerprints made from silicone, gelatin, or other materials commonly used in spoofing attacks. The reader also captures high-resolution fingerprint images and analyses multiple fingerprint features to improve identification accuracy while reducing the risk of false acceptance.

The system also addresses bypass threats through layered security architecture. The fingerprint reader communicates with the Aperio wireless locking system using encrypted communication protocols, protecting authentication data from interception or manipulation. The lock hardware includes tamper protection mechanisms that detect attempts to open the device casing or interfere with internal electronics. In addition, the system can be configured with multi-factor authentication, requiring both a fingerprint and a card credential or PIN before access is granted. This significantly reduces the risk that a single compromised biometric credential could allow unauthorised entry.

**Impact on Locksmithing Practices**

For locksmiths and security professionals, understanding spoofing and bypass techniques is essential when selecting, installing, and maintaining biometric locks. A biometric system may appear secure on the surface, but vulnerabilities can exist in sensor technology, installation methods, or supporting infrastructure. Knowledge of these attack methods allows locksmiths to recommend systems with strong anti-spoofing features, ensure that hardware is installed in ways that minimise bypass opportunities, and advise clients on appropriate security configurations. This awareness helps ensure that biometric access control systems provide genuine protection rather than a false sense of security.

## Integration with Traditional Locks

Biometric security systems are increasingly integrated with traditional mechanical locking systems to combine the convenience of biometric authentication with the proven reliability of mechanical security. Rather than completely replacing traditional locks, many biometric systems operate as electronic control mechanisms that activate or release a conventional locking device, such as a deadbolt, mortise lock, or electric strike. This hybrid approach allows buildings to maintain familiar mechanical hardware while adding advanced identity verification technologies such as fingerprint, facial recognition, or iris scanning. For locksmiths and security professionals, understanding how biometric systems interface with traditional locking components is essential for ensuring both secure installation and reliable operation.

In many installations, the biometric device functions as an authentication reader, while the actual locking mechanism remains a conventional mechanical lock adapted for electronic control. For example, a fingerprint reader mounted beside a door may connect to an electric strike installed in the door frame. When the biometric system verifies an authorised user, it sends an electrical signal to release the strike, allowing the door to open while the mechanical latch or deadbolt remains physically secure when the system is locked. In other configurations, biometric readers may control magnetic locks, motorised deadbolts, or electronic mortise locks, depending on the security requirements of the building.

Another common integration method involves biometric smart locks that replace the exterior portion of a traditional lock while retaining a mechanical key override. Many residential biometric locks, for example, incorporate a fingerprint scanner or keypad in the exterior housing while still containing a conventional deadbolt mechanism. Inside the lock, an electric motor or solenoid retracts the bolt when the biometric authentication succeeds. The mechanical key override ensures that access remains possible if the biometric system fails due to power loss, hardware malfunction, or other issues. This approach maintains a balance between modern electronic convenience and the reliability of mechanical locking systems.

Installation of biometric systems integrated with traditional locks requires careful consideration of door hardware, power supply, wiring, and device positioning. The biometric sensor must be installed in a location that is easily accessible for users while being protected from environmental factors such as moisture, direct sunlight, or physical damage. For fingerprint scanners, installers must ensure that the sensor is mounted at a comfortable height and angle for consistent finger placement. Facial recognition devices require correct positioning to capture faces clearly, which often involves adjusting camera height and ensuring adequate lighting. Iris scanners require even more precise alignment to allow accurate imaging of the user's eye.

In wired installations, biometric readers are typically connected to door controllers, access control panels, or power supplies through concealed wiring. These systems may operate using low-voltage cabling that connects the reader to the electronic locking device and central management system. Proper cable routing, secure mounting, and protection of wiring channels are essential to prevent tampering or accidental damage. In wireless systems, biometric devices may communicate with locks or controllers using Bluetooth, Wi-Fi, or other wireless protocols, reducing installation complexity but requiring careful configuration of network security.

Integration also requires attention to fail-safe and fail-secure operation modes, which determine how the lock behaves during power loss. For example, magnetic locks used in emergency exits are often configured as fail-safe so that doors unlock automatically when power is lost, allowing safe evacuation. Motorised deadbolts and electric strikes may be configured as fail-secure to remain locked during power interruptions. Locksmiths must ensure that the chosen configuration complies with building safety regulations and the intended security design.

Testing and commissioning are important final steps in installation. After the biometric system is installed, locksmiths must enrol authorised users, test authentication processes, and confirm that the locking mechanism operates correctly under different conditions. This includes verifying that the biometric reader communicates properly with the locking hardware, ensuring that mechanical overrides function correctly, and confirming that the system logs access events where applicable. Proper installation and configuration ensure that biometric systems enhance security rather than introducing vulnerabilities.

**Case Study Scenario: Integrating a Fingerprint Access System with a Traditional Door Lock in a Medical Clinic**

A private medical clinic upgrades its security system to better control staff access to sensitive areas such as the medication storage room and patient records office. Previously, these doors were secured using traditional mechanical mortise locks with standard keys, which created several operational issues. Keys were sometimes misplaced, staff members occasionally shared keys, and it was difficult for management to monitor who accessed restricted rooms. To improve accountability and convenience while maintaining the reliability of existing hardware, the clinic decides to integrate a biometric fingerprint access control system with the existing mechanical locks.

A professional locksmith is contracted to design and install the new system. After assessing the doors and existing lock hardware, the locksmith determines that the most appropriate solution is to install a fingerprint reader connected to an electric strike mounted in the door frame. The original mortise lock remains in place so the door can still latch and secure mechanically. When a staff member places their finger on the reader, the biometric system verifies the fingerprint against the stored database of authorised users. If the identity is confirmed, the reader sends a signal to the electric strike, temporarily releasing the latch so the door can be opened.

During installation, the locksmith carefully considers the positioning of the fingerprint reader to ensure consistent and reliable operation. The reader is mounted beside the door at an accessible height where staff can comfortably place their finger. Because the clinic corridor receives significant foot traffic, the device is installed in a recessed protective housing to reduce the risk of accidental damage. Low-voltage wiring is run through the wall cavity to connect the reader to the door controller and power supply, ensuring that cables are concealed and protected from tampering.

The locksmith also integrates a mechanical key override as a backup access method. Although the biometric system controls normal access, the original mortise lock and key cylinder remain functional. This ensures that authorised personnel can still open the door if there is a power outage, equipment failure, or emergency maintenance requirement. The system is configured as fail-secure, meaning the door remains locked during power loss, which is appropriate for protecting medication storage areas.

After installation, the locksmith performs testing and commissioning procedures. Staff members enrol their fingerprints into the system, and multiple test entries are conducted to ensure reliable recognition. The locksmith verifies that the electric strike releases correctly, that the mechanical key override operates properly, and that all access events are recorded in the system logs. The clinic management is also trained on how to enrol new users and revoke access permissions when staff roles change.

This installation demonstrates how biometric authentication can be successfully integrated with traditional mechanical locking hardware. The clinic benefits from improved access control, the ability to monitor entry events, and the convenience of keyless access, while still retaining the reliability of conventional locks. For locksmiths, the project illustrates the practical considerations involved in integrating biometric readers with existing door hardware, including sensor placement, wiring protection, backup access methods, and compliance with security and safety requirements.

**Impact on Locksmithing Practices**

Understanding the integration of biometric security systems with traditional locking hardware is highly relevant for anyone interested in locksmithing because modern security installations increasingly combine mechanical locks with electronic access control technologies. Traditional locksmithing historically focused on the design, installation, and maintenance of purely mechanical devices such as deadbolts, mortise locks, and cylinders. However, many contemporary security systems now rely on hybrid solutions, where biometric readers act as authentication devices that control mechanical locking hardware. For locksmiths, this means that effective practice now requires knowledge of both mechanical locking mechanisms and electronic access control systems, including how biometric devices interface with traditional locks.

This integration influences locksmithing practices by expanding the technical skill set required in the profession. Locksmiths installing biometric systems must understand how fingerprint readers, facial recognition devices, or iris scanners communicate with locking hardware such as electric strikes, motorised deadbolts, or electronic mortise locks. In practical terms, this requires knowledge of low-voltage wiring, electronic control panels, power supplies, and communication protocols, in addition to traditional door hardware installation. Locksmiths must also ensure that mechanical components remain properly aligned and secure while electronic systems control the locking mechanism.

The topic also highlights the importance of correct installation techniques in modern locksmithing. Biometric sensors must be positioned carefully so that users can interact with them consistently and comfortably. Improper placement can lead to frequent authentication failures, reducing the effectiveness of the system. At the same time, wiring and electronic components must be concealed and protected to prevent tampering or accidental damage. Locksmiths therefore play a key role in ensuring that biometric systems are physically secure, ergonomically accessible, and integrated correctly with door hardware.

Another important influence on locksmithing practice is the need to maintain redundancy and reliability. Many biometric locks retain mechanical key overrides or conventional locking mechanisms to ensure that access is still possible if the electronic system fails. Locksmiths must therefore understand how to maintain and service both systems together. This hybrid approach reinforces the continuing importance of traditional locksmithing skills such as cylinder servicing, key control, and mechanical lock maintenance, even when advanced biometric technologies are used.

Integration of biometric systems also introduces new responsibilities related to system configuration, testing, and user management. Locksmiths may need to enrol authorised users, test biometric authentication accuracy, configure fail-safe or fail-secure operation, and verify that the locking hardware responds correctly to electronic signals. This shifts part of the locksmith's role from purely mechanical work toward security system commissioning and access control management, requiring greater familiarity with electronic security systems and operational procedures.

Finally, this knowledge influences how locksmiths advise clients and design security solutions. Understanding how biometric readers interact with traditional locks allows locksmiths to recommend appropriate hardware combinations based on the building's security requirements, safety regulations, and operational needs. In environments such as medical clinics, offices, or secure facilities, locksmiths must ensure that biometric systems improve security without compromising reliability or emergency access requirements.

## Ethical and Legal Considerations

Ethical and legal considerations are critical when implementing biometric security systems because these systems collect and use unique biological characteristics such as fingerprints, facial features, and iris patterns to identify individuals. Unlike passwords or key cards, biometric data is inherently linked to a person's identity and cannot easily be changed if it is compromised. For this reason, organisations and security professionals must ensure that biometric systems are used responsibly, legally, and in a manner that respects individual rights. Locksmiths and security installers involved in deploying biometric locks must therefore understand not only the technical aspects of the systems but also the ethical and legal responsibilities associated with handling biometric information.

One important ethical consideration is informed consent. Individuals whose biometric data is collected should clearly understand why the data is being collected, how it will be used, and who will have access to it. In workplaces or secure facilities where biometric access control systems are installed, employees or authorised users should be informed about the purpose of the system and the nature of the data being stored. Ethical practice requires transparency so that individuals are not unknowingly subjected to biometric monitoring or identification systems. Providing clear information and obtaining consent helps maintain trust between organisations and the people who use the system.

Another key issue is data privacy and protection. Biometric information is highly sensitive because it represents permanent personal identifiers. If biometric data were stolen or misused, the affected individual could face long-term security risks because biometric characteristics cannot be replaced in the same way that passwords or identification cards can be changed. To address this risk, most modern biometric systems store biometric information as encrypted templates rather than raw images, reducing the possibility that the original biometric pattern could be reconstructed. Security professionals must ensure that biometric databases are protected using strong encryption, restricted access controls, and secure data transmission methods.

Legal frameworks in many countries regulate the collection, storage, and processing of biometric data. These laws are designed to protect personal information and ensure that organisations handle biometric identifiers responsibly. For example, privacy legislation may require organisations to explain why biometric data is collected, limit its use to specific purposes, and ensure that it is stored securely. In Australia, biometric data is generally treated as sensitive personal information under privacy regulations, meaning organisations must follow strict requirements when collecting and managing it. Locksmiths involved in installing biometric access systems should therefore ensure that their installations comply with relevant privacy and data protection laws.

There are also ethical concerns related to fairness and discrimination in biometric systems. Some biometric technologies have historically demonstrated differences in accuracy across demographic groups due to variations in training data or sensor design. For example, facial recognition systems may perform differently depending on lighting conditions, skin tone, or facial characteristics. Ethical deployment of biometric systems requires careful evaluation to ensure that the technology does not unfairly exclude or disadvantage particular groups of people. Organisations must select systems that have been tested for fairness and reliability across diverse populations.

Another legal and ethical issue involves data retention and system management. Biometric data should only be stored for as long as it is necessary for the intended security purpose. Organisations must establish clear policies for deleting biometric templates when they are no longer required, such as when an employee leaves the organisation or a user no longer requires access. Secure deletion procedures help prevent unnecessary long-term storage of sensitive personal data and reduce the risk of misuse or unauthorised access.

### Impact on Locksmithing Practices

Security professionals must also consider the risk of surveillance and misuse. Biometric technologies can potentially be used to monitor individuals' movements or activities if implemented without appropriate safeguards. Ethical use requires limiting biometric systems to their intended purpose, such as controlling access to secure areas, rather than using them for excessive monitoring or tracking without justification. Clear governance policies and access controls should be implemented to prevent misuse of biometric systems.

## Chapter 11 Review Questions

**Question 1**

A residential property installs a fingerprint-enabled smart lock. When enrolling users, the system converts each fingerprint image into a digital template rather than storing the full fingerprint image.

Why is this approach used?

A. It reduces battery consumption
B. It improves privacy and security by avoiding storage of full biometric images
C. It increases door lock strength
D. It eliminates the need for encryption

**Question 2**

A factory installs fingerprint scanners at employee entrances. Workers frequently arrive with wet or dirty hands, and the scanner sometimes fails to recognise authorised users.

What issue does this illustrate?

A. False acceptance errors
B. Poor network security
C. Reduced reliability due to environmental conditions
D. Incorrect lock installation

**Question 3**

A high-security research laboratory chooses iris recognition instead of fingerprint authentication.

What is the primary reason for selecting iris scanning?

A. Iris patterns change frequently
B. Iris recognition has extremely low false acceptance rates and high identification accuracy
C. Iris scanners require no cameras
D. Iris scanning does not require any electronic components

**Question 4**

An attacker attempts to unlock a fingerprint reader using a silicone replica of an authorised user's fingerprint.

What type of attack is this?

A. Mechanical bypass
B. Biometric spoofing
C. Wireless replay attack
D. Door frame manipulation

**Question 5**

A commercial office installs a facial recognition entry system. The device occasionally fails to recognise users when the lighting in the lobby changes significantly.

What factor is affecting system performance?

A. Mechanical bolt alignment
B. Environmental conditions influencing recognition accuracy
C. Key duplication errors
D. Lock cylinder wear

**Question 6**

A security consultant recommends combining fingerprint recognition with a PIN code for access to a restricted laboratory.

What security strategy does this represent?

A. Single-factor authentication
B. Multi-factor authentication
C. Mechanical redundancy
D. Door reinforcement

**Question 7**

A medical clinic installs fingerprint readers connected to electric strikes while keeping the existing mechanical mortise locks in place as backups.

What advantage does this design provide?

A. It eliminates the need for wiring
B. It combines biometric authentication with reliable mechanical backup access
C. It removes the need for authentication
D. It prevents any electronic failure

**Question 8**

During installation of a facial recognition system, a locksmith carefully positions the camera at an appropriate height and angle.

Why is correct sensor positioning important?

A. It ensures consistent biometric capture and reliable recognition
B. It reduces electricity usage
C. It prevents users from touching the door
D. It increases mechanical strength of the lock

**Question 9**

A company installs a biometric access system but fails to encrypt stored biometric templates.

What risk does this create?

A. Increased lock wear
B. Exposure of sensitive biometric data if the system is compromised
C. Reduced door strength
D. Slower door closing speed

**Question 10**

A government facility installs an iris recognition system with near-infrared illumination.

What is the purpose of using near-infrared lighting?

A. To heat the camera sensor
B. To improve visibility of iris patterns regardless of eye colour or lighting conditions
C. To unlock the door faster
D. To eliminate biometric templates

# Chapter 12

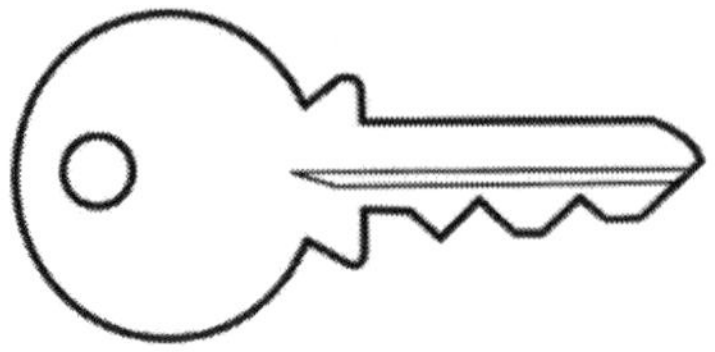

# Access Control in Critical Environments

This chapter examines the principles and technologies used to control and manage access within critical environments, where security failures can have serious consequences for safety, operations, and infrastructure. Facilities such as airports, hospitals, and data centres operate under demanding conditions that require reliable systems to regulate who can enter specific areas while ensuring that authorised personnel can move efficiently during routine operations and emergencies. As these environments often contain sensitive assets, critical infrastructure, and restricted operational zones, access-control systems must combine strong physical security with advanced electronic authentication and monitoring capabilities.

In modern facilities, access control rarely relies on a single locking device or security mechanism. Instead, it is implemented through integrated security systems that combine mechanical locks, electronic access control, biometric authentication, surveillance systems, and centralised monitoring platforms. These technologies work together to create layered security architectures that protect sensitive areas while providing traceability through event logging and real-time monitoring. Understanding how these systems interact is essential for security professionals and locksmiths responsible for installing, maintaining, and troubleshooting access-control infrastructure in high-risk environments.

The chapter explores how access-control systems operate within airports, hospitals, and data centres, highlighting the unique operational requirements of each environment. It examines how security zoning, multi-factor authentication, and integrated security management systems are used to balance strict protection of sensitive areas with the need for efficient operational workflows. The discussion also introduces key design concepts such as multi-layered security systems, fail-safe and fail-secure locking mechanisms, and emergency override systems, all of which play an important role in maintaining both safety and security.

## Airports, Hospitals, and Data Centres

Access control in high-stakes environments is essential for security, safety, and continuous operations. In airports, hospitals, and data centres, facilities house sensitive infrastructure, restricted zones, and high-value assets that demand robust access management to prevent unauthorized entry while enabling efficient movement of authorized personnel. Across these sectors, locksmiths and security professionals must understand how modern access-control ecosystems function within complex, high-security settings, including how systems integrate with surveillance, alarm, and security-management platforms to support situational awareness and rapid response to incidents [293-296] The literature consistently highlights that multimodal approaches, combining something the user has (cards/tokens), something the user is (biometrics), and something the user knows (PINs/passwords), enhance resilience against credential loss or theft and support safer operations in critical facilities [293, 295]. In aviation and healthcare contexts, there is particular emphasis on minimizing delays to legitimate activity while maintaining tight control over sensitive areas, a tension that drives adoption of layered, context-aware solutions rather than single-factor systems [294, 297, 298].

Airports constitute one of the most controlled environments due to the need to safeguard passengers, aircraft, and aviation infrastructure. Security zoning in airports typically stratifies spaces into public areas, staff-only zones, airside operational sectors, and highly restricted locations such as control towers and baggage-handling facilities. Access control systems at airports are designed to restrict entry to restricted zones to authorized personnel, and they are frequently integrated with centralized management platforms that monitor entry and exit events, support event correlation, and feed into broader security operations centres. In these facilities, electronic access cards and biometrics, such as fingerprint and facial recognition, are commonly deployed alongside centralised access-control software that correlates access events with CCTV and alarm data to enforce strict movement controls throughout the facility [294, 298-300]. High-security locking hardware and emergency override mechanisms are routinely required to comply with aviation security regulations, and locksmiths operating in airport environments must stay current with regulatory expectations while ensuring minimal disruption to critical operations during maintenance or incident response [294, 298]. The integrated security posture, combining access control with video surveillance and real-time monitoring, supports continuous operations and rapid threat detection, underscoring the need for reliable, scalable, and interoperable solutions suitable for demanding airport environments [293, 294, 298].

Hospitals present a parallel but distinct set of access-control requirements, balancing security with the imperative for rapid access during emergencies. Healthcare facilities contain sensitive zones such as operating theatres, pharmaceutical storage rooms, laboratories, and patient-records departments, where access controls must prevent unauthorized entry while not impeding urgent medical care. Hospital access control typically employs electronic key-card systems, biometric authentication, keypad entries, and automated door controls to limit access to restricted areas, with additional security measures often applied to high-risk areas such as maternity units or psychiatric wards. The design challenge in hospitals is to avoid hindering clinical workflows or emergency response while providing robust audit trails and access governance. Locksmiths and security professionals in this sector must select locking hardware and electronic-control configurations that integrate seamlessly with healthcare IT systems, preserve patient safety, and maintain operability during power outages or hospital-wide emergencies [293, 294, 298, 301]. Multimodal biometric approaches are increasingly considered to improve security guarantees in hospital contexts, given the sensitive nature of data and assets and the need to reduce impersonation risks without compromising rapid patient care [293].

Data centres represent another archetype of high-security environments where access control must tightly regulate who can reach critical infrastructure while supporting operational continuity and governance across complex physical and digital assets. Data-centre security typically employs multiple layers of protection, including perimeter fencing, security checkpoints, electronic access control, and biometric authentication (e.g., fingerprint or iris recognition). Many data centres also incorporate mantrap entry systems to enforce controlled progression into secure areas, ensuring that only authenticated personnel can access server rooms and other sensitive spaces. Because data centres store highly valuable information and support essential services across sectors, the access-control design emphasizes high reliability, integration with alarm and surveillance systems, and resilience to adverse conditions or disruptions. In this context, locksmiths must install high-security locking hardware and synergize physical-access systems with the broader security architecture to safeguard both physical assets and digital information [293, 294, 298, 302].

Across all three environments, access-control systems are increasingly embedded within broader security-management ecosystems that include surveillance cameras, alarm systems, and real-time monitoring software. This integration enables security operators to track access events, detect abnormal patterns, and respond swiftly to potential threats. The emphasis on continuous operation often necessitates redundancy, failover capabilities, and robust maintenance practices to ensure that access control remains functional under demanding conditions. The literature indicates that integrated platforms provide a holistic view of security, enabling corroboration between access events and surveillance data, and supporting incident response and regulatory compliance requirements in high-stakes facilities [293, 294, 302]. In practice, the strongest security outcomes arise from systems designed for layered defence, interoperability, and alignment with regulatory frameworks governing sensitive environments such as aviation, healthcare, and data processing facilities [293, 294, 298].

When considering technology choices, there is substantial evidence supporting the efficacy of multimodal and credential-agnostic approaches in critical environments. Studies and project reports discuss a range of modalities, from RFID and smart cards to keypad-based, biometric, and IoT-enabled solutions, with recurring emphasis on redundancy, auditability, and ease of integration with existing infrastructure. For instance, RFID-based door locks integrated with smart-card credentials are described as practical for large facilities that require scalable, layered access control, with configurations including solenoid-driven locks, keypad inputs, and networked controllers to support centralized management and remote administration [301, 303, 304] . Similarly, research on NFC-enabled and biometric-enabled systems highlights the vulnerabilities of single-factor keys and emphasizes the value of combining card-based access with biometric verification to strengthen security in high-risk settings such as hospitals and data centres [293, 294, 305]. Several studies also discuss IoT-enabled door locks and cloud-connected management platforms, illustrating how remote monitoring, logging, and analytics contribute to improved situational awareness and rapid enforcement of access policies in distributed critical environments [296, 298, 304, 306]. While not all sources focus on the strict regulatory context of aviation or hospital security, the consensus across the literature is that robust access control in critical environments benefits from multimodal credentials, device-hardening, and tight integration with security-management systems to achieve reliable protection of people, assets, and operations [293, 294, 298, 302].

In practice, locksmiths working in airports, hospitals, and data centres must balance security, usability, and regulatory compliance. This entails selecting locking hardware and electronic-control architectures that can withstand high use, support emergency egress, and operate reliably during contingencies, while ensuring that access events are accurately logged and auditable. It also requires understanding how to implement and maintain layered access-control approaches, including credential management, biometric enrolment, physical anti-tamper measures, and integration with surveillance and alarm systems. The body of work reviewed here indicates that the most effective solutions in critical environments are those that blend physical security with intelligent, integrated management tools, enabling authorized personnel to move efficiently when needed, while maintaining tight control over sensitive zones and assets. Through a synthesis of the cited sources, it is evident that access-control success in airports, hospitals, and data centres hinges on layered, multimodal authentication, robust locking hardware, and cohesive integration with security-management platforms to support continuous operations and resilience in the face of evolving threats [153, 293, 294, 298, 301, 302].

### Impact on Locksmithing Practices

Understanding access control in critical environments such as airports, hospitals, and data centres is highly relevant to someone interested in locksmithing because these facilities represent some of the most demanding and complex security environments in which locking and access-control systems are used. Unlike standard residential or small commercial installations, critical infrastructure requires systems that can reliably control access to sensitive areas while still allowing authorised personnel to move efficiently during daily operations or emergencies. For locksmiths, this means that modern practice increasingly involves understanding how mechanical locks, electronic access systems, and integrated security platforms work together to protect people, infrastructure, and information.

One of the key influences on locksmithing practice is the growing use of integrated security ecosystems. In environments such as airports and data centres, door locks are rarely standalone devices. Instead, they are connected to access-control controllers, surveillance systems, alarm networks, and central security-management software that monitors activity across an entire facility. Locksmiths must therefore understand how door hardware interfaces with electronic systems that log entry events, trigger alarms, and integrate with CCTV monitoring platforms. This requires locksmiths to move beyond purely mechanical work and develop familiarity with networked security systems, credential management, and system configuration.

Access control in critical environments also highlights the importance of layered security design, which significantly influences locksmithing decisions. Facilities such as airports and data centres often implement multiple authentication factors, including cards, biometric identifiers, and PIN codes, to reduce the risk of credential theft or impersonation. Locksmiths involved in these installations must select hardware that supports these technologies and ensure that door locks can interface with biometric readers, card readers, and electronic access controllers. As a result, locksmithing practice increasingly involves designing and installing multi-factor authentication systems rather than relying solely on traditional key-based access.

Another important consideration is the need to balance security with operational efficiency, particularly in environments such as hospitals. In healthcare facilities, access control systems must protect restricted areas such as pharmaceutical storage rooms and laboratories while still allowing medical staff to respond rapidly to emergencies. Locksmiths must therefore select and install hardware that supports quick and reliable access while maintaining strong security controls. This may involve installing electronic locks with emergency override features, fail-safe exit systems, or automatic door controls that comply with healthcare safety standards.

Working in critical environments also requires locksmiths to understand regulatory compliance and safety requirements. Airports must comply with strict aviation security regulations, hospitals must meet healthcare safety standards, and data centres must protect sensitive information infrastructure. Locksmiths installing or maintaining locking systems in these environments must ensure that the hardware, access-control configuration, and emergency release mechanisms meet these regulatory obligations. This includes ensuring proper fail-safe or fail-secure operation, emergency egress compliance, and tamper-resistant installations.

The complexity of these environments also places greater emphasis on reliability, durability, and redundancy in locking systems. Doors in airports, hospitals, and data centres often operate continuously and may be used thousands of times per day. Locksmiths must therefore select high-quality locking hardware designed for heavy use and ensure that systems remain operational even during power failures or network disruptions. Backup access methods, redundant power supplies, and regular maintenance procedures become essential components of locksmithing practice in these settings.

Understanding access control in critical environments allows locksmiths to play a broader role as security system advisors and integrators. Rather than simply installing locks, locksmiths may assist organisations in designing comprehensive access-control strategies that protect sensitive zones while supporting operational workflows. By understanding the security challenges faced by airports, hospitals, and data centres, locksmiths can recommend appropriate hardware, authentication technologies, and system configurations that support both security and efficiency.

## Multi-Layered Security Systems

Building a multi-layered security system involves designing a series of coordinated security measures that work together to protect a facility from unauthorised access, theft, or sabotage. Rather than relying on a single lock or access device, a layered system introduces several levels of protection so that if one security measure fails, other layers continue to provide protection. This approach is commonly described as defence in depth, where each layer acts as a barrier that slows, detects, or stops an intruder. For locksmiths and security professionals, building such a system requires careful planning, risk assessment, and integration of both physical and electronic security technologies.

The first step in building a multi-layered security system is conducting a security risk assessment. This process involves identifying what assets need protection, such as people, equipment, sensitive data, or restricted areas. The assessment also examines potential threats, including theft, vandalism, unauthorised entry, or internal security risks. By analysing the facility layout, operational requirements, and likely threats, security professionals can determine which areas require stronger protection and what types of security controls are appropriate. This assessment forms the foundation for designing an effective layered security strategy.

The next step is establishing perimeter security, which forms the outermost layer of protection. Perimeter controls are designed to prevent unauthorised individuals from entering the property in the first place. Typical measures include fencing, gates, vehicle barriers, lighting, and surveillance cameras that monitor the facility boundary. Controlled entry points may be installed where authorised personnel must present credentials such as access cards or biometric identification. By limiting entry points and monitoring activity around the facility, perimeter security reduces the risk of intrusion and allows security staff to detect suspicious activity early.

After perimeter protection is established, the system must address building entry control. This layer focuses on controlling access to the building itself through secure doors and entry points. Electronic locks, keycard readers, keypad entry systems, and biometric devices such as fingerprint or facial recognition scanners are often used to authenticate authorised users. These systems may be connected to central access-control software that records entry events, manages user permissions, and allows administrators to control access remotely. Locksmiths play a key role in installing door hardware such as electric strikes, magnetic locks, motorised deadbolts, and high-security cylinders that support these systems.

The next layer involves internal access zoning, which restricts movement within the building. Not all employees require access to every area of a facility, so security systems divide buildings into different zones with varying access levels. Sensitive areas such as server rooms, laboratories, medication storage areas, or financial records offices may require additional authentication measures beyond the main building entrance. This might include secondary card readers, biometric verification, or mantrap entry systems that allow only one authorised individual to pass through at a time. Zoning ensures that individuals can only enter areas relevant to their roles.

Another important component is surveillance and monitoring systems. Security cameras, motion detectors, and alarm systems provide real-time monitoring and detection of suspicious activity. These systems are often integrated with access control systems so that security personnel can view camera footage when a door is accessed or an alarm is triggered. Centralised security management software allows operators to monitor events, track access logs, and respond quickly to potential security incidents. Continuous monitoring adds an important detection layer that helps prevent or respond to breaches.

A well-designed layered system must also include redundancy and emergency planning. Security systems must continue to function even during power failures, equipment malfunctions, or emergency situations. Backup power supplies such as batteries or uninterruptible power systems (UPS) ensure that electronic locks and access control devices remain operational. Mechanical key overrides or emergency release mechanisms allow authorised personnel to access doors if electronic systems fail. Locksmiths must ensure that these backup systems are properly installed and comply with safety and building regulations.

Finally, building a multi-layered security system requires ongoing management and maintenance. Access permissions must be regularly reviewed, especially when staff roles change or employees leave the organisation. Hardware components such as locks, sensors, and readers require periodic inspection and servicing to ensure reliable operation. Security audits and system testing help identify weaknesses and ensure that all layers of the system continue to function as intended.

### Impact on Locksmithing Practices

Understanding how multi-layered security systems are designed is highly relevant to locksmiths because modern security rarely relies on a single lock or device. Locksmithing has evolved from primarily installing and servicing mechanical locks to designing integrated security solutions that combine physical hardware with electronic access control, surveillance, and monitoring systems. A locksmith working in commercial, institutional, or high-security environments must understand how locks fit within a broader "defence in depth" strategy, where multiple security layers such as perimeter protection, controlled entry points, and internal zoning work together to protect a facility. This knowledge allows locksmiths to recommend appropriate locking hardware and ensure that it supports the overall security design rather than functioning as an isolated component.

The concept of multi-layered security also influences how locksmiths approach installation and system integration. Modern locksmithing frequently involves installing electronic locking devices such as electric strikes, magnetic locks, motorised deadbolts, and biometric readers that operate within access control systems. Locksmiths must understand how these devices interact with door hardware, building infrastructure, and security management software. In addition, they must consider factors such as fail-safe versus fail-secure operation, emergency egress requirements, wiring routes, and compatibility with surveillance or alarm systems. As security technology advances, locksmiths increasingly act as security integrators who ensure that mechanical and electronic systems operate reliably together.

Multi-layered security affects the ongoing responsibilities and professional skills required in locksmithing practice. Locksmiths must be capable of conducting basic security risk assessments, advising clients on appropriate hardware for different security zones, and maintaining systems to ensure reliable long-term operation. They may also be responsible for tasks such as managing access credentials, servicing electronic locking components, and testing backup systems such as key overrides or battery power supplies. By understanding layered security principles, locksmiths can provide more comprehensive security solutions, helping organisations protect people, assets, and sensitive areas while maintaining safe and efficient access for authorised users.

## Fail-Safe Vs Fail-Secure Design

Fail-safe and fail-secure design are two fundamental concepts used in electronic locking and access control systems to determine how a lock behaves during a power failure or system malfunction. These design principles ensure that doors respond appropriately to emergencies while still maintaining security where required. The choice between fail-safe and fail-secure operation depends on the type of facility, the level of security required, and applicable safety regulations such as fire safety and emergency egress requirements.

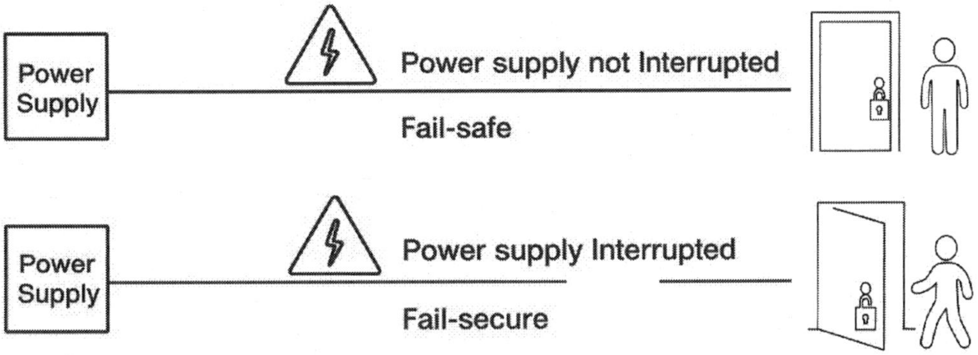


*Figure 80: Fail Safe vs Fail Secure.*

Fail-safe locks are designed to unlock automatically when power is lost. In normal operation, electricity is required to keep the lock engaged. If the power supply is interrupted, the locking mechanism releases and the door becomes unlocked. This design is commonly used in locations where life safety is the priority, such as emergency exits, stairwell doors, and fire escape routes. Magnetic locks (maglocks) are a common example of fail-safe devices because they rely on continuous electrical power to hold the door closed. If power is interrupted during an emergency, the lock releases so occupants can exit the building safely. Fail-safe systems are therefore preferred in areas where building evacuation must never be obstructed.

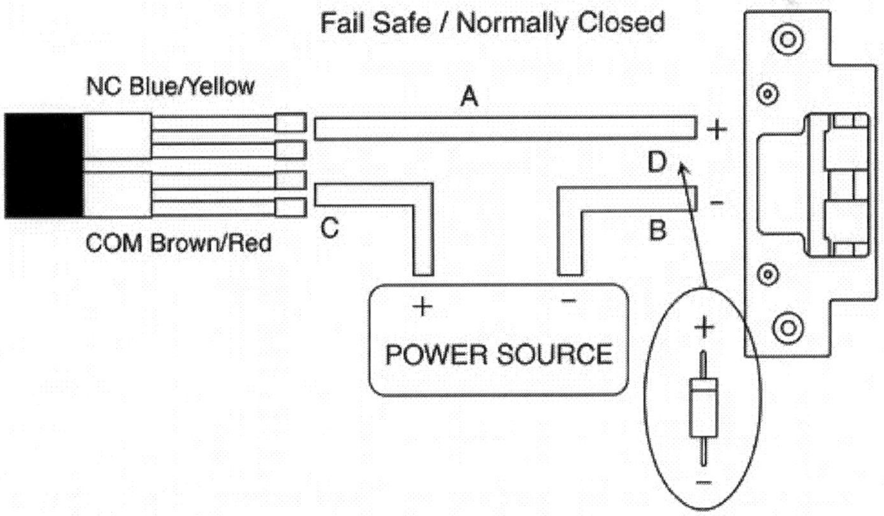


*Figure 81: Sample Fail Safw Lock wiring.*

In contrast, fail-secure locks remain locked when power is lost. These locks require electrical power to unlock the door, but the locking mechanism stays engaged if power fails. Fail-secure designs are typically used in locations where maintaining security is more important than automatic unlocking during a power outage. Examples include secure storage rooms, data centres, laboratories, and restricted access areas where unauthorised entry must be prevented even during power interruptions. Electric strikes and motorised deadbolts are often configured as fail-secure devices because the mechanical locking mechanism remains engaged unless an electrical signal releases it.

Selecting between fail-safe and fail-secure operation requires careful consideration of building safety codes, emergency procedures, and security risks. For example, doors used as emergency exits must generally be fail-safe or equipped with mechanical exit devices to ensure occupants can leave quickly during fires or other emergencies. However, sensitive areas such as medication storage, server rooms, or financial records offices often require fail-secure locking to prevent unauthorised access during power failures. In many buildings, both systems are used together depending on the function of each door.

**Impact on Locksmithing Practices**

For locksmiths and security professionals, understanding fail-safe and fail-secure design is essential when installing electronic locking systems. Proper configuration ensures that the locking hardware meets safety regulations, supports emergency evacuation procedures, and still provides effective protection for restricted areas. Incorrect selection or installation of these systems can create serious safety risks or security vulnerabilities, making careful planning and compliance with building codes a critical part of modern locksmithing practice.

## Emergency Override Systems

Emergency override systems are safety mechanisms built into locking and access-control systems that allow doors or secured areas to be opened during emergencies, even if normal access restrictions are in place. These systems are designed to ensure that people can exit a building safely or that emergency responders can gain rapid access when required. In environments where electronic access control, biometric systems, or high-security locks are used, emergency override functions provide a critical safeguard so that security measures do not prevent evacuation or delay emergency response.

One of the most common forms of emergency override is the mechanical override, which allows a door to be unlocked using a traditional key even if the electronic system fails. Many electronic locks and biometric access systems include a concealed key cylinder or mechanical lock as a backup. This ensures that authorised personnel, such as building managers or security staff, can still gain access if there is a power failure, electronic malfunction, or system outage. Mechanical overrides are particularly important in security systems that rely on electronic components such as electric strikes, magnetic locks, or motorised deadbolts.

Another type of emergency override involves emergency exit hardware, often referred to as panic bars or crash bars. These devices allow occupants to quickly open a door by pushing a horizontal bar, automatically releasing the locking mechanism. Panic hardware is commonly installed on exit doors in public buildings such as schools, hospitals, theatres, and commercial facilities. The purpose is to ensure that people can exit immediately during emergencies such as fires, evacuations, or other hazardous situations without needing keys, codes, or electronic authentication.

Emergency override systems may also include fire alarm integration and automatic door release mechanisms. In many buildings, electronic locks are connected to the fire alarm system so that doors automatically unlock when the alarm is activated. For example, magnetic locks on emergency exits are often configured to release when the fire alarm is triggered or when power is interrupted. This design ensures that escape routes remain accessible during emergencies and helps comply with building safety codes and fire regulations.

In some high-security environments, authorised emergency override credentials may be used. These credentials allow designated personnel such as emergency responders, security supervisors, or facility managers to bypass normal access restrictions. For example, special emergency access cards or override codes may unlock multiple doors within a building to allow rapid movement during a crisis. These systems are carefully controlled to ensure that override privileges are only available to authorised individuals.

**Impact on Locksmithing Practices**

From a locksmithing perspective, emergency override systems are essential because locksmiths are responsible for ensuring that security systems balance safety and security requirements. When installing or servicing electronic locking systems, locksmiths must ensure that emergency release mechanisms function correctly, comply with building and fire regulations, and remain reliable under real-world conditions. Proper design and testing of emergency override systems help ensure that buildings remain secure during normal operation while still allowing safe evacuation and emergency access when needed.

## Case Study: Access Control Failures in High-Risk Facilities

### Case Study (Hypothetical): Access Control Failures in a Pharmaceutical Research Facility

A large pharmaceutical research facility operated several secure laboratories where sensitive drug research and proprietary data were stored. Access to these laboratories was controlled through an electronic keycard system integrated with magnetic locks and surveillance cameras. Staff members were issued individual access cards that allowed entry to specific laboratories based on their roles and clearance levels. The facility relied heavily on the electronic access control system, assuming that the card-based authentication and monitoring software provided sufficient protection for restricted areas.

Over time, several weaknesses developed in the access control system. One of the primary issues was poor credential management. When staff members changed departments or left the organisation, their access permissions were not always updated or removed promptly. In some cases, access cards remained active long after an employee had left the facility. In addition, several employees began sharing access cards to allow colleagues to enter laboratories when they forgot their credentials. Because the system relied only on single-factor authentication (the access card), it was difficult to determine whether the person using the card was actually the authorised employee.

The failure became evident when an internal security audit revealed that an unauthorised contractor had entered a restricted laboratory using a borrowed access card. Although the contractor had legitimate permission to work elsewhere in the building, they did not have clearance for the laboratory. The breach occurred because the cardholder allowed the contractor to use their card to save time, bypassing security procedures. The access-control system recorded the entry as a valid event because the card belonged to an authorised employee, highlighting a weakness in the authentication process.

Further investigation revealed additional design flaws in the facility's access control system. The magnetic locks used on laboratory doors were configured as fail-safe, meaning the doors automatically unlocked during power interruptions. While this configuration met fire safety requirements, it created a security vulnerability because backup power systems had not been properly maintained. During a brief electrical fault earlier in the year, several laboratory doors automatically unlocked, leaving sensitive areas temporarily unsecured.

Following the incident, the facility implemented several improvements to its security system. The organisation introduced multi-factor authentication, combining access cards with biometric fingerprint verification for entry into high-security laboratories. Access privileges were reviewed and updated regularly to ensure that only authorised personnel retained entry rights. Staff members also received training on security procedures, including strict policies prohibiting the sharing of access credentials.

This case study highlights the importance of layered security, proper system configuration, and regular security audits. By combining physical security hardware, electronic authentication, and clear operational procedures, organisations can significantly reduce the risk of access control failures in high-risk facilities.

**Impact on Locksmithing Practices**

From a locksmithing and security professional perspective, this case demonstrates how access control failures often result from a combination of technical, procedural, and human factors. Even when advanced electronic systems are installed, weaknesses in system configuration, maintenance, and credential management can undermine security. Locksmiths involved in designing and maintaining such systems must ensure that locking hardware, authentication methods, and backup mechanisms work together to provide both safety and security.

## Chapter 12 Review Questions

### Question 1

A security team at a large airport installs access-control readers at entrances to staff-only and airside zones. Each entry event is automatically logged and linked with CCTV footage.

What is the primary purpose of integrating access control with surveillance systems?

A. To eliminate the need for physical locks
B. To provide real-time monitoring and correlate entry events with video evidence
C. To increase door opening speed
D. To reduce the number of employees required at the airport

### Question 2

A hospital installs electronic locks on pharmaceutical storage rooms but also ensures that authorised staff can gain rapid access during medical emergencies.

What key design requirement is being addressed?

A. Maximising lock weight
B. Limiting staff movement
C. Balancing strong security with operational efficiency
D. Eliminating electronic systems

### Question 3

A data centre installs a mantrap entry system before its main server room. Staff must authenticate at two doors before entering the secure area.

What is the primary purpose of this design?

A. To reduce electricity usage
B. To prevent unauthorised tailgating and ensure controlled entry
C. To eliminate the need for authentication
D. To speed up staff movement

**Question 4**

A security consultant recommends combining RFID cards, fingerprint scanners, and PIN codes for access to a high-security research facility.

What security principle does this approach represent?

A. Single-factor authentication
B. Multi-factor authentication
C. Mechanical redundancy
D. Lock reinforcement

**Question 5**

A locksmith installs electronic locks in a hospital emergency exit corridor that automatically unlock when power is lost to allow safe evacuation.

What type of system is being used?

A. Fail-secure locking
B. Manual override locking
C. Fail-safe locking
D. Mechanical deadbolt locking

**Question 6**

A server room door uses an electric strike configured so that the door remains locked during a power outage to protect sensitive data infrastructure.

What type of locking design is this?

A. Fail-safe
B. Fail-secure
C. Magnetic release
D. Mechanical bypass

**Question 7**

A corporate facility installs perimeter fencing, monitored entry gates, card-controlled building entrances, and biometric readers for restricted internal areas.

What type of security strategy does this represent?

A. Multi-layered defence system
B. Single-device protection
C. Passive monitoring system
D. Mechanical-only security

**Question 8**

During a fire alarm in a large office building, several magnetic locks automatically release so that occupants can exit through emergency doors.

What system integration allowed this to occur?

A. CCTV integration
B. Fire alarm integration with electronic locks
C. Key duplication systems
D. Door alignment mechanisms

**Question 9**

A pharmaceutical research facility discovers that former employees still have active access cards and that some workers have been sharing credentials.

What was the primary weakness in the security system?

A. Poor door construction
B. Weak key cutting procedures
C. Poor credential management and single-factor authentication
D. Faulty surveillance cameras

**Question 10**

A security consultant advises installing backup batteries and mechanical key overrides for an electronic access control system in a data centre.

What is the primary purpose of these features?

A. To increase network speed
B. To provide redundancy and maintain access during system failures
C. To reduce system cost
D. To eliminate electronic authentication

# Part 7

# Locksmithing in Unusual and Extreme Contexts

# Chapter 13

# Locks in Harsh Environments

Locking systems are often associated with residential doors, offices, or commercial buildings, but in many industries locks must operate under conditions far more demanding than those found in everyday environments. Facilities such as ships, offshore oil platforms, mining operations, and remote industrial installations expose locking hardware to corrosive saltwater, dust contamination, vibration, extreme temperatures, and heavy mechanical stress. In these settings, locks are not simply security devices—they are critical components that contribute to operational safety, equipment protection, and regulatory compliance. Selecting the wrong type of lock for these environments can lead to rapid deterioration, system failure, or serious safety risks.

This chapter explores how locks are designed, selected, and maintained for use in harsh environments, particularly in marine, mining, and offshore industrial applications. It examines the environmental challenges that affect lock performance and explains the specialised materials, mechanisms, and protective features required to ensure reliability under demanding conditions. The chapter also presents examples of marine-grade padlocks, weatherproof industrial locks, explosion-proof hardware, watertight door locks, and electronic access systems used in industrial facilities. Understanding these specialised locking solutions highlights the expanding role of modern locksmiths, who must increasingly integrate knowledge of materials science, environmental engineering, and industrial safety standards when designing and maintaining security systems in extreme environments.

## Marine, Mining, and Offshore Applications

Locks used in marine, mining, and offshore environments must operate reliably in some of the harshest conditions encountered in industrial security. These environments expose locking hardware to saltwater corrosion, high humidity, dust, vibration, extreme temperatures, and mechanical stress. In addition, many of these facilities operate continuously and require locking systems that maintain both security and operational reliability under demanding conditions. For locksmiths and security professionals, selecting appropriate locks for these environments requires understanding specialised materials, sealing technologies, and locking mechanisms designed specifically for harsh industrial applications.

**Environmental Challenges in Marine, Mining, and Offshore Settings**

Before selecting locks for these environments, it is important to understand the operational conditions that influence lock performance.

Marine environments, such as ships, ports, and coastal infrastructure, expose locking hardware to saltwater spray, high humidity, and constant corrosion risk. Salt deposits can quickly degrade standard steel components, leading to seized mechanisms or weakened structural integrity. Marine locks must therefore be made from corrosion-resistant materials such as marine-grade stainless steel (316), brass, bronze, or specialised coated alloys.

Mining environments present different challenges. Underground mines expose locks to dust, abrasive particles, vibration from heavy machinery, and sometimes explosive atmospheres. Surface mining operations also introduce exposure to mud, chemicals, and weather extremes. Locks in mining facilities must therefore be dust-resistant, impact-resistant, and capable of operating despite heavy contamination.

Offshore oil and gas platforms combine many of the hazards found in both marine and industrial settings. Locks may be exposed to salt spray, high winds, mechanical vibration, hazardous chemicals, and strict safety regulations related to fire and explosion risk. Equipment used in offshore facilities must often comply with hazardous-area standards, such as explosion-proof design requirements.

**Lock Types Suitable for Harsh Industrial Environments**

<u>1. Marine-Grade Padlocks</u>

Marine-grade padlocks are specifically designed to resist corrosion and environmental damage. These locks are commonly used on ships, docks, marinas, and offshore platforms.

Typical features include:

- 316 stainless steel or brass bodies
- Corrosion-resistant shackles
- Weatherproof seals to prevent water ingress
- Non-rusting internal springs and components
- Drainage holes to prevent water accumulation

Examples of applications include:

- Securing deck equipment
- Locking storage lockers on vessels
- Securing chain barriers on docks
- Protecting maintenance tool cabinets

Because marine environments are highly corrosive, stainless steel padlocks are often preferred over hardened steel locks that may rust rapidly.

Several widely used commercial padlocks illustrate the types of locking hardware commonly selected for marine and offshore applications. These locks are designed to withstand highly corrosive environments such as saltwater exposure, humidity, and constant outdoor use. Manufacturers address these challenges by using corrosion-resistant materials such as brass and marine-grade stainless steel, sealed lock cylinders, and internal components designed to resist rust and contamination.

One of the most recognised manufacturers of corrosion-resistant padlocks is ABUS, which produces several locks specifically designed for marine environments. The ABUS 70IB Marine Padlock features a solid brass lock body combined with a stainless steel shackle, corrosion-resistant internal components, and a sealed keyway that prevents salt and dirt from entering the locking mechanism. This model is widely used for securing boats, docks, and outdoor storage areas. Another example is the ABUS 75IB Marine Grade Stainless Steel Padlock, which uses both a stainless steel body and shackle to provide high resistance to saltwater corrosion. These locks are commonly used to secure boat deck lockers, marina storage cages, dock equipment, and outdoor toolboxes that are frequently exposed to seawater.

Another widely used manufacturer is Master Lock, which produces a range of marine-grade padlocks designed for harsh coastal environments. The Master Lock 640D Marine Padlock features a solid brass body and a hardened steel shackle with a rust-resistant coating, along with a dual ball locking mechanism that improves resistance to forced entry. The Master Lock 40 Marine Padlock incorporates a brass body with stainless steel internal components and protective seals to improve weather resistance. These locks are typically used to secure dock gates, storage lockers on vessels, chain barriers at marinas, and maintenance cabinets exposed to outdoor conditions.

The British manufacturer Squire also produces high-security padlocks widely used in industrial marine environments. The Squire Stronghold SS50 Marine Padlock features a stainless steel lock body and corrosion-resistant shackle designed to operate reliably in harsh outdoor conditions. Another example is the Squire Marine Grade Brass Padlock, which uses a solid brass body combined with a stainless steel shackle and weatherproof internal components. These locks are frequently installed on offshore platform storage compartments, dockside equipment enclosures, port facility gates, and shipboard storage areas where long-term corrosion resistance is essential.

American Lock manufactures heavy-duty industrial padlocks designed for harsh operational environments, including marine and coastal infrastructure. The American Lock A1106 Marine Padlock features a solid brass body with a stainless steel shackle and weather-resistant construction suitable for demanding outdoor use. Another example is the American Lock A5200 Marine Padlock, which incorporates a laminated steel body with corrosion-resistant coatings and a hardened shackle for improved durability. These locks are commonly used to secure industrial dock gates, port container storage areas, and marine maintenance lockers where strong and durable locking hardware is required.

In high-security environments, locks produced by ABLOY are often selected because of their advanced disc-detainer locking mechanisms, which perform reliably even when exposed to dirt, moisture, and corrosion. The ABLOY PL321 Marine Padlock features a hardened steel body with corrosion protection and a disc-detainer cylinder that is resistant to contamination and picking. The ABLOY PL362 Marine Padlock is a heavy-duty model designed for extreme environments and provides resistance to both corrosion and physical attack. These locks are commonly used in highly sensitive locations such as offshore oil platforms, port infrastructure, marine industrial facilities, and critical infrastructure storage areas where both environmental durability and high security are required.

2. Weatherproof Industrial Padlocks

Industrial weatherproof padlocks are widely used in mining operations, construction sites, and processing facilities.

These locks are typically designed with:

- Protective rubber or polymer outer covers
- Dust seals to prevent contamination
- Heavy-duty internal mechanisms
- High shackle strength to resist cutting
- Lockout compatibility for safety systems

Mining operations frequently use these locks to secure:

- Equipment cabinets
- Electrical panels
- Conveyor system access points
- Safety lockout points during maintenance

Some industrial padlocks are specifically designed to function even when covered in dust, mud, or grease.

Industrial weatherproof padlocks are widely used in mining operations, construction sites, processing plants, and heavy industrial facilities where locks must continue operating despite exposure to dust, mud, vibration, and extreme weather conditions. Manufacturers design these locks with protective rubber or polymer covers, sealed keyways, reinforced shackles, and corrosion-resistant internal mechanisms so they can withstand contamination and heavy daily use. Many industrial padlocks are also designed to support lockout–tagout (LOTO) safety systems, which are critical for isolating machinery during maintenance and preventing accidental equipment activation.

One commonly used example is the Master Lock 4400 Series weatherproof padlock, manufactured by Master Lock. This lock features a Zenex thermoplastic body with a protective outer coating, making it highly resistant to chemicals, dust, and corrosion. The lock includes a stainless steel shackle and sealed internal components to prevent contamination. It is frequently used in mining and industrial environments for securing electrical isolation points, safety lockout stations, and equipment maintenance panels, particularly where lockout–tagout procedures are required.

Another widely used industrial padlock is the ABUS 83WP Weatherproof Padlock from ABUS. This padlock includes a protective weatherproof jacket that shields the lock body from dust, water, and debris, along with a hardened steel shackle designed to resist cutting and forced entry. The sealed cylinder helps prevent dirt and grease from entering the mechanism, allowing the lock to function reliably even in harsh industrial conditions. These locks are often installed on conveyor system access gates, outdoor machinery cabinets, and equipment enclosures in mining sites.

Heavy-duty industrial operations frequently use padlocks produced by American Lock, particularly the American Lock A700 Series weatherproof padlocks. These locks feature laminated steel bodies with corrosion-resistant coatings and weather-sealed cylinders that help protect internal components from contamination. The hardened boron alloy shackles provide high resistance to cutting or prying, making them suitable for high-security industrial environments. In mining facilities, these locks are commonly used to secure electrical panels, control boxes, and tool storage areas where both durability and security are required.

Industrial lockout systems often rely on specialised padlocks produced by Brady, such as the Brady Safety Lockout Padlock Series. These padlocks are specifically designed for lockout–tagout procedures in hazardous industrial environments. They feature lightweight yet durable polymer bodies, corrosion-resistant shackles, and non-conductive materials that improve electrical safety. Brady lockout padlocks are widely used in mining and processing plants to secure machine isolation points, maintenance lockout stations, and safety interlocks during equipment servicing.

Another manufacturer known for robust industrial locking hardware is Squire, which produces heavy-duty weather-resistant padlocks such as the Squire Stronghold Series. These locks are designed with hardened steel bodies, protective coatings, and high-strength shackles capable of withstanding extreme industrial use. They are commonly installed on industrial gates, mining facility equipment cages, and heavy machinery storage areas where strong physical security is required alongside environmental durability.

Across mining, construction, and processing facilities, these industrial weatherproof padlocks play a critical role in protecting equipment and maintaining workplace safety. Their sealed mechanisms, corrosion-resistant materials, and heavy-duty construction allow them to function reliably even when exposed to dust, mud, grease, vibration, and extreme weather, conditions that would quickly damage standard commercial padlocks. As a result, selecting appropriate industrial-grade locking hardware is essential for maintaining both security and operational reliability in harsh industrial environments.

3. Explosion-Proof and Hazardous-Area Locks

In offshore oil platforms and some mining environments, equipment must be designed to prevent sparks that could ignite flammable gases or vapours.

Locks used in these environments may include:

- ATEX-certified locking hardware
- Non-sparking materials such as brass or bronze
- Sealed mechanisms to prevent ignition sources
- Anti-static materials

These locks are commonly installed on:

- Hazardous chemical storage cabinets
- Offshore drilling control rooms
- Fuel storage access points
- Gas monitoring equipment enclosures

Compliance with hazardous-area safety regulations is essential in these applications.

ATEX-certified locking hardware refers to locks, latches, and access-control components that are designed and approved for use in explosive or hazardous environments where flammable gases, vapours, or combustible dust may be present. The term ATEX originates from the French phrase *"ATmosphères EXplosibles"* and refers to a set of European Union directives that regulate equipment used in potentially explosive atmospheres. These directives ensure that equipment installed in hazardous locations is designed in a way that prevents ignition sources such as sparks, excessive heat, or electrostatic discharge.

In hazardous industrial environments—such as offshore oil platforms, petrochemical refineries, chemical plants, gas storage facilities, and certain mining operations—even a small spark from metal friction, electrical discharge, or static electricity could ignite an explosive atmosphere. ATEX-certified locking hardware is therefore designed to eliminate or minimise these risks. Locks used in these environments are typically constructed from non-sparking materials such as brass, bronze, or specialised stainless steel alloys that reduce the likelihood of producing sparks when struck or manipulated. In addition, these locks often incorporate anti-static finishes and sealed internal mechanisms to prevent the accumulation of combustible dust or vapours within the lock body.

ATEX-certified locking hardware must also meet strict engineering and testing requirements to ensure safe operation. Components such as cylinders, shackles, latches, and housings are designed to withstand harsh environmental conditions without producing heat or mechanical friction that could trigger ignition. Many ATEX locks use sealed enclosures and compression latching systems to prevent the entry of flammable gases or dust into electrical cabinets and equipment housings. This is particularly important for locks installed on control panels, instrumentation cabinets, or monitoring equipment where electrical components are present.

ATEX certification is based on two primary European directives: ATEX 2014/34/EU, which governs equipment and protective systems intended for use in explosive atmospheres, and ATEX 1999/92/EC, which outlines workplace safety requirements for environments where explosive atmospheres may occur. Equipment certified under these directives is labelled according to specific hazard zones, such as Zone 0, Zone 1, or Zone 2 for gases and vapours, and Zone 20, Zone 21, or Zone 22 for combustible dust environments. Locking hardware used in these areas must meet the safety requirements appropriate for the specific hazard classification.

In practice, ATEX-certified locks are commonly installed on hazardous chemical storage cabinets, offshore drilling control rooms, fuel storage access points, gas monitoring equipment enclosures, and explosion-proof electrical cabinets. These locks often take the form of quarter-turn latches, compression locks, industrial cabinet locks, or heavy-duty padlocks, depending on the application.

The goal is to provide secure access control while ensuring that the locking mechanism itself does not introduce any ignition risk.

For locksmiths and industrial security professionals, understanding ATEX-certified locking hardware is important when working in oil and gas, mining, petrochemical, and industrial processing environments. Installing standard locking hardware in hazardous zones could create serious safety risks or violate regulatory requirements. Therefore, locksmiths operating in these sectors must ensure that locks, latching mechanisms, and access-control components are certified for hazardous environments, constructed from appropriate materials, and installed in accordance with safety standards. This ensures that facilities remain both secure and compliant with regulations governing explosive atmospheres.

Locks used in explosion-prone or hazardous environments must meet strict safety requirements to prevent ignition of flammable gases, vapours, or combustible dust. Facilities such as offshore oil platforms, petrochemical plants, refineries, and certain mining environments operate under regulations that require equipment capable of functioning safely in potentially explosive atmospheres. In these environments, locking hardware is often manufactured from non-sparking materials such as brass, bronze, or stainless steel, and may include ATEX-certified components, sealed locking mechanisms, and anti-static finishes to minimise ignition risks. These locks are commonly installed on hazardous chemical storage cabinets, offshore drilling control rooms, fuel storage areas, and gas monitoring equipment enclosures.

One manufacturer known for hazardous-environment locking solutions is ABLOY, which produces high-security industrial locks suitable for extreme environments. The ABLOY PL362 padlock, for example, features a hardened steel body with corrosion-resistant coatings and a sealed disc-detainer locking mechanism that performs reliably in environments exposed to moisture, dirt, and industrial contaminants. Because the mechanism does not rely on springs that could corrode or jam, it is particularly suitable for offshore installations and petrochemical facilities. These locks are often used on fuel storage access points, offshore equipment cabinets, and restricted areas within oil and gas platforms.

Another well-known supplier is Squire, which manufactures heavy-duty industrial padlocks used in hazardous environments. The Squire Stronghold SS80CS stainless steel padlock is constructed entirely from corrosion-resistant stainless steel and includes sealed internal components to prevent contamination. Its robust design and high resistance to environmental damage make it suitable for installations in offshore drilling platforms, hazardous material storage areas, and industrial processing plants where equipment must withstand both corrosion and harsh operating conditions.

Industrial hazardous-area equipment often incorporates locks manufactured by Emka, a company specialising in locking systems for industrial enclosures and control cabinets. EMKA produces ATEX-compliant quarter-turn locks and compression latches designed for electrical cabinets and equipment housings located in explosive atmospheres. These locking systems are commonly installed on gas monitoring equipment enclosures, hazardous electrical control cabinets, and offshore process control systems, where sealed and anti-spark designs are essential for compliance with hazardous-area safety regulations.

Another manufacturer supplying locking systems for hazardous industrial environments is Dirak, which produces specialised enclosure locks and latches for industrial and hazardous-area installations. DIRAK locking systems include explosion-proof quarter-turn latches and compression locks designed for electrical panels and machinery enclosures operating in potentially explosive atmospheres. These locks are frequently used in chemical processing plants, refinery control cabinets, and offshore instrumentation enclosures, where maintaining secure access while preventing spark generation is critical.

Heavy industrial facilities also utilise locking hardware from Master Lock, particularly brass padlocks designed for hazardous environments. Models such as the Master Lock 6835 brass safety padlock use non-sparking brass construction and corrosion-resistant components, making them suitable for use around flammable materials and fuel storage areas. These locks are commonly installed on chemical storage cabinets, fuel storage cages, and industrial safety lockout points where both environmental durability and ignition prevention are required.

Across offshore oil platforms, mining operations, and chemical processing facilities, explosion-resistant locking hardware plays a critical role in maintaining safety. By using non-sparking materials, sealed mechanisms, anti-static finishes, and certified hazardous-area designs, these locks help prevent ignition risks while ensuring that restricted equipment and hazardous materials remain securely protected. For locksmiths and industrial security professionals, selecting locks that comply with hazardous-area standards is essential to ensure both operational safety and regulatory compliance in environments where even a small spark could have serious consequences.

### 4. Watertight Marine Door Locks

Ships and offshore structures often use watertight doors and hatches to maintain vessel integrity. These doors require specialised locking systems capable of maintaining pressure seals.

Features typically include:

- Heavy stainless-steel construction
- Compression handles that secure watertight seals
- Corrosion-resistant internal components
- High-strength locking cams

Applications include:

- Ship bulkhead doors
- Engine room access
- Offshore platform maintenance compartments
- Submersible equipment housings

These locks are essential for maintaining safety and structural integrity in maritime environments.

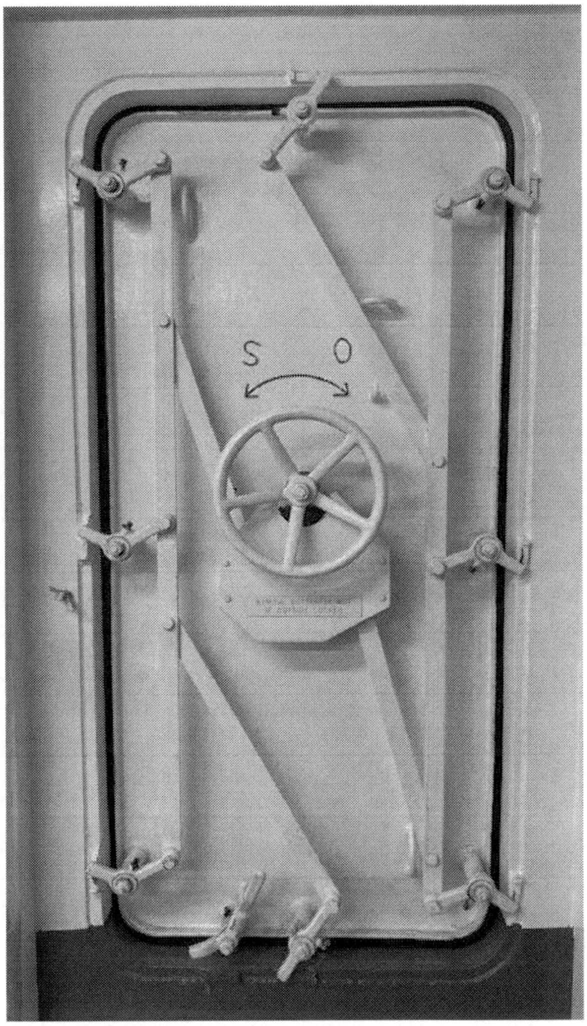


*Figure 82: Watertight Marine Door. Capt. NaZri, Public Domain, via Pexels.*

5. Electronic Access Locks for Offshore and Industrial Sites

Modern offshore platforms and large mining operations increasingly use electronic access control systems to manage personnel movement.

These locks may include:

- Weatherproof electronic keypads
- RFID card readers
- Biometric fingerprint scanners
- Motorised locking bolts

Industrial electronic locks are typically designed with:

- IP-rated waterproof housings
- corrosion-resistant materials
- sealed electronic circuits
- backup mechanical override keys

Examples of usage include:

- Offshore control rooms
- equipment monitoring stations
- hazardous material storage areas
- restricted engineering zones

Electronic access locks used in offshore platforms, mining operations, refineries, and heavy industrial facilities are designed to withstand extreme environmental conditions while providing secure and controlled access to restricted areas. These systems often combine weatherproof electronic keypads, RFID card readers, biometric authentication devices, and motorised locking mechanisms to regulate personnel movement and maintain detailed access records. Because these installations are frequently exposed to moisture, dust, vibration, and corrosive environments, industrial electronic locks typically feature IP-rated waterproof housings, corrosion-resistant materials, sealed electronic components, and mechanical override keys to ensure reliability even during power failures or system malfunctions.

One widely used manufacturer of industrial access-control hardware is ASSA ABLOY, which produces robust electronic locking systems for demanding environments. Products such as the ASSA ABLOY Aperio electronic locking platform are commonly installed in industrial and offshore facilities. Aperio locks integrate wireless communication with RFID card readers, allowing doors to connect to centralised access-control systems while reducing wiring requirements. These locks are frequently installed in offshore control rooms, maintenance workshops, and restricted engineering areas, where authorised personnel must be authenticated before entry.

Another major provider of industrial electronic locking technology is Salto Systems. The Salto XS4 electronic locking system is designed for high-use environments and supports RFID cards, mobile credentials, and biometric authentication integration. Many models feature weather-resistant housings and sealed electronic components, making them suitable for industrial outdoor installations. These locks are commonly used in equipment monitoring stations, industrial laboratories, and engineering access points within large mining or offshore facilities.

Electronic access locks manufactured by dormakaba are also widely deployed in industrial environments. Systems such as the dormakaba exos access control platform combine electronic door hardware with centralised security management software. These systems often incorporate RFID readers, keypad authentication, and biometric verification options, enabling facility managers to monitor personnel movement across complex industrial sites. Dormakaba locks are frequently installed in hazardous material storage areas, industrial processing rooms, and data control centres within mining and energy facilities.

Another manufacturer specialising in rugged industrial locking solutions is Southco, which produces electronic rotary latches and access-control systems designed for equipment enclosures and heavy machinery compartments. Southco electronic locks often include motorised latching mechanisms, sealed housings, and corrosion-resistant construction, allowing them to operate reliably in environments exposed to dust, moisture, and vibration. These systems are commonly installed on industrial control cabinets, remote monitoring equipment housings, and offshore instrumentation panels.

Biometric access control is increasingly used in high-security industrial environments, and companies such as Suprema provide advanced biometric authentication systems suitable for these installations. Devices such as the Suprema BioEntry series combine fingerprint scanning with RFID card authentication and networked access-control software. These systems are typically installed at restricted engineering zones, critical infrastructure control rooms, and offshore operational centres, where higher levels of identity verification are required.

Across offshore platforms, mining operations, and industrial processing plants, these electronic access locks play a vital role in controlling personnel access to sensitive areas. Their rugged design, environmental protection features, and integration with centralised security systems allow organisations to maintain secure operations while ensuring that authorised personnel can move efficiently within complex industrial facilities.

### Materials Used in Harsh Environment Locks

The durability of locks used in marine, mining, and offshore environments depends heavily on the materials used in their construction. These industries expose locking hardware to extreme conditions such as saltwater corrosion, moisture, dust, chemicals, and heavy mechanical wear. As a result, standard residential lock materials are often unsuitable because they can corrode, seize, or degrade quickly when exposed to such environments. To ensure reliable long-term performance, manufacturers use specialised materials that provide enhanced resistance to corrosion, contamination, and mechanical stress.

One of the most commonly used materials in harsh environments is 316 stainless steel, which offers excellent resistance to corrosion, particularly in saltwater and marine atmospheres. This material is widely used for lock bodies, shackles, and external components on ships, docks, offshore platforms, and coastal facilities. Brass is another commonly used material because it is naturally corrosion resistant and produces very few sparks when struck, making it suitable for hazardous environments such as fuel storage areas and chemical processing facilities. Bronze alloys are also frequently used in marine locking hardware due to their durability and strong resistance to saltwater corrosion.

Additional materials are often used to protect internal lock components and extend the operational life of the locking mechanism. Nickel-plated steel is commonly applied to internal parts such as springs, pins, and cylinders to provide corrosion protection while maintaining mechanical strength. Many industrial locks also include polymer or protective coatings on the outer surfaces of the lock body to shield against chemicals, moisture, dust, and physical abrasion. Together, these specialised materials significantly extend the service life of locks used in harsh environments compared to standard residential locking hardware.

### Maintenance Considerations

Locks operating in marine, mining, and offshore environments are exposed to extreme conditions that can accelerate wear, corrosion, and mechanical failure. For this reason, regular inspection and maintenance are essential to ensure that locking hardware continues to operate reliably and securely. Harsh environmental factors such as saltwater exposure, dust, vibration, and chemical contamination can gradually degrade lock components if they are not properly maintained.

Typical maintenance procedures involve several routine checks and preventative actions. Locks exposed to marine environments should be rinsed with fresh water to remove salt deposits that can cause corrosion or interfere with the locking mechanism. Corrosion-resistant lubricants should be applied periodically to internal components to maintain smooth operation and protect against rust. Maintenance inspections should also include checking protective seals and covers to ensure they remain intact and continue to prevent moisture, dust, or debris from entering the lock body.

In addition, technicians should regularly inspect shackles, locking bolts, and other load-bearing components for signs of wear, deformation, or corrosion that could weaken the lock or reduce its security. Any worn or damaged parts should be replaced before they fail to prevent operational disruptions or security breaches. Preventative maintenance is especially important in remote offshore platforms, mining sites, or industrial facilities, where equipment failure can cause delays, increase safety risks, or interrupt critical operations.

### Impact on Locksmithing Practices

For locksmiths and security professionals, working with marine, mining, and offshore locks requires specialised knowledge beyond conventional lock installation. Professionals must understand environmental risks, corrosion protection, hazardous-area safety standards, and industrial security requirements. Selecting inappropriate locking hardware for these environments can lead to rapid corrosion, lock failure, or safety hazards.

As industrial facilities increasingly integrate electronic access control with traditional locking hardware, locksmiths must also be able to install weather-resistant electronic locking systems while ensuring mechanical reliability. By understanding the demands of harsh industrial environments, locksmiths can recommend and install locking systems that maintain security, safety, and operational reliability under extreme conditions.

## Corrosion, Pressure, and Extreme Temperatures

Locks used in harsh environments must withstand several extreme environmental conditions that can significantly affect their performance and lifespan. Among the most important of these conditions are corrosion, pressure, and extreme temperatures. These factors can damage locking mechanisms, weaken materials, and interfere with the smooth operation of lock components. Understanding how these environmental forces affect locks is essential for selecting appropriate locking hardware and ensuring reliable operation in harsh industrial settings.

Corrosion is one of the most significant threats to locks used in marine and offshore environments. Corrosion occurs when metal components react chemically with oxygen, moisture, or salt, causing the material to gradually deteriorate. In marine environments, saltwater and salt-laden air accelerate corrosion, which can cause lock bodies, shackles, springs, and internal pins to rust or seize. This process can weaken structural components and prevent the lock from functioning properly. To reduce corrosion risks, locks used in these environments are often constructed from materials such as 316 stainless steel, brass, or bronze, and may include protective coatings or sealed cylinders that prevent moisture and contaminants from entering the mechanism.

Pressure can also influence the performance of locks, particularly in offshore, subsea, and deep mining environments. In underwater installations or pressurised environments, external pressure can place mechanical stress on lock bodies, seals, and internal components. High pressure can cause seals to deform, allow water ingress, or interfere with the movement of internal locking parts. In some offshore installations, equipment enclosures are designed with compression seals and pressure-resistant housings to maintain internal integrity. Locks used in these systems must therefore be robust enough to maintain their structural integrity and continue functioning despite mechanical stress and environmental pressure.

Extreme temperatures present another challenge for locks used in industrial environments. Very high temperatures, such as those found near industrial machinery, processing plants, or desert mining operations, can cause metal components to expand, lubricants to break down, and electronic locking systems to malfunction. Conversely, extremely cold environments—such as offshore platforms in polar regions or high-altitude mining sites—can cause materials to contract and lubricants to thicken or freeze, making locks difficult to operate. Temperature fluctuations can also lead to condensation inside the lock, increasing the risk of corrosion. To address these challenges, industrial locks are often designed with temperature-resistant materials, specialised lubricants, and sealed housings that help maintain reliable operation across a wide temperature range.

**Impact on Locksmithing Practices**

For locksmiths and security professionals working in these environments, understanding the effects of corrosion, pressure, and extreme temperatures is essential when selecting and maintaining locking hardware. Choosing locks constructed from appropriate materials, ensuring proper sealing and lubrication, and conducting regular maintenance inspections can significantly reduce the risk of lock failure in harsh environmental conditions.

## Dust, Vibration, and Mechanical Fatigue

Locks installed in harsh industrial environments such as mining operations, construction sites, offshore platforms, and heavy manufacturing facilities are frequently exposed to dust, vibration, and mechanical fatigue. These environmental factors can significantly affect the reliability and lifespan of locking hardware. If not properly managed, they can cause locks to malfunction, become difficult to operate, or fail completely. For locksmiths and security professionals working in these environments, understanding how these forces affect lock performance is essential for selecting appropriate locking systems and implementing effective maintenance strategies.

Dust contamination is a major issue in environments such as mines, quarries, cement plants, and construction sites where airborne particles are constantly present. Dust can enter the keyway or internal components of a lock and accumulate inside the cylinder, interfering with the movement of pins, springs, and other internal mechanisms. Over time, this buildup can cause locks to become stiff, jammed, or completely inoperable. Fine abrasive particles can also accelerate wear on internal components, reducing the lock's lifespan. Dust-related problems can be mitigated by using locks with sealed cylinders, protective keyway covers, or dust-resistant housings. Regular maintenance, including cleaning and the use of dry lubricants such as graphite or PTFE, can also help prevent dust from binding internal parts.

Vibration is another significant factor affecting locks in harsh industrial settings. Machinery, engines, conveyor systems, and heavy equipment can produce continuous vibrations that gradually loosen lock components or cause internal parts to shift out of alignment. Repeated vibration may also cause fasteners, mounting hardware, or locking bolts to loosen over time. In some cases, vibration can cause keys to become difficult to insert or turn due to misalignment within the locking mechanism. To mitigate vibration-related problems, locks used in these environments are often designed with reinforced bodies, heavy-duty internal components, and secure mounting systems. Using locking hardware with anti-vibration fasteners, thread-locking compounds, or vibration-resistant cylinders can also help maintain stability.

*Figure 83: A lockout device applied to a hoist brake. Wtshymanski, CC BY-SA 4.0, via Wikimedia Commons.*

Mechanical fatigue occurs when lock components are subjected to repeated stress over time. Industrial locks are often used thousands of times in high-traffic areas such as equipment rooms, access gates, and machinery cabinets. Continuous operation, combined with environmental stresses such as vibration and temperature fluctuations, can gradually weaken components such as shackles, springs, locking bolts, and internal mechanisms. Mechanical fatigue can eventually lead to cracked components, weakened shackles, or reduced locking strength. Mitigation strategies include selecting locks constructed from high-strength materials such as hardened steel, stainless steel, or industrial-grade alloys, which are designed to withstand heavy use. Regular inspection and preventative replacement of worn parts are also important to ensure that locks continue to function safely and reliably.

In harsh industrial environments, the combined effects of dust, vibration, and mechanical fatigue can significantly shorten the lifespan of locking hardware if not properly managed. Effective mitigation requires a combination of robust lock design, appropriate material selection, protective sealing, and routine maintenance procedures. By selecting locks specifically designed for industrial conditions and performing regular inspections, locksmiths and facility managers can ensure that locking systems remain reliable and secure even in the most demanding operational environments.

**Impact on Locksmithing Practices**

Understanding the effects of dust, vibration, and mechanical fatigue is highly relevant for anyone involved in locksmithing because these environmental factors directly influence the performance, durability, and reliability of locking systems used in industrial settings. Locksmiths who work in sectors such as mining, manufacturing, construction, and offshore operations must recognise that standard residential or commercial locks are often unsuitable for these demanding environments. Instead, they must select locking hardware designed to withstand contamination, constant movement, and repeated mechanical stress. This knowledge allows locksmiths to recommend appropriate solutions such as sealed cylinders, heavy-duty padlocks, corrosion-resistant materials, and vibration-resistant mounting systems. By understanding the environmental risks that affect locks, locksmiths can design and install security systems that remain functional even under harsh operational conditions.

This understanding also influences everyday locksmithing practices, particularly in maintenance, inspection, and preventative servicing. Locks installed in industrial environments require more frequent checks to identify wear, contamination, or loosening components before failures occur. Locksmiths may implement specialised maintenance procedures such as cleaning contaminated cylinders, applying suitable lubricants, tightening mounting hardware, and replacing worn components. They must also advise facility operators on selecting durable locking systems and establishing maintenance schedules that reduce the risk of operational disruption. As a result, knowledge of environmental impacts on locking hardware enables locksmiths to move beyond simple lock installation and become essential contributors to the long-term reliability and safety of industrial security systems.

## Designing Locks for Remote Locations

Designing locks for remote locations requires careful consideration of reliability, environmental resistance, ease of maintenance, and operational continuity. Remote sites such as mining operations, offshore platforms, telecommunications towers, pipelines, and isolated infrastructure facilities often operate far from technical support or replacement parts. In these environments, locks must function reliably for long periods without servicing while also providing effective protection against unauthorized access. Unlike locks used in residential or urban settings, remote-location locks must be engineered to tolerate harsh environmental conditions, limited maintenance access, and the potential for operational disruptions if a lock fails.

The first consideration when designing locks for remote environments is durability and environmental resistance. Remote sites are frequently exposed to extreme weather conditions including rain, dust, salt spray, humidity, freezing temperatures, and intense heat. Locks installed in these locations must therefore be constructed from materials that resist corrosion and environmental degradation. Common materials used include 316 stainless steel, brass, bronze alloys, and corrosion-resistant coatings that protect both the external body and internal components. Sealed lock cylinders, weatherproof housings, and protective keyway covers help prevent water, dirt, and debris from entering the locking mechanism. These design features ensure that the lock continues to operate even after prolonged exposure to challenging environmental conditions.

Another important design requirement is mechanical simplicity and reliability. Remote locations often lack immediate access to locksmiths or maintenance technicians, meaning that locks must operate reliably with minimal servicing. Simple mechanical locking systems with fewer moving parts are generally preferred because they reduce the likelihood of failure. High-quality springs, hardened locking components, and reinforced shackles improve durability and reduce wear over time. In some cases, locks are designed with oversized components or heavy-duty internal mechanisms to ensure consistent operation even after thousands of use cycles.

Redundancy and emergency access are also critical considerations when designing locks for remote installations. If a lock fails or becomes damaged, it may prevent access to critical equipment such as communication infrastructure, power systems, or safety equipment. For this reason, many remote locking systems incorporate backup access methods, such as master key systems, emergency override keys, or dual-locking mechanisms. In electronic access-control systems, backup power supplies and mechanical override cylinders are often included so that authorised personnel can still gain entry if electronic systems fail.

Another key factor is ease of maintenance and servicing. Since remote locations may be difficult to access, locks should be designed to require minimal maintenance and allow simple servicing when necessary. Modular lock components that can be quickly replaced without specialised tools are particularly useful in these situations. Drainage channels, sealed bearings, and self-lubricating components can also reduce maintenance requirements. In some industrial designs, locks include protective dust covers or weather shields that extend the service life of the mechanism.

Security considerations must also be balanced with operational practicality. Remote facilities often rely on a small number of authorised personnel who require reliable access to equipment and infrastructure. Key management systems, restricted keyways, or electronic credential systems may be used to ensure that only authorised individuals can access sensitive areas. However, these systems must remain practical for field operations, particularly when workers are wearing gloves, working in low-light conditions, or operating in hazardous environments. Locks designed for remote sites often feature large keyways, ergonomic handles, or simplified access systems to improve usability in demanding working conditions.

Finally, modern remote security solutions increasingly integrate electronic monitoring and access management. Some advanced locking systems incorporate RFID, keypad entry, or remote access monitoring to track who enters a facility and when. In extremely isolated environments, locks may be integrated with satellite communication systems or remote security platforms that allow administrators to manage access permissions without physically visiting the site. However, these systems must still include mechanical backup access to ensure reliability if electronic systems fail.

**Example: Lock Design for an Offshore Wind Farm**

An offshore wind farm is a typical example of a remote infrastructure environment where locking systems must be carefully designed to ensure both security and operational reliability. Wind turbines are located many kilometres offshore and are exposed to saltwater spray, high winds, heavy rain, temperature fluctuations, and constant vibration from turbine operation. Access to these turbines is usually limited to scheduled maintenance visits by specialised technicians, meaning that locks installed on turbine towers, control cabinets, and service platforms must remain operational for extended periods without maintenance. Because a malfunctioning lock could prevent technicians from accessing critical systems or emergency equipment, reliability is a primary design consideration.

*Figure 84: Vestas V90-3MW wind turbine of the Kentish Flats Offshore Wind Farm, Thames Estuary. Phil Hollman from London, UK, CC BY 2.0, via Wikimedia Commons.*

The external access doors at the base of each wind turbine tower are typically fitted with marine-grade locking systems constructed from 316 stainless steel or corrosion-resistant alloys to withstand saltwater corrosion. These locks are often designed with sealed cylinders and weatherproof housings to prevent moisture, salt deposits, and debris from entering the internal mechanism. Protective keyway covers may be used to shield the cylinder from sea spray and airborne contaminants. Inside the tower, additional locks may secure electrical control cabinets, safety equipment storage areas, and ladder access points. These internal locks must also resist corrosion and vibration, while remaining easy to operate in confined working spaces.

Because offshore wind turbines are remote and costly to access, mechanical reliability and redundancy are essential. Locks used in these installations are typically simple, heavy-duty mechanical systems with minimal internal complexity to reduce the risk of failure. Many wind farms also implement master key systems that allow authorised maintenance teams to access multiple turbines using controlled key hierarchies, while restricting access to sensitive areas such as power electronics compartments. In addition, emergency override mechanisms are often incorporated so that authorised personnel can gain access if a primary locking system becomes damaged or jammed due to environmental conditions.

Maintenance considerations are also critical in offshore wind farms. Lock systems must be designed to minimise servicing requirements, as technicians may only visit turbines periodically during scheduled maintenance cycles. Locks may include drainage channels, corrosion-resistant springs, and sealed internal components to reduce wear and contamination. During routine turbine inspections, maintenance crews typically check locking mechanisms for corrosion, stiffness, or salt buildup and apply marine-grade lubricants to ensure smooth operation.

In modern offshore wind installations, physical locks may also be integrated with electronic access-control systems to track personnel access to turbine towers and control rooms. RFID cards, key fobs, or digital credentials can be used to record entry events and restrict access to authorised technicians. These systems may be connected to central management platforms onshore, allowing operators to monitor who has accessed specific turbines and when. However, due to the remote nature of offshore wind farms, electronic systems are normally paired with mechanical backup locks so that access remains possible even if power or network connectivity is disrupted.

This example demonstrates how lock design for offshore wind farms must combine corrosion resistance, mechanical reliability, redundancy, and operational practicality. By carefully selecting materials, simplifying mechanisms, and incorporating both mechanical and electronic access solutions, security professionals and locksmiths can ensure that locking systems continue to function effectively in the demanding offshore environment while supporting safe and efficient maintenance operations.

### Impact on Locksmithing Practices

Understanding how locks are designed for remote and harsh environments is highly relevant to locksmithing because it expands the role of the locksmith beyond simple installation or repair of residential locks. Locksmiths working in industries such as renewable energy, telecommunications, mining, and offshore infrastructure must understand how environmental conditions and operational constraints affect lock performance. In locations such as offshore wind farms, locks must withstand corrosion, vibration, moisture, and long maintenance intervals while still providing reliable security. This requires locksmiths to select specialised hardware, such as marine-grade stainless steel locks, sealed cylinders, and heavy-duty mechanical systems that are specifically designed for remote industrial environments. Knowledge of these requirements allows locksmiths to design access systems that remain functional even when servicing opportunities are limited.

This understanding also influences professional locksmithing practices, particularly in areas such as lock selection, system design, and maintenance planning. Locksmiths involved in infrastructure projects must consider factors such as master key systems for large facilities, emergency override access for critical equipment, and compatibility with electronic access-control systems. They may also advise operators on preventative maintenance schedules, corrosion protection methods, and the use of redundant locking systems to ensure operational continuity. In complex environments like offshore wind farms, locksmiths contribute not only to security but also to safety, operational reliability, and infrastructure resilience, demonstrating how modern locksmithing increasingly intersects with engineering, industrial security, and critical infrastructure management.

## Chapter 13 Review Questions

### Question 1

A marina installs padlocks on storage cages located on the dock. Within six months, several locks seize due to corrosion caused by saltwater exposure.

What would have been the most appropriate lock type for this environment?

A. Laminated steel padlocks designed for indoor storage
B. Marine-grade padlocks made from corrosion-resistant materials
C. Plastic padlocks designed for temporary use
D. Standard brass desk drawer locks

### Question 2

An underground mining facility installs locks on electrical isolation panels used during equipment maintenance. The locks must resist dust contamination and support lockout–tagout safety procedures.

Which type of lock would best meet this requirement?

A. Weatherproof industrial padlocks with sealed cylinders
B. Standard residential deadbolts
C. Decorative gate padlocks
D. Magnetic cabinet locks

### Question 3

A petrochemical refinery installs locks on cabinets containing gas monitoring equipment located in an area where flammable vapours may be present.

What is the most important requirement for these locks?

A. Bright colour identification
B. Non-sparking materials and hazardous-area certification
C. Lightweight construction
D. Plastic lock bodies

**Question 4**

A ship uses watertight doors between compartments to maintain vessel integrity during emergencies.

Which feature is most important in locks used on these doors?

A. Decorative finishes
B. Compression handles that maintain watertight seals
C. Lightweight aluminium components
D. Plastic latch mechanisms

**Question 5**

An offshore drilling platform installs electronic locks on control rooms to manage access for authorised technicians. These locks must withstand moisture, vibration, and salt spray.

Which design feature is most critical for reliable operation?

A. Chrome plating
B. Key-only access systems
C. IP-rated waterproof housings and sealed electronics
D. Glass control panels

**Question 6**

A locksmith installs standard indoor padlocks on conveyor equipment access gates in a quarry. Within months, the locks begin jamming due to dust entering the cylinders.

What feature would best prevent this problem?

A. Longer keys
B. Sealed keyways or dust-resistant lock housings
C. Decorative covers
D. Plastic shackles

**Question 7**

A mining facility reports that locks installed on machinery cabinets frequently loosen or fail due to constant vibration from nearby equipment.

What solution would best address this issue?

A. Installing lighter locks
B. Using vibration-resistant mounting systems and heavy-duty lock hardware
C. Reducing the size of the lock body
D. Switching to electronic keypad locks only

**Question 8**

Technicians maintaining offshore wind turbines report that locks occasionally fail due to salt deposits and corrosion.

What maintenance practice would best reduce this problem?

A. Painting the lock body annually
B. Rinsing locks with fresh water and applying corrosion-resistant lubricant
C. Replacing keys frequently
D. Increasing lock weight

**Question 9**

A telecommunications tower located in a remote desert region requires a lock that must function reliably for long periods without maintenance.

Which design principle is most appropriate for this environment?

A. Highly complex locking systems with many internal components
B. Simple mechanical locks with durable materials and minimal moving parts
C. Locks made entirely from plastic
D. Systems that require frequent servicing

**Question 10**

A security consultant designs a locking system for a remote wind farm where access to turbine towers must still be possible even if electronic systems fail.

What design feature should be included?

A. Multiple decorative locks
B. Electronic locks without backup access
C. Mechanical override keys or backup access methods
D. Magnetic-only door locks

# Chapter 14

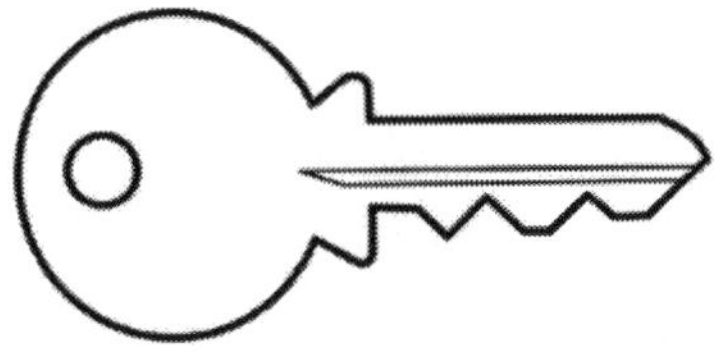

# Vehicle and Transport Security Evolution

Advances in automotive electronics, connectivity, and software-defined vehicle architecture are rapidly changing how vehicles are secured, accessed, and controlled. For security professionals and locksmiths, this transition represents a shift from purely mechanical lock systems toward integrated cyber-physical security platforms that combine traditional locking mechanisms with electronic control units, wireless communication technologies, and cloud-based management systems.

This chapter explores the evolution of vehicle and transport security, beginning with the transition from conventional mechanical keys to electronic access systems such as Remote Keyless Entry (RKE), Passive Keyless Entry and Start (PKES), smartphone-based digital keys, and cloud-connected vehicle access platforms. It examines how these technologies operate, how they improve convenience and fleet management capabilities, and how they introduce new technical challenges related to authentication, encryption, and cybersecurity. As vehicles become increasingly connected and automated, understanding the design, operation, and vulnerabilities of modern access-control systems is essential for anyone involved in automotive security, locksmithing, fleet management, or vehicle maintenance.

The chapter also considers the practical implications of these technologies for modern locksmithing practice. As vehicle security systems become more electronically driven, locksmiths must develop expertise not only in traditional lock mechanisms but also in radio-frequency communication, digital key programming, electronic control systems, and secure diagnostic procedures. By examining the evolution of vehicle access technologies, this chapter provides the foundation needed to understand both the opportunities and challenges associated with securing modern vehicles in an increasingly connected transport environment.

## Beyond Car Keys: Digital Vehicle Access

For more than a century, vehicles have relied on mechanical keys as the primary method of access and ignition. Traditional metal keys physically interact with the vehicle's lock cylinders and ignition switch to allow entry and start the engine. However, advances in electronics, wireless communication, and smartphone technology have led to the development of digital vehicle access systems, which are increasingly replacing or supplementing traditional car keys. These systems allow drivers to unlock, start, and manage their vehicles using electronic credentials rather than a physical key.

One of the earliest steps beyond mechanical keys was the remote keyless entry (RKE) system, which allows drivers to unlock or lock their vehicle using a wireless key fob. These systems typically operate using radio frequency signals transmitted between the key fob and a receiver inside the vehicle. When the correct signal is received, the vehicle's central locking system activates and unlocks the doors. Modern systems often include additional features such as remote trunk release, alarm activation, and panic functions. Although RKE systems still rely on a physical key fob, they represent an important transition toward fully digital vehicle access.

*Figure 85: Remote central locking. James086, CC BY-SA 4.0, via Wikimedia Commons.*

A more advanced system is passive keyless entry and start (PKES), sometimes referred to as a "smart key" system. In this configuration, the driver does not need to press a button on the key fob. Instead, the vehicle automatically detects the presence of the authorised key through low-frequency radio signals and encrypted communication. When the driver approaches the vehicle, proximity sensors trigger the door locks to unlock. Once inside the vehicle, the driver can start the engine using a push-button ignition. This technology improves convenience while maintaining secure authentication between the vehicle and the key device.

*Figure 86: BMW E65 remote key fob (BMW E65 key fob can support passive keyless entry and start, but only if the vehicle is equipped with BMW's Comfort Access option). Hugo venter at English Wikipedia, CC BY-SA 3.0, via Wikimedia Commons.*

The next stage in digital vehicle access is the use of smartphone-based digital keys. Many modern vehicles now allow a smartphone to function as a virtual car key using technologies such as Bluetooth, Near Field Communication (NFC), or Ultra-Wideband (UWB). In these systems, the driver's phone communicates securely with the vehicle to unlock doors and enable ignition. Some systems allow the phone to remain in the user's pocket while the vehicle automatically recognises the authorised device. Smartphone-based digital keys also allow owners to share temporary access with other drivers through mobile apps, making it possible to grant or revoke vehicle access remotely.

Another development is cloud-connected vehicle access systems, which integrate digital keys with online vehicle management platforms. Through manufacturer mobile applications, vehicle owners can unlock doors remotely, track vehicle location, or grant temporary access to other users such as family members, delivery drivers, or maintenance personnel. This technology is commonly used in car-sharing services, fleet management systems, and ride-sharing platforms, where multiple users may require controlled access to the same vehicle.

Digital vehicle access systems also incorporate advanced security features to protect against unauthorized entry. Encryption protocols, rolling codes, and secure authentication methods ensure that communication between the vehicle and digital credential cannot easily be intercepted or duplicated. Ultra-Wideband technology is increasingly used because it can accurately measure the distance between the vehicle and the authorised device, reducing the risk of relay attacks in which criminals extend wireless signals to trick the vehicle into unlocking.

Despite the convenience of digital access systems, most vehicles still include mechanical backup systems. Hidden emergency key blades or mechanical override locks are typically integrated into the vehicle design so that the owner can access the vehicle if the digital system fails due to battery loss, electronic malfunction, or network failure. This redundancy ensures that drivers are not locked out of their vehicles in situations where electronic access systems are unavailable.

**Remote Keyless Entry (RKE) System**

A Remote Keyless Entry (RKE) system is an electronic access-control system used in modern vehicles that allows drivers to lock or unlock their car doors remotely using a handheld transmitter, commonly known as a key fob. Instead of inserting a physical key into a door lock, the driver presses a button on the key fob, which sends a wireless signal to the vehicle. When the vehicle recognises and authenticates the signal, it activates the central locking system to perform the requested function. These functions may include unlocking or locking the doors, opening the trunk, or activating a panic alarm.

An RKE system consists of several interconnected electronic components that work together to provide secure wireless access to the vehicle. One of the most important components is the key fob transmitter. The key fob is a small handheld device that contains a microcontroller, a radio-frequency (RF) transmitter, a battery, one or more push buttons, and an encryption chip. When the driver presses a button on the fob, the microcontroller generates a digital signal that includes an identification code and a command such as lock or unlock. The RF transmitter then sends this signal as a radio wave to the vehicle. Most RKE systems operate on radio frequencies such as 315 MHz, commonly used in North America, or 433.92 MHz, which is widely used in Europe and many other regions. These frequencies allow reliable short-range communication between the key fob and the vehicle.

Inside the vehicle is a receiver module, which is typically integrated into the body control module (BCM) or central locking controller. This receiver constantly listens for signals transmitted by authorised key fobs. When a signal is received, the system first detects the radio transmission and then decodes the digital information contained within the signal. The system verifies that the signal is coming from a recognised key fob before taking any action. If the transmitted code is valid, the receiver sends instructions to the vehicle's central locking system.

Once the signal has been authenticated, the central locking system activates the door lock actuators. These actuators are small electric motors located inside each door that mechanically move the locking mechanism to either lock or unlock the door. Depending on the command transmitted from the key fob, the system may also unlock the trunk, flash the vehicle's hazard lights, sound the horn, or activate interior lighting. These responses provide confirmation to the driver that the command has been successfully executed.

Early remote locking systems used fixed digital codes, meaning that the same code was transmitted every time the key fob button was pressed. This approach created a security weakness because attackers could intercept the signal and replay it later to unlock the vehicle. To address this vulnerability, modern RKE systems use rolling code encryption, also known as code hopping. In a rolling code system, a new encrypted code is generated each time the button is pressed. The vehicle and the key fob share a synchronised encryption algorithm that allows the vehicle to predict the next valid codes in the sequence. If the transmitted code matches one of the expected codes, the system accepts it and updates its internal counter. This process prevents attackers from reusing intercepted signals.

The signal transmission process typically occurs very quickly. When the driver presses the unlock button on the key fob, the microcontroller generates an encrypted digital message that contains a device identification number, a command such as lock or unlock, and a rolling code value. The RF transmitter broadcasts this signal, and the vehicle's receiver detects and analyses it. After verifying the code through its encryption system, the vehicle activates the central locking actuators. The doors unlock and the vehicle may flash its lights to confirm the action. This entire sequence generally takes less than a second.

The effective operating range of an RKE system usually falls between 5 and 20 metres, although this distance can vary depending on environmental conditions, the strength of the key fob battery, and the design of the system. Obstacles such as buildings, metal structures, or electromagnetic interference can reduce signal range. A weak battery inside the key fob may also limit the distance over which the signal can be transmitted.

Modern RKE systems often include a range of additional features beyond simple door locking and unlocking. These features may include remote trunk release, panic alarm activation, remote window control, vehicle locator functions that flash the lights in large parking areas, and remote engine start systems. These capabilities enhance both convenience and security for vehicle owners.

Although RKE systems are generally secure, they can still be vulnerable to certain types of attack. These may include signal replay attacks, relay attacks that extend the wireless signal, radio jamming, or cloning of key fobs in older systems. To address these risks, vehicle manufacturers incorporate advanced encryption, rolling code algorithms, and improved signal authentication technologies.

For automotive locksmiths, understanding how RKE systems operate is essential when diagnosing remote locking faults or programming replacement key fobs. Locksmiths must often use specialised diagnostic equipment, programming tools, and manufacturer-specific procedures to synchronise new remotes with the vehicle's electronic control units. As vehicles increasingly rely on electronic access systems rather than purely mechanical locks, modern locksmithing continues to evolve, combining traditional mechanical skills with expertise in radio-frequency communication, electronic security systems, and digital access technologies.

When a vehicle has lost its Remote Keyless Entry (RKE) keys, access must be regained through legitimate and lawful methods that protect both the vehicle and the owner's security. In professional practice, automotive locksmiths and authorised technicians follow structured procedures that verify ownership, minimise damage, and restore proper key functionality. The process generally involves vehicle access, key replacement, and electronic reprogramming.

The first and most important step is verification of ownership and authority to access the vehicle. Professional locksmiths and vehicle technicians will normally request identification and proof that the person requesting access owns or is authorised to operate the vehicle. This may include registration documents, driver identification, or other official records. This step is essential to prevent unauthorised access and to comply with legal and professional standards within the locksmithing industry.

Once authority has been verified, the next step is gaining entry to the vehicle without causing damage. Many vehicles still retain a mechanical keyway in the door lock, even if the vehicle is primarily operated using a remote key fob. In these situations, locksmiths can create or obtain a replacement mechanical key that will operate the door lock. Automotive locksmiths may use vehicle identification information, key codes, or manufacturer databases to cut a new key that matches the vehicle's lock configuration.

In vehicles where the mechanical key is unavailable or the key code is unknown, professional locksmiths may use specialised vehicle entry tools designed to safely unlock doors without damaging the locking mechanism, window seals, or vehicle body. These tools are specifically designed for legitimate service work and require training and experience to use correctly. The goal is to gain entry while avoiding damage to the vehicle's locking system or electronic components.

Automotive locksmiths and roadside assistance technicians use specialised vehicle entry tools to unlock vehicles when keys have been lost, locked inside, or when electronic key systems fail. These tools are designed to allow access without damaging the vehicle's locking mechanisms, body panels, windows, or electronic systems. Because modern vehicles contain sensitive wiring, airbag systems, and complex door mechanisms, these tools must be used carefully and only by trained professionals who understand vehicle construction and locking systems.

One of the most widely used tools is the inflatable air wedge. An air wedge is a small reinforced bladder attached to a hand pump. It is inserted carefully between the vehicle door frame and the body of the car, usually at the upper corner of the door near the window. When the technician gently inflates the wedge, it creates a small gap between the door and the frame without bending or damaging the metal. This gap provides enough space to insert other tools that can reach the internal locking controls. The air wedge must be inflated slowly and carefully to avoid deforming the door frame or damaging weather seals.

Once a small opening has been created, technicians often use a long-reach tool, sometimes called a lockout rod. This tool is a long, thin rod made from steel or reinforced flexible materials that can be shaped to reach the interior door controls. Through the gap created by the air wedge, the long-reach tool can be manoeuvred to press the door unlock button, pull the interior door handle, or lift the lock knob. Many of these tools include protective coatings or plastic tips to prevent scratching the vehicle interior. Successful use of the long-reach tool requires knowledge of where the internal controls are located and careful manipulation to avoid damaging wiring or trim components.

Another commonly used tool is the slim jim, which is a thin flat strip of metal designed to slide down between the window glass and the outer weather stripping of the door. The purpose of the slim jim is to reach the internal linkage rods that connect the lock cylinder to the door latch mechanism. By hooking and lifting the correct linkage rod, the technician can unlock the door. However, modern vehicles often contain side-impact airbags, electronic wiring harnesses, and plastic shield plates inside the door, which makes slim jim use more complex and potentially risky. For this reason, slim jims are now used less frequently and only on certain vehicles where the door design permits safe access to the linkage.

A more advanced approach uses lockout kits that include a range of specialised tools designed for different vehicle types. These kits may contain wedges, long-reach tools, grabbers, flexible probes, and specialised hooks that allow the technician to reach different types of locking controls. Many modern lockout kits are designed to accommodate vehicles with electronic lock buttons, recessed handles, or complex door panel layouts.

Some technicians also use plastic wedges or protective wedges in combination with air wedges. Plastic wedges are placed between the door frame and the body of the vehicle to stabilise the gap created by the inflatable wedge. They help distribute pressure and protect the vehicle's paint and weather seals. These wedges are often made from durable plastic or nylon to prevent scratching the vehicle.

The process of using these tools follows a careful sequence. First, the technician selects a safe point on the door frame where a small gap can be created. An air wedge or protective wedge is inserted and inflated slowly to create a controlled opening. Next, a long-reach tool or similar instrument is inserted through the gap. The technician then carefully manipulates the tool to activate the interior unlocking mechanism, such as pressing the unlock button or pulling the handle. Once the door unlocks, the wedge and tools are removed and the door frame is checked to ensure that it has returned to its normal position.

In some situations, locksmiths may also use vehicle-specific entry tools designed for particular makes and models. These tools take into account differences in door structure, lock placement, and interior controls. Professional training and experience are important because improper tool use can damage the door structure, affect window alignment, or interfere with electrical systems.

For automotive locksmiths, the correct use of vehicle entry tools is an essential skill. Modern vehicle lockout procedures require not only mechanical skill but also an understanding of vehicle design, door construction, electronic systems, and safety considerations. When used correctly, specialised vehicle entry tools allow technicians to regain access quickly and safely while preserving the integrity of the vehicle's locking and security systems.

After gaining access to the vehicle, the next stage is replacing or programming a new key or key fob. Modern vehicles with RKE systems typically require the new remote key to be electronically paired with the vehicle's security system. This process involves using vehicle diagnostic equipment or manufacturer-approved programming tools to register the new key with the vehicle's immobiliser or body control module. During this procedure, previously lost or stolen keys can often be erased from the system to prevent unauthorised future access.

In some cases, the process may involve retrieving the vehicle's key code from the manufacturer or dealer database using the vehicle identification number (VIN). This code allows a locksmith or dealer technician to cut a new mechanical key that matches the vehicle's lock system. Once the key is cut, it can then be programmed to operate the remote entry system and immobiliser.

For vehicles equipped with advanced keyless systems such as passive keyless entry and push-button start, the replacement process can be more complex. These systems use encrypted communication between the vehicle and the key fob, which means that programming often requires specialised diagnostic equipment and secure access to manufacturer software. In such cases, the vehicle may need to be serviced by an authorised dealer or a locksmith equipped with advanced automotive programming tools.

For professional locksmiths, understanding these procedures is essential because modern vehicle security systems combine mechanical locks, electronic access control, and immobiliser technology. Proper handling of lost RKE keys ensures that vehicle security is maintained while restoring authorised access. The objective is always to regain entry and functionality safely, legally, and without compromising the vehicle's security system.

Key Programming

Programming or pairing Remote Keyless Entry (RKE) keys to a vehicle is the process of electronically registering a key fob with the vehicle's onboard control systems so that the remote transmitter is recognised and authorised to control functions such as door locking, trunk release, alarm activation, and in some cases immobiliser functions. Because modern vehicles use encrypted electronic security systems, pairing a key fob involves synchronising the transmitter with the vehicle's Body Control Module (BCM), immobiliser system, or central locking controller.

The first step in the process is confirming the correct key fob for the vehicle. Key fobs are designed to operate on specific frequencies and communication protocols that vary by manufacturer and model. If an incorrect transmitter is used, the vehicle will not recognise the signal. Automotive locksmiths and technicians typically confirm compatibility using the vehicle's make, model, year, and VIN (Vehicle Identification Number). Many manufacturers also require the replacement key fob to match a specific part number.

Before programming begins, the technician normally checks that the vehicle battery and key fob battery are fully charged. A weak battery in either the vehicle or the remote can interrupt the programming process. The technician may also ensure that all doors are closed, the vehicle is in park, and any existing keys are available, since some programming procedures require all authorised keys to be present during the pairing process.

There are several different methods used to program RKE keys depending on the manufacturer and vehicle design. The simplest method is known as manual programming mode, which is available on some vehicles, particularly older models. In these systems, the technician places the vehicle into programming mode using a sequence of actions such as inserting and removing the key from the ignition, turning the ignition on and off multiple times, or opening and closing the door locks in a specific pattern. Once the vehicle enters programming mode, the technician presses a button on the key fob. The vehicle then stores the unique code transmitted by the fob and confirms successful programming by cycling the door locks or flashing the lights.

Many modern vehicles require the use of diagnostic programming tools rather than manual procedures. Automotive locksmiths and technicians use specialised scan tools that connect to the vehicle's onboard diagnostic port (OBD-II port), usually located under the dashboard. These diagnostic devices communicate directly with the vehicle's electronic control modules. Using the programming interface, the technician can access the vehicle's security system and register a new remote key.

During diagnostic programming, the scan tool sends commands to the BCM or immobiliser system to place the vehicle into a key learning mode. The technician then activates the key fob so that its unique digital identification code is transmitted to the vehicle. The control module stores this code in its memory, allowing the vehicle to recognise the key in the future. Many vehicles allow multiple key fobs to be registered, and the programming process may include adding new keys while removing lost or stolen ones.

Modern RKE systems often incorporate rolling code encryption to enhance security. When a key fob is programmed, the vehicle and the key share a synchronised encryption algorithm that generates a new code every time the remote is used. This prevents intercepted signals from being reused by attackers. During programming, the encryption system establishes a secure link between the transmitter and the vehicle's control module.

Some vehicles also combine the RKE system with the immobiliser or transponder system. In these cases, programming the remote key may also require programming the embedded transponder chip that allows the engine to start. This process may involve separate steps for pairing the remote locking functions and registering the transponder with the engine control unit (ECU). Without this pairing, the vehicle may unlock but the engine will remain immobilised.

Another important aspect of programming is erasing previously stored keys. When keys have been lost or stolen, technicians may delete all existing keys from the vehicle's memory and then reprogram only the keys that are currently available. This ensures that lost remotes cannot be used to access the vehicle in the future.

Advanced vehicles equipped with smart keys or passive keyless entry systems use more complex communication methods that include proximity detection and encrypted authentication. Programming these systems may require manufacturer-specific software, security access codes, or online authentication through the vehicle manufacturer's database. In these cases, the process may only be performed by authorised dealers or locksmiths equipped with advanced programming equipment.

For automotive locksmiths and technicians, the ability to program RKE keys is an essential skill because modern vehicle security systems rely heavily on electronic access control. Proper programming ensures that the vehicle recognises authorised keys while maintaining the integrity of the vehicle's anti-theft system. As vehicles continue to integrate more advanced digital security technologies, automotive locksmithing increasingly requires knowledge of vehicle electronics, encryption systems, diagnostic tools, and manufacturer-specific programming procedures in addition to traditional mechanical key cutting skills.

### Passive Keyless Entry and Start (PKES)

A Passive Keyless Entry and Start (PKES) system is an advanced vehicle access and ignition technology that allows a driver to unlock, lock, and start a vehicle without physically pressing buttons on a key fob or inserting a key into the ignition. Instead, the system detects the presence of an authorised key automatically when it is within a short range of the vehicle. The driver can simply approach the vehicle, touch the door handle, and start the engine using a push-button start system. PKES systems are now widely used in modern vehicles because they improve convenience while maintaining electronic security through encrypted communication.

A PKES system works through a coordinated interaction between the vehicle, the key fob (often called a smart key), and several antennas located around the vehicle. The vehicle continuously monitors for authorised keys by transmitting low-frequency signals from antennas located in areas such as the door handles, inside the cabin, and sometimes in the trunk area. These antennas operate at very low radio frequencies, typically around 125 kHz, which are designed to operate over a short range of about one to two metres.

When a person carrying a smart key approaches the vehicle and touches the door handle or pulls on it, the system activates one of the vehicle's low-frequency antennas. This antenna sends out a short-range challenge signal that essentially asks whether an authorised key is nearby. Because the signal is low frequency, it does not travel very far, which helps ensure that only keys located close to the vehicle respond.

If a smart key is within range, it receives this low-frequency challenge signal through a receiver coil built into the key fob. The smart key then wakes from its low-power standby mode and responds by transmitting an encrypted identification signal back to the vehicle. This response typically uses a higher-frequency radio transmission, commonly around 315 MHz or 433 MHz, similar to the frequencies used in standard remote keyless entry systems.

Once the vehicle receives the response from the smart key, it verifies the authenticity of the signal using its stored security credentials. Modern PKES systems rely on encrypted challenge–response authentication protocols. The vehicle sends a random challenge code to the key, and the key uses an internal encryption algorithm to calculate the correct response. The vehicle performs the same calculation using the stored key data. If the calculated response matches the one transmitted by the key, the vehicle confirms that the key is legitimate.

After authentication is successful, the vehicle unlocks the door and allows entry. The system may also provide confirmation through actions such as unlocking the door locks, illuminating exterior lights, or enabling the door handles. Because the process occurs automatically when the driver touches the handle, the system is described as passive, meaning that the user does not need to press any buttons on the key fob.

Once inside the vehicle, the PKES system continues to monitor the presence of the authorised key. Interior antennas located within the cabin detect whether the smart key is inside the vehicle. When the driver presses the push-button start switch, the system again performs an authentication check. If the authorised key is detected inside the vehicle and the correct encryption response is received, the system disables the immobiliser and allows the engine to start.

The PKES system must also ensure that the key is located inside the vehicle rather than outside, because otherwise someone standing outside with a key could start the vehicle. To do this, the system compares signals received by different antennas inside and outside the vehicle. By analysing the signal strength and antenna location, the system determines whether the key is inside the passenger compartment.

Another feature of PKES systems is passive locking. When the driver exits the vehicle and walks away with the smart key, the system detects that the key is no longer within range. Some vehicles automatically lock the doors when the key moves a certain distance away, while others require the driver to touch a sensor on the door handle. This provides both convenience and security.

The smart key itself is designed to conserve battery power. Most of the time the key remains in a low-power sleep mode and only activates when it detects the vehicle's low-frequency challenge signal. This allows the key fob battery to last for several years under normal use.

Although PKES systems are designed with strong encryption and authentication mechanisms, they have also introduced new types of security risks. One of the most widely discussed vulnerabilities is the relay attack, in which attackers use radio equipment to extend the communication range between the vehicle and the key. In such attacks, a device near the vehicle relays signals to another device located near the key inside a building or house. This can trick the vehicle into believing the key is nearby. To counter this risk, modern PKES systems increasingly incorporate additional security measures such as motion sensors in key fobs, ultra-wideband (UWB) distance measurement, improved encryption protocols, and automatic key deactivation after periods of inactivity.

For automotive locksmiths and vehicle security specialists, understanding PKES systems is important because these technologies combine radio-frequency communication, encryption algorithms, immobiliser systems, and vehicle control modules. Diagnosing faults, replacing smart keys, or reprogramming PKES systems requires specialised diagnostic tools and manufacturer-specific procedures. As vehicle access technologies continue to evolve, locksmithing is expanding beyond traditional mechanical locks to include advanced electronic security systems and digital authentication technologies.

### Key Programming

Programming or pairing a Passive Keyless Entry and Start (PKES) system involves electronically registering a smart key with the vehicle's security system so that the vehicle can recognise and authenticate the key for both passive entry and engine start functions. Unlike conventional Remote Keyless Entry (RKE) systems, PKES systems integrate several security components, including the Body Control Module (BCM), immobiliser system, engine control unit (ECU), and multiple antenna modules. Because of the higher security level involved, PKES key programming usually requires specialised diagnostic tools and controlled procedures.

The first step in the process is verifying that the correct smart key is being used for the vehicle. PKES systems use manufacturer-specific frequencies, encryption protocols, and key identifiers. Technicians confirm compatibility using the vehicle's make, model, production year, and Vehicle Identification Number (VIN). The replacement key must match the correct electronic specification and part number. If the incorrect key type is used, the vehicle will not accept the key during the programming process.

Before programming begins, technicians normally perform basic system checks. The vehicle battery and the smart key battery must be sufficiently charged to ensure stable communication during programming. All vehicle doors are typically closed, and the vehicle must be placed in a specific state such as ignition-off or accessory mode. In some vehicles, all previously authorised keys must be present during the programming procedure because the system may erase existing keys during the pairing process.

Modern PKES systems are generally programmed using vehicle diagnostic equipment connected to the onboard diagnostic port (OBD-II). Automotive locksmiths or technicians connect a diagnostic scanner or manufacturer-approved programming device to the vehicle. The programming tool communicates with the vehicle's electronic control modules, particularly the BCM and immobiliser controller. Through the diagnostic interface, the technician accesses the security menu and selects the function for key learning or smart key registration.

Because PKES systems are part of the vehicle's anti-theft system, many manufacturers require security authentication before programming can begin. The diagnostic tool may request a security code or a time-based security access procedure. In some vehicles this involves entering a manufacturer-provided PIN code, while others require secure online authentication through the manufacturer's service portal. This security step prevents unauthorised programming of keys.

Once the system is placed into key learning mode, the technician introduces the new smart key to the vehicle. The vehicle activates its internal low-frequency antennas, typically operating around 125 kHz, which detect the presence of the smart key inside the vehicle cabin. The key receives the low-frequency challenge signal and responds using its radio transmitter, usually operating at 315 MHz or 433 MHz depending on the region.

The vehicle then performs an encrypted challenge–response authentication process. During this process, the vehicle sends a random challenge value to the key. The key uses its internal cryptographic algorithm and stored secret key to generate a response. The vehicle performs the same calculation using its stored data. If the responses match, the system recognises the key as legitimate and stores its identification data in the vehicle's memory.

Once the smart key is successfully authenticated, the vehicle registers the key within its secure key database. The BCM stores the unique identifier of the key and synchronises encryption parameters used for future communication. The system may confirm successful programming by displaying a message on the instrument panel, flashing indicator lights, or cycling the door locks.

In many vehicles the programming procedure also allows technicians to erase lost or stolen keys from the system. When keys are deleted, the immobiliser memory is cleared and only the newly programmed keys are stored. This is an important security step because it prevents previously lost keys from being used to gain access or start the vehicle.

After programming is complete, technicians normally perform functional verification tests. These tests confirm that passive entry works correctly at each door, that the push-button start system recognises the key inside the cabin, and that remote locking and unlocking functions operate normally. The technician may also check the system's ability to detect when the key is outside the vehicle to prevent unauthorised starting.

Some advanced PKES systems use additional technologies such as Ultra-Wideband (UWB) positioning, motion sensors in the key fob, or encrypted proximity verification. These systems may require manufacturer-specific diagnostic software and secure online access to program keys. In certain cases, programming can only be performed by authorised dealers or certified locksmiths equipped with approved equipment.

## Smartphone-Based Digital Keys

A Smartphone-Based Digital Key system allows a driver to unlock, lock, and start a vehicle using a smartphone instead of a traditional key or key fob. These systems are becoming increasingly common in modern vehicles as manufacturers integrate mobile technology with vehicle access systems. Digital keys operate through secure wireless communication between the smartphone and the vehicle, typically using technologies such as Bluetooth Low Energy (BLE), Near Field Communication (NFC), or Ultra-Wideband (UWB). The system works by storing a secure digital credential on the user's smartphone that functions as the electronic equivalent of a vehicle key.

At the centre of the system is the digital key credential, which is securely stored in the smartphone. This credential is typically stored within the phone's secure element, trusted execution environment, or secure hardware module. These secure storage areas are designed to protect sensitive information such as payment cards and biometric authentication data. When a digital key is issued, the vehicle manufacturer's system generates a unique encrypted credential and transfers it to the smartphone through a secure pairing process using the manufacturer's mobile application.

The initial setup process usually begins when the vehicle owner uses the manufacturer's mobile app to link the vehicle with their smartphone. This may involve scanning a QR code displayed on the vehicle's infotainment screen, entering a pairing code, or logging into the manufacturer's connected vehicle service account. During this process, the vehicle and smartphone establish a trusted connection, and the digital key credential is installed on the phone. Once installed, the smartphone effectively becomes an authorised key for the vehicle.

Different communication technologies are used depending on the design of the digital key system. Bluetooth Low Energy (BLE) is commonly used for passive access systems. BLE allows the vehicle to detect when the smartphone is nearby, typically within a few metres. When the driver approaches the vehicle, BLE antennas in the vehicle communicate with the phone to confirm the presence of a valid digital key. If authentication is successful, the vehicle unlocks automatically when the driver touches the door handle.

Near Field Communication (NFC) is another method used by many digital key systems. NFC requires the smartphone to be placed very close to the vehicle, usually by touching the phone to a designated reader on the door handle or dashboard. Because NFC operates over extremely short distances, typically a few centimetres, it provides strong security against wireless relay attacks. NFC is often used as a backup access method if the smartphone battery is low or if the wireless passive entry system is unavailable.

More advanced digital key systems use Ultra-Wideband (UWB) technology. UWB allows the vehicle to measure the precise distance and direction of the smartphone. By calculating the exact position of the phone relative to the vehicle, UWB systems can prevent certain types of attacks such as relay attacks, where signals are artificially extended by attackers. UWB can determine whether the phone is actually next to the vehicle rather than several metres away.

When the driver approaches the vehicle carrying the smartphone, the vehicle's access control system sends a wireless request asking for authentication. The smartphone receives this request and uses its stored digital key credential to generate an encrypted response. The vehicle verifies the response using its internal security system. If the authentication is valid, the vehicle unlocks and allows entry.

Once inside the vehicle, the smartphone continues to communicate with the vehicle to enable the engine start function. Similar to passive keyless entry systems, the vehicle checks that the authorised digital key is located inside the cabin before allowing the engine to start. The driver can then start the vehicle using the push-button start system. If the phone leaves the vehicle, the system will detect its absence and prevent the vehicle from restarting.

Smartphone digital key systems often integrate biometric authentication for additional security. Before transmitting the digital key credential, the smartphone may require fingerprint recognition, facial recognition, or a device passcode. This ensures that only the authorised phone user can operate the vehicle.

Another important feature of digital key systems is the ability to share access electronically. Vehicle owners can send temporary or permanent digital keys to other people through a mobile application. For example, a driver can send a digital key to a family member or friend through a secure link. The recipient installs the key on their own smartphone and can then access the vehicle according to the permissions granted. Access can also be revoked remotely at any time.

These systems rely heavily on advanced encryption and authentication protocols. Communication between the smartphone and vehicle is encrypted to prevent interception. Digital keys follow standards developed by organisations such as the Car Connectivity Consortium (CCC), which has established the Digital Key specification used by many manufacturers. These standards ensure compatibility between smartphones and vehicles while maintaining strong security.

From a vehicle security and locksmithing perspective, smartphone digital keys represent a major shift in how vehicles are accessed and controlled. Instead of relying on physical keys or radio transmitters, access credentials are stored digitally and managed through mobile devices and cloud services. This means that vehicle access systems increasingly involve software security, encrypted communications, mobile device integration, and digital credential management.

As this technology becomes more widespread, automotive locksmiths and vehicle security professionals must expand their expertise beyond traditional mechanical locks and key programming. Modern vehicle access systems now require knowledge of mobile device integration, wireless communication protocols, cryptographic authentication, and manufacturer-specific digital key platforms. These developments represent the ongoing evolution of vehicle security from purely mechanical systems to highly sophisticated digital access technologies.

### Key Programming

Programming or pairing Smartphone-Based Digital Keys involves securely linking a smartphone to a vehicle's electronic access-control system so that the phone can function as a digital vehicle key. Unlike traditional key programming, which typically involves registering a physical key or key fob with the vehicle, digital key pairing requires coordination between the vehicle's onboard systems, the smartphone, and the manufacturer's connected vehicle platform or mobile application. The process establishes a secure digital credential on the smartphone that the vehicle can recognise and authenticate.

The first stage of the process is vehicle and user authentication. Before a digital key can be issued, the vehicle owner must verify their identity through the manufacturer's connected vehicle service. This is usually done through the manufacturer's official mobile application, where the user logs into their vehicle account using registered credentials. The system confirms that the user is authorised to access and control the vehicle. In many cases, the vehicle must also be physically present, and at least one existing authorised key may be required during the setup process to prevent unauthorised pairing.

Once the user's identity has been confirmed, the system begins the initial pairing process between the smartphone and the vehicle. The vehicle typically enters a digital key setup mode through the infotainment system or through the manufacturer's mobile app. During this stage, the smartphone and vehicle establish a secure communication channel. This may involve scanning a QR code displayed on the vehicle's screen, entering a pairing code, or placing the smartphone near a designated reader inside the vehicle. This step ensures that the pairing process occurs only when the phone is physically near the vehicle.

After the connection is established, the manufacturer's system generates a digital key credential. This credential functions as the electronic equivalent of a traditional vehicle key. The credential is encrypted and transmitted to the smartphone using a secure communication protocol. On the smartphone, the credential is stored in a secure hardware environment, such as the device's secure element, trusted execution environment, or digital wallet platform. This protected storage prevents unauthorised applications or users from accessing the digital key data.

Different wireless communication technologies may be used during and after the pairing process. Many systems rely on Near Field Communication (NFC) during initial pairing because NFC requires the smartphone to be placed very close to the vehicle's reader, providing a secure environment for credential transfer. Once the digital key is installed on the phone, the system may use Bluetooth Low Energy (BLE) or Ultra-Wideband (UWB) for ongoing communication between the vehicle and the smartphone during normal operation.

After the digital key credential has been installed, the vehicle stores the smartphone's digital identifier within its vehicle access-control system, typically managed by the Body Control Module (BCM) or central security controller. The vehicle now recognises the smartphone as an authorised key. The pairing process may be confirmed through visual messages on the infotainment display, notifications within the mobile app, or by automatically unlocking the vehicle when the phone is detected nearby.

Once pairing is complete, the technician or vehicle owner usually performs functional testing. This involves verifying that the smartphone can unlock the doors, lock the vehicle, and enable the engine start system when the phone is inside the cabin. If the system supports passive entry, the vehicle should automatically detect the phone when it approaches. The engine start system will also confirm that the digital key is inside the vehicle before allowing the push-button start system to operate.

Many smartphone digital key systems also allow the primary user to share digital keys with other users. Through the mobile application, the vehicle owner can send a digital key to another smartphone user. The recipient receives an invitation and installs the digital key on their device. The vehicle's access control system then recognises the additional authorised user. Permissions can be customised to allow full access or restricted access, such as limiting vehicle speed or access times.

If a smartphone is lost or replaced, the digital key can usually be revoked remotely through the manufacturer's account system. This prevents the lost phone from being used to access the vehicle. A new smartphone can then be paired using the same secure pairing procedure.

From a vehicle security perspective, the pairing process relies heavily on encryption, secure hardware storage, and authentication protocols. Many modern systems follow the Car Connectivity Consortium (CCC) Digital Key standard, which defines how vehicles and smartphones securely exchange credentials. These standards ensure that digital keys can operate safely while protecting against unauthorised access.

For automotive technicians and locksmiths, smartphone digital key systems represent a significant shift in vehicle access technology. Programming and pairing these systems now involves understanding mobile device security, wireless communication protocols, cloud-based authentication systems, and vehicle electronic control modules. As vehicles continue to adopt digital access technologies, locksmithing increasingly requires expertise not only in mechanical locks and key programming but also in secure digital credential management and smartphone integration with vehicle security systems.

### Cloud-Connected Vehicle Access Systems

A Cloud-Connected Vehicle Access System is a modern vehicle access and security technology that allows vehicles to be locked, unlocked, started, monitored, and managed through internet-connected services. Instead of relying solely on physical keys or local wireless signals, these systems integrate the vehicle with cloud-based platforms, mobile applications, and remote servers. This allows authorised users to control and monitor the vehicle from virtually anywhere using a smartphone, tablet, or web interface.

At the centre of the system is the vehicle telematics control unit (TCU). The TCU is an onboard electronic module that connects the vehicle to external communication networks, typically using cellular data (such as 4G or 5G), satellite communication, or sometimes Wi-Fi. The TCU continuously communicates with the vehicle's internal electronic systems, including the body control module (BCM), immobiliser, central locking system, alarm system, and other security-related components. Through this connection, the vehicle can send information to cloud servers and receive commands from authorised users.

The cloud platform serves as the central management system for vehicle access and communication. Vehicle manufacturers operate secure cloud servers that store vehicle data, authentication credentials, and access permissions. When a user interacts with a vehicle through a mobile application, the command is first transmitted to the cloud platform. The cloud system verifies the user's identity and determines whether the user has permission to perform the requested action, such as unlocking the vehicle or starting the engine.

A typical interaction begins when a user opens the vehicle manufacturer's mobile application on their smartphone. After logging into their account, the user selects a command such as unlocking the vehicle. The smartphone sends this command to the manufacturer's cloud servers through the internet. The cloud platform then verifies the user's credentials using authentication systems such as passwords, biometric verification, or multi-factor authentication.

Once the user is authenticated, the cloud server sends a secure command through the cellular network to the vehicle's telematics control unit. The TCU receives the message and passes the command to the appropriate vehicle control module, such as the body control module responsible for door locking. The BCM then activates the door lock actuators to unlock the vehicle. The vehicle may send confirmation back to the cloud system, which then updates the mobile application to confirm that the command was successfully executed.

Cloud-connected vehicle systems also support remote engine start functions. In these systems, the same authentication and command process occurs. However, before allowing the engine to start, the vehicle performs several safety checks. These checks may include verifying that the vehicle is in park, confirming that the doors are locked, and ensuring that immobiliser authentication requirements are met. Once these conditions are satisfied, the vehicle allows the engine to start remotely.

Another important feature of cloud-connected systems is digital access management. Vehicle owners can grant temporary or permanent access to other users through the cloud platform. For example, a vehicle owner can send a digital key to another person through the mobile application. The recipient installs the digital key on their smartphone and can access the vehicle according to the permissions granted. Access rights can include full vehicle operation or limited permissions such as unlocking doors only.

Cloud-connected vehicle systems also support vehicle monitoring and telemetry functions. The vehicle can transmit information such as location, battery status, fuel levels, security alerts, and maintenance data to the cloud servers. The owner can view this information in real time through the mobile application. In fleet management environments, these systems allow companies to monitor multiple vehicles, manage driver access, and track vehicle usage remotely.

Security is a critical component of cloud-connected vehicle access systems. Communication between the smartphone, cloud servers, and the vehicle is protected using advanced encryption protocols such as Transport Layer Security (TLS). These encryption methods prevent attackers from intercepting or modifying commands. Authentication systems also ensure that only authorised users can access the vehicle. In addition, many systems log all access events so that vehicle owners or administrators can review activity history.

Another security feature is the ability to revoke or disable access remotely. If a smartphone is lost or if a user's access privileges need to be removed, the vehicle owner can deactivate the digital key through the cloud platform. The next time the vehicle communicates with the cloud server, the revoked key will no longer be recognised. This provides a significant advantage over traditional keys, which often require physical replacement if lost.

Cloud-connected systems also support software updates and security improvements through over-the-air (OTA) updates. Manufacturers can remotely update the vehicle's software to improve functionality, fix vulnerabilities, or add new features. These updates are downloaded through the telematics system and installed directly on the vehicle's electronic control units.

From a vehicle security and locksmithing perspective, cloud-connected vehicle access systems represent a major technological evolution. Traditional vehicle access relied entirely on mechanical locks and locally transmitted radio signals, but modern vehicles increasingly depend on networked systems, digital credentials, encrypted communication, and remote management platforms. Automotive locksmiths and vehicle security professionals must therefore understand how these digital access systems operate, including their communication technologies, authentication protocols, and integration with vehicle electronic control systems.

As the automotive industry continues to move toward connected and autonomous vehicles, cloud-connected access systems are expected to become standard across many vehicle platforms. These systems enable new levels of convenience, remote control, and fleet management while maintaining strong security through encryption, authentication, and centralised access management.

**Impact on Locksmithing Practices**

For locksmiths and automotive security professionals, the shift toward digital vehicle access represents a major evolution in the field. Modern automotive locksmithing increasingly requires knowledge of electronic security systems, wireless communication technologies, digital credential management, and vehicle programming tools. While traditional lock-picking and key-cutting skills remain important, professionals must now understand how to diagnose electronic access issues, program replacement key fobs, and manage digital key systems. As vehicles continue to integrate more advanced electronic technologies, locksmithing is evolving into a field that combines mechanical expertise with electronic and cybersecurity knowledge.

## CAN Bus Vulnerabilities and Relay Attacks

Modern vehicles rely on complex internal networks to coordinate the communication of numerous subsystems, and the Controller Area Network (CAN) bus is among the most critical of these infrastructures. CAN enables ECUs such as engine control units, body control modules, braking systems, transmission controllers, dashboards, and door locking mechanisms to exchange messages over a shared data network, thereby reducing wiring complexity and enabling coordinated functionality across vehicle systems [307-309].

However, CAN was designed with reliability and efficiency as primary goals, not security, and many implementations transmit messages without encryption or authentication. This fundamental design choice creates a security-oriented vulnerability landscape in which an attacker who gains access to the CAN network can inject or spoof messages that appear legitimate to other ECUs, potentially enabling unauthorized control of vehicle functions [307, 308, 310, 311]. Consequently, modern vehicle security research and practice emphasize understanding CAN bus vulnerabilities and countermeasures, including gateway devices, encrypted protocols, and authentication mechanisms to mitigate unauthorized frame injection and control signal spoofing [307, 312-314].

The CAN bus provides interconnections among multiple electronic modules that coordinate in-vehicle functions via broadcast frames on a shared medium. In practice, these modules include the engine control unit (ECU), body control module (BCM), braking system, transmission controller, dashboard systems, and door locking mechanisms, all of which rely on timely reception and processing of CAN frames. The core strength of CAN, its low latency and broadcast nature, also underpins its vulnerability: many CAN implementations lack built-in source authentication or message encryption, so ECUs may accept and act on malicious frames if they resemble legitimate traffic, enabling an attacker to influence critical subsystems through spoofed commands [307, 308, 311]

Analyses of these vulnerabilities emphasize that attackers can exploit physical access to the CAN network, often via the OBD-II diagnostic port, to connect a device that injects frames onto the bus. Because the messages are not cryptographically protected and the network typically treats all messages as trusted, compromised nodes can propagate unauthorized commands across the bus, undermining safety-critical functions and increasing the risk of loss of control [307, 309, 313]. A related concern is the susceptibility of CAN to DoS-style disruptions via bus-off conditions and error-handling mechanisms; compromising such mechanisms can yield widespread ECU disconnects and degraded vehicle operation, underscoring how in-vehicle network design choices can create security blind spots if not complemented by defensive controls [307, 308]. Given these vulnerabilities, researchers and practitioners advocate for architectural defences such as gateway modules that enforce authentication and filter CAN traffic, as well as encrypted or authenticated vehicle-network protocols to prevent unauthorized message injection and command execution [307, 308, 312].

The security relevance of CAN is best understood as an access problem: if an attacker can (a) join the bus electrically or logically, and (b) transmit plausible frames, CAN's trust model often does the rest. Research repeatedly shows that "one compromised ECU can influence others" unless segmentation and authentication are engineered into the architecture.

A particularly salient class of security concerns for modern vehicles involves relay attacks in the context of Passive Keyless Entry and Start (PKES) systems. In a relay attack, two actors cooperate to extend the effective reach of a key's wireless credentials: one attacker near the vehicle captures the vehicle's challenge and relays it to the key holder located remotely, while the second attacker near the key relays the key's response back to the vehicle. The net effect is that the vehicle "believes" the authorised key is nearby, enabling wireless entry and potentially enabling starting and immobilization bypass, depending on the system design. This relay mechanism exploits the PKES's reliance on proximity signals rather than secure cryptographic containment and has been demonstrated in practice as a potent method for defeating PKES protections [307, 315, 316].

Mitigations include Ultra-Wideband (UWB) based distance measurement, which uses precise time-of-flight to determine proximity rather than signal strength, rendering relay-based extensions of range more difficult, and adding motion sensors to keys that place the key in a sleep mode when stationary to prevent active relay operation. Secure gateway modules and encrypted intra-vehicle communications are also described as complementary defences that can prevent unauthorized message flow into the CAN network and protect access control signals from relay abuse [307, 311, 315].

**Paths onto CAN and adjacent trust zones**

**OBD-II / diagnostics** remain a primary physical exposure. The 2011 automotive attack-surface analysis calls OBD-II "the most significant automotive interface" and notes that it is the principal means for diagnostics and ECU updates; in many vehicles it can reach multiple internal bus segments. The same work notes the mandate for SAE J2534 "PassThru," which standardizes programmatic access to vehicle buses via reprogramming/diagnostic tools—useful for repair, risky if misused or exposed.

**Physical tapping / exposed harnesses** are a growing real-world concern. CVE-2023-29389 (as described by ) documents a case where physically proximate attackers accessed a control CAN bus via externally reachable connectors and sent forged authorization messages to enable driving, with exploitation noted "in the wild." Independent technical writeups tie this class of theft to "CAN Injection" and emphasize that the underlying weakness is architectural trust of internal messages.

**Compromised ECUs and wireless gateways** provide scalable, remote reach. The 2011 attack-surface paper demonstrated that remote exploitation can be feasible over multiple channels (e.g., Bluetooth, cellular) and can enable "long distance vehicle control" depending on architecture and vulnerabilities. Modern examples include Bluetooth-stack vulnerabilities in infotainment platforms (e.g., PerfektBlue CVE chain in OpenSynergy BlueSDK), which can provide code execution in IVI (in-vehicle infotainment) and potentially enable lateral movement depending on internal segmentation and additional weaknesses.

**Aftermarket and third-party devices** (diagnostic dongles, fleet telematics, insurance trackers) are a recurring weak link in security assessments because they can (1) bridge CAN to wireless networks and (2) vary widely in engineering quality and patch practices. Academic analysis of vehicle attack surfaces explicitly calls out the risk of externally connected tools and boundary interfaces.

**What attackers do once on CAN**

The attack patterns below are repeatedly observed in both experimentation and field exploitation. While details vary by OEM and vehicle architecture, the common failure mode is that receiving ECUs treat bus messages as inherently trustworthy.

*Table 8: Comparative table of attack types, access, and mitigations.*

| Attack type | Required access level | Typical mechanism (high-level) | Likely impact | Representative documentation | Mitigations (most effective first) |
|---|---|---|---|---|---|
| CAN message spoofing / injection | Physical (OBD-II or harness) or logical (compromised ECU) | Send forged frames with expected IDs/data to trigger actions on victim ECUs | Unauthorized unlock/start, sensor tampering, safety feature manipulation | CAN lacks authenticator/source IDs and is broadcast ; exploited-in-the-wild CAN injection | Message authentication + freshness (e.g., AUTOSAR SecOC) ; secure gateways/allowlists |

| Attack type | Required access level | Typical mechanism (high-level) | Likely impact | Representative documentation | Mitigations (most effective first) |
|---|---|---|---|---|---|
| | | | (vehicle-dependent) | case documented as CVE-2023-29389 | between domains ; segmentation; secure ECU design |
| CAN replay | Physical/logical bus access; prior observation | Record valid traffic and replay at a later time (works when higher-layer checks absent or weak) | Re-trigger actions, bypass weak state checks | CAN broadcast enables snooping and injection | Freshness counters/timestamps with MAC (SecOC-style) ; state-aware checks; IDS tuned for repeats |
| CAN DoS / bus starvation | Physical/logical bus access | Flooding, priority abuse, or bus-state manipulation to starve other nodes | Loss of functionality; limp mode; safety degradation | "Fragility to DoS" noted as inherent challenge in automotive CAN ; DoS fault injection literature | Gateway rate limiting & arbitration-aware filtering ; resilient design; IDS for timing anomalies |
| Diagnostic abuse / ECU reflash | Physical diagnostic access or compromised reflashing path | Use diagnostic services to read/write memory or reflash firmware (credentialing varies) | Persistent compromise; new "trusted" behavior; key/immobilizer tampering (vehicle-dependent) | Firmware extraction possible via CAN in many ECUs examined | Secure boot + signed firmware; restrict reflashing to least-exposed bus segments ; protect diagnostics with authenticated sessions |
| Wireless-to-CAN pivot (IVI/telematics compromise) | Remote proximity (Bluetooth/Wi-Fi) or internet/cellular (telematics) depending on bug | Compromise externally reachable ECU, then pivot laterally to bus(ses) | Scalable attacks; privacy and potentially safety impacts | Uconnect CAN access via infotainment compromise described in CVE-2015-5611 ; PerfektBlue chain enables IVI RCE and may enable lateral movement depending on architecture | Domain isolation + authenticated gateways ; harden wireless stacks and patch supply chain; monitoring and incident response |
| Keyless relay (PKES) | Proximity to car and proximity | Relay request/response to defeat | Unauthorized entry and | Foundational relay demonstration across multiple car | UWB secure ranging / distance bounding ; motion- |

| Attack type | Required access level | Typical mechanism (high-level) | Likely impact | Representative documentation | Mitigations (most effective first) |
|---|---|---|---|---|---|
| | (often indirect) to key's radio environment | proximity assumption | potentially start/drive | models ; large-scale vulnerability testing by ADAC | sensor "sleeping" fobs ; user-configurable PKES disablement |
| RKE rolling-code exploitation (e.g., jam/capture, resync abuse) | Proximity during capture; later proximity for replay | Abuse receiver resynchronizati on windows or capture-and-replay patterns | Unauthorized unlock (start is vehicle-dependent) | RollBack paper (Black Hat) explains replay/resync attack families | Stronger resync policies; cryptographic redesign; limit acceptance windows; anomaly detection for repeated patterns |

Industry and academic work on securing automotive networks identifies several concurrent approaches to mitigate CAN vulnerabilities and relay-based attacks. One line of defence involves integrating gateway modules that enforce authentication and filterCAN traffic across network boundaries, thereby preventing unauthorized frames from reaching sensitive ECUs and reducing the blast radius of a compromised module [307, 308, 312]. Another approach emphasizes the adoption of encrypted or authenticated communication protocols within vehicle networks to safeguard against message injection and spoofing, especially for critical control signals. In the realm of PKES-related relay risks, UWB-based distance measurement provides a robust proximity test that is much harder to spoof than conventional signal-strength approaches, while motion-sensing keys reduce exposure by forcing an idle key to a non-responsive state when unmoved for a period [307, 311, 315].

Additional research has explored intrusion detection systems (IDS) tailored to CAN traffic, employing ensemble learning and real-time analysis to identify anomalous CAN frames and distinguish legitimate operation from spoofed or injected frames, thereby enabling timely detection and response to attacks without imposing prohibitive latency on vehicle operation [311]. Related studies address broader network vulnerability assessment methodologies, incorporating risk-based indices and scenario analyses to quantify and compare the potential impact of different attack vectors in automotive and related cyber-physical domains [312, 317-319].

### Relay attacks against PKES and RKE

Relay attacks are best viewed as distance fraud: the attacker does not break cryptography; they extend the communication range so the car concludes the authorized key is "nearby." The NDSS PKES relay paper describes a two-device setup with one device near the vehicle and another near the key, relaying messages between them. The authors emphasize that the relay can be transparent to modulation and cryptographic strength because it forwards signals rather than forging responses.

PKES systems commonly use a low-frequency (LF) wakeup / challenge from vehicle to key and a higher-frequency response channel from key to vehicle. The NDSS paper's access-control summary shows LF involvement for passive modes (outside/inside) and UHF for key responses in normal operation, illustrating why relaying the "car ⇒ key" leg can be sufficient.

Large-scale testing suggests the issue is widespread across vendors and model years. The 2019 ADAC document states that, among "way over 300 models," only four were "better protected" in their testing context, and it lists many affected vehicles (with illegal unlocking and/or engine start possible). similarly drove industry response through consumer security ratings and explicitly notes the introduction of "sleeping fobs" (motion-sensor-enabled) as a countermeasure to frustrate relay thieves.

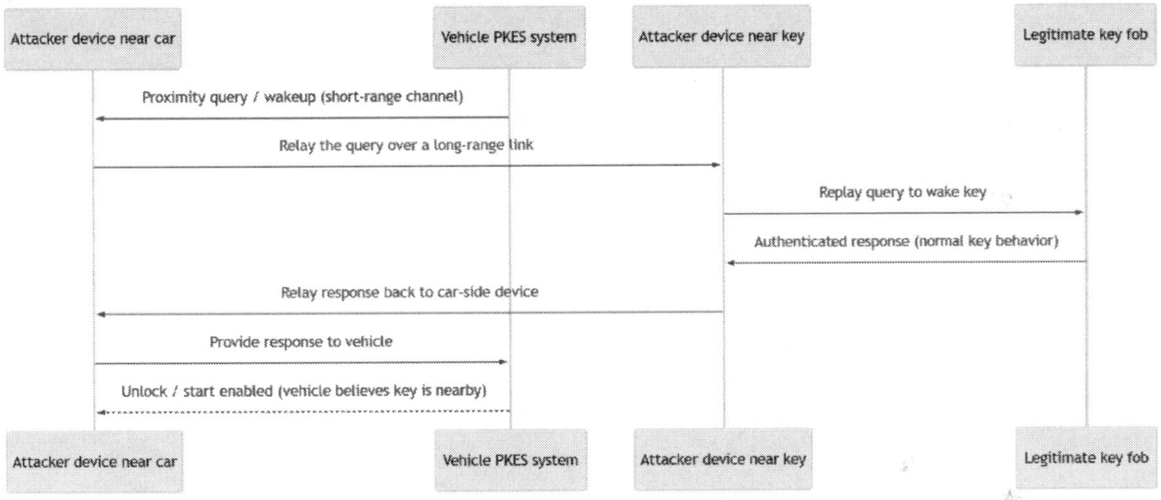


*Figure 87: Relay attack flow diagram (sequence).*

### RKE attacks: beyond relay to replay/resynchronization abuse

Remote Keyless Entry (RKE) typically requires an explicit button press, so "classic relay" is less central than in PKES. However, attack families exist that defeat RKE's rolling-code defenses through capture/abuse and receiver resynchronization behavior.

The RollBack work (documented in a Black Hat technical paper and later ACM publication line) describes a time-agnostic replay-and-resynchronize attack: instead of repeatedly jamming until the moment of theft, the attacker captures signals once and later replays a sequence to trigger rollback-like resynchronization in the receiver, restoring acceptance of previously used codes. This matters for real-world use cases like rentals/car-sharing (brief access to fobs) and for post-event forensic timelines.

Earlier "capture + jam" concepts are publicly associated with the "RollJam" class of attacks (widely discussed in security journalism and referenced by follow-on academic work); RollBack explicitly positions itself as removing the need for continuous jamming.

**Interaction between relay attacks and CAN exploitation**

From a theft and unauthorized-use perspective, relay and CAN exploitation often appear as parts of multi-stage chains:

**Entry / authorization bypass**: relay (PKES) or RKE exploitation yields door unlock and sometimes start authorization.

**Persistence / scaling**: once inside, attackers may seek persistence by reprogramming keys, compromising modules, or manipulating diagnostic flows (vehicle-dependent).

**Drive-away enablement**: CAN injection can bypass key validation entirely by injecting forged "key validated / immobilizer release" type messages on trusted internal networks (as described in CVE-2023-29389 and technical analyses).

These chains are enabled by the same systemic property: internal trust of bus messages. The CAN injection narrative stresses that thieves "don't attack the hard part" (cryptographic key challenge-response) but instead bypass the smart key system by injecting internal messages that ECUs accept as authoritative when not protected. The 2010 vehicle security analysis similarly observes that a compromised component can inject new packets to induce actions because CAN is broadcast and lacks authenticators.

In remote compromise scenarios, the pivot can be: wireless flaw → IVI/telematics code execution → gateway traversal → CAN influence. The 2015 Uconnect case is a canonical example where infotainment compromise enabled CAN-bus access and vehicle control implications, triggering a recall and additional mitigations including software updates and carrier-level exposure reduction. More recent middleware disclosures (PerfektBlue) explicitly discuss IVI as an entry point and note lateral movement possibilities, contingent on vehicle architecture and additional vulnerabilities.

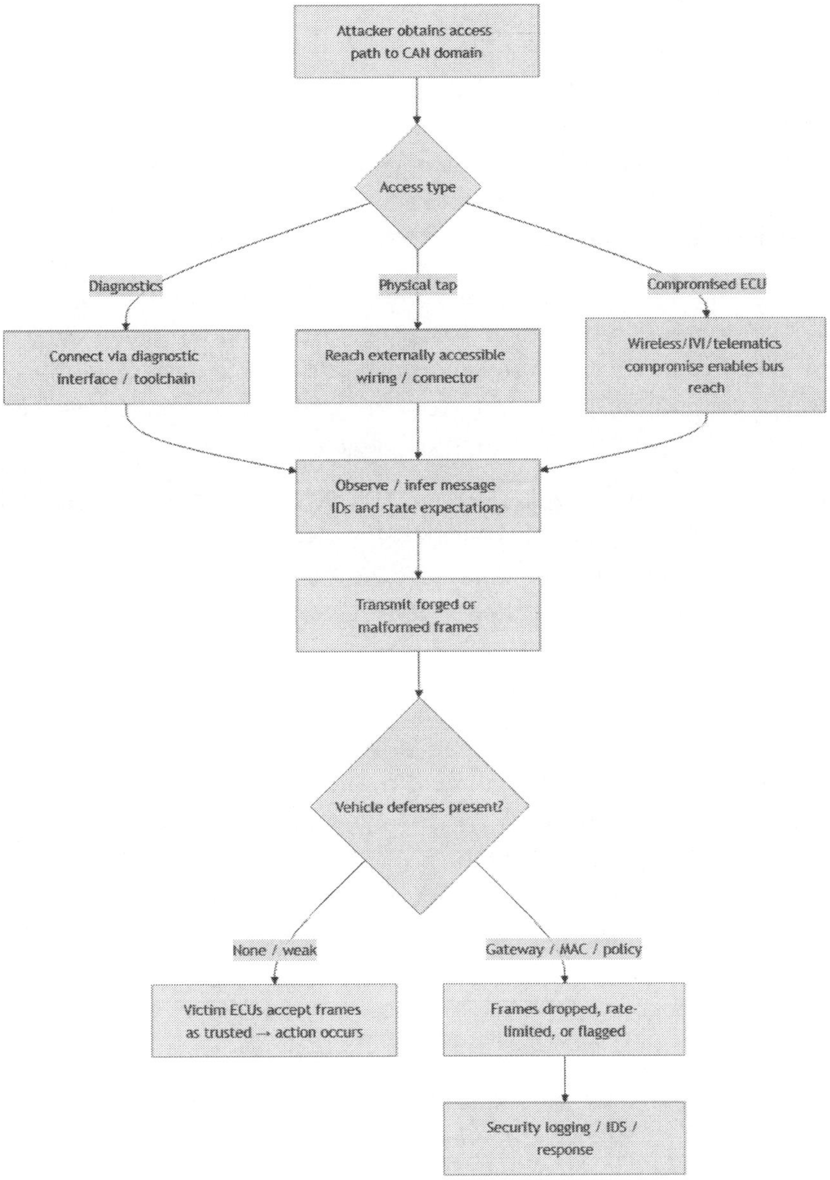


*Figure 88: CAN injection flow diagram (high-level).*

### Detection and mitigation strategies

Mitigation must be layered because threats span physical access, radio proximity fraud, and remote software compromise. The most effective programs treat vehicle cybersecurity as an engineering discipline spanning lifecycle governance, architecture, cryptography, and update pipelines.

Hardening CAN and in-vehicle network architecture

Authenticated communication (message integrity + freshness) is the direct countermeasure to spoofing/replay. SecOC is an industry-standardized approach to append authentication information (e.g., MAC) and an optional freshness value; receiver-side verification checks both authenticity and freshness to defeat spoofing and replay. This approach is specifically designed as a practicable add-on for legacy communication stacks, but it creates deployment realities: key provisioning, payload overhead (tight in classic CAN), and compatibility constraints.

Secure gateways and segmentation reduce blast radius. Research and industry guidance repeatedly recommend isolating infotainment/telematics from safety-critical domains, with gateways enforcing allowlists/rate limits and preventing diagnostic abuse. The 2011 attack-surface paper specifically recommends restricting reflashing paths to the bus with the smallest external attack surface, acknowledging that multi-bus ECUs and gateway architectures complicate this.

Emerging protocol-level security for next-generation CAN is developing. CAN XL is standardized in ISO 11898-1:2024 and introduces larger payloads; notes that CANsec is under development as a data-link-layer security protocol for CAN XL frames, including freshness and authentication tags. This is promising for future platforms but does not remediate the existing fleet where classic CAN remains prevalent.

Detecting CAN attacks with IDS and telemetry

IDS approaches typically model timing, ID sequences, payload distributions, or physical characteristics to detect anomalies. Surveys show significant research investment across heuristic and ML approaches. However, the 2026 adversarial evaluation of CAN IDS warns that ML-based IDS can suffer increased missed attacks under protocol-compliant adversarial perturbations, underscoring that IDS must be validated against adaptive adversaries rather than only offline benchmarks. This suggests practical IDS deployment should combine: (1) multi-signal telemetry (not only CAN payload), (2) gateway-enforced policy controls, and (3) safety-aware failover strategies to avoid either silent compromise or denial-of-service via false alarms.

Mitigating relay attacks and keyless weaknesses

UWB secure ranging / distance bounding is the primary technical path to defeating classical relays while preserving hands-free convenience. The states that Digital Key Release 3 uses BLE for authentication and establishes a secure ranging session with UWB to enable accurate distance measurement and mitigate relay attacks, while maintaining NFC as a backup. A CCC whitepaper explicitly says UWB secure ranging "protect[s] against relay attacks" and ties ranging to secure element storage and short-lived ranging keys.

Motion-sensor "sleeping" fobs reduce exposure when keys are stationary at home. Thatcham reports that manufacturers introduced sleeping fobs and that tested models passed relay-attack testing within their rating program, though they also caution that longer-term fixes are needed.

Pragmatic user mitigations (Faraday pouches/boxes, disabling passive entry, storing keys away from entry points) remain effective but shift burden to users and are inconsistent with the original convenience goal. The foundational relay paper itself noted that immediate mitigations often reduce or disable PKES convenience, reflecting this usability-security trade-off.

Important nuance: even UWB is not a universal silver bullet. Security research on UWB ranging highlights subtle "distance reduction" avenues that can arise from implementation and clock behaviour, demonstrating why security claims must be validated end-to-end (standard + implementation + environment). The practical takeaway for engineers is not "UWB is unsafe," but that secure ranging must be treated as a cryptographic/PHY system requiring rigorous validation and strong receiver design assumptions.

Secure boot, secure updates, and lifecycle controls

Technical mitigations fail without patch delivery. Regulatory and standards frameworks increasingly mandate cybersecurity management and update governance:

- / ISO/SAE 21434 defines lifecycle cybersecurity risk management processes for vehicle E/E systems.
- UN Regulations R155 (cybersecurity management) and R156 (software updates) establish performance/audit requirements and an update management regime, reinforcing secure boot and signed-update expectations as industry baseline practice (even if technical mechanisms vary by OEM).

**Practical Recommendations**

Practical recommendations and research gaps continue to emerge as vehicle access systems become increasingly digital and interconnected. While modern vehicles provide convenience through keyless entry, wireless communication, and integrated diagnostics, these same features introduce new security risks. Both vehicle owners and automotive security professionals must therefore adopt updated practices that reflect the evolving cyber-physical nature of modern vehicle systems.

For vehicle owners, several practical steps can significantly reduce the risk of unauthorized vehicle access without requiring detailed technical knowledge. One of the most effective measures is reducing exposure to keyless relay attacks. Owners can do this by using "sleeping" key fobs, where supported by the manufacturer, or by storing keys in radio-frequency (RF) shielding containers when they are not in use. These containers prevent the wireless signals from being intercepted and relayed by attackers. Such measures are widely recommended by automotive security organisations and have been validated through independent testing programs.

Vehicle owners should also ensure that their vehicle's infotainment and telematics systems remain updated whenever manufacturers provide software updates. Modern vehicles rely heavily on networked software, and vulnerabilities discovered after a vehicle is released can often be addressed through firmware or system updates. Historical incidents have shown that unpatched software can provide pathways for attackers to gain remote access to internal vehicle networks, including the Controller Area Network (CAN) bus, which governs many critical vehicle functions. Regular updates therefore play an important role in maintaining the security of the vehicle's electronic architecture.

Another important consideration for owners is the treatment of aftermarket diagnostic devices, tracking systems, or OBD-port accessories. These devices should be regarded as cyber-physical components because they interact directly with the vehicle's internal electronic systems. Owners should only install products from reputable manufacturers that provide ongoing software updates and technical support. Diagnostic interfaces are extremely powerful by design, and poorly secured devices can unintentionally introduce vulnerabilities that allow unauthorized access to vehicle systems.

For locksmiths and automotive security practitioners, the professional environment is rapidly evolving. Traditional locksmithing roles that focused primarily on mechanical key cutting and replacement are increasingly expanding into system security maintenance and electronic credential management. Modern vehicles incorporate immobilizers, encrypted key authentication, and networked access control systems, meaning that locksmiths must operate with processes that ensure both technical accuracy and procedural accountability.

One essential practice for security professionals is the implementation of strong authorization and audit controls. This includes verifying customer identity before performing key programming, documenting work performed on vehicle access systems, and maintaining clear service records. These procedures help ensure that key enrolment, immobilizer changes, or diagnostic access cannot be misused. Because many modern service procedures involve modifying electronic security states within the vehicle, maintaining traceable records is an important professional safeguard.

Automotive security practitioners must also remain aware of the rapid evolution of vehicle security technologies. Manufacturers are introducing systems such as gateway filtering, secured diagnostic channels, encrypted communications, and "sleeping" key fobs designed to mitigate relay attacks. Professionals should stay informed about these developments and provide guidance to customers regarding both risks and protective measures. Independent test bodies, industry standards, and security research reports can serve as valuable reference points for maintaining up-to-date knowledge.

Finally, unexplained anomalies in immobilizer behaviour, key authorization, or vehicle electronic systems should be treated as potentially security-relevant events. Modern vehicle theft methods increasingly involve electronic intrusion techniques such as CAN bus injection, which may leave few or no visible signs of forced entry. Locksmiths and automotive security specialists must therefore approach unusual system behaviour with a forensic mindset, recognising that electronic evidence may indicate attempted or successful unauthorized access.

**Impact on Locksmithing Practices**

For automotive locksmiths, security engineers, and vehicle security specialists, understanding CAN vulnerabilities and relay attacks is indispensable for diagnosing risks, informing owners, and guiding the development of more secure access and control systems. The cybersecurity landscape surrounding in-vehicle networks intersects with mechanical and functional aspects of vehicle design, wireless communications, and gateway security architectures. As vehicle technology evolves toward increased connectivity and automation, professionals must consider CAN security not only in terms of protecting physical access and immobilization signals, but also in terms of safeguarding the integrity and authenticity of intravehicular communications and access-control protocols. The adoption of gateway devices, encrypted CAN messaging, and proximity-based authentication improvements underscores a shift toward defence-in-depth in automotive security [307, 308, 311, 312].

## Heavy Vehicle and Fleet Security

Heavy vehicle and fleet security has become an increasingly important issue in modern transport operations, particularly in industries such as logistics, construction, mining, and public transport. Unlike passenger vehicles, heavy vehicles, such as trucks, buses, and specialized transport equipment, often operate within complex fleets, carry high-value cargo, and travel long distances across multiple jurisdictions. Because of these factors, they present attractive targets for theft, cargo hijacking, unauthorized use, and cyber intrusion. Effective security therefore requires a combination of physical protection, electronic systems, operational procedures, and fleet management strategies.

One of the fundamental aspects of heavy vehicle security is physical vehicle protection. Heavy vehicles are often parked in depots, roadside rest areas, freight terminals, or customer facilities where access control may vary significantly. Physical security measures may include secure depots with controlled entry points, fencing, lighting, surveillance cameras, and monitored parking areas. Vehicles themselves may incorporate reinforced door locks, ignition protection systems, steering wheel locks, and tamper-resistant compartments. For high-value cargo operations, sealed trailers, lockable container systems, and tamper-evident seals are commonly used to detect unauthorized access during transit.

Another critical component of fleet security is electronic vehicle protection systems. Modern heavy vehicles increasingly incorporate electronic immobilisers, keyless access systems, and telematics units that communicate with fleet management platforms. Immobilisers prevent the engine from starting unless an authorized key, transponder, or electronic credential is present. In many modern fleets, drivers authenticate themselves through RFID cards, electronic keys, or driver identification systems integrated into the vehicle's ignition or onboard computer. These systems help ensure that only authorized drivers can operate fleet vehicles.

Closely related to electronic security is the use of GPS tracking and telematics systems. Most commercial fleets now rely on telematics platforms that provide real-time location tracking, driver behaviour monitoring, and operational data such as speed, fuel consumption, and engine status. From a security perspective, GPS tracking allows fleet managers to monitor vehicle location, detect route deviations, and quickly respond if a vehicle is stolen or diverted. Some systems also support geofencing, which triggers alerts if a vehicle enters or leaves predefined geographic areas. This capability is particularly valuable for high-value cargo transport or restricted operational zones.

Fleet operators also rely heavily on cargo security measures, especially in logistics and freight transport. Cargo theft is a major issue globally, and criminals often target vehicles carrying electronics, pharmaceuticals, alcohol, or other high-value goods. To reduce risk, fleets may use electronic cargo seals, trailer door sensors, and load-monitoring systems that alert managers when trailers are opened unexpectedly. Some advanced systems integrate sensors that detect trailer movement, door opening, or cargo weight changes, providing early warning of potential theft.

Driver authentication and personnel security are another key element of heavy vehicle fleet protection. Because drivers have direct control over the vehicle and cargo, companies must ensure that only properly authorized and trained personnel have access to fleet vehicles. This may involve background checks, driver identification cards, biometric authentication systems, or secure digital driver profiles within fleet management software. Many modern fleets require drivers to log in to the vehicle's onboard computer before starting a shift, which allows operators to track who is driving each vehicle and helps ensure regulatory compliance with work-hours and safety requirements.

Cybersecurity has also emerged as a growing concern for heavy vehicle fleets. Modern trucks and buses increasingly rely on networked electronic control systems, including the Controller Area Network (CAN bus), telematics units, and wireless communications systems. These technologies improve efficiency and maintenance monitoring, but they also introduce potential vulnerabilities. Unauthorized access through telematics systems, diagnostic ports, or wireless interfaces could theoretically allow attackers to manipulate vehicle systems or extract sensitive operational data. As a result, fleet operators are beginning to adopt cybersecurity practices such as secured diagnostics, software updates, encrypted communications, and restricted access to onboard systems.

Operational procedures and training also play a major role in maintaining fleet security. Drivers and dispatch staff must be trained to recognise suspicious behaviour, follow secure parking procedures, and report incidents quickly. Many companies implement policies such as avoiding predictable routes, limiting disclosure of cargo information, and requiring check-ins at designated intervals during long journeys. Incident reporting systems and security audits help organisations identify weaknesses and improve their security practices over time.

Finally, regulatory and insurance requirements influence heavy vehicle security practices. In many jurisdictions, transport companies must comply with chain-of-responsibility laws, safety management systems, and cargo protection requirements. Insurance providers may also require specific security measures, such as GPS tracking, secure parking, or cargo monitoring, before covering high-value shipments. These regulatory and contractual obligations encourage fleet operators to implement structured security systems rather than relying on informal or ad-hoc practices.

Overall, heavy vehicle and fleet security is no longer limited to simple mechanical locks or basic tracking devices. Modern fleet protection involves an integrated approach combining physical security, electronic monitoring, cybersecurity measures, operational procedures, and driver accountability. As transport systems become increasingly connected and automated, effective fleet security will continue to evolve to address both traditional theft risks and emerging cyber-physical threats.

Two commercially available examples illustrate how modern heavy vehicle and fleet security systems integrate physical protection, electronic monitoring, and telematics technology to protect vehicles, drivers, and cargo. These systems are widely used in logistics, freight transport, and construction fleets to reduce theft risk, monitor vehicle use, and maintain operational control.

Geotab – GO Fleet Telematics Platform

The Geotab GO device is a widely used commercial fleet management and vehicle security platform designed for heavy vehicles such as trucks, buses, and construction equipment. The system consists of a small telematics unit installed in the vehicle's OBD or diagnostic port, which connects to the vehicle's internal electronic network and transmits operational data to a cloud-based management platform.

From a security perspective, the system provides real-time GPS vehicle tracking, allowing fleet operators to see the exact location of vehicles at any time. The platform also supports geofencing, which enables operators to create virtual geographic boundaries. If a vehicle enters or leaves a designated area—such as a depot, construction site, or approved route—the system automatically generates an alert. This capability is particularly valuable for preventing unauthorized vehicle use or detecting theft.

The Geotab platform also monitors driver identification, engine activity, and vehicle diagnostics, helping fleet operators verify who is using the vehicle and how it is being operated. Integration with driver authentication systems can ensure that vehicles only operate when a registered driver credential is present. In addition, the platform records vehicle events such as ignition activation, harsh braking, or unexpected movement, which can indicate security incidents or misuse.

For locksmiths and vehicle security technicians, systems like Geotab demonstrate how vehicle access and immobiliser systems interact with telematics and diagnostic interfaces. When replacing keys or programming immobilisers on fleet vehicles, technicians must understand how these electronic systems connect to the vehicle network, as unauthorized access to diagnostic ports could potentially allow security systems to be altered.

ABLOY – ABLOY PROTEC2 CLIQ High-Security Locking System

The ABLOY PROTEC2 CLIQ system is a high-security mechanical and electronic locking solution widely used in transport fleets, infrastructure systems, and high-value cargo operations. The system combines mechanical high-security lock cylinders with programmable electronic keys, providing strong resistance to picking, drilling, and unauthorized duplication while allowing fleet managers to control and audit access.

Each CLIQ key contains an electronic chip that communicates with the lock cylinder when inserted. Access rights are programmed into the key, meaning that specific drivers, technicians, or depot personnel can be given permission to open only certain locks—such as a vehicle door, trailer compartment, fuel tank, or equipment cabinet. Access permissions can be updated electronically through management software, which allows administrators to quickly revoke access if a key is lost or a driver leaves the organisation.

For heavy vehicle fleets, the CLIQ system is commonly used to secure trailer doors, cargo compartments, fuel tanks, and service panels. Because the keys record usage data, fleet managers can track who accessed a lock and when, providing an audit trail that can assist in investigating cargo theft or unauthorized vehicle access.

From a locksmithing perspective, systems like ABLOY CLIQ represent the shift toward integrated mechanical–electronic access control systems. Locksmiths installing or servicing these systems must understand both the mechanical security features of the lock cylinder and the electronic credential management software used to program keys. This reflects the broader trend in modern locksmithing where security professionals increasingly manage access control systems rather than just physical locks.

**Impact on Locksmithing Practices**

Heavy vehicle and fleet security is relevant to someone interested in locksmithing because modern vehicle security has evolved far beyond traditional mechanical locks. Trucks, buses, and other heavy vehicles often operate within large fleets, carry valuable cargo, and travel long distances, making them attractive targets for theft, unauthorized use, and cargo hijacking. As a result, security systems used in these vehicles typically combine mechanical locks, immobilisers, electronic keys, driver authentication systems, and monitoring technologies. For locksmiths, this means that servicing heavy vehicles increasingly involves understanding both physical locking systems and electronic access controls. Locksmiths may be required to install or maintain reinforced locks, trailer locking devices, ignition protection systems, and tamper-resistant compartments, while also working with electronic key programming, immobiliser systems, and driver identification credentials used to control who can operate fleet vehicles.

The development of telematics, GPS tracking, and networked vehicle electronics has also influenced locksmithing practices by expanding the profession into areas related to system security and access control management. Modern fleets often integrate electronic driver authentication, cargo monitoring systems, and telematics platforms that record vehicle access and usage. Locksmiths working in this environment must therefore operate within stricter authorization procedures, ensuring that key replacement, immobiliser programming, and access credential management are carefully documented and verified. Additionally, awareness of emerging cyber-physical risks, such as vulnerabilities in diagnostic ports or electronic vehicle networks, is becoming increasingly important. As heavy vehicle security systems continue to integrate mechanical, electronic, and digital components, locksmiths must adapt their knowledge and practices to support the broader security needs of fleet operators and transport industries.

## Aviation and Maritime Locking Systems

Aviation and maritime locking systems constitute specialized security mechanisms that must deliver high levels of security, reliability, and safety in environments that impose severe environmental and operational constraints. The literature consistently emphasizes that these environments, characterized by vibration, corrosion, weather exposure, and safety-critical operations, drive locking systems to operate under stringent technical standards, with corrosion-resistant materials, tamper-resistant designs, and robust key control or electronic access schemes as core requirements [320-323]. Across aviation and maritime domains, locks secure a broad spectrum of components and facilities, from cockpit doors and maintenance access panels to cargo holds, engine rooms, cargo securing points, and port facilities, while also supporting integration with broader access-control and monitoring infrastructure to meet regulatory and safety imperatives [320, 324-326].

### Aviation locking systems: scope, requirements, and regulatory alignment

Aviation security architecture encompasses locking solutions deployed on aircraft structures and airport infrastructure. Aircraft feature numerous lockable components such as cockpit doors, service panels, cargo holds, avionics compartments, and maintenance access hatches. The design objective for these locks is to reconcile security with rapid authorized access, particularly in emergencies, which leads to the use of specialized quarter-turn latches, cam locks, or key-operated access panels configured for quick, maintenance personnel–friendly operation while maintaining flight safety in normal operation [320].

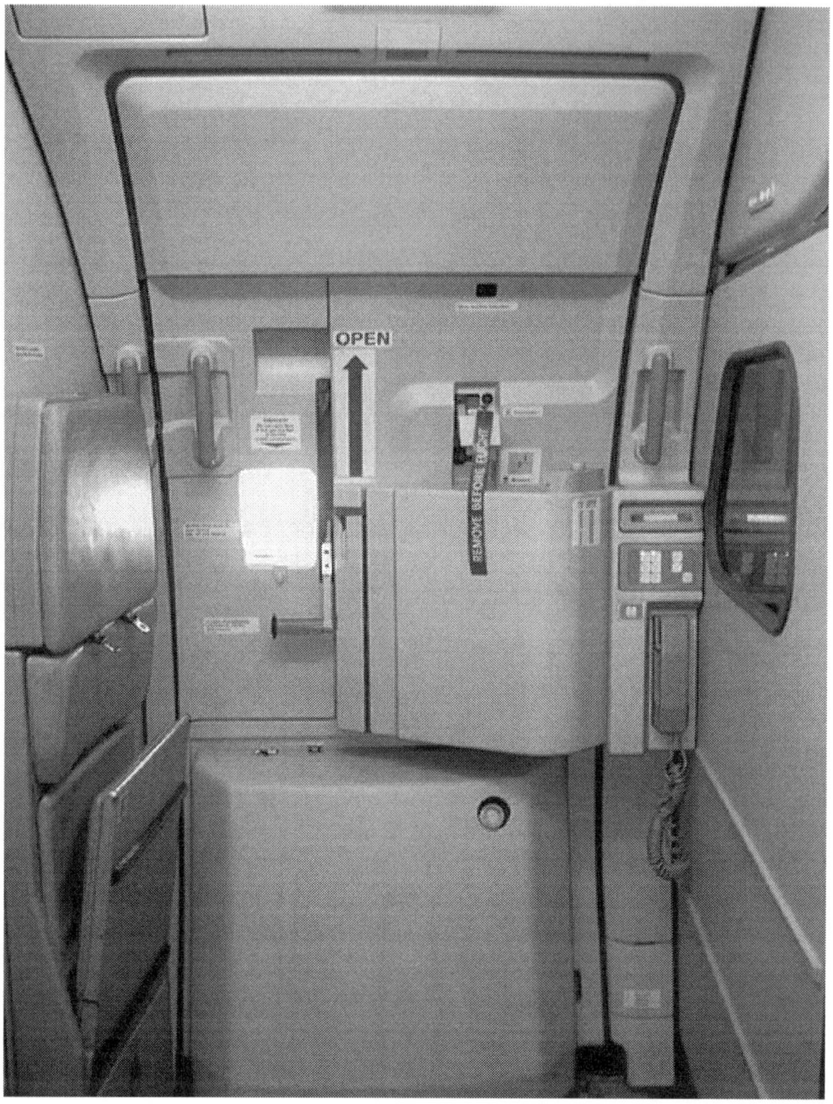


*Figure 89: Rear door of an Airbus A319 aircraft. Christopher Doyle, CC BY-SA 2.0, via Wikimedia Commons.*

Cockpit door locking systems in commercial aircraft are especially critical; modern systems employ reinforced doors with electronic locking mechanisms controlled from inside the flight deck and monitored via the aircraft security system, illustrating the convergence of physical tamper-resistance with electronic access monitoring and regulatory oversight [320, 323, 324].

Airport infrastructure expands the locking and access-control envelope to hangars, fuel storage areas, ground-support equipment, and airside zones, where access must be tightly controlled to prevent sabotage, theft, or unauthorized entry. Aviation locksmithing thus often involves high-security cylinders, restricted-key systems, and electronic access-control devices designed to meet aviation security regulations and to avoid interference with maintenance or emergency procedures. The tight regulatory environment, governed by national aviation authorities and international safety agencies, necessitates that locking systems comply with stringent safety and operational standards, ensuring access control that does not impede maintenance workflows or safety protocols [320, 323, 324].

Corrosion, material selection, and environmental durability in aviation systems have parallel implications for locking hardware. While aviation locks commonly use corrosion-resistant materials and tamper-resistant designs to withstand operational environments, there is ongoing emphasis on compatibility with safety-critical procedures and documentation requirements in service and key issuance, reflecting the intersection of mechanical security with regulatory compliance [320].

### Maritime locking systems: environmental resilience, security, and cargo/safety roles

Maritime locking systems face extreme environmental exposures, saltwater corrosion, humidity, vibration, and continuous duty cycles, that drive material choices toward corrosion-resistant alloys and finishes (e.g., stainless steel, brass, specialized marine alloys) and into lock designs that resist tampering while enabling reliable operation in harsh conditions. Maritime locks secure cabin doors, engine rooms, cargo holds, storage compartments, and safety equipment lockers, with reliability and access control being essential not only for theft prevention but also for protecting critical ship systems and restricting access to hazardous areas for authorized crew members [321, 323, 325]

In commercial shipping and cargo operations, locking systems extend to cargo-hold doors, container-locking points, and hazardous-material storage areas, where tamper-resistant locks and sealing systems enhance cargo integrity and enable tamper detection during voyages that can span weeks or months. Some modern vessels integrate electronic access-control systems that log crew entries to restricted areas such as bridge controls, navigation rooms, and engineering spaces, highlighting the convergence of physical security with digital access-control and monitoring systems in the maritime domain [321, 323, 325].

Ports and maritime infrastructure likewise rely on a blend of traditional locks and electronic access-control technologies (key cards, biometrics, monitored gates) to secure dockside warehouses, container terminals, fuel depots, and maintenance facilities. Maritime locksmiths typically service high-security padlocks, container locks, gate locks, and specialized marine locking devices designed to resist corrosion and physical attack, reflecting the broader security architecture required at scale in port environments [321, 323, 325].

### Materials, corrosion resistance, and long-term durability in aviation and maritime locking hardware

Across aviation and maritime contexts, locking hardware is typically manufactured from corrosion-resistant materials and finished to withstand aggressive environments, including marine atmospheres and aircraft-exposure conditions. Stainless steel, brass, marine alloys, and protective coatings are commonly referenced materials, chosen to extend service life and maintain functional integrity under salt spray, humidity, and vibration. The literature emphasizes that durable lock systems in these sectors must balance security with reliability and regulatory compliance, particularly in safety-critical locations like cockpit access points and engineered shipboard spaces [321, 323, 325].

Environmental durability has broader implications for maintenance practices and lifecycle considerations. For example, corrosion-diagnostic knowledge and durability testing are integral in both aviation- and maritime-adjacent contexts, where advanced coatings, sealants, and protective treatments are discussed as means to sustain performance in corrosive environments. Studies and reviews that address corrosion management strategies, materials selection, protective coatings, and long-term durability in marine and coastal environments underpin the rational choice of locking components and ancillary protective measures in aviation and maritime settings [321-323].

**Regulatory compliance, authorization procedures, and security governance**

Locking systems in aviation and maritime sectors operate within complex governance regimes and documentation requirements. The need to ensure access control without compromising emergency procedures or maintenance workflows drives the development of controlled-access schemes, restricted-key systems, and tamper-resistant hardware, paired with electronic access-control platforms and logging capabilities. In aviation, regulatory compliance with national authorities and international aviation security standards shapes not only hardware design but also key control and provisioning practices. In maritime contexts, security regimes at ports and onboard installations similarly require integrated access-control approaches and documented authorization processes for servicing and key issuance, reflecting the intersection of mechanical security with regulatory compliance and safety management [320, 323-325] .

Taken together, the aviation and maritime locking landscape demonstrates how locking technology adapts to high-stakes, harsh environments where safety, reliability, and security converge. Locks in these domains are engineered for high durability and corrosion resistance, with tamper-resistant features and controlled-key or electronic access systems, designed to integrate with broader security and safety-regulatory ecosystems. The locksmithing practice in such environments demands specialized knowledge of aircraft and vessel operations, environmental exposures, and regulatory requirements, including documentation and authorization workflows for servicing and key issuance [320, 323-325].

**Impact on Locksmithing Practices**

Aviation and maritime locking systems are highly relevant to someone interested in locksmithing because they demonstrate how locking technology must be adapted for specialized, safety-critical environments. Unlike standard residential or commercial locks, systems used in aircraft and marine vessels must operate reliably under conditions of vibration, corrosion, temperature variation, and heavy operational use. These environments also demand strict regulatory compliance and integration with broader safety systems. For locksmiths, this highlights the importance of understanding material selection, corrosion resistance, tamper-resistant design, and controlled key systems. Locksmiths working in aviation or maritime contexts may service specialized hardware such as quarter-turn latches, cam locks, restricted cylinders, and electronic access-control systems used on cockpit doors, maintenance panels, cargo holds, engine rooms, and port infrastructure. The need to balance strong security with rapid authorized access, particularly during emergencies or maintenance operations, also influences how these locking systems are designed and serviced.

The influence on locksmithing practice is significant because these sectors require locksmiths to operate within tightly regulated security and safety frameworks. In aviation and maritime environments, lock servicing, key issuance, and access control must often follow documented authorization procedures and compliance requirements set by national and international regulatory bodies. Locksmiths must therefore combine traditional mechanical skills with knowledge of electronic access systems, restricted key management, and integration with security monitoring platforms. Additionally, the harsh environmental conditions present in maritime and aviation operations require locksmiths to understand corrosion-resistant materials, protective coatings, and maintenance practices that ensure long-term durability of locking hardware. As a result, locksmithing in these industries becomes a specialized discipline that blends mechanical security, electronic access control, environmental engineering considerations, and strict procedural governance.

## Future of Autonomous Vehicle Security

The trajectory of autonomous vehicle security is intrinsically linked to the rapid progress of self-driving technologies, artificial intelligence, and the expanding ecosystem of connected vehicle networks. Contemporary autonomous vehicles rely on a layered architecture that integrates sensors, onboard processing, vehicle control systems, wireless communications, and cloud-based data services to enable safe operation. Core sensing and networking technologies such as radar, lidar, cameras, GPS, and vehicle-to-everything (V2X) communications empower interaction with other vehicles, road infrastructure, and traffic management systems, thereby enabling advanced navigation and automation while simultaneously broadening the attack surface inherent to a complex digital ecosystem [327-329]. The resulting security imperative must address both traditional physical access risks and sophisticated cyber threats that exploit software-defined and networked vehicle components [24, 327, 330].

A central security challenge in autonomous vehicles lies in protecting the electronic architecture and internal networks, notably the Controller Area Network (CAN) and newer Automotive Ethernet, which interconnect critical subsystems such as steering, braking, acceleration, and sensor inputs. Compromise of these networks could enable adversaries to alter vehicle behaviour or disable safety mechanisms; consequently, manufacturers are pursuing more secure network architectures that incorporate gateway controllers, network segmentation, intrusion detection, and encrypted communications to restrict access to vehicle control domains [327, 328, 330]. Secure gateway design is essential for authenticating connected ECUs and safeguarding data flowing across network boundaries, with secure key management and firewalling forming part of the defence-in-depth approach to in-vehicle networks [328, 330]. This view is reinforced by studies that emphasize secure gateway functionality, authentication of peripheral devices, and protection against wiretapping and spoofing within automotive networks [328, 330] .

Software integrity and over-the-air (OTA) update mechanisms are foundational to maintaining vehicle safety and resilience. Autonomous vehicles depend on frequent software updates to improve performance and remediate vulnerabilities; therefore, secure OTA processes must ensure that only authenticated, verified software is installed, and that transmissions are protected against alteration or injection of malicious code.

Cryptographic verification methods serve to confirm the provenance and integrity of updates, enabling manufacturers to prevent unauthorized modification of control software and ensuring a trustworthy update lifecycle [327, 328, 330]. The literature consistently underscores the need for robust code signing, secure boot procedures, and authenticated update channels as core components of vehicle cybersecurity strategies [327, 328, 330].

The wireless fabric supporting autonomous driving, particularly V2V and V2I communications, offers notable benefits for traffic efficiency and safety but introduces vulnerabilities related to interception, spoofing, and manipulation of messages. Security frameworks are being developed to mitigate these risks through encryption, authentication protocols, and digital certificates to verify the legitimacy of messages between vehicles and infrastructure, as well as to ensure message integrity and non-repudiation in the V2X domain [327-329, 331]. While encryption and authentication improve resilience, researchers also highlight the need for scalable, privacy-preserving approaches given the data-intensive and pervasive nature of V2X systems [329, 331].

Autonomous vehicles collect and transmit extensive data associated with location, driving behaviour, sensor streams, and occupant information. Protecting this data is essential for safety and privacy, necessitating secure data storage, minimization or anonymization techniques, and strict access controls to prevent unauthorized exposure or misuse of sensitive information [327, 329, 332, 333]. Privacy-preserving design choices, such as data minimization, secure enclaves, and privacy-aware data sharing policies, are increasingly treated as integral to the security architecture of autonomous mobility systems [327, 329, 333].

The evolution of autonomous and connected vehicles extends implications for physical security disciplines, including locksmithing and access control. Traditional mechanical keys are being superseded by digital access credentials, smartphone-based keys, biometric authentication, and cloud-managed access platforms. As vehicles become more automated and connected, locksmiths and automotive security professionals must broaden expertise to include electronic credential systems, secure software interfaces, and cyber-physical access protocols, alongside conventional lock mechanisms. This shift indicates a convergence of physical security and cybersecurity disciplines, with digital keys, encrypted interlocks, and vehicle authentication platforms forming essential competencies for modern automotive security practitioners [334-337].

Policy and regulatory developments are increasingly shaping autonomous vehicle cybersecurity by mandating governance structures for risk management, incident reporting, and continuous monitoring of vehicle systems. Governmental bodies and international standards organizations are articulating guidelines and regulations designed to ensure that manufacturers implement cybersecurity risk management across the vehicle lifecycle and respond rapidly to vulnerabilities or breaches. This regulatory impulse supports ongoing risk assessment, supply-chain security, and incident response planning, reinforcing a multidisciplinary security posture that integrates technical measures with governance, compliance, and oversight [327, 332, 333, 338].

The future of autonomous vehicle security envisions a multidisciplinary ecosystem that weaves together cybersecurity, physical access control, software integrity, data governance, and regulatory oversight. As vehicles become more autonomous and connected, security will depend on coordinated efforts among manufacturers, cybersecurity professionals, fleet operators, and security practitioners. For locksmiths and vehicle security technicians, this evolving landscape expands the scope of expertise beyond traditional lock mechanisms toward digital access systems, electronic vehicle architectures, secure communication protocols, and cybersecurity awareness. This reframing underscores the need for ongoing education and cross-disciplinary collaboration to ensure security practices keep pace with the rapid technological transformation characterizing modern transport systems [327, 331, 332, 335, 339].

**Impact on Locksmithing Practices**

The evolution of autonomous vehicle security is highly relevant to locksmithing because vehicle access and authorization systems are shifting from traditional mechanical keys to digitally managed credentials integrated with complex electronic vehicle architectures. Autonomous vehicles rely on interconnected systems such as the Controller Area Network (CAN), Automotive Ethernet, wireless communication interfaces, and cloud-based platforms. These systems govern vehicle control functions, authentication mechanisms, and communication with infrastructure and other vehicles. As a result, vehicle access is increasingly controlled through encrypted digital keys, smartphone credentials, biometric authentication, and cloud-managed authorization platforms. For locksmiths, this transformation expands the scope of the profession beyond mechanical key cutting and lock servicing toward understanding electronic control systems, secure gateways, credential provisioning, and the interaction between vehicle access systems and internal vehicle networks.

This technological shift significantly influences locksmithing practices by requiring practitioners to integrate cybersecurity awareness and digital access management into their skill sets. Tasks such as programming digital keys, managing vehicle authentication credentials, and interacting with diagnostic interfaces must be performed within secure frameworks that protect against unauthorized access to vehicle networks or software systems. Locksmiths must also understand secure software update processes, encrypted communications, and the role of authentication protocols used in connected vehicles. As regulatory frameworks increasingly mandate cybersecurity risk management and secure access governance across the vehicle lifecycle, locksmiths and automotive security technicians will need to follow stricter authorization procedures, documentation practices, and compliance requirements. Consequently, the future of locksmithing in automotive contexts reflects a convergence between traditional physical security expertise and emerging cyber-physical security disciplines.

## Chapter 14 Review Questions

### Question 1

A vehicle owner uses a smartphone app to unlock their car remotely through a cloud-connected access system. The command travels from the smartphone to cloud servers and then to the vehicle.

What security mechanism is most important to prevent interception of the command?

A. Mechanical key duplication control
B. Encrypted communication protocols such as TLS
C. Steel lock housings
D. Plastic key fobs

**Question 2**

A fleet management company allows drivers to access vehicles using digital keys issued through a mobile application. A driver leaves the company and their access must be removed immediately.

What feature of cloud-connected vehicle systems allows this to occur?

A. Remote credential revocation through the cloud platform
B. Replacing the vehicle ignition lock
C. Reprogramming the vehicle's engine
D. Disabling the vehicle battery

**Question 3**

A locksmith is diagnosing a modern vehicle where the owner's digital key no longer unlocks the doors after a phone was replaced.

What is the most likely cause?

A. Mechanical wear inside the door lock
B. The digital key credentials were not transferred or re-authenticated
C. The engine control unit is damaged
D. The vehicle tires need replacement

**Question 4**

A vehicle manufacturer pushes a software patch to improve vehicle security after a vulnerability is discovered in its digital key system.

What technology allows this update to be installed remotely?

A. Mechanical key cutting
B. Over-the-air (OTA) software updates
C. Door hinge adjustments
D. Manual lock cylinder replacement

**Question 5**

A locksmith working on a modern vehicle must program a replacement key fob after the owner loses their original one.

What additional knowledge is required compared with traditional automotive locksmithing?

A. Woodworking skills
B. Vehicle electronics and digital key programming systems
C. Metal forging techniques
D. Paint restoration skills

**Question 6**

A hacker attempts to intercept commands sent between a smartphone and a vehicle's cloud server.

Which security method primarily protects this communication?

A. Magnetic locking systems
B. Network encryption and authentication protocols
C. Manual door bolts
D. Rubber key covers

**Question 7**

A vehicle security engineer discovers that the Controller Area Network (CAN) bus inside a vehicle can be exploited if an attacker gains access to the internal wiring.

Why does this vulnerability exist?

A. CAN systems were originally designed for reliability rather than strong security
B. CAN systems are encrypted by default
C. CAN networks cannot transmit data
D. CAN systems are only used in aircraft

**Question 8**

A locksmith working on fleet vehicles must diagnose issues related to remote unlocking commands not reaching vehicles parked in remote locations.

Which factor could most likely cause this problem?

A. Broken windshield wipers
B. Poor network connectivity between the vehicle and cloud service
C. Incorrect tire pressure
D. Engine oil contamination

**Question 9**

A company wants to track vehicle usage, location, and driver activity across a large fleet.

Which feature of cloud-connected vehicle systems supports this function?

A. Mechanical lock cylinders
B. Vehicle telemetry and remote monitoring systems
C. Manual odometer recording
D. Physical key duplication

**Question 10**

A locksmith specialising in modern vehicles is asked to advise a client about security risks associated with digital vehicle access systems.

Which skill is now increasingly necessary for automotive locksmiths?

A. Knowledge of basic carpentry
B. Understanding of cybersecurity and digital authentication systems
C. Welding skills
D. Knowledge of plumbing systems

# Part 8

# Business, Legal, and Ethical Practice

# Chapter 15

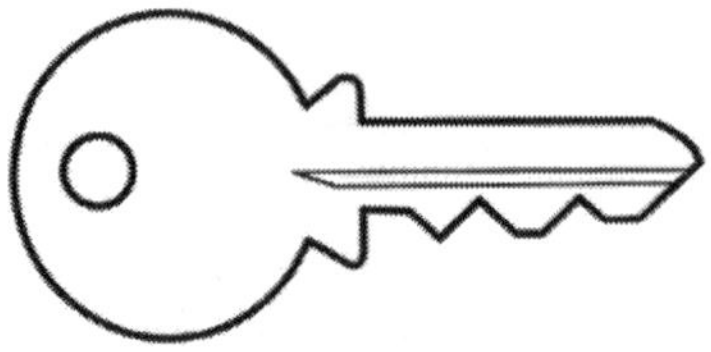

# The Modern Locksmith Business Model

Before exploring the specific components of the modern locksmith business model, it is important to understand how the locksmithing profession itself has evolved. Historically, locksmiths were primarily viewed as tradespeople who worked with mechanical locks, keys, safes, and door hardware. Their work focused on practical, hands-on tasks such as installing locks, cutting keys, repairing damaged mechanisms, and assisting customers during lockouts. While these services remain a fundamental part of the profession, the security environment surrounding buildings, vehicles, and infrastructure has changed significantly in recent decades.

Advances in technology have transformed how access to property and assets is controlled. Modern security systems increasingly combine traditional mechanical locking hardware with electronic access control, digital credentials, biometric authentication, and network-connected security platforms. As a result, the role of the locksmith has expanded from a narrow mechanical trade into a broader security profession that requires knowledge of both physical and electronic security systems. Locksmiths now work across a wide range of sectors, including residential security, commercial access control, automotive security technologies, and integrated building security systems.

This chapter examines how these developments have reshaped the locksmith profession and the way locksmith businesses operate. It explores the transition from traditional lock repair services to a modern security-focused business model that includes consulting, system integration, risk assessment, client education, and specialised market services. By understanding these changes, readers can see how modern locksmiths combine technical craftsmanship with analytical security planning to protect people, property, and assets in increasingly complex environments.

## Transition from Tradesperson to Security Consultant

The locksmithing profession has historically been viewed as a skilled trade centred on mechanical locks, key cutting, lock installation, and emergency lockout services. Traditional locksmiths were primarily reactive service providers who responded to immediate problems such as lost keys, broken locks, or lock replacements. Their expertise focused on physical hardware, including pin tumbler locks, safe mechanisms, door hardware, and mechanical access systems. While these skills remain important, the modern security environment has expanded far beyond mechanical locking systems. Advances in electronic security, smart access control, vehicle electronics, and integrated building security systems have transformed locksmithing from a narrowly defined trade into a broader security profession.

*Figure 90: Traditional locksmithing. muhammed diler, Public Domain, via Pexels.*

The transition from tradesperson to security consultant reflects the increasing complexity of security systems used in residential, commercial, industrial, and transport environments. Modern buildings and vehicles rarely rely on a single type of lock or security device. Instead, they use layered security architectures that combine mechanical locks, electronic access control, surveillance systems, alarms, biometric authentication, and digital credential management. As a result, locksmiths are increasingly expected to evaluate entire security environments rather than simply replace individual locks. This requires a more analytical and advisory approach in which the locksmith assesses risks, recommends appropriate security solutions, and designs integrated protection strategies.

A security consultant in the locksmithing context performs a much broader role than a traditional locksmith. Rather than only installing hardware, the consultant examines how people interact with a building, facility, or vehicle system and identifies potential vulnerabilities. This may involve reviewing door hardware, key control policies, electronic access logs, alarm systems, CCTV coverage, and physical access procedures. For example, in a commercial facility the locksmith may evaluate whether restricted key systems are required, whether electronic card readers should replace mechanical locks, or whether certain doors should include delayed egress hardware for safety compliance. In fleet operations or vehicle security, the consultant might examine immobiliser systems, telematics access controls, or digital key management policies.

Another major aspect of this transition is the shift toward risk assessment and security planning. Modern locksmith businesses often perform security audits for clients. A security audit involves analysing potential threats, evaluating the existing protective measures, and recommending improvements. This process may include examining physical entry points, reviewing access-control procedures, assessing key management practices, and considering emerging threats such as electronic intrusion or credential misuse. By providing structured risk assessments, locksmiths move beyond simple service work and become strategic advisors in the protection of assets and facilities.

Technology has been a primary driver of this transformation. Electronic access control systems, smart locks, wireless credentials, biometric readers, and cloud-managed security platforms are now common in many environments. Locksmiths must therefore understand networking principles, software configuration, wireless communication protocols, and cybersecurity considerations. For example, modern locksmiths may install and configure keyless entry systems, integrate door hardware with building management systems, or manage digital credentials for multiple users. In automotive contexts, locksmiths may program transponder keys, digital vehicle credentials, and smartphone-based access systems. This blending of mechanical and digital expertise reinforces the consultant role because clients often rely on locksmiths to interpret complex technology and recommend suitable solutions.

The business model of locksmithing also changes as this transition occurs. Traditional locksmith services often relied heavily on emergency work such as lockouts and urgent repairs. While these services still exist, many modern locksmith businesses generate significant revenue from planned security installations, system upgrades, maintenance contracts, and consulting services. Consulting work may involve designing master key systems, developing access control strategies for new buildings, or advising organisations on security policy and compliance requirements. These services create longer-term relationships with clients and position locksmiths as trusted security advisors rather than occasional repair technicians.

Professional credibility and education play an important role in supporting this shift. As locksmiths move toward a consultancy model, they must demonstrate expertise not only in lock mechanisms but also in security standards, building codes, and regulatory requirements. This may include understanding fire safety regulations related to exit hardware, compliance requirements for electronic access control systems, and standards governing vehicle security or fleet management. Continuing education, professional certifications, and training in electronic security technologies are increasingly important for locksmiths who wish to operate in this expanded professional role.

Communication and client engagement also become more significant. A locksmith acting as a security consultant must be able to explain complex security concepts in practical terms that clients can understand. This includes presenting risk assessments, outlining security improvement plans, and helping clients balance convenience, cost, and protection. Effective consultants also develop long-term relationships with clients by providing ongoing support, maintenance, and periodic security reviews. In this way, the locksmith's role evolves from simply fixing locks to actively guiding clients in managing their security systems.

Another factor influencing this transition is the convergence of physical security with cybersecurity. Many modern security systems rely on networked devices, cloud platforms, and digital credentials. This means that vulnerabilities may arise not only from physical lock failures but also from weak authentication systems, poorly managed digital keys, or unsecured communication networks. Locksmiths acting as security consultants must therefore be aware of cyber-physical security risks and understand how physical security devices interact with digital systems.

**Impact on Locksmithing Practices**

The transition from tradesperson to security consultant is highly relevant to someone interested in locksmithing because it reflects how the profession is evolving in response to modern security technologies and increasingly complex access-control environments. Traditional locksmithing focused primarily on mechanical skills such as cutting keys, repairing locks, and responding to lockouts. However, modern buildings, vehicles, and infrastructure now rely on integrated security systems that combine mechanical locks with electronic access control, surveillance systems, digital credentials, and networked monitoring platforms. For someone entering the locksmithing field, this means that the profession is no longer limited to servicing physical hardware. Instead, locksmiths must understand how different security components interact and be capable of evaluating the overall security environment to identify weaknesses and recommend appropriate solutions.

This evolution has significantly influenced locksmithing practices by expanding the role of the locksmith into risk assessment, system design, and security consulting. Modern locksmiths may conduct security audits, develop master key and access control strategies, advise clients on electronic locking systems, and manage digital credentials or vehicle access technologies. As a result, locksmithing now requires broader technical knowledge that includes electronic security systems, wireless communication technologies, and the integration of physical and digital security platforms. In practice, locksmiths increasingly act as advisors who help clients design layered security systems rather than simply replacing locks. This shift also encourages ongoing professional education, stronger documentation procedures, and closer collaboration with other security professionals, reflecting the growing convergence between physical security, electronic access control, and cybersecurity in modern locksmithing.

## Integrating Physical and Digital Services

The modern locksmith business model increasingly relies on the integration of physical security services with digital and electronic security technologies. Historically, locksmithing focused almost entirely on mechanical security devices such as door locks, safes, padlocks, and key systems. The locksmith's primary tasks involved installing locks, cutting keys, repairing mechanical mechanisms, and responding to emergency lockouts. However, advances in technology have transformed security systems across residential, commercial, industrial, and automotive environments. Many modern access-control systems now combine traditional mechanical locking devices with electronic authentication, wireless communication, cloud-based management platforms, and digital credentials. As a result, successful locksmith businesses must provide services that address both the physical and digital aspects of security.

One of the key drivers of this integration is the widespread adoption of electronic access control systems. These systems replace or supplement traditional keys with electronic credentials such as key cards, RFID tags, biometric identifiers, PIN codes, and smartphone-based digital keys. Electronic access control allows property owners and organisations to manage entry permissions electronically, track access activity, and quickly modify or revoke credentials when necessary. Locksmiths installing these systems must understand not only door hardware and locking mechanisms but also the electronic components that control access, including readers, controllers, power supplies, and network interfaces. Integrating these systems with existing door hardware requires careful planning to ensure that the mechanical and electronic components function reliably together.

Another area where physical and digital services intersect is smart lock technology. Smart locks are commonly used in homes, rental properties, and commercial environments where users want the convenience of remote access and digital credential management. These devices allow doors to be unlocked using smartphone applications, keypad codes, or wireless credentials instead of traditional keys. Many smart locks also connect to home automation systems or cloud-based management platforms that allow property owners to monitor and control access remotely. Locksmiths working with smart locks must therefore understand wireless communication protocols such as Bluetooth, Wi-Fi, and Zigbee, as well as the security implications of cloud-based credential management. They may also provide services such as configuring user permissions, integrating smart locks with other building systems, and troubleshooting connectivity or authentication issues.

The integration of physical and digital services also extends to vehicle security systems. Modern vehicles incorporate electronic immobilisers, remote keyless entry systems, passive keyless entry technologies, and smartphone-based digital keys. Automotive locksmiths must therefore combine traditional mechanical skills with expertise in electronic programming and vehicle diagnostics. Tasks such as programming transponder keys, pairing remote key fobs, and managing digital vehicle credentials require specialised diagnostic tools and knowledge of vehicle electronic control systems. In this context, locksmiths are not only servicing locks but also interacting with the vehicle's electronic architecture and digital authentication processes.

Commercial security systems often demonstrate the most advanced integration of physical and digital technologies. Businesses, government facilities, and industrial sites frequently use comprehensive security platforms that combine electronic door locks, surveillance cameras, alarm systems, biometric readers, and building management systems. In these environments, locksmiths may be responsible for installing high-security mechanical locks while also configuring electronic access control systems that communicate with networked security software. Integration may involve linking door hardware with card access systems, configuring audit trails that record entry events, and ensuring that emergency exit hardware complies with safety regulations. This requires locksmiths to understand how mechanical door hardware interacts with electronic locking mechanisms and how those systems connect to digital management platforms.

The shift toward integrated security services has also changed how locksmith businesses operate commercially. Instead of providing only one-time repair services, many locksmith companies now offer security system design, installation, maintenance, and consulting services. Clients may require assistance designing an access-control strategy for a new building, upgrading security hardware in an existing facility, or managing access credentials for a large workforce. By integrating physical and digital services, locksmiths can offer comprehensive security solutions that address multiple layers of protection. This approach not only improves security outcomes for clients but also creates long-term service relationships through system maintenance, software updates, and ongoing security reviews.

Another important consideration is the growing importance of cybersecurity awareness in locksmithing practice. Digital access-control systems and cloud-managed security platforms can introduce vulnerabilities if they are improperly configured or maintained. Weak passwords, outdated firmware, unsecured wireless networks, or poorly managed user credentials can undermine the effectiveness of even the strongest physical locking hardware. Modern locksmiths must therefore understand how to secure digital systems by implementing strong authentication procedures, managing encryption keys, and ensuring that software updates are applied regularly. This reflects the increasing convergence between traditional physical security and information security disciplines.

Training and professional development play a critical role in enabling locksmiths to integrate physical and digital services effectively. Modern locksmiths must remain familiar with evolving technologies such as biometric authentication systems, mobile credential platforms, vehicle security programming tools, and network-connected security devices. Many locksmiths now receive training in electronic security installation, networking basics, and security system integration. These skills allow them to design and implement security solutions that combine traditional mechanical protection with modern digital access control.

## Impact on Locksmithing Practices

This development is highly relevant to someone interested in locksmithing because it reflects the major technological shift occurring within the security industry. Locksmithing is no longer limited to servicing mechanical locks and cutting keys; modern security systems increasingly combine traditional locking hardware with electronic access control, wireless communication, and digital credential management. As buildings, vehicles, and infrastructure adopt technologies such as RFID cards, biometric authentication, smartphone-based keys, and cloud-managed access platforms, locksmiths must understand how these systems operate alongside conventional locks. For someone entering the profession, this means developing knowledge that spans both physical security hardware and digital access technologies. Understanding how mechanical door hardware integrates with electronic controllers, networked systems, and authentication platforms is essential for working effectively in modern residential, commercial, and automotive security environments.

The influence on locksmithing practices is significant because it changes both the technical skills required and the services locksmiths provide. Modern locksmiths are increasingly involved in installing electronic access control systems, configuring smart locks, programming vehicle keys, managing digital credentials, and integrating security hardware with networked management systems. This expands the role of the locksmith from a reactive technician who repairs locks to a security professional who designs, installs, and manages integrated security solutions. It also requires greater awareness of cybersecurity risks, such as protecting wireless communications, securing cloud-based access platforms, and maintaining software updates for connected devices. As a result, locksmithing practice now involves a blend of mechanical expertise, electronic system knowledge, and digital security awareness, reflecting the broader convergence of physical and information security technologies.

## Risk Assessments and Security Audits

In the modern locksmith business model, risk assessments and security audits have become essential professional services that extend locksmithing beyond traditional lock installation and repair. Historically, locksmiths were primarily called upon to fix specific problems such as broken locks, lost keys, or damaged door hardware. However, modern security environments are far more complex, and organisations now require systematic evaluations of their entire access-control infrastructure. Risk assessments and security audits allow locksmiths to examine how effectively a property, facility, or vehicle system is protected and to identify weaknesses that could lead to unauthorized access, theft, or safety incidents.

A security risk assessment is a structured process used to identify potential threats, vulnerabilities, and the possible consequences of security failures. In locksmithing practice, this involves examining how people access and move through a building or facility and determining where security weaknesses may exist. Locksmiths may evaluate door hardware, lock quality, key control procedures, access points, and emergency exit systems. They also consider external factors such as lighting, building layout, surveillance coverage, and the presence of alarm systems. By analysing these elements, locksmiths can determine whether existing security measures are appropriate for the risks associated with the property.

Security audits typically involve a detailed inspection of physical access-control systems. Locksmiths examine all entry points such as doors, gates, windows, and service access panels to determine whether they meet appropriate security standards. This may include checking whether locks are resistant to picking, drilling, or forced entry, verifying that door frames and strike plates are reinforced, and confirming that emergency exit hardware complies with fire safety regulations. Locksmiths may also inspect safes, secure storage cabinets, and restricted access areas to ensure that the locking mechanisms provide adequate protection for valuable assets.

Another important component of security auditing is key control management. In many organisations, poor key management practices create significant security risks. Locksmiths conducting audits may review how keys are issued, tracked, and stored. They may examine whether a restricted key system is required, whether master key systems are properly structured, and whether lost keys are promptly reported and replaced. In facilities with large numbers of employees, contractors, or service personnel, locksmiths may recommend improved key management procedures or the adoption of electronic access control systems to reduce the risks associated with uncontrolled key duplication.

Modern security audits also consider electronic access control systems. Many buildings now use electronic door locks, card readers, biometric scanners, or smartphone-based credentials instead of traditional keys. Locksmiths evaluating these systems must assess both the physical hardware and the electronic components that control access. This may include verifying that electronic locks are properly installed, ensuring that power supplies and backup systems are reliable, and reviewing access permissions stored within the system. Locksmiths may also examine audit logs to determine whether unusual or unauthorized access attempts have occurred.

Risk assessments in the modern locksmith business model increasingly include vehicle and fleet security evaluations. In automotive and transport contexts, locksmiths may assess vehicle immobiliser systems, keyless entry technologies, telematics security, and fleet access control procedures. For example, they may determine whether vehicles are vulnerable to relay attacks, whether diagnostic ports are adequately protected, or whether key management procedures for drivers are secure. Fleet operators often rely on locksmiths and security specialists to recommend improvements that reduce theft risks and improve vehicle access accountability.

Another important element of professional security assessments is compliance with regulations and safety standards. Locksmiths must ensure that access-control systems do not interfere with emergency evacuation procedures or building safety requirements. For example, exit doors must comply with fire safety regulations and allow rapid egress during emergencies. Locksmiths performing security audits often review door hardware to confirm that panic bars, emergency exit devices, and delayed egress systems meet applicable building codes and safety standards.

Security audits also involve developing security improvement strategies. After identifying vulnerabilities, locksmiths typically provide clients with detailed recommendations for improving their security systems. These recommendations may include upgrading locks to high-security cylinders, installing restricted key systems, adding electronic access control devices, improving door hardware reinforcement, or implementing new key management policies. In larger facilities, locksmiths may design master key systems or integrated access-control solutions that provide controlled access for different user groups.

The modern locksmith business model benefits significantly from offering risk assessments and security audits because these services create long-term relationships with clients. Instead of only responding to emergency repairs, locksmiths can provide ongoing security management services. Businesses, institutions, and property managers often require periodic security reviews to ensure that their systems remain effective as personnel, technology, and operational requirements change. This positions locksmiths as trusted security advisors who contribute to strategic planning rather than simply providing technical repairs.

Risk assessments and security audits highlight the growing role of locksmiths in the broader field of security consulting. Modern locksmiths must combine technical knowledge of locks and access-control hardware with analytical skills that allow them to evaluate security risks and recommend integrated solutions. By offering professional assessments, locksmiths help clients understand their vulnerabilities and implement appropriate security measures that protect property, assets, and people.

**Case Study Scenario: Security Risk Assessment and Audit for a Logistics Distribution Centre**

A large logistics company operates a regional distribution centre that stores and dispatches high-value consumer electronics. The facility includes a warehouse, administration offices, loading docks, a secure storage room for high-value inventory, and a fleet parking area for delivery vehicles. Over several months, the company experienced several minor security incidents, including misplaced keys, a delivery door left unlocked overnight, and unexplained access to the warehouse outside of normal operating hours. Although no major theft had occurred, management became concerned that their security procedures might contain vulnerabilities. Instead of waiting for a serious incident, the company engaged a professional locksmith business to conduct a comprehensive security risk assessment and security audit of the facility.

The locksmith began by conducting a site inspection and risk assessment to understand how employees, contractors, and delivery personnel moved through the facility. The assessment identified multiple access points, including staff entrances, warehouse roller doors, emergency exits, and vehicle gates. During the inspection, the locksmith discovered that several exterior doors were fitted with standard commercial locks that were more than ten years old and vulnerable to forced entry. In addition, the strike plates and door frames on several doors showed signs of wear and were not properly reinforced. Lighting around the rear loading dock was also inadequate, creating areas where unauthorized access could occur without being easily observed by security cameras.

During the security audit of physical locking systems, the locksmith examined all entry points and internal access areas. The audit revealed that the high-value storage room used a basic key-operated lock that could potentially be duplicated without authorization. Furthermore, the warehouse emergency exits were fitted with push-bar hardware that complied with fire safety requirements but lacked alarmed exit devices, meaning the doors could be opened without alerting staff or management. The locksmith also inspected the safe used to store company documents and discovered that its locking mechanism had not been serviced in several years, increasing the risk of malfunction.

A significant issue was discovered in the key management system. The company had issued multiple copies of facility keys to supervisors, maintenance contractors, and temporary staff over time, but there was no formal record of who held each key. Some keys had been lost but never reported, and several employees were using privately duplicated keys. The locksmith identified this as a major vulnerability, as uncontrolled key duplication could allow unauthorized individuals to enter the facility without detection.

The locksmith also evaluated the facility's electronic access control and surveillance systems. The administration office used a keypad lock system that required a four-digit code shared by several employees. Because the code had not been changed for several years, former employees and contractors might still know the access code. The locksmith recommended replacing the shared code system with an electronic card access system that would allow management to assign unique credentials to each employee and revoke them if staff left the organisation.

As part of the audit, the locksmith also examined the fleet vehicle security practices. The company operated a fleet of delivery trucks parked overnight in a fenced yard. The assessment revealed that spare vehicle keys were stored in an unlocked cabinet in the dispatch office, and drivers occasionally left remote key fobs inside the vehicles during loading operations. The locksmith advised that these practices increased the risk of vehicle theft and recommended a secure key cabinet with access logging, along with policies requiring drivers to retain keys at all times during shifts.

After completing the inspection, the locksmith prepared a detailed security audit report outlining the identified vulnerabilities and recommending improvements. Key recommendations included installing high-security restricted key cylinders on all external doors, implementing a structured master key system, reinforcing door frames and strike plates, upgrading lighting around the loading dock, and installing alarmed exit devices on emergency doors. The report also recommended replacing the shared keypad entry system with electronic card access control and implementing a formal key management policy with documented key issuance and tracking.

The company implemented many of these recommendations over the following months. The locksmith business was contracted to design the master key system, install the new high-security cylinders, and integrate the electronic access control system with the building's alarm and surveillance systems. The locksmith also trained management staff in key control procedures and scheduled annual security audits to ensure the system remained effective as the organisation expanded.

This scenario demonstrates how modern locksmiths operate not only as lock technicians but also as security consultants who assess risks, identify vulnerabilities, and design integrated security solutions. By performing a comprehensive security audit rather than simply replacing individual locks, the locksmith helped the organisation strengthen its security infrastructure, improve access control, and reduce the risk of theft or unauthorized entry.

## Impact on Locksmithing Practices

Understanding risk assessments and security audits is highly relevant to someone interested in locksmithing because it illustrates how the profession has expanded beyond simply repairing or installing locks. Modern security environments are far more complex, involving multiple access points, electronic systems, vehicle security technologies, and organisational procedures that influence how security is managed. A locksmith who understands how to conduct structured risk assessments can evaluate an entire facility or system rather than focusing only on individual locks. This allows the locksmith to identify weaknesses such as poor key management, outdated locking hardware, inadequate lighting, or poorly controlled access systems. For someone entering the locksmithing field, this highlights the importance of analytical thinking and the ability to evaluate security risks across buildings, vehicles, and operational environments rather than only performing mechanical repairs.

The influence on locksmithing practice is significant because it shifts the role of the locksmith toward security consulting and strategic planning. Modern locksmiths increasingly conduct detailed inspections of access points, key control systems, electronic access technologies, and compliance with safety regulations. They may also review vehicle security procedures, access logs, and fleet key management policies. Instead of responding only to emergencies such as lockouts or broken locks, locksmiths now provide proactive services such as security audits, vulnerability assessments, and recommendations for improving security infrastructure. This often leads to long-term relationships with clients through system upgrades, master key system design, electronic access control installation, and ongoing security reviews. As demonstrated in the logistics distribution centre case study, locksmiths who can perform professional risk assessments become trusted advisors who help organisations strengthen security systems, manage access more effectively, and reduce the likelihood of theft, unauthorized entry, or safety incidents.

## Client Education as a Service

Client education as a service refers to the practice of locksmiths providing guidance, training, and information to customers so they can better understand and manage their own security systems. Traditionally, locksmiths were called only when a lock failed, a key was lost, or a security device needed installation. Once the work was completed, the interaction between the locksmith and the client typically ended. In the modern security environment, however, locksmiths increasingly act as long-term security advisors. Educating clients about security risks, proper system use, and best practices helps prevent problems before they occur and allows customers to make informed decisions about protecting their property, assets, and personnel.

A key part of client education involves explaining how security systems function and how they should be used correctly. Many people install locks, electronic access systems, or smart security devices without fully understanding their capabilities or limitations. Locksmiths may therefore provide guidance on topics such as how restricted key systems prevent unauthorized duplication, how electronic access credentials can be issued or revoked, or how to correctly maintain locks and door hardware. By explaining these concepts clearly, locksmiths help clients understand why certain security measures are recommended and how they contribute to overall protection.

Client education is particularly important in environments where multiple people interact with security systems, such as commercial buildings, schools, healthcare facilities, and logistics operations. In these settings, employees, contractors, and visitors may have varying levels of access to different parts of a facility. Locksmiths can educate managers and staff about key control policies, credential management procedures, and access-control protocols. For example, organisations may be advised on how to track issued keys, how to respond when keys are lost, and how to limit duplication through restricted key systems. Training staff to follow consistent procedures reduces the likelihood of security breaches caused by poor key management.

Another important aspect of client education relates to electronic and digital security systems. As buildings and vehicles increasingly adopt electronic access control, smart locks, and digital credentials, many users require guidance on how these systems operate. Locksmiths may teach clients how to manage access permissions, update passwords or PIN codes, review access logs, and maintain the security of wireless or cloud-connected systems. In this way, locksmiths help ensure that advanced security technologies are used effectively and that users do not unintentionally create vulnerabilities through poor configuration or weak credential management.

Client education also plays a role in preventive security practices. Many security breaches occur not because locks fail but because users adopt unsafe habits, such as leaving doors unlocked, sharing access codes, or failing to report lost keys. Locksmiths can advise clients on practical measures such as maintaining secure storage for spare keys, regularly updating access credentials, performing routine maintenance on locks, and ensuring that security hardware is inspected periodically. By promoting these practices, locksmiths help reduce the likelihood of incidents that could otherwise require emergency repairs or system replacements.

In some cases, client education may involve formal training sessions or written security guidelines. Locksmith businesses may provide security briefings for facility managers, written procedures for key control, or instructional materials for operating electronic access systems. For example, a locksmith installing a master key system may provide documentation explaining how the system is structured, which keys provide access to specific areas, and how key issuance should be recorded. This ensures that the security system continues to function effectively after installation and that the client understands their responsibilities in maintaining it.

Client education also contributes to long-term business relationships. When locksmiths take the time to educate clients, they build trust and position themselves as knowledgeable professionals rather than simple service technicians. Clients who understand their security systems are more likely to seek ongoing advice, maintenance services, and system upgrades from the locksmith who originally assisted them. This approach shifts the locksmith business model toward proactive security management and consulting rather than purely reactive repair work.

From a broader perspective, client education reflects the evolving role of locksmiths within the integrated security industry. Modern security systems often combine mechanical locks, electronic access control, surveillance systems, alarms, and digital credential management. Clients may not have the technical knowledge to manage these systems effectively without guidance. By educating clients about security principles, proper system operation, and emerging risks, locksmiths help ensure that the security solutions they install remain effective over time.

**Sample Scenario: Client Education for a Commercial Office Building**

A locksmith company installs a new electronic access control system and restricted master key system for a medium-sized commercial office building that houses several businesses. The system includes electronic card readers on main entry points, restricted mechanical locks on internal offices, and a master key hierarchy for facility management. After installation, the building manager assumes that the system will operate automatically and does not initially plan to provide staff with instructions on how to use or manage it.

Recognising the risk of misuse or poor security practices, the locksmith offers client education as part of the service. The locksmith explains to the building manager that even the best security systems can fail if users do not understand how to operate them properly. For example, employees might share access cards, fail to report lost credentials, or prop doors open during busy periods. To prevent these issues, the locksmith schedules a training session with the building manager and selected staff members responsible for facility management and security.

During the training session, the locksmith explains how the restricted key system works and why keys cannot be duplicated without authorisation. The manager is shown how the master key hierarchy controls access to different areas of the building and how key issuance must be recorded to maintain accountability. The locksmith also explains how to respond if a key is lost, including rekeying procedures and updating key control records.

The locksmith then demonstrates how to manage the electronic access control system. Staff are trained on how to issue new access cards, deactivate lost or stolen cards, and adjust access permissions for employees or contractors. They are also shown how to review access logs to identify unusual activity, such as attempts to enter restricted areas outside of normal working hours. The locksmith emphasises the importance of regularly updating access credentials and removing access for employees who leave the organisation.

The session also covers preventive security practices. The locksmith explains that propped doors, shared access cards, or weak PIN codes can compromise the entire system. Staff are advised to ensure doors close properly, report lost credentials immediately, and avoid sharing access cards or security codes. Maintenance advice is also provided, including periodic inspection of door closers, lock mechanisms, and electronic readers.

At the conclusion of the session, the locksmith provides written security guidelines and operating procedures. These include a key control register template, instructions for managing electronic credentials, and recommended procedures for responding to security incidents. The building manager is also advised to schedule an annual security review to ensure the system remains effective as staff and tenants change.

Over time, the building management team follows the recommended procedures and maintains accurate records of key issuance and electronic credentials. Because employees understand how the system works, they are more likely to report lost cards promptly and follow proper security practices. As a result, the building experiences fewer security incidents, and the locksmith company continues to provide maintenance services and periodic security audits.

<u>Training Approach Used by the Locksmith</u>

The locksmith adopts a structured client education approach consisting of several stages:

**1. Initial briefing:** The locksmith explains the purpose of the installed security systems and why proper management is essential. This establishes the importance of security awareness among staff.

**2. Practical demonstration:** Staff are shown how to operate the access control system, issue credentials, review logs, and manage keys. Hands-on demonstrations help ensure that participants understand the system's functions.

**3. Policy guidance:** The locksmith provides recommendations for key control policies, credential management procedures, and reporting protocols for lost keys or cards.

**4. Written documentation:** Clear written instructions, system diagrams, and policy templates are provided so the organisation can maintain consistent procedures over time.

**5. Ongoing support and periodic reviews:** The locksmith offers follow-up support, maintenance, and annual security audits to ensure the system remains effective as the organisation grows or changes.

This scenario demonstrates how client education becomes a professional service within the modern locksmith business model. Instead of only installing locks or electronic systems, the locksmith helps the client understand how to manage and maintain security effectively. By educating the client, the locksmith reduces the likelihood of misuse, improves the overall effectiveness of the security system, and establishes a long-term professional relationship with the organisation.

### Impact on Locksmithing Practices

Client education as a service is highly relevant to someone interested in locksmithing because it reflects how the profession is evolving from a purely technical trade into a broader security advisory role. Traditionally, locksmiths focused on installing or repairing locks and cutting keys, with little involvement once the job was completed. However, modern security systems are increasingly complex and involve electronic access control, smart locks, digital credentials, and integrated security platforms. Without proper understanding, clients may misuse these systems by sharing access cards, failing to track issued keys, or neglecting maintenance procedures. By educating clients about how security systems function and how they should be managed, locksmiths help ensure that the systems they install are used correctly and remain effective over time. For someone entering the locksmithing profession, this highlights the importance of communication skills, the ability to explain technical systems clearly, and the role of the locksmith as a trusted advisor rather than simply a repair technician.

The influence on locksmithing practices is significant because it changes how locksmiths interact with clients and deliver services. Modern locksmiths may conduct training sessions, provide written security guidelines, demonstrate how to manage electronic access systems, and advise clients on best practices for key control and credential management. As illustrated in the commercial office building scenario, locksmiths can teach facility managers and staff how to issue and revoke access credentials, track keys, review access logs, and respond to lost credentials or security incidents. This proactive approach helps prevent security failures that could otherwise occur due to poor procedures or lack of awareness. It also strengthens long-term professional relationships with clients, as organisations often rely on locksmiths for ongoing support, system maintenance, and periodic security reviews. In this way, client education becomes an integral part of modern locksmithing practice, reinforcing the profession's shift toward integrated security management and consulting.

## Emerging Markets and Niches

Emerging markets and specialised niches are redefining the professional scope and growth potential of the modern locksmith business. Historically rooted in essential, hands-on tasks such as lock installation, key cutting, and lockout response, the locksmith discipline now encompasses a spectrum of high-value, technology-enabled services. This evolution is driven by rapid advancements in access-control technologies, digital credentials, and integrated security ecosystems that combine physical hardware with cyber-aware management platforms. The shift toward electronic and smart access control, alongside specialized sectors such as automotive security, fleet protection, and high-security installations, creates opportunities for locksmiths to deploy deep technical expertise beyond traditional emergency work and routine maintenance [340, 341].

The electronic and smart access-control market represents a primary growth axis for locksmiths, with homes, offices, and rental properties increasingly adopting digital locks, cloud-managed credentials, and networked security solutions. Locksmiths entering this space must extend their competencies to include networking concepts, wireless communication protocols, and cybersecurity best practices to configure, troubleshoot, and maintain these systems. Beyond initial installation, ongoing services such as managing user permissions, firmware updates, remote monitoring, and integration with home automation platforms are becoming routine demand drivers. This trend is corroborated by analyses of modern access-control deployments, which emphasize the need for locksmiths to operate at the intersection of physical hardware and digital credentialing infrastructures [340, 342]. While some sources focus on the technical intricacies of automotive and digital-key ecosystems, the implications for locksmith practice in electronic access control are consistent across residential, commercial, and multi-tenant contexts, underscoring a broad market expansion for skilled practitioners [342].

Automotive and vehicle security services constitute another substantial and rapidly expanding niche. Modern vehicles employ sophisticated immobilisers, transponder keys, remote keyless entry systems, and passive keyless entry technologies that require specialized diagnostic tools and programming capabilities. Locksmiths in this arena increasingly perform electronic key programming, remote fob pairing, and immobiliser resets, as well as advise on evolving digital-key and smartphone-based access paradigms. The literature documenting automotive security technologies, ranging from transponder-based immobilisers to advanced passive entry systems, highlights the growing complexity of the automotive security stack and the corresponding need for locksmiths to maintain up-to-date technical expertise and compliant security practices [342, 343]. Some discussions also emphasize historical and technical vulnerabilities associated with passive keyless-entry implementations and the importance of scrutinizing cipher strength and authentication mechanisms, which has direct relevance for professional practice in automotive security consulting and diagnostic services [342].

Commercial and institutional security consulting is likewise expanding as organisations seek integrated access-control solutions that blend mechanical locks with electronic credentials and monitoring technologies. Locksmiths operating in this niche may design comprehensive master-key systems, implement electronic access devices, conduct risk and security audits, and advise on organisational key-management policies. These engagements often culminate in long-term maintenance arrangements and periodic security reviews, contributing to stable revenue streams beyond episodic installation or emergency work. The expanding demand for security consulting aligns with broader industry movements toward holistic security architectures that integrate physical locking hardware with cyber-enabled credential management and comprehensive monitoring capabilities [341, 344, 345]. The long-term, project-based nature of these engagements is frequently cited as a differentiator that enables locksmiths to move from reactive service provision to proactive security stewardship [341].

Fleet and transport security represents a niche with strong growth potential, particularly as logistics and delivery operations invest in secure key management, vehicle immobiliser assessment, and telematics-enabled access controls. Locksmiths who understand both mechanical locking systems and electronic access technologies can contribute to safeguarding large vehicle fleets, supporting secure key management across multiple assets, and advising on immobiliser and digital credential strategies. The literature on fleet security underlines the convergence of mechanical and electronic systems in modern vehicles and the resulting demand for specialized expertise in immobilisers, digital keys, and connected access solutions.

This niche also benefits from cross-cutting knowledge in cyber-physical security, as vehicle access systems increasingly rely on networked platforms and data-driven monitoring to protect operations [341, 342, 346].

The hospitality and short-term rental sector has emerged as a notable market for locksmith services, driven by the deployment of electronic locks and digital credentialing to manage access across numerous units and locations. Locksmiths in this field may be responsible for installing smart lock fleets, integrating them with property-management and booking platforms, and performing maintenance at scale for large residential or commercial portfolios. The ability to remotely change access codes, audit entry activity, and maintain reliability across many properties places locksmiths who serve this sector in a position to offer scalable, high-value services that extend beyond traditional lock-related tasks. This trend is supported by evidence of widespread adoption of electronic and networked access-control solutions in hospitality and short-term rental properties, highlighting the strategic value of locksmith expertise in deployment, integration, and ongoing system stewardship [341, 347].

High-security and restricted-key systems constitute another specialized niche with growing importance for locksmith professionals serving banks, laboratories, data centres, and government facilities. These environments require locks with advanced resistance to picking and manipulation, patented key systems, and rigorous key control procedures. Locksmiths operating in this domain must possess detailed knowledge of high-security hardware, strict key-control protocols, and compliance frameworks governing sensitive facilities. The literature on high-security installations consistently identifies the need for specialized training and credentialing, as well as the ability to manage complex master-key architectures and secure key-issuance processes. This niche, while technically demanding, offers opportunities for high-margin projects and enduring client relationships due to the security-critical nature of the work [341, 348].

Biometric and identity-based access-control systems are an emerging niche reflecting broader security trends toward identity-centric protection. The integration of biometric devices, fingerprint, facial recognition, iris scanning. with door hardware and electronic access control expands the locksmith's role into ensuring proper installation, integration, and ongoing maintenance of identity-based systems. This shift requires locksmiths to understand how biometric authentication interacts with secure access control platforms, as well as the associated privacy and data-security considerations. The scholarly discourse surrounding biometric-based access control and identity verification emphasizes the growing relevance of locksmith expertise in delivering end-to-end, identity-centric security solutions within physical premises [341, 348, 349].

Cyber-physical security integration represents a cross-cutting domain in which locksmiths can contribute by bridging physical security hardware with digital monitoring platforms, alarms, surveillance systems, and cloud-based access-management. Mastery of how physical devices connect to networks, how digital credentials are managed, and how integrated systems are secured against cyber threats positions locksmiths as collaborators with IT and security professionals in delivering comprehensive, resilient security solutions. The cyber-physical convergence literature stresses the importance of holistic design thinking that unites hardware, software, and network considerations to mitigate threats across both cyber and physical domains, a framing that aligns closely with the evolving role of locksmiths in integrated security ecosystems [345, 350-352].

From a business strategy perspective, specialization in emerging markets allows locksmiths to diversify revenue streams away from traditional emergency work and to cultivate high-value, long-term engagements.

Engaging in electronic access control, automotive security, fleet protection, high-security installations, and integrated security consulting supports the development of advanced expertise that differentiates services in a competitive market. The literature on cyber-physical security and integrated systems further indicates that such diversification aligns with trends toward cross-disciplinary collaboration and the deployment of holistic security solutions that transcend conventional lock- and key-focused offerings. Collectively, these sources suggest that locksmiths who invest in training, certification, and ongoing education in these domains can secure resilient business growth and strategic partnerships with clients seeking comprehensive protection that spans physical and digital domains [341, 345, 350, 351].

Emerging markets and niches are reshaping the locksmith profession by expanding service types and client sectors beyond traditional lock-and-key operations. The convergence of electronic access-control technologies, automotive security innovations, and integrated security architectures creates substantial opportunities for locksmiths to develop specialized competencies, engage in longer-term contracts, and partner with organizations seeking robust, cyber-aware physical security solutions. By embracing these developments and committing to ongoing professional development, locksmiths can position themselves at the forefront of a modern security industry that blends mechanical expertise with digital and cyber-physical capabilities [341, 345, 350, 351].

## A Typical Day in the Life of a Modern Locksmith

A typical day for a modern locksmith reflects the profession's evolution from traditional lock-and-key services to a broader role that combines mechanical expertise, electronic security knowledge, and client advisory work. Rather than spending the entire day cutting keys or responding to emergency lockouts, many locksmiths now divide their time between installations, system configuration, client consultations, and ongoing security maintenance across multiple sectors.

The day often begins with planning and job preparation. Early in the morning, the locksmith reviews scheduled appointments, which may include residential smart lock installations, commercial access-control system maintenance, automotive key programming, or a security audit for a business client. Equipment and tools must be prepared carefully because modern locksmith work may require both traditional mechanical tools and specialised electronic devices such as diagnostic key programmers, access-control configuration software, or networking equipment. For example, if the locksmith is scheduled to program transponder keys for a vehicle, they will ensure that the appropriate programming interface and manufacturer-specific software are available before leaving the workshop.

The first field job of the day might involve electronic access control work at a commercial building. In this scenario, the locksmith visits an office complex where new employees require access credentials. The locksmith connects to the building's access-control system and programs new RFID cards while also reviewing the system logs to confirm that access permissions are correctly configured. They may also inspect door hardware to ensure electronic strikes, magnetic locks, and door closers are functioning correctly. If the building uses cloud-based access management, the locksmith may adjust user permissions remotely and provide brief guidance to the building manager on managing staff access.

Later in the morning, the locksmith might travel to a residential property to install a smart lock system. The homeowner wants the ability to control access remotely using a smartphone application. After installing the smart lock hardware, the locksmith connects it to the home's wireless network and demonstrates how to create digital keys for family members. The locksmith explains how to change access codes, update firmware, and review activity logs to ensure the homeowner understands how to manage the system securely. This interaction reflects the growing role of locksmiths in client education, ensuring customers can operate modern security technology effectively.

In the early afternoon, the locksmith may shift to the automotive security niche. A customer has lost the remote key for a modern vehicle equipped with a transponder-based immobiliser. Using a diagnostic programming device connected to the vehicle's onboard system, the locksmith programs a new key and synchronises the remote fob with the vehicle's electronic control unit. In addition to restoring access, the locksmith may delete previously registered keys from the vehicle's system to ensure that lost or stolen keys cannot be used to start the vehicle. This type of work requires specialised knowledge of automotive security technologies and access to manufacturer-compatible programming tools.

Later in the day, the locksmith may conduct a security risk assessment for a business client, such as a warehouse or logistics facility. The locksmith inspects entry points, evaluates lock quality, reviews key management procedures, and checks whether electronic access systems are functioning correctly. They may also examine lighting, camera coverage, and emergency exit hardware to ensure compliance with safety regulations. After the inspection, the locksmith prepares recommendations that might include upgrading locks to high-security cylinders, implementing a restricted key system, or introducing electronic access credentials to improve accountability.

Fleet security work can also form part of the day, particularly for locksmiths serving transport or logistics companies. During such visits, the locksmith may inspect vehicle key management procedures, review immobiliser systems, and advise on protecting vehicles against theft techniques such as relay attacks on keyless entry systems. Recommendations may include secure key storage systems, driver credential management policies, or telematics-based vehicle access monitoring.

Back at the workshop later in the afternoon, the locksmith may spend time performing administrative and technical tasks. These tasks might include preparing quotes for upcoming security installations, ordering specialised locking hardware, updating firmware for smart lock systems, or preparing reports from security audits conducted earlier in the day. Documentation is increasingly important in modern locksmithing, especially when dealing with restricted key systems, electronic access records, and compliance with security policies.

The day may conclude with an emergency service call, which remains a traditional aspect of locksmithing. A customer might be locked out of their home or office, requiring the locksmith to use non-destructive entry techniques to restore access. Even in these situations, modern locksmiths often encounter electronic locks or smart security systems that require knowledge beyond mechanical lock picking.

Overall, a typical day for a modern locksmith involves a diverse mix of mechanical work, electronic system management, consulting, and client education. Locksmiths move between different sectors—residential, commercial, automotive, and industrial, while applying both traditional craftsmanship and advanced technical knowledge. This diversity reflects the profession's transformation into a multidisciplinary security field where mechanical expertise is integrated with digital technology, risk assessment, and long-term security management.

**Impact on Locksmithing Practices**

Understanding emerging markets and specialised niches is highly relevant to someone interested in locksmithing because it illustrates how the profession is evolving beyond traditional lock-and-key services into a broader security technology field. Historically, locksmiths focused primarily on mechanical tasks such as lock installation, key cutting, and lockout response. However, modern security systems increasingly incorporate electronic access control, digital credentials, smart locks, vehicle security technologies, and integrated monitoring platforms. For someone entering the profession, this means that locksmithing now involves working with both physical security hardware and digital security systems. Developing knowledge in areas such as networking, electronic authentication systems, and vehicle security technologies allows locksmiths to participate in growing sectors such as smart home security, automotive key programming, and commercial access-control systems.

These emerging niches also influence locksmithing practices by expanding the range of services that locksmiths provide to clients. Modern locksmiths may install and configure electronic access systems, program automotive transponder keys, design master-key systems for commercial facilities, and conduct security risk assessments for businesses. As illustrated by the example of a typical day in the life of a locksmith, practitioners may work across several sectors in a single day, moving between residential smart lock installations, automotive key programming, and commercial security consulting. This diversity requires locksmiths to maintain a broader skill set that includes mechanical expertise, electronic diagnostics, and the ability to advise clients on security strategies and best practices.

The growth of specialised markets also affects the business model and long-term professional development of locksmiths. Rather than relying primarily on emergency lockout services, locksmiths can develop stable business opportunities through ongoing security maintenance, electronic access-control management, and long-term consulting arrangements with organisations. Specialising in areas such as fleet security, high-security restricted key systems, or biometric access technologies allows locksmiths to differentiate themselves in a competitive industry and provide higher-value services. Consequently, modern locksmithing increasingly requires continuous learning, professional certification, and collaboration with other security and technology professionals to meet the evolving demands of integrated physical and digital security environments.

## Chapter 15 Review Questions

### Question 1

A commercial property manager hires a locksmith not only to replace door locks but also to evaluate access control procedures, surveillance coverage, and key management policies across the entire facility.

What role is the locksmith performing?

A. Emergency technician
B. Security consultant
C. Key duplication specialist
D. Building inspector

**Question 2**

A locksmith business begins offering security audits to companies. During one inspection, the locksmith reviews door hardware, lighting conditions, key control procedures, and surveillance coverage.

What is the primary goal of this service?

A. To determine how many locks are installed in the building
B. To reduce the cost of security equipment
C. To identify vulnerabilities and recommend improvements
D. To increase the number of keys issued to employees

**Question 3**

A logistics company wants better control over employee access to restricted warehouse areas. The locksmith recommends replacing traditional keys with electronic credentials that can be revoked instantly.

What security benefit does this provide?

A. Easier duplication of credentials
B. Reduced need for door hardware
C. Improved control and monitoring of access permissions
D. Automatic door opening during emergencies

**Question 4**

A locksmith installs a smart lock system in a residential property and connects it to the homeowner's Wi-Fi network so it can be controlled through a smartphone app.

Which knowledge area is required beyond traditional mechanical locksmithing?

A. Advanced welding techniques
B. Networking and wireless communication protocols
C. Architectural design skills
D. Plumbing systems

**Question 5**

A warehouse security audit reveals that multiple employees possess privately duplicated keys that were never officially recorded.

What vulnerability does this situation represent?

A. Poor lighting conditions
B. Weak surveillance coverage
C. Uncontrolled key duplication and poor key management
D. Faulty door hinges

**Question 6**

A locksmith working with a fleet of delivery trucks inspects immobiliser systems and advises management on protecting vehicles against relay attacks on keyless entry systems.

Which niche area of locksmithing does this work represent?

A. Automotive and fleet security
B. Residential key cutting
C. Cabinet lock repair
D. Safe painting

**Question 7**

A property manager wants to install a security system where entry to the building is recorded and access permissions can be changed remotely.

Which technology best supports this requirement?

A. Mechanical padlocks
B. Master key duplication services
C. Electronic access control systems
D. Manual sign-in sheets

**Question 8**

A locksmith company begins offering training sessions to building managers on how to track issued keys, revoke electronic credentials, and review access logs.

What service is the locksmith providing?

A. Emergency lockout response
B. Client education and security training

C. Safe cracking services
D. Key stamping services

**Question 9**

A locksmith inspecting a commercial building notes that emergency exit doors comply with fire safety requirements but do not trigger alarms when opened.

What improvement might the locksmith recommend?

A. Removing the exit devices
B. Installing alarmed exit hardware
C. Reducing door width
D. Replacing door handles with knobs

**Question 10**

A locksmith business expands its services to include electronic access control installations, vehicle key programming, and security consulting rather than relying mainly on emergency lockouts.

What business strategy does this represent?

A. Limiting services to mechanical lock repairs
B. Increasing dependence on emergency work
C. Diversifying into modern security technologies and consulting
D. Eliminating electronic security services

# Chapter 16

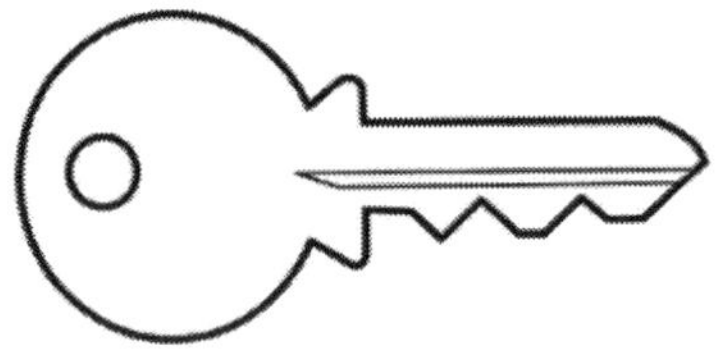

# Legal Frameworks and Compliance

Locksmithing is widely recognised as a technical trade, but it is also a profession that operates within a complex legal and regulatory environment. Because locksmiths possess the specialised skills and tools required to bypass locks, duplicate keys, and gain access to secured property, their work carries significant legal and ethical responsibilities. Governments, industry bodies, and professional associations around the world therefore establish regulatory frameworks designed to ensure that locksmiths operate with integrity, competence, and accountability. Understanding these frameworks is essential for anyone entering the profession, as compliance with licensing requirements, privacy laws, professional standards, and ethical obligations forms the foundation of responsible locksmith practice.

This chapter examines the legal frameworks that influence locksmithing practice and explains how regulatory requirements shape the way locksmiths operate in modern security environments. It explores topics such as licensing systems and regulatory models used in different countries, the ethical and legal responsibilities associated with working with security systems, and the importance of maintaining professional standards when handling sensitive access credentials and security information. The chapter also discusses how evolving technologies—such as electronic access control, biometric authentication, and cloud-managed security platforms—introduce new legal considerations related to privacy, cybersecurity, and professional liability.

By examining these issues, the chapter highlights how locksmithing has evolved from a purely mechanical trade into a regulated security profession that intersects with law enforcement, data protection, and building safety regulations. For aspiring and practicing locksmiths, a clear understanding of legal compliance is essential not only for operating lawfully but also for protecting clients, maintaining professional credibility, and ensuring that security systems perform their intended role in safeguarding people, property, and critical infrastructure.

## Licensing Requirements and Regulations

Licensing requirements and regulations for locksmiths vary widely around the world, reflecting differences in legal systems, public safety concerns, and the role locksmiths play within national security and consumer protection frameworks. In many jurisdictions, locksmiths are considered security professionals because they have the knowledge and tools to bypass locks, create keys, and access restricted areas. As a result, governments often regulate the profession to prevent misuse of locksmith skills for criminal purposes and to ensure that practitioners meet minimum standards of competence, integrity, and accountability.

### Purpose of Locksmith Licensing and Regulation

The primary purpose of locksmith licensing systems is to protect public safety and maintain trust in the security industry. Locksmiths frequently gain access to private homes, commercial premises, safes, vehicles, and security systems. Without regulatory oversight, individuals with criminal intent could misuse locksmith tools or knowledge to facilitate theft, fraud, or unlawful entry.

Licensing frameworks therefore typically require locksmiths to undergo background checks, criminal history screening, and sometimes fingerprinting before they are permitted to practice. These checks are designed to prevent individuals with serious criminal convictions—particularly those involving theft, burglary, fraud, or violence—from entering the profession.

Regulation also ensures that locksmiths possess appropriate technical skills and knowledge. Some jurisdictions require locksmiths to complete training programs, apprenticeships, or competency assessments before obtaining a license. In other regions, licensing may focus primarily on character and identity verification rather than formal technical education. In both cases, the goal is to ensure that practitioners are capable of performing locksmith work safely and responsibly.

Another important purpose of regulation is consumer protection. Licensing systems allow governments to enforce professional standards, investigate complaints, and discipline practitioners who engage in misconduct. Consumers can therefore verify that a locksmith is licensed, insured, and accountable to a regulatory authority before allowing them access to secure premises or sensitive property.

### Types of Locksmith Licensing Systems

Locksmith regulation around the world generally follows three broad models: government licensing systems, registration and background-check systems, and voluntary certification or industry self-regulation. These models differ in how strictly the profession is controlled and how governments or industry bodies ensure that locksmiths operate ethically and competently. The specific approach used in any country often reflects local legal traditions, security concerns, and the level of government oversight applied to security-related occupations.

Government licensing systems represent the most formal and regulated approach. In jurisdictions using this model, locksmiths must obtain an official government-issued license before they can legally operate a locksmith business or perform locksmith services. These licenses are typically administered by government departments responsible for consumer protection, police licensing, or security industry regulation. To obtain a license, applicants usually must meet several requirements designed to ensure both competence and trustworthiness. These requirements commonly include passing criminal background checks, providing fingerprints or identity verification, demonstrating technical competence or completing recognised training, registering their business operations, maintaining appropriate insurance or bonding, and periodically renewing their licenses. Operating without the required license in these jurisdictions can lead to serious consequences, including fines, business closure, or criminal penalties.

A second regulatory model involves registration and background-check systems. In these jurisdictions, locksmiths may not need a full professional license, but they are still required to register with a government authority or undergo police background checks before operating. These systems focus primarily on verifying the identity and criminal history of locksmiths rather than assessing their technical skills. The goal is to ensure that individuals who have the ability to bypass locks or access secure property are properly vetted. Registration systems typically require locksmiths to submit identification documents, undergo police vetting or criminal record checks, register business details with local authorities, and display identification when performing locksmith work. Although these systems impose fewer technical requirements than full licensing schemes, they still provide a level of accountability and oversight that helps protect consumers and property owners.

The third model is voluntary certification and industry self-regulation. In many countries, locksmithing is not directly regulated by government licensing laws. Instead, professional standards are maintained by industry associations and professional organisations that offer certification programs. These programs usually require locksmiths to demonstrate technical competency through examinations, adhere to professional codes of ethics, and participate in continuing professional development to keep their skills current. Membership in recognised industry bodies often signals professionalism and commitment to quality service. While these certifications are not always legally required, they can significantly enhance a locksmith's credibility and reputation. In many cases, customers, insurers, and commercial clients prefer to work with locksmiths who hold recognised certifications because they provide reassurance that the locksmith meets established professional standards.

**Locksmith Licensing in Australia**

In Australia, locksmiths are generally regulated under security industry legislation, and licensing requirements vary between states and territories. Locksmiths are typically classified as security equipment installers or security advisers.

Common requirements include:

- Police background checks
- Fingerprinting in some states
- Security industry license application
- Proof of training or experience
- Business registration and insurance

For example:

<u>New South Wales</u>

Locksmiths are regulated under the Security Industry Act 1997 and must obtain a Class 2C Security Equipment Specialist license if they install or maintain security equipment.

Requirements include:

- Criminal history check
- Approved training or experience
- Licensing through the NSW Police Force Security Licensing & Enforcement Directorate

Victoria

Locksmiths may require a Private Security License if they install or maintain electronic security equipment or alarms. Licensing is administered by Victoria Police Licensing & Regulation Division.

Queensland

In Queensland, locksmiths performing certain security work may require licensing under the Security Providers Act 1993.

Although traditional lock repair may not always require licensing, installing electronic security systems or working in security consulting often does.

### Locksmith Licensing in the United States

The United States has a highly fragmented regulatory framework because locksmith licensing is controlled at the state level rather than federally. As a result, requirements differ widely between states.

States With Strict Licensing

Some states require locksmiths to obtain government licenses.

Examples include:

- California
- Texas
- Illinois
- New Jersey
- Nevada

Typical requirements include:

- Criminal background checks
- Fingerprinting
- Licensing exams
- Business registration
- Surety bonds or insurance

For example, in Texas, locksmiths must obtain a license from the Texas Department of Public Safety Private Security Bureau, complete training, and pass a criminal history check.

States With Minimal Regulation

Other states have little or no direct locksmith licensing, although business registration laws still apply.

In these states, locksmiths may rely on voluntary certification from organisations such as:

- Associated Locksmiths of America (ALOA)
- Safe and Vault Technicians Association (SAVTA)

ALOA offers well-recognised certifications such as:

- Registered Locksmith (RL)
- Certified Registered Locksmith (CRL)
- Certified Professional Locksmith (CPL)
- Certified Master Locksmith (CML)

These certifications demonstrate technical competence but are not always legally required.

**Locksmith Licensing in the United Kingdom**

In the United Kingdom, locksmithing is largely unregulated by government licensing. Anyone can technically operate as a locksmith without a formal license.

However, professional standards are maintained through voluntary industry bodies such as:

- Master Locksmiths Association (MLA)

The MLA operates a rigorous accreditation system in which locksmiths must:

- Pass technical competency exams
- Undergo criminal record checks
- Demonstrate professional insurance
- Adhere to a code of conduct

The MLA Approved Company Scheme is widely recognised as a benchmark for professional locksmith services in the UK.

**Locksmith Licensing in Europe**

Across Europe, locksmith regulations vary significantly.

Germany

Germany often requires locksmith businesses to register under trade regulations (Handwerksordnung). Certain security-related activities may require trade certification or professional qualifications.

France

Locksmiths may operate under general trade registration but must comply with consumer protection laws. Professional training programs are available through vocational education.

Italy and Spain

Locksmiths generally operate as registered tradespeople, although security installation work may require additional certification or registration.

In many European countries, locksmiths working with electronic security systems must comply with broader security installer regulations.

### Locksmith Licensing in Canada

Canada regulates locksmiths primarily at the provincial level, and requirements vary across provinces.

Ontario

Locksmiths must hold a Private Security and Investigative Services License issued by the Ministry of the Solicitor General.

Requirements include:

- Criminal record check
- Security training course
- Licensing exam

British Columbia

Locksmiths may require a Security Worker License if performing certain security services.

### Ethical and Legal Responsibilities

Beyond meeting licensing requirements, locksmiths around the world are expected to adhere to strict ethical and legal standards due to the sensitive nature of their work. Locksmiths frequently gain access to private homes, businesses, vehicles, and secure facilities, which means that clients must be able to trust that the locksmith will act responsibly and professionally. One of the most important ethical responsibilities is verifying the identity and authority of a client before unlocking property or providing access to restricted areas. Locksmiths must ensure that the person requesting access has legitimate ownership or permission to enter the premises, vehicle, or container.

Maintaining confidentiality is another critical ethical obligation. Locksmiths often become aware of security system details, lock types, keying structures, or vulnerabilities during their work. This information must be treated as confidential and never disclosed to unauthorized individuals, as doing so could compromise the security of the client's property. Similarly, locksmiths must avoid unauthorized duplication of restricted or patented keys, as these systems are specifically designed to limit key copying and maintain strict access control.

Locksmiths are also responsible for protecting sensitive customer information. In modern security environments, locksmiths may work with digital access credentials, electronic access logs, alarm system codes, and other confidential data. Safeguarding this information is essential to prevent security breaches or misuse. Additionally, locksmiths must follow all relevant legal procedures when accessing locked property, ensuring that their actions comply with local laws and professional standards. Failure to follow these ethical and legal responsibilities can have serious consequences, including criminal charges, revocation of professional licenses, loss of business reputation, or civil liability for damages.

**Continuing Professional Development**

Because security technologies evolve rapidly, continuing professional development is an essential part of modern locksmith practice. Many licensing systems require locksmiths to complete ongoing training or renew their licenses periodically to ensure that their knowledge and skills remain current. Even in regions where formal licensing requirements are limited, professional locksmiths are encouraged to participate in training programs, workshops, and industry certification courses to keep up with technological advancements.

Training programs often cover emerging areas of locksmithing such as electronic access control systems, automotive key programming, cyber-physical security systems, and advanced safe and vault technologies. Locksmiths may also receive instruction in conducting security risk assessments, installing smart lock systems, and integrating mechanical security hardware with electronic access management platforms. As security systems increasingly combine physical hardware with digital technologies, locksmiths must understand how these systems operate and how to configure them securely.

Ongoing professional development helps ensure that locksmiths remain competent in working with modern security technologies such as biometric authentication systems, cloud-managed access control platforms, smartphone-based digital keys, and smart locks. By continually updating their skills and knowledge, locksmiths are better equipped to provide reliable services, adapt to new technologies, and maintain professional credibility within the evolving security industry.

**Impact on Locksmithing Practices**

Understanding licensing requirements and regulatory frameworks is highly relevant to anyone interested in locksmithing because it highlights that the profession is not simply a technical trade but also a regulated security occupation in many parts of the world. Locksmiths possess specialised knowledge and tools that allow them to bypass locks, duplicate keys, and gain access to secured property. Because these capabilities can be misused if placed in the wrong hands, many governments treat locksmiths as security professionals and impose regulatory requirements to ensure that practitioners are trustworthy and competent. For someone entering the locksmithing field, this means that professional practice often involves meeting legal obligations such as background checks, licensing applications, and compliance with security industry legislation. Understanding these requirements early helps prospective locksmiths plan their training, business registration, and career pathway within the legal framework of the jurisdiction in which they intend to operate.

The influence of licensing and regulation on locksmithing practices is significant because it establishes professional standards and accountability within the industry. Licensing systems require locksmiths to demonstrate integrity, technical competence, and compliance with legal and safety requirements before they are allowed to provide services to the public. In practical terms, this means locksmiths must verify the identity and authority of clients before unlocking property, protect confidential security information, and follow strict procedures when handling restricted keys or access credentials. These regulatory expectations shape how locksmiths interact with clients, document their work, and maintain records of key systems or electronic access credentials. As a result, locksmithing practice increasingly includes administrative responsibilities, compliance with regulatory bodies, and adherence to professional codes of conduct.

Regulation also influences how locksmith businesses operate and develop professionally. Many licensing systems require continuing professional development, meaning locksmiths must regularly update their knowledge to keep pace with evolving security technologies such as electronic access control systems, automotive immobilisers, biometric authentication devices, and smart locks. This encourages locksmiths to pursue ongoing education, professional certification, and specialised training throughout their careers. In turn, these requirements help elevate locksmithing from a purely mechanical trade into a recognised security profession that combines technical expertise, ethical responsibility, and regulatory compliance. For individuals entering the field, understanding licensing and regulatory expectations helps ensure that they can operate legally, build professional credibility, and maintain the trust required to work in an industry responsible for protecting people, property, and critical infrastructure.

## Privacy Laws and Access Control Data

Privacy laws and access control data management are increasingly important considerations in modern locksmithing, particularly as security systems evolve from purely mechanical locks to integrated electronic and digital access control platforms. Traditional locksmithing involved physical keys and mechanical lock mechanisms, which generally did not store or process personal data. However, modern security systems often record detailed information about who enters a building, when access occurs, and which credentials were used. As a result, locksmiths who install, configure, or maintain these systems may encounter sensitive personal information and must comply with privacy and data protection laws in the jurisdictions where they operate.

Access control systems commonly collect and store various forms of data. Electronic access systems may record user credentials such as employee identification numbers, RFID card identifiers, biometric data, PIN codes, or smartphone-based digital credentials. These systems also generate access logs that record the time and location of each entry attempt, whether access was granted or denied, and which credential was used. In commercial environments, this data may be used to monitor building security, track employee access to restricted areas, investigate security incidents, or comply with regulatory requirements. Because this information can identify individuals and reveal patterns of movement within a facility, it is often classified as personal data under privacy legislation.

Locksmiths working with electronic security systems may therefore have responsibilities related to the protection of personal information. When installing or maintaining access control systems, locksmiths may configure user databases, manage digital credentials, retrieve access logs, or reset system administrators. In doing so, they may temporarily access sensitive information such as employee names, access permissions, identification numbers, or biometric templates. Professional locksmiths must treat this information confidentially and ensure that it is handled in accordance with applicable privacy laws. This typically includes limiting access to authorised personnel, preventing unauthorized copying or disclosure of data, and ensuring that system security settings protect stored information.

Around the world, privacy and data protection laws increasingly regulate how personal information collected by security systems must be handled. In the European Union, for example, the General Data Protection Regulation (GDPR) establishes strict rules governing the collection, processing, storage, and transfer of personal data. Organisations using electronic access control systems must ensure that the data collected is necessary for legitimate security purposes, stored securely, and retained only for appropriate periods. Locksmiths who install or configure such systems may need to ensure that system settings comply with GDPR requirements, such as enabling secure authentication, limiting administrative access, and ensuring encryption of stored data where appropriate.

In Australia, privacy obligations are governed primarily by the Privacy Act 1988 and the Australian Privacy Principles (APPs). These laws regulate how organisations collect, store, and manage personal information. If access control systems store identifiable user data or access logs linked to individuals, organisations operating those systems must ensure that the information is collected for legitimate purposes, stored securely, and accessible only to authorised personnel. Locksmiths involved in configuring or maintaining these systems must understand that access control data may constitute personal information and should therefore be handled carefully and confidentially.

In the United States, privacy regulation is more fragmented and often sector-specific. Although there is no single national privacy law equivalent to GDPR, various federal and state laws regulate the handling of personal information. Some states, such as California, have enacted comprehensive privacy laws like the California Consumer Privacy Act (CCPA), which governs how businesses collect and use personal data. Additionally, industries such as healthcare and finance are subject to strict privacy requirements that may affect how access control systems store and manage user information. Locksmiths working with electronic security systems in these environments must ensure that the systems they install support secure credential management and appropriate data protection practices.

Biometric access control systems introduce additional privacy considerations because they collect highly sensitive biometric identifiers such as fingerprints, facial recognition data, or iris scans. In some jurisdictions, biometric data is subject to enhanced legal protections because it is unique to an individual and cannot easily be changed if compromised. Laws governing biometric data often require explicit consent from individuals before their biometric information can be collected and may impose strict requirements on storage security and data retention. Locksmiths who install or maintain biometric access control devices must therefore ensure that systems are configured to protect biometric templates and that clients understand their legal obligations regarding biometric data management.

Cloud-based access control systems also create new privacy challenges. Many modern security platforms store user credentials and access logs on remote servers managed by third-party providers. This means that access control data may be transmitted across networks or stored in different countries, potentially subjecting it to multiple legal jurisdictions. Locksmiths who deploy cloud-managed access systems should understand how data is transmitted, stored, and protected by the platform. They may also need to advise clients about issues such as data residency, system security settings, and the importance of strong authentication for system administrators.

Another key privacy concern in locksmithing relates to access logs and surveillance integration. Many electronic access control systems integrate with video surveillance systems, allowing security personnel to correlate access events with recorded footage. While this can improve security monitoring, it also increases the amount of personal information collected about individuals entering or leaving a facility. Privacy laws in many countries require organisations to inform individuals when such monitoring occurs and to ensure that surveillance data is used only for legitimate security purposes. Locksmiths involved in installing integrated security systems should be aware of these requirements and may need to configure systems to ensure compliance with privacy regulations.

Ethical responsibilities also play an important role in protecting privacy in locksmithing practice. Even when privacy laws do not explicitly regulate every aspect of access control data, locksmiths are expected to maintain strict confidentiality regarding any security information they encounter. This includes protecting system passwords, not disclosing details about security configurations, and ensuring that sensitive information is not shared with unauthorized individuals. Ethical locksmith practice also requires avoiding unnecessary access to data and ensuring that client systems are configured to minimise privacy risks.

### Impact on Locksmithing Practices

Understanding privacy laws and access control data management is highly relevant to someone interested in locksmithing because it reflects the significant technological transformation occurring within the security industry. Traditional locksmithing primarily involved mechanical locks and physical keys, which rarely involved the collection or storage of personal information. Modern security systems, however, increasingly rely on electronic access control platforms that record and store detailed information about users and access events. These systems may collect credentials such as RFID card identifiers, PIN codes, biometric templates, or smartphone-based digital keys, as well as maintain access logs showing when individuals enter or attempt to enter specific areas. For someone entering the locksmithing profession, this means that the work now extends beyond mechanical hardware into systems that process sensitive personal data, requiring an understanding of privacy regulations and responsible data management practices.

The influence on locksmithing practices is substantial because locksmiths who install, configure, or maintain electronic access control systems may have direct or indirect access to confidential information. This may include employee identification records, access permissions, biometric identifiers, or detailed entry logs that reveal patterns of movement within a facility. As a result, locksmiths must handle this information with strict confidentiality and ensure that systems are configured to protect personal data. In practical terms, this means implementing secure authentication procedures, restricting administrative access, protecting system passwords, and ensuring that sensitive data is not copied or disclosed to unauthorized individuals. Locksmiths may also need to ensure that the systems they install comply with applicable privacy laws in their jurisdiction, such as the European Union's General Data Protection Regulation (GDPR), Australia's Privacy Act 1988 and Australian Privacy Principles, or various state-level privacy laws in the United States.

Privacy regulations also influence how locksmiths design, configure, and advise clients on security systems. When installing electronic or cloud-based access control systems, locksmiths must consider issues such as secure storage of access logs, protection of biometric data, encryption of communications, and proper management of digital credentials. In environments where surveillance systems integrate with access control platforms, locksmiths may also need to ensure that monitoring systems comply with legal requirements regarding data collection and notification. These responsibilities encourage locksmiths to develop broader knowledge in cybersecurity awareness, digital credential management, and secure system configuration. Consequently, privacy law considerations reinforce the evolving role of locksmiths as security professionals who manage not only physical access but also the protection of sensitive digital information associated with modern security systems.

## Liability and Professional Responsibility

Liability and professional responsibility are critical aspects of modern locksmithing because locksmiths work directly with systems designed to protect property, assets, and personal safety. Clients rely on locksmiths to install, repair, and maintain security systems that prevent unauthorized access to homes, businesses, vehicles, and sensitive facilities. If a locksmith performs work negligently, provides incorrect advice, or fails to follow legal and professional standards, the consequences can include property loss, theft, injury, or financial damage. As a result, locksmiths must understand their legal responsibilities and the potential liabilities associated with their work.

Professional liability in locksmithing often arises from negligence or failure to exercise reasonable care when performing services. For example, if a locksmith installs a lock incorrectly and it fails to secure a door properly, the locksmith may be held responsible if a burglary occurs due to that failure. Similarly, if a locksmith improperly programs an electronic access control system or fails to secure administrative access credentials, unauthorized individuals may gain access to restricted areas. In such cases, clients may seek compensation for damages through civil legal action. To reduce this risk, locksmiths are expected to follow accepted industry standards, manufacturer installation guidelines, and applicable building and safety regulations.

Another important aspect of liability involves verifying the authority of individuals requesting locksmith services. Locksmiths frequently receive requests to unlock homes, vehicles, safes, or commercial premises. If a locksmith fails to confirm that the person requesting access is the legitimate owner or authorized user of the property, they could inadvertently assist in unauthorized entry or theft. This could result in both criminal and civil liability. Professional locksmith practice therefore requires careful verification of identity and authorization before granting access to locked property.

Locksmiths also carry responsibilities related to the protection of confidential security information. During their work, locksmiths may gain knowledge of key systems, alarm codes, access credentials, or vulnerabilities within a client's security infrastructure. Disclosing or mishandling this information could compromise the security of the client's property and expose the locksmith to legal consequences. Professional responsibility therefore requires maintaining strict confidentiality and ensuring that sensitive information is not shared with unauthorized individuals.

Insurance and risk management play an important role in managing liability in locksmith businesses. Many professional locksmiths carry liability insurance to protect against claims resulting from accidental damage, faulty installations, or professional errors. Insurance may cover legal costs, property damage, or compensation claims if a client suffers loss due to the locksmith's work. In addition to insurance, locksmith businesses often implement clear service agreements, documentation procedures, and client authorization forms to reduce legal risks and clarify responsibilities.

Professional responsibility also extends to compliance with licensing laws, safety regulations, and industry standards. In jurisdictions where locksmith licensing is required, operating without the appropriate license can result in penalties, fines, or business closure. Locksmiths must also ensure that the hardware they install complies with fire safety regulations, building codes, and emergency exit requirements. For example, exit doors in commercial buildings must allow rapid evacuation during emergencies, and improperly installed locking hardware could create serious safety hazards.

As locksmithing evolves to include electronic access control systems, biometric authentication devices, and cloud-based security platforms, professional responsibility increasingly includes cybersecurity awareness. Locksmiths configuring electronic security systems must ensure that digital credentials, administrative accounts, and network connections are properly secured. Weak passwords, unsecured wireless networks, or poorly configured access permissions could create vulnerabilities that expose clients to security breaches.

Ethical conduct is closely linked to professional responsibility in locksmithing. Locksmiths are expected to operate with honesty, integrity, and respect for client privacy. This includes refusing to perform work that appears unlawful, avoiding unnecessary damage when gaining entry to locked property, and providing accurate advice about security systems. Maintaining ethical standards not only protects clients but also helps preserve public trust in the locksmith profession.

**Case Study Scenario: Liability and Professional Responsibility in a Commercial Security Installation**

A medium-sized accounting firm located in a metropolitan office building decides to upgrade its security systems after experiencing several incidents involving misplaced keys and unauthorized after-hours entry. The firm hires a locksmith company to replace existing mechanical locks with a modern electronic access control system that uses key cards and PIN codes. The locksmith is responsible for installing electronic door locks, configuring access permissions for employees, and setting up the administrative controls for the system.

During installation, the locksmith successfully installs the electronic locks and programs access cards for staff members. However, in order to speed up the setup process, the locksmith uses a simple default administrator password for the system's management software and does not advise the client to change it immediately. The locksmith also neglects to configure proper access restrictions within the system, leaving multiple staff members with administrative privileges that allow them to alter access permissions and view access logs. Although the system appears to function correctly, these configuration oversights create a hidden vulnerability in the firm's security system.

Several weeks later, a former employee who still knows the default administrator password gains remote access to the system's management interface and reactivates their access card. The individual enters the office building after hours and removes confidential financial documents from the premises. The company later discovers the breach during an internal audit and reviews the access logs. The investigation reveals that the system's administrative password was never changed from the default configuration and that the former employee had been able to exploit this weakness to regain access.

The firm files a complaint against the locksmith company, claiming that the system was improperly configured and that the locksmith failed to follow professional installation practices. Security experts reviewing the case determine that the locksmith should have followed the manufacturer's configuration guidelines, which require changing default administrator credentials during installation and restricting administrative access to authorised management personnel. Because the locksmith did not implement these safeguards or advise the client about the importance of doing so, the company argues that the locksmith's negligence contributed to the security breach.

The case highlights the legal liability that locksmiths may face when professional standards are not followed. The accounting firm seeks compensation for the financial losses associated with the stolen documents, as well as the costs of investigating the incident and upgrading the security system. Fortunately, the locksmith business carries professional liability insurance, which covers a portion of the legal expenses and damages associated with the claim. However, the incident still damages the locksmith's professional reputation and results in stricter internal procedures for future installations.

Following the incident, the locksmith company implements new professional practices designed to reduce liability risks. These include mandatory system security checks during installation, documented client authorization forms, written guidance for clients on changing default passwords, and detailed system configuration records. The company also introduces additional staff training on cybersecurity practices related to electronic access control systems.

This scenario demonstrates how liability and professional responsibility influence locksmithing practices in modern security environments. Locksmiths are responsible not only for installing physical hardware but also for ensuring that electronic systems are configured securely and in accordance with industry standards. By exercising reasonable care, following manufacturer guidelines, verifying client authorization, and maintaining proper documentation, locksmiths can reduce the risk of security failures while protecting both their clients and their own professional reputation.

## Impact on Locksmithing Practices

Understanding liability and professional responsibility is highly relevant to someone interested in locksmithing because the profession directly affects the safety, security, and legal rights of clients. Locksmiths are entrusted with systems that protect homes, businesses, vehicles, and valuable assets, which means their work carries significant responsibility. Unlike many trades, locksmiths are often granted temporary access to secured property and sensitive information about security systems. As a result, individuals entering the profession must understand that their actions can have legal consequences if work is performed negligently, if security systems are improperly installed or configured, or if client information is mishandled. Awareness of liability helps prospective locksmiths recognise that the role involves not only technical skill but also careful judgement, ethical behaviour, and compliance with legal obligations.

Liability considerations strongly influence locksmithing practices by encouraging adherence to professional standards and careful work procedures. Locksmiths must follow manufacturer installation guidelines, building codes, and security industry standards to ensure that the systems they install function correctly and provide the intended level of protection. This includes correctly installing mechanical locks, properly configuring electronic access control systems, and ensuring that security devices comply with fire safety and emergency exit regulations. In practical terms, locksmiths must document their work, verify client authorization before providing access to locked property, and ensure that any system credentials or administrative accounts are securely configured. These procedures reduce the risk of security failures and protect both the client and the locksmith from potential legal disputes.

The case study scenario also illustrates how professional responsibility increasingly extends into areas such as electronic security and cybersecurity. Modern locksmiths frequently work with systems that include digital credentials, access management software, and cloud-based security platforms. Improper configuration, such as leaving default passwords unchanged or granting excessive administrative access, can create vulnerabilities that allow unauthorized entry. For someone entering the locksmith profession, this demonstrates that technical competence now includes understanding how to secure digital access systems as well as mechanical locks. Consequently, locksmithing practices now emphasize risk management, system testing, and proper security configuration to ensure that installed systems are resilient against misuse or exploitation.

Liability considerations also encourage locksmiths to adopt structured business practices such as service agreements, identity verification procedures, and professional liability insurance. These measures help clarify responsibilities, protect against potential claims, and demonstrate professionalism to clients. Ethical conduct also plays a key role in reducing liability, as locksmiths must maintain confidentiality regarding security systems and refuse requests that appear unlawful or suspicious. By understanding liability and professional responsibility, individuals entering the locksmithing field can appreciate the importance of careful workmanship, ethical behaviour, and ongoing professional development in maintaining both legal compliance and public trust in the profession.

## Working with Law Enforcement

Working with law enforcement is an important aspect of professional locksmith practice because locksmiths possess specialised skills and tools that allow them to access secured property, vehicles, and safes. Police and other law enforcement agencies occasionally require the assistance of locksmiths when they encounter situations involving locked premises, vehicles, or secure containers during investigations, emergency responses, or lawful entry operations. In these situations, locksmiths may be asked to provide technical expertise in gaining access without unnecessary damage, securing premises after police operations, or assisting with forensic and investigative tasks related to locks and security systems.

One of the most common ways locksmiths work with law enforcement is by assisting with lawful entry to locked premises. When police officers have legal authority to enter a building, such as through a search warrant, court order, or emergency powers, they may call a locksmith to unlock doors or bypass locking mechanisms. Locksmiths can often gain access using specialised tools and techniques that minimize damage to doors, locks, and frames. This is particularly useful when police need to preserve property, maintain evidence integrity, or avoid unnecessary destruction of a building's structure. In many cases, the locksmith may also assist with temporarily securing the property after the operation, for example by rekeying locks or installing temporary locking hardware.

Locksmiths may also assist law enforcement with vehicle access and automotive security issues. During investigations, police may need to gain entry to locked vehicles that contain evidence, stolen property, or individuals who are in danger. Locksmiths with expertise in automotive locking systems and electronic vehicle access technologies can help officers unlock vehicles safely and without damaging the locking mechanism or vehicle electronics. In some cases, locksmiths may also assist in recovering vehicles that have been stolen or abandoned by helping to disable compromised locks or reprogram vehicle keys once ownership has been verified.

Another area where locksmiths may assist law enforcement is in forensic examination and security analysis. Locksmiths who have specialized knowledge of lock mechanisms may be consulted to determine whether a lock has been picked, forced, manipulated, or tampered with. This type of expertise can be valuable in burglary investigations, insurance claims, or cases involving unauthorized entry. By examining lock cylinders, strike plates, and door hardware, locksmiths may be able to identify evidence of forced entry or determine whether a security device failed or was bypassed through specific techniques.

Locksmiths also sometimes support law enforcement by helping secure crime scenes or sensitive facilities. After a burglary, domestic incident, or other criminal investigation, property owners may need their locks replaced or rekeyed quickly to prevent further unauthorized access. Police may recommend or request the involvement of a locksmith to ensure the premises are secured once the investigation has concluded.

Similarly, locksmiths may assist in securing seized properties, vehicles, or evidence storage areas by installing new locks or restricted key systems.

Because locksmiths work closely with security systems, their interaction with law enforcement is governed by strict legal and ethical considerations. Locksmiths must ensure that they only provide assistance when law enforcement officers can demonstrate lawful authority to access the property in question. This typically involves verifying official identification and confirming that a warrant, court order, or other legal authorization exists. Locksmiths must avoid participating in any activity that could be considered unlawful entry or violation of property rights. Maintaining proper documentation of such interactions is also considered good professional practice.

Confidentiality and professional conduct are also essential when working with law enforcement. Locksmiths may become aware of sensitive information related to criminal investigations, security vulnerabilities, or evidence handling. This information must be treated confidentially and should not be disclosed to unauthorized individuals. In some cases, locksmiths may be asked to provide written statements or expert testimony regarding the condition of locks or security systems involved in a criminal investigation.

Working with law enforcement also requires locksmiths to maintain high professional standards and technical competence. Modern security systems include electronic access control, biometric authentication, and advanced automotive immobiliser technologies, and law enforcement may rely on locksmith expertise when dealing with these systems. As a result, locksmiths who stay up to date with evolving security technologies are better equipped to assist authorities in situations where specialized access or security knowledge is required.

**Case Study Scenario: Locksmith Assisting Law Enforcement in a Lawful Entry Operation**

Police in a regional city receive credible intelligence that a residential property is being used to store stolen electronics linked to a series of commercial burglaries. After presenting the evidence to a magistrate, investigators obtain a lawful search warrant authorizing entry to the property. When officers arrive at the house early the next morning, they find that the front door is secured with a high-security deadbolt and reinforced door hardware. Because forced entry could damage the door and potentially disturb evidence inside the premises, the police request the assistance of a licensed locksmith to perform a controlled entry.

The locksmith arrives at the location and first verifies the authority of the attending officers by confirming their identification and reviewing the search warrant documentation. Once satisfied that the officers have lawful authority to enter the premises, the locksmith assesses the locking hardware. The door is fitted with a high-security cylinder and a reinforced strike plate, which would require significant force to break. Using specialized non-destructive entry techniques and professional lock-picking tools, the locksmith carefully manipulates the lock and successfully opens the door without damaging the lock or door frame. This allows police officers to enter the premises while preserving the physical condition of the door and lock, which may later be examined as part of the investigation.

Inside the property, officers discover a storage room secured with a heavy-duty padlock and several locked cabinets containing electronic equipment suspected to be stolen goods. Again, the locksmith assists by opening the padlocks using controlled techniques that avoid unnecessary destruction. Because the condition of the locks may provide forensic information about whether they were previously tampered with, the locksmith carefully documents the process and ensures that the locks remain intact for later examination by investigators.

During the search, police also locate a locked vehicle parked in the property's garage. Investigators believe the vehicle may have been used to transport stolen items between locations. The locksmith is asked to assist with vehicle access so officers can inspect the interior. Using automotive locksmith tools and knowledge of modern vehicle locking systems, the locksmith safely unlocks the vehicle without damaging the door or triggering the alarm system. This allows police to recover several boxes of stolen electronics and related documentation from the vehicle.

After the search is completed and evidence is collected, the police request that the locksmith secure the property to prevent unauthorized access while the investigation continues. The locksmith installs temporary locks on the main entrance and provides the police with new keys that will remain under official control until the legal process determines ownership of the property. The locksmith also documents the work performed and provides a brief written statement describing the condition of the locks and the entry methods used during the operation.

Several weeks later, investigators contact the locksmith again to provide expert insight regarding the condition of one of the locks recovered from the property. The locksmith explains that the lock shows no signs of forced entry, suggesting that individuals involved in the operation likely possessed legitimate keys or duplicate keys. This information becomes relevant to the investigation because it indicates that the property may have been accessed by someone with authorized access or inside knowledge.

This case study demonstrates how locksmiths can play a valuable role in supporting law enforcement operations. By providing technical expertise in non-destructive entry, vehicle access, lock analysis, and property security, the locksmith helped police conduct a lawful search efficiently while preserving evidence and minimizing property damage. The scenario also highlights the importance of verifying legal authority, maintaining confidentiality, documenting work performed, and adhering to professional and ethical standards when working alongside law enforcement agencies.

**Impact on Locksmithing Practices**

Working with law enforcement is highly relevant to anyone interested in locksmithing because it illustrates the broader role locksmiths play within public safety and the justice system. Locksmiths are not only service providers for residential or commercial clients but are also skilled security professionals whose expertise may be required during police investigations, emergency responses, and lawful entry operations. Law enforcement agencies often encounter situations where locked doors, safes, or vehicles must be accessed legally and efficiently, and locksmiths possess the specialised tools and technical knowledge needed to perform these tasks without causing unnecessary damage. Understanding this relationship helps aspiring locksmiths recognise that their skills may contribute directly to criminal investigations, evidence preservation, and the protection of property and people.

This relationship influences locksmithing practices by reinforcing the importance of legal awareness and strict professional procedures. Locksmiths must be able to recognise when law enforcement has lawful authority to request assistance, such as through a search warrant, court order, or emergency powers. Before performing any work, locksmiths must verify official identification and confirm the legitimacy of the request to avoid participating in unlawful entry. As a result, professional locksmith practice includes careful documentation, identity verification procedures, and adherence to legal requirements. These practices protect both the locksmith and the client while ensuring that locksmith services are not misused for unauthorized access.

Working with law enforcement also shapes the technical standards expected of locksmiths. Police operations often require non-destructive entry techniques so that locks, doors, and other physical evidence remain intact for forensic examination. Locksmiths must therefore develop advanced skills in lock manipulation, bypass techniques, and controlled entry methods that preserve evidence while allowing lawful access. In addition, locksmiths may be asked to analyse locks after a crime to determine whether they were picked, forced, or tampered with. This investigative role requires a deep understanding of lock mechanisms, wear patterns, and evidence preservation, further elevating the locksmith's role from a simple tradesperson to a technical security specialist.

Confidentiality and professional conduct are also strongly influenced by interactions with law enforcement. Locksmiths assisting with investigations may encounter sensitive information related to criminal activity, security vulnerabilities, or evidence handling. Ethical locksmith practice requires maintaining strict confidentiality and avoiding the disclosure of any details that could compromise an investigation or a client's security. These responsibilities reinforce the importance of integrity, discretion, and trustworthiness within the locksmith profession.

The case study demonstrates how locksmiths can contribute to law enforcement operations by providing non-destructive entry, assisting with vehicle access, analysing lock conditions, and helping secure premises after police operations. For someone interested in locksmithing, this example highlights the real-world impact of locksmith skills and the importance of maintaining high professional standards. It shows that locksmiths must combine technical expertise, legal awareness, and ethical conduct to support both their clients and broader public safety objectives.

Ultimately, understanding how locksmiths interact with law enforcement influences the way locksmiths approach their daily work. It encourages the development of careful verification procedures, detailed record keeping, and adherence to professional standards when accessing secured property. It also emphasizes the need for ongoing technical training, particularly in modern security technologies such as electronic access control systems, automotive immobilisers, and high-security locking devices. By maintaining these competencies, locksmiths can operate responsibly, support lawful investigations when required, and uphold the trust placed in them as security professionals.

## Standards and Codes of Practice

Standards and codes of practice form an important part of the legal and professional framework governing locksmithing and the wider security industry. While laws and regulations establish the legal requirements that locksmiths must follow, standards and codes of practice provide detailed guidance on how work should be performed safely, professionally, and consistently. These documents are typically developed by standards organisations, government agencies, or professional industry bodies and outline recommended procedures, performance benchmarks, and technical specifications for security products and services. For locksmiths, understanding and applying these standards helps ensure that installations and services meet recognised levels of quality, reliability, and safety.

Technical standards are particularly important when locksmiths install or maintain locking hardware and security systems. Standards may specify the performance characteristics that locks must meet, including resistance to picking, drilling, forced entry, and environmental conditions. For example, many countries adopt national or international lock standards that classify locks based on their security performance. These classifications allow locksmiths to recommend appropriate hardware for different applications, such as residential doors, commercial premises, or high-security environments. By following recognised standards, locksmiths help ensure that the locks and security devices they install provide the level of protection expected by clients and regulators.

Codes of practice complement technical standards by outlining professional and ethical expectations for locksmiths and security practitioners. These codes often address issues such as verifying customer authority before unlocking property, maintaining confidentiality about security systems, using non-destructive entry techniques whenever possible, and providing accurate advice about security risks. Professional locksmith associations frequently require members to follow a formal code of ethics that emphasises honesty, integrity, and respect for client privacy. Adhering to these codes helps maintain public trust in the profession and reduces the risk that locksmith skills could be misused.

Building and fire safety regulations also influence locksmith practices through standards governing door hardware and emergency exit systems. In commercial and public buildings, locksmiths must ensure that locks and door hardware comply with safety requirements that allow occupants to exit quickly during emergencies. For example, panic bars, emergency exit devices, and fire-rated door hardware must meet specific performance standards to ensure they function correctly during evacuations. Installing non-compliant locking hardware could create serious safety hazards and expose locksmiths to legal liability if an emergency occurs.

Electronic security systems introduce additional standards related to system reliability, cybersecurity, and integration with other building systems. Locksmiths installing electronic access control systems must often follow manufacturer guidelines, network security recommendations, and industry standards that address credential management, encryption, and secure system configuration. These standards help ensure that digital access systems protect both physical security and sensitive user data. As locksmithing increasingly involves cyber-physical security technologies, understanding these standards becomes essential for professional practice.

Many countries also maintain national standards bodies that publish formal security and building standards relevant to locksmith work. For example, Australia uses standards developed by Standards Australia, while the United States relies on organisations such as ANSI and UL to establish security hardware standards. In Europe, standards are often issued through the European Committee for Standardization (CEN). These standards help create consistency across the security industry by defining how products should be tested, classified, and installed.

Standards and codes of practice provide locksmiths with structured guidance on how to design, install, and maintain security systems safely and effectively. Around the world, a number of well-recognised technical standards and professional codes influence locksmith practice. These standards establish measurable performance requirements for locks, door hardware, access control systems, and security procedures, while codes of practice outline the ethical and operational expectations of security professionals. Understanding these specific standards allows locksmiths to ensure that their work complies with recognised industry benchmarks and legal expectations.

One important group of standards relates to the performance and testing of locks and cylinders. In Australia, locksmiths commonly encounter AS 4145 – Locksets and Hardware for Doors, which specifies requirements for durability, security, corrosion resistance, and operational performance of door hardware. This standard includes classifications for different levels of security and is frequently referenced in commercial building specifications. In Europe, similar requirements are outlined in EN 1303 – Cylinders for Locks, which establishes security grades based on resistance to drilling, picking, and key control features. In the United States, the ANSI/BHMA A156 series standards govern door hardware performance, including locksets, deadbolts, and exit devices. These standards classify locks into grades (Grade 1, Grade 2, and Grade 3) based on durability and security performance. Locksmiths rely on these classifications when recommending appropriate locking hardware for residential, commercial, or high-security environments.

Another important category involves high-security locks and burglary resistance standards. For example, UL 437 – Standard for High Security Cylinders in the United States establishes rigorous testing requirements for resistance to picking, drilling, and other forms of manipulation. Products certified to UL 437 are commonly used in sensitive facilities such as government buildings, banks, and data centres. In Europe, EN 1627 – Burglary Resistance for Doors and Windows provides classifications for physical security resistance levels (RC1–RC6). Locksmiths working with high-security installations often rely on these standards to ensure that the door hardware and locking systems provide appropriate protection against forced entry.

Building safety standards also play a critical role in locksmith work, particularly in commercial and public buildings. For example, in Australia the National Construction Code (NCC) sets requirements for building safety, including emergency egress and fire safety provisions. Locksmiths must ensure that locking hardware installed on exit doors allows occupants to escape quickly during emergencies. Standards such as AS 1905.1 – Fire-Resistant Doorsets and AS 1428 – Design for Access and Mobility influence the selection and installation of door hardware to ensure compliance with fire safety and accessibility requirements. Similarly, in the United States the NFPA 101 – Life Safety Code establishes requirements for exit hardware, panic bars, and emergency door operation. Locksmiths must ensure that security hardware does not interfere with evacuation procedures or accessibility requirements.

Electronic security and access control systems introduce additional standards governing system integration and cybersecurity. One widely recognised international standard is ISO/IEC 27001 – Information Security Management Systems, which outlines best practices for protecting digital information and access control data. While locksmiths may not always be responsible for full compliance with this standard, it influences how electronic access systems are configured and managed. Another relevant framework is ISO/IEC 27002, which provides guidelines for managing user authentication, access permissions, and security monitoring. Locksmiths installing electronic access systems must often work with IT professionals to ensure that digital credentials and system data are protected against cyber threats.

There are also standards related specifically to electronic access control and alarm systems. For example, in Australia the AS/NZS 2201 series covers intrusion detection and alarm systems, including system design, installation, and maintenance requirements. Locksmiths working with integrated security systems may need to ensure that electronic locks, alarms, and access control devices comply with these standards. In Europe, electronic access systems are often governed by EN 60839 – Alarm and Electronic Security Systems, which provides guidelines for system performance and reliability. Compliance with these standards ensures that electronic security devices operate reliably and integrate properly with other security infrastructure.

In addition to technical standards, locksmiths must also follow professional codes of ethics and industry codes of practice. For example, the Master Locksmiths Association (MLA) Code of Practice in the United Kingdom requires locksmiths to verify customer authority before performing entry work, maintain confidentiality regarding security systems, and operate with honesty and integrity. In the United States, the Associated Locksmiths of America (ALOA) Code of Ethics similarly emphasises professional conduct, customer verification, and responsible handling of security information. These codes are designed to prevent misuse of locksmith skills and to protect public trust in the profession.

Standards and codes of practice also influence insurance requirements and legal liability. Many insurers require security installations to comply with recognised standards before providing coverage for buildings or security systems. If a burglary occurs and the installed locks do not meet recognised standards, insurance claims may be denied or reduced. In legal disputes, courts often assess whether a locksmith followed accepted industry standards when performing work. Compliance with recognised standards therefore provides an important defence against negligence claims and demonstrates that the locksmith acted according to professional best practice.

Compliance with standards and codes of practice also plays a role in insurance, legal liability, and professional reputation. Insurance providers may require security installations to meet recognised standards before providing coverage for a building or security system. Similarly, in the event of a dispute or legal claim, courts may assess whether a locksmith followed accepted industry standards when performing work. Demonstrating compliance with recognised standards can therefore help protect locksmiths from claims of negligence and show that they acted according to professional best practice.

In modern locksmithing, standards and codes of practice therefore serve as a bridge between legal regulation, technical performance, and professional ethics. They guide locksmiths in selecting appropriate hardware, installing security systems correctly, protecting client information, and maintaining safe building access. For individuals entering the locksmith profession, familiarity with relevant standards and professional codes is essential for delivering reliable security services, complying with legal obligations, and maintaining the credibility of the locksmith industry.

**Impact on Locksmithing Practices**

Understanding standards and codes of practice is highly relevant for anyone interested in locksmithing because these frameworks define how locksmith work should be carried out safely, professionally, and in compliance with recognised industry expectations. Locksmiths are responsible for installing and maintaining systems that protect property, assets, and personal safety, and clients rely on these systems to function reliably in both normal and emergency situations. Standards provide clear benchmarks for lock performance, installation procedures, and security system configuration, allowing locksmiths to recommend appropriate hardware for different environments such as homes, commercial buildings, and high-security facilities. By following recognised standards—such as those governing lock durability, burglary resistance, fire-rated door hardware, or electronic access control—locksmiths ensure that their work meets the quality, safety, and reliability levels expected by regulators, insurers, and clients.

These standards and codes directly influence everyday locksmithing practices by shaping how locksmiths select products, install hardware, manage electronic security systems, and interact with customers. Professional codes of practice reinforce ethical behaviour, requiring locksmiths to verify customer authority before granting access, protect confidential security information, and use non-destructive entry techniques whenever possible. Technical standards also guide compliance with building and fire safety regulations, ensuring that locks do not interfere with emergency exits or accessibility requirements. As modern locksmithing increasingly includes electronic access control and cyber-physical security systems, standards related to cybersecurity and data protection also become relevant. By understanding and applying these frameworks, locksmiths not only deliver higher-quality security solutions but also reduce legal risk, support insurance compliance, and maintain the professional credibility and trust essential to the security industry.

## Chapter 16 Review Questions

### Question 1

A locksmith arrives at a residential property where a person claims to be locked out of their house and asks the locksmith to unlock the door immediately. The locksmith asks the person to show identification and proof of residence before performing the work.

Why is the locksmith taking this step?

A. To verify the client's authority before granting access
B. To determine the type of lock installed on the door
C. To record the brand of the locking hardware
D. To calculate the service fee

**Question 2**

A locksmith business in New South Wales installs and maintains electronic access control systems for commercial buildings.

Under Australian security legislation, what license is typically required for this type of work?

A. Class 2C Security Equipment Specialist license
B. Electrical contractor license only
C. Residential building license
D. Fire safety inspector license

**Question 3**

A locksmith installs a biometric access control system in an office building that stores fingerprint templates for employees.

What additional responsibility must the locksmith consider when configuring the system?

A. Colour matching of door hardware
B. Compliance with privacy and data protection laws
C. Replacement of door hinges
D. Recording the building height

**Question 4**

A locksmith installs electronic door locks for a company but leaves the default administrator password unchanged.

Several weeks later, an unauthorized person uses the default password to gain access to the system.

What professional issue does this situation illustrate?

A. Equipment durability problems
B. Software compatibility issues
C. Professional liability due to negligent configuration
D. Incorrect key cutting procedures

**Question 5**

During a security installation in a commercial building, the locksmith installs a lock on an emergency exit door that prevents the door from opening quickly during an evacuation.

Which requirement has the locksmith failed to consider?

A. Energy efficiency standards
B. Fire and emergency egress regulations
C. Window glazing standards
D. Soundproofing regulations

**Question 6**

A locksmith installs a cloud-based access control system that records when employees enter and exit a building.

Which type of information must be protected under many privacy regulations?

A. Door frame dimensions
B. Access logs linked to identifiable individuals
C. Paint colour of doors
D. Manufacturer serial numbers of locks

**Question 7**

A locksmith assists police officers who have a legal search warrant to enter a locked property.

Before performing the entry, the locksmith verifies the officers' identification and the warrant documentation.

Why is this verification necessary?

A. To confirm lawful authority for entry
B. To determine the type of lock installed
C. To record the time of entry
D. To estimate the cost of repairs

**Question 8**

A locksmith installs high-security cylinders in a government building and ensures that the locks meet recognised testing standards for resistance to picking and drilling.

Why are such standards important?

A. They improve lock colour consistency
B. They reduce the number of keys required
C. They ensure locks meet recognised security performance benchmarks
D. They simplify lock manufacturing

**Question 9**

A locksmith provides installation services for electronic access control systems but does not keep documentation of system configuration or client authorization.

Which risk does this create for the locksmith business?

A. Difficulty identifying door hardware brands
B. Increased legal liability and compliance risk
C. Lower installation speed
D. Reduced key cutting accuracy

**Question 10**

A locksmith joins a professional association that requires members to follow a formal code of ethics, including verifying client authority and protecting confidential security information.

What is the purpose of such codes of practice?

A. To standardise lock colours
B. To regulate the cost of locksmith services
C. To guide professional conduct and ethical behaviour
D. To determine the number of locks used in buildings

# Part 9

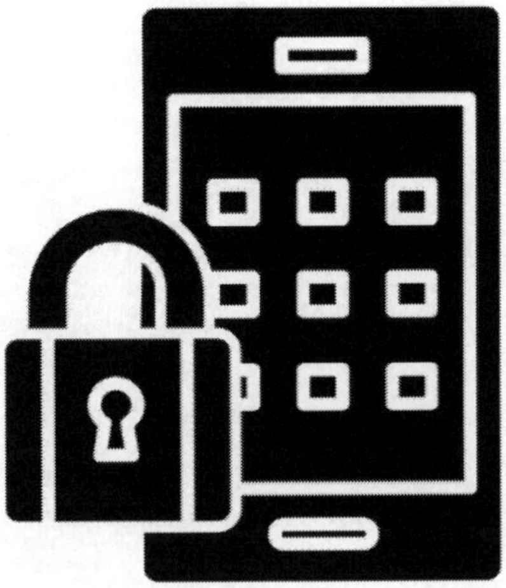

# Innovation and the Future of Locksmithing

# Chapter 17

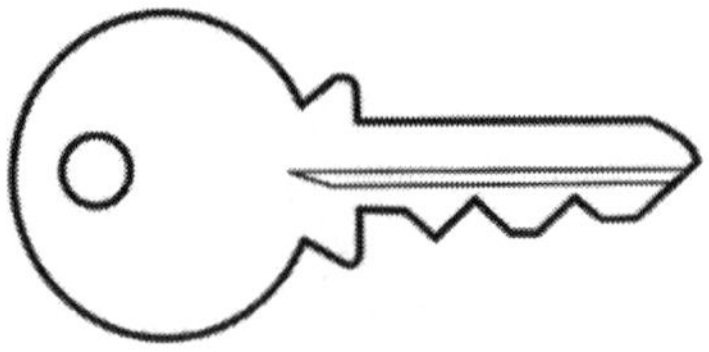

# The Death of the Key?

For thousands of years, the key has been the defining symbol of security. From the earliest wooden pin locks used in ancient civilizations to the precision-engineered cylinders found in modern buildings, the basic idea has remained remarkably consistent: possession of a physical key grants access to a locked space. Locksmiths built their craft around this principle, mastering the design, installation, and manipulation of mechanical locks and keys that protected homes, businesses, vehicles, and valuables.

Today, however, that long-standing relationship between locks and keys is beginning to change. Increasingly, doors open without a metal key ever being inserted into a lock. Smartphones unlock offices, fingerprint scanners grant entry to secure facilities, and vehicles start automatically when a digital credential is detected nearby. In many environments, smart homes, hotels, offices, and shared workspaces, the traditional key is quietly being replaced by electronic credentials, wireless authentication, and cloud-managed access systems. This transformation raises a provocative question for the security industry: are we witnessing the death of the key, or simply the evolution of access control into a new digital form?

The shift toward keyless systems has been driven by rapid advances in electronics, wireless communications, and digital identity technologies. Electronic access control systems can now issue, modify, and revoke credentials instantly without physically changing locks. Mobile devices can act as digital keys, biometric systems can authenticate users without any token at all, and cloud platforms allow administrators to control building access from anywhere in the world. These capabilities provide unprecedented convenience and flexibility, but they also introduce new risks, including cybersecurity vulnerabilities, digital credential theft, and system reliability concerns.

For locksmiths, these developments represent both a challenge and an opportunity. The profession that once revolved almost entirely around mechanical hardware is evolving into a multidisciplinary security field that blends traditional lock expertise with knowledge of electronics, networking, and cybersecurity. Modern locksmiths increasingly install smart locks, configure access control systems, manage digital credentials, and advise clients on emerging security threats that did not exist in purely mechanical environments.

This chapter explores the technological revolution reshaping access control. It examines the rise of electronic access control, digital identity, mobile credentials, and blockchain-based security systems, while also addressing the risks associated with fully digital locking ecosystems. As the boundaries between physical security and digital infrastructure continue to blur, understanding these technologies is essential for anyone seeking to understand the future of locksmithing.

## Keyless Systems and Digital Identity

The emergence of keyless systems and digital identity marks a watershed in locksmithing and physical security, shifting the focus from physical tokens to electronic and cyberphysical authentication. Early forms of access control in buildings relied on mechanical keys and locks, but the late twentieth century saw the adoption of electronic access control (EAC) using magnetic stripe cards, proximity cards, and keypad codes. This transition enabled centralized management of credentials, with the ability to issue, modify, or revoke access without physically changing locks or keys, thereby introducing dynamic and scalable security management that superseded purely static mechanical systems [346, 353, 354]. In parallel, the building security domain increasingly integrated with broader security infrastructures such as alarms and surveillance, enabling event logging and policy enforcement across offices, hospitals, universities, hotels, and multi-unit residences where traditional keys would be impractical at scale [353-355].

In the automotive sector, the first widely adopted keyless technology took the form of remote keyless entry (RKE) systems introduced in the 1980s and gaining ubiquity in the 1990s. RKE systems rely on a handheld transmitter or key fob that communicates a coded RF signal to the vehicle to unlock doors and, in many cases, enable engine control features, thereby delivering substantial convenience while expanding the attack surface through wireless communications [342, 356, 357]. The evolution progressed to passive keyless entry and start (PKES) systems, where the vehicle detects an authorized key fob in proximity via low-frequency signalling and authenticates the user automatically upon approach, unlocking doors when touched and enabling push-button ignition without a physical key insertion. This convenience however introduced new security challenges, notably relay attacks that exploit wireless links between the key and vehicle, and the literature documents a spectrum of vulnerabilities and countermeasures in this domain, including cryptographic weaknesses, protocol design flaws, and the need for stronger authentication architectures [342, 357-359].

Beyond the automobile, the broader security landscape experienced parallel growth in electronic access control within buildings through RFID cards, smart cards, and keypad-based systems. These technologies facilitated granular access control, time-restricted entry, and integration with security ecosystems encompassing alarms and surveillance, while enabling centralized administration of large user populations. The transition from mechanical to electronic credentials allowed detailed audit trails and policy enforcement, and the modern trend has continued toward digital identities that reside on devices or in cloud platforms, enabling flexible management of permissions and rapid adaptation to organizational changes [353, 355, 356, 360]. Digital identity expands the concept of authorization from possession of a physical credential to possession and verification of electronic credentials, which can be stored on smartphones or wearable devices and managed through centralized identity management infrastructures, often leveraging mobile keys, encrypted digital certificates, or biometric attestations as part of the authentication process [355, 358, 361, 362].

The convergence of digital identity with access control systems is often realized through mobile-based digital keys that interact with locks via Bluetooth, NFC, or cloud-based platforms. Smartphone-based digital keys have become a prominent archetype, enabling users to unlock doors or vehicles using mobile applications that communicate with locks and access control systems, thereby centralizing management of permissions and reducing reliance on physical tokens. In residential contexts, smart locks allow homeowners to issue digital access codes or mobile keys to family members and guests, while commercial settings employ digital identity platforms that link with enterprise identity management to enforce policy-based access such as role-based or time-bound permissions. This evolution is tightly coupled to the broader digitization of identity management in organizations and the deployment of cloud-based access control systems, which permit remote credential provisioning and revocation, real-time activity monitoring, and integration with building management systems. However, this reliance on wireless and networked credentials introduces cyber risks, including eavesdropping, cloning, replay, and provisioning attacks, underscoring the need for strong encryption, secure authentication protocols, regular updates, and robust lifecycle management to mitigate risks in the digital identity-enabled security landscape [342, 353, 355, 356, 361].

Biometric authentication represents a notable branch within digital identity for access control. By leveraging inherent physiological or behavioural traits such as fingerprints, facial recognition, iris patterns, or voice characteristics, biometric systems offer passwordless verification and the potential to eliminate the need for physical credentials entirely. While biometrics can enhance security and user convenience, they also raise privacy and data protection concerns because biometric identifiers are highly sensitive personal data and, once compromised, cannot be reissued in the same way as a password or token. The literature addresses both the security benefits and the privacy implications of biometric-enabled access control, highlighting the need for careful data protection, secure storage of biometric templates, and compliance with relevant privacy frameworks when deploying biometric systems as part of digital identity-enabled access control [355, 358, 362].

Cloud-based access control systems constitute a further major development in the digital identity trajectory, enabling credential management, policy enforcement, and monitoring through centralized cloud platforms rather than local hardware controllers. Cloud-enabled access systems facilitate remote provisioning and revocation, real-time audit logging, and seamless integration with enterprise identity management and security information and event management (SIEM) ecosystems. The shift to cloud-managed credentials supports scalable administration in large buildings, residential complexes, and rental properties, while also introducing dependencies on network connectivity and potential exposure to cloud-centric threat models. The literature notes that cloud-based architectures should be designed with strong encryption, authentication, authorization controls, and continuous monitoring to mitigate cyber risks inherent in connected access control environments [342, 353, 355, 361] .

The cybersecurity dimensions of keyless and digital identity-enabled access control are prominent components of the broader security paradigm, given that digital credentials traverse wireless links and internet-connected infrastructures. Attack surfaces include relay and replay attacks, credential cloning, and network intrusions, necessitating robust cryptographic schemes, secure rolling codes or equivalent cryptographic primitives, mutual authentication, and resistance to side-channel and forged credential threats. Scholarly work demonstrates both the feasibility of such attacks on real-world systems and the corresponding countermeasures, including the development of hardening strategies, protocol improvements, and hardware-forward security considerations. This evolving landscape has expanded the professional remit of locksmiths and security technicians from traditional mechanical concerns to multidisciplinary cybersecurity awareness, credential lifecycle management, and integrated system design that encompasses electronics, networking, and digital identity considerations [342, 346, 355-357, 363-365].

In professional practice, locksmiths and security technicians increasingly engage in activities that span installation and configuration of electronic locks, programming of digital credentials, integration with building and identity management networks, and advisory services regarding cybersecurity risk. The profession is described as widening from a purely mechanical craft to a multidisciplinary security field that fuses mechanical know-how with electronics, networking, and digital identity governance. This contemporary scope reflects both the technical maturation of electronic access systems and the essential need for cybersecurity literacy among practitioners who install, manage, and service modern keyless and digital identity–driven access control solutions [342, 346, 355, 356, 363].

### Impact on Locksmithing Practices

The emergence of keyless systems and digital identity is highly relevant to anyone interested in locksmithing because it demonstrates how the profession is evolving from a purely mechanical trade into a technologically advanced security discipline. Historically, locksmiths worked primarily with physical keys, cylinders, and mechanical locking hardware. However, the widespread adoption of electronic access control systems, digital credentials, and wireless authentication technologies means that modern locksmiths increasingly work with systems that combine physical security with electronics, software, and networked infrastructure. Understanding these technologies is essential for individuals entering the profession because many modern security installations, particularly in commercial buildings, vehicles, and residential smart homes, now rely on electronic credentials such as RFID cards, mobile keys, biometric identifiers, or cloud-managed access systems rather than traditional keys alone.

This technological shift significantly influences locksmithing practices by expanding the technical competencies required of professionals in the field. Locksmiths are now frequently responsible for installing electronic locks, configuring access control systems, programming digital credentials, and integrating security hardware with broader building or identity management systems. In addition to traditional mechanical skills, locksmiths must understand wireless communication technologies such as RFID, Bluetooth, and NFC, as well as cybersecurity principles that protect digital access credentials from threats such as cloning, replay attacks, or network intrusion. These responsibilities require locksmiths to adopt new practices such as secure credential management, system configuration auditing, firmware updates, and collaboration with IT or security professionals when deploying integrated access control solutions.

The emergence of digital identity also influences how locksmiths advise clients on security strategies. Instead of simply recommending stronger locks or restricted key systems, locksmiths may now evaluate access control policies, digital credential lifecycles, and authentication methods within an organisation. For example, locksmiths may help businesses implement role-based access permissions, time-restricted credentials, or biometric authentication systems that improve accountability and monitoring. In residential environments, locksmiths increasingly assist homeowners in installing and configuring smart locks that allow remote access management through smartphone applications or cloud platforms.

These developments also introduce new ethical and legal responsibilities for locksmiths. Because digital access systems often store personal data such as access logs, user identities, or biometric templates, locksmiths must understand privacy considerations and ensure that systems are configured to protect sensitive information. Additionally, the cybersecurity risks associated with wireless and networked systems require locksmiths to remain informed about vulnerabilities and countermeasures in electronic access technologies. Consequently, the emergence of keyless systems and digital identity has expanded locksmithing into a multidisciplinary security profession that combines traditional craftsmanship with knowledge of electronics, networking, cybersecurity, and digital identity management.

## Blockchain-Based Access Control

Blockchain-based access control is an emerging approach to managing physical and digital access permissions using distributed ledger technology (DLT). In traditional access control systems, permissions are typically stored and managed in a centralized database controlled by a building management system, security server, or cloud platform. While these centralized systems can be effective, they may also create single points of failure, making them vulnerable to hacking, system outages, or unauthorized administrative changes. Blockchain technology offers an alternative architecture in which access permissions and authentication records are stored in a decentralized, tamper-resistant ledger distributed across multiple network participants.

A blockchain is essentially a chronologically ordered chain of digital records (blocks) that are cryptographically linked together. Each block contains transaction data that has been validated by the network and permanently recorded. Once a block is added to the chain, altering the information within it becomes extremely difficult because any modification would require changing every subsequent block and gaining consensus from the network. This immutability makes blockchain particularly attractive for applications that require secure record-keeping and trusted verification of transactions, including access control events.

In a blockchain-based access control system, permissions and credentials can be represented as cryptographic tokens or digital identities stored on the blockchain. Instead of relying on a centralized server to verify whether a user is authorized to access a building, door, or digital system, the verification process can be performed through the blockchain network. When a user attempts to access a secured location, the access control device checks the blockchain ledger to confirm that the user's credential is valid and that the required permissions exist.

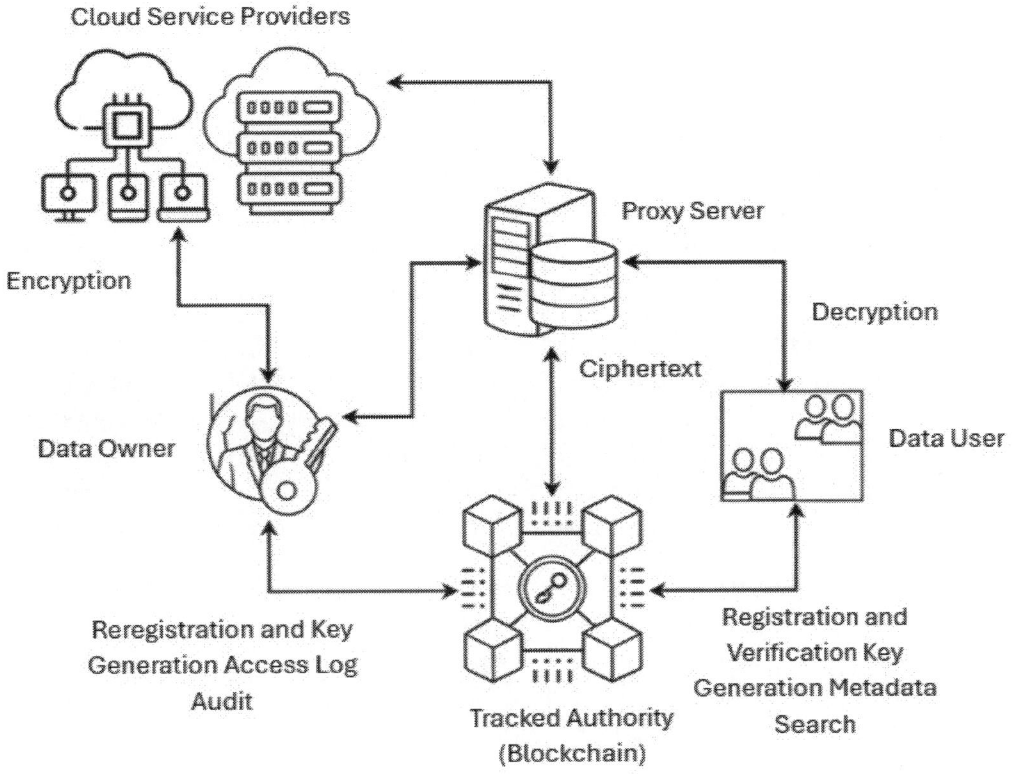


*Figure 91: Applying Access Control Enabled Blockchain [366].*

One key feature of blockchain access control systems is the use of smart contracts, which are self-executing pieces of code stored on the blockchain. Smart contracts automatically enforce predefined rules governing access permissions. For example, a smart contract could specify that an employee is permitted to enter a laboratory only between certain hours, or that a contractor's access expires automatically after a project deadline. When a user presents their credential, the smart contract verifies whether the access conditions are satisfied and authorizes or denies entry accordingly.

Blockchain systems can also provide secure identity management through decentralized identity (DID) frameworks. In these systems, individuals control their own digital identity credentials rather than relying entirely on centralized identity providers. A user may possess a digital identity stored in a secure wallet on a smartphone or secure hardware device. When authentication is required, the user proves their identity through cryptographic verification rather than submitting personal information to a central database. This approach can enhance privacy because sensitive identity data does not need to be stored in a single vulnerable database.

Another important advantage of blockchain-based access control is auditability and transparency. Every access event recorded on the blockchain becomes part of a permanent and verifiable log. Security administrators can review access records to determine who accessed a facility, when access occurred, and which credentials were used. Because these records cannot easily be altered or deleted, blockchain provides a highly reliable audit trail for security investigations, regulatory compliance, and incident analysis.

Blockchain access control systems can also support distributed access management across multiple organisations or facilities. In traditional systems, each building or organisation may maintain its own access control database. Blockchain enables a shared security infrastructure in which multiple organisations can trust the same distributed ledger. For example, a group of companies operating in a shared office complex could use a blockchain system to manage employee credentials across all buildings without relying on a single central authority.

Despite its advantages, blockchain-based access control also presents technical and operational challenges. Blockchain networks may require significant computational resources and network connectivity, which can introduce latency or complexity in real-time access decisions. Additionally, integrating blockchain systems with existing door hardware, access readers, and building management systems can be technically challenging. For these reasons, many blockchain access control implementations currently operate as hybrid systems, combining blockchain identity verification with conventional local access control hardware.

Security considerations are also important when implementing blockchain-based access control. Although blockchain itself provides strong protection against data tampering, vulnerabilities can still exist in the surrounding system components, such as mobile credential storage, authentication devices, or network communication channels. Effective security architecture must therefore include encryption, secure device authentication, and protection against credential theft or unauthorized access to digital wallets.

Blockchain-based locking systems represent an emerging form of access control technology that integrates smart locks, digital identity, and distributed ledger technology to regulate who can access a physical space. Unlike traditional systems that rely on mechanical keys or centralized access control servers, these systems use a blockchain network to verify credentials and store access permissions within a secure, tamper-resistant digital ledger. Because the ledger is distributed across multiple network participants, the system does not rely on a single controlling authority. This decentralized architecture allows access permissions to be issued, verified, and revoked securely while maintaining a permanent and verifiable record of access activity.

A blockchain-based locking system typically consists of several interconnected components that work together to authenticate users and enforce access permissions. One of the most important components is the smart lock hardware installed on the door or secured access point. These locks are usually electronic devices equipped with a microcontroller, wireless communication capabilities such as Bluetooth, Wi-Fi, NFC, or Zigbee, an electronic actuator that physically engages or disengages the locking mechanism, and a secure authentication module that manages credential verification. The lock communicates with a smartphone, credential device, or network gateway to determine whether a user requesting entry has the appropriate authorization.

Another key component of the system is the user's digital identity credential. Instead of carrying a traditional metal key, the user possesses a digital credential that proves their identity through cryptographic authentication. These credentials may be stored in a variety of secure formats, including smartphone digital wallets, smart cards, hardware security tokens, or wearable devices. Each credential contains cryptographic keys that allow the system to confirm that the person requesting access is the legitimate credential holder.

The blockchain network itself functions as the system's distributed database. It stores information such as user identities, assigned access permissions, access event logs, and the rules governing how access is granted or denied. Because blockchain data is stored across many independent nodes in the network, it is extremely difficult for unauthorized parties to alter or manipulate the stored records. This distributed architecture enhances system security and creates a transparent audit trail of all access events.

Another critical element of blockchain-based locking systems is the use of smart contracts, which are automated programs stored on the blockchain that enforce access control rules. Smart contracts define the conditions under which access will be granted. These rules may specify who is allowed to access a particular door, when access is permitted, how long a credential remains valid, and whether additional security measures such as multi-factor authentication are required. When a user attempts to unlock a door, the smart contract automatically checks whether the request meets the defined conditions. If the rules are satisfied, the system authorizes the lock to open; if not, access is denied. This automated decision-making process allows access control to operate securely without relying on a centralized authority.

**Step-by-Step Operation of a Blockchain Locking System**

**Step 1: User Identity Creation**

A user is first issued a digital identity credential.

This process typically involves:

1. Creating a cryptographic key pair

- Private key (held by the user)
- Public key (registered on the blockchain)

2. The user's public key is linked to their access permissions.

For example:

User Public Key → Access Permission

- Door A → Allowed
- Door B → Denied

This information is written to the blockchain as a transaction.

**Step 2: Access Permission Assignment**

An administrator assigns permissions through a **smart contract**.

For example:

- Alice can access Building Door 3
- Only between 8:00 AM and 6:00 PM
- Until 30 June

These rules are encoded into the smart contract and stored on the blockchain ledger.

Once recorded, the permission becomes tamper-resistant.

**Step 3: User Attempts to Unlock the Door**

When the user approaches the door:

1. The user's smartphone or credential device communicates with the smart lock using:

- Bluetooth
- NFC
- Wi-Fi
- QR code authentication

2. The device sends a cryptographic authentication message proving ownership of the private key.

This proves the user is the legitimate credential holder.

**Step 4: Blockchain Verification**

The lock or gateway then verifies authorization.

This can happen in two ways:

**Local Verification**

Some systems store relevant blockchain data locally on the lock.

The lock checks:

- The user's public key
- Access permissions
- Smart contract rules

**Network Verification**

Other systems query a blockchain node or gateway.

The node verifies the access request using the blockchain ledger.

**Step 5: Smart Contract Execution**

The smart contract checks the access rules:

Examples:

- Is this user authorized?
- Is the time within the allowed window?
- Has the credential expired?
- Has access been revoked?

If all conditions are satisfied, the contract returns authorization = true.

**Step 6: Door Unlocks**

The smart lock receives the authorization result.

If valid:

- The lock activates its motor
- The bolt retracts
- The door unlocks

If invalid:

- The request is rejected
- Access is denied.

**Step 7: Access Event Recorded**

The system records the access attempt.

A transaction may be written to the blockchain containing:

- User identity
- Door identifier
- Timestamp
- Result (granted or denied)

Because blockchain records are immutable, the log cannot easily be altered.

This provides a highly reliable audit trail.

One of the major advantages of blockchain-based locking systems is the ability to revoke or change access permissions instantly. In traditional mechanical locking systems, removing a user's access often requires physically collecting keys or replacing locks, which can be time-consuming and costly. In a blockchain-based system, access permissions are controlled digitally. If a user leaves an organisation or no longer requires access, an administrator can submit a transaction to the blockchain that revokes the user's credentials. The system's smart contract automatically updates the permission status, and all locks connected to the system will immediately recognize that the credential is no longer valid. As a result, the user's access is denied without the need to physically change locks or reissue keys.

Many blockchain locking systems operate using offline or hybrid architectures to ensure reliable performance. Because blockchain networks can sometimes introduce delays when processing transactions, relying on real-time blockchain verification for every door access request could slow system response times. To address this issue, some systems store local snapshots of access permissions directly on the smart lock or within a nearby gateway device. These local copies are periodically synchronized with the blockchain ledger to ensure that permission updates remain accurate. In some configurations, a gateway server acts as an intermediary between the lock hardware and the blockchain network, helping to manage communication and verification processes. This hybrid approach improves the speed, reliability, and practicality of blockchain-based access control in real-world environments.

Blockchain locking systems also provide several important security advantages. One of the most significant benefits is the presence of tamper-resistant records. Access logs stored on a blockchain cannot easily be altered or deleted, creating a reliable and permanent audit trail of access events. The decentralized nature of blockchain networks also improves security because there is no single central server that attackers can compromise in order to manipulate permissions or system records. User authentication is typically based on cryptographic key pairs, allowing individuals to prove their identity securely. In addition, blockchain systems can provide transparent audit trails that allow administrators to verify when access occurred and which credentials were used. These systems can also support secure credential sharing by issuing temporary digital access tokens that expire automatically after a defined period.

A wide range of real-world applications can benefit from blockchain-based locking technology. In smart buildings, these systems can manage access to offices, shared workspaces, and apartment complexes, allowing administrators to issue and revoke digital credentials for occupants or staff members. In short-term rental environments, such as vacation rentals or serviced apartments, digital guest keys can be issued to visitors and automatically expire at the end of their stay. Supply chain and logistics facilities may use blockchain locks to control warehouse access or secure cargo containers during transport. Vehicle-sharing platforms can also employ blockchain-based digital keys to manage access to shared cars or fleet vehicles. Additionally, secure storage environments such as data centres or equipment lockers can use blockchain locking systems to maintain precise records of who accessed sensitive infrastructure and when access occurred.

As access control technologies continue to evolve, the role of locksmiths is expanding beyond traditional mechanical lock systems into more advanced electronic and digital security environments. Modern locksmiths increasingly work with smart lock hardware, electronic credential systems, digital identity platforms, and networked access control infrastructure. These technologies allow access permissions to be managed electronically rather than through physical keys, enabling greater flexibility and control over who can enter secure spaces. As a result, locksmiths must develop broader technical knowledge that includes not only mechanical security devices but also electronic systems and digital authentication methods.

Blockchain-based locking systems represent a further extension of this technological shift by combining physical security hardware with decentralized digital identity and authentication systems. In this environment, locksmiths may be involved in installing and configuring smart locks that interact with blockchain networks, ensuring that the hardware operates correctly within the broader access control architecture. Their work may also include integrating locks with digital identity systems that manage cryptographic credentials, allowing authorized users to access buildings or facilities using secure digital keys rather than traditional metal keys.

In addition to installation and integration tasks, locksmiths may also assist with managing access permissions within blockchain-based access control systems. This may involve configuring smart contracts, issuing or revoking digital credentials, and ensuring that user permissions are properly synchronized across the network. Because these systems often operate within connected digital environments, locksmiths may also play an advisory role by helping clients understand potential cybersecurity risks associated with electronic and blockchain-enabled locking systems. By combining traditional lock expertise with knowledge of digital security technologies, modern locksmiths are increasingly positioned as specialists in integrated cyber-physical security solutions.

**Impact on Locksmithing Practices**

For locksmiths and security professionals, the development of blockchain-based access control represents a further expansion of the profession into cyber-physical security integration. Professionals involved in installing and managing modern access control systems may encounter blockchain-enabled identity platforms, decentralized credential systems, and smart-contract-based authorization mechanisms. Understanding these technologies allows security practitioners to participate in the design, installation, and maintenance of next-generation access control infrastructures.

## Mobile Credentials and Cloud-Based Systems

Mobile credentials are digital access keys stored on a user's smartphone or mobile device, enabling doors and access control readers to be unlocked via a mobile application rather than a physical key or plastic access card [367-369]. The communication between the mobile device and the locking system can occur through several technologies, including Bluetooth Low Energy (BLE), Near Field Communication (NFC), Wi-Fi, or QR codes, with the credential transmitted to the reader in an encrypted form for authorization verification before the lock is enabled [367, 370]. In addition to possession factors, mobile credentials frequently integrate on-device biometric verification such as fingerprint or facial recognition, which adds an extra layer of security beyond traditional keys [367, 371]. The capability of mobile credentials to combine device-based authentication with hardware-based access control has been recognized as expanding the reach of electronic access control beyond conventional keys [367, 368, 370].

A central feature of mobile credentials is the ability to issue and revoke access remotely through a cloud-based management platform, enabling instant changes to permissions without distributing physical credentials and supporting dynamic provisioning for new or departing personnel [367, 372-374]. This dynamic management is particularly advantageous in environments with frequent personnel changes, such as offices, hotels, coworking spaces, and rental properties, where temporary or time-bound credentials can be issued to visitors or contractors and programmed to expire after a defined period, reducing logistical burdens associated with traditional key control [367, 372, 374].

Cloud-based access control systems extend management capabilities beyond the local premises, allowing administrators to configure doors, credentials, and permissions from any location with internet access, and to monitor activity, generate reports, and modify permissions through web interfaces or mobile apps that communicate with devices in buildings [367, 372, 373]. The cloud infrastructure supports comprehensive audit trails by logging events such as door openings, failed authentications, and system alerts, storing these records securely to support incident investigation, workforce movement monitoring, and regulatory compliance, while enabling integration with other building technologies for a coordinated security posture [367, 373, 375]. When viewed through a locksmithing lens, these systems extend the practitioner's role from mechanical installation to configuring smart locks and credential readers, integrating devices with network infrastructure, and aligning access policies within cloud-based management platforms [367, 376].

Mobile credentials and cloud-based access systems rely on encrypted communications and network connectivity, necessitating sound configuration of secure authentication protocols, robust management of administrative accounts, and protections against unauthorized access. Weak passwords, unsecured networks, or poorly configured access permissions can create vulnerabilities that undermine building security, making cybersecurity considerations a central obligation for locksmiths and security professionals who implement these systems [367, 369, 377]. The contemporary security paradigm for cloud and mobile-enabled access includes data integrity, confidentiality, and auditable logging, often leveraging cryptographic techniques and tamper-evident records to support forensics and compliance. Several lines of work emphasize audit trails, transparency in logging, and cryptographic enforcement to ensure secure, auditable access in cloud environments, including discussions of secure logging, encryption, hashing, and anomaly detection within cloud-auditing frameworks [370, 372, 378, 379]. In practice, this translates into combining device-level security features (biometric verification, secure storage of credentials) with cloud-based governance (RBAC/ABAC access control, policy-driven permissions, continuous monitoring) to enact defence-in-depth in access control [370, 372, 373]. The literature also notes that the evolving security landscape necessitates collaboration with information technology professionals and the application of cybersecurity principles, reflecting the cross-disciplinary nature of modern locksmith practice in electronically enabled facilities [369, 372, 377].

The adoption of mobile credentials and cloud-based access systems broadens the set of services that locksmith businesses offer. In addition to installation and maintenance of mechanical hardware, locksmiths increasingly engage in configuring smart locks, credential readers, and networked access control devices, as well as assisting clients with user account setup, access policy configuration, and security policy implementation within cloud management platforms [367, 376]. The ongoing management dimension often takes the form of subscription-based service models that cover system administration, remote credential management, software updates, and security audits, reflecting a shift from one-time installations to continuous security monitoring and support [367, 376]. Cloud-based systems enable real-time visibility into door events and security posture, supporting regulatory compliance regimes that require robust audit trails and incident reporting [372, 375]. The governance aspect is reinforced by cloud-native capabilities such as centralized logging, data retention policies, and role-based access management, which together facilitate least-privilege access and traceability across the organizational security ecosystem [373, 375, 379].

### Impact on Locksmithing Practices

The emergence of mobile credentials and cloud-based access control systems is highly relevant to anyone interested in locksmithing because it reflects the broader transformation of the security industry from purely mechanical locks toward digitally managed access systems. Traditionally, locksmiths worked primarily with physical keys, lock cylinders, and mechanical door hardware. However, modern access control increasingly relies on smartphones, encrypted digital credentials, and cloud-managed platforms to grant or restrict entry to buildings and secure areas. Understanding how mobile credentials communicate with locks through technologies such as Bluetooth, NFC, or Wi-Fi allows locksmiths to install, configure, and maintain these systems effectively. As organizations move away from physical keys toward digital identity-based access, locksmiths who understand these technologies are better equipped to meet modern security requirements and remain relevant in a rapidly evolving profession.

These technologies also significantly influence locksmithing practices by expanding the role of the locksmith beyond hardware installation into areas that include system configuration, digital credential management, and cybersecurity awareness. Locksmiths increasingly assist clients with setting up smart locks, integrating access control readers with network infrastructure, and managing permissions through cloud-based platforms. Because mobile and cloud systems rely on encrypted communications and secure authentication protocols, locksmiths must also understand potential cybersecurity risks such as weak passwords, unsecured networks, or improperly configured permissions that could compromise building security. As a result, the profession is evolving into a multidisciplinary field that combines mechanical security expertise with knowledge of electronic systems, networking, and digital identity management, positioning locksmiths as specialists in modern integrated security solutions.

## Risks of Fully Digital Locking Ecosystems

Fully digital locking ecosystems represent a significant evolution in access control technology, replacing traditional mechanical keys with electronic credentials, network-connected locks, and centralized or cloud-based management platforms. While these systems provide greater convenience, flexibility, and scalability, they also introduce a range of technical, operational, and cybersecurity risks that must be carefully managed. Understanding these risks is important for security professionals and locksmiths because digital locking environments operate very differently from purely mechanical lock systems and may create new vulnerabilities if not properly designed and maintained.

One of the most significant risks associated with fully digital locking systems is cybersecurity vulnerability. Unlike mechanical locks, which require physical manipulation to bypass, digital locks rely on electronic communication, software, and network connectivity. Attackers may attempt to exploit weaknesses in wireless communication protocols such as Bluetooth, Wi-Fi, or radio frequency systems used by smart locks and mobile credentials. If encryption protocols are weak or improperly implemented, malicious actors could intercept authentication signals, clone credentials, or perform replay attacks to gain unauthorized access. Additionally, poorly secured cloud platforms or access control servers may become targets for hacking attempts, potentially allowing attackers to manipulate permissions or unlock doors remotely.

Another important risk involves system outages and technical failures. Digital locking systems depend on electronic components, power supplies, and network connectivity to operate correctly. If the system loses power, experiences a network failure, or encounters software errors, users may be unable to access secured areas. While many electronic locks include battery backups or fail-safe modes, system reliability remains a critical consideration, especially in environments where continuous access is essential, such as hospitals, data centres, or emergency facilities. Mechanical locks, by contrast, can often continue functioning without electricity or network infrastructure.

Credential management and identity security also present challenges in digital locking ecosystems. In traditional lock systems, access control depends on possession of a physical key. In digital systems, access permissions are linked to electronic credentials such as smartphones, RFID cards, PIN codes, or biometric identifiers. If these credentials are compromised, stolen, or improperly managed, unauthorized individuals may gain access to secure areas. For example, a lost smartphone containing a digital key or an employee account that has not been properly deactivated could create security vulnerabilities. Effective credential lifecycle management, including timely revocation of access permissions and strong authentication policies, is essential to reduce these risks.

Fully digital systems also create privacy and data protection concerns. Many electronic access control systems collect and store detailed records of user activity, including entry times, locations, and credential identifiers. While these audit logs can improve security monitoring and incident investigation, they also represent sensitive personal data that must be protected. Unauthorized access to access-control databases could reveal patterns of movement within a facility or expose personal identity information. Compliance with privacy regulations and secure data storage practices therefore becomes an important part of managing digital locking systems.

Another risk involves system complexity and integration challenges. Modern digital access systems often integrate multiple technologies, including electronic locks, credential readers, cloud services, mobile applications, alarm systems, and surveillance platforms. As system complexity increases, the potential for configuration errors also increases. Incorrectly configured permissions, poorly secured administrator accounts, or software compatibility issues between devices can create vulnerabilities that undermine the intended security of the system. Proper installation, testing, and ongoing maintenance are therefore critical to ensure that digital locking ecosystems function securely and reliably.

Digital locking systems may also introduce supply chain and software dependency risks. Many modern smart locks rely on proprietary software platforms, firmware updates, and cloud services maintained by manufacturers or service providers. If a vendor discontinues support for a product, experiences technical failures, or suffers a cybersecurity breach, the entire access control system may be affected. Organizations that rely heavily on digital locking infrastructure must therefore consider long-term support, vendor reliability, and update management when selecting security systems.

The risk of overreliance on automation is another concern in fully digital environments. Automated access control systems often depend on software rules, algorithms, and centralized management policies. While automation improves efficiency, errors in system configuration or software logic could unintentionally grant or deny access to users. For example, a misconfigured permission rule might allow unauthorized access to restricted areas or prevent legitimate personnel from entering critical facilities. Regular system audits and manual oversight are necessary to ensure that automated systems continue to operate as intended.

Finally, the transition to fully digital locking ecosystems changes the professional responsibilities of locksmiths and security technicians. Professionals working with these systems must understand both physical security hardware and digital security principles. This includes knowledge of network security, encryption, credential management, and system integration. Locksmiths must also be prepared to diagnose electronic faults, manage firmware updates, and address cybersecurity risks associated with connected access control devices. As a result, the profession is increasingly evolving into a multidisciplinary field that combines traditional locksmithing skills with electronic security and information technology expertise.

**Case Study Scenario: Security Breach in a Fully Digital Locking Ecosystem**

A large technology company operates from a multi-level office building that uses a fully digital access control system instead of traditional mechanical keys. All entry points, including office doors, meeting rooms, and equipment storage areas, are fitted with smart locks connected to a centralized cloud-based access control platform. Employees use mobile credentials stored on their smartphones to unlock doors through Bluetooth communication with the electronic locks. The system allows the company's security administrators to remotely manage access permissions, track entry logs, and integrate door activity with the building's surveillance and alarm systems.

During the initial installation of the system, a locksmith company was responsible for installing the smart locks, configuring credential readers, and connecting the devices to the building's network infrastructure. The system appeared to function correctly, and employees enjoyed the convenience of unlocking doors with their smartphones instead of carrying physical keys or access cards. However, the security configuration of the system was not thoroughly reviewed after installation, and several default settings were left unchanged in the cloud-based access management platform.

Several months later, a cybersecurity researcher discovered that the wireless communication protocol used by the smart locks could be intercepted under certain conditions. An attacker positioned near the building was able to capture encrypted authentication signals transmitted between a legitimate employee's smartphone and the door lock. Because the encryption protocol was not configured with the strongest available security settings, the attacker was able to perform a replay attack, transmitting the captured signal to the lock and gaining unauthorized entry to the building after business hours.

At the same time, another issue emerged related to credential management. A former contractor who had worked on the building's IT infrastructure still had an active digital access credential stored on their mobile device because their access had not been properly revoked when the contract ended. The individual was able to enter the building and access restricted areas without triggering suspicion because the access control system still recognized their credential as valid.

The company eventually detected unusual activity in the system's access logs during a routine security review. Investigators noticed that doors had been unlocked outside normal operating hours and that access attempts were occurring from credentials that should have been deactivated. The security team temporarily disabled the mobile credential system and requested assistance from the locksmith company that had originally installed the locks.

The locksmith technicians conducted a detailed inspection of the digital locking ecosystem. They identified several issues that contributed to the security breach, including weak wireless encryption settings, outdated firmware on several smart locks, and inadequate procedures for managing digital credentials when employees or contractors left the organization. In addition, the locksmiths discovered that some administrative accounts in the cloud management system had overly broad permissions, increasing the risk of accidental configuration errors.

To address these vulnerabilities, the locksmith team implemented several corrective measures. The wireless communication settings on the locks were updated to use stronger encryption protocols, and all devices were upgraded to the latest firmware versions. The company also introduced stricter credential lifecycle management procedures to ensure that access permissions are automatically revoked when employees leave the organization. Multi-factor authentication was enabled for system administrators, and access permissions were reconfigured according to role-based security policies.

This case study demonstrates the risks associated with fully digital locking ecosystems and the importance of proper system configuration, cybersecurity awareness, and credential management. Although digital access systems offer convenience and flexibility, they can introduce new vulnerabilities if security controls are not carefully implemented. The scenario also highlights the evolving role of locksmiths, who must now understand not only physical locking hardware but also electronic systems, network security, and digital access management in order to maintain secure modern access control environments.

### Impact on Locksmithing Practices

The risks associated with fully digital locking ecosystems are highly relevant to anyone interested in locksmithing because they highlight how the profession is evolving beyond traditional mechanical lock systems into complex electronic and networked security environments. In the past, locksmithing primarily involved physical locks, keys, and mechanical hardware where security vulnerabilities were mostly related to physical manipulation such as picking, drilling, or forced entry. Modern digital locking systems, however, rely on wireless communication, cloud platforms, electronic credentials, and software-based management. This means that security weaknesses may now arise from cybersecurity vulnerabilities, communication protocols, credential management errors, or software misconfigurations, rather than purely mechanical flaws. For someone entering the locksmith profession, understanding these risks is essential because the effectiveness of modern security systems depends not only on the physical lock hardware but also on how the electronic and digital components are configured and protected.

These risks also influence locksmithing practices by expanding the knowledge and responsibilities expected of modern locksmiths. Professionals working with digital access systems must be able to install and maintain electronic locks, configure network-connected access control systems, manage digital credentials, and understand basic cybersecurity principles such as encryption, authentication, and secure system configuration. Locksmiths may also need to troubleshoot system failures, manage firmware updates, and ensure that access permissions are correctly configured to prevent unauthorized entry. In addition, locksmiths must consider issues such as system reliability, privacy protection, and compliance with security standards when advising clients on digital locking solutions. As a result, locksmithing is increasingly becoming a multidisciplinary security profession that integrates traditional mechanical expertise with electronic security, networking, and cybersecurity knowledge to manage the risks associated with modern digital access control systems.

## Chapter 17 Review Questions

### Question 1

A large university replaces traditional metal keys with RFID access cards that allow administrators to grant or revoke building access remotely without replacing door locks.

What advantage does this electronic access control system provide?

A. It reduces the need for electricity in the building
B. It allows centralized management of user access permissions
C. It prevents all forms of unauthorized entry
D. It eliminates the need for door hardware

**Question 2**

A modern vehicle automatically unlocks when the driver approaches with the key fob in their pocket and starts with a push-button ignition.

Which technology does this system represent?

A. Remote Keyless Entry (RKE)
B. Passive Keyless Entry and Start (PKES)
C. RFID building access control
D. Mechanical immobiliser

**Question 3**

A cybersecurity researcher demonstrates that a thief can extend the wireless signal between a car and its key fob to unlock the vehicle even though the owner is inside their home.

What type of attack is being demonstrated?

A. Brute-force encryption attack
B. Replay attack
C. Relay attack
D. Physical lock manipulation

**Question 4**

A residential smart lock allows a homeowner to send temporary digital access codes to guests through a smartphone application. The code automatically expires after the guest leaves.

What capability of digital access systems does this illustrate?

A. Mechanical key duplication
B. Static credential management
C. Time-restricted digital credential control
D. Manual door unlocking

**Question 5**

A company installs fingerprint scanners at secure laboratory entrances so employees no longer need access cards.

Which authentication method is being used?

A. Password-based authentication
B. Biometric authentication

C. RFID authentication
D. Mechanical verification

**Question 6**

A large office complex installs a cloud-managed access control system that allows administrators to manage building entry permissions from a web dashboard.

What major benefit does cloud-based access control provide?

A. Automatic replacement of door locks
B. Remote administration of access credentials and permissions
C. Elimination of electronic locks
D. Removal of authentication requirements

**Question 7**

A new access control system stores permission records across many computers in a distributed network so that the access logs cannot easily be altered.

Which technology is most likely being used?

A. Centralized password database
B. Magnetic stripe card system
C. Blockchain-based access control
D. Mechanical master key system

**Question 8**

A building security system stores door entry events permanently so administrators can later verify exactly who entered a facility and when.

What feature of blockchain-based systems enables this?

A. Wireless credential broadcasting
B. Immutable and tamper-resistant records
C. Mechanical lock redundancy
D. Manual key tracking

**Question 9**

A hotel installs a system where guests unlock doors using a smartphone app communicating with the lock via Bluetooth.

What type of credential is being used?

A. Mechanical key credential
B. Magnetic stripe credential
C. Mobile digital credential
D. Manual access token

**Question 10**

A security consultant warns a company that their fully digital locking system could fail if network connectivity or power is lost.

Which risk of digital locking ecosystems is being described?

A. Key duplication risk
B. Lock cylinder wear
C. System outages and technical failures
D. Door frame damage

# Chapter 18

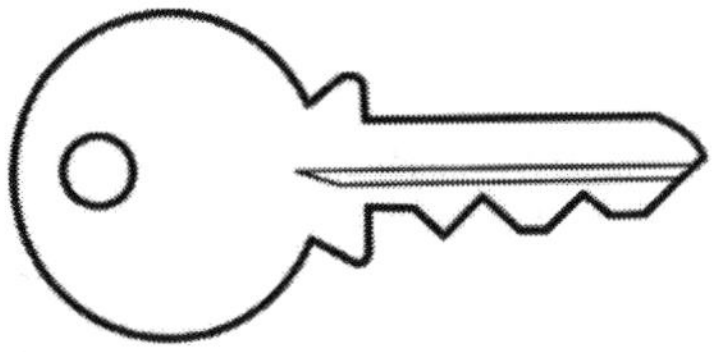

# AI, Automation, and Predictive Security

This chapter explores how emerging technologies are redefining modern security infrastructures. Artificial intelligence now enables surveillance systems to interpret video streams, detect unusual behaviour, and assist in identifying potential threats before they escalate. Predictive maintenance tools allow security systems to monitor the health of locks and door hardware, identifying signs of wear or malfunction before failures occur. Automated threat detection platforms integrate data from cameras, electronic locks, sensors, and access logs to generate rapid alerts and coordinated responses across entire facilities. Together, these technologies represent a shift from reactive security practices toward predictive and adaptive protection strategies.

At the same time, these powerful technologies introduce new responsibilities and challenges. As AI-driven surveillance and automated decision-making become more common, organizations must carefully consider issues such as privacy, data protection, algorithmic bias, transparency, and human oversight. Security professionals must ensure that technological innovation strengthens security without undermining civil liberties or public trust. Responsible deployment of intelligent security systems therefore requires not only technical expertise but also ethical awareness and sound governance.

For locksmiths and security practitioners, the rise of AI-enabled security technologies marks an important turning point in the profession. Modern locksmithing increasingly extends beyond the installation of mechanical locks to include electronic access control systems, biometric authentication devices, networked security platforms, and integrated building security infrastructure. Understanding how automation, predictive analytics, and AI-driven monitoring operate is becoming essential for professionals responsible for designing, installing, and maintaining secure entry systems. As security technologies continue to evolve, locksmiths are increasingly positioned at the intersection of physical security, digital identity, and intelligent infrastructure, playing a critical role in shaping the future of access control.

## AI In Surveillance and Access Control

Artificial intelligence (AI) is increasingly embedded in modern surveillance and access control systems, reshaping how security infrastructures detect threats, manage identities, and control entry to secure spaces. The shift from human-centric monitoring and fixed-credential access to AI-enabled decision-making enhances efficiency, responsiveness, and resilience of security operations. Foundationally, traditional surveillance relied on manual observation by operators and rule-based interpretation of feeds, while access control depended on static credentials such as keys, cards, or PINs. AI introduces automatic data analysis, pattern recognition, anomaly detection, and real-time decision-making, enabling more efficient operations, faster threat responses, and reduced reliance on continuous human oversight [380, 381]. The integration of AI thus marks a move from passive recording toward proactive, autonomous security postures that harmonize multiple data streams across CCTV, access logs, and IoT-enabled sensors [382, 383].

A core AI-enabled capability in surveillance is computer vision, which permits cameras to analyse video streams without constant human attention. AI-powered systems can detect objects, identify faces, flag unusual behaviours, and track movement across camera networks. They can alert security personnel when activity deviates from expected patterns, such as entry into restricted zones, known security concerns associated with a person, loitering, unauthorized access attempts, or abandoned objects, thereby enabling rapid responses and reducing the burden of perpetual manual review [380, 384, 385]. The ability to correlate detections across multiple camera views supports more robust situational awareness and incident reconstruction, ultimately improving response times and investigative efficiency [382, 384, 385]. In parallel, AI-driven video analytics enhance forensic capabilities by enabling rapid search over large archives for criteria like clothing, facial matches, or vehicle plates, leading to more efficient incident investigations [381, 384].

Facial recognition constitutes a prominent AI-enabled mechanism within access control systems. By analysing distinctive facial features and matching them against biometric templates, authorized individuals may gain automatic entry without carrying physical credentials. This modality is increasingly deployed in high-demand environments such as airports, corporate facilities, government buildings, and other secure venues where swift identity verification is valued [386-388]. Nevertheless, facial recognition raises substantial privacy and ethical concerns related to data storage, consent, and potential misuse; these issues require careful governance, transparency, and compliance with privacy protections as part of responsible deployment [381, 386, 387]. The literature consistently notes the dual-edged nature of facial biometrics: offering convenience and security while presenting risks around data handling, bias, and potential misidentification, emphasizing the need for human oversight in high-stakes decisions [383, 387, 388].

Beyond facial recognition, AI enables behavioural analysis and anomaly detection in surveillance systems. Rather than merely flagging the presence of individuals, AI can interpret behavioural patterns to identify unusual or suspicious activities. Examples include detection of individuals operating outside normal hours in restricted zones, patterns consistent with attempted intrusions, or atypical crowd movements signalling potential threats. The strength of these approaches lies in leveraging temporal data to distinguish normal from risky activity with greater accuracy than traditional rule-based monitoring [383, 384, 387]. In access control contexts, AI also supports decision-making by combining multiple authentication factors, facial recognition, fingerprint scanning, mobile credentials, and behavioural cues, to assess the likelihood that a granted entry should be authorized. Contextual factors such as location, time, and past user behaviour inform the risk assessment, allowing adaptive patterns such as requesting additional verification when access occurs unusually or from unfamiliar devices [383, 384, 389, 390].

Automated incident detection and response is a critical application area for AI in integrated security ecosystems. AI can automatically generate alerts when rules are violated (for example, door forced events, repeated failed access attempts, or detections of unauthorized individuals in restricted zones) and trigger predefined responses such as door locking, alarm activation, or escalation to security personnel. In cohesive security architectures, AI can coordinate responses across cameras, alarms, and building management systems to synchronize protection measures and optimize incident containment [383, 384, 390]. AI also powers automated forensic workflows by enabling rapid retrieval of relevant footage given criteria like appearance, movements, or vehicle features, thereby reducing manual review time and supporting faster containment and investigation of security breaches [381, 384].

The broader security value proposition of AI in surveillance and access control extends to system optimization and resilience. By automating routine monitoring and decision-making, AI reduces labour intensity, enables real-time risk assessment, and supports scalable surveillance across large facilities and networks. However, the deployment of AI-based security systems must grapple with privacy, ethics, and governance considerations; comprehensive policies, transparency, and appropriate oversight mechanisms are essential to address concerns about biometric data collection, retention, and potential for misuse [381, 386, 387]. Additionally, algorithmic biases and accuracy limitations, particularly in facial recognition, necessitate robust validation, diverse training data, and human-in-the-loop strategies to minimize false positives and false negatives that could result in wrongful denial of entry or unwarranted security responses [381, 383, 390].

In professional practice, the growing adoption of AI in surveillance and access control broadens the scope of locksmiths and security professionals who must understand how AI-enabled hardware integrates with existing infrastructure. Modern security installations increasingly incorporate biometric readers, AI-enabled cameras, and integrated platforms; professionals must ensure correct installation, secure integration, and proper configuration to balance security, usability, and privacy. This expansion of the profession reflects the converging trends of digital security technologies with traditional access control expertise, reinforcing the need for interdisciplinary knowledge across hardware, software, and policy dimensions [380, 390, 391].

When evaluating references, the literature consistently emphasizes both capabilities and caveats of AI-enabled surveillance and access control. Computer vision and facial recognition deliver rapid, scalable identity verification and threat detection but carry privacy and bias concerns requiring governance and human oversight [381, 383, 386, 387]. Behavioural analytics and anomaly detection offer nuanced threat assessment beyond fixed rules but also face reliability challenges and potential bias, particularly under changing user behaviour or environmental conditions [384, 387]. Automated incident response and cross-system coordination provide powerful resilience but demand careful design to avoid unintended consequences and to ensure privacy-preserving operation [381, 383, 384, 390]. Taken together, AI-enhanced security systems offer substantial improvements in threat detection, identity management, and entry control while demanding deliberate attention to privacy, ethics, bias, and human-in-the-loop safeguards to ensure responsible and effective deployment in real-world settings [381, 383, 384, 386, 387, 390].

Below are some well-known commercial examples that demonstrate how artificial intelligence is being applied in surveillance and access control systems in real-world security environments.

**Example 1: Avigilon (Motorola Solutions) AI Video Analytics and Access Control**

One widely used commercial system that incorporates artificial intelligence into surveillance and access control is the Avigilon Unity platform developed by Motorola Solutions. Avigilon combines AI-enabled video analytics with access control management to create an integrated security ecosystem used in airports, stadiums, government facilities, and large corporate campuses.

The system uses AI-powered computer vision to analyse video feeds from CCTV cameras in real time. Its analytics can automatically detect and classify objects such as people and vehicles, recognize unusual behaviour, and identify security events such as loitering, intrusion into restricted areas, or abandoned objects. The platform can also perform facial recognition matching, comparing faces captured by cameras against a watchlist database. When a match occurs, the system can automatically alert security staff or trigger predefined responses such as locking doors or activating alarms.

Avigilon's Appearance Search™ technology is another example of AI functionality. Instead of manually reviewing hours of video footage, security personnel can search recorded video using physical characteristics such as clothing colour or object type. The AI system scans video archives to locate matching individuals across multiple cameras. In access control contexts, the system can integrate with credential readers and door controllers so that events such as forced doors, repeated access failures, or unauthorized entries trigger AI-driven alerts and camera tracking. Locksmiths installing these systems must therefore understand not only door hardware and electronic locks but also camera placement, system integration, and network connectivity.

**Example 2: Hikvision DeepinMind AI Surveillance Cameras**

Another commercial example is Hikvision's DeepinMind AI camera series, which incorporates artificial intelligence into video surveillance and access control environments. These cameras are commonly used in retail environments, office buildings, transportation hubs, and industrial facilities.

DeepinMind cameras use deep learning algorithms to analyse video streams and detect specific events or behaviours. For example, the system can perform facial recognition for access control gates, identify license plates at vehicle entry points, and detect abnormal activities such as trespassing or crowd formation. The AI engine can distinguish between humans, vehicles, and other moving objects, significantly reducing false alarms that are common with traditional motion detection systems.

In access control scenarios, these cameras can integrate with facial recognition terminals and smart door controllers. When an individual approaches an entrance, the camera scans the face and compares it to a database of authorized users. If the match is successful, the system automatically unlocks the door. If the person is not recognized, the system may deny access, notify security personnel, or record the event for investigation. AI systems can also track individuals across multiple cameras throughout a facility, allowing security teams to reconstruct movements during incident investigations.

For locksmiths and security technicians, installing such systems requires knowledge of biometric readers, electronic locks, network configuration, and AI-enabled video analytics platforms, illustrating how the locksmith profession increasingly intersects with advanced digital security technologies.

### Example 3: Ring and AI-Based Residential Security (Optional Additional Example)

In the residential sector, Ring security systems provide another example of AI in surveillance and access control. Ring doorbells and cameras incorporate AI-powered motion detection and person recognition features. The system can distinguish between people, animals, vehicles, and packages, sending targeted alerts to homeowners via mobile applications.

When integrated with smart locks, such as Ring-compatible electronic deadbolts, homeowners can remotely unlock doors through the mobile application after verifying visitors via the camera feed. AI-based alerts can detect unusual activity around entry points, improving residential security monitoring. While simpler than enterprise systems, these solutions illustrate how AI-powered surveillance and access control technologies are now entering everyday homes, expanding the scope of locksmith work into smart home security installation and configuration.

## Impact on Locksmithing Practices

The increasing use of artificial intelligence in surveillance and access control is highly relevant to someone interested in locksmithing because it illustrates how the security industry is evolving from traditional mechanical protection toward intelligent, data-driven security systems. Historically, locksmiths focused on installing and servicing physical locks, keys, and mechanical door hardware. Modern security environments, however, increasingly rely on integrated electronic systems that combine smart locks, biometric authentication, surveillance cameras, and AI-based monitoring platforms. These systems do not simply restrict access based on possession of a key or card; they analyse identity, behaviour, and contextual data to determine whether entry should be granted. Understanding how AI-enabled surveillance and access control systems operate allows locksmiths to participate effectively in modern security installations that incorporate facial recognition readers, intelligent video analytics, and automated incident detection.

The integration of AI also significantly influences locksmithing practices by expanding the technical responsibilities of professionals working in the field. Locksmiths who install or maintain electronic access control systems must ensure that biometric readers, credential devices, and AI-enabled cameras are correctly installed and integrated with broader security infrastructure such as network systems, cloud platforms, and building management systems. They may also assist clients in configuring authentication policies, managing credentials, and ensuring that systems comply with privacy and cybersecurity requirements. As a result, locksmithing is increasingly becoming a multidisciplinary security profession, requiring knowledge of electronic hardware, digital identity systems, and cybersecurity principles in addition to traditional lock and key expertise. Understanding AI-driven security technologies therefore enables locksmiths to design and maintain modern access control systems that balance security, efficiency, and ethical considerations in increasingly automated security environments.

## Predictive Maintenance of Lock Systems

Predictive maintenance of lock systems refers to the use of data monitoring, sensors, and analytics to anticipate when locks, access control devices, or door hardware are likely to fail or require servicing before a malfunction occurs. Traditional locksmith maintenance typically follows either a reactive approach, where repairs occur only after a lock fails, or a scheduled preventative approach, where locks are inspected and serviced at regular intervals regardless of their actual condition. Predictive maintenance represents a more advanced strategy that uses real-time data from electronic locks, access control systems, and monitoring software to identify patterns that indicate wear, malfunction, or security risks. By analysing this data, maintenance can be scheduled precisely when it is needed, reducing unexpected failures and improving the reliability of security systems.

In modern electronic locking systems, predictive maintenance is often enabled through sensors and integrated monitoring software. Smart locks and access control devices may record operational data such as the number of door cycles, battery levels, actuator performance, motor resistance, communication errors, and failed access attempts. This data is transmitted to a central management system or cloud platform where analytics tools evaluate performance trends. For example, if a smart lock's motor begins requiring more power to actuate the locking mechanism, the system may interpret this as early evidence of mechanical wear, friction in the bolt assembly, or misalignment between the lock and strike plate. The system can then generate an alert recommending inspection or servicing before the lock fails completely.

Predictive maintenance is particularly valuable in environments where locks are used frequently or where access reliability is critical. Facilities such as hospitals, airports, office buildings, data centres, and hotels may have hundreds or thousands of electronically controlled doors that operate continuously. In these environments, a failed lock could disrupt operations, compromise security, or create safety hazards during emergency evacuations. Predictive maintenance systems allow facility managers and security technicians to monitor the health of door hardware remotely and schedule maintenance activities in advance, reducing downtime and improving operational continuity.

Machine learning and artificial intelligence technologies are increasingly being used to enhance predictive maintenance capabilities. These systems analyse large datasets from lock operations across many devices and locations to identify patterns associated with equipment failure. For example, an AI-based maintenance system might learn that a particular type of electronic lock typically begins showing increased actuator resistance after a certain number of operating cycles. By comparing real-time operational data with historical failure patterns, the system can predict when a specific lock is approaching the end of its reliable service life and notify technicians to replace components such as motors, batteries, or electronic control modules.

Predictive maintenance also applies to mechanical locks and door hardware when integrated into modern building management systems. Sensors placed in doors or locking mechanisms can monitor door alignment, vibration, usage frequency, and environmental conditions such as humidity or temperature. These factors can affect lock performance and longevity. For example, a door that is slightly misaligned may place excessive stress on a deadbolt or latch, leading to accelerated wear. Early detection of these conditions allows locksmiths to correct the problem before the lock becomes difficult to operate or fails entirely.

Another advantage of predictive maintenance is improved lifecycle management of security equipment. By collecting data about how long locks and components operate before requiring servicing, organizations can better plan replacement schedules and inventory management. Instead of replacing hardware at arbitrary intervals, facility managers can base decisions on actual usage patterns and performance data. This approach reduces maintenance costs, extends the service life of equipment, and ensures that critical security infrastructure remains reliable.

For locksmiths and security professionals, predictive maintenance represents an important shift in the way lock systems are serviced and managed. Modern locksmiths increasingly work with networked access control systems that provide diagnostic data and remote monitoring capabilities. Their responsibilities may include configuring monitoring systems, interpreting maintenance alerts, and performing proactive servicing based on predictive analytics. This requires knowledge not only of mechanical lock components but also of electronic hardware, network connectivity, and system management software.

Ultimately, predictive maintenance helps transform security systems from passive infrastructure into actively monitored assets. By identifying potential failures before they occur, organizations can maintain consistent security performance while minimizing disruptions. For locksmiths, this approach expands the profession beyond traditional repair work toward proactive system management, long-term security planning, and integration with intelligent building technologies.

## Impact on Locksmithing Practices

Predictive maintenance is highly relevant to someone interested in locksmithing because it illustrates how the profession is evolving from reactive repair work to proactive security system management. Traditionally, locksmiths were called when a lock failed, became difficult to operate, or required replacement. With modern electronic and smart lock systems, however, operational data can be monitored continuously to detect early signs of wear or malfunction. Understanding predictive maintenance allows locksmiths to identify potential problems before they lead to lock failure, security breaches, or access disruptions. This capability is particularly important in environments such as hospitals, airports, office complexes, and data centres, where a malfunctioning lock can create operational disruptions or safety risks. For individuals entering the locksmith profession, knowledge of predictive maintenance systems helps them provide higher-value services that focus on reliability, system performance, and long-term security planning rather than simply responding to faults after they occur.

The influence of predictive maintenance on locksmithing practices is significant because it expands the technical scope of the profession. Locksmiths increasingly work with network-connected access control systems, smart locks, and monitoring platforms that generate operational data about lock usage, battery condition, actuator performance, and door alignment. Instead of only installing and repairing locks, locksmiths may now interpret system alerts, analyse performance data, and perform targeted maintenance based on predictive analytics. This requires new competencies in electronic security systems, software platforms, and networked infrastructure alongside traditional mechanical skills. As buildings become more technologically integrated and security systems become more intelligent, predictive maintenance enables locksmiths to act as security system specialists who manage and optimise access control infrastructure, helping clients maintain reliable and efficient locking systems while reducing unexpected failures and maintenance costs.

## Automated Threat Detection

Automated threat detection encompasses the deployment of advanced technologies such as artificial intelligence (AI), machine learning (ML), video analytics, and sensor integration to automatically identify potential security threats within surveillance, access control, and broader security ecosystems. Rather than relying solely on human operators to monitor cameras or review logs, automated threat detection systems analyse large volumes of data in real time to identify unusual behaviour, unauthorized access attempts, or suspicious activities, generating alerts or triggering automated responses to enhance speed and accuracy while reducing the workload on security personnel [392-394]. The integration of multimodal data streams. from video, access control logs, alarms, motion sensors, and building management systems, enables unified threat detection platforms that correlate disparate events and improve situational awareness, response times, and incident containment [395-397].

A central enabler of automated threat detection is AI-driven video analytics. Modern surveillance cameras equipped with computer vision algorithms can continuously analyse video feeds to identify objects, movements, and behaviours, distinguishing between people, vehicles, and animals; detecting entry into restricted areas; and recognizing behaviours such as loitering near secured entrances [395, 398]. Beyond simple event detection, these systems can identify unusual activities such as abandoned objects, tailgating through access-controlled doors, or attempts to force open doors, and they can automatically alert security personnel or activate countermeasures when threat criteria are met [395, 398]. The literature also emphasizes the importance of explainability and robust performance under real-world conditions, given the wide deployment contexts from corporate campuses to critical infrastructure [395].

Automated threat detection is widely utilized in electronic access control systems, which generate detailed logs of credential usage, including door openings, failed authentication attempts, and unusual patterns. Automated analysis of these logs can reveal suspicious behaviour such as repeated failed attempts to access restricted doors, credentials used at two locations simultaneously, or access attempts outside normal hours; upon detection, the system can issue alerts, temporarily suspend credentials, or notify administrators for investigation [396]. This stream of activity naturally aligns with research on anomaly detection in dynamic access environments, where ML models learn normal usage patterns and identify deviations indicative of compromise or misuse [396]. The combination of access-event analytics with video and sensor data strengthens threat assessment by providing corroborating evidence for a given anomaly [396].

Behavioural analytics extend threat detection from single-event signalling to patterns over time. Systems analyse movement and activity trajectories to determine whether an activity deviates from established norms, such as an employee who consistently accesses a building during daytime hours attempting entry late at night, which may trigger suspicion [396, 399]. Additionally, tracking movement across multiple cameras can help identify individuals who show suspicious patterns, such as surveilling behaviour or repeated proximity to restricted doors [396]. This longitudinal perspective complements instantaneous detection by enabling context-rich risk scoring and more accurate threat classification [396, 399].

Automated threat detection systems frequently integrate multiple security technologies into a unified platform, where video surveillance, access control, alarms, motion sensors, and building management data feed a central security management system. AI algorithms analyse the combined data streams to identify threats more accurately than when systems operate in isolation. For example, correlating a door-forced alarm with concurrent near-door surveillance can classify the situation as a likely intrusion attempt, enabling faster and more reliable responses [398]. Such integrated platforms support automated responses, including door locking, alarm activation, staff notification, and evidence capture for investigation, with dynamic policy adjustments such as additional authentication requirements or area restrictions during incidents [398]. The convergence of data sources and automated decision-making is particularly valuable in large facilities (airports, hospitals, transportation hubs, corporate campuses, data centres, and government installations) where manual monitoring would be infeasible or inefficient [398, 400].

Automated threat detection systems often incorporate response mechanisms that are triggered automatically when threats are detected. These responses can include locking doors, sounding alarms, notifying security personnel, recording surveillance footage, or adjusting access controls (e.g., temporary credential suspension or area-wide access restrictions). Some systems even adjust security policies dynamically in response to risk indicators, thereby enhancing containment and reducing exposure to evolving threats [400]. The capability to enact rapid, automated responses is an important practical benefit in safeguarding high-risk facilities and large-scale deployments [398].

The value proposition of automated threat detection is particularly pronounced in large or complex facilities where manual monitoring would be impractical. Airports, hospitals, transportation hubs, corporate campuses, data centres, and government facilities frequently deploy these systems to monitor thousands of cameras and access points simultaneously, enabling continuous monitoring and rapid detection of security incidents. This supports enhanced situational awareness, faster incident response, and improved resilience in critical infrastructure contexts [394, 398, 400]. The literature also discusses AI-enabled Zero Trust and adaptive IAM as complementary approaches, underscoring the role of behavioural analytics in continuous verification and risk-based access decisions within dynamic security ecosystems [396, 397].

Some commercially available examples of automated threat detection systems used in modern surveillance and access control environments follow. These systems illustrate how artificial intelligence, video analytics, and integrated security platforms are used in real-world facilities such as airports, corporate campuses, retail environments, and critical infrastructure.

**1. Verkada AI Security Platform**

One prominent commercial example of automated threat detection is the Verkada AI-powered security platform, which integrates cloud-managed surveillance cameras, access control devices, sensors, and environmental monitoring systems.

Verkada cameras use AI-based computer vision analytics to automatically detect people, vehicles, license plates, and unusual behaviour within monitored environments. The system continuously analyses video streams and can trigger alerts when events such as loitering, unauthorized entry into restricted areas, or suspicious movement patterns are detected. Verkada's software platform also correlates data from multiple sources, including door access logs and camera footage, enabling security teams to quickly identify incidents such as tailgating through access-controlled doors or forced entry events.

Another capability of Verkada's platform is Person of Interest (POI) detection, where facial recognition can identify individuals on a watchlist and notify security personnel when they appear within camera coverage. The platform can automatically link surveillance footage with access control records, allowing investigators to reconstruct security incidents rapidly. Verkada systems are widely deployed in schools, corporate campuses, healthcare facilities, and retail chains, demonstrating how automated threat detection improves situational awareness in large security environments.

**2. Avigilon (Motorola Solutions) AI Video Analytics**

Avigilon, a Motorola Solutions company, provides an enterprise-grade surveillance and access control platform incorporating AI-powered video analytics and automated threat detection.

Avigilon cameras use deep learning algorithms to perform object classification, behaviour analysis, and unusual activity detection. The system can detect events such as perimeter breaches, unauthorized entry, crowd formation, or suspicious loitering. When an anomaly is detected, the system automatically alerts security staff and can trigger predefined responses such as recording video, activating alarms, or locking doors.

A key feature of the platform is Appearance Search™, which allows security personnel to locate individuals across a network of cameras based on characteristics such as clothing colour, gender, or object possession. In integrated deployments, Avigilon systems correlate surveillance analytics with access control systems, enabling automated alerts when suspicious activity is detected at controlled entry points. These systems are commonly used in airports, transportation networks, stadiums, and government facilities, where monitoring large areas manually would be impractical.

**3. Genetec Security Center with AI Analytics**

The Genetec Security Center platform is another widely used integrated security management system that combines video surveillance, access control, license plate recognition, and analytics into a single automated threat detection environment.

Genetec uses AI-based analytics to analyse video streams and detect security incidents such as unauthorized perimeter access, suspicious vehicle behaviour, or abnormal movement patterns within facilities. The platform can correlate events from multiple systems—for example, linking a forced-door alarm with camera footage and access logs to determine whether an intrusion attempt is occurring.

The platform also includes AutoVu license plate recognition, which automatically identifies vehicles entering restricted areas and checks them against watchlists or authorized vehicle databases. This capability is commonly used in airports, logistics hubs, city surveillance networks, and critical infrastructure facilities. Automated threat detection within Genetec systems allows rapid identification of security risks while reducing the burden on human operators.

**4. Hikvision DeepinMind AI Cameras**

The Hikvision DeepinMind series provides another example of AI-driven threat detection within surveillance systems. These cameras use deep learning algorithms embedded directly within the camera hardware to analyse video streams in real time.

DeepinMind cameras can detect and classify people, vehicles, faces, and behaviours, enabling automated alerts for events such as perimeter intrusion, unattended objects, loitering, or unauthorized access attempts. The system can also integrate with facial recognition terminals and electronic door controllers to create automated access control responses.

For example, if an individual attempts to access a secured area without authorization, the system can automatically generate alerts, record the event, and activate additional security measures. These systems are widely used in transport hubs, commercial buildings, and industrial facilities, where automated monitoring improves detection of suspicious activity.

**5. Honeywell Pro-Watch Integrated Security System**

The Honeywell Pro-Watch platform is a large-scale security management system used in airports, government facilities, and enterprise environments. Pro-Watch integrates video surveillance, access control, intrusion detection, and identity management into a unified platform capable of automated threat detection.

The system uses analytics to identify anomalies such as repeated failed credential attempts, unauthorized door access, or unusual movement patterns captured by cameras. Pro-Watch can correlate events from multiple systems to classify potential security incidents and automatically trigger responses such as door lockdowns, alarm activation, or alerts to security staff.

Honeywell systems are often deployed in critical infrastructure and high-security environments, where rapid automated threat detection and response are essential.

For locksmiths and security professionals, automated threat detection represents a growing area of integration with access control systems and building security infrastructure. Modern locksmithing increasingly intersects with AI-driven surveillance, cybersecurity, and integrated security management, as intelligent platforms monitor events, detect anomalies, and coordinate responses across electronic locks, readers, and door hardware connected to unified security ecosystems. Understanding automated threat detection mechanisms informs the design and configuration of access control installations that support advanced monitoring capabilities, and helps professionals implement interoperable solutions that leverage data from cameras, sensors, and management platforms [396, 401].

The surveyed literature situates automated threat detection within broader AI-driven cybersecurity trends, including anomaly detection, UEBA, and predictive analytics, and notes that AI-enabled threat detection is increasingly deployed beyond traditional IT networks to cloud, IoT, and core infrastructure contexts. Several works discuss applications in cloud security, network security, and critical infrastructure, pointing to trends such as SOAR integration, adaptive policy enforcement, and post-quantum considerations as part of a forward-looking security architecture. Across these domains, the core message is that AI-enabled threat detection improves speed and accuracy, supports real-time responses, and reduces manual workload, provided that instruments for privacy, explainability, and governance are adequately addressed [392, 393, 395-397, 402-405].

Automated threat detection, underpinned by AI-based video analytics, ML-driven anomaly detection, and integrated security architectures, offers substantial improvements in the speed, accuracy, and scalability of threat identification and response within surveillance and access control systems. By analysing multimodal data in real time, identifying patterns that signal risk, and automating alerts and responses, these systems alleviate operator workload and bolster situational awareness in diverse environments—from corporate campuses to critical infrastructure. The effectiveness of these systems, however, depends on careful attention to privacy, bias, governance, and lifecycle management, ensuring that automated capabilities augment human operators rather than supplant essential oversight. Ongoing advances in behavioural analytics, adaptive IAM, and AI-enabled Zero Trust architectures will continue to influence the design and deployment of automated threat detection in increasingly complex security landscapes [392-399].

**Impact on Locksmithing Practices**

Automated threat detection is highly relevant to someone interested in locksmithing because it demonstrates how modern security systems are evolving beyond traditional mechanical protection toward intelligent, integrated security environments. Historically, locksmiths focused primarily on installing and maintaining physical locks and keys, with security decisions largely based on mechanical barriers. However, contemporary access control systems increasingly operate within larger digital ecosystems that include AI-enabled surveillance cameras, electronic locks, credential readers, sensors, and centralized management platforms. Automated threat detection systems analyse data from these components to identify suspicious behaviour, unauthorized access attempts, or anomalies in credential usage. Understanding how these systems operate enables locksmiths to design and install access control solutions that integrate effectively with surveillance systems, alarm networks, and building security platforms, ensuring that physical entry points function correctly within the broader automated security infrastructure.

The influence of automated threat detection on locksmithing practices is significant because it expands the professional role of locksmiths from mechanical technicians to integrated security specialists.

| |
|---|
| Locksmiths working with modern electronic access systems must understand how door hardware, electronic locks, and credential readers interact with surveillance systems, analytics platforms, and automated response mechanisms. For example, an access-controlled door may be linked to video analytics capable of detecting tailgating, forced entry attempts, or unusual access patterns. In such systems, locksmiths may be involved in configuring door hardware to support security policies, integrating locks with centralized monitoring platforms, and ensuring that entry points operate correctly during automated security responses such as lockdowns or alarm events. As automated threat detection technologies continue to develop, locksmiths increasingly require knowledge of electronic security systems, data integration, and cybersecurity principles to effectively support modern intelligent access control environments. |

## Ethical Concerns of AI-Driven Security

A core ethical concern is privacy erosion in AI-enabled security environments. Systems that monitor individuals through video, facial recognition, behaviour tracking, and data logging collect extensive personal data such as images, biometric templates, access records, and movement patterns. While such data can enhance threat detection and incident reconstruction, it also raises risks of excessive surveillance, data misuse, improper storage, and unauthorized access. Several studies frame privacy as a primary risk when AI-driven security scales within buildings, campuses, and public spaces, emphasizing the need for privacy-preserving design and governance to mitigate surveillance overreach [406-409]. In addition, scholarly work on AI-enabled surveillance stresses that citizens' awareness of continuous data analysis is often limited, underscoring the importance of transparent data governance and rights-respecting deployment practices [409, 410].

Algorithmic bias and fairness Bias and fairness in AI-based security decisions, such as facial recognition, identity verification, or behavioural analysis, are repeatedly identified as ethical challenges. Training data biases and insufficient demographic representation can yield disparate performance, with higher error rates for certain groups potentially leading to wrongful denial of access or intensified security scrutiny for specific populations. Addressing these biases requires careful dataset selection, rigorous testing, ongoing monitoring, and transparent reporting of performance across demographics to ensure fair outcomes [406, 408, 411, 412]. While some literature documents historical disparities in facial recognition accuracy, there is broad consensus on the necessity for fairness audits and bias mitigation throughout system lifecycles [406, 408, 411].

Transparency and explainability are repeatedly identified as critical ethical dimensions, given the opaque nature of many AI models used in security decisions. The lack of understandable explanations for automated access denial, alarm triggering, or suspicion labelling challenges accountability and the ability of affected individuals to challenge outcomes. Several sources advocate for mechanisms that provide intelligible explanations and enable human oversight, aligning with the broader call for accountable AI deployment in security contexts [406, 409, 411, 412]. The literature also discusses the ambiguity of responsibility when AI-driven security decisions cause harm, highlighting the need for clear accountability frameworks that allocate responsibility among developers, vendors, operators, and configurators [406, 409, 411, 412].

The ethical management of sensitive security data, biometric templates, facial images, access logs, and behavioural analytics, entails robust data protection measures. Data breaches or mismanagement can have severe consequences because biometric data is typically non-replaceable. Strong encryption, access controls, data governance, and compliance with privacy regulations are repeatedly recommended to mitigate these risks and uphold privacy and security objectives simultaneously [407, 409]. Some literature also emphasizes privacy-preserving AI techniques (e.g., differential privacy, federated learning) as practical means to balance threat detection with privacy protection [407, 409, 412].

Over-automation and reduced human oversight present ethical concerns, as fully autonomous security decisions may lead to unintended harms if systems err or operate outside intended conditions. A common design principle recommended is a human-in-the-loop approach where AI assists rather than replaces human judgment, thereby preserving human accountability and contextual understanding in high-stakes security settings [406-408, 412].

Societal and regulatory dimensions Beyond technical issues, the ethical discourse addresses the balance between safety and civil liberties at a societal level. The deployment of AI surveillance in public spaces, workplaces, and transport raises concerns about mass surveillance, potential power abuses, and long-term impacts on individual freedoms. These broader concerns motivate calls for ethical frameworks, regulatory oversight, and transparent governance to ensure responsible AI use in security technologies [406, 409, 411]. Some analyses discuss governance at organizational and policy levels, arguing for comprehensive regulatory regimes and international norms to guide responsible AI deployment in security and counterterrorism contexts [409, 413].

**Impact on Locksmithing Practices**

For security practitioners such as locksmiths integrating with electronic access control, biometric authentication, and AI-enabled platforms, the ethical implications are particularly salient. The collection and processing of sensitive identity data necessitate privacy-by-design principles, robust security controls, and adherence to regulatory requirements while delivering reliable access control solutions [406, 408, 412]. This perspective aligns with broader calls for responsible deployment that respects rights and maintains effective physical security.

## Chapter 18 Review Questions

### Question 1

A large airport installs an AI-powered surveillance system that automatically analyses CCTV footage and alerts security staff when someone enters a restricted area or leaves a suspicious package unattended.

Which AI capability enables this functionality?

A. Computer vision and automated video analytics
B. Manual camera monitoring
C. Mechanical lock control
D. Traditional motion detection sensors

**Question 2**

A corporate office installs facial recognition terminals at entry gates. Employees simply walk toward the entrance and the system compares their facial features with stored biometric templates to grant access.

What type of authentication system is being used?

A. PIN-based authentication
B. RFID credential authentication
C. Biometric facial recognition authentication
D. Mechanical key authentication

**Question 3**

A security platform analyses access logs and identifies that an employee who normally enters the building during daytime hours is attempting access late at night from an unfamiliar location.

What type of AI capability is detecting this behaviour?

A. Behavioural analysis and anomaly detection
B. Static credential verification
C. Physical lock inspection
D. Manual security review

**Question 4**

A hospital uses a smart access control system that automatically locks doors, activates alarms, and notifies security personnel when multiple failed access attempts occur at a restricted door.

What feature of AI-enabled security systems does this demonstrate?

A. Automated incident detection and response
B. Manual alarm activation
C. Lock cylinder replacement
D. Static access logging

**Question 5**

A building security system tracks the performance of electronic locks and sends an alert when a lock motor begins requiring more power to operate, indicating possible mechanical wear.

What maintenance strategy does this represent?

A. Reactive maintenance
B. Preventive maintenance
C. Predictive maintenance
D. Emergency repair

**Question 6**

A data centre installs sensors in doors that monitor the number of lock cycles and detect door misalignment. The system schedules maintenance before the lock fails.

What advantage does predictive maintenance provide in this scenario?

A. Eliminates all maintenance costs
B. Prevents unexpected equipment failure
C. Removes the need for locksmiths
D. Reduces door usage

**Question 7**

A large corporate campus installs an AI security platform that combines surveillance cameras, motion sensors, and access control logs into a single system that detects suspicious behaviour automatically.

What is the primary benefit of integrating multiple data sources in automated threat detection?

A. Simplifies key cutting procedures
B. Improves situational awareness and threat detection accuracy
C. Eliminates electronic locks
D. Reduces the number of cameras required

**Question 8**

A security system detects that a person repeatedly attempts to access multiple restricted doors using different credentials. The system automatically disables the credential and alerts administrators.

Which technology enables this automated detection?

A. Mechanical key control
B. Machine-learning anomaly detection
C. Traditional door alarms
D. Manual security review

**Question 9**

A smart building installs AI-powered cameras that can search recorded footage for a person wearing a red jacket across hundreds of cameras within seconds.

What AI capability enables this search?

A. Automated forensic video analytics
B. Mechanical surveillance monitoring
C. Static video recording
D. Manual footage review

**Question 10**

A company deploys an AI-based facial recognition system but later discovers that the system incorrectly denies access more frequently for certain demographic groups.

What ethical issue does this scenario illustrate?

A. Hardware malfunction
B. Algorithmic bias in AI systems
C. Mechanical lock failure
D. Encryption vulnerability

# Chapter 19

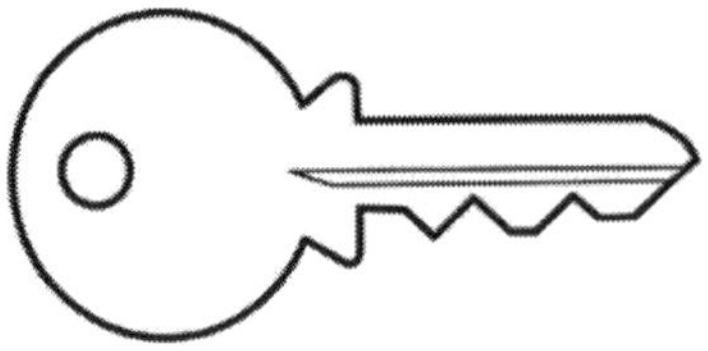

# Designing the Next Generation of Locks

Modern locking systems must address challenges that traditional locks were never designed to solve. Buildings now contain interconnected security technologies including surveillance systems, access control software, mobile credentials, and building automation platforms. Users expect seamless access through smartphones, biometric authentication, or digital credentials while still maintaining strong protection against unauthorized entry. At the same time, organizations must manage large numbers of users, monitor access events in real time, and adapt security permissions as operational needs change. As a result, the design of locking systems increasingly involves software, electronics, materials science, and human-factors engineering alongside traditional mechanical security design.

Another major influence on the next generation of locks is the growing recognition that security must work in harmony with human behaviour. Systems that are difficult to use, confusing to operate, or inconvenient in everyday situations often lead users to bypass security procedures entirely. Doors are propped open, codes are shared, and access credentials are mismanaged when security systems fail to account for the needs and behaviours of real users. Modern lock design therefore places increasing emphasis on usability, accessibility, and adaptability, ensuring that security systems are both effective and practical in real-world environments.

Sustainability and lifecycle considerations are also reshaping the development of locking hardware. Manufacturers are exploring new materials, modular architectures, and environmentally responsible production methods that reduce waste while maintaining durability and security performance. At the same time, rapid advances in digital technologies mean that locks must now be designed to evolve over time. Systems must be upgradeable, interoperable with other security technologies, and capable of adapting to new authentication methods without requiring complete replacement.

These changes mark a significant shift in how locks are conceived, manufactured, and deployed. Rather than being isolated mechanical devices, modern locks are increasingly part of integrated security infrastructures that combine physical hardware, digital networks, and intelligent management systems.

This chapter explores the key principles shaping the next generation of lock design, including user-centred security systems, sustainable materials and manufacturing innovations, modular and upgradeable lock architectures, and the evolving balance between open and proprietary security ecosystems. Understanding these developments is essential for locksmiths, engineers, security professionals, and designers who must navigate a rapidly changing landscape in which the lock is no longer just a piece of hardware, but a dynamic component of modern security infrastructure.

Together, these concepts illustrate how the future of locking technology is being defined not only by stronger materials and more complex mechanisms, but by the integration of technology, usability, sustainability, and system interoperability.

## User-Centred Lock Design

User-centred lock design refers to the approach of designing locking systems and access control technologies with a primary focus on the needs, abilities, and behaviours of the people who will use them. Rather than concentrating solely on technical security features such as resistance to picking or forced entry, user-centred design considers how individuals interact with locks in everyday situations. The goal is to create locking systems that are secure, intuitive, accessible, and convenient to operate while still maintaining strong protection against unauthorized access. In modern security engineering, user-centred design has become increasingly important as locking technologies expand beyond mechanical keys to include electronic credentials, mobile devices, biometric systems, and integrated access control platforms.

User-centred lock design appeals to a wide range of clients because it focuses on making security systems secure, practical, and easy to use in real-world environments. The following groups are particularly likely to benefit from or demand user-centred locking solutions.

**Residential Homeowners:** Homeowners represent one of the largest client groups for user-centred lock design. Many residents want security systems that are simple to operate, convenient, and adaptable to everyday life. Smart locks, mobile credentials, and keypad systems allow homeowners to avoid carrying physical keys while still maintaining strong security.

User-centred features such as remote locking, temporary digital guest keys, and clear lock status notifications are particularly attractive in residential environments. For example, a homeowner may provide temporary access to family members, cleaners, or delivery personnel without distributing physical keys. Locksmiths who understand user-centred design can recommend systems that balance convenience and safety for households.

**Property Managers and Apartment Complexes:** Managers of multi-tenant residential buildings are another major client group for user-centred lock systems. Apartment complexes often require flexible access control for residents, guests, maintenance staff, and service providers.

User-centred access systems allow administrators to:

- Issue and revoke digital credentials remotely
- Provide temporary access for contractors or guests
- Manage large numbers of users without physical key management
- Monitor access events across the building

These capabilities reduce operational complexity while improving security for residents.

**Commercial Offices and Corporate Facilities:** Businesses and corporate offices benefit significantly from user-centred access control because employees need secure but convenient entry to buildings and workspaces. In large organizations, traditional key management can be inefficient and difficult to control.

User-centred systems allow companies to implement:

- Keycard or smartphone access credentials
- Biometric authentication for secure areas
- Role-based access permissions
- Remote credential management for employees and contractors

These systems improve productivity while maintaining security policies across office environments.

**Healthcare Facilities:** Hospitals and healthcare facilities have particularly complex access requirements. These environments must balance rapid access for staff during emergencies with strict security for sensitive areas such as medication storage, laboratories, and patient records rooms.

User-centred lock design in healthcare often includes:

- Hands-free or biometric access for medical staff
- Quick emergency override mechanisms
- Controlled access to restricted clinical areas
- Systems that function reliably during high-stress situations

Lock systems must also be accessible to staff wearing gloves or carrying equipment.

**Elderly Care and Assisted Living Facilities:** Facilities that support elderly or mobility-impaired individuals require lock systems designed with accessibility and ease of use in mind. Traditional locks may be difficult for residents with reduced dexterity, arthritis, or visual impairment.

User-centred solutions may include:

- Large keypads or touchless entry systems
- Voice-assisted or mobile-based access
- Simplified door hardware such as lever handles
- Emergency access features for caregivers

These features allow residents to maintain independence while ensuring safety.

**Educational Institutions:** Schools, universities, and research campuses manage large populations of students, staff, and visitors. These environments benefit from flexible and scalable access control systems.

User-centred lock systems allow institutions to:

- Issue digital credentials to students and staff
- Restrict access to laboratories or administrative areas

- Monitor access to buildings after hours
- Rapidly update permissions when individuals leave the institution

These systems simplify access management across large campuses.

Hospitality Industry (Hotels and Short-Term Rentals): Hotels, resorts, and short-term rental providers require locking systems that are both secure and easy for guests to use. User-centred design in hospitality focuses on frictionless guest experiences.

Examples include:

- Mobile room keys on smartphones
- Temporary digital credentials that expire at checkout
- Keycard systems that are simple for guests to operate
- Integration with booking and property management systems

These features reduce front-desk workload while improving guest convenience.

**Industrial and Critical Infrastructure Facilities:** Industrial sites, manufacturing plants, data centres, and critical infrastructure facilities require high levels of security but must also allow workers to move efficiently between areas.

User-centred access systems in these environments may include:

- Multi-factor authentication for restricted zones
- Access credentials integrated with employee identification systems
- Systems that operate reliably in harsh environments
- Lock hardware designed for frequent or heavy use

This ensures security does not interfere with operational workflows.

**Government and High-Security Facilities:** Government buildings, defence installations, and research laboratories often require strict access control combined with reliable user interaction.

User-centred lock design in these settings may incorporate:

- Biometric authentication
- Multi-factor identity verification
- Secure audit trails of access events
- Systems that are intuitive for authorized personnel to operate under pressure

**Locksmith and Security Integration Clients:** Many clients seeking advanced locking systems rely on locksmiths and security consultants to design solutions that are both secure and usable. Locksmiths often serve as advisors to businesses, property developers, and facility managers.

In these cases, user-centred design helps locksmiths recommend solutions that:

- Reduce user frustration
- Prevent unsafe security workarounds
- Improve system reliability
- Align security technology with the daily behaviour of users

A fundamental principle of user-centred lock design is usability. Locks must be easy for authorized users to operate correctly while remaining difficult for unauthorized individuals to bypass. Poorly designed locking systems can create confusion, lead to improper use, or encourage users to adopt insecure behaviours. For example, if a door lock is difficult to operate, users may prop doors open, share access credentials, or bypass security controls entirely. User-centred design seeks to prevent these behaviours by ensuring that locks function in a straightforward and reliable manner. This may involve intuitive interfaces, clear feedback mechanisms such as audible clicks or indicator lights, and simplified authentication processes.

Another important aspect of user-centred lock design is accessibility and inclusivity. Lock systems must accommodate a wide range of users, including individuals with disabilities, limited mobility, reduced hand strength, or visual impairments. Traditional mechanical locks may be difficult for some individuals to operate due to small keyways or stiff mechanisms. Modern design approaches therefore incorporate features such as lever handles instead of round knobs, large and clearly marked keypads, voice prompts, or smartphone-based access systems that reduce the need for fine motor control. Compliance with accessibility standards in buildings also influences lock design, particularly in public spaces where all users must be able to enter and exit safely.

User-centred lock design also considers context and environment. Locks are used in many different situations, including residential homes, commercial buildings, hospitals, transportation hubs, and industrial facilities. Each environment presents different operational requirements. For example, locks used in healthcare facilities may need to support rapid emergency access while still protecting restricted areas such as medication storage rooms. Locks used in apartment buildings may require flexible credential management for residents, guests, and maintenance staff. Understanding how users interact with doors and security systems within specific environments allows designers to tailor locking solutions to real-world operational needs.

Another key element is error tolerance and fail-safe operation. Users may occasionally make mistakes when interacting with locks, such as entering incorrect PIN codes or attempting access with an expired credential. User-centred systems are designed to handle these situations without causing unnecessary frustration or security risks. For example, electronic locks may provide feedback when a credential is invalid, allow a limited number of retry attempts, or offer alternative authentication methods. In emergency situations, such as fires or evacuations, locks must also allow occupants to exit quickly without requiring specialized knowledge or credentials. Designing locks that function safely during emergencies is an essential component of user-centred security design.

Modern user-centred lock design also integrates convenience and flexibility, particularly in digital access control systems. Technologies such as smartphone credentials, biometric authentication, and temporary digital keys allow users to access secured spaces without carrying physical keys. For example, a homeowner may grant temporary digital access to a guest or service provider, or an office administrator may adjust employee access permissions remotely. These features improve convenience while maintaining security, demonstrating how user-centred design balances usability with protection.

Security feedback and situational awareness are also important considerations. Users should be able to clearly understand the status of a lock, whether it is locked, unlocked, or malfunctioning. Visual indicators, audible alerts, and mobile notifications can provide immediate feedback, helping users interact confidently with the system. For instance, many smart locks display LED indicators or send smartphone notifications confirming that a door has been successfully locked. This feedback reduces uncertainty and encourages proper use of security systems.

User-centred lock design also takes into account human behaviour and security psychology. People often prioritise convenience over strict security practices. If security systems are overly complex or inconvenient, users may find ways to bypass them. For example, employees may share access codes or leave doors unlocked to avoid repeated authentication. By understanding typical user behaviour, designers can create security systems that encourage secure habits while minimizing frustration. This approach recognizes that effective security depends not only on technical strength but also on how people interact with security technologies.

User-centred lock design is a structured process that places the needs, behaviours, and limitations of real users at the centre of the design and implementation of locking systems. Instead of beginning purely with technical specifications such as resistance to attack or mechanical complexity, designers start by understanding how people interact with doors, access control systems, and security technologies in everyday environments. The objective is to ensure that the resulting lock system is not only secure but also intuitive, reliable, and practical for those who use it. Implementing user-centred lock design involves several systematic stages.

The first step is understanding the users and their environment. Designers must identify who will interact with the lock system and how they will use it. This may involve homeowners, office workers, healthcare staff, maintenance personnel, visitors, or emergency responders. Each group may have different requirements. For example, hospital staff may need rapid access to certain areas, while visitors may require restricted and temporary access. Designers gather information through observation, interviews, surveys, and analysis of existing security procedures. Understanding how doors are used, how frequently they are accessed, and what problems users encounter with existing locks provides critical insight for designing improved systems.

The next stage involves defining functional and usability requirements. Based on the user research, designers establish what the lock system must accomplish. These requirements may include ease of operation, accessibility for people with disabilities, rapid emergency egress, secure authentication methods, and clear user feedback. For example, a user-centred design might require that a lock provide visual or audible confirmation when it engages, or that it allow entry using multiple authentication methods such as keys, PIN codes, or mobile credentials. Accessibility considerations may require lever handles instead of round knobs, large keypad buttons, or smartphone access for individuals with limited hand strength.

Once requirements are defined, designers proceed to concept development and system design. At this stage, potential solutions are developed that address both security and usability needs. Designers may create mechanical, electronic, or hybrid locking systems depending on the environment. The design must balance convenience with protection against unauthorized access. For instance, a system may include biometric authentication for secure areas while also maintaining mechanical override capabilities for emergencies. Designers also consider how the lock integrates with broader systems such as alarms, surveillance cameras, and building management platforms.

The fourth stage involves prototyping and usability testing. Prototype lock systems or access control interfaces are tested with representative users in realistic environments. Observing how individuals interact with the system reveals potential usability problems. For example, users may struggle to read keypad displays in low lighting or may not receive clear confirmation when the door locks. Testing allows designers to identify these issues early and refine the system before full deployment. Iterative testing and refinement are key components of user-centred design, ensuring that the system works reliably for real users rather than only under ideal laboratory conditions.

Another important aspect of user-centred lock design is error tolerance and safety planning. Designers must anticipate how users may make mistakes and ensure that the system handles these errors safely. For example, a user may forget a PIN code, lose a mobile credential, or accidentally lock themselves out. A well-designed system provides recovery mechanisms such as secondary authentication methods, administrative override options, or emergency unlocking procedures. Safety features are also critical, particularly in public buildings where doors must allow occupants to exit quickly during emergencies.

The design process must also consider human behaviour and security practices. Many security failures occur because users find systems inconvenient or confusing and attempt to bypass them. For example, employees may share access codes or prop doors open if authentication procedures are too cumbersome. User-centred lock design seeks to minimise these behaviours by making secure actions easier than insecure alternatives. If entering a secure area is quick and intuitive, users are more likely to follow proper security procedures.

Another step in implementing user-centred lock design is integrating technology in ways that simplify interaction. Modern locks may incorporate smartphone credentials, biometric recognition, or proximity authentication. These technologies can reduce the cognitive effort required to operate locks while maintaining strong security. For example, a smartphone-based credential system allows administrators to issue and revoke access permissions remotely while allowing users to unlock doors without carrying physical keys.

After deployment, designers and security professionals must perform ongoing monitoring and improvement. User-centred design is not a one-time process but an ongoing cycle. Feedback from users, operational data from access control systems, and incident reports can reveal areas where the system can be improved. For example, access logs may show frequent failed authentication attempts at certain doors, indicating usability issues or unclear instructions. Continuous improvement ensures that the lock system evolves alongside user needs and technological advancements.

Finally, successful user-centred lock design requires collaboration between multiple professionals, including locksmiths, security engineers, architects, IT specialists, and facility managers. Locksmiths contribute practical knowledge about lock hardware, installation constraints, and real-world security vulnerabilities. Architects and facility designers ensure that door hardware integrates with building layouts and safety regulations. IT professionals manage digital access credentials and networked security infrastructure. By combining expertise from these disciplines, designers can create lock systems that provide effective security while remaining easy and intuitive for users to operate.

**Impact on Locksmithing Practices**

For locksmiths and security professionals, user-centred design is becoming increasingly relevant as access control systems grow more complex. Locksmiths may be involved in selecting lock hardware, configuring electronic access systems, and advising clients on security solutions that balance usability with protection. Understanding user-centred principles helps locksmiths recommend locking systems that meet both security requirements and practical operational needs. As the next generation of locks continues to incorporate digital credentials, biometrics, and integrated building systems, user-centred design will play a crucial role in ensuring that security technologies remain both effective and practical for the people who rely on them.

## Sustainability and Materials Innovation

Sustainability and materials innovation are becoming important considerations in the design of next-generation locking systems. Traditionally, locks have been manufactured using durable but resource-intensive materials such as brass, steel, and zinc alloys, with design priorities focused primarily on strength, durability, and resistance to physical attack. While these properties remain essential, modern product design increasingly considers the environmental impact of manufacturing processes, material sourcing, product lifespan, and end-of-life disposal. As global industries move toward more sustainable manufacturing practices, lock designers and security hardware manufacturers are exploring new materials, production methods, and lifecycle strategies that reduce environmental impact while maintaining high security standards.

One important area of innovation involves the use of recycled and environmentally responsible metals. Brass and steel remain common materials for lock bodies, cylinders, and internal components because of their strength, corrosion resistance, and machinability. However, many manufacturers are now incorporating recycled metal alloys into production processes. Recycling metals significantly reduces energy consumption compared with producing new metal from raw ore. For example, recycled brass or aluminium can retain most of the mechanical properties of newly produced materials while reducing the environmental footprint of lock manufacturing. By increasing the proportion of recycled content in lock hardware, manufacturers can lower energy use, reduce mining impacts, and support circular material economies.

Another area of development involves advanced alloys and lightweight materials that improve durability while reducing resource consumption. Modern metallurgical techniques allow engineers to create alloys with improved strength-to-weight ratios, corrosion resistance, and wear characteristics. These materials can extend the service life of locks and door hardware, reducing the frequency with which components need to be replaced. In some cases, high-strength aluminium alloys or stainless steel components can provide sufficient strength while using less material overall. Reducing material mass without sacrificing performance contributes to more efficient manufacturing and transportation.

One widely used material in modern locks is stainless steel, particularly grades such as AISI 304 and AISI 316. Stainless steel is valued for its exceptional corrosion resistance, strength, and durability. Grade 304 stainless steel is commonly used in door hardware, lock bodies, and strike plates where resistance to rust and environmental exposure is important. Grade 316 stainless steel offers even greater resistance to corrosion because it contains molybdenum, making it particularly suitable for marine environments, coastal installations, and outdoor security hardware. Stainless steel components are often used in high-security padlocks, commercial door locks, and access control hardware because they resist both environmental degradation and physical attack.

Another important group of materials used in lock manufacturing is high-strength aluminium alloys, such as 6061 and 7075 aluminium. Aluminium alloys provide an excellent strength-to-weight ratio and are widely used in aerospace and structural applications. In lock systems, aluminium alloys may be used for housings, electronic lock enclosures, or structural components where weight reduction is beneficial. Alloy 6061 offers good corrosion resistance and machinability, while alloy 7075 provides significantly higher strength due to the presence of zinc as a primary alloying element. These materials allow designers to produce strong components that are lighter than steel, which is particularly useful in portable security devices or large electronic access systems.

Zinc alloys, particularly Zamak alloys, remain very common in lock manufacturing. Zamak is a family of zinc-based die-casting alloys containing aluminium, magnesium, and copper. These alloys are widely used for lock housings, knobs, handles, and other structural components because they can be produced efficiently through die casting. Zamak alloys provide good dimensional accuracy, corrosion resistance, and moderate mechanical strength while allowing manufacturers to create complex shapes with minimal machining. Although zinc alloys are not typically used for the most critical security components such as locking bolts, they remain essential materials in many lock assemblies due to their cost-effectiveness and manufacturing efficiency.

In higher-security lock systems, manufacturers often use hardened alloy steels for critical components such as locking bolts, shackles, and anti-drill inserts. Hardened steels may include chromium-molybdenum alloys or other heat-treated steels designed to resist cutting, drilling, and physical attack. These materials are commonly used in high-security padlocks, safes, and industrial locking mechanisms. For example, hardened boron steel or chrome-moly steel may be used in padlock shackles to resist bolt cutters and saw attacks. Heat treatment processes significantly increase hardness and wear resistance, improving the lock's ability to withstand mechanical tampering.

Another class of advanced materials used in lock systems includes titanium alloys, particularly in specialized high-performance or lightweight applications. Titanium alloys offer extremely high strength-to-weight ratios and excellent corrosion resistance. While titanium is more expensive than steel or aluminium, it may be used in high-end security products, marine hardware, or specialized locking devices where weight reduction and corrosion resistance are critical. Titanium components also exhibit strong resistance to fatigue and environmental degradation, making them suitable for long-term use in harsh conditions.

Modern lock designs also incorporate engineering polymers and reinforced composite materials in selected components. Materials such as glass-filled nylon, polycarbonate, and fibre-reinforced polymers are used for electronic lock housings, internal carriers, keypad assemblies, and other non-load-bearing components. Glass-filled nylon, for example, combines lightweight properties with improved stiffness and durability compared with standard plastics. These materials resist corrosion, reduce manufacturing weight, and can be moulded into complex shapes using efficient manufacturing processes. In electronic locks and smart lock systems, polymer housings protect sensitive electronics while providing durability and insulation.

Another innovative material used in some high-security locks is ceramic or carbide-based materials, particularly for anti-drill protection. Tungsten carbide inserts are often embedded within lock cylinders or protective plates to resist drilling attacks. Carbide is extremely hard and difficult to machine, which helps prevent attackers from drilling through the lock to manipulate internal components. These materials are typically used in combination with hardened steel components to create layered defence against mechanical attack.

Surface treatments and coatings also play a critical role in enhancing material performance. Physical vapour deposition (PVD) coatings, anodized aluminium finishes, and corrosion-resistant plating can significantly improve wear resistance and environmental durability. For example, anodizing aluminium creates a protective oxide layer that increases corrosion resistance, while PVD coatings provide highly durable finishes used in architectural hardware and premium lock products.

In addition to metal innovation, designers are exploring the use of engineering polymers and composite materials in non-critical lock components. High-performance polymers such as reinforced nylon, polycarbonate, or fibre-reinforced composites can replace metal parts in certain applications, including housings, internal carriers, and electronic lock enclosures. These materials are lighter, corrosion-resistant, and often easier to manufacture using moulding processes that produce less waste than traditional machining. While critical security components such as locking bolts or cylinders typically remain metal, selective use of advanced polymers can improve manufacturing efficiency and reduce environmental impact.

One commonly used material is glass-filled nylon, often referred to as glass-reinforced polyamide (PA-GF). Nylon by itself is a strong engineering polymer with good wear resistance and low friction. When reinforced with short glass fibres—typically 15–40 percent by weight—its strength, stiffness, and dimensional stability increase significantly. Glass-filled nylon is widely used for internal lock carriers, actuator mounts, electronic lock housings, and battery compartments in smart locks. It performs well under repeated mechanical loading, resists temperature changes, and maintains its shape under stress. Because it can be injection moulded into complex shapes, manufacturers can integrate multiple functions into a single component, reducing part count and assembly complexity.

Another important engineering polymer is polycarbonate (PC). Polycarbonate is known for its exceptional impact resistance and optical clarity, which makes it useful for protective covers, keypad faces, indicator windows, and electronic enclosures. It is significantly tougher than many other plastics and can withstand physical impacts without cracking. In electronic locks and access control readers, polycarbonate is often used to protect internal electronics from environmental damage while allowing LED indicators or displays to remain visible. Polycarbonate can also be blended with other polymers, such as ABS, to produce materials that combine strength, impact resistance, and improved moulding properties.

A widely used blended polymer in lock systems is ABS (acrylonitrile butadiene styrene) or PC-ABS blends. ABS offers good strength, toughness, and surface finish quality, making it suitable for housings, keypad panels, and trim components in residential smart locks and electronic access control devices. When combined with polycarbonate, the resulting PC-ABS blend provides improved heat resistance and impact strength while maintaining ease of moulding. These materials are commonly used in consumer-grade smart locks, card readers, and digital keypad assemblies where both durability and aesthetic appearance are important.

Another engineering polymer sometimes used in lock components is acetal, also known as polyoxymethylene (POM). Acetal has excellent dimensional stability, low friction, and high wear resistance. These properties make it suitable for moving internal components such as gear wheels, cams, bushings, and sliding mechanisms within electronic locks. Because acetal has very low moisture absorption and maintains precise dimensions over time, it is frequently used in precision mechanical parts where smooth movement and consistent performance are required.

In more demanding environments, designers may use polyamide-imide (PAI) or polyetheretherketone (PEEK), which are high-performance engineering polymers with exceptional strength, chemical resistance, and thermal stability. Although these materials are more expensive, they can be used in specialized locking systems exposed to extreme temperatures, harsh chemicals, or heavy wear. For example, industrial locking systems used in chemical processing plants or high-temperature environments may incorporate PEEK components for their durability and resistance to degradation.

Composite materials also play a role in lock design. Fibre-reinforced composites, such as carbon fibre reinforced polymers (CFRP) or glass fibre composites, can be used in structural housings or protective covers where both strength and light weight are desirable. While these materials are not typically used in small residential locks due to cost considerations, they may appear in specialized security equipment, high-end safes, or advanced electronic lock systems where durability and reduced weight are priorities.

Another class of materials used in modern lock systems is thermoplastic elastomers (TPE). These materials combine the elasticity of rubber with the processing advantages of plastics. TPE materials are often used for seals, gaskets, keypad overlays, and protective covers that improve weather resistance and user comfort. For example, outdoor smart locks may use TPE seals to prevent water ingress and protect internal electronics from moisture.

Surface treatments and additives further enhance the performance of polymer components. Many engineering polymers incorporate UV stabilizers, flame retardants, antimicrobial additives, or wear-resistant fillers depending on the intended environment. UV stabilizers help prevent degradation in outdoor installations, while flame-retardant grades may be required for electronic lock components used in commercial buildings. These additives ensure that polymer components maintain their performance over long service lives.

Sustainability in lock design also involves product longevity and durability. A lock that functions reliably for decades is inherently more sustainable than one that requires frequent replacement. High-quality materials, corrosion-resistant finishes, and robust internal mechanisms extend the operational life of locking systems, reducing material waste and maintenance costs. In environments such as commercial buildings, hotels, and industrial facilities, durable lock systems can remain operational through many years of heavy use. Designing locks that resist environmental degradation, mechanical wear, and tampering contributes to both sustainability and security performance.

Another important consideration is modular design and serviceability. Traditional locks often require full replacement when internal components fail or when access control requirements change. Next-generation designs increasingly incorporate modular components that can be repaired, upgraded, or replaced individually. For example, an electronic lock may allow replacement of the electronic control module without removing the entire mechanical assembly. Similarly, modular cylinder systems allow rekeying or cylinder replacement without replacing the entire lock body. This approach reduces waste, lowers maintenance costs, and extends the useful life of the hardware.

The growth of electronic and smart locking systems also raises sustainability considerations related to energy consumption and battery use. Designers are increasingly focusing on low-power electronics, efficient wireless communication protocols, and energy-harvesting technologies that reduce battery requirements. Some electronic locks incorporate energy-efficient motors, sleep modes, or near-field communication systems that minimize power consumption. In certain cases, locks may harvest small amounts of energy from door movement or user interaction to supplement battery power. Improving energy efficiency reduces maintenance requirements and environmental impact.

Manufacturing processes are also evolving to support sustainability goals. Modern production techniques such as precision casting, additive manufacturing, and automated machining can reduce material waste and improve manufacturing efficiency. Additive manufacturing, commonly known as 3D printing, allows designers to produce complex components with minimal material waste while enabling rapid prototyping and customization. These technologies allow manufacturers to optimize designs for both strength and material efficiency, supporting more sustainable production methods.

Another emerging area is environmentally responsible coatings and finishes. Locks often require protective coatings to resist corrosion, wear, and environmental exposure. Traditional finishing processes sometimes involve chemical treatments that can have environmental impacts. Newer coating technologies aim to reduce hazardous chemicals while maintaining durability. For example, powder coatings, advanced anodizing processes, and environmentally safer plating methods can provide corrosion protection while reducing environmental contamination during manufacturing.

A widely used environmentally responsible finishing method is powder coating. Powder coating involves applying a dry powder composed of polymer resins, pigments, and additives to a metal surface using electrostatic charge. The coated component is then heated in an oven where the powder melts and forms a continuous protective layer. Unlike traditional liquid paints, powder coatings contain little or no solvents, which significantly reduces emissions of volatile organic compounds. Powder coatings also produce minimal waste because unused powder can often be collected and reused. The resulting finish is highly durable, resistant to scratches, and capable of protecting lock hardware from corrosion and weather exposure, making it suitable for architectural door hardware and outdoor locks.

Another environmentally responsible finish commonly used on aluminium components is anodizing. Anodizing is an electrochemical process that converts the surface of aluminium into a thick, protective oxide layer. This oxide layer improves corrosion resistance, surface hardness, and wear resistance while allowing for decorative colouring. Because anodizing transforms the metal surface itself rather than applying a separate coating, it produces a durable finish that does not peel or flake. Modern anodizing processes are designed to reduce chemical waste and water consumption compared with older methods. Anodized finishes are often used on aluminium lock housings, electronic lock enclosures, and architectural door hardware where durability and appearance are important.

Physical Vapour Deposition (PVD) coatings represent another advanced and environmentally responsible finishing technology. In a PVD process, metal compounds such as titanium nitride or zirconium nitride are vaporized in a vacuum chamber and deposited as an extremely thin but hard coating on the surface of the lock component. PVD coatings are highly resistant to scratching, tarnishing, and corrosion, and they can produce decorative finishes such as gold, bronze, or black without the use of hazardous plating chemicals. Because the process occurs in a controlled vacuum environment and uses minimal chemical waste, PVD coatings are considered more environmentally friendly than many traditional electroplating techniques. PVD finishes are commonly used on high-end door hardware and luxury lock products.

Another environmentally improved finishing method is trivalent chromium plating, which has largely replaced older hexavalent chromium plating processes. Hexavalent chromium coatings were historically used to produce corrosion-resistant chrome finishes but are highly toxic and tightly regulated due to environmental and health concerns. Trivalent chromium plating provides similar corrosion resistance and appearance while using less hazardous chemicals. Many modern lock manufacturers have transitioned to trivalent chromium systems to comply with environmental regulations and reduce hazardous waste during production.

Some manufacturers also use electroless nickel plating, which provides uniform corrosion-resistant coatings without the need for electrical current during plating. Electroless nickel can produce a smooth, durable surface that protects lock components from wear and environmental damage. Modern plating systems are designed to minimize chemical waste and improve recycling of plating solutions, making them more environmentally responsible than older plating technologies.

In addition to metal-based coatings, manufacturers sometimes use eco-friendly organic coatings and water-based finishes. Water-based coatings replace solvent-heavy paints with formulations that significantly reduce emissions of volatile organic compounds. These finishes are often applied to decorative lock hardware, door handles, and architectural components where aesthetic appearance is important. Although water-based coatings may require more precise application conditions, they reduce air pollution and improve worker safety in manufacturing facilities.

Surface treatments may also include anti-corrosion conversion coatings, such as environmentally safer zinc or phosphate treatments that prepare metal surfaces for painting or powder coating. These treatments improve adhesion and corrosion resistance while using chemical formulations designed to minimize hazardous by-products.

For locksmiths and security professionals, sustainability and materials innovation are becoming increasingly relevant as clients seek environmentally responsible building technologies. Architects, facility managers, and government agencies are increasingly incorporating sustainability standards into building design and procurement processes. Locksmiths may therefore encounter lock systems that use recycled materials, modular components, or energy-efficient electronic technologies. Understanding these innovations enables locksmiths to recommend products that align with both security requirements and sustainability objectives.

Ultimately, sustainability and materials innovation represent an important dimension in the design of next-generation locks. By combining durable materials, efficient manufacturing processes, modular design strategies, and energy-efficient electronics, modern lock systems can reduce environmental impact while maintaining strong physical security. As security technologies continue to evolve alongside broader environmental priorities, sustainable design principles will play an increasingly significant role in shaping the future of locking hardware and access control systems.

## Impact on Locksmithing Practices

Sustainability and materials innovation are increasingly relevant to someone interested in locksmithing because they influence the types of locking hardware that professionals install, maintain, and recommend to clients. Historically, locksmithing focused primarily on mechanical performance, durability, and resistance to forced entry. However, modern construction projects, government procurement policies, and corporate sustainability initiatives increasingly require building components, including door hardware and access control systems, to meet environmental performance standards. As a result, locksmiths may encounter lock systems manufactured from recycled metals, lightweight alloys, advanced polymers, and environmentally responsible coatings. Understanding these materials allows locksmiths to select products that provide appropriate durability, corrosion resistance, and security performance while also meeting sustainability requirements in modern buildings.

Materials innovation also affects how locks are serviced, repaired, and maintained in professional practice. New materials such as stainless steel alloys, reinforced polymers, and advanced coatings can significantly improve the longevity of locking hardware, reducing corrosion, wear, and environmental degradation. This can extend maintenance intervals and reduce the need for frequent replacement of lock components. At the same time, modular design approaches, enabled by modern materials and manufacturing methods, allow locksmiths to repair or upgrade individual parts rather than replacing entire lock assemblies. For example, electronic lock modules, cylinders, or actuator components may be replaced independently, reducing material waste and maintenance costs. Locksmiths who understand the properties of these materials can diagnose wear patterns, select compatible replacement parts, and ensure that locking systems continue to operate reliably over long service lives.

The increasing use of electronic locks and smart access control systems further connects sustainability considerations with locksmithing practice. Energy-efficient electronics, low-power wireless communication technologies, and improved battery management systems are being incorporated into modern lock designs to reduce power consumption and maintenance requirements. Locksmiths who install and maintain these systems must understand how electronic components interact with mechanical locking mechanisms and how to service devices designed for long operational lifespans. In addition, environmentally responsible coatings such as powder coatings, anodized finishes, and PVD treatments influence the durability of lock hardware exposed to outdoor or high-traffic environments. Knowledge of these finishes helps locksmiths select appropriate products for different environmental conditions, such as coastal areas, industrial facilities, or high-use commercial buildings.

Finally, sustainability and materials innovation influence the advisory role of locksmiths when working with architects, builders, and facility managers. Many construction projects now incorporate sustainability certification frameworks or environmental design standards, which may encourage the use of long-lasting materials, recyclable components, and energy-efficient technologies. Locksmiths who understand modern materials and sustainable design principles can provide informed recommendations about locking hardware that balances environmental responsibility with high security performance. As the security industry continues to evolve alongside broader sustainability goals, locksmiths increasingly operate not only as technicians but also as security consultants who help clients select durable, efficient, and environmentally responsible locking systems for modern buildings.

## Modular and Upgradeable Lock Systems

Modular and upgradeable lock systems represent a forward-looking paradigm in security hardware that emphasizes interchangeability, serviceability, and adaptability. The shift from traditional, monolithic lock assemblies toward modular architectures is motivated by several practical and environmental considerations. First, modular designs enable targeted replacement rather than complete unit disposal when components wear, fail, or require functionality changes. This translates into reduced maintenance downtime, lower material waste, and extended service life for the core locking platform, since only the failed or outdated modules need replacement rather than the entire mechanism [414-416]. The ability to reconfigure locks by swapping modules, such as updating credentials readers, control electronics, or power systems, underpins ongoing alignment with evolving security requirements without forcing lock removal or full demolition in the field, a pattern echoed in broader modular design literature that links modularity to improved maintainability and lifecycle efficiency [416-418].

A canonical instance of modular design in locking systems is the interchangeable core cylinder, widely used in commercial and institutional environments. The cylinder module, containing the key pins and locking mechanism, can be rapidly removed and replaced with a control key, enabling quick rekeying or permission updates across a facility without dismounting the lock body itself. This modular cylinder approach materially reduces maintenance time and material waste because the lock body remains in place while only the core is exchanged [414-416]. Beyond mechanical cores, electronic and smart locks embody modularity through discrete, replaceable subassemblies: the latch or deadbolt assembly, the electronic control unit, credential readers, wireless transmitters, and power sources are designed as separate modules. When a component becomes obsolete or fails, it can be upgraded or swapped without replacing the whole unit, enabling transition from legacy credentialing methods (for example, RFID) to newer technologies such as mobile credentials or biometrics by upgrading only the reader module while preserving the core hardware [414, 415, 419, 420].

The modular unlock paradigm also supports future-proofing against rapid technological change in access control. As authentication technologies evolve—mobile credentials, biometrics, cloud-managed access—the modular approach allows organizations to adopt new methods by installing compatible modules within the existing system framework rather than deploying an entirely new hardware platform. This concept is corroborated by discussions of upgrading paths in modular lock ecosystems, which emphasize the ability to integrate new readers and authentication modules without wholesale hardware replacement [414, 417, 421, 422]. In practice, a building that begins with keypad-based access can upgrade to biometric or Bluetooth-based credentials simply by integrating updated reader modules into the same mechanical platform, preserving door preparation and mechanical standards while expanding security capabilities [414, 417, 419, 422].

Modularity in lock systems yields notable maintenance and serviceability advantages, especially in large facilities with extensive lock deployments. With modular designs, technicians can diagnose faults and perform component-level replacements without complete lock disassembly, shortening repair times and reducing labour costs. This is particularly advantageous in environments with high lock density, such as hospitals, universities, hotels, and large office campuses, where component failures are common and downtime carries substantial operational impact. The literature consistently notes that modularity enables targeted interventions, quicker fault isolation, and lower total maintenance expenditure by enabling module-level replacements rather than end-to-end lock refurbishment [414-416, 419].

Sustainability considerations strongly favour modular lock architectures. By enabling repair, refurbishment, and selective part replacement rather than scrapping entire locks, modular designs align with broader sustainability aims within building design and product manufacturing, including waste reduction and longer system lifespans. Several analyses of modular systems in related domains emphasize that longevity, reparability, and the capacity to upgrade components contribute to reduced material waste and better resource utilization over the product lifecycle, supporting circular economy objectives in security hardware [414, 417-419]. The environmental benefits are particularly salient when considering large-scale deployments where lifecycle waste from monolithic units can be substantial; modularity offers a practical path to decoupling durability from obsolescence [417, 418, 421].

Modularity also facilitates cross-environment adaptability, enabling lock platforms to span residential, commercial, industrial, and outdoor installations. A shared mechanical foundation with configurable authentication technologies or environmental protections allows manufacturers to produce families of products that can be tailored to specific settings without redesigning the entire system. For example, outdoor-ready modules can be weather-sealed or integrated with building management systems as needed, without altering the core functional architecture of the lock. The literature supports this adaptability as a key advantage of modular lock platforms, where common mechanical cores support diverse environmental and integration requirements through plug-in modules [414, 417, 419, 422, 423].

In terms of design and ecosystem implications, modularity has implications beyond individual products to maintenance strategy and lifecycle planning. Integrated modular design methodologies emphasize early inclusion of maintenance performance considerations, structure-based module identification, and optimization of module boundaries to minimize maintenance costs while maximizing availability and repairability. Several studies in modular design for maintenance-oriented systems show how decomposition into high-cohesion, low-coupling modules can improve maintenance effectiveness and facilitate rapid upgrades or replacements without compromising overall system performance. Although these studies focus on broader engineering contexts, their insights are transferable to lock systems, reinforcing the claim that module boundaries should align with maintenance workflows and anticipated upgrade paths to maximize uptime and minimize lifecycle costs [416, 417, 424].

**Impact on Locksmithing Practices**

For locksmiths and security professionals, modular and upgradeable lock systems significantly influence professional practice. Instead of replacing entire locks, locksmiths may focus on diagnosing individual modules, upgrading access technologies, or reconfiguring credential systems to meet new security requirements. This requires an understanding of both mechanical lock components and electronic access control technologies. Locksmiths may also assist clients in planning upgrades that extend the life of existing hardware while incorporating new authentication methods or management systems.

## Open Vs Proprietary Security Systems

Open versus proprietary security systems represent an important design consideration in the development of next-generation locking technologies. As locks increasingly integrate electronic components, network connectivity, and digital identity systems, decisions about whether a security platform should be open or proprietary significantly influence interoperability, security management, long-term maintainability, and customer flexibility. Understanding the distinction between these two approaches is essential for designers, manufacturers, and locksmiths involved in modern access control systems.

An open security system is designed using publicly documented standards, communication protocols, and interfaces that allow components from different manufacturers to work together. Open systems are typically based on widely adopted technical standards such as standard communication protocols, open APIs, or interoperable credential technologies. In an open lock ecosystem, a building's access control infrastructure may include locks, readers, controllers, and software platforms from different vendors that communicate through shared standards. This interoperability allows organizations to select hardware and software based on performance, cost, or functionality rather than being limited to a single manufacturer's product line.

A widely recognized example of an open access control standard is Open Supervised Device Protocol (OSDP). OSDP is an open communication protocol developed by the Security Industry Association (SIA) for communication between access control panels and peripheral devices such as card readers, biometric scanners, and keypads. Unlike older protocols, OSDP supports encrypted communication and secure device supervision, improving system security while allowing devices from different manufacturers to interoperate. Many modern electronic lock and reader manufacturers design their products to support OSDP, allowing system integrators to deploy equipment from multiple vendors within a single access control network.

Another important open standard used in modern security systems is ONVIF (Open Network Video Interface Forum). ONVIF is an international standard that allows video surveillance cameras, video management software, and network devices from different manufacturers to communicate with one another. While ONVIF primarily relates to video surveillance, it often integrates closely with access control systems. For example, a door access event recorded by an electronic lock can trigger a connected ONVIF-compatible camera to capture and store footage of the event. By using ONVIF-compliant devices, security systems can integrate surveillance and access control components without requiring equipment from a single manufacturer.

A well-known example of an open architecture security platform is Genetec Security Center. Genetec is a unified security platform that integrates access control, video surveillance, automatic license plate recognition, and intrusion detection systems. The platform supports numerous open communication standards and integrates hardware from many different manufacturers. This allows organizations to deploy electronic locks, card readers, cameras, and sensors from multiple vendors while managing them through a single software platform. Because the system supports open interfaces and software development kits (SDKs), organizations can also integrate the platform with building management systems or custom security applications.

Another example is HID Global's Seos credential technology, which is designed to support open and interoperable digital credential systems. Seos credentials can be stored on smart cards, smartphones, or wearable devices and used across different access control platforms. The system allows credentials to be securely managed through open standards, enabling integration with multiple access control systems and mobile credential platforms. This approach supports flexible deployment in environments such as corporate offices, universities, and government facilities where access control systems may evolve over time.

The KNX building automation standard also provides an example of an open system that can integrate with electronic locking and access control technologies. KNX is an open standard for building automation used worldwide in smart buildings. It allows devices such as lighting systems, HVAC controls, security sensors, and access control hardware to communicate over a shared network. Electronic locks integrated into KNX-based systems can communicate with other building automation devices, enabling coordinated security functions such as automatically locking doors during alarms or adjusting access permissions during emergencies.

Another widely used open security ecosystem involves Bluetooth Low Energy (BLE) mobile credential systems. Many modern smart locks support BLE communication standards that allow smartphones to act as access credentials. Because BLE is an open wireless communication protocol, compatible locks, readers, and mobile devices can communicate without relying on a single proprietary platform. This enables mobile credential systems that operate across different manufacturers' hardware and software environments.

In the smart home sector, Matter (formerly Project CHIP) represents a newer open standard designed to unify smart devices across different manufacturers. Matter aims to allow smart locks, lighting systems, sensors, and other smart home devices to operate seamlessly across multiple ecosystems such as Apple Home, Google Home, and Amazon Alexa. Although still evolving, Matter represents an effort to create a unified open architecture for connected devices, including smart locking systems.

One key advantage of open systems is flexibility and interoperability. Because open standards allow multiple manufacturers to produce compatible devices, organizations can integrate locks, credential readers, sensors, and management software from different suppliers. For example, a facility might use electronic locks from one manufacturer, card readers from another, and access management software from a third vendor, all connected through a common protocol. This approach reduces vendor lock-in and allows organizations to adapt their security systems as technology evolves without replacing the entire infrastructure.

Open security systems also support scalability and long-term adaptability. As buildings expand or security requirements change, new devices can often be added without major redesign. For example, a facility that initially uses keycard-based access control may later integrate biometric authentication or smartphone credentials if the underlying platform supports open communication standards. Because open systems are designed for compatibility, upgrades and expansions are typically easier to implement.

However, open systems can present security and integration challenges if not properly managed. Interoperability between devices from multiple manufacturers requires careful configuration and testing to ensure that security features function correctly across the system. Differences in firmware, software updates, or device capabilities may create compatibility issues if standards are implemented inconsistently. Security professionals must therefore ensure that open systems are configured securely and that software updates are maintained across all integrated devices.

In contrast, a proprietary security system is designed and controlled by a single manufacturer or vendor. Proprietary systems use communication protocols, software platforms, and hardware components that are unique to that manufacturer. In these systems, locks, readers, controllers, and management software are typically designed to function only within the vendor's own ecosystem. Access to system documentation, programming tools, or firmware updates may also be restricted to authorized partners or technicians.

One prominent example of a proprietary security system is HID Global's iCLASS access control platform. The iCLASS system uses proprietary smart card technology and encrypted communication protocols designed specifically by HID. The readers, cards, credential management software, and access controllers are intended to operate together within the HID ecosystem. Organizations that deploy iCLASS systems typically use HID readers and credential technologies across their facilities. The proprietary encryption and authentication mechanisms used in the system are designed to enhance security and prevent credential cloning, but the system is generally optimized for hardware produced by the same manufacturer.

Another example is Salto Systems' electronic access control platform, which is widely used in hotels, commercial buildings, and institutional facilities. Salto's technology combines electronic locks, access control management software, and proprietary communication methods. One of the key features of Salto's system is its Virtual Network technology, which allows electronic locks to communicate with a central management system through encoded data written to user keycards. Although the system integrates advanced features such as mobile access credentials and cloud management, many of its core technologies remain proprietary and are designed to operate within the Salto product ecosystem.

A widely recognized proprietary system in the hospitality industry is Assa Abloy's VingCard electronic lock system. VingCard systems are commonly used in hotels and resorts worldwide to manage guest room access using electronic keycards or mobile credentials. The locks, card encoders, management software, and communication infrastructure are designed to function together as a single integrated platform. Because of the specialized requirements of hospitality environments—such as automated check-in systems, guest credential expiration, and integration with property management software—many components of the VingCard platform are proprietary to ensure compatibility and reliability across the system.

Another example is Schlage's ENGAGE access control platform, developed by Allegion. The ENGAGE system integrates electronic locks, wireless communication devices, credential readers, and centralized management software. While the platform can integrate with certain third-party systems through approved interfaces, the core communication technologies, lock hardware, and system management tools are designed primarily for use within the Allegion product ecosystem. This approach allows Allegion to tightly control system functionality, firmware updates, and device compatibility.

A further example is Brivo's cloud-based access control system. Brivo provides a fully integrated platform that combines proprietary access controllers, cloud-based management software, credential readers, and mobile access credentials. Because the system is cloud-managed, the hardware and software infrastructure are designed to operate together under Brivo's controlled environment. While the system supports integration with some external services, many of the platform's key features rely on proprietary hardware and software components designed specifically for the Brivo ecosystem.

Another proprietary system commonly used in enterprise environments is Honeywell's Pro-Watch security management platform. Pro-Watch integrates access control, video surveillance, and intrusion detection systems under a centralized management framework. Although Honeywell systems can integrate with some external devices, the platform is largely built around Honeywell's own hardware and software technologies. The system is frequently used in high-security environments such as airports, government facilities, and large corporate campuses where centralized control and tightly integrated infrastructure are important.

Proprietary security systems are also common in the consumer smart lock market. For example, systems such as August Smart Lock, Yale Assure Lock, and Level Lock often operate within proprietary smartphone applications and cloud management platforms. While these devices may support certain open communication standards such as Bluetooth or Wi-Fi, many of their advanced features—including remote access management, firmware updates, and integration with home automation systems—are controlled through the manufacturer's own software ecosystem.

The main advantage of proprietary systems is tight integration and simplified system management. Because all components are developed by the same manufacturer, the system is usually optimized to work together with minimal compatibility issues. Hardware and software are designed as a unified platform, which can simplify installation, configuration, and maintenance. Manufacturers can also control security updates and system behaviour more directly, reducing the risk of misconfiguration between incompatible devices.

Proprietary systems may also offer enhanced security control in certain environments. Since the communication protocols and internal system architecture are not publicly available, it may be more difficult for attackers to analyse system vulnerabilities. Manufacturers can tightly control access to system programming and updates, limiting the ability of unauthorized parties to modify the system.

However, proprietary systems can also create vendor dependency, sometimes referred to as vendor lock-in. Organizations that adopt a proprietary platform may find it difficult to integrate hardware from other manufacturers or migrate to alternative systems in the future. If the vendor discontinues support for the platform, raises licensing costs, or ceases operations, the organization may face significant challenges upgrading or maintaining its access control infrastructure. In some cases, upgrading the system may require replacing large portions of the hardware rather than simply updating individual components.

From a design perspective, many next-generation locking systems attempt to balance the strengths of both approaches through hybrid architectures. These systems may use proprietary hardware designs while supporting open communication standards that allow integration with broader building management systems or security platforms. For example, a manufacturer might develop proprietary lock hardware but allow the system to communicate with external management platforms using standardized network protocols.

Hybrid security architectures combine elements of open systems and proprietary systems to balance interoperability, flexibility, and vendor control. In a hybrid architecture, some components of the system follow open standards—allowing integration with other platforms—while other components remain proprietary to the manufacturer. This approach allows organizations to benefit from the reliability and optimized integration of proprietary hardware while still maintaining compatibility with broader security ecosystems such as building management systems, surveillance platforms, or third-party software. Hybrid architectures are increasingly common in modern access control and smart lock environments where security infrastructure must integrate with multiple technologies.

One well-known example of a hybrid architecture is the HID Global access control ecosystem. HID produces proprietary credential technologies such as iCLASS and Seos, along with proprietary reader hardware. However, these readers and controllers often support open communication standards such as OSDP (Open Supervised Device Protocol) and integrate with a wide range of third-party access control software platforms. This allows organizations to use HID credentials and hardware while still connecting the system to external security management software or building management systems. In this model, the credential technology remains proprietary, but the communication framework supports open interoperability.

Another example is Genetec Security Center, which functions as a hybrid security architecture combining open integration capabilities with proprietary system components. Genetec provides its own security management platform, analytics tools, and system architecture, but it supports integration with numerous third-party cameras, electronic locks, and access control devices through open standards and application programming interfaces (APIs). For example, a facility might use Genetec software to manage a system that includes cameras from multiple manufacturers, electronic locks from different vendors, and building automation systems integrated through open protocols. In this architecture, the management platform remains proprietary while the connected hardware ecosystem remains flexible.

Schlage's ENGAGE platform, developed by Allegion, provides another hybrid example. ENGAGE locks use proprietary wireless communication technology and lock management features, but they can also integrate with open access control platforms through standardized communication interfaces. For instance, ENGAGE locks may connect to third-party access control systems using standard credential technologies such as RFID or mobile credentials. This enables organizations to integrate Schlage electronic locks into larger security infrastructures while still benefiting from Allegion's proprietary device management tools.

A similar hybrid model is used by Assa Abloy's Aperio wireless lock technology. Aperio locks use proprietary wireless communication between locks and hubs, ensuring secure and reliable communication within the system. However, the hubs connect to existing access control systems using open integration standards. This allows Aperio locks to be integrated into access control systems from other manufacturers without requiring complete replacement of the existing infrastructure. In practice, this means that a facility may retain its existing access control platform while adding Aperio wireless locks to doors that were previously difficult to connect using wired systems.

Another example of hybrid architecture appears in smart home ecosystems such as those supported by Apple Home, Google Home, and Amazon Alexa. Many smart locks use proprietary hardware and firmware designed by the manufacturer, but they support open wireless communication standards such as Bluetooth, Wi-Fi, Zigbee, Z-Wave, or the emerging Matter standard. This allows locks from different manufacturers to interact with broader home automation systems while maintaining proprietary device management features within the manufacturer's application environment.

Hybrid architectures are also common in enterprise cloud-based access control platforms. Systems such as Brivo or Kisi use proprietary cloud management software and hardware controllers but often support open APIs that allow integration with other enterprise systems such as identity management platforms, building automation systems, and video surveillance systems. Through these APIs, organizations can connect access control events with broader operational workflows, such as automatically granting building access when an employee account is activated in an identity management system.

**Impact on Locksmithing Practices**

For locksmiths and security professionals, understanding the distinction between open and proprietary systems is increasingly important as access control technologies become more digitally integrated. Locksmiths may be responsible not only for installing physical locks but also for configuring electronic access systems, integrating credential technologies, and advising clients on system selection. Clients such as commercial buildings, hospitals, universities, and government facilities often need guidance on whether an open or proprietary system better suits their operational needs.

## Chapter 19 Review Questions

### Question 1

A homeowner installs a smart lock that allows them to grant temporary smartphone access to a cleaner and receive notifications when the door is locked or unlocked. The system is designed to be easy to use while maintaining security.

What design principle is primarily demonstrated in this situation?

A. Mechanical security hardening
B. User-centred lock design

C. Industrial lock standardisation
D. Lock material sustainability

**Question 2**

A hospital installs door locks that allow medical staff to open restricted doors quickly using biometric authentication while still protecting medication storage rooms from unauthorized access.

Why is this design approach important in healthcare environments?

A. It reduces lock manufacturing costs
B. It improves lock aesthetics
C. It balances rapid emergency access with security control
D. It eliminates the need for electronic access systems

**Question 3**

A building manager installs a keypad lock system, but users frequently leave doors open because the lock interface is confusing and difficult to operate.

Which design principle was most likely neglected during system development?

A. Environmental corrosion resistance
B. Usability in user-centred design
C. Lock manufacturing efficiency
D. Surface coating durability

**Question 4**

An assisted-living facility replaces traditional round doorknobs with lever handles and installs large keypad entry systems to accommodate residents with arthritis and limited hand strength.

What design concept best explains this change?

A. Modular lock design
B. Mechanical lock redundancy
C. Access credential encryption
D. Accessibility and inclusive lock design

**Question 5**

A large university campus installs electronic locks that allow administrators to issue digital access credentials to students and revoke them when they graduate.

Which advantage of user-centred access control systems does this scenario demonstrate?

A. Reduction of mechanical wear in locks
B. Elimination of all security risks
C. Flexible management of user credentials
D. Improved lock corrosion resistance

**Question 6**

A manufacturer develops lock bodies using recycled brass instead of newly mined metal, reducing energy consumption in the production process.

Which design objective is being addressed?

A. Mechanical tamper resistance
B. Environmental sustainability in lock manufacturing
C. Increased lock complexity
D. Credential encryption strength

**Question 7**

A high-security padlock uses hardened steel shackles and tungsten carbide inserts to resist drilling and cutting attacks.

What is the primary reason for using these materials?

A. They increase decorative appearance
B. They improve resistance to physical attacks
C. They simplify electronic programming
D. They reduce manufacturing costs

**Question 8**

An electronic lock is designed so that the keypad reader, battery pack, wireless module, and mechanical latch can all be replaced individually without removing the entire lock.

What type of design architecture does this represent?

A. Monolithic lock design
B. Integrated lock assembly
C. Modular lock system
D. Traditional mechanical lock design

**Question 9**

A commercial building installs an access control system where locks from one manufacturer, readers from another, and management software from a third company communicate using a shared protocol such as OSDP.

What type of system architecture is being used?

A. Proprietary security system
B. Closed security ecosystem
C. Open security system
D. Offline mechanical system

**Question 10**

A hotel installs electronic door locks that operate only with software and hardware produced by the same manufacturer, preventing integration with other brands.

Which system type does this represent?

A. Hybrid security architecture
B. Proprietary security system
C. Open security platform
D. Distributed authentication system

# Part 10

# Professional Mastery and Reflection

# Chapter 20

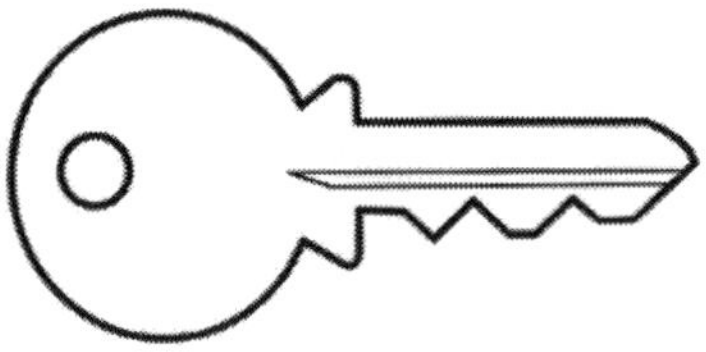

# Becoming a Security Thinker, Not Just a Locksmith

Modern buildings rarely rely on a single protective measure. Instead, they incorporate multiple layers of protection that include physical locks, electronic access control systems, surveillance technologies, alarms, lighting, administrative policies, and human behaviour. In such environments, the effectiveness of any single lock cannot be evaluated in isolation. A door may be fitted with a high-security cylinder, yet remain vulnerable if key control procedures are weak, if staff routinely bypass access controls, or if door hardware and monitoring systems are poorly maintained. Understanding how these elements interact requires a shift in perspective, from thinking purely in terms of hardware to thinking in terms of systems.

This chapter introduces that shift in mindset. It explores how locksmiths can develop the habits and perspectives required to operate not only as technicians but also as security thinkers. Systems thinking allows locksmiths to analyse how people, technology, policies, and physical infrastructure interact to influence real-world security outcomes. Rather than asking only which lock should be installed on a door, the security thinker asks broader questions about risk, workflow, access management, and resilience within the environment being protected.

In addition to systems thinking, the chapter also examines the importance of continuous learning, reflective practice, and strategic planning in modern locksmithing. Security technologies, threats, and operational requirements continue to evolve rapidly. Locksmiths who remain effective in this environment must continually update their knowledge, learn from real-world experience, and adapt their recommendations to changing conditions. By developing these habits, locksmiths expand their professional role from hardware installers to trusted advisors who can design and maintain comprehensive security solutions.

Ultimately, becoming a security thinker does not replace the traditional skills of locksmithing, it builds upon them. Mechanical knowledge, installation expertise, and practical problem-solving remain essential. What changes is the scope of thinking. The locksmith begins to see locks not as isolated devices, but as components within larger security systems that must function reliably alongside people, procedures, and technology. This chapter explores how adopting this broader perspective strengthens both the effectiveness of security systems and the professional role of the locksmith in modern society.

## Systems Thinking in Security

## Beyond the Key

Systems thinking in security is the practice of understanding security not as a collection of separate devices, locks, cameras, alarms, credentials, guards, policies, but as an interconnected whole in which each part influences the others. For a locksmith moving toward the mindset of a security thinker, this is a major shift. A traditional locksmith might look at a door and ask, "What lock should be fitted here?" A security thinker asks broader questions: Who uses this door? What are they protecting? What other controls exist around it? How might people bypass the lock? What happens if power fails? How does this entry point relate to alarms, surveillance, evacuation, maintenance access, and organizational policy? Systems thinking therefore moves security practice from hardware selection alone to the analysis of relationships, dependencies, and consequences across the whole environment.

At its core, systems thinking recognizes that security problems rarely arise from a single weak component. A strong lock on a poorly reinforced door may fail because the frame gives way. A high-security cylinder may be undermined by poor key control. An advanced biometric reader may be defeated if staff prop the door open for convenience. A well-monitored site may still be vulnerable if access privileges are not revoked when staff leave. In each of these cases, the failure is not purely mechanical or purely digital; it emerges from the interaction between physical hardware, human behaviour, procedures, technology, and the surrounding environment. Systems thinking helps security professionals see these interactions and understand that the effectiveness of a security measure depends on how well it functions within the broader system.

One of the central ideas in systems thinking is that every security environment contains multiple layers. Physical barriers, electronic access control, surveillance, intrusion detection, lighting, staffing, policy, training, and incident response all contribute to protection. None of these elements should be considered in isolation. A locksmith who adopts systems thinking understands that a lock is only one layer in a layered defence strategy. For example, the right solution for a restricted office may not simply be a stronger cylinder. It may involve appropriate door hardware, access logs, after-hours alarms, camera coverage, controlled key issuance, and staff procedures for visitors. The lock remains important, but it is understood as one component within a larger defensive design.

Systems thinking also requires attention to flows of people, credentials, information, authority, and risk. In security, people move through spaces, credentials are issued and revoked, information about access events is generated and stored, and risks shift over time depending on use patterns and operational changes. A security thinker studies these flows. In a hospital, for example, staff, patients, visitors, contractors, and emergency responders all move through the same building in different ways and with different authorization needs. The locksmith who thinks systemically does not just fit locks to doors; they consider traffic patterns, emergency overrides, restricted medication areas, accessibility requirements, and the consequences of delays or misuse. The question becomes not only "Can this area be locked?" but "How should access to this area function safely and effectively within the daily operation of the facility?"

Another important aspect of systems thinking is recognizing interdependence. In modern security, one decision often affects many others. Changing from mechanical keys to electronic credentials affects access policies, staff training, maintenance requirements, privacy obligations, and cybersecurity considerations. Adding smart locks may improve flexibility, but it may also create dependence on batteries, firmware updates, network infrastructure, and vendor support. Installing high-security door hardware may improve resistance to attack, but it may also affect emergency egress, accessibility compliance, and user convenience. A security thinker anticipates these connected effects. Rather than solving one problem narrowly and creating three others, they aim to understand the broader consequences of design choices.

Feedback is another systems-thinking principle with strong relevance to security. Security systems constantly generate feedback, whether through failed access attempts, maintenance faults, alarm activations, audit logs, user complaints, or incident reports. A locksmith working only as a technician may treat each issue as a separate repair task. A security thinker asks what pattern the feedback reveals. Are users frequently making mistakes with a keypad because the interface is poor? Are doors being forced because access rights are too restrictive during legitimate workflows? Are locks wearing out quickly because of door misalignment or heavy traffic? Feedback is not just evidence of failure; it is information that helps improve the overall system. Systems thinking encourages continuous review, learning, and redesign rather than simple fault correction.

Human behaviour sits at the centre of systems thinking in security. Many security breakdowns occur not because devices fail technically, but because people adapt around them. Staff share access codes, leave keys unsecured, wedge doors open, ignore alarm signals, or choose convenience over procedure. A systems thinker does not dismiss this as carelessness alone. Instead, they ask why the system encourages such behaviour. Is the process too slow? Is the lock difficult to use? Are policies unrealistic? Is training inadequate? In this way, systems thinking avoids blaming users too quickly and instead examines how design, workflow, and culture interact. This is particularly important for locksmiths moving into broader security roles, because the best hardware solution is often the one that real people will use correctly and consistently.

Systems thinking also emphasizes context. The same lock or access method may be appropriate in one environment and completely unsuitable in another. A keypad lock may work well for a short-term rental property, but not for a high-security laboratory. A mobile credential system may suit a corporate office, but not a remote industrial site with unreliable network coverage. A high-security mechanical cylinder may be ideal for an external plant room, but insufficient on its own for a controlled records archive. The security thinker therefore avoids one-size-fits-all recommendations. They assess threats, user needs, building function, environmental exposure, maintenance capacity, legal requirements, and organizational priorities before deciding how the system should be designed.

Risk assessment becomes far more meaningful when viewed through systems thinking. Instead of asking only, "What attack might defeat this lock?", the broader question becomes, "How could this system fail, be bypassed, misused, or become ineffective?" This includes forced entry, credential misuse, social engineering, poor maintenance, policy failures, insider threats, emergency conditions, and technology obsolescence. It also includes what happens after an incident: how it is detected, how access is contained, how evidence is preserved, and how the site recovers. A locksmith who thinks this way starts to operate less like a tradesperson responding to isolated tasks and more like a security consultant designing for resilience.

Resilience is a particularly important systems-thinking concept. Good security is not just about preventing every possible incident; it is also about ensuring that when something goes wrong, the system can still function, recover, and protect people safely. For example, if an electronic access control system fails, is there a secure override? If a credential is compromised, can permissions be revoked quickly? If a door is damaged, can the area be temporarily secured without disrupting emergency access? If power or network connectivity is lost, what happens to critical entry points? Systems thinking encourages the design of security measures that are robust, repairable, and adaptable rather than brittle and overdependent on one technology or process.

For locksmiths, adopting systems thinking changes professional practice in several ways. It means surveying more than just the lock and door. It means asking better questions during client consultations. It means understanding how physical security interacts with electronic systems, with user behaviour, with policy, and with the operational purpose of the space. It means being able to explain to clients that the strongest lock is not always the best solution if it disrupts workflow, creates unsafe egress conditions, or leads people to bypass the control altogether. It also means seeing opportunities for better solutions, integrating door hardware with credential management, recommending key control policies, identifying dependencies in smart lock deployments, and noticing where maintenance or training weaknesses could undermine security.

By means of a practical example, imagine a regional private hospital engaged a locksmith after several security concerns emerged around its medication storage and after-hours staff access. Management initially believed the problem was simple: the lock on the pharmacy storeroom door needed to be upgraded because there had been two incidents in which unauthorized staff entered the area. A traditional lock-focused response might have been to install a stronger high-security cylinder or an electronic keypad. However, the locksmith approached the site using systems thinking and began by looking beyond the door itself.

During the site assessment, the locksmith observed that the pharmacy corridor was used by pharmacists, nurses, cleaners, maintenance staff, contractors, and emergency personnel at different times of day. The storeroom door already had a relatively good lock, but the door frame was worn, the self-closing mechanism was inconsistent, and the strike plate alignment was poor. More importantly, staff workflow created security weaknesses. Nurses frequently needed urgent after-hours access to nearby treatment rooms, and when the area became busy, the pharmacy corridor door was sometimes wedged open for convenience. CCTV coverage of the corridor was incomplete, and cleaning contractors were using a shared after-hours access card that also allowed access to adjacent controlled zones. The issue was therefore not just the pharmacy lock; it was the interaction between physical hardware, access permissions, user behaviour, and operational pressure.

The locksmith then mapped the whole security system surrounding the area. This included the outer corridor door, the internal pharmacy storeroom door, the existing card access system, CCTV placement, alarm coverage, door-closing hardware, lighting, staff access policies, emergency override procedures, and key control arrangements. Interviews with staff revealed that the pharmacy keypad code had been shared informally because some legitimate users found the access process too slow during urgent patient care tasks. Audit logs also showed repeated failed access attempts late at night, but these were not intrusion attempts; they were staff using expired temporary credentials because access permissions had not been updated promptly after role changes. What first appeared to be a weak lock was actually a system design problem involving workflow, credential management, monitoring gaps, and hardware wear.

Rather than recommending only a stronger lock, the locksmith proposed a layered solution. The pharmacy storeroom received an upgraded electronic lock linked to individualized staff credentials rather than a shared code. The door frame was reinforced, the strike plate was realigned, and the closer was replaced so the door would reliably latch every time. Access rights were restructured so that cleaning contractors could enter only the external service corridor and not the pharmacy zone. CCTV coverage was adjusted so that both the corridor approach and the pharmacy entrance were visible. A forced-door alarm was added to the pharmacy entry, and the hospital introduced a policy requiring immediate review of access permissions when staff roles changed. For emergency situations, an override process was retained for authorized senior staff, but it required logging and review.

The locksmith also addressed the human side of the problem. Staff were trained on why shared codes and wedged doors created risk, but the training was paired with practical redesign so the secure process was not unnecessarily burdensome. Nurses who needed legitimate rapid access to nearby treatment areas were given appropriate credentials so they no longer felt pressure to bypass controls. The hospital's facilities team was also instructed to include door alignment, closer performance, and latch operation in routine maintenance checks, because the locksmith recognized that small mechanical faults could gradually undermine the whole system.

Several months later, the hospital reported a significant improvement. Unauthorized access incidents stopped, doors were no longer being wedged open, and audit records became more reliable and meaningful. Importantly, staff no longer viewed the security system as an obstacle to patient care because it had been redesigned around actual workflow. The locksmith's value to the client was not simply that a lock had been changed, but that the broader relationships among hardware, people, policy, monitoring, and operational needs had been understood and improved.

This case study shows how systems thinking changes locksmith practice. The problem was not solved by treating the lock as an isolated object. Instead, the locksmith examined the full security environment: the physical door condition, the electronic credential structure, the behaviour of users, the visibility provided by surveillance, the access needs created by healthcare operations, and the policies governing after-hours movement. By doing so, the locksmith acted not merely as a tradesperson fitting hardware, but as a security thinker designing a resilient and workable system.

In practical terms, systems thinking in security often involves looking at several domains at once: the physical domain of doors, frames, barriers, and hardware; the technical domain of readers, controllers, networks, and software; the human domain of users, habits, and errors; the procedural domain of policies, authorization, and response plans; and the environmental domain of layout, lighting, weather, and occupancy patterns. Security becomes strongest when these domains support one another rather than conflict. A locksmith becoming a security thinker learns to read all of them together.

Ultimately, systems thinking in security is about seeing relationships instead of isolated parts, causes instead of symptoms, and long-term function instead of one-off fixes. It encourages the locksmith to move beyond the narrow question of what hardware fits a door and toward the larger question of how protection actually works in the real world. That is the mindset shift at the heart of becoming a security thinker: understanding that locks matter greatly, but they only achieve their full value when designed, selected, installed, and managed as part of an integrated security system.

## Impact on Locksmithing Practices

Understanding systems thinking in security is highly relevant to someone interested in locksmithing because the profession has evolved far beyond simply installing or repairing mechanical locks. Modern locksmiths operate in environments where physical security hardware interacts with electronic access control, surveillance systems, building management technologies, and organizational procedures. A locksmith who understands systems thinking recognises that a lock is only one element within a much larger security ecosystem. This broader perspective enables locksmiths to design solutions that address real-world risks rather than focusing narrowly on hardware alone.

Traditionally, locksmithing centred on the mechanical aspects of security. A locksmith might evaluate the quality of a cylinder, install a stronger deadbolt, or repair a faulty latch. While these skills remain essential, modern buildings rely on integrated security systems where door hardware interacts with electronic credentials, alarms, cameras, and administrative policies. Systems thinking helps locksmiths understand how these components influence one another. For example, installing a high-security cylinder may appear to improve protection, but if key control procedures are weak or doors are frequently left open due to workflow pressures, the security benefit may be minimal. By examining the whole system, locksmiths can identify where real vulnerabilities exist.

This approach significantly influences locksmithing practice because it changes how security problems are diagnosed and solved. Instead of assuming that a lock failure or security incident is caused solely by inadequate hardware, a systems-thinking locksmith investigates multiple factors. These include the physical condition of doors and frames, the way people use the space, the management of keys or credentials, surveillance coverage, alarm integration, and operational procedures. This broader investigation often reveals that security failures arise from interactions between these elements rather than from a single defective component.

Systems thinking also influences how locksmiths assess and recommend security solutions. When evaluating an entry point, a locksmith using systems thinking considers several questions beyond the lock itself. These may include who requires access to the space, how often the door is used, whether the area is monitored by surveillance systems, how access permissions are managed, and how emergency evacuation procedures operate. By understanding these relationships, the locksmith can recommend solutions that support both security and operational needs. For example, a secure door in a hospital must allow rapid access during emergencies while still protecting restricted areas such as medication storage.

Another important influence on locksmithing practice is the concept of layered security. Systems thinking emphasises that effective protection rarely depends on a single barrier. Instead, security is achieved through multiple layers that support one another. A lock may be supported by reinforced doors and frames, electronic access control systems, surveillance cameras, alarms, lighting, and staff procedures. Locksmiths who understand this layered approach can contribute more effectively to comprehensive security planning by ensuring that door hardware integrates properly with other protective measures.

Human behaviour is another factor that systems thinking brings into locksmithing practice. Security technologies are used by people, and the way individuals interact with locks often determines whether security measures succeed or fail. If a lock is difficult to operate, people may prop doors open or share access codes. If key management procedures are inconvenient, users may duplicate keys or store them insecurely. A locksmith who understands systems thinking recognises that security solutions must account for human behaviour and practical workflows. In many cases, improving usability and access management can strengthen security more effectively than simply installing stronger hardware.

The increasing integration of electronic access control systems further highlights the relevance of systems thinking for locksmiths. Many modern locks are connected to digital credential systems, cloud-based management platforms, or building automation networks. These technologies introduce new considerations such as network reliability, power supply, cybersecurity, and software management. Locksmiths who understand systems thinking can better evaluate how these technologies interact with mechanical locking components and how system failures might affect building access during emergencies or outages.

Systems thinking also encourages locksmiths to consider long-term resilience rather than short-term fixes. Instead of focusing only on preventing forced entry, security professionals examine how systems will respond if something goes wrong. For example, electronic locks must have secure override procedures if power fails. Access credentials must be revocable if lost or compromised. Doors must remain secure even when maintenance issues arise. By considering these scenarios, locksmiths can design security systems that remain functional and safe under a range of conditions.

The hospital case study illustrates how this mindset influences practical locksmithing work. Initially, management believed the solution to unauthorized access was simply replacing a lock. However, the locksmith's systems-thinking approach revealed multiple contributing factors, including worn door hardware, incomplete surveillance coverage, shared access codes, and workflow pressures that encouraged staff to bypass controls. By addressing these interconnected elements rather than focusing only on the lock itself, the locksmith helped design a solution that improved security while supporting hospital operations.

Adopting systems thinking therefore transforms the role of the locksmith from a hardware technician into a broader security advisor. Locksmiths who understand the relationships between physical security devices, user behaviour, and organizational policies are better equipped to diagnose vulnerabilities, recommend appropriate technologies, and design effective access control strategies. This broader capability increases the professional value of locksmiths within modern security environments.

## Continuous Learning and Adaptation

Continuous learning and adaptation are central to becoming a security thinker rather than remaining only a locksmith because security is not a fixed field. Locks, credentials, attack methods, building technologies, user expectations, regulations, and business risks all change over time. A locksmith can become highly competent at fitting hardware, rekeying cylinders, opening safes, and servicing doors, but a security thinker recognizes that yesterday's good solution may be inadequate tomorrow. Continuous learning therefore means more than attending training occasionally; it is the ongoing habit of updating knowledge, questioning assumptions, reviewing failures, and adjusting practice as threats and technologies evolve.

In traditional locksmithing, skill was often built through repetition and mastery of proven mechanisms. Mechanical principles remain important, but the security environment has expanded. Today, locksmiths encounter electronic access control, smart locks, mobile credentials, cloud-managed platforms, biometrics, audit logs, wireless communication, privacy obligations, and cybersecurity dependencies. At the same time, criminals and intruders adapt as well. Attack methods evolve from picking and forced entry to credential theft, relay attacks, social engineering, and exploitation of poor system configuration. Continuous learning is therefore essential because the modern security professional must understand not only how locks work, but how entire systems fail, how users behave, and how attackers adapt.

A security thinker treats learning as part of daily professional practice. This includes learning from formal sources such as manufacturer training, industry standards, certification programs, technical manuals, trade publications, and professional associations. It also includes learning from informal sources such as field experience, client feedback, maintenance records, incident reports, and conversations with other specialists. A door that repeatedly fails to latch properly, a credential system that generates frequent user complaints, or a site that keeps experiencing workarounds are all opportunities to learn. Continuous learning means being alert to what the field is teaching in real time, not only what was learned in past training.

Adaptation is the practical partner of learning. It is not enough to collect information if professional practice stays the same. A security thinker takes new knowledge and applies it. If repeated service calls show that a certain electronic lock struggles in harsh outdoor environments, future recommendations should change. If audit logs reveal that users frequently share keypad codes, then the access method or policy should be reconsidered. If a manufacturer releases a firmware update to fix a known vulnerability, the security thinker understands that updating the system is part of responsible security maintenance. Adaptation means changing methods, recommendations, and designs in response to evidence rather than simply repeating familiar habits.

One important dimension of continuous learning is technical breadth. To think like a security professional, locksmiths increasingly need knowledge that crosses traditional boundaries. Mechanical lock knowledge is still foundational, but it now sits alongside door hardware, fire and egress requirements, electronic access systems, wireless credentials, surveillance integration, network dependencies, and cybersecurity basics. A locksmith who learns continuously becomes more capable of understanding how these domains affect one another. For example, a decision to install wireless smart locks is no longer only about convenience and door preparation; it may also involve battery maintenance, network reliability, user credential management, privacy considerations, and fallback access in the event of system failure.

Another dimension is learning from incidents and near misses. Security thinkers do not view problems merely as isolated faults to be repaired and forgotten. They ask what each incident reveals about the wider system. A forced door may indicate poor frame reinforcement, but it may also indicate unrealistic access restrictions, bad traffic flow, or insufficient after-hours procedures. A shared access code may point not only to poor staff discipline, but also to a cumbersome system design that encourages workarounds. Continuous learning means examining incidents for patterns, causes, and lessons, then adjusting future designs accordingly. This reflective habit is one of the clearest differences between a technician who restores function and a security thinker who improves systems.

Continuous learning also involves keeping pace with changes in standards, regulation, and client expectations. Security hardware and access control systems do not operate in a vacuum. They are shaped by building codes, accessibility requirements, privacy law, fire safety obligations, procurement rules, and industry-specific compliance demands. Hospitals, schools, government facilities, transport sites, and commercial offices all have different legal and operational frameworks. A locksmith who does not keep learning may fit technically sound hardware that creates compliance problems, unsafe egress conditions, or poor user outcomes. A security thinker stays aware that lawful, safe, and usable security often depends on more than the lock itself.

Human behaviour is another area where learning and adaptation matter greatly. Security systems are used by real people under real pressures, and those pressures change. Staff turnover, new workflows, organizational growth, remote work arrangements, contractor access needs, and emergency procedures all affect how security functions day to day. A security thinker learns not just from hardware performance but from user behaviour. If a system is routinely bypassed, misunderstood, or resisted, then something in the design, policy, or training may need to change. Continuous learning in this sense includes observing how people actually use security and adapting systems so that secure behaviour becomes easier and more realistic.

Professional humility is an important part of continuous learning. Becoming a security thinker requires accepting that no one knows everything and that expertise can become outdated. The more complex security systems become, the more important it is to ask questions, consult other specialists, and be willing to revise one's view. Locksmiths may need to work with IT personnel, architects, fire safety consultants, facilities managers, and security integrators. Adaptation often happens through collaboration, where each professional brings a different perspective to a shared problem. The security thinker is willing to learn from those disciplines rather than defending a narrow view of security as only lock hardware.

Continuous learning also supports resilience in business practice. Clients increasingly expect security professionals to explain options, justify recommendations, and support long-term system performance. A locksmith who keeps learning can speak more confidently about lifecycle planning, modular upgrades, open versus proprietary systems, sustainability, predictive maintenance, and risk trade-offs. This strengthens professional credibility and shifts the practitioner's role from repair provider to trusted advisor. In this way, continuous learning does not only improve technical competence; it changes the value that the locksmith brings to clients.

A practical example can be seen in the evolution from mechanical master key systems to electronic credential management. A locksmith who stops learning may remain highly skilled in designing key hierarchies and rekeying buildings, but may struggle when a client wants time-based access permissions, audit trails, mobile credentials, or rapid credential revocation. A security thinker continues learning so that they can advise the client on when mechanical systems remain appropriate, when electronic systems offer better control, and how to manage the risks introduced by each. The point is not to abandon old knowledge, but to build on it so that practice evolves with the environment.

Adaptation also means revisiting old assumptions. For many years, stronger hardware was often treated as the primary answer to better security. In some environments that remains true, but not always. Continuous learning may reveal that in a particular site, the real weakness lies in key control, user training, poor door maintenance, bad lighting, weak surveillance integration, or unmanaged contractor access. The security thinker is willing to move beyond the instinct to solve every problem with heavier hardware and instead adapt recommendations to the actual risk picture.

**Impact on Locksmithing Practices**

Continuous learning and adaptation are highly relevant to someone interested in locksmithing because the security industry is constantly evolving. While traditional locksmithing focused primarily on mechanical locks and key systems, modern security environments now include electronic access control, smart locks, mobile credentials, biometrics, and cloud-based management platforms. At the same time, criminal methods and security threats are also changing, moving beyond traditional lock picking and forced entry to include credential theft, social engineering, and electronic attacks. For this reason, locksmiths must continually update their knowledge to remain effective. Continuous learning ensures that locksmiths understand both traditional mechanical security and the newer digital systems that are increasingly used in modern buildings.

This ongoing learning directly influences locksmithing practice by expanding the range of skills and knowledge required in the profession. A modern locksmith must understand not only lock mechanisms and door hardware but also access control software, wireless communication technologies, security policies, and building safety requirements. Continuous learning allows locksmiths to interpret audit logs, manage electronic credentials, understand system vulnerabilities, and recommend appropriate technologies for different environments. It also encourages locksmiths to learn from real-world experiences such as repeated service issues, user behaviour, and security incidents. By analysing these situations, locksmiths can adapt their recommendations and improve the overall reliability and effectiveness of security systems.

Continuous learning also changes the role of the locksmith from a technician who installs hardware to a professional who provides broader security advice. Clients increasingly expect security professionals to explain options, assess risks, and recommend systems that will remain effective over time. Locksmiths who continually update their knowledge can advise on topics such as electronic credential systems, system upgrades, key control procedures, and integrated security solutions. This ability to adapt and evolve strengthens professional credibility and allows locksmiths to design security systems that respond to changing technologies, operational needs, and emerging risks rather than relying solely on traditional lock hardware solutions.

## Reflective Practice in Locksmithing

Reflective practice in locksmithing refers to the habit of regularly thinking critically about one's work in order to improve professional judgment, technical skills, and security outcomes. Rather than simply completing a job and moving on to the next task, a reflective locksmith evaluates what happened during the job, what worked well, what problems occurred, and what could be improved in the future. This process encourages locksmiths to learn from their experiences, refine their decision-making, and continuously develop their expertise. Reflective practice is particularly important in security work because every installation, repair, or security incident provides insight into how locks and access systems perform in real-world environments.

In practical locksmithing, reflective practice often occurs after completing installations, service calls, or responding to security issues. For example, after installing a locking system, a locksmith may reflect on whether the chosen hardware was appropriate for the door construction, whether the installation process revealed any weaknesses in the door frame or surrounding structure, and whether the system integrates effectively with other security measures such as alarms or surveillance. If a lock repeatedly fails, becomes difficult to operate, or is bypassed by users, reflective practice encourages the locksmith to analyse why this occurred. The issue may relate to hardware quality, environmental conditions, incorrect installation, user behaviour, or operational pressures within the client's workplace.

Reflective practice also involves considering the broader context in which security systems operate. Locksmiths frequently work in environments such as hospitals, offices, schools, industrial facilities, or residential properties, each with different operational needs. A reflective locksmith asks questions such as: Did the security solution support the way people actually use the space? Were emergency exit requirements and accessibility considerations properly addressed? Did the chosen system encourage secure behaviour, or did it create frustration that might lead users to bypass it? By asking these questions, locksmiths learn to design solutions that balance security with usability and safety.

Another important aspect of reflective practice is learning from security incidents or near misses. If a building experiences unauthorized access, forced entry, or repeated lock failures, a reflective locksmith examines the underlying causes rather than simply repairing the immediate problem. For instance, if a high-security lock is installed but keys are poorly controlled, the real vulnerability lies in key management rather than the lock mechanism. Similarly, if staff repeatedly prop open a secure door, the locksmith might reflect on whether the access procedure is too inconvenient or whether additional door hardware or access control measures are required. Reflective practice encourages locksmiths to look beyond symptoms and identify the deeper causes of security weaknesses.

Reflective practice also supports professional growth by encouraging locksmiths to seek new knowledge and refine their skills. Through reflection, locksmiths may identify areas where additional training or research would be beneficial, such as electronic access control systems, cybersecurity considerations, or new locking technologies. This mindset helps locksmiths remain adaptable in an industry where technologies and threats continue to evolve. By reflecting on their work and actively seeking improvement, locksmiths strengthen both their technical competence and their ability to provide informed security advice to clients.

**Case Study Scenario Example**

A commercial office building hired a locksmith to improve security for a records storage room that contained sensitive financial documents. The building manager was concerned that several employees had recently accessed the room without authorization, and the initial request was simply to install a stronger lock. The locksmith installed a high-security cylinder and issued restricted keys to a small number of authorized staff. At the time of installation, the locksmith confirmed that the door hardware functioned correctly and that the lock met the client's immediate security requirements.

Several weeks later, the building manager contacted the locksmith again because the door was frequently being found unlocked during the day. When investigating the issue, the locksmith discovered that employees who required temporary access to nearby filing cabinets often asked authorized staff to unlock the room and leave the door open while they worked. In some cases, the door closer was weak, which prevented the door from fully latching after use. The locksmith repaired the closer and reminded staff to keep the door closed, but after completing the service call, the locksmith reflected on the situation and realized that the problem might not be purely mechanical.

Through reflective practice, the locksmith began to analyse the broader situation. The locksmith asked whether the original security solution had properly considered how the room was used during daily operations. Employees frequently needed short-term access to documents, and the restricted key system required them to locate one of the few authorized key holders each time access was required. This process was inconvenient, which encouraged staff to leave the door open or share access informally. The locksmith realised that although the high-security cylinder provided strong mechanical protection, it did not fully address the workflow and access needs of the workplace.

After discussing these observations with the building manager, the locksmith recommended modifying the access approach rather than simply relying on stronger hardware. The locksmith suggested installing an electronic access control lock that allowed authorized employees to use individual PIN codes or access cards. This would provide an audit trail showing who accessed the room and when, while still allowing flexible permission management. The door closer was also upgraded to ensure consistent latching, and staff received guidance on appropriate access procedures.

Several months later, the building manager reported that unauthorized access incidents had stopped and that employees were using the new system correctly. The locksmith reflected on the experience and recognised that the initial installation had focused mainly on physical hardware rather than operational behaviour. By analysing the situation after the service calls and adjusting the design, the locksmith improved both the security and usability of the system.

This case study demonstrates reflective practice in locksmithing. Instead of simply repairing faults and moving on, the locksmith evaluated the outcome of the original work, examined how users interacted with the security system, and identified the underlying causes of the problem. Through reflection, the locksmith improved future recommendations and developed a deeper understanding of how security systems function in real-world environments.

## Impact on Locksmithing Practices

Reflective practice is highly relevant to someone interested in locksmithing because the profession involves far more than installing or repairing lock hardware. Every security job provides information about how locks, doors, and access systems function in real environments with real users. By reflecting on each installation or service call, locksmiths can identify what worked well and what could be improved. This habit helps develop professional judgment and allows locksmiths to refine their decision-making over time. Instead of simply repeating the same methods on every job, reflective locksmiths gradually build deeper expertise by learning from both successful outcomes and problems encountered in the field.

Reflective practice also influences locksmithing by encouraging professionals to look beyond the immediate technical task and consider the broader security environment. Many security failures occur not because a lock is poorly designed, but because of factors such as user behaviour, poor key control, misaligned doors, or inconvenient access procedures. By reflecting on these factors, locksmiths can identify underlying causes of security weaknesses rather than only fixing visible symptoms. This leads to better recommendations for clients, such as improving door hardware, adjusting access policies, introducing electronic access systems, or modifying workflows so that security measures are easier to follow.

Finally, reflective practice supports long-term professional development and adaptability within the locksmithing industry. Security technologies, building requirements, and client expectations continue to evolve, particularly with the growth of electronic access control, smart locks, and integrated security systems. A locksmith who regularly reflects on their work becomes more aware of knowledge gaps and opportunities for improvement. This encourages ongoing learning and helps the locksmith transition from being purely a tradesperson to becoming a trusted security advisor who can design solutions that balance safety, usability, and effective protection.

## The Locksmith as a Security Strategist

The evolution of the locksmith from a primarily technical specialist to a broader security strategist is well documented across historical and contemporary sources. Traditional depictions emphasize hands-on expertise with mechanical mechanisms—installing, repairing, and bypassing locks to control entry to buildings, safes, and other secured spaces. This foundational view remains acknowledged in modern discussions of physical security, where core competencies in lock construction, door hardware, and access mechanisms form the baseline of professional practice [425-427]. However, this mechanical skill set is now situated within increasingly complex security ecosystems that integrate electronic, networked, and organizational dimensions. Contemporary perspectives frame the locksmith not merely as a technician repairing or defeating physical devices but as a strategic advisor who helps design and sustain comprehensive protection in built environments [427-429].

A key feature of the security-strategy role is holistic security planning that transcends individual devices. A storefront, hospital, data centre, or residential complex often requires coordinated interactions among multiple protective layers: physical barriers (doors, frames, locks), electronic access control (card readers, keypads, biometrics), surveillance systems, lighting, alarm networks, and the governance processes that govern how these elements are used. The strategic locksmith evaluates how these components complement each other rather than optimizing a single component in isolation. This approach aligns with broader security literature that emphasizes layered defence and system-level thinking, wherein the effectiveness of any one control depends on the context and integration with other measures [427, 428, 430].

Risk assessment and site-specific planning are central strategic duties for locksmiths. The decision-making process considers assets at risk, user populations, threat profiles, and the day-to-day movement patterns within a space. Different facilities demand distinct risk profiles and corresponding security architectures; for instance, a data centre has different confidentiality and availability concerns than a retail store or a private residence. The strategic locksmith uses context to guide choices about locking technologies, credential structures, and monitoring technologies so that security measures are proportionate, practical, and maintainable over time. This aligns with risk-focused perspectives that treat security as an adaptive process rather than a static installation [427, 429, 431].

Access control management is another defining strategic responsibility. Beyond preventing forced entry, many organizations require precise control over who can move where and when. The locksmith-turned-strategist designs and administers master key systems, hybrid mechanical-digital configurations, and electronic credential frameworks that reflect organizational roles, shifts, contractor access, and emergency procedures. A robust access strategy balances convenience for authorized users with restrictions on unnecessary access, and it requires ongoing governance to adapt to workforce changes. This emphasis on structured access control is consistent with contemporary practice that treats access management as a governance problem as much as a technical one [427-429].

Lifecycle planning and future flexibility are integral to strategic locksmithing. Security deployments are dynamic, with evolving technologies and shifting risks. A strategic locksmith anticipates maintenance needs, upgrades, and the potential for future integration of new credentials or smart-lock capabilities. Selecting hardware and platforms designed for interoperability reduces future disruption and cost while enabling scalable security solutions over time. This forward-looking planning is a hallmark of strategic thinking in security management and is repeatedly highlighted as essential for maintaining effectiveness in changing environments [427, 428, 432].

Education and advisory roles further distinguish the locksmith as a security strategist. Clients often require guidance on how different protective measures interact, how to implement key control policies, and how to train users and plan incident response. The advisory function helps build trust and positions the locksmith as a long-term partner in security management rather than a one-off service provider. This educational dimension is repeatedly emphasized as a core function of the strategist who integrates technical know-how with policy and operational considerations [427, 428, 433].

Technology integration, including cloud-based access platforms, mobile credentials, and interconnected building security systems, expands the strategist remit into networking and cybersecurity domains. Strategic locksmiths assess privacy, reliability, and workflow implications of technology adoption, balancing the benefits of flexible permissions and remote management with the risks of network exposure and software updates. The literature on security strategy increasingly underscores the need for such systems thinking that spans physical devices and digital infrastructure, reflecting the broader trend toward integrated, architecture-level security design [427, 428, 434].

Human behaviour and usability are recognized as critical determinants of security outcomes. Even the most sophisticated hardware and software can fail if systems are inconvenient or poorly understood by users. A locksmith acting as a strategist designs systems that encourage secure behaviour by aligning access processes with user workflows, adjusting schedules, or simplifying authentication methods to minimize workarounds that compromise security. This behavioural emphasis is consistent with risk-aware security planning that seeks to reduce opportunities for human error or intentional circumvention through intuitive design and operational alignment [428, 433].

In sum, the locksmith's shift toward strategic security practice entails moving from reactive, device-centred repair work to proactive, planning-oriented security management. The strategist integrates physical security devices with electronic controls, surveillance, organizational procedures, and lifecycle management. This synthesis enables a more resilient security posture that can adapt to evolving threats, organizational needs, and technological innovations. The literature consistently portrays the locksmith as a central figure in modern security management precisely because of this capacity to bridging technical expertise with strategic risk assessment, policy development, and system-wide design considerations [427-430, 433].

**Impact on Locksmithing Practices**

Understanding the locksmith as a security strategist is highly relevant for anyone interested in locksmithing because it reflects how the profession is evolving beyond purely mechanical work. While traditional locksmith skills such as installing locks, rekeying cylinders, and repairing door hardware remain fundamental, modern security environments require a broader perspective. Buildings increasingly rely on integrated systems that combine physical locks with electronic access control, surveillance, alarms, and organizational policies. Someone entering the locksmithing profession therefore benefits from understanding how these elements interact. This broader perspective allows locksmiths to move beyond simply fitting hardware to advising clients on how to design effective security systems that address real risks, operational needs, and user behaviour.

This shift also influences everyday locksmithing practices by encouraging professionals to approach jobs with a more analytical and planning-oriented mindset. Instead of focusing only on the immediate technical task—such as replacing a lock—the locksmith evaluates the wider security context: how people use the space, how access permissions are managed, how doors interact with alarms or surveillance, and how systems will be maintained over time. As a result, locksmiths increasingly provide recommendations on key control policies, access credential management, system upgrades, and long-term security planning. This strategic approach improves the effectiveness of security solutions and elevates the locksmith's role from a repair technician to a trusted security advisor capable of designing resilient, integrated protection systems.

## Effective Locksmithing Practices

Based on the "Impact on Locksmithing Practices" sections throughout *the book*, the following key points summarise the major practical implications for effective modern locksmithing practice. These points reflect the recurring themes across the chapters: integration of technology, human behaviour, ethics, systems thinking, and the expanding role of the locksmith as a security professional.

**Key Points for Effective Locksmithing Practice**

**Beyond the Key**

Foundations of Modern Locksmithing

- Understand that locksmithing now extends beyond mechanical locks into integrated physical–digital security systems.
- Maintain strong mechanical fundamentals, including lock mechanisms, door hardware, and physical attack resistance.
- Adopt a systems-thinking mindset, recognising that locks are one component within larger security ecosystems.
- Balance security and convenience, ensuring systems are secure enough to resist threats while remaining usable for legitimate users.

Technology and Digital Integration

- Develop competence in electronic access control systems, including keypads, cards, biometrics, and mobile credentials.
- Understand cybersecurity risks associated with smart locks and networked access systems.
- Evaluate both physical and digital vulnerabilities when recommending security solutions.
- Select systems that allow credential management, revocation, and audit logging.
- Plan for integration with surveillance, alarms, and building management systems.
- Consider device security, network reliability, and cloud dependencies when deploying digital locks.

Human-Centred Security Design

- Design systems around real user behaviour, not idealised procedures.
- Ensure locks are intuitive and easy to operate to prevent misuse or bypassing.
- Provide clear user feedback, such as audible clicks, indicator lights, or system notifications.
- Recognise that perceived security influences behaviour, not just actual technical strength.
- Design solutions that align with everyday workflows, especially in workplaces and shared environments.

Addressing Human Behaviour Risks

- Recognise that most security failures arise from human behaviour rather than mechanical faults.
- Minimise opportunities for user error through automation, such as self-locking doors.
- Implement layered security controls that do not rely on a single user action.
- Encourage secure habits through system design and training.
- Consider social and organisational factors, such as workplace culture and policy enforcement.

Professional Role and Client Advisory

- Transition from technician to security advisor, helping clients design effective access strategies.
- Conduct risk assessments that consider assets, threats, user needs, and operational environments.
- Recommend layered security solutions rather than single-device fixes.
- Educate clients about system capabilities, limitations, and correct usage.
- Plan for lifecycle management, including upgrades, maintenance, and system scalability.

Ethics, Privacy, and Legal Compliance

- Verify authorised access before performing locksmith services.
- Maintain strict confidentiality regarding keys, codes, and security vulnerabilities.
- Comply with privacy laws and data protection regulations when installing digital security systems.
- Provide transparent pricing, honest recommendations, and ethical service delivery.
- Commit to continuous professional development to remain competent in evolving technologies and threats.

**Overall principle**

Effective modern locksmithing is no longer defined solely by the ability to manipulate locks. It requires combining mechanical expertise, digital security knowledge, human-centred design, ethical practice, and strategic thinking to create security systems that work reliably in real-world environments.

## Chapter 20 Review Questions

### Question 1

A locksmith is asked to install a stronger lock on a warehouse door. Instead of immediately recommending hardware, the locksmith evaluates who uses the door, the surveillance coverage, alarm systems, and staff access procedures before proposing a solution.

Which concept best describes this approach?

A. Mechanical lock optimization
B. Systems thinking in security
C. Emergency access planning
D. Hardware standardisation

**Question 2**

A retail store installs a high-security cylinder, but staff frequently leave the door unlocked because retrieving the key from the manager slows down workflow.

What security issue does this scenario most clearly demonstrate?

A. Security failures often result from user behaviour rather than hardware weakness
B. Locks should always be replaced with electronic systems
C. Mechanical locks are obsolete in commercial environments
D. Stronger locks automatically prevent misuse

**Question 3**

A university campus uses electronic card readers, surveillance cameras, security lighting, and reinforced doors to protect laboratory areas. Each element works together to protect the facility.

Which security principle does this illustrate?

A. Key duplication management
B. Single-point security control
C. Layered security strategy
D. Mechanical redundancy

**Question 4**

A hospital installs electronic access control for medication storage. When the system loses power, authorized staff can still enter using a secure mechanical override.

Which systems-thinking principle does this design reflect?

A. Fail-safe and resilience planning
B. Simplified credential management
C. Surveillance integration
D. Door hardware standardisation

**Question 5**

A locksmith notices that a keypad lock frequently records incorrect PIN attempts. Investigation shows that the keypad display is difficult to read in low lighting.

How should a security thinker interpret this situation?

A. As a mechanical lock failure
B. As a human-interface usability issue

C. As a credential theft attempt
D. As an electronic lock programming fault

**Question 6**

A security professional regularly reviews alarm logs, user complaints, and maintenance reports to identify patterns that might indicate system weaknesses.

Which systems-thinking concept is being applied?

A. Credential revocation
B. Hardware reinforcement
C. Security feedback analysis
D. Lock replacement planning

**Question 7**

A locksmith attends manufacturer training courses and studies new technologies such as mobile credentials, biometric authentication, and smart locks in order to stay current in the profession.

Which professional principle does this behaviour represent?

A. Continuous learning and adaptation
B. Access credential auditing
C. Security hardware specialization
D. Incident recovery planning

**Question 8**

A locksmith installs a strong deadbolt for a records room but later reflects on the job and realises the system failed to address how employees frequently need temporary access. The locksmith recommends upgrading to an electronic access system with individual credentials.

Which professional practice does this example demonstrate?

A. Credential encryption design
B. Security policy enforcement
C. Reflective practice in locksmithing
D. Hardware durability testing

**Question 9**

A security consultant evaluates a building's doors, access control systems, surveillance coverage, staff workflows, and emergency procedures before designing a security plan.

Which professional role does this scenario best represent?

A. Locksmith as a mechanical technician
B. Locksmith as a security strategist
C. Locksmith as a building contractor
D. Locksmith as a maintenance supervisor

**Question 10**

A facility manager asks a locksmith to improve security for a building. Instead of recommending a single stronger lock, the locksmith proposes reinforced doors, access control credentials, CCTV coverage, and improved key management policies.

What principle of effective locksmithing practice is demonstrated?

A. Using a single high-security device
B. Prioritizing aesthetic door hardware
C. Implementing layered and integrated security solutions
D. Eliminating electronic security systems

# Chapter Review Questions Answer Key

Chapter 1

1. C; 2. B; 3. C; 4. C; 5. B; 6. C; 7. A; 8. B; 9. B; 10. B

Chapter 2

1. B; 2. C; 3. A; 4. B; 5. B; 6. B; 7. B; 8. B; 9. B; 10. B

Chapter 3

1. A; 2. C; 3. B; 4. B; 5. A; 6. A; 7. B; 8. B; 9. A; 10. B

Chapter 4

1. C; 2. B; 3. C; 4. B; 5. B; 6. B; 7. B; 8. C; 9. A; 10. B

Chapter 5

1. B; 2. C; 3. B; 4. B; 5. B; 6. B; 7. C; 8. B; 9. A; 10. B

Chapter 6

1. B; 2. B; 3. B; 4. B; 5. B; 6. A; 7. B; 8. B; 9. B; 10. B

Chapter 7

1. B; 2. C; 3. B; 4. B; 5. C; 6. B; 7. B; 8. B; 9. B; 10. B

Chapter 8

1. A; 2. D; 3. C; 4. D; 5. D; 6. D; 7. A; 8. D; 9. C; 10. D

Chapter 9

1. C; 2. A; 3. B; 4. A; 5. D; 6. B; 7. C; 8. A; 9. B; 10. C

Chapter 10

1. A; 2. B; 3. C; 4. A; 5. C; 6. B; 7. A; 8. B; 9. C; 10. A

Chapter 11

1. B; 2. C; 3. B; 4. B; 5. B; 6. B; 7. B; 8. A; 9. B; 10. B

Chapter 12

1. B; 2. C; 3. B; 4. B; 5. C; 6. B; 7. A; 8. B; 9. C; 10. B

Chapter 13

1. B; 2. A; 3. B; 4. B; 5. C; 6. B; 7. B; 8. B; 9. B; 10. C

Chapter 14

1. B; 2. A; 3. B; 4. B; 5. B; 6. B; 7. A; 8. B; 9. B; 10. B

Chapter 15

1. B; 2. C; 3. C; 4. B; 5. C; 6. A; 7. C; 8. B; 9. B; 10. C

Chapter 16

1. A; 2. A; 3. B; 4. C; 5. B; 6. B; 7. A; 8. C; 9. B; 10. C

Chapter 17

1. B; 2. B; 3. C; 4. C; 5. B; 6. B; 7. C; 8. B; 9. C; 10. C

Chapter 18

1. A; 2. C; 3. A; 4. A; 5. C; 6. B; 7. B; 8. B; 9. A; 10. B

Chapter 19

1. B; 2. C; 3. B; 4. D; 5. C; 6. B; 7. B; 8. C; 9. C; 10. B

Chapter 20

1. B; 2. A; 3. C; 4. A; 5. B; 6. C; 7. A; 8. C; 9. B; 10. C

# References


1. Gao, Q., *Biometric Authentication in Smart Grid.* 2012: p. 1-5.
2. Sai, K.N., T.D. Sunil, and M.N. Eshwarappa, *A Comprehensive Review of Door Lock Security Systems.* International Journal of Circuit Computing and Networking, 2024. **5**(1): p. 12-17.
3. Lindorfer, M., *Untitled.* 2020.
4. El-Khatib, K., et al., *Workshop on Cybersecurity.* 2009: p. 366.
5. Srinivas, C., *Smart Contact Lock: A Standalone App for Advanced Mobile Security Against Hacking.* International Journal on Science and Technology, 2025. **16**(1).
6. Chouraik, C., R. El-founir, and K. Taybi, *Comprehensive Review of Smart City Cybersecurity Strategies.* International Journal of Applied Research in Social Sciences, 2024. **6**(9): p. 2112-2120.
7. Oliha, J.S., P.W. Biu, and O. Chimezie, *Securing the Smart City: A Review of Cybersecurity Challenges and Strategies.* Open Access Research Journal of Multidisciplinary Studies, 2024. **7**(1): p. 094-101.
8. Botto-Tobar, M.B., S. Rehan, and R.P. Verma, *Protecting Smart Home From Cybersecurity Threats Strategies for Homeowners.* 2023. **12**(2): p. 83-98.
9. Angulo, J. and E. Wästlund, *Exploring Touch-Screen Biometrics for User Identification on Smart Phones.* 2012: p. 130-143.
10. Koong, C.-S., T.-I. Yang, and C.C. Tseng, *A User Authentication Scheme Using Physiological and Behavioral Biometrics for Multitouch Devices.* The Scientific World Journal, 2014. **2014**: p. 1-12.
11. Tait, B.L., *Secure Cloud-Based Biometric Authentication Utilising Smart Devices for Electronic Transactions.* International Journal of Electronic Security and Digital Forensics, 2014. **6**(1): p. 52.
12. Hussain, M. and U. Jain, *Simple and Secure Device Authentication Mechanism for Smart Environments Using Internet of Things Devices.* International Journal of Digital & Analog Cabled Systems, 2020. **33**(16).
13. Aziz, M.A., et al., *Vulnerability Analysis of Smart Lock Using NIST SP 800-115 Method.* Jurnal Penelitian Pendidikan Ipa, 2025. **11**(8): p. 264-272.
14. Fauzi, R., P.I.D.C. Wulan, and D.P. Perdana, *Systematic Literature Review on Cybersecurity Issues in the Smart Home Environment.* Compiler, 2024. **13**(2): p. 143.
15. Ghouzali, S., et al., *Trace Attack Against Biometric Mobile Applications.* Mobile Information Systems, 2016. **2016**: p. 1-15.
16. Wiercioch, A., S. Teufel, and B. Teufel, *The Authentication Dilemma.* Journal of Software, 2018. **13**(5): p. 277-286.
17. Choi, J., H. Cho, and J.H. Yi, *Personal Information Leaks With Automatic Login in Mobile Social Network Services.* Entropy, 2015. **17**(6): p. 3947-3962.
18. Teimoory, P., *Towards Robust Security in Smart Payment Systems: Challenges and Solutions.* Smart Cities and Regional Development (Scrd) Journal, 2025. **9**(3): p. 29-38.
19. Alam, T., *CMI Computing: A Cloud, MANET and Internet of Things Integration for Future Internet.* SSRN Electronic Journal, 2020.
20. Elkhameesy, N. and H.A.R. Mohamed, *A Proposed Model for Datacenter in -Depth Defense to Enhance Continual Security(Applied Study to ENR Datacenter – Egyptian National Railways).* International Journal of Information Technology and Computer Science, 2013. **5**(4): p. 55-67.
21. Qahtani, E.A., Y. Javed, and M. Shehab, *User Perceptions of Gmail's Confidential Mode.* Proceedings on Privacy Enhancing Technologies, 2021. **2022**(1): p. 187-206.

22. Aljeaid, D., et al., *Assessment of End-User Susceptibility to Cybersecurity Threats in Saudi Arabia by Simulating Phishing Attacks*. Information, 2020. **11**(12): p. 547.
23. Savitha, M., *A Unique Secure Multimodal Biometrics-Based User Authenticated Key Exchange Protocol for Generic HIoT Networks*. International Journal of Emerging Trends in Engineering Research, 2020. **8**(5): p. 1610-1619.
24. Sethuraman, S.C., et al., *Loki: A Physical Security Key Compatible IoT Based Lock for Protecting Physical Assets*. Ieee Access, 2022. **10**: p. 112721-112730.
25. Teague, C., L. Green, and D. Leith, *An Ambience of Power? Challenges Inherent in the Role of the Public Transport Transit Officer*. M/C Journal, 2010. **13**(2).
26. Stotz, T., et al., *Evaluating the Perceived Efficacy of Randomized Security Measures at Airports*. Risk Analysis, 2020. **40**(7): p. 1469-1480.
27. Nyemezu, C.O. and A.C. Opurum, *Breaking Through Barriers: Exploring the Connection Between Crime Obstructers and Security of Electronic Information Resources*. Alexandria the Journal of National and International Library and Information Issues, 2025.
28. Steinbruner, J., *The Cybernetic Theory of Decision*. 2021.
29. Życzkowski, M., et al., *Module Multisensor System for Strategic Objects Protection*. 2009.
30. Turtiainen, H., et al., *CCTV-FullyAware: Toward End-to-End Feasible Privacy-Enhancing and CCTV Forensics Applications*. 2022.
31. Nano, C.R., et al., *Perceptions of Safety and Crime Deterrence: The Impact of CCTV Cameras in Business Establishments*. International Journal of Research and Innovation in Social Science, 2025. **9**(11): p. 6123-6136.
32. Nte, N.D., G. Gande, and M. Uzorka, *The Challenges and Prospects of ICTs in Crime Prevention and Management in Nigeria: A Review of CCTV Cameras in Abuja*. Jurnal Pengabdian Hukum Indonesia (Indonesian Journal of Legal Community Engagement) Jphi, 2020. **5**(1): p. 75-100.
33. Agboola, T.O., J. Adegede, and J. Jacob, *Balancing Usability and Security in Secure System Design: A Comprehensive Study on Principles, Implementation, and Impact on Usability*. International Journal of Computing Sciences Research, 2024. **8**: p. 2995-3009.
34. Ardito, C., et al., *Handling Security, Usability, User Experience and Reliability in User-Centered Development Processes*. 2019: p. 759-762.
35. Liu, W., et al., *Center-Less Single Sign-on With Privacy-Preserving Remote Biometric-Based ID-MAKA Scheme for Mobile Cloud Computing Services*. Ieee Access, 2019. **7**: p. 137770-137783.
36. Ferreira, A., *Log Analysis of Human Computer Interactions Regarding Break the Glass Accesses to Genetic Reports*. 2013.
37. Ophoff, J., G. Johnson, and K. Renaud, *Cognitive Function vs. Accessible Authentication*. 2021: p. 1-5.
38. Shava, F.B. and D.v. Greunen, *Factors Affecting User Experience With Security Features: A Case Study of an Academic Institution in Namibia*. 2013: p. 1-8.
39. Godwin, Z.I., *How Human-Centered Cybersecurity (HCC) Training Fosters Organizational Cybersecurity Cultures (CSC)*. 2024: p. 187-204.
40. Lutz, O.H.-M., et al., *That Password Doesn't Sound Right*. 2020: p. 206-213.
41. Kazanskaia, A.N., *Confidentiality and Data Protection in Non-Profit Consulting: Legal, Ethical, and Practical Dimensions*. 2025.
42. Weichert, T., *Gesundheitsdatenschutz In vernetzten Zeiten*. Bundesgesundheitsblatt - Gesundheitsforschung - Gesundheitsschutz, 2018. **61**(3): p. 285-290.
43. Itani, W., A. Kayssi, and A. Chehab, *Hardware-Based Security for Ensuring Data Privacy in the Cloud*. 2013: p. 147-170.

44. Shafira Nur Lailiyah Maghfiroh, N., *Etika Dalam Asesmen Psikologi : Strategi Menjaga Kerahasiaan Dan Objektivitas Di Bimbingan Dan Konseling.* Harmoni Jurnal Ilmu Komunikasi Dan Sosial, 2025. **3**(2): p. 150-156.
45. Johan, E.S., *Peran Otoritas Jasa Keuangan Terhadap Perlindungan Kemanan Dan Kerahasiaan Data Pribadi Nasabah Pada Produk Perbankan Di Era Digital.* Mahalini Journal of Business Law, 2025. **2**(1): p. 24-38.
46. Rapôso, C.F.L., *Challenges and Compliance of Cloud Computing With the Lgpd: A Systematic Review.* 2025.
47. Vitla, S., *Securing the Physical and Digital Frontier: Leveraging Identity and Access Management (IAM) to Address the Lack of Controls on Physical Access to Sensitive Systems.* International Journal of Science and Research Archive, 2022. **6**(2): p. 108-125.
48. Sudarsan, S.V., O. Schelén, and U. Bodin, *Survey on Delegated and Self-Contained Authorization Techniques in CPS and IoT.* Ieee Access, 2021. **9**: p. 98169-98184.
49. Ovsianko, D. and E. Nyemkova, *Smart Contracts as Privacy-Preserving Mechanisms in Distributed Digital Twin Systems.* Cybersecurity Education Science Technique, 2025. **1**(29): p. 107-123.
50. Mohamed, S.I., et al., *ABACS: Attribute-Based Access Control System Using Digital Keys.* 2024.
51. Nkenyereye, L., et al., *Fog-Based and Secure Framework for Personal Health Records Systems.* Computers Materials & Continua, 2021. **66**(2): p. 1937-1948.
52. Garcia, R.D., et al., *Towards a Decentralized and Privacy-Preserving Quality of Experience System for 6G Networks.* Distributed Ledger Technologies Research and Practice, 2025.
53. Ikhalea, N., et al., *A Conceptual Framework for Enhancing Healthcare Data Security Using Blockchain and AI.* International Journal of Advanced Multidisciplinary Research and Studies, 2024. **4**(6): p. 1573-1590.
54. Wankhade, S., *Smart Lock System Using Iot and Solana Blockchain.* International Journal of Scientific Research in Engineering and Management, 2024. **08**(05): p. 1-5.
55. Zhang, J. and A. Datta, *Blockchain-Enabled Data Governance for Privacy-Preserved Sharing of Confidential Data.* Peerj Computer Science, 2024. **10**: p. e2581.
56. Li, M., et al., *Eunomia: Anonymous and Secure Vehicular Digital Forensics Based on Blockchain.* Ieee Transactions on Dependable and Secure Computing, 2023. **20**(1): p. 225-241.
57. Okoth, P.K., *Security Challenges in Civil Registration: Safeguarding Vital Information in an Evolving Landscape.* World Journal of Advanced Research and Reviews, 2023. **19**(1): p. 1051-1071.
58. Brown, C.L., *Health-Care Data Protection and Biometric Authentication Policies: Comparative Culture and Technology Acceptance in China and in the United States.* Review of Policy Research, 2012. **29**(1): p. 141-159.
59. Tsoukalas, I.A. and P.D. Siozos, *Privacy and Anonymity in the Information Society – Challenges for the European Union.* The Scientific World Journal, 2011. **11**: p. 458-462.
60. Veiga, A.d., *An Information Privacy Culture Instrument to Measure Consumer Privacy Expectations and Confidence.* Information and Computer Security, 2018. **26**(3): p. 338-364.
61. Huynh, T., *Everyone Is Safe Now: Constructing the Meaning of Data Privacy Regulation in Vietnam.* Asian Journal of Law and Society, 2024. **11**(4): p. 530-558.
62. Yang, L., Y. Lin, and B. Chen, *Practice and Prospect of Regulating Personal Data Protection in China.* Laws, 2024. **13**(6): p. 78.
63. Arun, K.D. and V. Varsha, *≪b>Artificial Intelligence in Personalisation and Its Impact on Consumer Trust</B>.* African Journal of Commercial Studies, 2026. **7**(1): p. 41-45.
64. Alturki, R., et al., *Analysis of an eHealth App: Privacy, Security and Usability.* International Journal of Advanced Computer Science and Applications, 2020. **11**(4).

65. Krasnova, H., N.F. Veltri, and W.E. Garah, *Effectiveness of Justice-Based Measures in Managing Trust and Privacy Concerns on Social Networking Sites: An Intercultural Perspective.* Communications of the Association for Information Systems, 2014. **35**.
66. Kumar, A. and D. Zhang, *Ethics and Policy of Biometrics*. 2010.
67. Liu, Y., et al., *PDPHE: Personal Data Protection for Trans-Border Transmission Based on Homomorphic Encryption*. Electronics, 2024. **13**(10): p. 1959.
68. Farrell, G. and N. Tilley, *Elegant Security: Concept, Evidence and Implications.* European Journal of Criminology, 2020. **19**(5): p. 932-953.
69. Cervi, A., *Keys and Locks*. 2017: p. 1-2.
70. Zhang, Y. and K.-H. Hsiao, *Traditional Chinese Yuekou Locks.* Mechanical Sciences, 2020. **11**(2): p. 411-423.
71. Wang, F., D. Yu, and Y. Jin, *Innovative Design of Pin Tumbler Lock Based on TRIZ.* Advanced Materials Research, 2012. **591-593**: p. 71-76.
72. Hanifullah, M.A. and K.D. Irianto, *Smart Door Lock System: A Mobile Application Development Based on IoT Technology.* International Journal Software Engineering and Computer Science (Ijsecs), 2025. **5**(2): p. 842-852.
73. Plumtree, W.G., *The Examination of Disc and Pin Tumbler Locks for Tool Marks Made by Lock Picks.* Journal of Forensic Sciences, 1975. **20**(4): p. 656-667.
74. Aronson, A., R. Pertsev, and A. Cohen, *A Novel Method for Extracting Forensic Evidence From Aluminum Foil That Has Been Used as Part of a "Foil Impressioning" Kit, Using Toolmarks Comparison Techniques.* Journal of Forensic Sciences, 2025. **71**(1): p. 211-221.
75. Meisel, L.O. and M. Weber, *Acid Attacks on Pin-tumbler Locks.* Journal of Forensic Sciences, 2024. **70**(1): p. 340-348.
76. Volkov, N., et al., *The Class Characteristic Mark of the H&M Mul-T-Lock Picking Tool in Toolmarks Examination.* Journal of Forensic Sciences, 2014. **59**(4): p. 1109-1112.
77. Boroš, M., et al., *Mechanical Characteristics of Cylinder Pin Tumbler Locks as They Relate to Resistance Testing.* Communications - Scientific Letters of the University of Zilina, 2018. **20**(2): p. 96-101.
78. Parker, A.E. and V. Parker, *Opto-Electronic Ignition Lock - The Ultimate Antitheft Device.* 1993. **1**.
79. Lad, A., *Operational Control of Keying Systems in Large High-Rise Building Projects.* International Journal of Innovative Research in Engineering & Multidisciplinary Physical Sciences, 2026. **14**(1).
80. Miller, D. and D.O. Petrov, *Research Specifics of Separate Types of Cylinder Locks Picked by Self Impressioning Master Keys.* Theory and Practice of Forensic Science and Criminalistics, 2018. **18**: p. 316-323.
81. Weiner, M.P., et al., *Security Analysis of a Widely Deployed Locking System.* 2013: p. 929-940.
82. Revathi, S., et al., *End to End Encrypted Smart Lock Using RSA Based on Opinion From Social Media Review Comments.* International Journal on Recent and Innovation Trends in Computing and Communication, 2023. **11**(11s): p. 282-290.
83. Blaze, M., *Toward a Broader View of Security Protocols.* 2006: p. 106-120.
84. Siegel, J. and S.E. Sarma, *Using Open Channels to Trigger the Invited, Unintended Consequences of the Internet of Things.* Ieee Security & Privacy, 2019. **17**(3): p. 49-55.
85. Pidnebennyi, R., *Comparative Analysis of Smart Lock Breaching Methods and Their Forensic Examination in Ukraine and the USA.* Theory and Practice of Forensic Science and Criminalistics, 2025. **38**(1): p. 182-199.
86. Lu, Y., *Research on Authentication Encryption Mechanism Based on Intelligent Door Lock Vulnerability Risk.* Matec Web of Conferences, 2021. **336**: p. 08009.
87. Strba, S., *Internet of Things Security: Ongoing Threats and Proposed Solutions.*

88. Wu, J., et al., *SPCL: A Smart Access Control System That Supports Blockchain.* Applied Sciences, 2024. **14**(7): p. 2978.
89. Putra, F.P.E., et al., *Service Quality Analysis of RFID-Based Smart Door Lock in Front One Azana Style Hotel Area.* Brilliance Research of Artificial Intelligence, 2024. **4**(1): p. 372-381.
90. Vardakis, G.E., et al., *Review of Smart-Home Security Using the Internet of Things.* Electronics, 2024. **13**(16): p. 3343.
91. Derbali, M., *Toward Secure Door Lock System: Development IoT Smart Door Lock Device.* 2023.
92. Naidu, G.H.S., *IoT Innovation in Hospitality: A Comprehensive Technical Analysis of Implementation and Impact.* International Journal for Multidisciplinary Research, 2024. **6**(6).
93. Zhang, T., et al., *Smart Service Interactional Experience for Family Travelers: Scale Development and Validation.* Journal of Hospitality & Tourism Research, 2024. **49**(4): p. 667-687.
94. Gad, A.A.A.S., *Evaluating Internet of Things Application in Sharm-El Sheik Hotels.* International Journal of Tourism Archaeology and Hospitality, 2022. **2**(3): p. 0-0.
95. Heartfield, R., et al., *A Taxonomy of Cyber-Physical Threats and Impact in the Smart Home.* Computers & Security, 2018. **78**: p. 398-428.
96. Barua, A., et al., *Security and Privacy Threats for Bluetooth Low Energy in IoT and Wearable Devices: A Comprehensive Survey.* Ieee Open Journal of the Communications Society, 2022. **3**: p. 251-281.
97. Syed, A., *Secure Integration of Oracle APEX Application With IoT Devices for the "My Smart Home" App.* International Journal of Leading Research Publication, 2023. **4**(11).
98. Muwanga, K.E. and E. Muwanguzi, *End User Security Using Smart Devices With Ability to Access IoT Services.* International Journal of Innovative Science and Research Technology, 2024: p. 2805-2810.
99. Asghar, I., et al., *Fortifying Smart Home Security: A Robust and Efficient User-Authentication Scheme to Counter Node Capture Attacks.* Sensors, 2023. **23**(16): p. 7268.
100. Satrya, G.B., Y.M. Agus, and A.B. Mnaouer, *A Comparative Study of Post-Quantum Cryptographic Algorithm Implementations for Secure and Efficient Energy Systems Monitoring.* Electronics, 2023. **12**(18): p. 3824.
101. Kosaraju, P. and V.M. Nadella, *Security and Privacy in IoT Ecosystems.* Universal Library of Engineering Technology, 2022: p. 69-78.
102. Purohit, A. and V. Sharan, *Enhancing Smart Home Security Using IoT-Based Intrusion Detection Systems.* International Journal of Scientific Research in Engineering and Management, 2024. **09**(01): p. 1-7.
103. Obaid, A.H., et al., *Dynamic Key Revocation and Recovery Framework for Smart City Authentication Systems.* International Journal of Innovative Research and Scientific Studies, 2025. **8**(4): p. 2112-2123.
104. Piedrahita-Solórzano, G., A. Gutierrez, and A.P. Gordillo, *Data Security Threats on Smart Devices at Home.* Arpha Conference Abstracts, 2023. **6**.
105. Rivas Bustos, D.F., J. Gutierrez, and S.J. Rueda, *Automated Vulnerability Scanning and Prioritisation for Domestic IoT Devices/Smart Homes: A Theoretical Framework.* 2025.
106. Pramod, T.C., et al., *CKMI: Comprehensive Key Management Infrastructure Design for Industrial Automation and Control Systems.* Future Internet, 2019. **11**(6): p. 126.
107. Li, T., et al., *Intelligent Lock Design Based on BLE.* Destech Transactions on Computer Science and Engineering, 2019(cscme).
108. Antonius, A. and H. Hugeng, *Automatic Door Lock Using Microcontroller.* International Journal of Application on Sciences Technology and Engineering, 2023. **1**(1): p. 361-364.
109. Lee, J., et al., *Securing Designs Against Scan-Based Side-Channel Attacks.* Ieee Transactions on Dependable and Secure Computing, 2007. **4**(4): p. 325-336.
110. Lee, J., et al., *Securing Scan Design Using Lock and Key Technique.* p. 51-62.

111. Lopez, K. and A. Rezaei, *K-Gate Lock: Multi-Key Logic Locking Using Input Encoding Against Oracle-Guided Attacks*. 2025: p. 794-800.
112. Su, Z., et al., *Design, Detection and Management of Mechanical Security Code Manhole Cover*. Matec Web of Conferences, 2018. **228**: p. 03007.
113. Pranata, S. and H.T. Nugroho, *2FYSH: Two-Factor Authentication You Should Have for Password Replacement*. Telkomnika (Telecommunication Computing Electronics and Control), 2019. **17**(2): p. 693.
114. Bennett, J., *Disrupting Prison Managerialism: Managing Prisons in an Age of Pandemic*. Incarceration an International Journal of Imprisonment Detention and Coercive Confinement, 2023. **4**.
115. Aziz, R.Z.A., A. Ikhsanudin, and M.S. Hasibuan, *GOVERNANCE EVALUATION ELECTRONIC SECURITY SYSTEM (ESS) (Case Study: ABC Central Bank)*. Sinkron, 2024. **8**(2): p. 713-726.
116. Ingale, K., *SecureShield: A Smart Security System*. Journal of Information Systems Engineering & Management, 2025. **10**(51s): p. 31-36.
117. *Fingerprint-Based Smart Door Locking System for Banks*. International Research Journal of Modernization in Engineering Technology and Science, 2024.
118. Chong, P.L., et al., *Designing an Autonomous Triggering Control System via Motion Detection for IoT Based Smart Home Surveillance CCTV Camera*. Malaysian Journal of Science and Advanced Technology, 2023: p. 80-88.
119. Fitzgerald, W.M., et al., *Anomaly Analysis for Physical Access Control Security Configuration*. 2012: p. 1-8.
120. Hamed, B., *Efficient Authorized Access Security System Control Using ATMEL 89C55 &Amp; Mobile Bluetooth*. International Journal of Computer Theory and Engineering, 2012: p. 112-118.
121. Malhotra, B., et al., *The Design, Development, and Implementation of a Web-Enabled Informatics Platform to Enhance the Well-Being of Individuals Aged 18-24 Years: Protocol for an Experimental Study (Preprint)*. 2022.
122. Chesher, C., *Mining Robotics and Media Change*. M/C Journal, 2013. **16**(2).
123. Jeong, S., *Design on Novel Door Lock Using Minimizing Physical Exposure and Fingerprint Recognition Technology*. Joiv International Journal on Informatics Visualization, 2022. **6**(1): p. 103.
124. Erwan, A.N.M., M.N.H.M. Alfian, and M.S.M. Adenan, *Smart Door Lock*. International Journal of Recent Technology and Applied Science (Ijortas), 2021. **3**(1): p. 1-15.
125. Adeosun, F.E. and A.E. Oke, *Examining the Awareness and Usage of Cyber Physical Systems for Construction Projects in Nigeria*. Journal of Engineering Design and Technology, 2022. **22**(1): p. 281-294.
126. Gebsombut, N., et al., *Smart Home Security: Factors Affecting Consumer Intention to Put Smart Home Security in Place*. Sage Open, 2025. **15**(3).
127. Simon, O.A., et al., *Electronic Doorbell System Using Keypad and GSM*. International Journal of Informatics and Communication Technology (Ij-Ict), 2020. **9**(3): p. 212.
128. Weise, M., et al., *OSSDIP: Open Source Secure Data Infrastructure and Processes Supporting Data Visiting*. Data Science Journal, 2022. **21**.
129. Manning, M., G.T.W. Wong, and M.W.S. Ip, *An Evaluation of an Australia-Based Home Burglary Prevention Program*. Security Journal, 2022. **36**(4): p. 593-614.
130. *Design and Development of Intelligent Access Control System Integrating Hi3861 and K210*. International Journal of Science and Engineering Applications, 2024.
131. Hasibuan, A., et al., *Performance of Secure Door Access System Using Identification Numbers, Fingerprints, and Facial Recognition*. Journal of Physics Conference Series, 2024. **2733**(1): p. 012003.

132. L, U.K.N., et al., *A Review on Bluetooth Low Energy Enabled Door Locking Systems*. 2023. **7**(9): p. 156-160.
133. Lively, C.P., *Social Isolation of Older Adults in Long Term Care as a Result of COVID-19 Mitigation Measures During the COVID-19 Pandemic*. Voices in Bioethics, 2021. **7**.
134. Champion, K., *A Risky Business? The Role of Incentives and Runaway Production in Securing a Screen Industries Production Base in Scotland*. M/C Journal, 2016. **19**(3).
135. Zegeye, A., *Fundamental and Key Concerns of the Shared Use Vehicle System*. Journal of Computing and Natural Science, 2023: p. 80-90.
136. Khurshid, A., et al., *Secure-CamFlow: A Device-oriented Security Model to Assist Information Flow Control Systems in Cloud Environments for IoTs*. Concurrency and Computation Practice and Experience, 2018. **31**(8).
137. Ahmad, W., et al., *Cyber Security in IoT-Based Cloud Computing: A Comprehensive Survey*. Electronics, 2021. **11**(1): p. 16.
138. Singh, N., R. Buyya, and H. Kim, *Securing Cloud-Based Internet of Things: Challenges and Mitigations*. Sensors, 2024. **25**(1): p. 79.
139. Surianarayanan, C. and P.R. Chelliah, *Integration of the Internet of Things and Cloud*. International Journal of Cloud Applications and Computing, 2023. **13**(1): p. 1-30.
140. Saleem, K., et al., *End-to-End Security Enabled Intelligent Remote IoT Monitoring System*. Frontiers in Physics, 2024. **12**.
141. Riad, K., *Robust Access Control for Secure IoT Outsourcing With Leakage Resilience*. Sensors, 2025. **25**(3): p. 625.
142. Mehmood, F., S. Ahmad, and D.H. Kim, *Design and Implementation of an Interworking IoT Platform and Marketplace in Cloud of Things*. Sustainability, 2019. **11**(21): p. 5952.
143. Shafagh, H., et al., *Towards Blockchain-Based Auditable Storage and Sharing of IoT Data*. 2017: p. 45-50.
144. Deng, Y., et al., *A Secure Subscription-Push Service Scheme Based on Blockchain and Edge Computing for IoT*. Ksii Transactions on Internet and Information Systems, 2022. **16**(2).
145. Vömel, C., et al., *The Secret Life of Keys: On the Calculation of Mechanical Lock Systems*. Siam Review, 2017. **59**(2): p. 393-422.
146. Reed, H., *A Beginner's Guide To Lock Impressioning*. 2018, United Locksmith: Lock Blog.
147. He, J., et al., *Optimization of Several Surface Treatment Processes for Alleviating Fretting Damage of a Locking Pin*. Friction, 2021. **10**(8): p. 1217-1233.
148. Zu, H., et al., *Effect of Metallic Coatings on the Wear Performance and Mechanism of 30CrMnSiNi2A Steel*. Materials, 2023. **16**(18): p. 6191.
149. None, N., *Safeguards Summary Event List (SSEL), January 1, 1990--December 31, 1996, Vol. 2, Rev. 5*. 1997.
150. Ayyubi, S., et al., *The System Monitoring Laboratorium Electrical Engineering With Internet of Things Based to Be Smaart Campus*. Procedia of Engineering and Life Science, 2021. **1**(2).
151. Hussain, S.M., et al., *Smart Security: An IoT-NFC Lock System for Efficient Access Management*. Malaysian Journal of Science and Advanced Technology, 2024: p. 322-329.
152. Nathiq, T.A., *Rancang Bangun Smart Doorlock Kamar Mesin Dan Anjungan Berbasis RFID Dan Arduino*. Venus Jurnal Publikasi Rumpun Ilmu Teknik, 2024. **2**(4): p. 14-40.
153. A, M.K., I.A. M, and R. Gowtham, *Revolutionizing Home Security: A Comprehensive Overview of an Advanced RFID Door Lock System for Keyless Access and Smart Home Protection*. Asian Journal of Applied Science and Technology, 2024. **08**(01): p. 01-13.
154. Alim, S.A. and T. Eseyin, *Development of a Fingerprint Lock Safe With Vibration Sensor*. Elektrika-Journal of Electrical Engineering, 2021. **20**(1): p. 9-14.

155. Duan, R., et al., *Communication and Remote Monitoring of Substation Access and Locks Control Based on Trusted WLAN Technology*. 2025: p. 122.
156. La Madrid, J., *Fingerprint Door Unlocking and Light Control System With Short Message Service (SMS) Notification Capability*. Isabela State University Linker Journal of Engineering Computing and Technology, 2025: p. 81.
157. Kim, Y.-i., et al., *Development of Guidelines for Physical Security Systems of Small-Scale Water Supplies*. Journal of the Korean Society of Water and Wastewater, 2013. **27**(6): p. 723-730.
158. Wahab, I.P.S.M.I., *Understanding the Implications of Non-Accused DNA at Crime Scenes for Criminal Defence Lawyers*. International Journal for Multidisciplinary Research, 2024. **6**(6).
159. Guleria, A., et al., *Global Adoption of Facial Recognition Technology With Special Reference to India—Present Status and Future Recommendations*. Medicine Science and the Law, 2024. **64**(3): p. 236-244.
160. datagram, *Lockpicking Forensics*. 2009.
161. DeMatteo, D., et al., *Legal and Ethical Requirements and Considerations Relevant to Forensic Practice*. 2025: p. 145-169.
162. Mulyk, Y., *Responsibility of a Foreign Expert in the Process of Economic Examination*. Eastern Europe Economy Business and Management, 2021(4(31)).
163. Triyanto Agung Praptono Wibowo, N., N. Rina Arum Prastyanti, and N. Zaenal Mustofa, *Kajian Yuridis Normatif Pertanggungjawaban Dokter Atas Kesalahan Diagnosis Dalam Pelayanan Medis Yang Menyebabkan Kematian Pasien*. Pemuliaan Keadilan, 2026. **3**(1): p. 29-36.
164. Qomariah, S.N. and H.D. Revani, *Analysis of Health Workers' Practice Negligence: Review of Ethical Codes and Legal Aspects*. Equalegum International Law Journal, 2025: p. 113-121.
165. Al-Azri, N.H., *Medical Liability of Healthcare Professionals Under Omani Law: A Primer*. Oman Medical Journal, 2020. **35**(5): p. e182-e182.
166. Belichenko, O. and A. Нємцев, *Medical Law in Action: Between Patient Protection and Doctor's Responsibility*. Visnik Nacional'nogo Universitetu «lvivska Politehnika» Seria Uridicni Nauki, 2025. **12**(46): p. 11-22.
167. Zawiślak-Białek, A.H., *Disciplinary Liability of Attorney-at-Law for Breach Professional Secrecy*. Ars Iuridica, 2023. **23**(1): p. 131-136.
168. Weinstock, R. and D. Weinstock, *Clinical Flexibility and Confidentiality: Effects of Reporting Laws*. Psychiatric Quarterly, 1989. **60**(3): p. 195-214.
169. Spector-Bagdady, K., et al., *Stemming the Standard-of-Care Sprawl*. The Hastings Center Report, 2017. **47**(6): p. 16-24.
170. Kamran, M. and S. Syahrul, *Professionalism and Medical Liability of Physician: Patient Protection and Civil Remedies*. Sign Jurnal Hukum, 2024. **6**(2): p. 206-218.
171. Steinert, T. and T. Henking, *Law and Psychiatry—current and Future Perspectives*. Frontiers in Public Health, 2022. **10**.
172. DeMatteo, D., et al., *Forensic Mental Health Practice and the Law*. 2025.
173. Batista, D.A., et al., *Exploring Blockchain Technology for Chain of Custody Control in Physical Evidence: A Systematic Literature Review*. Journal of Risk and Financial Management, 2023. **16**(8): p. 360.
174. Grigollo, F.V. and R.F. Fernandes, *Study on the Adoption of Historical Record and Chain of Custody by Brazilian Judicial Experts From the Collection to the Disposal of Evidence: A Survey Questionnaire*. Revista De Gestão Social E Ambiental, 2024. **18**(8): p. e08431.
175. D'Anna, T., et al., *The Chain of Custody in the Era of Modern Forensics: From the Classic Procedures for Gathering Evidence to the New Challenges Related to Digital Data*. Healthcare, 2023. **11**(5): p. 634.

176. Prayudi, Y. and A. Sn, *Digital Chain of Custody: State of the Art.* International Journal of Computer Applications, 2015. **114**(5): p. 1-9.
177. Onyeashie, B.I., et al., *A Bibliometric Analysis and Systematic Review of a Blockchain-Based Chain of Custody for Digital Evidence.* 2024: p. 112-131.
178. Ratnasari, D., Y. Prayudi, and B. Sugiantoro, *XML Approach for the Solution of Chain of Custody of Digital Evidence.* International Journal of Computer Applications, 2018. **179**(23): p. 20-25.
179. Jawad, T.M., *Adoption of Chain of Custody Improves Digital Forensic Investigation Process.* Iraqi Journal of Information & Communications Technology, 2018. **1**(2): p. 13-23.
180. Prayudi, Y., A. Ashari, and T.K. Priyambodo, *Digital Evidence Cabinets: A Proposed Framework for Handling Digital Chain of Custody.* International Journal of Computer Applications, 2014. **107**(9): p. 30-36.
181. Zaid, S.M.A. and B.M. Sabbar, *A Design of Digital Chain Custody for Republic of Iraq.* Iraqi Journal of Information & Communications Technology, 2020. **3**(3): p. 24-32.
182. Alketbi, S.K., *Maintaining the Chain of Custody: Anti-Contamination Measures for Trace DNA Evidence.* International Journal of Science and Research Archive, 2023. **8**(2): p. 457-461.
183. Zeresh Mtonga, L. and M. Mupeta, *Design and Development of a Web-Based Digital Evidence Management System for Cybercrime Investigations.* International Journal of Advanced Multidisciplinary Research and Studies, 2026. **6**(1): p. 1534-1539.
184. Grabner, G., A. Ahmed, and N. Baghaei, *Using Blockchain to Preserve Chain of Custody: Cloud Forensics Analysis (S).* 2023. **2023**: p. 380-385.
185. Azimuddin, S.S., et al., *Robust and Secure Evidence Management in Digital Forensics Investigations Using Blockchain Technology.* 2023: p. 214-243.
186. B, A., et al., *A Survey on Forensic Evidence Management Under AWS-S3 Service.* International Journal of Advanced Research in Science Communication and Technology, 2024: p. 101-105.
187. Singh, S., *Evidence Vault.* International Journal of Scientific Research in Engineering and Management, 2024. **08**(05): p. 1-5.
188. Appuhamy, P., *Challenges in Forensic Sampling in the Pre-Analytical Phase While Maintaining Chain of Custody in Sri Lanka and System Improvement in the Medico-Legal Unit of District General Hospital, Matale, Sri Lanka.* Medico-Legal Journal of Sri Lanka, 2022. **10**(2): p. 12.
189. PickPals, *Lock Picking Guide – How to pick a lock*. 2026, PickPals.
190. Ralph, T.C., *How To Pick Locks: The Definitive Guide To Lock Picking*, in *Lock Blog: A resource for consumers, locksmiths, and security professionals*. 2017, United Locksmith.
191. Grim Workshop, *How to Pick a Lock - Lock Picking Basics*. 2026, Grim Workshop,.
192. Chris, *Lock Picking Guides: Types of Locks and How to Pick them*. 2026, UK Bumpkeys.
193. Kurniawan, B. and H.T. Saputra, *Telegram Implementation on Security and Monitoring of Home Door Keys Based on Wemos and Internet of Things.* Journal of Applied Engineering and Technological Science (Jaets), 2022. **4**(1): p. 244-250.
194. Barrera, D., I. Molloy, and H. Huang, *Standardizing IoT Network Security Policy Enforcement.* 2018.
195. Jia, X., et al., *A2 Chain: A Blockchain-Based Decentralized Authentication Scheme for 5g-Enabled IoT.* Mobile Information Systems, 2020. **2020**: p. 1-19.
196. Dakić, V., et al., *Analysis of Azure Zero Trust Architecture Implementation for Mid-Size Organizations.* 2024.
197. Iglezakis, I., *Protecting Identity Without Comprising Privacy.* 2009: p. 62-88.
198. Behringer, F. and P. Baumann, *Integrating Identity and Access Management and Privileged Access Management for Enhanced Identity Security in Financial Institutions: A Zero-Trust Approach.* Cyber Security a Peer-Reviewed Journal, 2025. **9**(2): p. 114.

199. Hardas, B.M., et al., *The Role of Digital Identity (E-Kyc) in Expanding Microfinance Access: Contribution to SDG 10 and SDG 16*. Enterprise Development and Microfinance, 2026. **35**(2).
200. Hashim, M.H. and A.A. Yassin, *An Efficient Mechanism to Prevent the Phishing Attacks*. Iraqi Journal for Electrical and Electronic Engineering, 2021. **17**(1): p. 1-11.
201. Xing, Q., et al., *BRT: An Efficient and Scalable Blockchain-Based Revocation Transparency System for TLS Connections*. Sensors, 2023. **23**(21): p. 8816.
202. Palen, W.D.S., et al., *Development of Smart Key Prototype Management System*. Sar Journal - Science and Research, 2025: p. 42-48.
203. Özdayi, M.S., M. Kantarcıoğlu, and B. Malin, *Leveraging Blockchain for Immutable Logging and Querying Across Multiple Sites*. 2020.
204. Atallah, M.J., et al., *Dynamic and Efficient Key Management for Access Hierarchies*. Acm Transactions on Information and System Security, 2009. **12**(3): p. 1-43.
205. Khan, S., et al., *Accountable and Transparent TLS Certificate Management: An Alternate Public-Key Infrastructure With Verifiable Trusted Parties*. Security and Communication Networks, 2018. **2018**: p. 1-16.
206. Naikwade, N., S. Pathan, and P. Dhotre, *Federated Time-Bound Blockchain-Based Revocation With Ciphertext-Policy Attribute-Based Encryption and Redundant Shard Reassembly*. Engineering Technology & Applied Science Research, 2026. **16**(1): p. 31022-31027.
207. Yilmaz, E., *Machine Identity Management in Modern Enterprise Security: Concepts, Challenges, and the Role of Privileged Access Management Systems*. Engineering Technology & Applied Science Research, 2026. **16**(1): p. 32369-32376.
208. Jeevarathinam, A. and B. Kalaiyarasi, *Account Detection: Automated System for Blocking Malicious Access*. International Scientific Journal of Engineering and Management, 2025. **04**(03): p. 1-7.
209. Nagarajan, M., et al., *Implementation of Cloud Security by Identity Based Encryption*. International Journal of Recent Technology and Engineering (Ijrte), 2019. **8**(4): p. 9812-9816.
210. Shekar, M., et al., *Multi Tenancy Cloud Data With a Shared Privacy Preserving Trusted Keyword Search*. International Journal of Information Technology and Computer Engineering, 2025. **13**(2): p. 1290-1295.
211. Adams, A. and M.A. Sasse, *Users are not the enemy*. Communications of the ACM, 1999. **42**(12): p. 40-46.
212. Herley, C. *So long, and no thanks for the externalities: The rational rejection of security advice by users*. in *Proceedings of the New Security Paradigms Workshop (NSPW)*. 2009.
213. Beautement, A., M.A. Sasse, and M. Wonham. *The compliance budget: Managing security behaviour in organisations*. in *Proceedings of the New Security Paradigms Workshop (NSPW)*. 2008.
214. Kahneman, D., *Thinking, fast and slow*. 2011: Farrar, Straus and Giroux.
215. Reason, J., *Human error*. 1990: Cambridge University Press.
216. Norman, D.A., *The design of everyday things*. 2013: Basic Books.
217. Shenton, F.C., et al., *The Durability, Functionality and Acceptability of Novel Screened Doors and Windows After 4 Years of Use in a Gambian Village: A Cross-Sectional Survey*. Malaria Journal, 2022. **21**(1).
218. Corfar, D.A. and K.D. Tsavdaridis, *A Comprehensive Review and Classification of Inter-Module Connections for Hot-Rolled Steel Modular Building Systems*. Journal of Building Engineering, 2022. **50**: p. 104006.
219. Saiyar, H., *Utilization of Bluetooth Connections for Arduino Based Door Lock Security Tools*. Jurnal Riset Informatika, 2020. **2**(4): p. 213-218.
220. Mäusezahl, M., F. Munkes, and R. Löw, *Tutorial on Laser Locking Techniques and the Manufacturing of Vapor Cells for Spectroscopy*. New Journal of Physics, 2024. **26**(10): p. 105002.

221. K, A.K.K., *Design and Implementation of Car Door Latch Testing Machine.* International Journal for Research in Applied Science and Engineering Technology, 2022. **10**(6): p. 2242-2247.
222. Pettenghi, H., M.J. Avedillo, and J.M. Quintana, *Using Multi-Threshold Threshold Gates in RTD-based Logic Design: A Case Study.* Microelectronics Journal, 2008. **39**(2): p. 241-247.
223. Hoepman, J.-H. and B. Jacobs, *Increased Security Through Open Source.* Communications of the Acm, 2007. **50**(1): p. 79-83.
224. Saltzer, J.H., *Protection and Control of Information Sharing in Multics.* Acm Sigops Operating Systems Review, 1973. **7**(4): p. 119.
225. Rivera-Villicana, J., et al., *Informing a BDI Player Model for an Interactive Narrative.* 2018: p. 417-428.
226. Santos, J.P.O., *From Iron to the Industrial Cloud.* Narodna Umjetnost, 2020. **57**(1): p. 93-107.
227. Janes, B., H. Crawford, and T.J. Oconnor, *Never Ending Story: Authentication and Access Control Design Flaws in Shared IoT Devices.* 2020.
228. Zeteh, K.G. and A.O. Alabi, *An Improved Data Warehouse Security System Using Hybrid Authentication Techniques.* Stem Essentials, 2026: p. 06.
229. Nwogu, C., M. Eze, and C. Okunbor, *Design and Implementation of Access Door Control With Mo-Bile Alert.* International Journal of Engineering & Technology, 2020. **9**(2): p. 480-486.
230. Albalooshi, A. and A. Almahmood, *Biometric Authentication Module to Enhance Cubesat Security.* International Journal of Engineering Science Technologies, 2025. **9**(5).
231. Kumar, A.A. and D. Tl, *Security Measures Implemented in RESTful API Development.* Open Access Research Journal of Engineering and Technology, 2024. **7**(1): p. 105-112.
232. Ayyagiri, A., S. Jain, and A. Aggarwal, *Innovations in Multi-Factor Authentication: Exploring OAuth for Enhanced Security.* Innovative Research Thoughts, 2023. **9**(4): p. 254-267.
233. Ford, W. and B.S. Kaliski, *Server-Assisted Generation of a Strong Secret From a Password.* p. 176-180.
234. Ruchitha, M., *Door Lock System Using Raspberrypi.* International Journal of Scientific Research in Engineering and Management, 2025. **09**(06): p. 1-9.
235. Taware, S., *A Survey on Graphical Password Strategy Authentication.* International Journal of Scientific Research in Engineering and Management, 2024. **08**(05): p. 1-5.
236. Keri, M. and M. Niklekaj, *Strengthening Web Application Security Through Email Verification and JWT Authentication.* Ingenious, 2025. **5**(2): p. 91-105.
237. Chinthalapudi, S.S., *Enhancing Security in ASP.NET Core Applications: Implementing Oauth, JWT, and Zero-Trust Models.* 2025: p. 2561-2575.
238. Chuma, K.G. and M. Msomi, *Multi-Factor Authentication for Secured Access Control to Electronic Health Records in South African Public Hospitals (Preprint).* 2024.
239. Nanduri, V.S., *Strengthening Security: Implementation of Multi-Factor Authentication (MFA) in Data Center Environment.* International Journal of Scientific Research in Engineering and Management, 2024. **08**(008): p. 1-14.
240. Gorave, S.B., *Credential Harvesting and Account Takeover: Mechanisms, Impacts, and Defenses.* International Journal for Research in Applied Science and Engineering Technology, 2026. **14**(1): p. 122-126.
241. Deepak, A., *A Smart Login Authentication System Using Deep Learning–Based Face Recognition.* International Journal for Research in Applied Science and Engineering Technology, 2025. **13**(9): p. 11-21.
242. Vijayanand, S., et al., *A Novel Biometric-Centric Encryption Framework for Secure Online Payments and Cloud Storage.* 2025.

243. Sundaramoorthy, K., G. Lokanya, and A.P.P. J, *Secured E-Banking Using Multi-Identity Based Fully Homomorphic Encryption*. Journal of Electronics and Informatics, 2025. **7**(2): p. 122-132.
244. Falas, S., C. Konstantinou, and M.K. Michael, *Hardware-Enabled Secure Firmware Updates in Embedded Systems*. 2020: p. 165-185.
245. Shashikant, S., et al., *Fortifying Connected Vehicles Based Cybersecurity Measures for Secure Over-the-Air Software Updates*. 2024. **14**(1): p. 125-140.
246. Olaoye, G., *Authentication and Integrity of UEFI Capsule Updates*. 2025.
247. Iseal, S., *Understanding UEFI Capsule Update Mechanisms*. 2025.
248. Kovačević, A. and N. Gligorić, *Enhancing Security of Automotive OTA Firmware Updates via Decentralized Identifiers and Distributed Ledger Technology*. Electronics, 2024. **13**(23): p. 4640.
249. Mathew, A., *Secure Over-the-Air (OTA) Update Mechanisms for ADAS*. International Research Journal of Innovations in Engineering and Technology, 2024. **08**(04): p. 34-38.
250. Olaoye, G., *Role of Secure Boot in Protecting UEFI Capsule Updates*. 2025.
251. Avasthi, A. and R.K. Yadav, *Secure Firmware Updates in IoT Devices Using Hash Based Signature Schemes*. 2025: p. 278-302.
252. Zhao, Q., et al., *Fine-Grained Forward Secure Firmware Update in Smart Home*. Mathematics, 2023. **11**(14): p. 3084.
253. Falas, S., C. Konstantinou, and M.K. Michael, *A Hardware-Based Framework for Secure Firmware Updates on Embedded Systems*. 2019: p. 198-203.
254. Hayati, N., *The Design of Security Framework for LoRaWAN FUOTA*. Journal of Electrical Technology Umy, 2024. **7**(2): p. 82-88.
255. Weimerskirch, A., *Secure Software Flashing*. Sae International Journal of Passenger Cars - Electronic and Electrical Systems, 2009. **02**(1): p. 83-86.
256. Vankayala, V.N.M.K., *Decentralized Integrity: Blockchain-Based Frameworks for Enhancing Software Update Security*. International Journal of Scientific Research in Computer Science Engineering and Information Technology, 2024. **10**(5): p. 404-409.
257. Aldweesh, A., *Blockchain-Based Secure Firmware Updates for Electric Vehicle Charging Stations in Web of Things Environments*. World Electric Vehicle Journal, 2025. **16**(4): p. 226.
258. Buchmann, J., A. May, and U. Vollmer, *Perspectives for Cryptographic Long-Term Security*. Communications of the Acm, 2006. **49**(9): p. 50-55.
259. Schneller, L., C.N. Porter, and A. Wakefield, *Implementing Converged Security Risk Management: Drivers, Barriers, and Facilitators*. Security Journal, 2022. **36**(2): p. 333-349.
260. Jain, D., *Tracing the Timeline of Network Security From the 1950s to the Present (2023)*. International Journal of Scientific Research in Engineering and Management, 2023. **07**(08).
261. Salim, A., S. Tripathi, and R. Tiwari, *Applying Geo-Encryption and Attribute Based Encryption to Implement Secure Access Control in the Cloud*. International Journal of Computer Networks & Communications, 2019. **11**(4): p. 121-135.
262. Baza, M., et al., *Blockchain-Based Firmware Update Scheme Tailored for Autonomous Vehicles*. 2019: p. 1-7.
263. Vijayarajan, J., B.E. Балан, and V.S. Benitha, *Security Issues in IoT: Perspective Review*. Computer Networks and Communications, 2025: p. 101-117.
264. Dandotiya, S.K.G.A.S., *Analysis and Development of Security Framework for IOT Device*. Tuijin Jishu/Journal of Propulsion Technology, 2023. **44**(4): p. 995-1008.
265. Wu, H., et al., *Research on Programmable Logic Controller Security*. Iop Conference Series Materials Science and Engineering, 2019. **569**(4): p. 042031.
266. Mtetwa, N.S., et al., *Blockchain-Based Security Model for LoRaWAN Firmware Updates*. Journal of Sensor and Actuator Networks, 2022. **11**(1): p. 5.

267. Regueiro, C., et al., *A Blockchain-Based Audit Trail Mechanism: Design and Implementation.* Algorithms, 2021. **14**(12): p. 341.
268. Chapple, M.J., N.V. Chawla, and A. Striegel, *Authentication Anomaly Detection.* 2007: p. 17-22.
269. Jonker, H., et al., *Shepherd: A Generic Approach to Automating Website Login.* 2020.
270. Rich, M.S., *Enhancing Microsoft 365 Security: Integrating Digital Forensics Analysis to Detect and Mitigate Adversarial Behavior Patterns.* Forensic Sciences, 2023. **3**(3): p. 394-425.
271. Studiawan, H., C. Payne, and F. Sohel, *Graph Clustering and Anomaly Detection of Access Control Log for Forensic Purposes.* Digital Investigation, 2017. **21**: p. 76-87.
272. *Blockchain-Integrated Access Control for Wireless Edge Networks.* Journal of Wireless Mobile Networks Ubiquitous Computing and Dependable Applications, 2025. **16**(2): p. 775-792.
273. Santoso, R.S. and I.R. Mukhlis, *Implementation of Risk Management in Library Information System at Surabaya City Library.* Jurnal Kajian Informasi & Perpustakaan, 2023. **11**(2): p. 177.
274. Ramadhani, Y. and M. Sholeh, *Penerapan Tools Fail2Ban Untuk Mencegah Serangan Brute Force Pada Web Server Online Learning Uhamka (OLU).* Jurnal Teknik Informatika Dan Komputer, 2022. **1**(2): p. 41-45.
275. Chuah, E., et al., *Enabling Dependability-Driven Resource Use and Message Log-Analysis for Cluster System Diagnosis.* 2017: p. 317-327.
276. Mustafa, G., et al., *Usability Evaluation of a Push-Based Passwordless Authentication Model Using Public-Key Cryptography.* Iet Biometrics, 2026. **2026**(1).
277. Simoens, K., P. Tuyls, and B. Preneel, *Privacy Weaknesses in Biometric Sketches.* 2009: p. 188-203.
278. Nautsch, A., et al., *Homomorphic Encryption for Speaker Recognition: Protection of Biometric Templates and Vendor Model Parameters.* 2018.
279. Yang, W., et al., *Secure Fingerprint Authentication With Homomorphic Encryption.* 2020: p. 1-6.
280. Barni, M., et al., *A Privacy-Compliant Fingerprint Recognition System Based on Homomorphic Encryption and Fingercode Templates.* 2010: p. 1-7.
281. Bauspieß, P., et al., *Privacy-Preserving Preselection for Protected Biometric Identification Using Public-Key Encryption With Keyword Search.* Ieee Transactions on Industrial Informatics, 2023. **19**(5): p. 6972-6981.
282. Elazm, L.A.A., et al., *Efficient Hardware Design of a Secure Cancellable Biometric Cryptosystem.* Intelligent Automation & Soft Computing, 2023. **36**(1): p. 929-955.
283. Zhou, K. and J. Ren, *PassBio: Privacy-Preserving User-Centric Biometric Authentication.* Ieee Transactions on Information Forensics and Security, 2018. **13**(12): p. 3050-3063.
284. Li, M., et al., *Minutiae Matching With Privacy Protection Based on the Combination of Garbled Circuit and Homomorphic Encryption.* The Scientific World Journal, 2014. **2014**: p. 1-13.
285. Jegede, A., et al., *State of the Art in Biometric Key Binding and Key Generation Schemes.* International Journal of Communication Networks and Information Security (Ijcnis), 2022. **9**(3).
286. Bassit, A., et al., *Hybrid Biometric Template Protection: Resolving the Agony of Choice Between Bloom Filters and Homomorphic Encryption.* Iet Biometrics, 2022. **11**(5): p. 430-444.
287. Bauspieß, P., et al., *HEBI: Homomorphically Encrypted Biometric Indexing.* 2023: p. 1-10.
288. Zhang, H., et al., *Finger Vein Recognition With Cancelable Template-Based Partial Mutual Encryption.* Journal of Electronic Imaging, 2023. **32**(05).
289. Droandi, G. and R. Lazzeretti, *SHE Based Non Interactive Privacy Preserving Biometric Authentication Protocols.* 2015: p. 1-6.
290. El-Shafai, W., et al., *Optical Ciphering Scheme for Cancellable Speaker Identification System.* Computer Systems Science and Engineering, 2023. **45**(1): p. 563-578.
291. NisarBhat, A. and S. Kaur, *Enhancement of Biometric Template Security in Multibiometric Systems.* International Journal of Computer Applications, 2013. **69**(10): p. 36-41.

292. Golec, M., et al., *BioSec: A Biometric Authentication Framework for Secure and Private Communication Among Edge Devices in IoT and Industry 4.0.* Ieee Consumer Electronics Magazine, 2022. **11**(2): p. 51-56.
293. Dhawale, C., *Review and Applications of Multimodal Biometrics for Secured Systems.* 2017: p. 251-266.
294. Güler, İ., R.M. Zengin, and M. Sönmez, *Smart Cards: A Specific Application in the Hospital.* Journal of Medical Systems, 1998. **22**(6): p. 405-419.
295. Bhati, K.K., et al., *Microcontroller-Based Door Locking System.* 2025: p. 136-144.
296. Khan, S.R., *Development of Low Cost Private Office Access Control System (OACS).* International Journal of Embedded Systems and Applications, 2012. **2**(2): p. 1-7.
297. Sutar, M.S., et al., *Multi-Factor Authentication Doorlocking System.* International Research Journal of Modernization in Engineering Technology and Science, 2023.
298. Samuel, D.R., O. OsuekeChristian, and E. Dickson, *Microcontroller Based Code Locking System With Alarm.* Iosr Journal of Electrical and Electronics Engineering, 2014. **9**(1): p. 09-17.
299. Eseosa, O., *Access Control Using Smartcard and Passcode.* Iosr Journal of Electrical and Electronics Engineering, 2013. **4**(5): p. 29-34.
300. Gurung, C., et al., *Smart Lock Using IoT.* International Research Journal of Modernization in Engineering Technology and Science, 2023.
301. Kasim, B., et al., *Performance Analysis of RFID-Based Smart Door Lock Controlled by Arduino.* Journal of Advanced Research in Applied Mechanics, 2024. **122**(1): p. 163-174.
302. Vardhini, P.A.H., et al., *Patternized Secret Knock Detecting Smart Door Lock System.* International Journal of Latest Technology in Engineering Management & Applied Science, 2025. **14**(5): p. 725-728.
303. Supekar, S., et al., *Office Access System Using RFID and Smart Lock.* International Journal of Advanced Research in Science Communication and Technology, 2024: p. 252-256.
304. Kadak, T. and S. Özdemir, *Security-Oriented Smart Door Lock À La Internet of Things.* European Journal of Science and Technology, 2021.
305. Muke, S., et al., *NFC Hardware Device Based Access Control System Using Information Hiding.* Ijireeice, 2015. **3**(10): p. 69-72.
306. Hidayah, D.S., et al., *Perancangan Sistem Pengunci Otomatis Berbasis Mikrokontroler Arduino Dengan Integrasi Keypad Dan LCD.* Intro Journal Informatika Dan Teknik Elektro, 2025. **3**(2): p. 86-90.
307. Cho, K.-T. and K.G. Shin, *Error Handling of in-Vehicle Networks Makes Them Vulnerable.* 2016: p. 1044-1055.
308. Gao, Y. and M. Xu, *Defense Against Software-Defined Network Topology Poisoning Attacks.* Tsinghua Science & Technology, 2023. **28**(1): p. 39-46.
309. Palanca, A., et al., *A Stealth, Selective, Link-Layer Denial-of-Service Attack Against Automotive Networks.* 2017: p. 185-206.
310. Zhu, S., et al., *LHAP: A Lightweight Network Access Control Protocol for Ad Hoc Networks.* Ad Hoc Networks, 2006. **4**(5): p. 567-585.
311. Alalwany, E. and I. Mahgoub, *An Effective Ensemble Learning-Based Real-Time Intrusion Detection Scheme for an in-Vehicle Network.* Electronics, 2024. **13**(5): p. 919.
312. Rajkumar, V.S., et al., *Cyber Attacks on Protective Relays in Digital Substations and Impact Analysis.* 2020: p. 1-6.
313. Khederzadeh, M., *Vulnerability Analysis of Transmission Lines Based on the Performance of Mho Relays and Series/Shunt Compensation Systems.* 2007.
314. Elrawy, M.F., et al., *Modelling and Analysing Security Threats Targeting Protective Relay Operations in Digital Substations.* 2023.

315. Cultice, T. and H. Thapliyal, *Vulnerabilities and Attacks on CAN-Based 3D Printing/Additive Manufacturing*. 2023.
316. Johnson, A., et al., *Avoiding the Man on the Wire: Improving Tor's Security With Trust-Aware Path Selection*. 2017.
317. Bolandi, T.G., M.R. Haghifam, and M. Khederzadeh, *Real‐time Monitoring of Zone 3 Vulnerable Distance Relays to Prevent Maloperation Under Load Encroachment Condition*. Iet Generation Transmission & Distribution, 2017. **11**(8): p. 1878-1888.
318. Sun, Y., et al., *Counter-Raptor: Safeguarding Tor Against Active Routing Attacks*. 2017: p. 977-992.
319. Guan, X., et al., *Power Grids Vulnerability Analysis Based on Combination of Degree and Betweenness*. 2014: p. 4829-4833.
320. Godja, N. and F.D. Munteanu, *Environmentally Friendly Solutions as Potential Alternatives to Chromium-Based Anodization and Chromate Sealing for Aeronautic Applications*. Coatings, 2025. **15**(4): p. 439.
321. Sarode, D.B., et al., *Corrosion-Resistant Materials for Ocean Structures: Innovations, Mechanisms, and Applications*. Sustainable Marine Structures, 2025.
322. Korolov, V.P., I.V. Kushchenko, and E.A. Бочарова, *Parameterization of Requirements of Corrosion Protectability of Structural Steel*. Building and Reconstruction, 2023. **108**(4): p. 33-46.
323. Korolov, V.P., et al., *Management of the Quality of Corrosion Protection of Structural Steel Based on Corrosion Risk Level*. Journal of Materials Science and Engineering B, 2013. **3**(11).
324. Zhang, P., et al., *A Blockchain-Based Authentication Scheme and Secure Architecture for IoT-Enabled Maritime Transportation Systems*. Ieee Transactions on Intelligent Transportation Systems, 2022: p. 1-10.
325. Kovalenko, O.V., *Improving the Operational Reliability of Hydraulic Facilities When Using Polymer and Polymer-Cement Composites*. Міжвідомчий Тематичний Науковий Збірник Меліорація І Водне Господарство, 2019(2): p. 217-230.
326. Steven, H., *Enhancing Offshore Asset Safety Through UAV-Based Ultrasonic Thickness Measurement Inspections*. 2025.
327. Kumar, A., A. Gholve, and K. Kotalwar, *Automotive Security Solution Using Hardware Security Module (HSM)*. 2024. **1**.
328. Luo, F. and Q. Hu, *Security Mechanisms Design for in-Vehicle Network Gateway*. 2018. **1**.
329. Idris, A., et al., *Design of Robotics Implementation of a Fingerprint Biometric System*. Journal of Software Engineering and Simulation, 2024. **10**(9): p. 16-23.
330. Schleiffer, C., et al., *Secure Key Management - A Key Feature for Modern Vehicle Electronics*. 2013. **1**.
331. S, D.P.V., *AI-Driven Facial Recognition System for Secure and Efficient Bike Starter Authentication Using Machine Learning*. International Journal of Scientific Research in Engineering and Management, 2025. **09**(05): p. 1-9.
332. Yang, B., *Stakeholder Perspectives on Biometrics-Based Multi-Factor Authentication for eIDAS Levels of Assurance: Insights on Usability, Security, and Privacy*. 2025. **199**.
333. Santos, N., B. Ghita, and G.L. Masala, *Medical Systems Data Security and Biometric Authentication in Public Cloud Servers*. Ieee Transactions on Emerging Topics in Computing, 2024. **12**(2): p. 572-582.
334. Lichtermann, J. and R. Pettit, *Automotive Application of Biometric Systems and Fingerprint*. 2000. **1**.
335. Ansari, A., et al., *Mechanism to Identify Legitimate Vehicle User in Remote Keyless Entry System*. 2022. **1**.

336. Geetha, D.J.S., *Enhance the Security of Data Storage and Retrieval in Cloud Computing Through Authentication Based Encryption*. International Journal of Recent Technology and Engineering (Ijrte), 2019. **8**(4): p. 11927-11931.
337. Anjis Sapto Nugroho, N., et al., *Design of a Web-Based Digital Library Management System With Biometric Authentication*. Journal of Electrical Computer and Informatics, 2025. **1**(2): p. 31-39.
338. Sanap, M.J.R., *Secure Your Entry Ways: Using Cloud Based Door Access System*. International Scientific Journal of Engineering and Management, 2025. **04**(05): p. 1-7.
339. Reddy, M., N.M.A. Chisty, and A. Bodepudi, *A Review of Cybersecurity and Biometric Authentication in Cloud Network*. Engineering International, 2022. **10**(1): p. 9-18.
340. Schmitz, S., J. Kruppa, and P. Crowhurst, *Safety and Security Considerations of New Closure Systems*. 2000. **1**.
341. James, J. and S.V. Suryakala, *Advanced Vehicle Security Control and Accident Alert System*. International Journal of Engineering & Technology, 2018. **7**(2.8): p. 404.
342. Wouters, L., et al., *Fast, Furious and Insecure: Passive Keyless Entry and Start Systems in Modern Supercars*. Iacr Transactions on Cryptographic Hardware and Embedded Systems, 2019: p. 66-85.
343. Hicks, C., F.D. Garcia, and D. Oswald, *Dismantling the AUT64 Automotive Cipher*. Iacr Transactions on Cryptographic Hardware and Embedded Systems, 2018: p. 46-69.
344. Borse, Y. and M.A. Saleh, *Cloud Based Cyber Physical Systems Security Issues: A Survey*. International Journal of Computer Applications, 2019. **177**(9): p. 38-40.
345. Fitzgerald, J. and C. Morisset, *Can We Develop Holistic Approaches to Delivering Cyber-Physical Systems Security?* Research Directions Cyber-Physical Systems, 2024. **2**.
346. Shaikh, I.A., et al., *Vulnerability Assessment of Vehicle Access System Using Software-Defined Radio (SDR): Attacks, Testing, Comparative Analysis and Future Countermeasures*. 2025.
347. Doyle, B., et al., *MAKER: Call a 3D Locksmith – How 3D Printing Can Defeat Physical Security*.
348. Choraś, M., et al., *Cyber Threats Impacting Critical Infrastructures*. 2016: p. 139-161.
349. Orojloo, H. and M.A. Azgomi, *A Method for Modeling and Evaluation of the Security of Cyber-Physical Systems*. 2014: p. 131-136.
350. Xie, J., A. Ştefanov, and C.C. Liu, *Physical and Cyber Security in a Smart Grid Environment*. Wiley Interdisciplinary Reviews Energy and Environment, 2016. **5**(5): p. 519-542.
351. Zhang, H., B. Liu, and H. Wu, *Smart Grid Cyber-Physical Attack and Defense: A Review*. Ieee Access, 2021. **9**: p. 29641-29659.
352. Achar, S., et al., *Human Activity Recognition-Based Cyber-Physical System Security Through IoT-Cloud Computing*. 2023.
353. Rabinovich, V., et al., *Small Printed Meander Symmetrical and Asymmetrical Antenna Performances, Including the RF Cable Effect, in the 315 MHz Frequency Band*. Microwave and Optical Technology Letters, 2006. **48**(9): p. 1828-1833.
354. Beek, S.v.d. and F. Leferink, *Vulnerability of Remote Keyless-Entry Systems Against Pulsed Electromagnetic Interference and Possible Improvements*. Ieee Transactions on Electromagnetic Compatibility, 2016. **58**(4): p. 1259-1265.
355. Liu, H., et al., *Practical Contactless Attacks on Hitag2-Based Immobilizer and RKE Systems*. Destech Transactions on Computer Science and Engineering, 2018(CCNT).
356. Glocker, T., T. Mantere, and M. Elmusrati, *A Protocol for a Secure Remote Keyless Entry System Applicable in Vehicles Using Symmetric-Key Cryptography*. 2017: p. 310-315.
357. Schmitz, S. and C. Roser, *A New State-of-the-Art Keyless Entry System*. 1998. **1**.
358. Umadevi, R. and G. Karthikeyaraj, *Enhancing Automotive Security: Biometric Authentication for Vehicle Access Control*. International Journal for Multidisciplinary Research, 2023. **5**(5).

359. Shin, S.U., W. Shin, and K.-H. Rhee, *All-or-Nothing Transform and Remotely Keyed Encryption Protocols*. 2000: p. 178-195.
360. Daza, V. and X. Salleras, *LASER: Lightweight and Secure Remote Keyless Entry Protocol*. 2019.
361. Magda, D. and B. Payne, *RFID Key Fobs in Vehicles: Unmasking Vulnerabilities and Strengthening Security*. 2023.
362. Dhande, K.R., *Fourth Industrial Revolution and the Role of Cyber Security*. 2024: p. 241-257.
363. Joo, K., W. Choi, and D.H. Lee, *Hold the Door! Fingerprinting Your Car Key to Prevent Keyless Entry Car Theft*. 2020.
364. Moradi, A. and T. Kasper, *A New Remote Keyless Entry System Resistant to Power Analysis Attacks*. 2009: p. 1-6.
365. Eisenbarth, T., et al., *On the Power of Power Analysis in the Real World: A Complete Break of the KeeLoq Code Hopping Scheme*. p. 203-220.
366. Alharbi, A., *Applying Access Control Enabled Blockchain (ACE-BC) Framework to Manage Data Security in the CIS System*. Sensors, 2023. **23**(6): p. 3020.
367. Prakash, B. and N. Nalini, *Improved Security of Audit Trail Logs in Multi-Tenant Cloud Using ABE Schemes*. International Journal of Advanced Computer Science and Applications, 2014. **5**(11).
368. Chen, X., et al., *HIPPA's Compliant Auditing System for Medical Imaging System*. 2005.
369. Swanzy, P.N., A.M. Abukari, and E.D. Ansong, *Data Security Framework for Protecting Data in Transit and Data at Rest in the Cloud*. Current Journal of Applied Science and Technology, 2024. **43**(6): p. 61-77.
370. Yushui, G. and S. Pang, *Data Security Monitoring Platform in Cloud for Enterprise*. International Journal of Security and Its Applications, 2013. **7**(6): p. 67-78.
371. Ali, S., et al., *A Robust Algorithm for Authenticated Health Data Access via Blockchain and Cloud Computing*. Plos One, 2024. **19**(9): p. e0307039.
372. Prasad, A. and P. Chakrabarti, *Extending Access Management to Maintain Audit Logs in Cloud Computing*. International Journal of Advanced Computer Science and Applications, 2014. **5**(3).
373. Liu, S., *Securing the Clouds*. 2016: p. 220-232.
374. Owobu, W.O., et al., *Conceptual Framework for Deploying Data Loss Prevention and Cloud Access Controls in Multi-Layered Security Environments*. International Journal of Multidisciplinary Research and Growth Evaluation, 2022. **3**(1): p. 850-860.
375. None, N., et al., *Cybersecure Documentation and Record-Keeping Protocols for Safeguarding Sensitive Financial Information Across Business Operations*. 2023. **04**(03): p. 117-152.
376. Singh, P.K., *Design Patterns for Secure Multi-Tenant Architecture in Financial Services*. International Journal on Science and Technology, 2019. **10**(4).
377. Verma, S., *Cybersecurity Compliance in the Age of Remote Work: Challenges and Solutions*. World Journal of Advanced Engineering Technology and Sciences, 2025. **15**(1): p. 1112-1120.
378. Ganapathy, V., *Audit Trail and Transparency in Cloud-Based Auditing*. Edumania-an International Multidisciplinary Journal, 2024. **02**(02): p. 118-145.
379. Goundar, S. and I. Gondal, *AI–Blockchain Integration for Real-Time Cyber Security - System Design and Evaluation*. 2025.
380. Bastina, A.A.M. and N. Rama, *Biometric Identification and Authentication Providence Using Fingerprint for Cloud Data Access*. International Journal of Electrical and Computer Engineering (Ijece), 2017. **7**(1): p. 408.
381. None, N. and K. Kumar Puipati, *Biometric and Behavioral Authentication in IAM: Security, Privacy, and Continuous Verification Trade-Offs*. Frontiers in Emerging Computer Science and Information Technology, 2026. **03**(01): p. 01-14.

382. Olesia, B., *System of Individual Multidimensional Biometric Authentication*. International Journal of Emerging Trends in Engineering Research, 2019. **7**(12): p. 812-817.
383. Παπαϊωάννου, M., et al., *A Survey on Quantitative Risk Estimation Approaches for Secure and Usable User Authentication on Smartphones*. Sensors, 2023. **23**(6): p. 2979.
384. Abuhamad, M., et al., *Sensor-Based Continuous Authentication of Smartphones' Users Using Behavioral Biometrics: A Contemporary Survey*. 2020.
385. Pahuja, S. and N. Goel, *Multimodal Biometric Authentication: A Review*. Ai Communications, 2024. **37**(4): p. 525-547.
386. Iyen, C., A. Jacob, and A. Oluwasegun, *Development of Biometric User Identification and Access Control System*. European Journal of Applied Science Engineering and Technology, 2024. **2**(3): p. 194-204.
387. Παπαϊωάννου, M., et al., *Behavioral Biometrics for Mobile User Authentication: Benefits and Limitations*. 2023: p. 1-6.
388. Ramisetty, S. and B. Ramesh, *Impregnable Fingerprint Biometric a Survey*. International Journal of Computer Application, 2018. **1**(8).
389. Lone, S.A. and A.H. Mir, *Smartphone-Based Biometric Authentication Scheme for Access Control Management in Client-Server Environment*. International Journal of Information Technology and Computer Science, 2022. **14**(4): p. 34-47.
390. Lim, M.-H., A.B.J. Teoh, and J. Kim, *Biometric Feature-Type Transformation: Making Templates Compatible for Secret Protection*. Ieee Signal Processing Magazine, 2015. **32**(5): p. 77-87.
391. Ferrag, M.A., Λ. Μαγλαράς, and A. Derhab, *Authentication and Authorization for Mobile IoT Devices Using Biofeatures: Recent Advances and Future Trends*. Security and Communication Networks, 2019. **2019**: p. 1-20.
392. Basak, S., et al., *Introduction to AI and ML Technologies and Their Potential Applications in Cybersecurity*. 2024: p. 277-309.
393. I Ogbole, J., et al., *Security Analytics and Digital Forensics for Enterprise Risk Management, Advances and Practical Implications*. International Journal of Advanced Multidisciplinary Research and Studies, 2025. **5**(6): p. 2017-2028.
394. Mahdi, A.J., *Machine Learning Applications of Network Security Enhancement: Review*. Computer Science & It Research Journal, 2024. **5**(10): p. 2283-2300.
395. Bondhala, S., *Cybersecurity in AI-Driven Data Centers: Reinventing Threat Detection*. International Journal of Advanced Research in Science Communication and Technology, 2025: p. 510-519.
396. Palakurti, N.R., *Comprehensive AI Strategies for Cybersecurity for Innovations, Applications, and Future Directions*. International Journal of Advanced Research in Science Communication and Technology, 2024: p. 671-688.
397. None, N., et al., *Zero Trust Security Architecture (ZTSA)*. Journal of Artificial Intelligence & Cloud Computing, 2024: p. 1-8.
398. Brandão, P.R., *Exploring the Role of Artificial Intelligence in Detecting Advanced Persistent Threats*. Computers, 2025. **14**(7): p. 245.
399. Sanjai, S., *Kinesis:Web-Based Log Analysis for Intrusion Detection System*. International Journal for Research in Applied Science and Engineering Technology, 2025. **13**(12): p. 416-427.
400. Pandya, S.K. and U. Pandya, *The Role of AI in Enhancing Business Analytics and Security: A Systematic Review*. International Journal for Multidisciplinary Research, 2025. **7**(4).
401. Phanireddy, S., *AI-Driven Identity Access Management (IAM)*. International Journal of Scientific Research in Engineering and Management, 2021. **05**(06): p. 1-9.
402. Kumar, T., M. Memoria, and K. Joshi, *Empowering Cloud Security Through Advanced Monitoring and Analytics Techniques*. 2025: p. 163-186.

403. Guntupalli, R., *AI-Driven Threat Detection and Mitigation in Cloud Infrastructure: Enhancing Security Through Machine Learning and Anomaly Detection.* Journal of Informatics Education and Research, 2023. **3**(2).

404. Ravikumar, C.P., et al., *Enhancing Cloud Security and Efficiency Through AI-Driven Intrusion Detection and Machine Learning-Based Resource Management.* 2025: p. 241-256.

405. Vashishth, T.K., et al., *AI-Driven Threat Detection and Incident Response.* 2025: p. 51-82.

406. Sontan, A.D. and S.V. Samuel, *The Intersection of Artificial Intelligence and Cybersecurity: Challenges and Opportunities.* World Journal of Advanced Research and Reviews, 2024. **21**(2): p. 1720-1736.

407. *Enhancing Cybersecurity With Safe and Reliable AI: Mitigating Threats While Ensuring Privacy Protection.* International Journal of Computer Applications Technology and Research, 2024.

408. Yousefi, Z., N. Saadati, and S.A. Saadati, *AI in Elderly Care: Understanding the Implications for Independence and Social Interaction.* 2023. **1**(4): p. 26-32.

409. Nadella, G.S., et al., *Privacy and Security.* 2024: p. 157-174.

410. Ezzeddine, Y., *Artificial Intelligence in Law Enforcement Surveillance: Citizen Perspectives, Resistance and Counterstrategies.*

411. López-González, A., et al., *Ethics in Artificial Intelligence: An Approach to Cybersecurity.* Inteligencia Artificial, 2024. **27**(73): p. 38-54.

412. Goel, P.K., *Ethical Considerations in Implementing Artificial Intelligence in Cybersecurity.* 2024: p. 73-92.

413. Hayat khan, A., *Ethical Implications of Artificial Intelligence in Countering Violent Extremism.* 2025: p. 207-252.

414. Rawat, M., B.K. Lad, and A. Sharma, *Simulation-Based Joint Optimization of Fleet System Modularity and Level of Repair Decisions Considering Different Failure Rates of Components.* Grey Systems Theory and Application, 2020. **10**(3): p. 377-390.

415. Hartranft, J.J., *Expansion of USN Gas Turbine Repair Into the Intermediate Level Repair Facility.* 1993.

416. Wang, W.L. and J.T. Zhang, *Product Improvement Design Method Facing Modularization Disassembly.* Advanced Materials Research, 2010. **160-162**: p. 564-567.

417. Zheng, H., et al., *An Integrated Modular Design Methodology Based on Maintenance Performance Consideration.* Proceedings of the Institution of Mechanical Engineers Part B Journal of Engineering Manufacture, 2016. **231**(2): p. 313-328.

418. Yan, J.H. and C. Feng, *Sustainability-Oriented Product Modular Design Using Design Structure Matrix (DSM) Method.* Applied Mechanics and Materials, 2011. **128-129**: p. 1468-1471.

419. Amend, C., et al., *User Experience With Smartphone Repair Activities and the Impact of Modular Product Design.* 2021.

420. Zhang, Z., J.T. Heron, and A. Pena-Francesch, *Adaptive Magnetoactive Soft Composites for Modular and Reconfigurable Actuators.* Advanced Functional Materials, 2023. **33**(26).

421. Cassiani, J.D., et al., *Sustainable Strategies for Sea Locks: An <scp>LCA</Scp> Study on Reinforcement Corrosion Prevention Methods.* Structural Concrete, 2024. **26**(2): p. 2182-2200.

422. Husnutdinov, A.N. and K.G. Bakullin, *Development of Modular and Easily Replaceable Battery Packs With Impact on Reducing Maintenance Costs and Secondary Market for Batteries.* Ekonomika I Upravlenie Problemy Resheniya, 2025. **4/2**(157): p. 118-126.

423. Kim, W.S., et al., *Joint Displacement Evaluation of a Modular Road System.* Advanced Materials Research, 2013. **831**: p. 407-410.

424. Xu, H., Y. Fang, and N. Fard, *Optimal Design for Resilient Load-sharing Systems With Nonidentical Components.* Quality and Reliability Engineering International, 2018. **34**(6): p. 1254-1270.

425. Ryabchikova, I.A. and A.V. Petyukova, *Assessment of Occupational Risks in the Aircraft Industry.* Iop Conference Series Materials Science and Engineering, 2020. **962**(4): p. 042021.
426. Gurbanova, M.H., *The Risk of Professional Incompetence in Employees of Azerbaijan Caspian Shipping Company: Retrospective Study.* Marine Medicine, 2023. **9**(4): p. 34-39.
427. Ahmad, A., S.B. Maynard, and S. Park, *Information Security Strategies: Towards an Organizational Multi-Strategy Perspective.* Journal of Intelligent Manufacturing, 2012. **25**(2): p. 357-370.
428. Briggs, C.M., *Climate Change and Hybrid Warfare Strategies.* Journal of Strategic Security, 2020. **13**(4): p. 45-57.
429. Zajac, A., *Urzędnicy Najemni I Służba W Miastach Wołynia W Drugiej Połowie XVI – Pierwszej Połowie XVII Wieku.* Krakowskie Pismo Kresowe, 2018. **10**: p. 25-49.
430. Churchill, D., *The Spectacle of Security: Lock-Picking Competitions and the Security Industry in Mid-Victorian Britain.* History Workshop Journal, 2015. **80**(1): p. 52-74.
431. Gimaeva, Z.F., et al., *Diagnostic Significance of Lipid Profile Indicators for Assessing Chemical Workers' Cardiovascular Risk.* Russian Journal of Occupational Health and Industrial Ecology, 2022. **61**(1): p. 19-28.
432. Гимранова, Г.Г., et al., *The Influence of Production Factors on the Formation of the Health of Workers in the Production of Synthetic Rubbers, Preventive Measures.* Sanitarnyj Vrač (Sanitary Doctor), 2024(3): p. 194-202.
433. Magalhães, A.M.M.d., et al., *Erasmus Syndrome: An Underrecognized Entity.* Acta Médica Portuguesa, 2022. **36**(2): p. 122-126.
434. Pratikakis, P., J.S. Foster, and M. Hicks, *Locksmith.* Acm Transactions on Programming Languages and Systems, 2011. **33**(1): p. 1-55.

# Index

## A

## B

## C

## D



## E

## F



## G



## H



## I

## K



## L

## M



## N



## O



## P

## Q



## R

## S

## T



## U



## V



## W

Made in United States
Orlando, FL
04 November 2026

90202666R00448